Lecture Notes in Computer Science 4297

Commenced Publication in 1973
Founding and Former Series Editors:
Gerhard Goos, Juris Hartmanis, and Jan van Leeuwen

T0180641

Yves Robert   Manish Parashar
Ramamurthy Badrinath
Viktor K. Prasanna (Eds.)

High Performance Computing – HiPC 2006

13th International Conference
Bangalore, India, December 18-21, 2006
Proceedings

 Springer

Volume Editors

Yves Robert
LIP, École Normale Supérieure de Lyon
46 allée d'Italie, 69364 Lyon Cedex 07, France
E-mail: yves.robert@ens-lyon.fr

Manish Parashar
Rutgers, The State University of New Jersey
Department of Electrical and Computer Engineering
94 Brett Road, Piscataway, NJ 08854, USA
E-mail: parashar@caip.rutgers.edu

Ramamurthy Badrinath
HP - India Software Operations
29, Cunningham Road, Bangalore 560 052, India
E-mail: badrinath@hp.com

Viktor K. Prasanna
University of Southern California
Department of Electrical Engineering
Los Angeles, CA 90089-2562, USA
E-mail: prasanna@usc.edu

Library of Congress Control Number: 2006938017

CR Subject Classification (1998): D.1-4, C.1-4, F.1.2, G.1-2

LNCS Sublibrary: SL 1 – Theoretical Computer Science and General Issues

ISSN   0302-9743
ISBN-10   3-540-68039-X Springer Berlin Heidelberg New York
ISBN-13   978-3-540-68039-0 Springer Berlin Heidelberg New York

Springer is a part of Springer Science+Business Media

springer.com

© Springer-Verlag Berlin Heidelberg 2006
Printed in Germany

Typesetting: Camera-ready by author, data conversion by Scientific Publishing Services, Chennai, India
Printed on acid-free paper   SPIN: 11945918   06/3142   5 4 3 2 1 0

Message from the Program Chair

Welcome to the proceedings of the 13th International Conference on High-Performance Computing!

This year, we were delighted to receive 335 submissions from 5 continents and 39 different countries. Eventually, 52 submissions from 19 different countries were selected for presentation at the conference and publication in the conference proceedings.

This large number of submissions led us to use the same two-phase selection process as for HiPC 2005. First, all submitted papers were carefully considered by the Program Chair and Vice-Chairs to check their consistency with the minimal requirements for acceptance. At the end of this first stage, we were left with 282 submissions, which were further considered by the Program Committee. Each of these papers was reviewed by three Program Committee members. As many as 837 reviews were collected (2.97 per paper on the average), and each paper was discussed at the on-line Program Committee meeting. Finally, 52 out of 282 papers(18.4%) were accepted for presentation and publication in the proceedings.

These figures show that the selection process was highly competitive. We congratulate all authors of accepted papers for their success. Also, we would like to thank all authors of submitted papers for their interest in the conference, and we strongly encourage them to submit their work to forthcoming issues of HiPC.

Two outstanding papers were selected as "Best Papers"; one in the *Algorithms and Applications* area ("Algorithmic Ramifications of Prefetching in Memory Hierarchy," by Akshat Verma and Sandeep Sen), and the other in the *Software* area ("A Cache-Partitioning Aware Replacement Policy for Chip Multiprocessors," by Haakon Dybdahl, Per Stenstrom, and Lasse Natvig). They were presented in a separate plenary session and each paper was awarded a prize sponsored by InfoSys.

In the conference program, we were pleased to accommodate ten parallel technical sessions of high-quality contributed papers, plus the special plenary "Best Papers" session. In addition, this year's conference also featured three Tutorials, six Keynote Addresses, a Panel Session, a Poster Session, Industrial Exhibits, six Workshops and a Mini-Symposium.

It was a pleasure putting together this program with the help of six excellent Program Vice-Chairs and their 89 Program Committee members. The hard work of all the Program Committee members is deeply appreciated. I especially wish to acknowledge the dedicated effort put forth by the Vice-Chairs: Srinivas Aluru (Algorithms), John Morrison (Applications), Bradley Kuszmaul (Architecture), Yuanyuan Yang (Communication Networks), Stéphane Ubéda (Mobile and Sensor Computing), and Sanjay Rajopadhye (Systems Software).

Without their help and timely work, the quality of this program would not be as high, nor would the process have run so smoothly.

I thank the other organizers who contributed in assembling this program, including those who organized the keynotes, tutorials, workshops, awards, poster session, industry exhibits, and those who performed the administrative functions that were essential to the success of this conference. The work of Sushil K. Prasad in putting together the conference proceedings is also acknowledged, as well as the support provided by the Cyber Co-chairs, Sumir Chandra and Vijay Bhat.

I express heart-felt thanks to our General Co-chairs, Manish Parashar and Ramamurthy Badrinath; Steering Chair, Viktor Prasanna; and to the Vice-General Co-chairs, Rajendra V. Boppana and Rajeev Muralidhar; for all their useful advice. Lastly, I thank the Steering Chair and General Co-chairs for allowing me to serve our community as the Program Chair of this high-quality international conference. It has been a rewarding experience, and a pleasure to correspond with so many of you.

December 2006   Yves Robert
                                   Ecole Normale Supérieure de Lyon, France

Message from the General Co-chairs and the Vice General Co-chairs

On behalf of the organizers of the 13th International Conference on High-Performance Computing (HiPC), it is our pleasure to welcome you to Bangalore. I do hope you will find these proceedings exciting and rewarding.

Several new events made HiPC 2006 a special and exciting meeting. The call for papers included, for the first time, a separate track on Mobile and Sensor Computing. HiPC 2006 also featured a mini-symposium on High-Performance Computing Technologies, Applications and Experience aimed at bringing together the users and providers of HPC. Finally the HiPC Student Challenge brought together student teams from Indian institutions in a competition to design and develop distributed solutions to respond to and effectively manage national emergencies. These new events were complemented by the keynotes presented by internationally renowned research, the posters session presenting hot off-the-press research results, and the industry and research exhibits. As always, the meeting was preceded by a set of tutorials and was followed by workshops highlighting new and emerging aspects of the field.

The HiPC call for papers, once again, received an overwhelming response, attracting 335 submissions from 39 countries and establishing another record. For this, we would like to specially thank Yves Robert, Program Chair, who with remarkable dedication put together an outstanding technical program consisting of the 52 papers that appear in these proceedings. We would also like to thank the Program Committee for their efforts in assembling such an excellent program and the authors who submitted the high-quality material from which that program was selected. Finally, we would like to thank the presenters of the keynotes, posters and tutorials, the organizers of the workshops, and all the participants.

Arranging an exciting meeting with a high-quality technical program is easy when one is working with an excellent and dedicated team and can build on the practices and levels of excellence established by a quality research community. HiPC 2006 would not have been possible without the tremendous efforts of the many volunteers. We would like to acknowledge the critical contributions of each one. We would especially like to thank Viktor Prasanna, Chair of the HiPC Steering Committee, for his leadership, sage guidance, and untiring dedication. We would also like to welcome our new volunteers to the team - your efforts are critical to the continued success of this conference. Finally, we would like to gratefully acknowledge our academic and industry sponsors including the IEEE Computer Society, ACM SIGARCH, EATCS, IFIP, NASSCOM, MAIT, Infosys, DELL, HP, IBM, Microsoft, Satyam, Sun, Intel, AMD and Cray.

Manish Parashar
Rutgers University, USA
Ramamurthy Badrinath, HP, India
General Co-chairs

Rajendra V. Boppana
University of Texas at San Antonio, USA
Rajeev Muralidhar, Intel, India
Vice General Co-chairs

Message from the Steering Chair

It is my pleasure to welcome you to the proceeding of the 13th International Conference on High-Performance Computing held in Bangalore, the IT capital of India.

This conference would not be possible without the dedicated effort of many volunteers over the past year. First, I would like to single out the contributions of Yves Robert, Program Chair, for his outstanding contributions in putting together an excellent technical program. I appreciate his efforts in handling a large number of submissions in a timely manner and his overall coordination to ensure a quality program. Ramamurthy Badrinath and Manish Parashar as General Co-chairs provided the leadership in resolving numerous meeting-related issues and putting together the overall meeting program including the workshops and tutorials. They were ably assisted by Rajendra Boppana and Rajeev Muralidhar, Vice General Co-chairs. The poster/presentation session was organized by Rajeev Thakur. The meeting offers scholarships for India-based students. These scholarships were administered by Anu Bourgeois and Madhusudhan Govindaraju.

We have several continuing as well as new workshops. These workshops were coordinated by Manimaran Govindarasu. Rama Govindaraju put together the tutorials. His efforts in securing additional funding for tutorial speakers is appreciated. The Web site was maintained by Viraj Bhat and Sumir Chandra. The local arrangements were handled by C. Kalyana Krishna. Sushil Prasad liased with the authors and Springer to bring out the proceedings. Rajeev Raje, Bo Hong, and Manisha Dhanke handled the publicity for us. Susamma Barua acted as the Registration Chair.

Ramamurthy Badrinath with assistance from Raghuram Tupuri of AMD put together the Mini-Symposium on High-Performance Computing Technologies, Applications and Experience. It expands our HPC Users meeting that was organized in 2005. The intent of the mini-symposium is to provide a forum for vendors as well as HPC users in India to present the technologies and user experiences. In addition, Vijay Mann and Raghuram Tupuri coordinated the industry exhibits.

I would like to thank all our volunteers for their tireless efforts. The meeting would not be possible without the enthusiastic commitment of these individuals.

Major financial support for the meeting was provided by several leading IT companies and multinationals operating in India. I would like to acknowledge the following individuals and their organizations for their support:

N. R. Narayana Murthy and Kris Gopalakrishnan, Infosys
Harish Grama, IBM India
P. Gopalakrishnan, IBM India Research Lab
Daniel Dias, IBM India Research Lab
H. P. Raghunandan, IBM India Software Lab
Pratap Pattnaik, IBM T.J. Watson Research Center
Manish Gupta, IBM T.J. Watson Research Center
June GL Ng, IBM T.J. Watson Research Center

Subramanian Kannan, IBM T.J. Watson Research Center
Raghuram Tupuri, AMD
Vittal Kini and Kiran Panesar, Intel Research, India
Dinkar Sitaram and Faisal Paul, HP India
Jaideep Sen, Microsoft
B. Rudramuni, Dell India
Reza Rooholamini, Dell
V. Sridhar, Satyam
Venkat Ramana, Hinditron Infosystems

Finally, I would like to thank Animesh Pathak at USC for his continued assistance and enthusiasm in organizing the meeting.

December 2006   Viktor K. Prasanna
                                            University of Southern California, USA

Organization

Conference Organization

General Co-chairs

Manish Parashar, Rutgers University, USA
Ramamurthy Badrinath, HP, India

Vice General Co-chairs

Rajendra V. Boppana, University of Texas at San Antonio, USA
Rajeev Muralidhar, Intel, India

Program Chair

Yves Robert, Ecole Normale Supérieure de Lyon (ENS Lyon), France

Vice Program Chairs

Algorithms
Srinivas Aluru, Iowa State University, USA
Applications
John Morrison, University College, Cork, Ireland
Architecture
Bradley Kuszmaul, Massachusetts Institute of Technology, USA
Communication Networks
Yuanyuan Yang, State University of New York at Stony Brook, USA
Mobile and Sensor Computing
Stéphane Ubéda , Institut National des Sciences Appliquées de Lyon (INSA Lyon),
    France
Systems Software
Sanjay Rajopadhye, Colorado State University, USA

Steering Chair

Viktor K. Prasanna, University of Southern California, USA

Workshops Chair

Manimaran Govindarasu, Iowa State University, USA

Poster/Presentation Chair

Rajeev Thakur, Argonne National Laboratory, USA

Tutorials Chair

Rama K. Govindaraju, IBM, USA

Industry Liaison Co-chairs

Vijay Mann, IBM India Research Lab, India
Raghuram Tupuri, Advanced Micro Devices (AMD), India

Cyber Chair

Sumir Chandra, Rutgers University, USA
Viraj Bhat, Rutgers University, USA

Finance Co-chairs

Ajay Gupta, Western Michigan University, USA
B. V. Ramachandran, Software Technology Park, Bangalore, India

Local Arrangements Chair

C. Kalyana Krishna, Dell Product Group, India

Publications Chair

Sushil K. Prasad, Georgia State University, USA

Publicity Co-chairs

Rajeev R. Raje, Indiana University, Purdue University, USA
Bo Hong, Drexel University, USA
Manisha Dhanke, IBM India Software Labs, India

Registration Chair

Susamma Barua, California State University, Fullerton, USA

Steering Committee

P. Anandan, Microsoft Research, India
David A. Bader, Georgia Institute of Technology, USA
Ramamurthy Badrinath, HP, India
Frank Baetke, HP, USA
R. Govindarajan, Indian Institute of Science, India
Harish Grama, IBM, India
Manish Gupta, India Systems and Technology Lab, IBM, India
Vittal Kini, Intel, India
N. S. Nagaraj, Infosys, India
Viktor K. Prasanna, University of Southern California, USA (Chair)

Venkat Ramana, Cray-Hinditron, India
Sartaj Sahni, University of Florida, USA
Dheeraj Sanghi, Indian Institute of Technology, Kanpur, India
V. Sridhar, Satyam Computer Services Ltd., India
Harrick M. Vin, Tata Research, Development & Design Center (TRDDC),
    Pune, India

Program Committee

Algorithms

Mikhail Atallah, Purdue University, USA
David A. Bader, Georgia Institute of Technology, USA
Robert Germain, IBM T.J. Watson Research Center, USA
Ananth Grama, Purdue University, USA
Phalguni Gupta, Indian Institute of Technology, Kanpur, India
Bruce Hendrickson, Sandia National Laboratories, USA
Alex Pothen, Old Dominion University, USA
Sanguthevar Rajasekaran, University of Connecticut, USA
Abhiram Ranade, Indian Institute of Technology, Mumbai, India
Andrew Rau-Chaplin, University of Dalhousie, Canada
Oliver Sinnen, University of Auckland, New Zealand
Pavlos Spirakis, University of Patras, Greece
Srikanta Tirthapura, Iowa State University, USA
Denis Trystram, Informatique et Distribution (ID)-IMAG Grenoble, France

Applications

Brian Coghlan, Trinity College, Dublin, Ireland
Simon Dobson, University College, Dublin, Ireland
James Doherty, University College, Cork, Ireland
Bernd Freisleben, University of Marburg, Germany
George Gravvanis, Democritus University of Thrace, Greece
Vipin Kumar, University of Minnesota, USA
Alexey Lastovetsky, University College, Dublin, Ireland
Hyunyoung Lee, University of Denver, USA
Ami Marowka, Shenkar College of Engineering and Design, Israel
Esmond G. Ng, Lawrence Berkeley National Laboratory, USA
Marcin Paprzycki, Oklahoma State University, Tulsa, USA
Adarsh Patil, University College Cork, Ireland
Dana Petcu, Western University of Timisoara, Romania
Omer Rana, Cardiff University, UK
Joel Saltz, Ohio State University, USA
Matthew Smith, University of Marburg, Germany
Marek Tudruj, Polish-Japanese Institute of Information Technology, Warsaw, Poland

Architecture

Srini Devadas, Massachusetts Institute of Technology, USA
Sandhya Dwarkadas, University of Rochester, USA
Kanad Ghose, State University of New York Binghamton, USA
Andy Glew, Intel, USA
Sebastien Hily, Intel, USA
Mahmut Kandemir, Pennsylvania State University, USA
Christos Kozyrakis, Stanford University, USA
Sanjeev Kumar, Intel, USA
Milo Martin, University of Pennsylvania, USA
Pierre Michaud, Institut de Recherche en Informatique et Systèmes Aléatoires
    (IRISA), France
Yanos Sazeides, University of Cyprus, Cyprus
Hans Vandierendonck, University of Ghent, Belgium
Craig Zilles, University of Illinois at Urbana-Champaign, USA

Communication Networks

Jiannong Cao, Hong Kong Polytechnic University, China
Kartik Gopalan, Florida State University, USA
Lisandro Granville, Federal University of Rio Grande do Sul, Brazil
Qianping Gu, Simon Fraser University, Canada
Mathieu Latapy, LIAFA Paris, France
Xing Li, Tsinghua University, China
Bin Liu, Tsinghua University, China
Olav Lysne, Simula Research Laboratory, University of Oslo, Norway
Yavuz Oruc, University of Maryland at College Park, USA
Mohamed Ould-Khaoua, University of Glasgow, UK
Andrzej Pelc, Université du Québec en Outaouais, Canada
Congduc Pham, Universite de Pau, France
Greg Plaxton, University of Texas at Austin, USA
Jang-Ping Sheu, National Central University, Taiwan
Jingyuan Zhang, University of Alabama, USA

Mobile and Sensor Computing

Demet Aksoy, University of California at Davis, USA
Sajal K. Das, University of Texas at Arlington, USA
Marcelo Dias de Amorim, Laboratoire Informatique de Paris 6 (LIP6), France
Erol Gelembe, Imperial College, London, UK
Sandeep K. S. Gupta, Arizona State University, USA
Fredrik Manne, University of Bergen, Norway
Rajeev Muralidhar, Intel, India
Michael Neufeld, BBN Technologies, USA
Sotiris Nikoletseas, Computer Technology Institute, Patras, Greece

Thomas Noel, Université de Strasbourg, France
Stephan Olariu, Old Dominion University, USA
Christian Prehofer, DoCoMo Euro-Labs, Munich, Germany
Pedro Ruiz, University of Murcia, Spain
Stefan Weber, Trinity College Dublin, Ireland
Yang Yu, Motorola Research Laboratory, USA

Systems Software

Rumen Andonov, Institut National de Recherche en Informatique (INRIA), France
Eduard Ayguade, Universitat Politecnica de Catalunya (UPC), Barcelona, Spain
Scott Baden, University of California at San Diego, USA
Pedro Diniz, University of Southern California, Information Sciences Institute
    (ISI), Marina Del Rey, USA
Ramaswamy Govindarajan, Indian Institute of Science (IISc), Bangalore, India
Manish Gupta, IBM Yorktown Heights, USA
Ben Juurlink, Delft University of Technology, Netherlands
Uday Khedker, Indian Institute of Technology, Mumbai, India
Mitsuhisa Sato, Tsukuba University, Japan
Siang Wun Song, São Paulo University, Brazil
Michelle Strout, Colorado State University, USA
Chau-Wen Tseng, University of Maryland at College Park, USA
Frédéric Vivien, Institut National de Recherche en Informatique (INRIA) and Ecole
    Normale Supérieure de Lyon (ENS Lyon), France
Cho-Li Wang, Hong Kong University, China
Jingling Xue, University of New South Wales, Sydney, Australia

Workshop Organizers

Workshop on New Horizons in Compilers

Co-chairs

R. Govindarajan, IISc, Bangalore, India
Uday Khedker, IIT Mumbai, India
Rahul Simha, The George Washington University, USA
Bhagi Narahari, GWU, Washington, USA

Workshop on Many Core Computing

Co-chairs

S.K. Nandy, IISc, Bangalore, India
Vittal Kini, Intel Research, Bangalore, India

Workshop on Global Environment for Network Innovations (GENI): India

Chair

Vishal Mistra, Columbia University, NY, USA

Workshop on High-Speed DSP Architectures

Co-chairs

Serene Banerjee, Texas Instruments, India
C. P. Ravikumar, Texas Instruments, India

Workshop on Cutting-Edge Computing

Co-chairs

Harish K. Grama, IBM Software Lab., India
Albee Jhoney, IBM Software Lab., India

Workshop on Next-Generation Wireless Networks

Co-chairs

C. Siva Ram Murthy, IIT Madras, India
B. S. Manoj, University of California, San Diego, USA

Tutorials

Getting Going with the Grid and Its Applications

Mark Baker, The University of Reading, UK

MPI - Portable Scalable Programming for High-Performance Computing

Ewing ("Rusty") Lusk, Argonne National Labs, USA

Cluster and Parallel File Systems for High-Performance Computing Clusters

Gautam Shah, IBM Systems Development Lab., Poughkeepsie, NY, USA

List of Reviewers

In addition to the 95 PC members, the following colleagues provided reviews for HiPC 2006 papers. Their help is gratefully acknowledged.

Afrand Agah
Vijay Agneeswaran
Habib Ammari
Marcos Assuno
Isabelle Augé-Blum
Rosa M. Badia
Marinho Barcellos
Marina Blanton
Juan Botia
Ioannis Chatzigiannakis
Guillaume Chelius
Yen-Wen Chen
Hyun Jung Choe
Li-Der Chou
Julita Corbalan
Toni Cortes
Patrick Crowley
Pradip De
Chen Ding
Pierre-François Dutot
Christine Eisenbeis
Keith Frikken
Edgar Gabriel
Li Gao
Robert Germain
Sergi Girona
Maria Gradinariu

Isabelle Guén-Lassous
Sanjay Ha
Jun-Won Ho
Wei-jen Hsu
Chung-Ming Huang
Michael Huang
Neil Hurley
Jehn-Ruey Jiang
Mahesh Kallahalla
Randeep Kapoor
Chung-Ta King
Sumantra Kundu
Stefan Lankes
Vasia Liagkou
Jean Lorchat
Cristian Lumezanu
Clarissa Marquezan
Matthew Merten
Geyong Min
Julien Montavont
Nicolas Montavont
Joanna Moulierac
Manish Nair
Dimitrios Nikolopoulos
Sourav Pal
Alexander Pelov
Vincent Poirriez

Lakshmi Ramachandran
Noemi Rodriguez
Valter Roesler
Philippe Roose
Nirmalya Roy
Frode Sandnes
Loren Schwiebert
Kai Shen
Shiann-Tsong Sheu
David Simplot-Ryl
Jaime Spacco
Daniel Stefankovic
Thomas Sødring
Theocharis Theocharides
Jordi Torres
Sathish Vadhiyar
Fabrice Valois
Jin Wang
Zhijun Wang
Stefan Weber
Hsiao-Kuang Wu
Hui Wu
Min Yang
Il-Chul Yoon
Wei Zhang
Zhenghao Zhang

Table of Contents

Session II – Architectures

Session III – Network and Distributed Algorithms

Session IV – Application Software

Session V – Network Services

Session VI – Applications

Session VII – Ad-Hoc Networks

Session VIII – Systems Software

Session IX – Sensor Networks and Performance Evaluation

Session X – Routing and Data Management Algorithms

Navigability of Small World Networks

Pierre Fraigniaud

University Paris Sud, France

Abstract. The "Small World Phenomenon" a.k.a. "Six Degree of Separation Between Individuals" was identified by Stanley Milgram at the end of the 60s. Milgram's experiment demonstrated that letters from arbitrary sources and bound to
an arbitrary target can be transmitted along short chains of closely related individuals based solely on some characteristics of the target (occupation, state of
leaving, etc.). In his seminal work, Jon Kleinberg modeled and analyzed this phenomenon in the framework of "augmented networks". A network is navigable if it
can be augmented by random links so that greedy routing performs in a polyloga-
rithmic expected number of steps between any pair of nodes. This talk will survey
the recent results in this field. In particular, the connections between navigability
and low doubling dimension will be described. The possible use of the concept of
navigable networks in the framework of Grid Computing and P2P networks will
also be discussed.

Biography: Pierre Fraigniaud received the M.Sc. in Mathematics from UJF-Grenoble
(1987), and the Ph.D. degree in Computer Science from ENS Lyon (1990). He is Di-
recteur de Recherches at CNRS, leading the research group "Graph Theory and Fundamental Aspects of Communications" (GraFComm) at the Computer Science Dep. of U.
Paris Sud. His research interests include several aspects of communication networks,
parallel and distributed systems, and telecommunication systems. He is particularly
interested in routing algorithms, information dissemination problems (e.g., multicasting, broadcasting, etc.), peer-to-peer networks, and the algorithmic for mobile agents
(e.g., exploration, gathering, rendezvous, etc.). He is currently member of the Editorial Board of Theory of Computing Systems (TOCS) and Journal of Interconnection
Networks (JOIN). He recently acted as Program Chair for the 19th Int. Symp. on Distributed Computing (DISC 2005) and the 13th ACM Symp. on Parallel Algorithms and
Architectures (SPAA 2001).

Y. Robert et al. (Eds.): HiPC 2006, LNCS 4297, p. 1, 2006.
© Springer-Verlag Berlin Heidelberg 2006

Opportunities and Challenges for Future Generation Grid Research

Dennis Gannon

Professor of Computer Science & Science Director - Pervasive Computing Labs
Indiana University

Abstract. Grid systems are now a standard approach to solving problems in large scale, multidisciplinary scientific endeavors. These research groups are often geographically distributed, and to conduct their research, they need to share access to physical resources such as supercomputers, large databases, on-line instruments and distributed applications. Grid infrastructure helps solve their problems because it can provide a layer of middleware that virtualizes the access to these resources. The users see a single coherent computer system instead of a complex network of distributed resources. In most cases, users enter the grid though a "gateway" portal, which may be a web portal or a "thick" desktop client. The gateway gives the user a way to browse metadata about computational experiments to access data products, to monitor active workflows and to run applications and share results. The user can focus on the problems of science and not computer systems. All of this is made possible because of the Service Oriented Architecture (SOA) that underlies the core Grid middleware.

In this presentation, we will look at several examples of successful Scientific Grids and Gateways. We will also describe the fundamentals of the web service SOAs that works best in Grid systems. We will illustrate these ideas with an example called LEAD which is a Grid designed to improve our ability to predict meso-scale weather events such as hurricanes, typhoons and tornadoes. We will also describe how this entire approach to service virtualization is now being used in industry to better use the resources of a single, but distributed business enterprise. While a great deal of progress has been made there are many exciting and unsolved problems. As we go through the talk, we will highlight these challenges and research opportunities.

Biography: Dr. Gannon is a professor of Computer Science in the School of Informatics at Indiana University. He is also Science Director for the Indiana Pervasive Technology Labs. He received his Ph.D. in Computer Science from the University of Illinois in 1980 and his Ph.D. in Mathematics from the University of California in 1974. From 1980 to 1985, he was on the faculty at Purdue University. From 1997-2004 he was Chairman of the Indiana Computer Science Department. His research interests include software tools for high performance parallel and distributed systems and problem solving environments for scientific computation. His current work includes the design of software component architectures for multi-core and distributed systems and web service architectures for Grids and Grid Portals. He has been program chair or general

Y. Robert et al. (Eds.): HiPC 2006, LNCS 4297, pp. 2–3, 2006.

chair of a number of conferences including the International Conference on Supercomputing, Frontiers of Massively Parallel Computing, PPoPP, HPDC, Java Grande and the International Grid Conference. He was a co-founder of the Java Grande Forum and a Steering Committee member of the Global Grid Forum where he co-chaired the Open Grid Computing Environments and Open Grid Service Architecture working groups. He is one of the original architects of the Common Component Architecture and a founder of the CCA Forum.

Software Challenges for Multicore Computing

Kenneth Kennedy

John and Ann Doerr University Professor of Computational Engineering, Department of
Computer Science
Rice University

Abstract. Current technological trends have led chip manufacturers to move to
designs that include multiple processors, or cores, on each chip. In some cases,
these processors are homogeneous (e.g., Intel, AMD) and in others they are heterogeneous (e.g., Cell). It is clear is that these designs represent the future of
computational chips and they will effect enormous changes in the way software
is designed and implemented to take advantage of their power. In this talk, I will
survey issues that will be critical to making systems, particularly HPC systems
based on multicore chips usable by application developers. The talk presents a
series of "big questions" (not to be confused with "grand challenges") about software, particularly compilers and programming tools, for multicore chips. Topics
include utilization of bandwidth (both on and off chip), on-chip memory hierarchy (shared versus separate caches), methods for exploitation of parallelism (data
parallelism versus pipelining), and inter-core synchronization mechanisms. The
talk will also address the special challenges presented by on-chip heterogeneous
parallelism such as that found on the IBM Cell chip and planned for future Intel
designs. I will conclude with a discussion of my own group's preliminary research on compilers and tools for systems based on multicore chips and future
research directions for the computer science community as a whole.

Biography: Ken Kennedy is the John and Ann Doerr University Professor of Computational Engineering and Director of the Center for High Performance Software Research (HiPerSoft) at Rice University. He is a fellow of the Institute of Electrical and
Electronics Engineers, the Association for Computing Machinery, and the American
Association for the Advancement of Science. He was elected to the National Academy
of Engineering in 1990 and to the American Academy of Arts and Sciences in 2005.
From 1997 to 1999, he served as co-chair of the President's Information Technology
Advisory Committee (PITAC). For his leadership in producing the PITAC report on
funding of information technology research, he received the Computing Research Association Distinguished Service Award (1999) and the RCI Seymour Cray HPCC Industry Recognition Award (1999). Prof. Kennedy has published two books and over two
hundred technical articles and supervised thirty-seven Ph.D. dissertations on programming support software for high-performance computer systems. His current research
focuses on programming languages tools to improve the productivity of scientists and
engineers developing technical applications for complex platforms, particularly scalable parallel computers and the Grid. In recognition of his contributions to software for
high performance computation, he received the 1995 W. Wallace McDowell Award, the
highest research award of the IEEE Computer Society. In 1999, he was named the third
recipient of the ACM SIGPLAN Programming Languages Achievement Award.

Y. Robert et al. (Eds.): HiPC 2006, LNCS 4297, p. 4, 2006.

Imaging-Based Systems Biology

Gene Myers

Howard Hughes Medical Institute
Maryland, USA

Abstract. Arguably the most significant contribution of the human genome project is that we can now build a recombinant construct of every gene and every promotor in C. elegans (worm), D. melanogaster (fly), M. musculus (mouse), and H. sapiens (human). These include fluorescent proteins and other markers that can be induced at controlled time points via a change in temperature, light, or chemistry. Combined with tremendous advances in light and electron microscopy in recent years, I believe we are now poised to visualize the meso-scale of the cell, and development and small organs (e.g. a fly's brain) at the resolution of individual cells. Toward this end, my group is working on a number of preliminary imaging projects along these lines. These include (a) studies of development and gene expression in worms and flies, (b) neural patterning in flies and mice, and (c) the interpretation of signals from a new sub-wavelength resolution light microscope. We describe preliminary results on limited data sets and extrapolate on what we might be able to infer from such data. We further speculate on the potential implications of such work for the future of molecular biology.

Biography: Gene Myers is one of the seven initial investigators to sign on as group leaders at the new Janelia Farm Research Campus of the Howard Hughes Medical Institute. Gene comes to the JFRC from UC Berkeley where he was on the faculty of Computer Science starting in 2003. He was formerly Vice President of Informatics Research at Celera Genomics for four years where he and his team determined the sequences of the Drosophila, Human, and Mouse genomes using the whole genome shotgun technique that he advocated in 1996. Prior to that Gene was on the faculty of the University of Arizona for 18 years and he received his Ph.D in Computer Science from the University of Colorado in 1981. His research interests include design of algorithms, pattern matching, computer graphics, and computational molecular biology. His most recent academic work has focused on algorithms for the central combinatorial problems involved in DNA sequencing, and on a wide range of sequence and pattern comparison problems. Among the tools he has developed are Blast – a widely used tool for protein similarity searches, FAKtory – a system to support DNA sequencing projects, Anrep – a pattern matching language for applications in molecular biology, and Mac- & PC-Molecule – a molecular visualization tool for Apple and Wintel computers He was awarded the IEEE 3rd Millenium Acheivement Award in 2000, the Newcomb Cleveland Best Paper in Science award in 2001, and the ACM Kanellakis Prize in 2002. He was voted the most influential in bioinformatics in 2001 by Genome Technology Magazine and was elected to the National Academy of Engineering in 2003. In 2004 he won the International Max-Planck Research Prize.

Y. Robert et al. (Eds.): HiPC 2006, LNCS 4297, p. 5, 2006.
© Springer-Verlag Berlin Heidelberg 2006

Advanced Scientific Computing: An Extraordinary Tool for Extraordinary Science

Jeffrey Wadsworth

Oak Ridge National Laboratory
USA

Abstract. Work is under way throughout the world to realize the promise of petascale computing and to complete the emergence of simulation as the third leg of science, joining theory and experiment. By the end of this decade, our ability to attack previously unsolvable problems will provide the basis for transformational advances in science and engineering that will enable us to address global challenges in energy, environment, and national security. In the United States, three government agencies are pursuing petascale initiatives: the Defense Advanced Research Projects Agency, the National Science Foundation, and the U.S. Department of Energy (DOE). I will describe the new capability for high-end science that is being fielded by DOE's Leadership Computing Facility at Oak Ridge National Laboratory and discuss how this capability will be applied to such computationally challenging problems as climate modeling and prediction, astrophysics, nuclear fusion, systems biology, and materials design at the nanoscale.

Biography: As Director of Oak Ridge National Laboratory, Jeffrey Wadsworth is responsible for the management of the U.S. Department of Energy's largest multi-purpose science and energy laboratory, with a staff of 4150 and an annual budget of more than $1 billion. Previously, he worked at Stanford University, Lockheed Missiles and Space Company, and Lawrence Livermore National Laboratory. He earned baccalaureate and doctoral degrees from the University of Sheffield in England, which has also awarded him a D. Met. for published work and an honorary D. Eng. Dr. Wadsworth has published more than 275 papers in the open scientific literature on a wide range of materials science and metallurgical topics. He is the author of Superplasticity in Metals and Ceramics (Cambridge, 1997) and holds four U.S. patents. He is an elected Fellow of the American Association for the Advancement of Science, ASM International, and The Minerals, Metals & Materials Society (TMS) and a member of the Materials Research Society and the American Ceramic Society. He holds honorary professorships at Central South University, Changsha, China, and the University of Science and Technology Beijing, China. In January 2005, he was elected to membership in the National Academy of Engineering.

Y. Robert et al. (Eds.): HiPC 2006, LNCS 4297, p. 6, 2006.
© Springer-Verlag Berlin Heidelberg 2006

High-Performance Computing for the Masses

Zhiwei Xu

Institute of Computing Technology, Chinese Academy of Sciences
Beijing, China

Abstract. In this talk, the speaker will review the history and trends of high-performance computing from the users' viewpoint. Evolutional milestones in workload, usage modes, programming models and systems architectures will be identified. Essential challenges and bottlenecks will be analyzed. He will highlight the newly formed e-Nation strategy for China of 2006-2020, and summarize R & D efforts in China to provide a high-performance computing infrastructure that could benefit half of the population of China.

Biography: Zhiwei Xu received his Ph.D. degree from University of Southern California in 1987. He is a professor and deputy director at Institute of Computing Technology, Chinese Academy of Sciences. His research areas include distributed computing, net-centric operating system, and high-performance computing architecture. He is on the editorial boards of Journal of Grid Computing, Journal of Parallel and Distributed Computing, and IEEE Transactions on Computers.

Y. Robert et al. (Eds.): HiPC 2006, LNCS 4297, p. 7, 2006.
© Springer-Verlag Berlin Heidelberg 2006

Conquering Complexity in Information Systems

Harrick Vin

Tata Research, Development and Design Center
Tata Consultancy Services
Pune, India

Abstract. The complexity of large-scale "information plants" - consisting of a number of hardware and software components - has been increasing rapidly and is fast approaching a barrier. I argue that continuous "evolution" is a key contributor to this complexity. Information plants evolve to accommodate new software functionalities, hardware technology, application and user requirements, as well as changes in operating conditions (workload, faults, etc.). Today, evolving information plants in a timely manner while maintaining desired levels of performance, stability, and security is an art; system evolution tasks are manual and intuition-based. In this talk, I will illustrate, through examples, the complexity resulting from evolution in modern information systems, and advocate a broad research agenda in computing to conquer this complexity through managed evolution.

Biography: Dr. Harrick Vin is the head of the Systems Research Lab (SRL) at the Tata Research, Development and Design Center (TRDDC) in Pune, India. TRDDC is an R&D division of Tata Consultancy Services (TCS). Harrick's research interests are in the areas of networks, operating systems, distributed systems, and multimedia systems. Harrick received his Ph.D. in Computer Science from the University of California at San Diego. He has co-authored more than 100 papers in leading journals and conferences. Harrick is a recipient of several awards including the Faculty Fellow in Computer Sciences, Dean's Fellowship, National Science Foundation CAREER award, IBM Faculty Development Award, Fellow of the IBM Austin Center for Advanced Studies, AT&T Foundation Award, National Science Foundation Research Initiation Award, IBM Doctoral Fellowship, NCR Innovation Award, and San Diego Supercomputer Center Creative Computing Award. He has served on the Editorial Board of ACM/Springer Multimedia Systems Journal, IEEE Transactions on Multimedia, and IEEE Multimedia. He has been a guest editor for IEEE Network. He has served as the program chair, the program co-chair, and a program committee member for several conferences.

Y. Robert et al. (Eds.): HiPC 2006, LNCS 4297, p. 8, 2006.
© Springer-Verlag Berlin Heidelberg 2006

Algorithmic Ramifications of Prefetching in Memory Hierarchy

Akshat Verma[1] and Sandeep Sen[2,*]

[1] IBM India Research Lab
akshatverma@in.ibm.com
[2] Dept of Computer Science and Engineering, IIT Delhi
ssen@cse.iitd.ernet.in

Abstract. External Memory models, most notable being the I-O Model
[3], capture the effects of memory hierarchy and aid in algorithm design.
More than a decade of architectural advancements have led to new features not captured in the I-O model - most notably the prefetching capability. We propose a relatively simple *Prefetch model* that incorporates
data prefetching in the traditional I-O models and show how to design
algorithms that can attain close to peak memory bandwidth. Unlike (the
inverse of) memory latency, the memory bandwidth is much closer to the
processing speed, thereby, intelligent use of prefetching can considerably
mitigate the I-O bottleneck. For some fundamental problems, our algorithms attain running times approaching that of the idealized Random
Access Machines under reasonable assumptions. Our work also explains
the significantly superior performance of the I-O efficient algorithms in
systems that support prefetching compared to ones that do not.

1 Introduction

Algorithm analysis and design are based on models of computation that must
achieve a balance between abstraction and fidelity. The incorporation of memory hierarchy issues in the traditional Random Access Machine (RAM) model
took some time [1,2,4,3,14,11], eventually culminating in the I-O model of Aggarwal and Vitter[3]. The I-O model derives wide acceptance from its simplicity.
It manages to redress the lack of distinction among the memory access times of
the different tiers of memory in the RAM model and has been used extensively
in the design of various external memory algorithms [3,18,9]. Further work in
this direction led to the Cache model of Sen, Chatterjee and Dumeer [20] that
addresses the algorithm design issues under the constraints of limited associativity in memory hierarchy. These results show an inherent gap in complexities
of several problems between the RAM and the I-O models.

* Part of the research done when the author was visiting University of Connecticut
and supported by NSF Grant ITR-0326155.

Y. Robert et al. (Eds.): HiPC 2006, LNCS 4297, pp. 9–21, 2006.
© Springer-Verlag Berlin Heidelberg 2006

1.1 Motivation

The I-O models report their results in terms of the number of I-Os thus making an implicit assumption that every I-O has same cost. Dementiev and Sanders [9] present an efficient sorting algorithm in terms of the I-O time thus moving to to a more practical metric. However, all these models assume that the cost of I-O (in terms of time) is a fixed constant. A close look at memory access reveals that I-O cost can be broken into latency (time spent in seeking to the right location) and the transfer time (time spent in actual transfer of the block). Hence, a latency L is incurred before the start of the transfer of a block. A large number of techniques (like increase in bus bandwidth, advances in semiconductor technology) have led us to a stage where primary memory bandwidth is approaching processor speed. Similarly, the disk transfer times have significantly improved over the years where packing density and disk rotation speeds have greatly increased. Techniques like using parallel disks have also been useful to ensure that I-O bandwidth approaches processor speed [23,9,6]. Unfortunately the access latencies for both primary and secondary memory have not reduced in tandem with increase in memory bandwidth and processor speed, and I-O bottleneck is dominated by it.

The traditional approach for speeding up memory access has been to minimize the number of I-O's to reduce the total latency and its parallelization on multiple disk architectures. Pipelining and Prefetching support in contemporary architectures (including Pentium IV) [7,10] present another possibility, namely, overlapping access latencies (For a survey on system-level prefetching support, refer to [19] and references therein). Similarly, read-ahead caches on disks prefetch data in advance to hide the latency component. Also, disk scheduling algorithms like SCAN hide latency of queued requests while serving a block. Because of the huge difference in magnitude between latency and transfer times, the potential savings in I-O times are immense. As an example, consider a scenario where we read $1,000,000$ 10KB blocks sequentially where each read has a latency of 10 ms and transfer time of 0.1 ms per block. In the traditional I-O model we would incur a latency for each read and the total I-O time would be approximately 3 hours, whereas, with prefetching the total time is less than 2 minutes as we incur the latency only for the first block. On current systems with system-level prefetching [19], such sequential reads take significantly less time than predicted by the I-O model (Section 5). Moreover, many recently proposed disk scheduling algorithms strive to compensate for the lack of prefetching-awareness in algorithms by *idle waiting* at a head position or waiting to build large batches of requests before sending them to the disk controller [15,17]. In [15], the controller waits for more contiguous requests to be issued after serving a request, thus introducing idling. If such requests are issued before time using prefetching, such idling is eliminated thus increasing disk efficiency directly. Hence, incorporating prefetching in the I-O model not only ensures that the I-O times predicted by the model are meaningful but may also improve disk efficiency.

Moreover, algorithms designed for single cost I-O models [3,12,5] may not translate to optimal algorithms in a 2-cost prefetch model. Note that algorithms in the I-O model do not specify the relative order in which blocks are fetched.

Hence, such algorithms need to find an ordering of I-Os that is prefetch efficient (formally defined in Section 4). More importantly, computing a prefetch-efficient order of blocks ahead of time for certain problems (e.g. sorting) necessitates devising new techniques. Our experimental studies in Section 5 confirm that algorithms optimal in traditional I-O models but not prefetch aware may perform very poorly as compared to prefetch-aware algorithms.

1.2   Relationship with Other Models

In order to take advantage of prefetching, we work with a two-level memory model where the I-O cost is in terms of two parameters - L is the time to access the memory location and B_M is the transfer time for a memory block of size B. The request for accessing a block from the slower memory can be sent out prior to its actual use and moreover several such requests can be pipelined. This model has some similarities to [2] where $L = f(x_\ell)$ is a (monotonic) function of the last address x_ℓ in a block transfer and additional cost 1 thereafter. The authors had derived bounds for different families of the function f. By choosing a step function L (left open by [2]), in conjunction with other parameters of the I-O model our algorithms exploit features hitherto not analyzed. We would like to note that some of the recent experimental studies of external sorting [9] make extensive use of parallel threads which may invoke prefetching at a system level. In another approach the authors [6] look at *oblivious* sorting algorithms on a multi-disk machine where prefetching could turn out to be extremely relevant.

The design of cache-oblivious [12] algorithms has drawn a lot of attention and it may be pertinent to mention that it does not automatically take care of the issue addressed by us. The basic algorithm should be inherently recursive in nature and must be **aware** of the size of the internal memory to pipeline memory access. This is to avoid latency for every block of a (sufficiently large) sub-problem that can fit inside the internal memory. It is an interesting question, if every cache-oblivious algorithm can be converted into a prefetch-efficient algorithm by adding an extra (cache-aware) pipelined memory transfer step. A related area of study that has attracted a great deal of attention is the design of efficient system-level prefetching techniques independent of the algorithm running on the system [19,16]. These techniques identify regular data access patterns amongst the I-O requests and prefetch data accordingly. However, if the algorithms running on such systems are unaware of prefetching, the system-level prefetchers may not be effectively used. Hence, we design algorithms that efficiently use such prefetching support to reduce I-O times.

2   The Prefetch Model and Some Preliminaries

Aggarwal and Vitter [3] proposed an I-O model for an input of size N that reads blocks of size B, can transfer D blocks concurrently and works with a fast memory of size M. We formalize an extension of the I-O models to capture prefetching by introducing the following additional parameters -

- B_M as the time[1] to transfer one block of memory, where $B_M/B \geq 1$.
- L as the normalized latency in transferring from slow memory to fast memory. We always use L to denote read latency unless otherwise stated. In cases where we deal with both read and write latency, L_r denotes read latency and L_w denotes write latency.
- There is an explicit prefetch instruction and the Prefetch Latency is L.

A large block-size does not reduce the *Prefetch Model* to I-O model as the algorithms for $B_M = L$ may not be optimal. To simplify our presentation, we initially ignore the parameter D from the prefetch model and propose optimal algorithms in a single disk prefetch model. In this paper we make the following assumptions that are consistent with the existing architectures. (i) $N > M > B$ (ii) $(M/B)B_M > L$ [2] (iii) N, M, B are of the form 2^i to simplify analysis - the asymptotic bounds are not affected. The fast memory size (be it cache or registers) M is typically much larger than the size of the cache line B. Moreover, the latency incurred, L, is typically much smaller than the time to load the internal memory completely ($= \frac{M}{B}B_M$). Prefetch latency is typically same as the memory latency L or may differ from it by one or two cycles.

Definition 1. *The latency l_i of block i is defined as the additional latency that is incurred because of block i. Hence, if reads for block $(i-1, i)$ are given at (t_{i-1}, t_i) and the blocks are available at times (e_{i-1}, e_i), then $l_i = e_i - \max\{t_i, e_{i-1}\}$, where e_i's are ordered.*

Note that for blocking reads, this definition of latency is the same as $l_i = e_i - t_i$, which is the one commonly used. We modify the usual definition in order to define cumulative latency of a $m - block$ L_m simply as sum of the latency of the m blocks, where an $m - block$ denotes a set of m consecutive blocks.

We make a note here that complete control over prefetch is not realistic and, in practice, prefetching is constrained by the number of prefetch buffers, limitations due to associativity (in a Cache Model) and a streaming behavior in data access required for most forms of prefetching. Our results can be extended in a model that includes (a) limited prefetch buffers (b) small associativity (c) limited streams support for prefetching and (d) parallel disks, for which the reader is referred to [21].

Running time
We analyze the algorithms in terms of the total time that includes computation time and the I-O time. This is normalized with respect to the instruction cycle that takes unit time. The only I-Os (reads/writes) that we consider are I-Os to slow memory. Access to fast memory is counted along with the number of I-O operations. Since memory bandwidth is now within a small constant factor (2 to 4) of the processor speed, the running times that we derive have a multiplicative factor of B_M/B, which is $O(1)$ when $B_M = cB$ for some constant c.

---

[1] All the timing parameters are normalized wrt to the instruction cycle.
[2] Many of the technical results revolve around this assumption - c.f. Section 4.

2.1 Lower Bounds in the Prefetch Model

In the prefetch model, a block that has not been prefetched takes time $B_M + L$, whereas a block that has been prefetched takes B_M time. It is easy to see that if k is the minimum number of I-Os needed to solve a problem A, then $kB_M + L$ is the lower bound on total time in the prefetch model. The bound is obtained by assuming that there exists a prefetch algorithm that prefetches all but the first block. Similarly, if there exists an algorithm that uses k I-Os, then $k(L + B_M)$ is the upper bound on the I-O time by multiplying the number of I-O's by the time to transfer each block without prefetching. This upper bound is same as the *lower bound* on I-O time in the traditional I-O models and differs from the lower bound of the prefetch model by a factor of L/B_M (a factor of 1000s for typical disks). This general lower bound and $(M/B)B_M > L$ combined with the bound on the number of I-Os for individual problems [3] yields the following bounds in our prefetch model. For D disks, the bounds are divided by D.

Theorem 1. *The worst case I-O time required to sort N records and to compute any N-input FFT digraph or an N-input permutation network is $\Omega\left(\frac{N\log(1+N/B)}{\log(1+M/B)}\frac{B_M}{B}\right)$.*

Theorem 2. *The worst case I-O time required to permute N records is $\Omega(\min\{NB_M, \frac{N\log(1+N/B)}{\log(1+M/B)}\frac{B_M}{B}\})$.*

Theorem 3. *The worst case I-O time required to transpose a matrix with p rows and q columns, stored in row major order under the assumption that $M > B^2$, is $\Omega(N\frac{B_M}{B})$.*

3 Prefetch Model and PDM Algorithms

We now investigate similarities between algorithms in a Parallel Disk Model (PDM) and algorithms in the proposed Prefetch Model. We observe that both class of algorithms exploit essentially the same features in memory access. One may note that if a prefetch model algorithm can perform M/B I-Os in a pipelined fashion and hide the latency of all but the first of these M/B blocks, it would be efficient, i.e., it would take $O(B_M)$ time to perform a block I-O (since $L < (M/B)B_M)$). Similarly, a PDM algorithm with $M = DB$ (D is number of disks) needs to perform M/B I-Os concurrently and hence needs to predict the next M/B blocks required. Moreover, in both of these models, if the algorithm performs the minimal number of I-Os possible while maintaining the $O(M/B)$ pipelining or parallelism respectively, the algorithm is optimal in the respective models. This general idea has also been proposed in [22] to design efficient serial algorithms from parallel versions. We now present an emulation scheme to generate *Prefetch Model* algorithms from *PDM* algorithms using this insight.

3.1 PDM Emulation

We restrict *PDM* algorithms to only those parallel disk algorithms that deal with the case $M = DB$. The emulation works in the following manner. The sequential algorithm with prefetching performs I-O in blocks of D. It emulates the D disks as contiguous locations in D zones of the single disk. For every parallel I-O p_i performed by the PDM algorithm, let S_i be the set of D I-Os that the PDM algorithm performs concurrently. The emulation algorithm starts the prefetch of all these $|S_i|$ blocks together. When all the $|S_i|$ blocks are available in the fast memory, the emulation algorithm starts prefetch of the blocks in S_{i+1} corresponding to the next parallel I-O p_{i+1}. We show the following result (all proofs are omitted for lack of space and the reader is referred to [21]).

Theorem 4. *If the PDM algorithm performs k parallel I-Os, the corresponding sequential prefetching algorithm takes an I-O time of $O(kDB_M)$.*

A similar emulation scheme is obtained for parallel disk prefetch algorithms with the number of parallel disks $D' < D$. In a parallel disk prefetch model, we make the additional assumption that the fast memory available per disk is large enough to hide the latency for that disk, i.e., $\frac{M}{D'B}B_M > L$. Each of the D' disks now emulate D/D' disks and we have the following result.

Theorem 5. *If the PDM algorithm performs k parallel I-Os, the corresponding parallel prefetching algorithm with D' disks takes an I-O time of $O(kD/D'B_M)$.*

The above emulation scheme allows us to convert existing optimal PDM algorithms to algorithm optimal in our prefetch (sequential or parallel disk) model. It is easy to see that if a PDM algorithm is optimal in the number of parallel as well as block I-Os (i.e., it performs the minimal number of parallel I-Os as well as the total number of block I-Os across all the disks is minimum), the corresponding emulated prefetch algorithm is optimal in the prefetch model. Since the lower bound for most common problems in a PDM model is a factor D' (number of disks used) less than the lower bound in a sequential I-O model, an optimal PDM algorithm is also typically optimal in the traditional single disk I-O model. Hence, in most likelihood, such optimal PDM algorithms can be directly used to generate an optimal prefetch algorithm.

A drawback of the emulation strategy described here is that it is not easy to design theoretically optimal PDM algorithms (for $D = \Omega(M/B)$). Further, direct design often leads to simpler algorithms and also allows overlapping computation with memory access (which could save up to a factor of two).

4   Designing Optimal Algorithms Directly

The different techniques (from *prediction sequence balancing* to *sequence preservation*) employed for direct design of algorithms have a common underlying strategy: perform minimal number of I-Os in a prefetch-efficient manner, i.e., hide latency for all blocks other than the first.

Definition 2. *An algorithm that performs k I-Os is prefetch efficient if it takes I-O time $O(L + kB_M)$.*

We make a note here that the assumption $L < (M/B)B_M$ dictates the techniques that we use in designing algorithms. Observe that if $L = lB_M$, then a prefetch-efficient algorithm needs to prefetch l blocks in advance. Our assumption of a large l $(= M/B)$ covers real systems but requires our algorithms to be intelligent enough to predict the next M/B blocks and start prefetching for them. On the other hand, consider the extreme (though unrealistic) case of $l = O(1)$, where an optimal algorithm does not need to prefetch any blocks and any existing I-O optimal algorithms are optimal in this model.

We have essentially devised three techniques for designing optimal algorithms. We prove a general result for a class of algorithms called *sequence-preserving algorithms* and use it to design optimal algorithms for all straight-line algorithms considered (e.g., matrix transpose, permutation and FFT). We have devised a technique for *dynamic re-balancing of prefetched data* for algorithms that merge constant number of sequences (2-way sorts) and *prediction sequence balancing* for algorithms that merge large number of sequences (M/B-way sorts). We first present results for sequence-preserving algorithms and show its applications.

4.1   Sequence Preserving Algorithms

We define a class of straight-line algorithms that we call sequence-preserving algorithms and prove that in this class of algorithms, prefetching can hide latency. We will show later that many straight-line algorithms fall in this class. We begin with a technical lemma and some definitions.

Lemma 1. *For any set of k pre-determined block reads, the total time needed is $O(L + kB_M)$.*

Definition 3. *An instruction I_i precedes I_j in an algorithm A (i.e. $I_i < I_j$), iff I_i is executed before I_j in A.*

We define $I^{w,i}$ and $I^{r,i}$ as the ordered sets consisting of all the instructions that write and read respectively from memory location s_i, where the order is based on their usage time in A.

Definition 4. *The neighbourhood set N_I is defined as a set containing all the tuples of the form $\{I_1, I_2\}$ such that $I_1, I_2 \in \{I^{w,i} \cup I^{r,i}, I^{w,j} \cup I^{r,j}\}$ for some i, j and $\nexists I_3 : I_3 \in (I^{w,i} \cup I^{r,i} \cup I^{r,j} \cup I^{w,j})$ and $I_1 < I_3 < I_2$ or $I_2 < I_3 < I_1$.*

The neighbourhood set of an algorithm A consists of tuples $\{I_1, I_2\}$ of instructions such that I_1 and I_2 access (read or write) memory locations s_i and s_j at times T_1 and T_2 respectively. Also, none of the instructions in A executed between T_1 and T_2 access either of the two memory locations. We also define for all instructions of the form $I_m \in I^{w,j}$, ${}^*I_m$ as the last instruction in $I^{r,j}$ s.t. ${}^*I_m < I_m$ and I_m^* as the first instruction in $I^{r,j}$ s.t. $I_m < I_m^*$.

Definition 5. *A straight-line algorithm A is sequence preserving iff for all I_1 and I_1^*, s.t.(i) $\exists \{{}^*I_2, I_1\} \in N_I$ or $\exists \{{}^*I_2, {}^*I_1\} \in N_I$ and (ii) ${}^*I_2 < I_1$; then (a)I_2 exists,   (b)$\{I_1, I_2\} \in N_I$ (c) $I_1 < I_2 \Leftrightarrow I_1^* < I_2^*$.*

Essentially, a sequence preserving algorithm reads data in the same order as it had last written them, if it had written them earlier. Moreover, before reading any data that had been written earlier, all the reads before that write should also be written back. For the cases where any of the $\{I_1^*, I_2^*\}$ or $\{I_1, I_2\}$ are not defined, the corresponding precedence relation is assumed to hold by default. Using a constructive proof ([21]) of the following lemma, we convert existing I-O optimal sequence-preserving algorithms to prefetch-efficient algorithms.

Lemma 2. *For any I-O optimal sequence-preserving straight-line algorithm A, there exists a sequence-preserving algorithm A' such that if a write of block s_i of slow memory is made at time t, the read to block s_i of slow memory is made after at least M/B block I-Os. Also, A' performs no more I-Os than A and hence is also I-O optimal.*

We now state the key result for straight-line algorithms.

Theorem 6. *Any sequence-preserving straight-line algorithm that uses k I-Os has an equivalent algorithm that takes I-O time $O(L + kB_M)$.*

Corollary 1. *A sequence of k pre-determined reads, $k \geq M/B$, takes time $O(kB_M)$.*

Corollary 2. *A sequence of k pre-determined reads and writes, $k \geq M/B$, such that no writes follow reads, i.e., there does not exist a pair $I_1, I_2 \in \{I^{w,i}, I^{r,i}\}$ s.t.$I_1 < I_2$ for some i, takes time $O(kB_M)$.*

We have now characterized a class of algorithms such that prefetching is able to hide the latency in reading the blocks. We now specify a writing order for various I-O optimal algorithms and use Theorem 6 to devise algorithms optimal in the prefetch model.

Matrix Transpose Algorithm: We show that the following transposition-by-blocks algorithm is sequence-preserving. The algorithm transposes sub-matrices with B rows and B columns. It transposes the B rows by taking M/B rows at a time and computing the partial transposes. While writing them back, the algorithm ensures that it writes them in the order they need to be read. It then iterates till the transposition is complete. After computing all the block transposes, it rearranges the blocks in the required order taking linear time. Our writing order immediately ensures that our algorithm is sequence-preserving. If $M/B > B$, then the algorithm needs only one pass of the data to compute the block transposes. This leads to the following corollary of Theorem 6.

Corollary 3. *The total time to transpose a matrix with p rows and q columns, stored in row major order, is $\Theta(\frac{B_M}{B} N)$*

Note that even in the case that $M < B^2$, the above algorithm is sequence preserving. Moreover, the number of I-Os required in that case matches the lower bound of Aggarwal and Vitter for the general case [3] . Hence, the algorithm runs in time equal to the lower bound for the problem.

General Permuting Algorithm: Note that permuting is a special case of sorting. The algorithm for permuting is thus based on the M/B-way merge-sort algorithm of [3]. The algorithm has two phases. In the first phase, we permute the elements within runs of size M. Later, we merge the permuted runs taking them M/B at a time. The difference from merge sort though is that the the next set of blocks needed is known a priori in this case. We have the following theorem for permuting.

Theorem 7. *The total time required to permute N records is $\Theta(\min\{NB_M,$* $\frac{N\log(1+N/B)}{\log(1+M/B)}\frac{B_M}{B}\})$

We have also shown that the algorithm of Cormen et al. [8] for bit-matrix-multiply/complement (BMMC) permutation is sequence preserving and hence optimal in I-O time. Similarly, the inner loop of the algorithm for FFT and Permutation network in [3] is sequence-preserving and hence optimal.

4.2   Dynamic Rebalancing: Merge Sort

We illustrate the technique of *dynamic rebalancing of prefetched data* (balancing the amount of data being prefetched across all runs) using 2-way merge sort and show that it matches the I-O time to the Compute Time, i.e., $O(N \log N)$.

Merge Sort Algorithm: The merge sort algorithm is identical to the standard 2-way merge sort. Our prefetching strategy is the one that achieves the bounds needed. We describe our prefetching algorithm for the merging procedure of merge-sort first. We define A_1 and A_2 with sizes n_1 and n_2 as the two sorted arrays that are to be merged. Without loss of generality, we assume that $n_1 = n_2$.

> **Case (i)** $n_1 > M/2$: The prefetch algorithm prefetches $M/(2B)$ blocks of both the arrays and labels them from 1 to $M/(2B)$. It then prefetches the next block from A_1, if the last element of block 1 of A_1 is smaller than the last element of block 1 of A_2. Otherwise, it prefetches the next block of A_2. If it prefetches from A_1, then it decrements the label on each block of A_1 by 1. Otherwise, it does the same for A_2. The prefetching evaluation is performed every B_M cycles and either of A_1 or A_2 is prefetched depending on the evaluation. If at any time there are no blocks of A_1 left to be prefetched, the next block to be prefetched is from A_2 and vice versa. If both A_1 and A_2 have no blocks left to be prefetched, case (ii) is followed.
> **Case (ii)** $n_1 \leq M/2$: The prefetch algorithm prefetches $(M/2B)$ blocks each from both the arrays and labels them from 1 to $M/2$. It then prefetches the next set of arrays as the blocks of A_1 or A_2 are written out to slow memory, i.e., at most once every B_M cycles.

Note that the data manipulation in merge-sort is done only in the merging procedure. Hence, the reads are done just prior to merging. The merge-sort is performed in this manner. We initially load M/B blocks of the array and merge-sort them. We do this for all the (N/M) $M/B - blocks$. Hence, after this step, we have (N/M) $M/B - blocks$ that are all sorted and have to be merged taken 2 at a time, with the size doubling in each iteration of merge-sort. The prefetching algorithm for merging described earlier is then used for the remaining iterations. We have the following optimality result for merge sort.

Theorem 8. *The total time required to sort N numbers using 2-way merge sort in the prefetch model is $O(\frac{B_M}{B} \cdot N \log N)$.*

4.3 Randomized Merge Sort with Prediction Sequence Balancing

Although the two way mergesort has $\Theta(N \log N)$ running time, it does performs more passes than an I-O optimal algorithm. It is easy to verify that the standard M/B-way Merge Sort [3] is unable to hide the latency for most blocks because an adversary may force it to prefetch blocks out of order of their use. Since it has only constant memory available per run (as opposed to 2-way Merge Sort that had $M/2$ memory available), it can hide latency only for a constant fraction of blocks. The strategy of using a prediction sequence [9] used for parallel disks works either for small N $(N/B < M)$ or requires the complication of forming large meta-blocks *a priori*, which additionally increases the constants. Similarly, Columnsort algorithm [6] uses some novel techniques to ensure that data access is deterministic but the algorithm is not defined for large N.

We also pursue the idea that if an algorithm A could predict the order in which blocks are needed in any merge phase of the merge sort algorithm, A would be prefetch efficient, i.e., A would take $O(kB_M)$ time to perform k I-Os. We show ([21]) that an $O(M)$ sized *sliding window sample* of the prediction sequence is *sufficient* for predicting, with high probability, the order in which blocks are needed, if (a) the prediction sequence is *balanced* across all the runs being merged and (b) the input is randomized. Using the above result, the algorithm maintains one prediction sequence block from each of the M/B runs being merged in memory and uses it to prefetch blocks in advance. Hence, the technique is in some sense, a refinement of balancing prefetched data, the difference being that instead of balancing data over runs (as in Sec. 4.2), we now balance the in-memory prediction sequence across runs. For further details of the optimal M/B-way merge sort (optimalSort) algorithm and proof of its optimality, please refer to [21].

Theorem 9. *optimalSort sorts N records in an I-O time of $O(\frac{N \log(1+N/B)}{\log(1+M/B)} \frac{B_M}{B})$*

5 Experimental Results

We conducted a large number of experiments to study the relative performance of algorithms optimal in traditional I-O models ([3]) but prefetch-unaware as compared to algorithms that are prefetch-efficient. Matrix transpose and merge sort were used as sample problems to demonstrate the importance of incorporating prefetching when designing the algorithms. Matrix transpose represents the class of problems where the prefetch-optimal algorithm is derived from the optimal algorithm in the I-O model by finding a prefetch-efficient ordering, whereas, the standard M/B-way merge sort does not lead to any prefetch-efficient ordering and other algorithms (e.g. *2-way sort with dynamic rebalancing*) need to be devised in prefetch model. We use the *disksim* simulation environment to study the performance of various algorithms [13]. *Disksim* has been used in a large number of studies and approximates the behavior of a modern disk closely. We chose the disk model of *Seagate cheetah4LP* disk that has been validated against the real disk and matches its average response time to within 0.8%. *Seagate Cheetah4LP* supports sequential prefetching using readahead buffers. *C-SCAN* was chosen as the scheduling algorithm, N was 640000, and B was 512 Bytes.

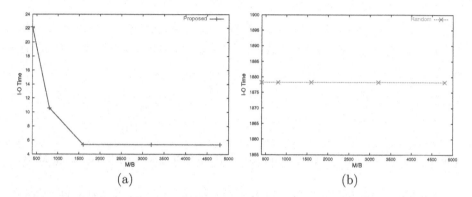

Fig. 1. Performance of (a) Prefetch-efficient and (b) Random Matrix Transpose with increasing M/B (Note that scales are different)

We performed three sets of experiments with the optimal I-O model matrix transpose. In the first set, prefetching was disabled and the algorithm picked the $B \times B$ sub-matrices in a random order. For the second set, the same algorithm was run with prefetching enabled. In the third set, the algorithm had a prefetch-efficient order (i.e. it was aware of the prefetching order and read the sub-matrices to match this order). We found that the first two sets showed no statistical difference with the second set performing marginally better in a few cases. Hence, we report only the second and third set of experiments. One may note that the random ordering of sub-matrices (Set 2), even though optimal in traditional I-O models ([3]), is not implemented in practice. We use such an ordering to demonstrate the inability of I-O models to differentiate between algorithms that have very different I-O times on real systems. Note that the I-O model predicts the running time of all the 3 sets as the running time of second set but it is clear (Fig. 1) that prefetch-efficiency makes a huge difference in performance. In fact, the performance improvement (ratio of Prefetch-Unaware disk I-O time to prefetch-efficient disk I-O time) is fairly close to the maximum achievable theoretically for this disk. The disk can prefetch up to 282 blocks ahead and hence the performance improvement due to prefetching alone is bounded by 282. However, random block accesses not only leads to more disk accesses but the cost of each disk access is also higher. We looked at the logs generated by

$DISKSIM$ and noticed that average positioning time for random block accesses is higher than that for prefetched access. Hence, we notice that the performance improvement even exceeds the bound of 282 for large m. One may also note that prefetch-unaware algorithms (Fig. 1 (b)) fails to improve the performance with additional memory as they do not use it for prefetching whereas we use the additional memory to hide the latency of more blocks.

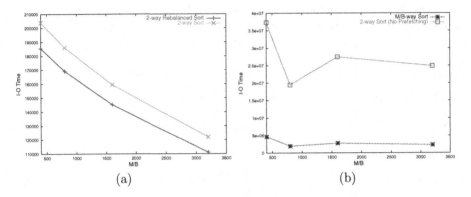

(a)   (b)

Fig. 2. Performance of (a) 2-way and (b) M/B-way Mergesort with increasing M/B

To evaluate the impact of prefetching on sorting, we studied the M/B-way mergesort that is optimal in the number of I-Os ([3]) but oblivious to prefetching. We compared it with the performance of our proposed 2-way mergesort that *dynamically rebalances data* and the standard 2-way mergesort. Prefetching was enabled for all the three algorithms. As a control experiment, we ran a *prefetch-disabled 2-way mergesort* (2-way sort with prefetching disabled). We observed that both the prefetch-enabled 2-way mergesorts comprehensively outperforms the M/B-way mergesort (Fig. 2). Note that the performance improvement for sorting does not approach the bound of 282. This is because 2-way mergesort performs more I-Os than the M/B-way mergesort. For the chosen value of parameters, a 2-way mergesort performs about 10 times more I-Os than M/B-way mergesort (evident from I-O time of Prefetch-disabled 2-way mergesort (Fig. 2(b)) as well). Hence, even though a 2-way mergesort has to perform a much larger number of I-Os, prefetching is not only able to compensate for it but allows it to outperform the prefetch-unaware algorithm by a significant margin. We also noticed that the average positioning time for the algorithms are almost same. Hence, prefetching alone attributes for the performance improvement of the 2-way sorting algorithm. We note that even the standard 2-way mergesort approaches the behavior of the *rebalanced sort* we propose and comprehensively outperforms the M/B-way mergesort. This is attributed to the fact that the simple 2-way mergesort naturally uses the sequential prefetching present in disks due to readahead caches. This may be an explanation as to why the naturally prefetch-efficient standard 2-way mergesort performs better than the more sophisticated prefetch-unaware M/B-way mergesort on many real systems.

References

1. A. Aggarwal, B. Alpern, A. Chandra, and M. Snir. A model for hierarchical memory. In *Proceedings of ACM Symposium on Theory of Computing*, 1987.
2. A. Aggarwal, A. Chandra, and M. Snir. Hierarchical memory with block transfer. In *Proceedings of IEEE Foundations of Computer Science*, pages 204–216, 1987.
3. A. Aggarwal and J. Vitter. The input/output complexity of sorting and related problems. *Communications of the ACM*, 31(9):1116–1127, 1988.
4. B. Alpern, L. Carter, E. Feig, and T. Selker. The uniform memory hierarchy model of computation. *Algorithmica*, 12(2):72–109, 1994.
5. G. S. Brodal and R. Fagerberg. On the limits of cache-obliviousness. In *Proceedings of STOC*, pages 307–315, 2003.
6. G. Chaudhry and T. H. Cormen. Getting more for out-of-core columnsort. In *Proceedings of ALENEX*, 2002.
7. T. Chen and J. Baer. Effective hardware-based data prefetching for high-performance processors. *IEEE Transactions on Computers*, 44(5):609–623, 1995.
8. T.H. Cormen, T. Sundquist, and L.F. Wisniewski. Asymptotically tight bounds for performing bmmc permutations on parallel disk systems. *SIAM Journal on Computing*, 28(1):105–136, 1999.
9. R. Dementiev and P. Sanders. Asynchronous parallel disk sorting. In *Proceedings of SPAA*, 2003.
10. N. R. Adiga et al. An overview of the bluegene/l supercomputer. In *Proceedings of Supercomputing (SC)*, 2002.
11. R. Floyd. Permuting information in idealized two-level storage. In *Complexity of Computer Computations*, pages 105–109. 1972.
12. M. Frigo, C. E. Leiserson, H. Prokop, and S. Ramachandran. Cache-oblivious algorithms. In *Proceedings of FOCS*, 1999.
13. B. Worthington G. Ganger and Y.Patt. The disksim simulation envirnoment (version 2.0). In *Available at http://www.ece.cmu.edu/ ganger/disksim/*.
14. J.-W. Hong and H. T. Kung. I/O complexity: The red-blue pebble game. In *Proceedings of the 13th Symposium on the Theory of Computing*, may 1981.
15. S. Iyer and P. Druschel. Anticipatory scheduling: A disk scheduling framework to overcome deceptive idleness in synchronous i/o. In *Proceedings of SOSP*, 2001.
16. M. Kallahalla and P. J. Varman. Optimal read-once parallel disk scheduling. In *Proceedings of IOPADS*, pages 68–77, 1999.
17. K. Lund and V. Goebel. Adaptive disk scheduling in a multimedia dbms. In *Proceedings of ACM Multimedia*, 2003.
18. U. Meyer and N. Zeh. I-o efficient undirected shortest paths. In *Proceedings of ESA*, pages 434–445, 2003.
19. K. J. Nesbit and J. E. Smith. Data cache prefetching using a global history buffer. In *Proceedings of HPCA*, pages 96–105, 2004.
20. S. Sen, S. Chatterjee, and N. Dumir. Towards a theory of cache-efficient algorithms. In *Journal of the ACM*, 2002.
21. A. Verma and S. Sen. Model and algorithms for prefetching in memory hierarchy. In *Working Draft, Available at http://www.research.ibm.com/people/a/ akshat_verma/akshat_verma.wip.html/$FILE/prefetch_main.ps*, 2005.
22. U. Vishkin. Can parallel algorithms enhance serial implementation? In *Communications of the ACM*, 1996.
23. J. Vitter and E. Shriver. Algorithms for parallel memory I: Two-level memories. *Algorithmica*, 12(2):110–147, 1994.

A Cache-Partitioning Aware Replacement Policy for Chip Multiprocessors*

Haakon Dybdahl[1], Per Stenström[2], and Lasse Natvig[1]

[1] Dept. of Computer and Information Science, Norwegian University of Science and Technology, N-7491 Trondheim, Norway
{dybdahl, lasse}@idi.ntnu.no
[2] Dept. of Computer Engineering, Dept. of Computer Engineering, Chalmers University of Technology, S-412 96 Goteborg, Sweden
per@ce.chalmers.se

Abstract. Chip multiprocessors (CMPs) usually employ shared, last-level caches to use on-chip memory resources effectively. Unfortunately, conventional replacement policies applied to shared caches fail to partition memory resources among cores to achieve an optimal execution throughput. This paper presents a novel replacement policy that dynamically estimates how many misses would be eliminated if one more block per set would be allocated to a certain processor taking into account the extra misses for some other processor. Our implementation makes novel use of *shadow tags* for the estimation. We show that it can yield 50% higher execution throughput on a 4-way CMP and in contrast to previously proposed schemes, we did not observe any noticeable degradation of performance for any application in the SPEC2000 we used.

1 Introduction

The first levels of cache in a chip multiprocessor (CMP) are often private to each processor whereas the last-level cache is usually shared. A shared cache can shield the long (and increasing) memory latency more effectively as it leverages the sharing of data among cores. However, conventional replacement policies, such as LRU, blindly victimize blocks regardless of which processor it belongs to. As the following experiment reveals, this can lead to lower than optimal global performance.

Figure 1 shows cache behavior for a limited sample period for some benchmark applications from SPEC2000. For the *mcf* application, the accesses either result in a cache hit in the most recently used (MRU) cache block in the set, or the accesses cause a cache miss. Even though the simulated cache is 16-way, all hits are to the MRU block, and hence the other 15 blocks per set are not needed. However, in a sharing situation, all the cache misses will evict cache blocks for other processors. Putting a constraint on the number of cache blocks per

* This work is partly sponsored by the HiPEAC Network of Excellence funded by EU under FP6.

Y. Robert et al. (Eds.): HiPC 2006, LNCS 4297, pp. 22–34, 2006.

processor for each set may prevent this from happening. *crafty* behaves similarly although it needs two cache blocks in each set. *gzip* and *gcc* do not need any constraints as there are few misses. Since they are reusing the cache blocks, they do not evict cache blocks for other processors.

An LRU algorithm augmented with balancing of cache partitions between processors can clearly improve performance as this can restrict one processor from using too much cache space. Partitioning cache resources across processors for shared caches have been studied before [1, 2] with algorithms based on monitoring the number of cache blocks allocated to each processor in the whole cache. Taking *mcf* and *crafty* as examples, this approach can erroneously victimize the most recently used blocks in a set. In fact, we have noticed that it provides mixed performance results.

In this paper, we take a radically different approach by monitoring and restricting the number of cache blocks allocated to a certain processor for each *set*. Our cache partitioning aware replacement policy associates with each block the ID of the processor that fetched the block into the cache. Additionally, a shadow tag per processor is associated with each set. When a block fetched by a processor is evicted, it will be kept track of by the shadow tag for that processor and set. With this basic infrastructure, the replacement policy dynamically monitors how many misses could be avoided if a processor had one more block by simply counting the hits to the LRU block and to the shadow tag for each processor. The new replacement algorithm victimizes the block associated with a processor that needs fewer blocks in favor of a processor that needs more blocks. Suh et al. [3, 4] also use counters for choosing a victim but do not use shadow tags to estimate what would be the potential gain of allocating more blocks to a processor.

Our detailed evaluation, based on most of the applications in SPEC2000, reveals that the new policy can increase the execution throughput of independent applications run on each processor core by up to 50%. Moreover, we did not see any noticeable degradation in performance with respect to any application.

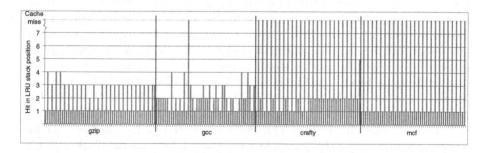

Fig. 1. The cache misses and hits in different LRU stack positions in the last-level cache for selected applications. Each access is either a hit in a position in the LRU stack or a cache miss (top line in the graph). Position 1 represents the most-recently-used (MRU) cache block in that set, position 2 is a hit in the second MRU position and so on.

We also found that making decisions on a per-set granularity usually works better than making it on a per-cache granularity as previous work suggested. Additionally, our scheme can be implemented with modest storage overhead.

The new scheme is described in Section 2. Sections 3 and 4 present the evaluation methodology and the results, respectively. Related work is discussed in Section 5 and we conclude in Section 6.

2   The New Scheme

Earlier work on partitioning cache resources in shared caches have used the total number of cache blocks per processor as a parameter for sharing [2] or other parameters for each processor such as miss rate and IPC rate [1]. Even though we also consider the total number of blocks allocated per processor, we noticed, and show in this paper, that this provides mixed results. Therefore, we present in this section a novel technique that uses the number of blocks per processor for each set as a basis for replacement decisions.

The new replacement policy is based on two guiding principles. First, we use the notion of overbooking of resources to increase utilization of all blocks in a set. Therefore, the number of blocks allocated to processors per set is larger than the number of blocks in the set. Second, the cache partitioning aware replacement policy aims at maximizing the total throughput, i.e. the number of instructions committed per time unit, by minimizing the total number of cache misses. If the total throughput is expected to increase by reducing the size of the partition for one processor and increasing it for a different processor, the change is done. In the next four sections, we present the infrastructure needed (Section 2.1), the replacement policy (Sections 2.2 and 2.3), and an analysis of the implementation costs (Section 2.4).

2.1   Structure

The hardware structures needed for the new scheme for a four core CMP are shown in Figure 2. Each cache block is extended with *processor identification* as shown in Figure 2(a). This field is updated with the value from the requesting processor every time a cache block is installed in the cache. When a cache block is evicted, the tag of the block is stored in the shadow tag table for the processor that fetched the block, see Figure 2(b). The last block that was evicted for processor 2 in set 1 is the block with tag *f*. Accesses that miss in the cache, but have a tag match in the shadow tag table, would hit in the cache if the partition had one more block in this set. This event is counted by the shadow tag's hit counter, see Figure 2(c). For example if processor 1 requests the block with tag *a* in set 0, the counter for the shadow tag's hits will increase from 10 to 11. The other counter, *hits in the LRU blocks*, is increased when a request hits in the LRU block for the involved processor in the cache. This number represents the increase in number of misses as a result of reducing the cache size by one block per set.

Index	Tag	LRU data	Cache line data	Processor ID
..	..	..	..	..

(a) Each cache block is extended to include processor ID.

Set number	Processor 1	Processor 2	Processor 3	Processor 4
0	a	b	c	d
1	e	f	g	h
..	..	..	..	..

(b) This is the structure with the shadow tags. Each cache set has one shadow tag per processor.

Counter	Processor 1	Processor 2	Processor 3	Processor 4
Hits in the LRU blocks	2	3	2	3
Hits in the shadow tags	10	11	9	2

(c) There are two global counters per processor.

Description	Processor 1	Processor 2	Processor 3	Processor 4
Max. no. of blocks in set	2	3	2	3

(d) The partitioning parameters (one per processor) used by the replacement policy.

Fig. 2. The extra storage requirements for the new scheme

A constraint is associated with each processor that limits the maximum number of blocks that can be in each set, see Figure 2(d). These values are used by the replacement algorithm to select a cache line for eviction.

2.2   The Partitioning Aware Replacement Policy

Algorithm 1 describes the new replacement policy for sharing cache space with the constraints from Figure 2(d). The search for a victim block starts at the bottom of the LRU stack (step 2) and steps the LRU stack towards the MRU block. If the processor that owns the cache block has too many cache blocks within the set (step 4), this block is chosen for eviction (step 5). If no block is found, the LRU cache block is evicted (step 8).

Algorithm 1. Pseudo code for finding a block for eviction. The function returns the position of the block to evict.

1: **function** FIND BLOCK TO EVICT
2:     **for** $LRU\_stack\_pos \leftarrow$ number of blocks per set, 1 **do**
3:         $proc\_id \leftarrow get\ processor\ that\ owns\ block(LRU\_stack\_pos)$
4:         **if** $max\ no\ of\ blocks\ in\ set[proc\_id] < no\ of\ blocks\ in\ set(proc\_id)$ **then**
5:             **return** LRU_stack_pos
6:         **end if**
7:     **end for**
8:     **return** $position\ of\ LRU\ block$
9: **end function**

2.3   Balancing the Cache Partition Sizes

The algorithm reevaluates the partition sizes per processor (see Figure 2(d)) on a regular basis. In our experiments we use 2000 cache misses in the last-level cache to trigger a reevaluation. The processor with the highest gain for increasing the cache size, i.e. the processor with most hits to its shadow tags (see Figure 2(c)), is compared to the processor with the lowest loss of decreasing cache size, i.e. the processor with the fewest hits to its LRU block. If the gain is higher than the loss, one cache block (per set) is given to the processor with the highest gain. The counters are reset after each reevaluation period.

In the initial partitioning, the total cache capacity is shared equally among all processors. Additionally, each processor receives two extra blocks per set. This *overbooking* of resources provides slack in case processors do not distribute their data uniformly.

A different approach used by Kim et al. [1] is changing the partitioning and measuring the performance difference, and then roll back if the performance was not increased. This might work for a two processor CMP, but complexity and uncertainty grow fast with increasing processor count since all processors influence the performance of each other. We have therefore not perused this technique.

2.4   Implementation Cost

The storage requirement for the new scheme used with the architecture presented in the evaluation section is 152 kbit. This is an increase of 0.5% in the storage requirement for the last-level cache. The storage is used for shadow tags (16%) and processor IDs in the blocks (84%). The evaluation section presents a CMP architecture with 4 processors, 4 MByte 16 way last-level cache with 4096 sets and 24 bits tag (we assume a 32 bit architecture). The shadow tags require $s*p*t$ bits where s is number of sets, p is number of processors, t is bits per tag, and for our architecture this is 384 kbit. However, shadow tags are not needed for all sets as shown in later in Section 4.4. We find that monitoring only 6% of the sets is sufficient for estimating cache sizes with almost no degradation of performance, and the number of bits is reduced downto 24 kbit. The field for the ID of the processor that fetched the block requires $\log_2 p$ bits per block, and this ID is required for every block. Our architecture with $4096 * 16$ blocks requires totally 128 kbit for this field. Finally, the storage for the two counters and one register per processor, sums up to 96 bits ($p * 3 * w$) if each register/counter (w) is 8 bit.

Most of the required logic is simple and can be allowed to be rather slow since accesses to main memory take hundreds of clock cycles. The fastest logic required is for indicating hits in the LRU block which requires access to the processor ID of all blocks in the set. The cycle time for this logic has to be as fast as a cache hit, but the logic can be pipelined (there is no need to immediately update the counter).

The physical placement of the shadow tags could be close to the cache tags. However, the latency requirement for these tags is more relaxed than for the cache tags. The shadow tags can therefore be moved further away from the processor and may operate at a lower voltage and hence consume less power.

3 Methodology

Simulation is used to compare the efficiency of the new scheme with a conventional LRU scheme and to earlier cache partitioning methods. The simulated architecture is shown in Figure 3. The baseline chip multiprocessor (CMP) architecture has four processors that share the last-level cache. We use a detailed

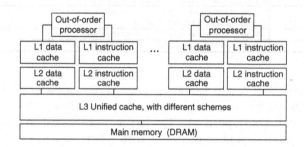

Fig. 3. Simulated architecture

pipeline-level clock cycle-accurate out-of-order execution model simulator with non-blocking caches to get statistics on improvements in instructions-per-cycle (IPC). The model is based on SimpleScalar version 3 [5], but is extended to simulate the new schemes and CMP configurations including congestion to main memory.

The baseline parameters for the simulator are shown in Table 1. All of the *SPEC2000* benchmark applications were used as workload with the reference data sets except for two. The simulator had compatibility problems with *vortex* and *sixtrack*, and they are not included in the experiments.

We create multiprogrammed workloads for our CMP architecture as follows. In each experiment, four randomly picked applications are run in parallel. Each application is randomly forwarded with cache system enabled between 0.5 and 1.5 billion instructions and then two hundred millions cycles are simulated.

4 Evaluation

Several of the SPEC2000 applications have a small working set which more or less fits into the L1 and L2 cache. These applications are not sensitive to enhancements of the last-level (L3) cache. The goal of our scheme is to improve performance for the applications which are sensitive to the performance of the last-level cache. Additionally, the scheme should be robust and should not degrade the performance for any application. We first classify the applications with

Table 1. Parameters for the simulated processors

Parameter	Value
Register Update Unit Size	128 instructions
Load Store Queue	64 instructions
Fetch queue size	4 instructions
Fetch, Decode, Issue and Commit width	4 instructions/cycle
Functional Units	4 INT ALUs, 4 FP ALUs, 1 INT Multiply/Divide, 1 FP Multiply/Divide
Branch Predictor	Comb., Bimodal 4K table, 2-Level 1K table, 10-bit history table, 4K Chooser
Branch Target Buffer	512-entry, 4 way
Mispredict Penalty	7 cycles
L1 Instruction/Data Cache	64K, 2-way (LRU), 64 B Blocks, 2/3 cycle latency
L2 Instruction/Data Cache	128/256K, 4-way (LRU), 64 B Blocks, 7/7 cycle latency
L3 Cache	4 MByte unified, 16-way (LRU), 64 B Blocks, 19 cycle latency
Main Memory	250 cycles first chunk, 4 cycles inter chunk. Chunk size 8 bytes. 9 GBytes/s teoretical limit for 4.5 GHz processor
I-TLB/D-TLB	128-entry, fully associative, 30 cycles miss penalty
Chip multiprocessor	4 processors sharing the L3 cache

respect to their sensitivity to last-level cache performance in Section 4.1. We then consider the performance improvements and robustness of the new scheme in Sections 4.2 and 4.3, respectively. The last part of the evaluation is concerned with sensitivity analysis and comparisons with related schemes.

4.1   Classification of Workloads

The number of cache accesses to the last-level cache is generated by forty experiments with random configurations of four applications. The numbers of accesses per application for all experiments are averaged and shown as a logarithmic plot in Figure 4. We classify the applications as (a) either being last-level cache intensive or (b) not depending on the last-level cache. The applications with more than 2 million last-level cache accesses are classified as last-level cache intensive. The number of cache misses could have been used to classify the applications as well, but then the applications that work very well for conventional LRU would not have been included in the evaluation of the new scheme. If the new scheme degraded performance for these applications it might not have been detected.

4.2   Throughput Improvements

This subsection evaluates the throughput improvements (i.e. total number of instructions committed) for the last-level cache intensive applications found in the previous section. The configuration is a four processor CMP. Each processor has 200 million clock cycles simulated (the clock is synchronized with the other processors) after warm up. Sixty different experiments were run with the last-level cache intensive applications. Each application is represented 18 times on average in these experiments. The total number of instructions committed for each application for the new scheme divided by the total number of instructions committed by the conventional scheme is shown in Figure 5. Positive speedups (i.e. > 1) are shown for all of the applications and the best numbers are for

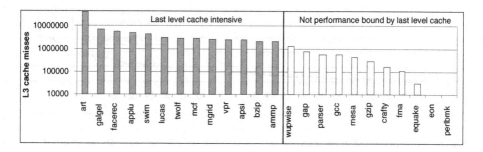

Fig. 4. Average number of last-level cache accesses for each application in a logarithmic plot

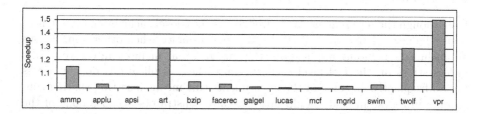

Fig. 5. Speedups for last-level cache intensive applications

vpr, *twolf* and *art* with speedups of 1.51, 1.30 and 1.29 respectively. Not all applications benefit from slightly larger caches which is one reason why not all applications run much faster with the new scheme.

Instead of considering improvements per application, improvement per experiment (that is four randomly picked applications run together) is shown in Figure 6 as *average speedup*. The total number of committed instructions for all experiments is increased by 7% for the new scheme compared to the conventional scheme. All experiments have a performance gain, i.e. the sum of committed instruction for the four benchmarks run in parallel is increased. This shows that the new scheme provides a robust and improved performance across different experiments.

A computer system is often bound by the slowest running application. In these cases, the harmonic mean is more important than the average mean [6]. The *harmonic speedup* in Figure 6 is the harmonic mean of the four applications with the new scheme divided by the harmonic mean of the same four applications with the conventional scheme. The highest speedup on the right hand side of the graph is 2.0. This shows that the new scheme is not only increasing performance for the fastest running applications, but also improves performance for the slowest running applications. This is not surprising since the goal of the scheme is to reduce the total number of cache misses and the slowest running application often has most cache misses.

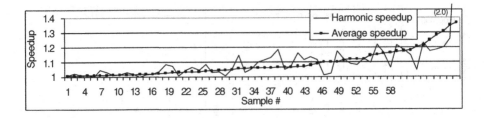

Fig. 6. The speedup for each experiment summing up all instructions committed per core in the CMP. The applications used are the last-level cache intensive category. The results are sorted on average speedup.

4.3   Robustness

Even though the new scheme works well for the last-level cache intensive applications, the new scheme should not degrade performance when other applications are included. Including applications that are not that memory intensive results in less stress on the last-level cache because these applications do not access the L3 frequently. In some of the experiments this means inserting sharing constraints in a system that should not have any constraints because none of the processors are using too much last-level cache space. However, as shown in Figure 7, the new scheme works impressively well in this setting. The figure shows speedups when combining both categories of the SPEC2000 benchmarks. Even though there are some degradation of performance of about 1% for *facerec* and 0.5% for *wupwise*, speedups of 13%, 10%, 5% and 4% are provided for *twolf, vpr, parser* and *art* respectively.

The increase in number of committed instructions per experiment is shown in Figure 8. Even though there are some experiments with fewer committed instructions, in most cases the performance is improved. The total number of committed instructions is 0.8% higher with the new scheme compared to a conventional architecture. By comparing this graph to the graph for the applications which are intensive to the last-level cache (Figure 6), we see that the performance gain is now reduced. This is due to the lower competition of the cache since many applications do not require much cache space in L3 and hence there is less room for improvement. The bandwidth required in and out of the chip is also reduced and the effect of this bottleneck becomes less significant.

4.4   Reducing Number of Shadow Tags

The experiments in the previous sections were done with four shadow tags for every set to predict marginal gains of increasing cache size for the processors. However, it is not necessary to implement shadow tags in all sets. Previous work has revealed that monitoring the sets with the lowest index works well and better than randomly generated subsets or subsets based on prime numbers [7]. The result of monitoring 1/16 of the sets with lowest index is shown in Figure 9. As shown, the median performance is slightly increased by only having shadow tags

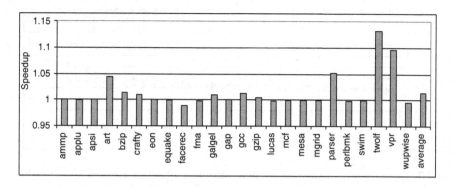

Fig. 7. Total number of instructions committed per application with the new scheme divided by the number of instructions committed per application with a conventional scheme

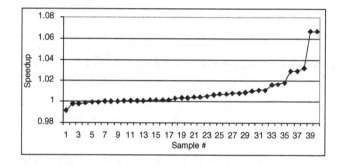

Fig. 8. The speedup for each experiment summing up all instructions committed per core in the CMP. Combination of both categories SPEC2000 applications are used.

in a subset of all cache sets. We conjecture that it is due to the higher contention and congestion in the sets with the lowest number. This makes monitoring these sets more important, and hence improves performance. There is one experiment however where reducing the number of shadow tags causes a penalty of 24%. The total number of committed instructions are reduced by 0.2% for all experiments when reducing the number of shadow tags. LRU hits are counted for in all sets,

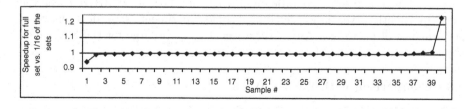

Fig. 9. The speedup of monitoring all sets vs. monitoring only 1/16 of the sets

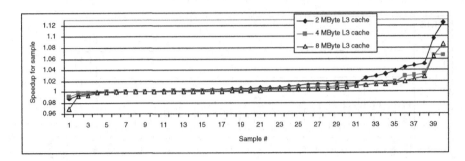

Fig. 10. The speedup for the new scheme with different L3 cache sizes. Combinations of both categories SPEC2000 applications are used as workload. Each line in the graph is sorted independently.

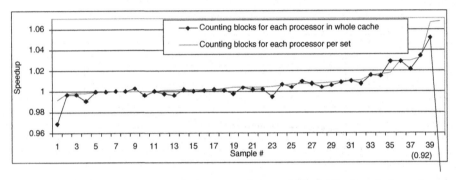

Fig. 11. Two different replacement policies: Counting blocks in the whole cache per processor vs. per set

but the numbers are normalized when compared to shadow tag hits. The cost of monitoring only $1/16 \approx 6\%$ of the sets is not very high and is discussed in Section 2.4.

4.5   Cache Size Sensitivity

The speedup of the new scheme when using both categories SPEC2000 applications for 2, 4 and 8 MByte L3 caches is shown in Figure 10. We see a higher gain for a 2 MByte than for a 4 MByte cache. This is due to more conflicts in a smaller cache and more room for improvements. For the 8 MByte cache the gains are slightly lower, as expected, and some experiments show degradation of performance of up to 3%. An 8 MByte cache is quite large for the SPEC2000 applications making constraints less effective.

4.6   Comparison with Earlier Work

Different Granularity of the Cache Constraints. Earlier attempts to partition the cache dynamically have been partitioning the cache globally and counted

the total number of blocks used per processor [1, 2]. In the new scheme we count
the number of blocks per processor *per set* when finding the block for eviction. In
Figure 11 the new scheme is compared to a modified new scheme where counting
is done globally. Lower performance is seen when counting globally for a configuration with a 16 way cache. Experiments with a 4 way cache result in lower
performance for the new scheme. This is because there is no room for constraints
with so few blocks in each set.

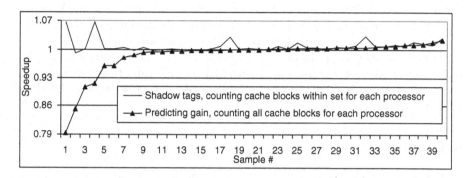

Fig. 12. Two different schemes: (a) Predicting the marginal gains with counting blocks
in the whole processor and (b) the new scheme (shadow tags and counting blocks within
set)

Not Using Shadow Tags. Earlier work did not use the extra shadow tags for
calculating gains. Suh and Rudolph have looked at dynamic cache partitioning
[2], and instead of shadow tags they used counters for hits to the LRU and
second LRU cache blocks, and a formula for estimating the marginal gains:
$M(p) = 2 * H(p) - Q(p)$, where $M(p)$ is the marginal gain for increasing the
cache size for processor p with one block per set, $H(p)$ is the number of hits in
the LRU cache blocks for all sets and $Q(p)$ is the number of hits in the second
LRU cache blocks in all sets. The results from this approximation combined with
counting globally are shown in Figure 12. Even though this approximation shows
speedup for some experiments, the overall result is a degradation of performance.
The figure includes a graph for the new scheme, and as shown the new scheme
results in a much more stable performance for the same experiments.

5   Related Work

Chishti et al. proposed a scheme where cache blocks evicted for one processor
can be stored in the cache storage for other processors if free space is available
[8]. Zhang and Asanovic [9], and Chang and Sohi [10] evaluate a combination
of shared and private caches. The goal is to have the size of the shared cache
available to all processors with the speed of the local cache. None of these studies
have provided constraints on capacity usage as in our scheme nor have they
considered the marginal gains/loss for different cache partitioning.

6   Conclusion

Conventional chip multiprocessors do not consider cache usage per processor when deciding which cache block to evict. The scheme proposed in this work establishes constraints on the number of cache blocks in each set that each processor can allocate. We have shown that our new scheme outperforms the LRU replacement policy in almost every case and hence increases performance compared to conventional architectures. Compared to previous work it has much better robustness. This is due to two contributions: (a) our use of shadow tags which is an accurate measure of increased performance for increased cache sizes and (b) putting constraints within each single set instead of across all sets.

Methods for increasing the overall performance can lead to starvation for the slowest processors. i.e. the processors with highest miss rate. However, since the goal of the scheme is to minimize the total number of cache misses, this does not happen with our scheme.

References

1. Kim, S., Chandra, D., Solihin, Y.: Fair cache sharing and partitioning on a chip multiprocessor architecture. PACT (2004)
2. Suh, G., Devadas, S., Rudolph, L.: Dynamic cache partitioning for simultaneous multithreading systems. IASTED Parallel and Dist. Computing Systems (2001)
3. Suh, G.E., Devadas, S., Rudolph, L.: A new memory monitoring scheme for memory-aware scheduling and partitioning. HPCA (2002)
4. Suh, G.E., Devadas, S., Rudolph, L.: Dynamic partitioning of shared cache memory. The Journal of Supercomputing, Vol 28, No 1 (2004)
5. Austin, T., Larson, E., Ernst, D.: SimpleScalar: An infrastructure for computer system modeling. IEEE Computer, Volume 35, Issue2 (2002)
6. Smith, J.E.: Characterizing computer performance with a single number. Communications of the ACM 31(10) (1988) 1202–1206
7. Dybdahl, H., Stenström, P.: Enhancing lower level cache performance by early miss determination and block bypassing. submitted to ICCD (2006)
8. Chishti, Z., Powell, M.D., Vijaykumar, T.N.: Optimizing replication, communication, and capacity allocation in CMPs. SIGARCH Comput. Arc. News 33(2) (2005)
9. Zhang, M., Asanovic, K.: Victim replication: Maximizing capacity while hiding wire delay in tiled chip multiprocessors. In: ISCA. (2005)
10. Chang, J., Sohi, G.S.: Cooperative caching for chip multiprocessors. ISCA (2006)

A Security-Oriented Task Scheduler for Heterogeneous Distributed Systems

Tao Xie[1] and Xiao Qin[2]

[1] Department of Computer Science, San Diego State University,
San Diego, CA 92182, USA
xie@cs.sdsu.edu
[2] Department of Computer Science, New Mexico Institute of Mining and Technology,
Socorro, NM 87801, USA
xqin@cs.nmt.edu

Abstract. High quality of security is increasingly critical for applications running on heterogeneous distributed systems, where processors operate at different speeds and communication channels have different bandwidths. Although there are a few scheduling algorithms in the literature for heterogeneous distributed systems, they generally do not take into account of security requirements of applications. In this paper, we propose a novel heuristic scheduling algorithm, or SATS, which is conducive to improving security of heterogeneous distributed systems. First, we formalize a concept of security heterogeneity for our scheduling model in the context of distributed systems. Next, we devise the SATS algorithm aiming at scheduling tasks to maximize the probability that all tasks are executed without any risk of being attacked. Empirical results demonstrate that with respect to security and performance, the proposed scheduling algorithm outperforms existing approaches for heterogeneous distributed systems.

Keywords: Security heterogeneity, heterogeneous distributed system, scheduling, degree of security deficiency, risk-free probability.

1 Introduction

Over the last decade, heterogeneous distributed systems have been emerging as popular computing platforms for computationally intensive applications with diverse computing needs [6]. To date they have been applied to security sensitive applications, such as banking systems and digital government, which require new approaches to security. Inherently, distributed systems are more vulnerable to threats than centralized systems, since it is difficult to control processing activities of the distributed systems and information can be accessed over networks. A variety of techniques like authentication [8] and access control [12] are widely used to secure distributed systems. Although these techniques can be applied to distributed systems, the conventional security techniques lack the ability to express heterogeneity in security services. Our study is intended to introduce a concept of security heterogeneity, which provides a means of measuring overhead incurred by security services in the context of heterogeneous distributed systems.

Y. Robert et al. (Eds.): HiPC 2006, LNCS 4297, pp. 35–46, 2006.

Scheduling algorithms play a key role in obtaining high performance in parallel and distributed systems [10][18]. The objective of scheduling algorithms is to map tasks onto sites and order their execution in a way to optimize overall performance. In this work we consider the issue of dynamic task scheduling. Nowadays, a wide variety of scheduling algorithms for distributed systems have been reported in the literature [1][3]. Peng and Shin proposed a new scheduling algorithm for tasks with precedence constraints in distributed systems [9]. Arpaci-Dusseau introduced two key mechanisms in implicit coscheduling for distributed systems [1]. The above algorithms were designed for homogeneous distributed systems.

In recent years, the issue of scheduling on heterogeneous distributed systems has been addressed and reported in the literature [5][16]. Ranaweera and Agrawal developed a scalable scheduling scheme called STDP for heterogeneous systems [11]. Srinivasan and Jha incorporated reliability cost, defined to be the product of processor failure rate and task execution time, into scheduling algorithms for tasks with precedence constraints [15]. A. Dogan and F. Özgüner studied reliable matching and scheduling for tasks with precedence constraints in heterogeneous distributed systems. Due to the lack of security awareness, these algorithms are not suitable for security-sensitive distributed computing applications. On the other hand, however, an increasing number of mission-critical applications with high demands of quality of security service have been emerging in various distributed systems [4][7][13][14]. For example, in a real-time stock quote update and trading web service system, each incoming request from business partners and each outgoing response from an enterprise's back-end application have deadlines and security requirements, which have to be dealt with by a server located between the business partners and enterprise backend applications [17]. Furthermore, the server can judiciously select a suitable security level from the range of security service levels, which are predefined combinations of transport and message security mechanisms. Typical security levels in a real-time quote and trading system are [17]: Routing + message security; Routing + SSL; Routing + SSL + message security; Routing + SSL + client authentication and Routing + SSL + message security + client authentication.

In our previous work, we proposed a family of dynamic security-aware scheduling algorithms for a cluster [18] and a Grid [19]. Unfortunately, these scheduling algorithms only support homogeneous computing applications, thus limiting their applicability to heterogeneous distributed systems. Hence, we are motivated in this study to formalize the security heterogeneity concept, and to propose a scheduling algorithm to improve security of heterogeneous distributed systems while minimizing computational overhead.

2 Modeling Tasks and Their Security Requirements

In this study, we consider a queuing architecture of an n-site distributed system in which n heterogeneous sites are connected via a network to process independent tasks submitted by m users. Let $M = \{M_1, M_2, ..., M_n\}$ denote the set of heterogeneous sites.

2.1 System Model

The system model, depicted in Figure 1, is composed of a task schedule queue, STAS task scheduler, and n local task queues. The function of STAS is intended to make a good task allocation decision for each arrival task to satisfy its security requirements and maintain an ideal performance in conventional performance metrics such as average response time.

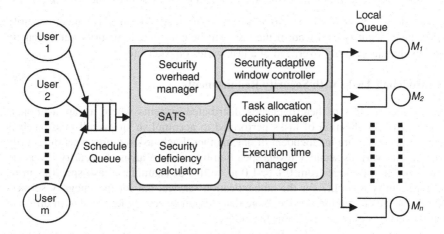

Fig. 1. System model of the SATS strategy

A schedule queue is used to accommodate incoming tasks. SATS scheduler then processes all arrival tasks in a First-Come First-Served (FCFS) manner. After being handled by SATS, the tasks are dispatched to one of the designated site $M_i \in M$ for execution. The sites, each of which maintains a local queue, can execute tasks in parallel. The main component of the system model above is SATS, which is composed of five modules: (1) Execution time manager; (2) Security overhead manager; (3) Degree of security deficiency (*DSD*) calculator; (4) Security-adaptive window controller; and (5) Task allocation decision maker. Since execution time of each task can be estimated by code profiling and statistical prediction [2], we assume that the execution time of each arrival task for each site is a prior and this information is managed in the execution time manager module. Similarly, we assume that the security overhead for each arrival task on each site is a prior, and this information is maintained in the security overhead manager module. The *DSD* calculator is used to calculate discrepancies between an arrival task's security level requirements and the security levels that each site offers. The function of security-adaptive window controller is to vary size of the window to discover a suitable site for the current arrived task so that (1) its security demands can be well met; (2) the total execution time can be as small as possible.

After retrieving information like execution time on each site, security overhead on each site, degree of security deficiency on each site and the size of security-adaptive window for the current task from the corresponding modules, the task allocation decision maker will decide which site will be assigned to the task.

Each site in the system model above is inherently heterogeneous in both computation and security. Computational heterogeneity means that for each task the execution time on different sites is distinctive. While each task has an array of security service requests, each site offers the security services with different levels. The level of a security service provided by a site is normalized in the range from 0.1 to 1.0. Suppose site M_j, offers q security services, $P_j = (p_j^1, p_j^2, ..., p_j^q)$, a vector of security levels, characterizes the security levels provided by the site. p_j^k is the security level of the kth security service provided by M_j. To meet security requirement, security overhead of the task will be considered. Security heterogeneity suggests that the security overhead of a task varies on each site.

2.2 Modeling Tasks with Security Requirements

We consider a class of heterogeneous distributed systems where an application is comprised of a collection of tasks performed to accomplish an overall mission. It is assumed that tasks are independent of one another. Each task requires a set of security services with various security levels specified by a user. Values of security levels are normalized in the range from 0.1 to 1.0 as well. For example, a task specifies in its request security level 0.7 for the authentication service, 0.3 for the integrity service, and 0.8 for the encryption service. Note that the same security level value in different security services may have various meanings.

Suppose there is a task T_i submitted by a user, T_i is modeled as a set of rational parameters, e.g., $T_i = (a_i, E_i, f_i, l_i, S_i)$, where a_i and f_i are the arrival and finish times, and l_i denotes the amount of data (measured in MB) to be protected. E_i is a vector of execution times for task T_i on each site in M, and $E_i = (e_i^1, e_i^2, ..., e_i^n)$. Suppose T_i requires q security services, $S_i = (s_i^1, s_i^2, ..., s_i^q)$, a vector of security levels, characterizes the security requirements of the task. s_i^k is the security level of the kth security service required by T_i. Please note that a way of quantitatively measuring security is still an open question to be answered. Still, we believe that the proposed security requirement model is of help in this research because of the following two reasons. First, although the measurements of security requirements and security levels are not completely objective, performance improvements of the SATS algorithm compared with the two existing approaches in terms of security (i.e., risk-free probability and degree of security deficiency) are still valid because the performance of the three algorithms is evaluated using an identical set of criteria under the same circumstance. Second, quantitatively modelling security requirements and security levels makes it possible for us to compare the security performance of different algorithms and perceive the performance discrepancies among them.

A security-aware scheduler has to make use of a function to measure the security benefits gained by each arrival task. In particular, the security benefit of task T_i is quantitatively modeled as a function of the discrepancy between security levels requested and the security levels offered. The security benefit function for task T_i on site M_j is denoted by $DSD: (S_i, P_j) \rightarrow \Re$, where $\Re$ is the set of non-negative real numbers:

$$DSD(s_i) = \sum_{k=1}^{q} w_i^k g(s_i^k, p_j^k) \, , \quad g(s_i^k, p_j^k) = \begin{cases} 0, \text{if } s_i^k \leq p_j^k \\ s_i^k - p_j^k, \text{otherwise} \end{cases}, \tag{1}$$

where $0 \leq w_i^k \leq 1$, $\sum_{k=1}^{q} w_i^k = 1$.

Note that w_i^k is the weight of the kth security service for task T_i. Users specify in their requests the weights to reflect relative priorities given to the required security services. *Degree of Security Deficiency*, or *DSD*, is defined to be a weighted sum of q discrepancy values between security levels requested by a task and the security levels offered by a site. For each task, a small *DSD* value means a high satisfaction degree. Zero *DSD* value implies that a task's security requirements can be perfectly met. That is, there exists at least one site M_j in M that can satisfy the following condition:

$$\forall k \in [1, q], s_i^k \leq p_j^j.$$

Let X_i be all possible schedule for task T_i, $x_i \in X_i$ be a scheduling decision of T_i. Given a task T_i, the *degree of security deficiency value* (*DSDV*) of T_i is expected to be minimized:

$$DSDV(x_i) = \min_{x_i \in X_i} \{DSD(x_i)\} = \min_{x_i \in X_i} \left\{ \sum_{k=1}^{q} w_i^k(g(s_i^k, p^k(x_i))) \right\} \tag{2}$$

where $p^k(x_i) = p_j^k$ if the task is allocated to site j. A security-aware scheduler strives to minimize the system's overall *DSDV* value defined as the sum of the degree of security deficiency of submitted tasks (See Equation 1). Thus, the following *DSDV* function needs to be minimized:

$$SDSD(x) = \min_{x \in X} \left\{ \sum_{T_i \in T} DSDV(x_i) \right\}. \tag{3}$$

where T is a set of submitted tasks. Substituting Equation 2 into 3 yields the following security objective function. Thus, our proposed SATS scheduling algorithm makes an effort to schedule tasks in a way to minimize Equation 4:

$$SDSD(x) = \min_{x \in X} \left\{ \sum_{T_i \in T} \left(\min_{x_i \in X_i} \left\{ \sum_{k=1}^{q} w_i^k(f(s_i^k, p^k(x_i))) \right\} \right) \right\}. \tag{4}$$

Since the degree of security deficiency for task T_i merely reflects the security service satisfaction degree experienced by the task, it is inadequate to measure quality of security for T_i during its execution. Therefore, we derive in this section the probability $P_{rf}(T_i, M_j)$ that T_i remains risk-free during the course of its execution.

The quality of security of a task T_i with respect to the kth security service is calculated as $\exp\left(-\lambda_i^k \left(e_i^j + \sum_{l=1}^{q} c_{ij}^l(s_i^l)\right)\right)$ where λ_i^k is the task's risk rate of the

kth security service, and $c_{ij}^l(s_i^l)$ is the security overhead experienced by the task on site j. The risk rate is expressed as:

$$\lambda_i^k = 1 - \exp\left(-\alpha(1 - s_i^k)\right)$$ (5)

Note that this model assumes that risk rate is a function of security levels, and the distribution of risk-free for any fixed time interval is approximated using a *Poisson* probability distribution. The risk rate model is just for illustration purpose only. Thus, the model can be replaced by any risk rate model with a reasonable parameter α.

The quality of security of task T_i on site M_j can be obtained below by considering all security services provided to the task. Consequently, we have:

$$P_{rf}\left(T_i, M_j\right) = \prod_{k=1}^{q} \exp\left(-\lambda_i^k\left(e_i^j + \sum_{l=1}^{q} c_{ij}^l(s_i^l)\right)\right) = \exp\left(-\left(e_i^j + \sum_{l=1}^{q} c_{ij}^l(s_i^l)\right)\sum_{k=1}^{q}\lambda_i^k\right).$$ (6)

Using equation 6, we obtain the overall quality of security of task T_i in the system as follows,

$$P_{rf}(T_i) = \sum_{j=1}^{n}\left\{P[x_i = j] \cdot P_{rf}(T_i, M_j)\right\} = \sum_{j=1}^{n}\left\{p_{ij} \cdot \exp\left(-\left(e_i^j + \sum_{l=1}^{q} c_{ij}^l(s_i^l)\right)\sum_{k=1}^{q}\lambda_i^k\right)\right\}$$ (7)

where p_{ij} is the probability that T_i is allocated to site M_j. Given a task set T, the probability that all tasks are free from being attacked during their executions is computed based on Equation 7. Thus,

$$P_{rf}(T) = \prod_{T_i \in T} P_{rf}(T_i).$$ (8)

By substituting the risk rate model into Equation 8, we finally obtain $P_{rf}(X)$ as shown below:

$$P_{rf}(X) = \prod_{T_i \in T}\left\{\sum_{j=1}^{n}\left\{p_{ij} \cdot \exp\left(-\left(e_i^j + \sum_{l=1}^{q} c_{ij}^l(s_i^l)\right)\sum_{k=1}^{q}\lambda_i^k\right)\right\}\right\}.$$ (9)

In summary, DSD values show us security service satisfaction degrees experienced by tasks, while risk-free probability measured by Equation 9 defines quality of security provided by a heterogeneous distributed system. In Section 5 these two metrics are used to evaluate security of distributed systems.

2.3 Heterogeneity Model

The computational weight of task T_i on site M_j (e.g., w_i^j) is defined as a ratio between its execution time on M_j and that on the fastest site in the system. That is, we have $w_i^j = e_i^j \big/ \min_{k=1}^{n}\left(e_i^k\right)$. The computational heterogeneity level of task T_i, referred to as

H_i^C, can be quantitatively measured by the standard deviation of the computational weights. Formally, H_i^C is expressed as:

$$H_i^C = \sqrt{\frac{1}{n}\sum_{j=1}^{n}\left(w_i^{avg} - w_i^j\right)^2} \text{ , where } w_i^{avg} = \left(\sum_{j=1}^{n} w_i^j\right)\Big/ n \cdot \qquad (10)$$

The computational heterogeneity of a task set T can be computed by

$$H^C = \frac{1}{|T|}\sum_{T_i \in T} H_i^C \cdot$$

There are three types of security heterogeneities. (i) Security heterogeneity of a particular task T_i indicates the difference of security requirements in the q security services requested by the task (see Equation 11). (ii) Security heterogeneity of a given security service provided by each site in a distributed system reflects the discrepancy of the offered security levels of the service in the system (see Equation 12). (iii) Security heterogeneity of a particular site M_j shows the deviation of the q security levels provided by the site (see Equation 13).

Given a task T_i and its security requirement $S_i = (s_i^1, s_i^2, ..., s_i^q)$, the heterogeneity of security requirement for T_i is measured by the standard deviation of the security levels in the vector. Thus,

$$H_i^S = \sqrt{\frac{1}{q}\sum_{j=1}^{q}\left(s_i^{avg} - s_i^j\right)^2} \text{ , where } s_i^{avg} = \left(\sum_{j=1}^{q} s_i^j\right)\Big/ n \cdot \qquad (11)$$

The security requirement heterogeneity of a task set T can be computed by

$$H^S = \frac{1}{|T|}\sum_{T_i \in T} H_i^S \cdot$$

The heterogeneity level of the kth security service in a distributed system is defined as:

$$H_k^V = \sqrt{\frac{1}{n}\sum_{i=1}^{n}\left(p_{avg}^k - p_i^k\right)^2} \text{ , where } p_{avg}^k = \left(\sum_{i=1}^{n} p_i^k\right)\Big/ n \cdot \qquad (12)$$

Using Equation 12, the heterogeneity of security services can be written as

$$H^V = \frac{1}{q}\sum_{k=1}^{q} H_k^V \cdot$$

Finally, the heterogeneity level of security services in site M_j is defined to be:

$$H_j^M = \sqrt{\frac{1}{q}\sum_{k=1}^{q}\left(p_j^{avg} - p_j^k\right)^2} \text{ , where } p_j^{avg} = \left(\sum_{k=1}^{q} p_j^k\right)\Big/ q \cdot \qquad (13)$$

2.4 Security Overhead Model

Now we consider security overhead incurred by security services. The following security overhead model includes three services, namely, encryption, integrity, and authentication [18]. The security overhead model can be easily extended to incorporated more security services.

Suppose task T_i requires q security services, which are provided in sequential order. Let s_i^k and $c_{ij}^k(s_i^k)$ be the security level and overhead of the kth security service, the security overhead c_{ij} experienced by T_i on site M_j can be computed using Equation 14. In particular, the security overhead of T_i with security requirements for the three services above is modelled by Equation 15.

$$c_{ij} = \sum_{k=1}^{q} c_{ij}^k(s_i^k), \text{ where } s_i^j \in S_i . \tag{14}$$

$$c_{ij} = \sum_{k \in \{a, e, g\}} c_{ij}^k(s_i^k), \text{ where } s_i^j \in S_i . \tag{15}$$

where $c_{ij}^e(s_i^e)$, $c_{ij}^g(s_i^g)$, and $c_{ij}^a(s_i^a)$ are overheads caused by the authentication, encryption, and integrity services [18]. Our security level assignment is reasonable because a security mechanism providing higher quality of security imposes higher overhead than mechanisms offering lower security.

The encryption overhead c_i^e of T_i on M_j is computed using Equation 16, where π_i^e is the CPU time spent in encrypting security sensitive data.

$$c_{ij}^e(s_i^e) = \pi_{ij}^e s_i^e, \text{ where } s_i^e \in S_i . \tag{16}$$

The integrity overhead can be calculated using the following equation, where l_i is the amount of security sensitive data, and $\mu^g(s_i^g)$ is a function mapping a security level into its corresponding integrity service performance.

$$c_{ij}^g(s_i^g) = l_i / \mu^g(s_i^g), \text{ where } s_i^g \in S_i . \tag{17}$$

The security level of a security mechanism can be quantitatively measured by the amount of cost needed to successfully break the mechanism. However, quantitatively measuring the security level of a security mechanism is a nontrivial research issue, and it is out of the scope of this work.

3 The SATS Algorithm

For task T_i, the earliest start time on site M_j is $es_j(T_i)$, which can be computed by Equation 18:

$$es_j(T_i) = r_j + \sum_{T_l \in W_j} \left(e_l^j + \sum_{k=1}^{q} c_{lj}^k(s_l^k) \right) \tag{18}$$

where r_j represents the remaining overall execution time of a task currently running on the jth site, and $e_l^j + \sum_{k=1}^{q} c_{lj}^k(s_l^k)$ is the overall execution time (security overhead is factored in) of waiting task T_l assigned to site M_j prior to the arrival of T_i. Thus, the earliest start time of T_l is a sum of the remaining overall execution time of the running task and the overall execution times of the tasks with earlier arrival on site $M_{j.}$ Therefore, the earliest completion time for task T_i on site M_j can be calculated as:

$$ec_j(T_i) = es_j(T_i) + e_i^j + \sum_{k=1}^{q} c_{ij}^k(s_i^k) = r_j + \sum_{T_i \in W_j}\left(e_l^j + \sum_{k=1}^{q} c_{lj}^k(s_l^k)\right) + e_i^j + \sum_{k=1}^{q} c_{ij}^k(s_i^k)$$ (19)

1.**for** each task T_i submitted to the schedule queue **do**
2.   **for** each site M_j in the system **do**
3.   Use **Eq.18** to compute $es_j(T_i)$, the earliest start time of T_i on site M_j;
4.   Use **Eq.19** to compute $ec_j(T_i)$, the earliest completion time of T_i on M_j;
5.   **end for**
6.   Sort all sites in earliest completion time in a non-decrease order
7.   **for** each site in the security-adaptive window **do**
8.   Use **Eq.1** to compute $DSD(s_i)$ /* Compute DSD for each site*/
9.   **end for**
10.   Select the site M_r that can offer the smallest DSD value and assign T_i on it
11.   Update site M_r's earliest available time es_j
12.   Use **Eq.6** to compute risk-free probability for task T_i
13.   Record start time and completion time for task T_i
14.**end for**

Fig. 2. The SATS algorithm

The SATS algorithm is outlined in Figure 2. The goal of the algorithm is to deliver optimal quality of security while maintaining high performance for tasks running on heterogeneous systems. To achieve the goal, SATS manages to minimize degree of security deficiency (see Equation 1) of each task (see Step 10 in Fig. 2) without performance deterioration. Before optimizing the degree of security deficiency of task T_i, SATS sorts all the sites in a non-decrease order in T_i's total execution time based on the information retrieved from the execution time manager and the security overhead manager (see Figure 1). Step 7 computes the degree of security deficiencies for the task on each site in the light of the security deficiency calculator. Combining the input from the security-adaptive window controller, the task allocation decision maker decides a site to which the task is allocated.

4   Simulations

Using extensive simulation experiments based on San Diego Supercomputer Center (SDSC) SP2 log, we evaluate in this section the potential benefits of the SATS algorithm. The real trace was sampled on a 128-node (66MHz) IBM SP2 from May 1998 through April 2000. To simplify our experiments, we utilized the first three months data with 6400 parallel tasks in simulation. Since the trace was sampled from a homogenous environment, to reflect the heterogeneity of the simulated distributed system, we translated the "execution time" of each task from a single value to a vector with n (number of sites) elements based on the heterogeneity model described in Section 2.3. In purpose of revealing the strength of SATS, we compared it with two well-known scheduling algorithms, namely, *Min-Min* and *Sufferage* [14]. *Min-Min* and *Sufferage* are non-preemptive task scheduling algorithms, which schedule a stream of independent tasks onto a heterogeneous distributed computing system. The two algorithms are briefly described below.

(1) MINMIN: For each submitted task, the site that offers the earliest completion time is tagged. Among all the mapped tasks, the one that has the minimal earliest completion time is chosen and then allocate to the tagged site.
(2) SUFFERAGE: Allocating a site to a submitted task that would "suffer" most in terms of completion time if that site is not allocated to it.

Table 1. Characteristics of system parameters

Parameter	Value (Fixed) - (Varied)
Number of tasks	(6400)
Number of sites	(16) – (8, 16, 32,64,128)
Task arrival rate	Decided by the trace
Size of security-adaptive window	(8) – (1, 2, 4, 8, 16)
Site security level (uniform dist.)	(0.1 – 1.0)
Task security level (uniform dist.)	(0.1 – 1.0)
Weights of security services	Authentication=0.2, Encryption=0.5, Integrity =0.3

4.1 Simulator and Simulation Parameters

The parameters of sites in the simulated distributed system are chosen to resemble real-world workstations like IBM SP2 nodes. Table 1 summarizes the key configuration parameters of the simulated distributed system used in our experiments.

We modified the trace by adding a block of data to be secured for each task in the trace. The size of the security-required data assigned to each task is controlled by a uniform distribution (see Table 1). Although "task number", "submit time", and "execution time" of tasks submitted to the system are taken directly from the trace, "size of data to be secured", "number of sites", "computational heterogeneity", and "security heterogeneity" are synthetically generated in accordance with the above model since they are not available in the trace. The performance metrics we used include: *risk-free probability* (see Equation 9), *degree of security deficiency* (see Equation 1), *site utilization* (defined as the percentage of total task running time out of total available time of a given site), *makespan* (the latest task completion time in the task set), *average response time* (the response time of a task is the time period between the task's arrival and its completion and the average response time is the average value of all tasks' response time), *slowdown ratio* (the slowdown of a task is the ratio of the task's response time to its service time and the slowdown ratio is the average value of all tasks' slowdowns).

4.2 Overall Performance Comparisons

The goal of this experiment is two fold: (1) to compare the proposed SATS algorithm against the two heuristics, and (2) to understand the sensitivity of SATS to the size of security-adaptive window.

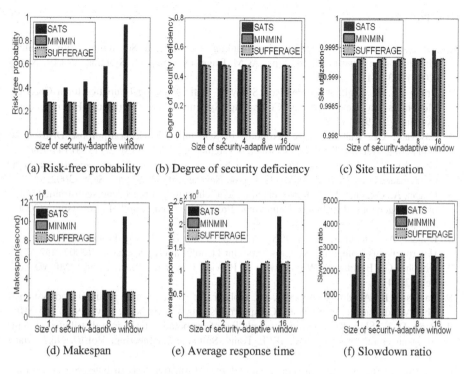

(a) Risk-free probability   (b) Degree of security deficiency   (c) Site utilization

(d) Makespan   (e) Average response time   (f) Slowdown ratio

Fig. 3. Performance impact of size of security-adaptive window

Figure 3 shows the simulation results for the three algorithms on a distributed system with 16 sites. Since MINMIN and SUFFERAGE do not have a security-adaptive window, their performance in all six metrics keeps constant. We observe from Figure 3a that SATS significantly outperforms the two heuristics in terms of risk-free probability, whereas MINMIN and SUFFERAGE algorithms exhibit similar performance.

5 Summary and Future Work

In this paper, we considered the security requirements of applications in the context of task scheduling in heterogeneous distributed systems because an increasing number of applications running on heterogeneous distributed systems requires not only descent scheduling performance but also high quality of security. To solve this problem, we proposed a security-adaptive scheduling heuristic that is based on the concept of security heterogeneity. Experimental results demonstrate that our strategy outperforms existing approaches in both security and performance. In future research, the heuristic will be extended to schedule parallel applications.

References

1. Arpaci-Dusseau, A.C.: Implicit Coscheduling: Coordinated Scheduling with Implicit Information in Distributed Systems. ACM Trans. on Computer Systems, Vol.19, No.3, (2001) 283-331
2. Braun , T. D. et al.: A Comparison Study of Static Mapping Heuristics for a Class of Meta-tasks on Heterogeneous Computing Systems. Proc. Workshop Heterogeneous Computing, (1999) 15-29
3. Casavant, T.L, Kuhl, J.G.: A Taxonomy of Scheduling in General-purpose Distributed Computing Systems. IEEE Trans. Software Engineering, Vol.14, No.2, (1988) 141-154
4. Connelly, K., Chien, A.A.: Breaking the barriers: high performance security for high performance computing. Proc. Work-shop on New security paradigms, Sept. (2002)
5. Dogan, A., Özgüner, F.: Reliable matching and scheduling of precedence-constrained tasks in heterogeneous distributed computing. Proc. Int'l Conf. Parallel Processing, (2000) 307-314
6. Dogan, A., Özgüner, F.: LDBS: A Duplication Based Scheduling Algorithm for Heterogeneous Computing Systems. Proc. Int'l Conf. Parallel Processing, (2002) 352-359
7. Donoho, G.: Building a Web Service to Provide Real-Time Stock Quotes. MCAD.Net, Feb. (2004)
8. Lampson, B., Abadi, M., Burrows, M., Wobber, E.: Authentication in distributed systems: Theory and practice. ACM Trans. Computer Systems, Vol.10, No.4, Nov. (1992) 265-310
9. Peng, D.-T., Shin, K.G.: Optimal scheduling of cooperative tasks in a distributed system using an enumerative method. IEEE Trans. Software Engineering, Vol.19, No.3, Mar. (1993) 253-267
10. Petrini F., Feng, W.-C.: Scheduling with Global Information in Distributed Systems. Proc. 20th Int'l Conf. Distributed Computing Systems, April (2000) 225 – 232
11. Ranaweera, S., Agrawal, D.P.: Scheduling of Periodic Time Critical Applications for Pipelined Execution on Heterogeneous Systems. Proc. Int'l Conf. Parallel Processing, Sept. (2001) 131-138
12. Sandhu, R.S. et al.: Role-Based Access Control Models. IEEE Computer, Vol.29,No.2, (1996) 38-47
13. Son, S.H., Mukkamala, R., David, R.: Integrating security and real-time requirements using covert channel capacity. IEEE Trans. Knowledge and Data Engineering. Vol.12, No.6, (2000) 865 – 879
14. Song, S., Kwok, Y.-K., Hwang, K.: Trusted Job Scheduling in Open Computational Grids: Security-Driven Heuristics and A Fast Genetic Algorithms. Proc. Int'l Symp. Parallel and Distributed Processing, (2005)
15. Srinivasn, S., Jha, N.K.: Safety and Reliability Driven Task Allocation in Distributed Systems. IEEE Trans. Parallel and Distributed Systems, Vol.10, No.3, Mar. (1999) 238-251
16. Topcuoglu, H., Hariri, S., Wu, M.-Y.: Performance-effective and Low-complexity Task Scheduling for Heterogeneous Computing. IEEE Trans. Parallel and Distributed Sys., Vol.13, No.3, Mar. (2002)
17. VeriSign Corp.: Simplifying Application and Web Services Security - VeriSign Trust Gateway. (2003)
18. Xie, T., Qin, X.: Scheduling Security-Critical Real-Time Applications on Clusters. IEEE Transactions on Computers, vol. 55, no. 7, July (2006) 864-879
19. Xie, T., Qin, X.: Enhancing Security of Real-Time Applications on Grids through Dynamic Scheduling. Proc. 11th Workshop Job Scheduling Strategies for Parallel Processing, MA, June (2005)

Minimizing Average Response Time for Scheduling Stochastic Workload in Heterogeneous Computational Grids

Jie Hu and Raymond Klefstad

Department of Electrical Engineering and Computer Science
University of California, Irvine, CA 92697, USA
{jieh, klefstad}@uci.edu

Abstract. Scheduling stochastic workloads is a difficult task. We analyze minimum average response time of computational grids composed of nodes with multiple processors when stochastic workloads are scheduled to the grids. We propose an algorithm to achieve minimum average response time of grids. We compare the minimum average response time of grids with the average response time of grids with load balancing scheduling in different cases. Specifically, we analyze the impact of differential processor speeds, the number of processors per node, and utilization rate of the grids on the difference between these two scheduling strategies. These analysis provide deeper understanding of average response time of grids, which will allow us to design more efficient algorithms for Grid workload scheduling.

1   Introduction

In order to take advantage of the idle nodes connected to the Internet and to meet the increasing demand of computation, especially in the field of scientific computation, Grid Computing was proposed in mid 1990s [1], [2], [3]. Availability of high performance network nodes and interconnects has enabled development of computational grids[4]. It is well known that computers on the Internet are often idle or lightly loaded. The goal of computational grids is to better utilize the computing power of all nodes over the Internet[1], [2], [3].

However, grid researchers still face challenges, such as security, resource allocation, and task scheduling[5]. Because there are numerous distributed heterogeneous computers and jobs in a grid, one important challenge is how to assign jobs to nodes efficiently. The scheduling problem involves organize, integrate, and manage the whole grid efficiently to maximize throughput or minimize completion time for the stream of jobs. The nature of variability of processing power in grid environments increases the difficulty of the scheduling problem.

Different schedulers may focus on different metrics, such as load balancing and average response time. From the point of the view of systems, load balancing is an important metric. On the other hand, from the view of users, average response time (ART) of grids is more important. However, we cannot achieve

Y. Robert et al. (Eds.): HiPC 2006, LNCS 4297, pp. 47–59, 2006.
© Springer-Verlag Berlin Heidelberg 2006

load balancing (LB) state and minimum average response time (MinART) state at the same time in heterogeneous grids. The ART of grids with LB scheduling may be poor in some cases. If ART of grids is poor, grids are less attractive because use of a grid may even slow down job execution. In addition, most grid designers utilize the relative processing speed of each node to do scheduling and to analyze the performance of schedulers[6], [7]. This method is good for LB scheduling, but it may make ART of grids worse when some nodes have multiple processors.

In order to be efficient, some grid schedulers must be given a time estimate of each submitted job. This assumption is not always realistic, as the duration of a job is rarely known before its completion. Indeed, this duration may depend on its parameters and input data. It also requires users to run or measure their jobs beforehand and provide durations to schedulers. In our work, we use stochastic processing to indicate that the duration of each job is not known but is given by a random variable with a fixed mean following the exponential distribution. Moreover, we do not assume that the submission times of jobs are known in advance. We suppose that the arrival time of each job is also drawn with a random variable following the Poisson distribution. This probabilistic approach is driven by the dynamic nature of grids, where it is not always possible to predict both the duration and arrival time of jobs.

In this paper, we study a randomized algorithm to obtain MinART by statically scheduling stochastic workload to heterogeneous grids. Unlike the work in [6] where jobs are allocated to a computing node proportional to its relative processing speed, we focus on MinART of grids. First we present how to obtain MinART by constructing an optimization problem. Then based on the result, the solution of the optimization problem is proposed. Finally, we compare the difference of ARTs of grids between MinART scheduling and LB scheduling from the work in [6]. In addition, we present impacts of differential processor speeds, the number of processors per node and utilization rate of grids on the difference. The contribution of this paper is an extensive analysis of the MinART behavior of heterogeneous computational grids.

The remainder of this paper is organized as follows. In section 2, we present our model and then propose a optimization model to obtain minimum average response time. Section 3 describes our theoretical analysis, solution, and case study. Section 4 presents initial simulation results from our theoretical analysis. Section 5 describes the related research. Section 6 gives our conclusions and plans for future work.

2 Problem Formulation

2.1 Model of Computation

Considering a heterogeneous grid as shown in Figure 1, we assume there are N nodes in the heterogeneous grid. To simplify the model, we assume network latencies and transfer times are negligible. In addition, we assume every job is independent from the others and can be processed only sequentially. In the grid, there is a central scheduler which can reach every node in the grid. Every job is initially submitted to the scheduler. The scheduler would randomly dispatch the submitted jobs to nodes, but with a distribution of probability based on some rules such as the capacity of the different nodes. We assume that the arrival rate of jobs follows the Poisson distribution and that the processing times of jobs are exponentially distributed. From Figure 1, we notice the whole system mixes up multiple $M/M/k$ models where $k \geq 1$. Because we assume that every node is locally homogeneous in our model, each node n_i has m ($m \geq 1$) processors each with relative processing speed s_i, which is the relative processing speed compared to a reference processor.

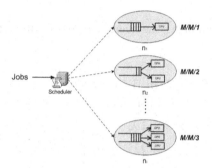

Fig. 1. A Heterogeneous Grid

Every job is nonpreemptive. Without loss of generality, we assume the job j has an execution time l_j, which is the time required by a reference processor for processing the job. Therefore, a processor with relative speed s_i would spend $\frac{l_j}{s_i}$ units of time on processing the job.

2.2   Mathematical Model

In this section, we construct an optimization problem to obtain the MinART of the model. We assume our model has N nodes. n_i ($i \in [1..N]$) represents the number of processors inside node i and s_i ($i \in [1..N]$) refers to the relative processing speed of a single processor inside node i. The total job arrival rate of the system is λ which follows the Poisson distribution. In our model, jobs are randomly dispatched, but with a distribution of probability to minimize the average response time of jobs. Therefore, the goal of the problem is to assign the appropriate stream of jobs to the node i with flow rate λ_i ($i \in [1..N]$) to obtain MinART, where $\lambda = \sum_{j=1}^{N} \lambda_j$. Because our model is a queueing model, without loss of generality, the response time of job k with execution time e_k is $r_k = e_k + w_k$. w_k is the waiting time of job k. Hence the average response time of m jobs is $W(m) = \frac{\sum_{k=1}^{m} r_k}{m}$. The average response time of the whole system is $W = \lim_{m \to +\infty} W(m)$. In our model, every node is an $M/M/n_i$ $(n_i \geq 1)$ system with arrival rate λ_i, and the processors each with the job service rate $\mu_i = \frac{s_i}{E(l)}$ (l is the random variable of the job execution time which follows the exponential distribution). Therefore, for node i, the average response time is as follows [8]:

$$W_i = \frac{(\frac{\lambda_i}{\mu_i})^{n_i} \cdot P_0}{\mu_i \cdot n_i \cdot n_i! (1 - (\frac{\lambda_i}{n_i \cdot \mu_i})^2)} + \frac{1}{\mu_i}, \text{ where } P_0 = \Big(\sum_{k=0}^{n_i-1} \frac{\lambda_i^k}{\mu_i^k \cdot k!} + \frac{\lambda_i^{n_i}}{\mu_i^{n_i} \cdot n_i! \cdot (1 - \frac{\lambda_i}{n_i \cdot \mu_i})} \Big)^{-1}$$

Usually the computational power of grids is greater than the workload. Hence we assume the utilization rate of grids is always less than 1, which means $\frac{\lambda}{\sum_{i=1}^{N} \mu_i \cdot n_i} <$ 1. The problem of minimizing the average response time of jobs can be written as follows:

$$Min \ W(\beta) = \frac{\sum_{i=1}^{N} \lambda_i \cdot W_i}{\lambda}$$
$$\text{where } \beta = (\lambda_1, \lambda_2, \cdots, \lambda_N), \text{ subject to}$$
$$\sum_{j=1}^{N} \lambda_j = \lambda, \qquad\qquad (1)$$
$$\lambda_i \geq 0, \ (i \in [1..N]) \qquad (2)$$
$$\frac{\lambda_i}{n_i \cdot \mu_i} < 1, \ (i \in [1..N]) \qquad (3)$$

Then a nonlinear optimization problem is shown as above. Inequation (3) describes $M/M/n$ $(n \geq 1)$ models must have the system utilization rate less than 1. Otherwise, average response times of the models are infinity. Inequation (2) illustrates that the job arrival rate of every node cannot be negative. Equation (1) means the total job arrival rate of the system should be equal to the sum of sub job arrival rates of all nodes.

3   Theoretical Analysis

In this section, we discuss properties of the optimization problem as shown in the previous section through theoretical analysis. Then we present the solution. Finally, we compare the ART difference between LB and MinART via case study and theoretical analysis.

3.1   Existence and Uniqueness of Solutions

Although, for $M/M/1$ queueing system models, overall optimization problems have been obtained[9], to the best of our knowledge, there are few solutions to $M/M/n$ models for $n > 1$, published [10]. Usually two main properties of nonlinear optimization problems are concerned: Does the nonlinear optimization problem have any optimal solution? Is the solution global and unique?

Lemma 1. *For* $0 \leq \frac{\lambda_i}{n_i \cdot \mu_i} < 1$, $(i \in [1..N])$, *the* $W(\beta)$ *is strictly convex.*

Proof. Because every term $\lambda_i \cdot W_i$ in the function $W(\beta)$ has the same format of the formula and is independent from the other terms, without loss of generality, we inspect the second derivative of $W(\beta)$ with respect to λ_i as follows:

$$\frac{\partial^2 W(\beta)}{\partial \lambda_i^2} = \frac{\partial^2 (\lambda_i \cdot W_i)}{\partial \lambda_i^2} = 2 \cdot \frac{\partial W_i}{\partial \lambda_i} + \lambda_i \cdot \frac{\partial^2 W_i}{\partial \lambda_i^2}$$

From [11], the average response time W_i is strictly increasing in ρ where $\rho = \frac{\lambda_i}{n_i \cdot \mu_i}$. Hence W_i is also strictly increasing in λ_i. Therefore, the first term is positive. In addition, from [12], we have $\frac{\partial^2 W_i}{\partial \lambda_i^2} > 0$ for $0 \leq \frac{\lambda_i}{n_i \cdot \mu_i} < 1$, $(i \in [1..N])$. Therefore, the second term is also positive. Finally,

$$\frac{\partial^2 W(\beta)}{\partial \lambda_i^2} = \frac{\partial^2 (\lambda_i \cdot W_i)}{\partial \lambda_i^2} = 2 \cdot \frac{\partial W_i}{\partial \lambda_i} + \lambda_i \cdot \frac{\partial^2 W_i}{\partial \lambda_i^2} > 0$$

We have $\frac{\partial^2 W(\beta)}{\partial \lambda_i^2} > 0$ $(i \in [1..N])$ which means $W(\beta)$ is strictly convex.

Theorem 1. *The nonlinear optimization problem* $Min(W(\beta))$ *has the unique global optimal solution.*

Proof. Since $W(\beta)$ is a strictly convex function, the unique global optimal solution to the minimizing problem on $W(\beta)$ exists. [13]

3.2 Solution

We use the Lagrange multiplier method to solve the problem. The problem can be rewritten to remove inequation constraints as follows:

$$Min \ W(\beta) = \sum_{i=1}^{N} \lambda_i \cdot W_i$$
$$\text{where } \beta = (\lambda_1, \lambda_2, \cdots, \lambda_N), \text{ subject to}$$
$$\lambda - \sum_{j=1}^{N} \lambda_j = 0, \qquad (4)$$
$$s_i^2 - \lambda_i = 0, \ (i \in [1..N]) \quad (5)$$

s_i, $(i \in [1..N])$ is a slack variable. The multiplier[14], $L(\beta, \omega, \eta)$ is defined as

$$L(\beta, \omega, \eta) = \sum_{i=1}^{N} \lambda_i \cdot W_i + \omega \cdot (\lambda - \sum_{j=1}^{N} \lambda_j) + \sum_{j=1}^{N} \eta_j \cdot (s_i^2 - \lambda_i)$$

and we must solve

$$\begin{cases} \frac{\partial L}{\partial \lambda_i} = \frac{\partial (\lambda_i \cdot W_i)}{\partial \lambda_i} - \omega - \eta_i = 0 & (6) \\ \frac{\partial L}{\partial s_i} = 2\eta_i \cdot \lambda_i \cdot s_i = 0 & (7) \\ \frac{\partial L}{\partial \omega} = \lambda - \sum_{j=1}^{N} \lambda_j = 0 & (8) \\ \frac{\partial L}{\partial \eta_i} = s_i^2 - \lambda_i = 0 & (9) \end{cases}$$

The equations (7) and (9) indicate that

$$\begin{cases} \lambda_i = s_i = 0, \eta_i \geq 0 \\ \lambda_i = s_i^2 \geq 0, \eta_i = 0 \end{cases}$$

From above constraints, we know the last term of the multiplier $L(\beta, \omega, \eta)$ ($\sum_{j=1}^{N} \eta_j \cdot (s_i^2 - \lambda_i)$) is always equal to zero in the optimal solution so that equation (6) can be rewritten as $\frac{\partial L}{\partial \lambda_i} = \frac{\partial(\lambda_i \cdot W_i)}{\partial \lambda_i} - \omega = 0$. And λ_i can be divided into two

---

1) Sort nodes by the relative speed of processors

Use a basic sorting algorithm to obtain an array $k_1 \ldots k_N$ where

$s_{k_1} \geq s_{k_2} \geq \cdots \geq s_{k_N}$

2) Group nodes with the same relative speed of processors

j=1

Create a group g_j and put s_{k_1} into g_j

for i=2 to N do

begin

   if $(s_{k_{i-1}} = s_{k_i})$

   Put s_{k_i} into group g_j

   else

   j=j+1

   Create a group g_j and put s_{k_i} to group g_j

end

3) Compute the new solution for the selected groups

for m=1 to j do

begin

   if $(\lambda < \sum_{i=1}^{m} (\sum_{p \in g_i} n_{k_p} \cdot \mu_{k_p}))$ (where $\mu_i = \frac{s_i}{E(l)}$)

   begin

   Find the solution by solving equations (6) and (8) with

   $\lambda_{k_p} > 0$ $(\forall p \in \bigcup_{t \in [1..m]} g_t)$ and $\lambda_{k_h} = 0$ $(h \notin \bigcup_{t \in [1..m]} g_t)$

   if $(m = j)$ or $(\frac{\partial(\lambda_{k_p} \cdot W_{k_p})}{\partial \lambda_{k_p}} \leq \frac{1}{\mu_{k_q}})$ $(\forall p \in \bigcup_{t \in [1..m]} g_t$ and $\forall q \in g_{m+1})$

   STOP, MinART is obtained

   end

end

Fig. 2. The MinART Algorithm

---

sets: $\Phi = \{\lambda_i | \lambda_i > 0, i \in [1...N]\}$ and $\Theta = \{\lambda_i | \lambda_i = 0, i \in [1...N]\}$. For $\lambda_i \in \Phi$, $\frac{\partial(\lambda_i \cdot W_i)}{\partial \lambda_i} = \omega$, which means all the first partial derivatives are equal to the same value. Therefore, with equation (8), we can solve the problem with a given Φ

and Θ if the inverse function of $\frac{\partial(\lambda_i \cdot W_i)}{\partial \lambda_i}$ is available. We can obtain the optimal solution by solving the problems with all possible Φ and Θ.

However, in the worst case, there are 2^N different Φ and Θ. We notice that $\lambda_k \cdot W_k$ is a convex function in $\lambda_k \geq 0$ so that $\frac{\partial(\lambda_k \cdot W_k)}{\partial \lambda_k}$ is also a monotonically increasing function. Therefore, the minimum value is at $\lambda_k = 0$ where $\frac{\partial(\lambda_k \cdot W_k)}{\partial \lambda_k}|_{\lambda_k=0} = W_k|_{\lambda_k=0} = \frac{1}{\mu_k}$. On the other hand, $\lambda_k = 0$ $(k \in \Theta)$ means $\frac{\partial(\lambda_k \cdot W_k)}{\partial \lambda_k}|_{\lambda_k=0} = \frac{1}{\mu_k} \leq \omega$ for node k. It indicates that if $s_k \leq s_j$ and $\lambda_k > 0$, we must have $\lambda_j > 0$. Hence, in the worst case, the number of the possible sets Φ and Θ can be reduced to N if all nodes are sorted by the relative speed of their processors.

Based on the above analysis, the proposed algorithm is shown as in Figure 2. In the algorithm, we have to find the inverse functions of $\frac{\partial(\lambda_i \cdot W_i)}{\partial \lambda_i}$ to solve the equations. However, the inverse functions may not have a closed-form expression when the number of processors inside one node is greater than 2. Hence it is hard to solve the equations (6) and (8). On the other hand, we notice that every temporary solution inside step 3) has $\lambda_{k_p} > 0$ $(p \in \bigcup_{t \in [1..m]} g_t)$ which means the temporary solution is a local minimum inside the feasible region of λ_{k_p} $(p \in \bigcup_{t \in [1..m]} g_t)$[14]. It implies that we can obtain the solution by Newton's method with feasible initial point instead of solving the equations for every loop inside step 3). To use Newton's method, the constraint (4) of minimization problem can be eliminated by expressing one λ_{k_p} $(p \in \bigcup_{t \in [1..m]} g_t)$ by the others.

Specifically, for each node i with an M/M/1 model, we have $\frac{\partial(\lambda_i \cdot W_i)}{\partial \lambda_i} = \frac{\mu_i}{(\mu_i - \lambda_i)^2}$. Because $\frac{\partial(\lambda_{k_i} \cdot W_{k_i})}{\partial \lambda_{k_i}} = \frac{\partial(\lambda_{k_p} \cdot W_{k_p})}{\partial \lambda_{k_p}}$ $(p, i \in \bigcup_{t \in [1..m]} g_t)$ in every solution of step 3), λ_i of each M/M/1 system with $\lambda_i > 0$ can be expressed by λ_p, where p is one of the most powerful M/M/1 nodes. Therefore, we can simply the optimization problem by expressing job flow variables of all M/M/1 nodes with a single variable.

3.3 Comparison Between Load Balancing and MinART

It is well known that the average response time is an important metric for grids. A lower ART means jobs are processed faster, implying that the utilization of grids is more efficient and grids exhibit more power. Different scheduling strategies may result in different ARTs on the same system. On the other hand, another important metric is load balancing. Although in homogeneous systems LB state and MinART state are the same state, they are different states in heterogeneous environments. In this section, we compare the difference of ARTs between LB state and MinART state through case study and theoretical analysis.

In some works [6], authors consider grids are utilized efficiently when LB state of heterogeneous grids is achieved. In general, LB state of grids is the state where every node has the same utilization rate. Usually LB scheduling is considered to be able to increase throughput of grids and decrease average response time of grids because workloads are dispatched into grids evenly. However, LB state is not the same as the MinART state in heterogeneous grids. The ART difference between two states is sometimes large. Especially in some cases with $M/M/n$ models where $n > 1$, the difference may become larger.

In our model, LB state can be expressed as follows:

$$\lambda_i = \lambda \cdot \frac{n_i \cdot s_i}{\sum\limits_{j=1}^{N} n_j \cdot s_j}, \; (i \in [1..N])$$

Therefore, $\rho_i = \frac{\lambda_i}{n_i \cdot \mu_i} = \frac{\lambda \cdot E(l)}{\sum\limits_{j=1}^{N} n_j \cdot s_j}$, (where $\mu_i = \frac{s_i}{E(l)}$) $(i \in [1..N])$

From the above formulas, we notice the utilization rate of every node is equal to the utilization rate of the whole system. Hence every node is utilized evenly. We inspect ARTs of grids at LB state and MinART state via the following case study.

In this case, we assume a grid has two nodes (n_1 and n_2) each with an M/M/1 model. Therefore, we have $\mu_1 = \frac{s_1}{E(l)}$ and $\mu_2 = \frac{s_2}{E(l)}$. For LB state, we obtain $\lambda_1 = \frac{\mu_1}{\mu_1 + \mu_2} \cdot \lambda$ and $\lambda_2 = \frac{\mu_2}{\mu_1 + \mu_2} \cdot \lambda$ according to $\frac{\lambda_1}{\mu_1} = \frac{\lambda_2}{\mu_2}$. The ART of the grid at LB state can be easily obtained as follows:

$$ART_{LB} = \frac{1}{\lambda} \cdot \left(\frac{\lambda_1}{\mu_1 - \lambda_1} + \frac{\lambda_2}{\mu_2 - \lambda_2} \right) = \frac{2}{\mu_1 + \mu_2 - \lambda}$$

Then in order to obtain MinART state, we construct an optimization problem first as follows:

$$Min \; W(\lambda_1) = \frac{1}{\lambda} \cdot \left(\frac{\lambda_1}{\mu_1 - \lambda_1} + \frac{\lambda - \lambda_1}{\mu_2 - (\lambda - \lambda_1)} \right)$$
$$\text{subject to } 0 \le \lambda_1 < \mu_1 \text{ and } 0 \le \lambda - \lambda_1 < \mu_2$$

Here we replace λ_2 with $\lambda - \lambda_1$ so that the equation constraint is eliminated in this case.

Therefore, we obtain the solution of the optimization problem as follows:

$$\frac{\partial W(\lambda_1)}{\partial \lambda_1} = \frac{\mu_1}{(\mu_1 - \lambda_1)^2} - \frac{\mu_2}{(\mu_2 + \lambda_1 - \lambda)^2} = 0$$
$$\Rightarrow \lambda_1 = \frac{\mu_1 \cdot \sqrt{\mu_2} + \sqrt{\mu_1} \cdot \lambda - \mu_2 \cdot \sqrt{\mu_1}}{\sqrt{\mu_1} + \sqrt{\mu_2}}$$

However, the solution must be inside the feasible region:

$$0 \le \lambda_1 < \mu_1 \text{ and } 0 \le \lambda_2 < \mu_2$$
$$\Rightarrow max \left\{ \mu_2 - \sqrt{\mu_1 \cdot \mu_2}, \mu_1 - \sqrt{\mu_2 \cdot \mu_1} \right\} < \lambda < \mu_1 + \mu_2$$

Without loss of generality, we assume $\mu_1 \le \mu_2$. Therefore, for $0 \le \lambda \le \mu_2 - \sqrt{\mu_1 \cdot \mu_2}$, the solution is $\lambda_1 = 0$ and $\lambda_2 = \lambda$. The MinART of the grid is

$$MinART = \frac{1}{\lambda} \cdot \left(\frac{2 \cdot \lambda - (\sqrt{\mu_1} - \sqrt{\mu_2})^2}{\mu_1 + \mu_2 - \lambda} \right), \; (\mu_2 - \sqrt{\mu_1 \cdot \mu_2} < \lambda < \mu_1 + \mu_2)$$
$$MinART = \frac{1}{\mu_2 - \lambda}, \; (0 < \lambda \le \mu_2 - \sqrt{\mu_1 \cdot \mu_2})$$

Therefore, the relative difference of ARTs between two states is

$$\frac{\Delta T}{ART_{LB}} = \frac{ART_{LB}-MinART}{ART_{LB}} = \frac{(\sqrt{\mu_1}-\sqrt{\mu_2})^2}{2\cdot\lambda}, \ (\mu_2 - \sqrt{\mu_1\cdot\mu_2} < \lambda < \mu_1+\mu_2)$$

$$\frac{\Delta T}{ART_{LB}} = \frac{ART_{LB}-MinART}{ART_{LB}} = \frac{\mu_2-\mu_1-\lambda}{2\cdot(\mu_2-\lambda)}, \ (0 < \lambda \le \mu_2 - \sqrt{\mu_1\cdot\mu_2})$$

From the above formulas, we obtain the range of the relative difference as follows:

$$\frac{(\sqrt{\mu_1}-\sqrt{\mu_2})^2}{2\cdot(\mu_1+\mu_2)} < \frac{\Delta T}{ART_{LB}} < \frac{\mu_2-\mu_1}{2\cdot\mu_2}$$

Therefore, we notice if $\mu_1 = \mu_2$, which means the grid is a homogeneous system, we obtain

$$\frac{ART_{LB}-MinART}{ART_{LB}} = 0 \text{ and } \lambda_{1LB} = \lambda_{1MinART} = \frac{\lambda}{2}$$

This indicates LB state and MinART state are the same state. However, in heterogeneous grids, the relative difference of ARTs between LB and MinART decreases gradually to the minimum extremum $\frac{(\sqrt{\mu_1}-\sqrt{\mu_2})^2}{2\cdot(\mu_1+\mu_2)}$ with the steady increase of the system load. On the other hand, when the system load decreases, the relative difference increases gradually to the maximum extremum $\frac{\mu_2-\mu_1}{2\cdot\mu_2}$. This implies that ART of LB state may be undesired when the difference of computational powers between two nodes is large or system load is not heavy.

4   Numerical Experiments

In previous sections, we proposed an algorithm to obtain the MinART. In addition, we have analyzed the difference of ARTs between LB and MinART state through theoretical analysis and a case study. In order to explore and validate the behavior of ART differences of heterogeneous grids between MinART and LB, we present more complicated cases and data in this section. In order to simplify experiments, we only inspect two nodes cases in this paper.

4.1   Two Nodes with M/M/1 Models

From the above section, we obtain formulas for two nodes cases with $M/M/1$ models. Therefore, we can compare ART differences among different cases according to these formulas.

Figure 3 illustrates the relative difference of ARTs between LB and MinART states ($\frac{ART_{LB}-MinART}{ART_{LB}}$). In this experiment, we fix the service rate of one node (node 1) to 1 and increase the service rate of the other node (node 2) from 2 to 5 gradually. From Figure 3 we notice, with the differential of the service rate between two nodes increasing, relative differences of ARTs are also enlarging. Although relative differences of ARTs go down with the increase of the system load, they would converge to different minimum extrema which are not equal to zero eventually. Before the point ($\rho = \frac{\lambda}{\mu_1+\mu_2} = \frac{\mu_2-\sqrt{\mu_1\cdot\mu_2}}{\mu_1+\mu_2}$), relative differences decrease quickly from different maximum extrema with the system load increasing. However, after the point, the decrease of relative differences is slow. Moreover, with the enlargement of differential of processor speeds, relative differences go down slower when the system load becomes heavy. We find out

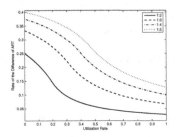   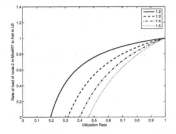

Fig. 3. Relative differences of ARTs in the cases of different system utilization rates and differential processor speeds

Fig. 4. Rate of load difference of node 1 between LB and MinART states with different system utilization rate and differential processor speeds

relative differences are very large in some cases such as the cases of 1:5 and 1:4 as shown in Figure 3. Figure 4 demonstrates the rate of the load of node 1 in MinART state to the load of node 1 in LB state($rate = \frac{\lambda_{1_{MinART}}}{\lambda_{1_{lb}}}$, node 1). With the increase of the system load, the relative load difference of node 1 between two states becomes smaller and would converge to the same value eventually. Because the utilization rate of every node should be less than 1 (otherwise the average response time would be infinite), loads of node 1 at LB state and MinART state are very close when the system load is heavy. Moreover, we notice that the relative load difference of node 1 becomes large with the increase of the differential of service rates between two nodes.

4.2 Two Nodes with M/M/n Models

Then in the next experiment, we inspect the impact of $M/M/n$ ($n > 1$) models. Similar to previous experiments we study only the case of two nodes. We let node 1 have only one processor ($M/M/1$), while node 2 has multiple processors ($M/M/n$, $n > 1$). We fix the service rate of node 1 to 8 and the total service rate of node 2 to 4.

Relative ART differences ($\frac{ART_{LB} - MinART}{ART_{LB}}$) in the cases of different numbers of processors inside node 2 are shown in Figure 5. With the number of processors increasing, relative differences are also rising. Relative differences are high when system load is light and moderate. Relative differences go down with the increase of the system load and would converge to the minimum extremum of $M/M/1$ models eventually.

From Figure 5, we notice that relative differences are very high when the system load is light. Relative differences in some $M/M/n$ models are over 30% in Figure 5. The reason for high relative differences is that LB scheduling ignores the power of a single processor inside nodes, which can deteriorate ARTs of grids in some cases. Because we fix the total service rate of node 2 to 4 and service rate of node 1 to 8, with the number of processors increasing, the power of a single

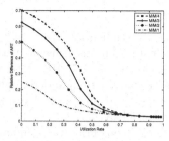 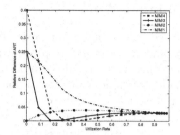

Fig. 5. Relative differences of ARTs in the cases of different system utilization rate and different numbers of processors inside node 2

Fig. 6. Relative differences of ARTs in the cases of different system utilization rate and different numbers of processors inside node 1

processor inside node 2 goes down. This leads the power difference between two processors from different nodes to increase, which results in relative differences of ARTs rising.

However, in some other cases, increasing the number of processors may decrease the difference of powers of processors from different nodes. Now we still fix the service rate of node 1 and node 2 to 8 and 4. But node 1 has multiple processors ($M/M/n$ model) and node 2 only has one processor ($M/M/1$ model). Because the power of node 1 is larger than node 2, increasing the number of processors inside node 1 may decrease the power difference of processors from different nodes.

Figure 6 demonstrates relative differences of ARTs between MinART and LB states. We notice some interesting phenomena from it. When node 1 has two processors, which means the power difference among all processors is zero, the relative difference starts from zero and increases gradually to the maximum extremum which is the minimum extremum of $M/M/1$ models. When the number of processors of node 1 is three and four which means the power difference of processors rises, relative differences of ARTs are very high in the light system load state. However, relative differences drop to zero rapidly with the system load increasing gradually. Then relative differences rise slowly and converge to the same value which is the minimum extremum of $M/M/1$ models eventually.

We think the reason for the phenomena is that the impact of the power difference of processors from different nodes counteracts the impact of the power difference of nodes. When system load is light, the impact of the power difference of processors dominates. However, the trend is reversed with the system load increasing. Therefore, relative differences of all cases would converge to the minimum extremum of the $M/M/1$ model eventually.

From above experiments, we notice that there are two impacts on scheduling stochastic workload in heterogeneous grids: the power difference of processors and power difference of nodes. These two impacts can counteract or enhance each other. Ignoring one of them may result in undesired ARTs of grids.

5   Related Work

The scheduling problem on grids has been widely studied[15], [16]. Different models and algorithms were proposed to handle heterogeneous grids [6]. Models of some other works are also similar to grid environments [7], [17], [18]. Although the models of some works may appear theoretical and far from actual applications, the models are realistic in some particular cases and starting with a simple model is the traditional approach to tackle hard problems[6].

In [6], the authors proposed a model of heterogeneous grids very similar to ours. In their model, nodes with more power should process more jobs. An algorithm was proposed to achieve the LB state. Then they showed several important metrics of the model such as queue length and ART. In addition, they analyzed the case where the utilization rate of grids is greater than 1. They also measured the algorithm with several different distributions. However, due to the heterogeneity of grids, load balancing schedulers may deteriorate the ART of grids. Therefore, the power of heterogeneous grids cannot be utilized efficiently.

Models of some other works are similar to heterogeneous grids. In [17], authors constructed a general model of heterogeneous distributed systems. And an algorithm based on the Lagrange multiplier to obtain MinART was proposed. However, their model missed the constraints for $M/M/1$ models, so that the global optimal solution may be inside infeasible regions. In [18], the model and algorithm based on Newton's method is similar to the work in [17], except for the goal of the algorithm, which is to minimize the sojourn time of each job. Compared to MinART, which is a global optimal state to maximize the system performance, it can be considered as an individual optimal state to maximize the individual performance. Nevertheless, such state cannot be stable and may even deteriorate the system performance [19]. In addition, Newton's method cannot handle any constraint so that the solution may be not available by this method.

6   Conclusions and Future Work

We analyzed how to obtain the minimum average response time of heterogeneous grids. Scheduling is one of the key services required for enabling performance on distributed and heterogeneous platforms. A goal of heterogeneous grids is to speed up the job processing time, which is related to ART. We compared the difference of ARTs between load balancing state and MinART state. Moreover, we explored the impacts of the heterogeneity of grids on ART difference through theoretical analysis and case study. We found that the ART of load balancing state may be far away from MinART in some case. And the difference of ARTs between two states in different cases was studied. The range of relative ART differences was found in some cases.

Our future works are directed toward more complex cases: decentralized scheduling, mixed dynamic and static scheduling, and estimating the load of nodes. These new constraints will more than likely make our analysis more difficult. We also want to extend our work toward fault tolerance and reliability by adding witness and replica mechanism.

References

1. Foster, I.: Internet Computing And The Emerging Grid. Nature Web Matters (2000)
2. F. Berman, G.F., Hey, T.: Grid Computing: Making the Global Infrastructure a Reality. John Wiley and Sons (2003)
3. Foster, I., Kesselman, C.: The Grid: Blueprint for a New Computing Infrastructure. Morgan Kaufman (1999)
4. OPTIPUTER: (http://www.optiputer.net/)
5. Foster, I.: The Challenges of Grid Computing. Condor Week Presentation at via Access Grid (2002)
6. Vandy Berten, J.l.G., Jeannot, E.: On the Distribution of Sequential Jobs in Random Brokering for Heterogeneous Computational Grids. IEEE TRANSACTIONS ON PARALLEL AND DISTRIBUTED SYSTEMS, VOL. 17, NO. 2 (2006)
7. Hui, C.C., Chanson, S.T.: Hydrodynamic Load Balancing. IEEE TRANSACTIONS ON PARALLEL AND DISTRIBUTED SYSTEMS, VOL. 10, NO. 11 (1999)
8. Ross, S.M.: Introduction to Probability Models. Harcourt Brace and Company (1993)
9. Douligeris C, M.R.: A game theoretic perspective to flow control in telecommunication networks. Journal of the Franklin Institute 329, pp. 383C402. (1992)
10. Atsushi Inoie., H.K., Touati, C.: A paradox in optimal flow control of $M/M/n$ queues. Computers and Operations Research 33 pp. 356C368 (2006)
11. Lee HL, C.M.: A note on the convexity of performance measures of $M/M/c$ queueing systems. Journal of Applied Probability 20 pp. 920C3. (1983)
12. Harel A, Z.P.: Strong convexity result for queueing systems. Operations Research 35 pp. 405C18 (1987)
13. Lee HL, C.M.: An interior-point ℓ_1-penalty method for nonlinear optimization. International Conference on Continuous Optimization (ICCOPT I) (2004)
14. E.MILLER, R.: Optimization: Foundation and Applications. John Wiley and Sons (2000)
15. A. Galstyan, K.C., Lerman, K.: Resource Allocation in the Grid Using Reinforcement Learning. Third International Joint Conference on Autonomous Agents and Multiagent Systems - Volume 3, pp. 1314-1315 (2004)
16. H. Casanova, J.H., Yang, Y.: Algorithms and Software to Schedule and Deploy Independent Tasks in Grid Environments. Proc. Workshop Distributed Computing, Metacomputing, and Resource Globalization (2002)
17. Li, J., Kameda, H.: Load Balancing Problems for Multiclass Jobs in Distributed/Parallel Computer Systems. IEEE TRANSACTIONS ON COMPUTERS, VOL. 47, NO. 3 (1998)
18. Zeng, Z., Veeravalli, B.: Rate-Based and Queue-Based Dynamic Load Balancing Algorithms in Distributed Systems. Proceedings of the Tenth International Conference on Parallel and Distributed Systems (ICPADS04) (2004)
19. Hassin, R., Haviv, M.: TO QUEUE OR NOT TO QUEUE: EQUILIBRIUM BEHAVIOR IN QUEUEING SYSTEMS. Kluwer Academic Publishers (2003)

Advanced Reservation-Based Scheduling
of Task Graphs on Clusters

Anthony Sulistio[1], Wolfram Schiffmann[2], and Rajkumar Buyya[1]

[1] Grid Computing and Distributed Systems Lab
Dept. of Computer Science and Software Engineering
The University of Melbourne, Australia
{anthony, raj}@csse.unimelb.edu.au
[2] Dept. of Mathematics and Computer Science
University of Hagen, Germany
Wolfram.Schiffmann@FernUni-Hagen.de

Abstract. A Task Graph (TG) is a model of a parallel program that consists of many subtasks that can be executed simultaneously on different processing elements. Subtasks exchange data via an interconnection network. The dependencies between subtasks are described by means of a Directed Acyclic Graph. Unfortunately, due to their characteristics, scheduling a TG requires dedicated or uninterruptible resources. Moreover, scheduling a TG by itself results in a low resource utilization because of the dependencies among the subtasks. Therefore, in order to solve the above problems, we propose a scheduling approach for TGs by using advance reservation in a cluster environment. In addition, to improve resource utilization, we also propose a scheduling solution by interweaving one or more TGs within the same reservation block and/or backfilling with independent jobs.

1   Introduction

A Task Graph (TG) is a model of a parallel program that consists of many subtasks that can be executed simultaneously on different processing elements (PEs). Subtasks exchange data via an interconnection network. The dependencies between subtasks are described by means of a Directed Acyclic Graph (DAG). Executing a TG is determined by two factors: a *node weight* that denotes the computation time of each subtask, and an *edge weight* that corresponds to the communication time between dependent subtasks [1]. Thus, to run these TGs, we need a target system that is tightly coupled by fast interconnection networks. Typically, cluster computers provide an appropriate infrastructure for running parallel programs.

Scheduling TGs in a cluster environment is a challenging process because of the following constraints: *Firstly*, a TG requires a fixed number of processors for execution. Hence, a user needs to reserve the exact number of PEs. *Secondly*, due to communication overhead between the subtasks on different PEs, each subtask must be completed within a specific time period. *Finally*, each subtask needs to

Y. Robert et al. (Eds.): HiPC 2006, LNCS 4297, pp. 60–71, 2006.

wait for its parent subtasks to finish executing in order to satisfy the required dependencies.

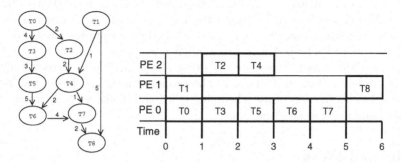

Fig. 1. Illustration of a task graph (left) and its schedule (right) on 3 PEs

Scheduling a TG on a resource can be visualized by a time-space diagram as shown in Figure 1. In this figure, a TG consists of 9 subtasks ($T0 - T8$), and as an example it was scheduled using 4 target processing elements (TPEs). Each subtask has a node weight of 1 time unit, and its edge weight is also shown on Figure 1 (left) in a number next to the arrow line. In order to minimize the schedule length (overall computation time) and the communication costs of a TG, its subtasks must be assigned to appropriate PEs and they must be started after their parent subtasks. In this example, $T6$ depends on $T4$ and $T5$, so it must wait for both subtasks to finish and it will be scheduled on $PE0$ in order to minimize the communication cost. However, this schedule does not make an efficient use of the given TPEs. Although this schedule assigned the subtasks to 3 PEs, only 2 PEs are actually needed. In general, the right number of schedule's PEs can not be determined in advance. Thus, the resulting schedules might not be able to make an efficient use of the available PEs. Therefore, in this paper, we will talk about how this problem can be improved upon by means of an advanced reservation-based scheduling.

If we consider DAGs with different node and edge weights, the general scheduling problem is NP-complete [2]. Thus, in practice, heuristics are most often used to compute optimized (but not optimal) schedules. Unfortunately, time-optimized algorithms do not make an efficient use of the given PEs. In this context, the *efficiency* is measured by the ratio of the *total node weight* in relation to the *overall available* processing time. As an example, in Figure 1, the efficiency of this TG schedule is 9/18 or 50%, which is quite low because $PE1$ and $PE2$ are mostly idling. If there are no idle PEs at all time, then the efficiency can be said to be optimal (100%).

In [3], a comprehensive test bench (comprised of 36,000 TGs with up to 250 nodes), is used to evaluate the schedule's efficiency of several popular heuristics, such as such as DLS [4], ETF [5], HLFET [6] and MCP [7]. Essentially, it reveals that the efficiency of the DAG-schedules is mostly below 60%, which means a lot of the provided computing power is wasted. The main reason is due to the constraints of the schedule as demonstrated in the previous example.

The contribution of this paper is as follows. We propose an approach to schedule TGs by using advance reservation in a cluster environment. Moreover, to improve the efficiency or to maximize the CPU utilization, we also propose a scheduling solution by interweaving one or more TGs within the same reservation block and/or backfilling with other independent jobs.

The rest of this paper is organized as follows. Section 2 mentions some related work in this area. Section 3 describes the proposed model, whereas Section 4 evaluates the effectiveness of the scheduling solution. Finally, Section 5 concludes the paper and gives some future work.

2   Related Work

Some systems are available for running DAG applications in the Grid or cluster computing environment, such as Condor [8,9], GrADS [10], Pegasus [11], Taverna [12] and ICENI [13]. However, only ICENI provides a reservation capability in its scheduler [14]. In comparison to our work, the scheduler inside ICENI does not consider backfilling other independent jobs with the reserved DAG applications. Hence, ICENI resource scheduler does not consider the efficiency of the reserved applications towards CPU utilization.

With regards to the efficiency analysis of functional parallel programs, i.e. executing two or more tasks concurrently, there are only few works done so far. In [15], the authors analyze the efficiency of TG schedules, such as ECPFD [16], DLS [4] and BSA [17] with respect of different Communication-to-Computation (CCR) values. The authors report that a resource efficiency drops down if the CCR value is increased and it also depends on the network topology. Moreover, they find that for coarse grained parallel programs (low CCR), the efficiency achieved is lower than 50%. In [15], the efficiency is defined as speedup of a TG schedule divided by number of processors, where the speedup denotes a ratio of measured parallel execution time to sequential execution time. However, it can be easily shown that this definition of efficiency is equivalent to the one already given in the previous section. Hence, the above findings are similar with [3] as mentioned earlier, except that in [15], the experiments were conducted on a real system because the model accuracy should be evaluated. Therefore, the main goal of our work is to increase the scheduling efficiency of these TGs.

3   Description of the Model

3.1   System Model

Figure 2 shows the open queueing network model of a resource applied for our work. In this model, there are two queues: one is reserved for TGs while the other one is for parallel and independent jobs. Each queue has a finite buffer with size S to store objects waiting to be processed by one of P independent CPUs or

PEs. All PEs are connected by a high-speed network. PEs in a resource can be homogeneous or heterogeneous. For this paper, we assume that a resource has homogeneous PEs, each having the same processing power.

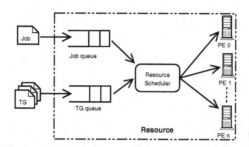

Fig. 2. Overall model where a user submits a set of task graphs on a resource

In this model, as shown in Figure 2, we assume that we have already known the optimal schedules for each TG in the queue and that their run times are also identified. With this assumption, the resource scheduler only needs to reserve and run these TGs. Moreover, the resource scheduler can perform futher optimization methods that will be discussed later on.

3.2   User Model

A user provides the following parameters during submission:

- $TG = \{T1, T2, ..., Tn\}$: Task Graph (TG) that consists of a set of dependent subtasks, where each subtasks has a node and edge weight.
- $List = \{TG_1, TG_2, ..., TG_k\}$: a collection of TGs.
- PE : number of CPUs requested.
- $start$: reservation start time.
- $finish$: reservation finish time.

A user needs to make a reservation by specifying a tuple $< PE, start, finish >$ to a resource. Once a reservation has been confirmed, then the user sends $List$ to the resource before the start time, otherwise the reservation will be cancelled. More details on the states of Advance Reservation can be found on [18].

3.3   Scheduling Model

The aim of our reservation-based scheduler is to improve the efficiency of each TG. Therefore, for executing TGs, we propose the following approaches:

1. **Rearranging subtasks:** This is done by rearranging all subtasks in a TG based on the total number of subtasks executed on each PE. For example,

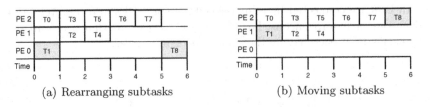

(a) Rearranging subtasks   (b) Moving subtasks

Fig. 3. Scheduling a task graph. The shaded subtasks denote the before (a) and after (b) a moving operation.

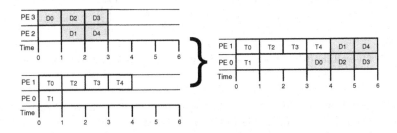

Fig. 4. Combining the execution of two TG by interweaving

we relocate all subtasks of $PE0$, $PE1$ and $PE2$ as depicted in Figure 1 to $PE2$, $PE0$ and $PE1$ respectively as shown in Figure 3(a). This fundamental step is required as a basis for the next step.

2. **Moving subtasks:** This is done by moving one or more subtasks from one PE to another as long as there are empty slots. For example, we move $T1$ and $T8$ as mentioned in Figure 3(a) from $PE0$ to $PE1$ and $PE2$ respectively as shown in Figure 3(b). With this approach, the best case scenario would result in the reduction of the schedule's PEs (SPEs). Hence, the available PEs can be used to run another TG by interweaving and/or backfilling with independent jobs as discussed in the next step.

3. **Interweaving TGs:** This can be done by combining two or more TGs from *List* and still keeping the original allocation and dependencies untouched.

For example, in Figure 4, two TGs that require the same number of PEs are interlocked. In general, the number of PEs do not matter. Each TG has an earliest task to start with. Without the loss of generality, this TG can be placed on a PE that will be availabe next. Due to the time relation in a schedule, we can now look if the second earliest TG "row" can be placed on another PE. If yes, we can proceed in this way until the second TG is completely placed. If there are no PEs available to fit the time relations, we delay all the previously placed task rows of that schedule appropriately. Of course this will create gaps of idle processor-cycles. But these gaps can be hopefully closed by the following backfilling step.

4. **Backfilling a TG or remaining gaps between interweaved TGs:** This can be done if there are smaller independent jobs that can be fit in and executed without delaying any of the subtasks of a TG.

In this step, we try to close the gaps by using (independent) jobs from another queue. In contrast to the interweaving step, the best fitting jobs should be selected out of this queue. We start with the first gap and look for the job that has an estimated schedule length lower or (best) equal to the gap's length. Jobs that can not be used to fill enough gaps must be scheduled after all the parallel programs are executed. As an example, there is enough gap on $PE0$ in Figure 4 to put 2 small independent jobs, each runs for 1 time unit.

4 Performance Evaluation

In order to evaluate the performance of our advanced reservation-based scheduler (AR), we compare it with two standard algorithms, i.e. First Come First Serve (FCFS) and EASY backfilling (Backfill) [19]. We use GridSim toolkit [18] to conduct the experiment with different parameters. We simulate the experiment with three different target systems that consist of clusters with varying number of processors, i.e. 16, 32 and 64 PEs. Then, we run the experiment by submitting both TGs and other jobs (taken from a workload trace) into these systems.

4.1 Experimental Setup

Test Bench Structure. In this experiment, we use the same test bench (created by a task graph generator), as discussed in [1] and [3], to evaluate the performance of our scheduler. Therefore, we briefly describe the structure of the test bench. More detailed explanation of the test bench can be found in [1].

TGs with various properties are synthesized by a graph generator whose input parameters are varied. The directory tree that represents the structure of test bench are shown in Figure 5. The total number of TGs at each level within a path of the tree is shown on the right side. The parameters of a TG is described as follows (from top to bottom level in Figure 5):

– Graph Size (GS): denotes the number of nodes or subtasks for each TG. In Figure 5, The parameters of a generated TG are grouped into three categories: 7 to 12 nodes (GS7_12), 13 to 18 nodes (GS13_18) and 19 to 24 nodes (GS19_24).
– Meshing Degree (MD) or Number of Sons (NoS): denotes the number of dependencies between the subtasks of each TG. When a TG has a low, medium

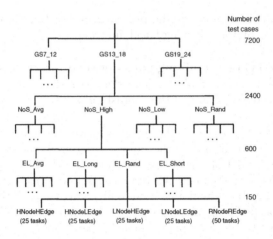

Fig. 5. Structure of the test bench

and strong meshing degree, the NoS in Figure 5 are NoS_Low, NoS_Avg and NoS_High respectively. TGs with random meshing degrees are represented as NoS_Rand.

– Edge Length (EL): denotes the distance between connected nodes. When a TG has a short, average and long edge length, Figure 5 depicts the notation as EL_Short, EL_Avg and EL_Long respectively. TGs with random edges are represented as EL_Rand.

– Node- and Edge-weight: denotes the Computation-to-Communication Ratio with a combination of heavy (H), light (L) and random (R) weightings for the node and edge.

From this test bench, we also use the optimal schedules for the branches of GS7_12 and GS13_18 for both 2 and 4 TPEs. Each branch contains 2,400 task graphs, hence the maximum number of task graphs that we use is 9,600. These optimal schedules were computed and cross-checked by two independent informed search algorithms (branch-and-bound and A*) [1]. Note that at the time of conducting this experiment, the optimal schedules of GS19_24 for 4 TPEs are not yet completed. Therefore, we do not incorporate the schedules of GS19_24 for 2 TPEs into the experiment for consistency.

Workload Trace. We also take two workload traces from the Parallel Workload Archive [20] for our experiment. We use the trace logs from DAS2 fs4 (Distributed ASCI Supercomputer-2 or DAS in short) cluster of Utrecht University, Netherlands and LPC (Laboratoire de Physique Corpusculaire) cluster of Universite Blaise-Pascal, Clermont-Ferrand, France. The DAS cluster has 64 CPUs with 33,795 jobs, whereas the LPC cluster has 140 CPUs with 244,821 jobs. The detailed analysis for DAS and LPC workload traces can be found in [21] and [22] respectively. Since both original logs recorded several months of run-time period with thousands of jobs, we limit the number of submitted jobs to be 1000, which is roughly a 5-days period from each log. If the job requires more than the total PEs of a resource, we set this job to the maximum number of PEs.

In order to submit 2,400 TGs within the 5-days period, a Poisson distribution is used. 4 TGs arrive in approximately 10 minutes for conducting the FCFS and Backfill experiments. When using the AR scheduler, we set the limit of each reservation slot to contain only 5 TGs from the same leaf of the test bench tree from Figure 5. Hence, only 480 reservations were created during the experiment, where every 30 minutes a new reservation is requested. If there are no available PEs, then the resource scheduler will reserve the next available ones.

4.2   Results

Figure 6 and 7 show a huge gain for using AR scheduler for the total completion time on 4 TPEs for both the DAS and LPC trace respectively, especially on a resource that has 16 PEs. Note that for 2 TPEs, the results are similar, hence they are being omitted in this paper.

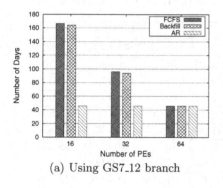

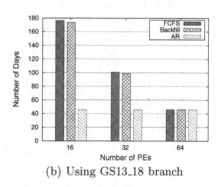

(a) Using GS7_12 branch   (b) Using GS13_18 branch

Fig. 6. Total completion time on the DAS trace with 4 TPEs (lower is better)

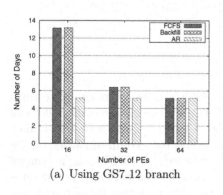

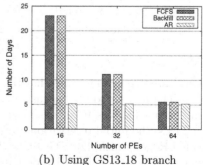

(a) Using GS7_12 branch   (b) Using GS13_18 branch

Fig. 7. Total completion time on the LPC trace with 4 TPEs (lower is better)

Table 1. Average percentage of reduction in a reservation duration time

Task Graph Parameters	2 TPEs (% reduction)			4 TPEs (% reduction)		
	GS7_12	GS13_18	Avg	GS7_12	GS13_18	Avg
MD Low	2.06	2.15	2.10	14.99	22.80	18.89
MD Avg	6.59	7.73	7.16	13.68	19.87	16.78
MD High	9.66	9.61	9.64	12.33	16.55	14.44
MD Rand	5.35	4.68	5.02	15.80	23.54	19.67
EL Long	0.21	0.00	0.11	9.52	11.85	10.69
EL Short	11.92	13.99	12.96	16.89	23.04	19.96
EL Avg	3.64	3.03	3.34	13.83	22.55	18.19
EL Rand	7.89	7.15	7.52	16.55	25.32	20.94
LNode LEdge	4.02	3.99	4.00	8.42	10.94	9.68
LNode HEdge	6.80	8.01	7.41	9.73	12.62	11.17
HNode LEdge	5.75	5.47	5.61	23.74	25.72	24.73
HNode HEdge	7.57	6.69	7.13	18.78	26.31	22.55
RNode REdge	5.67	6.05	5.86	12.26	24.60	18.43

Table 2. Average of total backfill time on the DAS trace (in seconds)

Task Graph Parameters	2 TPEs			4 TPEs		
	GS7_12	GS13_18	Avg	GS7_12	GS13_18	Avg
MD Low	1,089.00	432.00	760.50	711.33	209.67	460.50
MD Avg	4,499.00	2,301.33	3,400.17	2,121.33	2,585.33	2,353.33
MD High	598.67	145.00	371.83	197.67	614.33	406.00
MD Rand	943.33	1,041.67	992.50	698.67	644.33	671.50
EL Long	2,834.67	1,627.33	2,231.00	1,574.33	491.33	1,032.83
EL Short	1,811.33	1,114.00	1,462.67	467.33	2,469.33	1,468.33
EL Avg	2,263.67	379.67	1,321.67	777.33	329.00	553.17
EL Rand	220.33	799.00	509.67	910.00	764.00	837.00
LNode LEdge	1,760.67	865.33	1,313.00	981.33	329.67	655.50
LNode HEdge	602.67	74.67	338.67	436.67	9.33	223.00
HNode LEdge	620.67	102.00	361.33	201.67	146.67	174.17
HNode HEdge	1,259.67	382.00	820.83	509.33	962.67	736.00
RNode REdge	2,886.33	2,496.00	2,691.17	1,600.00	2,605.33	2,102.67

There are two main reasons that the AR scheduler manages to complete much earlier. The first reason is because a set of TGs in a single reservation slot can be interweaved successfully, as shown in Table 1. For TGs on a GS7_12 branch fot 4 TPEs, the initial reservation duration time is reduced up to 23.74% on the HNode LEdge branch. For TGs on a GS13_18 branch for 4 TPEs, the maximum reduction is 26.31% on the HNode HEdge branch. In constrast, the reduction is much smaller for 2 TPEs on the same branches. The reduction in the reservation duration time can also be referred to as an increase in the efficiency of scheduling

Table 3. Average of total backfill time on the LPC trace (in seconds)

Task Graph Parameters	2 TPEs			4 TPEs		
	GS7_12	GS13_18	Avg	GS7_12	GS13_18	Avg
MD Low	2,451.67	1,640.67	2,046.17	1,136.00	815.67	975.83
MD Avg	883.00	474.00	678.50	718.00	2,874.33	1,796.17
MD High	1,902.33	1,916.67	1,909.50	2,334.00	678.00	1,506.00
MD Rand	2,474.67	1,698.67	2,086.67	2,172.00	1,020.33	1,596.17
EL Long	2,018.67	1,611.33	1,815.00	1,889.00	1,419.33	1,654.17
EL Short	1,830.67	1,835.00	1,832.83	1,610.00	1,846.33	1,728.17
EL Avg	2,469.00	1,213.67	1,841.33	1,218.33	455.00	836.67
EL Rand	1,393.33	1,070.00	1,231.67	1,642.67	1,667.67	1,655.17
LNode LEdge	1,578.33	978.00	1,278.17	1,459.33	1,419.00	1,439.17
LNode HEdge	1,126.33	1,051.33	1,088.83	1,387.67	541.67	964.67
HNode LEdge	2,114.33	683.00	1,398.67	828.00	940.33	884.17
HNode HEdge	1,121.67	1,529.33	1,325.50	838.00	1,011.00	924.50
RNode REdge	1,771.00	1,488.33	1,629.67	1,847.00	1,476.33	1,661.67

TGs in this experiment. Overall, these results show that the achievable reduction depends on the size of the TGs and their graph properties as well.

The second reason is because there are many small independent jobs that can be used to fill in the "gaps" within a reservation slot, as depicted in Table 2 and 3. However, on average, the AR scheduler manages to backfill more jobs from the LPC trace into the reservation slot compare to the DAS trace. This is due to the characteristics of workload jobs themselves. The first 1000 jobs from the LPC trace are primarily independent jobs that require only 1 PE with an average runtime of 23.11 seconds. In contrast, the first 1000 jobs from the DAS trace contain a mixture of independent and parallel jobs that require on average 9.15 PEs with an average runtime of 3676.70 seconds. These phenomena also explain why the total completion time on the DAS trace took much longer than the LPC one.

An interesting observation to note from Figure 6 and 7 is that, the total completion time for the AR scheduler is the same for a resource with 16, 32 and 64 PEs. The FCFS and Backfill algorithm only manage to finish within the same time as the AR scheduler when a resource has 64 PEs. Hence, the AR scheduler executes these jobs and TGs more efficiently.

5   Conclusion and Future Work

In this paper, we have presented a novel approach to schedule TGs by using advance reservation in a cluster environment. In addition, to improve resource utilization, we proposed a scheduling solution (AR scheduler) by interweaving one or more TGs within the same reservation block and/or backfilling with other independent jobs.

The results showed that the AR scheduler performs better than the standard FCFS and Easy backfilling algorithms for reducing both the reservation duration time and the total completion time. The AR scheduler managed to interweave a set of task graphs with a reduction of up to 23.74% and 26.31% on 7–12 nodes and 13–18 nodes with 4 target processing elements (TPEs) respectively. However, much smaller reduction is noticed for 2 TPEs on same nodes. These results also showed that the achievable reduction depends on the size of the task graphs and their graph properties as well. Finally, the results showed that when there are many small independent jobs, the AR scheduler accomplished to fill these jobs into the reservation blocks.

An extension to this work is to consider scheduling task graphs with an economy model in order to see how efficient the AR scheduler is in terms of resource profits and user costs. Moreover, the AR scheduler can be extended to interweave task graphs from different reservation slots within a specified time block.

Acknowledgement. We would like to thank Chee Shin Yeo, Uros Cibej and anonymous reviewers for their comments on the paper.

References

1. Hoenig, U., Schiffmann, W.: A comprehensive test bench for the evaluation of scheduling heuristics. In: Proc. of the 16th International Conference on Parallel and Distributed Computing and Systems (PDCS'04), Cambridge, USA (2004)
2. Coffman, E.G., ed.: Computer and Job-Shop Scheduling Theory. Wiley (1976)
3. Hoenig, U., Schiffmann, W.: Improving the efficiency of functional parallelism by means of hyper-scheduling. In: Proc. of the the 35th International Conference on Parallel Processing (ICPP - in print), Ohio, USA (2006)
4. Sih, G.C., Lee, E.A.: A compile-time scheduling heuristic for interconnection-constrained heterogeneous processor architectures. IEEE Transactions on Parallel and Distributed Systems $4(2)$ (1993) 75–87
5. Hwang, J.J., Chow, Y.C., Anger, F.D., Lee, C.Y.: Scheduling precedence graphs in systems with interprocessor communication times. SIAM Journal on Computing $18(2)$ (1989) 244–257
6. Adam, T.L., Chandy, K.M., Dickson, J.: A comparison of list scheduling for parallel processing systems. Communications of the ACM 17 (1974) 685–690
7. Wu, M.Y., Gayski, D.D.: Hypertool: A programming aid for message-passing systems. IEEE Transactions on Parallel and Distributed Systems $1(3)$ (1990) 330–343
8. Tannenbaum, T., Wright, D., Miller, K., Livny, M.: Condor – a distributed job scheduler. In Sterling, T., ed.: Beowulf Cluster Computing with Linux. MIT Press (2001)
9. Thain, D., Tannenbaum, T., Livny, M.: Distributed computing in practice: the Condor experience. Concurrency – Practice and Experience $17(2\text{-}4)$ (2005) 323–356
10. Berman, F., Chien, A., Cooper, K., Dongarra, J., Foster, I., Gannon, D., Johnsson, L., Kennedy, K., Kesselman, C., Mellor-Crummey, J., Reed, D., Torczon, L., Wolski, R.: The GrADS project: Software support for high-level grid application development. International Journal of High Performance Computing Applications $15(4)$ (2001) 327–344

11. Deelman, E., Blythe, J., Gil, Y., Kesselman, C., Mehta, G., Patil, S., Su, M.H., Vahi, K., Livny, M.: Pegasus: Mapping scientific workflow onto the grid. In: Across Grids Conference 2004, Nicosia, Cyprus (2004)
12. Oinn, T., Addis, M., Ferris, J., Marvin, D., Senger, M., Greenwood, M., Carver, T., Glover, K., Pocock, M., Wipat, A., Li, P.: Taverna: a tool for the composition and enactment of bioinformatics workflows. Bioinformatics **20**(17) (2004) 3045–3054
13. McGough, S., Young, L., Afzal, A., Newhouse, S., Darlington, J.: Performance architecture within ICENI. UK e-Science All Hands Meeting (2004) 906–911
14. McGough, S., Young, L., Afzal, A., Newhouse, S., Darlington, J.: Workflow enactment in ICENI. UK e-Science All Hands Meeting (2004) 894–900
15. Sinnen, O., Sousa, L.: On task scheduling accuracy: Evaluation methodology and results. Journal of Supercomputing **27**(2) (2004) 177–194
16. Ahmad, I., Kwok, Y.K.: On exploiting task duplication in parallel program scheduling. IEEE Transactions on Parallel and Distributed Systems **9**(8) (1998) 872–892
17. Kwok, Y.K., Ahmad, I.: Link contention-constrained scheduling and mapping of tasks and messages to a network of heterogeneous processors. Cluster Computing: The Journal of Networks, Software Tools, and Applications **3**(2) (2000) 113–124
18. Sulistio, A., Buyya, R.: A grid simulation infrastructure supporting advance reservation. In: Proc. of the 16th International Conference on Parallel and Distributed Computing and Systems (PDCS'04), Cambridge, USA (2004)
19. Mu'alem, A.W., Feitelson, D.G.: Utilization, predictability, workloads, and user runtime estimates in scheduling the ibm sp2 with backfilling. IEEE Transactions on Parallel and Distributed Systems **12**(6) (2001) 529–543
20. Feitelson, D.: Parallel workloads archive. http://www.cs.huji.ac.il/labs/parallel/workload (2006)
21. Li, H., Groep, D., Walters, L.: Workload characteristics of a multi-cluster supercomputer. In Feitelson, D.G., Rudolph, L., Schwiegelshohn, U., eds.: Job Scheduling Strategies for Parallel Processing. Springer-Verlag (2004) 176–193 Lect. Notes Comput. Sci. vol. 3277.
22. Medernach, E.: Workload analysis of a cluster in a grid environment. In Feitelson, D.G., Rudolph, L., Schwiegelshohn, U., eds.: Job Scheduling Strategies for Parallel Processing. Springer-Verlag (2005) 36–61

Estimation Based Load Balancing Algorithm for Data-Intensive Heterogeneous Grid Environments

Ruchir Shah[1], Bharadwaj Veeravalli[2], and Manoj Misra[1]

[1] Department of Electronics and Computer Engineering, Indian Institute of
Technology, Roorkee, India 247667
r4_ruchir@yahoo.co.in, manojfec@iitr.ernet.in
[2] Computer Networks and Distributed Systems Laboratory, Department of Electrical
and Computer Engineering, National University of Singapore, Singapore 117576
elebv@nus.edu.sg

Abstract. Grid computing holds the great promise to effectively share
geographically distributed heterogeneous resources to solve large-scale
complex scientific problems. One of the distinct characteristics of the
Grid system is resource heterogeneity. The effective use of the Grid requires an approach to manage the heterogeneity of the involved resources
that can include computers, data, network, etc. In this paper, we proposed a de-centralized and adaptive load balancing algorithm for heterogeneous Grid environment. Our algorithm estimates different system
parameters (such as job arrival rate, CPU processing rate, load at processor) and effectively performs load balancing by considering all necessary affecting criteria. Simulation results demonstrate that our algorithm
outperforms conventional approaches in the event of heterogeneous environment and when communication overhead is significant.

Keywords: Grid systems, Heterogeneous environment, Load balancing,
Average response time, Communication overhead, Migration.

1 Introduction

One of the primary goals of the Grid computing [6,8] is to share access to geographically distributed heterogeneous resources in a transparent manner. With
its multitude of resources, a proper scheduling and efficient load balancing across
the Grid can lead to improved overall system performance and a lower turnaround time for individual jobs. Many load balancing algorithms have been proposed for traditional distributed systems. However, little work has been done
to cater the following unique characteristics for the Grid computing environment: heterogeneous resources, considerable communication delay and dynamic
network topology. The present work is targeted to the Grid model where heterogeneous resources are connected through arbitrary topology and network bandwidth also varies from link to link.

Y. Robert et al. (Eds.): HiPC 2006, LNCS 4297, pp. 72–83, 2006.

Load balancing involves assigning to each processor work proportional to its performance, thereby minimizing the execution time of a job. But there are a wide variety of issues that need to be considered for heterogeneous Grid environment. For example, capacities (in terms of processor speed) of the machines differ because of processor heterogeneity. Also their usable capacities vary from moment to moment according to load imposed upon them. Further, in Grid computing, as resources are distributed in multiple domains in the Internet, not only the computational nodes but also the underlying network connecting them is heterogeneous. The heterogeneity results in different capabilities for job processing and data access. For instance, typical bandwidth in Local Area Network (LAN) vary from $10Mbps$ to $1Gbps$, whereas it is few $kbps$ to $Mbps$ for Wide Area Network (WAN). Also network bandwidth across resources varies from link to link for large-scale Grid environments. Further, network topology among resources is also not fixed due to dynamic nature of the Grid. The focus of our work is to present load balancing algorithm adapted to heterogeneous Grid computing environment which considers all necessary factors such as load at processor, processor heterogeneity and job migration cost for load balancing. The conventional parallel processing scheduling and load balancing methods can not be applied directly to the Grid system due to its unique characteristics. In this paper, we attempt to formulate an adaptive, decentralized, sender-initiated load balancing algorithm for heterogeneous Grid environments which is based on ELISA [7]. Below we highlight some key contributions from the literature that are relevant to the context of our paper.

1.1 Related Works

One problem that is critical to effective utilization of computational grids is the efficient scheduling of jobs. Many job scheduling algorithms [1,3,4,5,9] have been proposed to deal with the heterogeneity and dynamic nature of distributed systems so as to optimize some figure of merit, for instance, minimize average job response time or better resource utilization. Martin [9] studied the effects of communication latency, overhead and bandwidth in cluster architecture to observe the impact on application performance. Shan et al. [3] considered heterogeneity in the system as well as in workload to optimize execution time in which sender processor collects status information about neighboring processors by communicating with them at every load balancing instant. This can lead to frequent message transfer. For large-scale Grid environment where communication latency is very large, status exchange at each load balancing instant can lead to large communication overhead. In our approach, the problem of frequent exchange of information is alleviated by estimating load, based on system state information received at sufficiently large interval of time. Arora et al. [5] proposed a decentralized load balancing algorithm for heterogeneous Grid environment. Although this work attempts to include communication latency between two nodes during triggering process on their model, it did not consider the actual cost for job transfer which is a significant factor in load balancing decision.

Our approach for load balancing in heterogeneous Grid environment is based on periodic status exchange concept of ELISA [7]. In ELISA, load balancing decision is taken based on queue length only. But due to unique characteristic of heterogeneous Grid environment, load balancing decision should consider all affecting factors which are current load at processor, processor heterogeneity and migration cost of a job. In the rest of the paper, we have used terms "Processor", "Resource" and "Node" interchangeably.

1.2 Our Contributions

In [2], it was pointed out that serious performance degradation will occur for slower networks such as for Grid environment if data migration cost is not considered when scheduling jobs. Further, to minimize total execution time of a job, load should be assigned to each processor proportional to its performance, taking into account processor heterogeneity in terms of its speed. In a heterogeneous Grid environment, performance of the system is largely affected by resource heterogeneity, considerable communication delay, dynamic changing environment etc. In this work, we have proposed a dynamic, de-centralized, sender-initiated load balancing algorithm which is applicable to heterogeneous Grid environments. One of the important characteristics of our algorithm is to estimate system parameters such as job arrival rate, processing rate and to perform proactive job migration. We study the effects of several influencing factors such as job size, arrival rate, number of migrations, and processor heterogeneity to show the behavior of our algorithm.

The paper is organized as follows: In Section 2, we present our Grid system model and problem definition. In Section 3, we provide design of our load balancing algorithm. In Section 4, we describe reference algorithms which are used to compare the results of our algorithms. Performance of our algorithm and discussion of results are presented in Section 5. Finally, Section 6 concludes the paper.

2 System Model and Problem Definition

Our Grid system model consists of a set of M heterogeneous resources, labeled as $P_1, P_2, ..., P_M$, connected by a communication network. The resources may be of different hardware architecture and processing speed can be different for different resources. There is no possibility of dropping of a job due to unavailability of buffer space as we assume that each resource has an infinite capacity buffer. For any resource P_i, jobs are assumed to arrive randomly at the processors, the inter-arrival time being exponentially distributed with average $1/\lambda_i$. The jobs are assumed to require service time that are exponentially distributed with mean $1/\mu_i$. All jobs are assumed to be mutually independent and can be executed on any node. Thus, each node is modeled as a $M|M|1$ Markov chain, with the number of jobs queued up for processing at each node representing the state of the system. Job size is assumed to have normal distribution with a given mean and variance. This job size includes both program and data size. As Grid is dynamic in nature, there is no fixed network topology. In our model, we consider arbitrary network topology to capture this constraint. Also data transfer rate is not same for each link connecting two resources. Nodes which are directly connected to a node constitute its *buddy set*. We also assume that each node has knowledge about its buddy nodes (in terms of processor speed and communication latency between them) and load balancing is carried out within buddy sets only. It may be noted that two neighboring buddy sets may have few nodes common to each set. Job arrival rates and service rates are such that for some node (say P_i), $\lambda_i > \mu_i$ (that is P_i is unstable), but whole system always remains stable, that is $\sum_{i=1}^{M} \lambda_i < \sum_{i=1}^{M} \mu_i$.

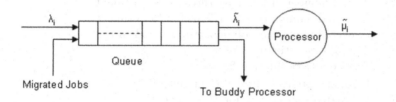

Fig. 1. System Model

Before presenting the exact problem statement, we first describe notations and terminology that are used throughout the paper below (refer to Table 1). We will now introduce certain key performance metrics of interest considered in this paper.

2.1 Performance Metrics

We have considered two performance metrics of relevance in our study. The response (or turnaround) time is probably the single most important measure for an individual submitting a job. If N jobs are processed by the system, then Average Response Time (ART) of the system can be calculated as follows:

$$ART = \frac{1}{N} \sum_{i=1}^{N} (FT_i - AT_i) \qquad (1)$$

where AT_i is the time at which the ith job arrives and FT_i is the time at which it leaves the system. The delay due to job transfer, waiting time in queue and processing time, together constitute the response time.

At the system level, we consider total execution time as performance metric to measure algorithm's efficiency. It indicates time at which all N jobs get finished.

Thus, our objective is to design an adaptive, de-centralized load balancing algorithm which minimizes the Average Response Time (ART) of the jobs for

Table 1. List of notations and terminology

M	Number of heterogeneous processors $(P_1, P_2, ..., P_M)$
N	Number of jobs to be processed
λ_i	Actual arrival rate for P_i (Poisson distribution)
$1/\mu_i$	Actual mean service time for P_i (Exponential distribution)
CST	Current System Time
T_s	Status exchange period
T_e	Load estimation period
ERT_i^j	Estimated Run Time of job j on P_i
EFT_i^j	Estimated Finish Time of job j on P_i
$A_i(t)$	Actual number of job arrivals for P_i in time t
$D_i(t)$	Actual number of job departures for P_i in time t
$EA_i(t)$	Expected number of job arrivals for P_i in time t
$ED_i(t)$	Expected number of job departures for P_i in time t
α	Arrival rate estimation factor
β	Service rate estimation factor
$\widetilde{\lambda}_i(T)$	Estimated arrival rate for P_i at time T
$\widetilde{\mu}_i(T)$	Estimated service rate for P_i at time T
$\widetilde{L}_i(T)$	Estimated load on P_i at time T
$\widetilde{L}_{k,i}(T)$	Estimated load on buddy processor P_k calculated by P_i at time T
$Q_i(T)$	Number of jobs waiting in queue for P_i at time T

heterogeneous Grid environments. Our algorithm is highly adaptive in nature in the sense that number of job migration for execution of N jobs triggers by available network bandwidth, processor heterogeneity and current load at processor.

3   Design of Load Balancing Algorithm

In any distributed systems, even simple load sharing policies yields significant improvements in performance over the no sharing case. But in heterogeneous Grid environment, where data size is very large or network bandwidth is low, it is critical to consider data transfer overhead when making job migration decision. Further, when resources are heterogeneous, we need to assign jobs to processors according to its performance. Our algorithm is based on ELISA and does parameter estimation and information exchange at regular intervals. We shall first describe how each P_i estimates its parameters and then will elaborate how load balancing is carried out in our algorithm.

As shown in Fig. 2, at each periodic interval of time T_s, called the status exchange interval, each P_i in the system calculates its status parameters which are estimated arrival rate, service rate and load on processor using following relationships:

$$\widetilde{\lambda}_i(T_{n-1}) = \alpha * \widetilde{\lambda}_i(T_{n-2}) + (1 - \alpha) * (A_i(T_s)/T_s) \tag{2}$$

$$\widetilde{\mu}_i(T_{n-1}) = \beta * \widetilde{\mu}_i(T_{n-2}) + (1 - \beta) * (D_i(T_s)/T_s) \tag{3}$$

$$\widetilde{L}_i(T_{n-1}) = Q_i(T_{n-1})/\widetilde{\mu}_i(T_{n-1}) \tag{4}$$

Thus, in the above relationship, by changing value of the parameter α ($0 \leq \alpha \leq 1$), one can vary the estimate. A value of 0.5 for α would mean that an equal weight has been considered for the current period and the previous estimate of the arrival rate. Similarly, we tune the parameter β ($0 \leq \beta \leq 1$) for service rate estimation.

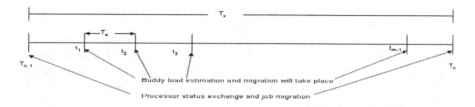

Fig. 2. Estimation and status exchange intervals

Each P_i in the system exchanges its status information with the processors in its buddy set. In Fig. 2, T_{n-1} and T_n represent the status exchange instant. Each status exchange period is further divided into equal subintervals called estimation interval T_e. These points are known as estimation instants. In Fig. 2, t_1, t_2,...,t_{m-1} represent estimation instants. As each processor balances the load within its buddy set, every processor estimates the load in the processors belonging to its buddy set at each estimation instants by following equations:

$$\widetilde{L}_{k,i}(T_{n-1} + t_i) = ((EA_k(T_e) - ED_k(T_e))/\widetilde{\mu}_k(T_{n-1})) + \widetilde{L}_{k,i}(T_{n-1} + t_{i-1}) \tag{5}$$

where $i = 1, 2, 3, ..., m - 1$ and
$EA_k(T_e) = a$ such that

$$\sum_{x=0}^{a} \frac{e^{(-\widetilde{\lambda}_k(T_{n-1})*T_e)} * (\widetilde{\lambda}_k(T_{n-1}) * T_e)^x}{x!} \cong 1 \tag{6}$$

$ED_k(T_e) = d^1$ such that

$$\sum_{x=0}^{d} \frac{e^{(-\widetilde{\mu}_k(T_{n-1})*T_e)} * (\widetilde{\mu}_k(T_{n-1}) * T_e)^x}{x!} \cong 1 \tag{7}$$

Depending on the accuracy required, computations of $EA_k(T_e)$ and $ED_k(T_e)$ can be terminated after computing a sufficiently large number of terms in (6)

---

[1] Note that number of job departures can not be greater than number of job arrivals. That is,

$$ED_k(T_e) \leq (EA_k(T_e) + \widetilde{L}_{k,i}(T_{n-1} + t_{i-1}) * \widetilde{\mu}_k(T_{n-1}))$$

and (7). The status exchange instants and the estimation instants together constitute the set of transfer instants $(T_{n-1}, t_1, t_2, ..., t_{m-1}, T_n)$ in Fig. 2. At the transfer instants, rescheduling of jobs is carried out. By making the interval between status exchange instants large, and by restricting the exchange of information to the buddy set, the communication overheads are kept at a minimum.

Based on this calculated buddy load, each processor calculates average load in its buddy set. P_i will take decision of job migration if its load is greater than an average load in its buddy set and will try to distribute its load such that load on all buddy processors get finished at almost same time taking into account node's heterogeneity in terms of processor speed. This average buddy load can be calculated using following relationships.

Let S_i denote the weight of a processor P_i which is a normalized measure of its speed. So a value of 2 for S_i means P_i will take half amount of time to execute job than time taken by reference[2] processor having value of 1 for S_i. Here, each P_i will calculate normalized buddy average load (NBL_{avg}) using value of $\widetilde{L}_{k,i}(T)$ and S_i by following equation:

$$NBL_{avg} = \frac{\sum_{k \epsilon buddyset_i} S_k * \widetilde{L}_{k,i}(T)}{\sum_{k \epsilon buddyset_i} S_k} \tag{8}$$

NBL_{avg} indicates average load for reference processor. P_i is considered as a sender processor, if $NBL_{avg} < S_i * \widetilde{L}_i(T)$. Now P_i will try to transfer its extra load to all receiver processors P_k such that they receive extra load based on their current load ($\widetilde{L}_{k,i}(T)$) and processor weight (S_k). After determining how much load P_i can transfer to P_k, as shown in Fig. 3, P_i will calculate expected finish time of job j on buddy processor (P_k) by estimating load on P_k at time $CST + t_c^j$ (where t_c^j is migration time for job j from P_i to P_k). Job will be migrated to P_k only if $EFT_k^j < EFT_i^j$,
where

$$EFT_i^j = Q_i(CST)/\widetilde{\mu}_i(T_{n-1}) + ERT_i^j \tag{9}$$

$$EFT_k^j = max(\ (\widetilde{L}_{k,i}(CST) + (EA_k(t_c^j) - ED_k(t_c^j))/\widetilde{\mu}_k(T_{n-1})),\ t_c^j\ ) + ERT_k^j \tag{10}$$

In (10), first term which is maximum of two values - approximate wait time of job j on P_k and job transfer time - indicates expected starting time of job j on P_k. We assume that these activities can be performed simultaneously. So job will be migrated only if its expected finish time on destination processor is less than expected finish time on source processor. Figure 4 shows complete working of our load balancing algorithm.

4   Reference Algorithms

We have used two algorithms, which are relevant to our context, as reference algorithms to compare results of our algorithm.

---

[2] This could be an abstract processor within the system.

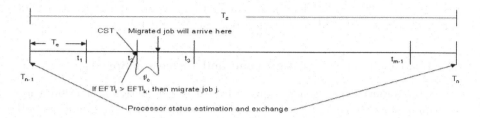

Fig. 3. Job migration decision

Main Algorithm

At the status exchange instant, for each processor:
1. Estimate the arrival rate, service rate and load on processor using equations (2), (3) and (4).
2. Communicate the status defined by a 3-tuple as: <estimated arrival rate, estimated service rate, estimated load> to all processors in the buddy set.
3. Call TRANSFER.

At the estimation instant, for each processor
1. Estimate the load for each processor in the buddy set using equation (5), (6) and (7).
2. Call TRANSFER.

Procedure TRANSFER by P_i, i=1,2,...,M.

1. Estimate an average normalized buddy load using equation (8).
2. If load of a processor is greater than average load (as computed in 1), then
   a) Construct active set as follows: if a processor in the buddy set has load less than the average normalized buddy load , include processor in active set.
   b) Determine how much load can be transferred to buddy processors in active set such that load on all processors gets finished at almost same time.
   c) Attempt to migrate the load in excess over average buddy load to all buddy processors in active set by calculating EFT_k^j on destination processor and migrating job only if $EFT_k^j < EFT_i^j$.

Fig. 4. Load balancing algorithm

4.1   Load Balancing Based on Queue Length

This approach is similar to ELISA [7]. In ELISA, each processor estimates queue length of its buddy processor at estimation instant using information exchanged at status exchange instant. Here, each processor will exchange estimate of arrival rate, service rate and queue length at each status exchange period T_s. At T_e, each processor will calculate queue length of its buddy processors by estimating number of job arrivals and departures. P_i will transfer its job to destination processor (P_k) if its queue length is greater than average queue length in its buddy set. Interested readers can refer to [7] for more details.

4.2 Load Balancing Based on Load and Processor Speed

Owing to resource heterogeneity, queue length is not always best criteria for determining load imbalance. Instead, product of average processing time of a job and queue length provides better load index for balancing load. In this approach, load balancing is done based on load (in terms of expected time to execute all jobs waiting in queue) rather than based on queue length. Here, P_i will take decision of job migration if its load is greater than an average load in its buddy set and will try to distribute its load such that load on all buddy processors get finished at almost same time on all buddy processors taking into account node's heterogeneity in terms of processor speed.

5 Performance Evaluation and Discussions

Here we present the results of our simulation study and compare the performance of our proposed algorithm with other algorithms discussed in Section 4. In our simulation model, we have considered 16 heterogeneous processors connected by communication channels assuming an arbitrary topology generated by a graph generator tool as shown in Fig. 5. Weight on each link indicates data transfer rate in $Mbps$. Various parameter values used for simulation are shown in Table 2. These parameter values are used for all cases unless otherwise stated explicitly.

We have used following notations for our algorithms for this section:

LBQL - Load Balancing based on Queue Length
LBLS - Load Balancing based on Load and processor Speed
LBALL - Load Balancing considering All criteria (Our proposed algorithm)

Table 2. Parameter values

Parameter	Value
Mean inter-arrival time	Exponentially distributed in $[1, 4]$
Mean service time	Exponentially distributed in $[1, 4]$
N	10000
T_s	20
T_e	5
α	0.5
β	0.5
Job Size	Normal distribution with μ=5MB and σ=1MB

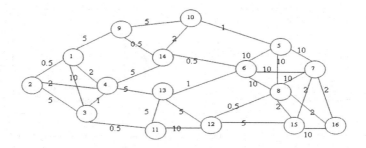

Fig. 5. Network topology and bandwidth across link

5.1   Performance for Heterogeneous Case

For heterogeneous case, we considered different data transfer rate for each link as shown in Fig. 5. We also considered resource heterogeneity by setting value of S_i to 2 for randomly half of processors. Our algorithm gives better performance for *ART* as can be seen from Fig. 6(a). As from Fig. 6(c), total number of migration performed by our algorithm is less than other reference algorithms as it is not always advisable to perform migration in event of low data transfer rate and resource heterogeneity.

5.2   Performance for Homogeneous Case

This is a special case to our heterogeneous environment. In this case, we have considered all nodes are homogeneous, that means S_i is set to 1 for all processors. Also, network bandwidth is fixed and it is $10Mbps$ for all links. As shown in Fig. 7(a) and 7(b), ART and total execution time are almost same for all algorithms. Even number of migration performed for execution of N jobs is also identical as can be seen from Fig. 7(c). This observation indicates that our algorithm gives as good performance as given by LBQL or LBLS in case of homogeneous case.

5.3   Effect of Job Size

Our algorithm also takes into account job migration cost and as job size largely affects job migration cost, it will be very interesting to measure performance of algorithms by varying job size. In this set of experiments, we varied job size from $5MB \pm 1MB$ to $50MB \pm 20MB$. From Fig. 8(a) and 8(b), as we increase job size, performance of our algorithm is far better than other referenced algorithms in terms of decrease in *ART* and total execution time. As from Fig. 8(c), for our algorithm, total number of migration gets decreased as we increase job size whereas it remains almost constant for LBQL and LBLS irrespective of large communication overhead due to increase in job size.

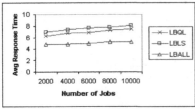

(a) ART comparison

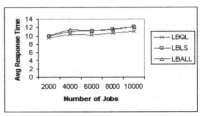

(a) ART comparison

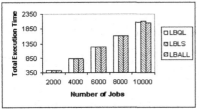

(b) Execution Time comparison

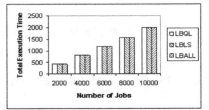

(b) Execution Time comparison

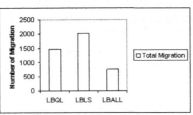

(c) Total migration comparison

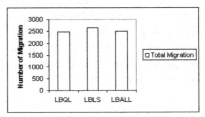

(c) Total migration comparison

Fig. 6. Heterogeneous case

Fig. 7. Homogeneous case

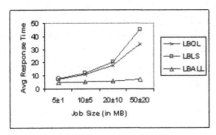

(a) ART comparison

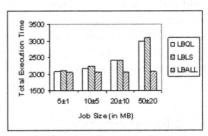

(b) Execution Time comparison

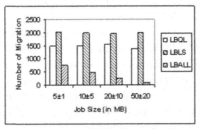

(c) Total migration comparison

Fig. 8. Effect of job size

6   Conclusions

Grid computing environment has many unique characteristics such as resource heterogeneity, communication overhead and dynamic nature that makes it different from the traditional distributed systems. These characteristics have a significant impact on the performance of load balancing. In this paper, we have proposed a de-centralized, distributed and adaptive load balancing algorithm to cater for these characteristics. Our algorithm estimates different types of system parameters and effectively does load balancing to improve performance of the system. Through simulation experiments, it is found that when communication overhead is significant and resources are heterogeneous, performance of our load balancing algorithm is far better than other reference algorithms.

Although we have considered resource heterogeneity in terms of processor speeds, it can be extended to heterogeneity in other dimensions such as, machine architecture, operating system, available storage space, etc. and can measure the performance of the algorithm.

References

1. Y.Murata, H.Takizawa, T.Inaba, and H.Kobayashi, "A distributed and cooperative load balancing mechanism for large-scale P2P systems", International Symposium on Applications and Internet Workshops (SAINT 06), pp. 126-129, Jan 2006.
2. H.Shan, L.Oliker, R.Biswas, and W.Smith, "Scheduling in heterogeneous grid environments: The effects of data migration", in Proceedings of ADCOM2004: International Conference on Advanced Computing and Communication, Ahmedabad, Gujarat, India, December, 2004.
3. H.Shan, L.Oliker, and R.Biswas, "Job superscheduler architecture and performance in computational grid environments", ACM/IEEE Conferece on Supercomputing, November, 2003.
4. V.Subramani, R.Kettimuthu, S.Srinivasan, and P.Sadayappan, "Distributed job scheduling on computational grid using multiple simultaneous requests", Proceedings of 11th IEEE Symposium on High Performance Distributed Computing (HPDC 2002), July 2002.
5. M.Arora, S.K.Das, and R.Biswas, "A De-centralized scheduling and load balancing algorithm for heterogeneous grid environments", Proceedings of the International Conference on Parallel Processing Workshops (ICPPW,2002), pp. 499-505, 2002.
6. I.Foster, C.Kesselman, and S.Tuecke, "The anatomy of the grid: Enabling scalable virtual organizations", International Journal of High Performance Computing Applications, Vol. 15, Issue 3, pp. 200-222, 2001.
7. L.Anand, D.Ghose, and V.Mani, "ELISA: An Estimated Load Information Scheduling Algorithm for distributed computing systems", International Journal on Computers and Mathematics with Applications, Vol.37, Issue 8, pp. 57-85, April 1999.
8. I.Foster and C.Kesselman, "The Grid : Blueprint for a future computing infrastructure", Morgan Kaufmann Publishers, USA, 1999.
9. R.Martin, A.Vahdat, D.Culler, and T.Anderson, "Effects of communication latency, overhead and bandwidth in a cluster architecture", Proceedings 24th Annual International Symposium on Computer Architecture, pp. 85-97, June 1997.

Improving the Flexibility of the Deficit Table Scheduler*

Raúl Martínez, Francisco J. Alfaro, and José L. Sánchez

Departamento de Sistemas Informáticos
Universidad de Castilla-La Mancha
02071 - Albacete, Spain
{raulmm, falfaro, jsanchez}@info-ab.uclm.es

Abstract. A key component for networks with Quality of Service (QoS) support is the egress link scheduler. The table-based schedulers are simple to implement and can offer good latency bounds. Some of the latest proposals of network technologies, like Advanced Switching and Infini-Band, define in their specifications one of these schedulers. However, these schedulers do not work properly with variable packet sizes and face the problem of bounding the bandwidth and latency assignments.

We have proposed a new table-based scheduler, the Deficit Table (DTable) scheduler, that works properly with variable packet sizes. Moreover, we have proposed a methodology to configure this table-based scheduler that partially decouples the bandwidth and latency assignments.

In this paper we propose a method to improve the flexibility of the decoupling methodology. Moreover, we compare the latency performance of this strategy with two well-known scheduling algorithms: the Self-Clocked Weighted Fair Queuing (SCFQ) and the Deficit Round Robin (DRR) algorithms.

1 Introduction

Current packet networks are required to carry not only traffic of applications such as e-mail or file transfer, which does not require pre-specified service guarantees, but also traffic of other applications that require different performance guarantees. The best-effort service model, though suitable for the first type of applications, is not so for applications of the other type [10]. A key component for networks with QoS support is the output scheduling algorithm, which selects the next packet to be sent and determines when it should be transmitted, on the basis of some expected performance metrics.

An ideal scheduling algorithm implemented in a high performance network with QoS support should satisfy two main properties: good end-to-end delay and simplicity. The design of a traffic scheduling algorithm involves an inevitable

---

* This work was partly supported by the Spanish CICYT under Grant TIC2003-08154-C06-02, by the Junta de Comunidades de Castilla-La Mancha under Grant PBC-05-005-1, and by the Spanish State Secretariat of Education and Universities under FPU grant.

Y. Robert et al. (Eds.): HiPC 2006, LNCS 4297, pp. 84–97, 2006.

trade-off among these properties. Many scheduling algorithms have been proposed. Among them, the "sorted-priority" family of algorithms are known to offer very good delay [13]. These algorithms assign each packet a service tag and transmit packets in an increasing order of service tag. However, their computational complexity is very high, making their implementation in high-speed networks rather difficult.

In order to avoid the complexity of the sorted-priority approach, the Deficit Round Robin (DRR) algorithm [11] has been proposed. In the DRR algorithm, a list of flow[1] *quantums* is visited sequentially, each quantum indicating the amount of data that can be transmitted from the flow in question. The sum of all the quantums is called the frame length. The main problem of this algorithm is that its delay and fairness depend on the frame length. Depending on the situation, the frame can be very long, and thus, the latency and fairness would be very bad.

On the other hand, in the table-based schedulers instead of serving packets of a flow in a single visit per frame, the service is distributed throughout the entire frame. This approach is followed in [3] and in two of the last high performance network interconnection proposals: Advanced Switching [1] and InfiniBand [5]. These table-based schedulers can provide a good latency performance with a low computational complexity [3, 9, 2]. However, these schedulers do not work properly with variable packet sizes, as is usually the case in current network technologies. Moreover, they face the problem of bounding the bandwidth and latency assignments [9, 2]. This may involve a waste of resources because the flows with the highest latency requirements are probably going to be assigned more bandwidth than they actually require.

In [7] we reviewed these problems and proposed a new table-based scheduler that works properly with variable packet sizes. Moreover, we proposed a methodology to configure this scheduler in such a way that it permits us to decouple partially the bounding between the bandwidth and latency assignments. We called this new scheduler Deficit Table scheduler, or just DTable scheduler. As we stated in [7], the latency performance of the DTable scheduler depends on the maximum amount of data that is allowed to be transmitted per table entry. The more information is allowed to be transmitted the worse latency performance we get. Therefore, this maximum amount of data should be kept as small as possible. However, one of the parameters that our configuration methodology uses to increase the decoupling between the bandwidth and the latency assignments is indeed this value.

In this paper we propose and evaluate a method to increase the decoupling between the bandwidth and the latency assignments without increasing too much the maximum amount of data that is allowed to be transmitted per table entry. Moreover, we compare the latency performance of the DTable scheduler with the latency performance of two well-known scheduling algorithms: the Self-Clocked Weighted Fair Queuing (SCFQ) [4] and the DRR algorithms. We have chosen

---

[1] In this paper we will use the term *flow* to refer both to a single flow or an aggregated of several flows with similar characteristics.

the SCFQ algorithm as an example of "sorted-priority" algorithm and the DRR algorithm as one of the simplest scheduling mechanism proposed in the literature.

The structure of the paper is as follows: In Section 2, we review the DTable scheduler and our methodology to decouple the bandwidth and latency assignments. In Section 3, we propose a method to improve the latency performance of the DTable scheduler. Details on the experimental platform and the performance evaluation are presented in Section 4. Finally, some conclusions are given.

2   The Deficit Table Scheduler

In [7] we proposed a new table-based scheduling algorithm that works properly with variable packet sizes. We called this algorithm Deficit Table scheduler, or just DTable scheduler, because it is a mix between the already proposed table-based schedulers and the DRR algorithm. The scheduler works in a similar way than the DRR algorithm but instead of serving packets of a flow in a single visit per frame, the service is distributed throughout the entire frame.

The DTable scheduler defines an arbitration table in which each table entry has associated a flow identifier and an *entry weight*. Moreover, each flow has assigned a *deficit counter* that is set to 0 at the start. When scheduling is needed, the table is cycled through sequentially until an entry assigned to an active flow is found. A flow is considered active when it stores at least one packet and the link-level flow control, if exists, allows that flow to transmit packets. When a table entry is selected, the *accumulated weight* is computed. The accumulated weight is equal to the sum of the deficit counter for the selected flow and the current entry weight. The scheduler transmits as many packets from the active flow as the accumulated weight allows. When a packet is transmitted, the accumulated weight is reduced by the packet size.

The next active table entry is selected if the flow becomes inactive or the accumulated weight becomes smaller than the size of the packet at the head of the queue. In the first case, the remaining accumulated weight is discarded and the deficit counter is set to zero. In the second case, the unused accumulated weight is saved in the deficit counter, representing the amount of weight that the scheduler owes the queue. Note that, if this scheduler is employed in a network with a credit-based link-level flow control, like Advanced Switching, the weights are usually expressed in flow control credits.

We set the minimum value that a table entry can have associated to the Maximum Transfer Unit (MTU) of the network. This is the smallest value that ensures that there will never be necessary to cycle through the entire table several times in order to gather enough weight for the transmission of a single packet. Note that this consideration is also made in the DRR algorithm definition [11].

In [7] we have also proposed a methodology to configure the DTable scheduler to decouple, at least partially, the bounding between the bandwidth and latency assignments. With this methodology we set the maximum distance between any consecutive pair of entries assigned to a flow depending on its latency requirement. By fixing this separation, it is possible to control the maximum latency of a network element crossing. This is because this distance determines the maximum time that a packet at the head of a flow queue is going to wait until being transmitted. Therefore, given a maximum number of hops, we can control the maximum end-to-end latency [2].

Moreover, we set the weights of the table entries assigned to a flow depending on its bandwidth requirement. With this methodology we can assign the flows with a bandwidth varying between a minimum and a maximum value that depends not only on the number of table entries assigned to each flow, but also on two table configuration parameters. We have called these parameters w and k.

Supposing an arbitration table with N entries in a network with a certain MTU, the w parameter determines the maximum weight M that can be assigned to a single table entry in function of the MTU: $M = MTU \times w$. The k parameter determines the total weight that can be distributed between all the table entries. We call this value the *bandwidth pool*: $pool = N \times MTU \times k$. The total number of weight units (as stated before, a weight unit is usually equivalent to a flow control credit) from the bandwidth pool that the table entries of a flow have assigned fixes the bandwidth that the flow has actually assigned.

Note that $k, w \geq 1$ because, as stated before, the minimum weight that can be assigned to a table entry is the MTU. Note also that $k \leq w$ because the bandwidth pool cannot be larger than the theoretical maximum weight among all the entries ($N \times M$).

The w and k parameters fix the minimum bandwidth $min\phi_i$ and the maximum bandwidth $max\phi_i$ that can be assigned to the ith flow depending on the number of table entries n_i that it has assigned:

$$min\phi_i = \frac{n_i \times MTU}{pool} = \frac{n_i \times MTU}{N \times MTU \times k} = \frac{n_i}{N} \times \frac{1}{k}$$

$$max\phi_i = \frac{n_i \times M}{pool} = \frac{n_i \times MTU \times w}{N \times MTU \times k} = \frac{n_i}{N} \times \frac{w}{k}$$

Summing up, the DTable scheduler is a table-based scheduler that is able to deal properly with variable packet sizes and considers the possibility of a link-level flow control mechanism. Moreover, with our configuration methodology we can provide a flow with latency and bandwidth requirements in a partially independent way.

3   Improving the Latency Performance of the DTable Scheduler

As stated before, using the DTable scheduler and our methodology, we can assign each flow a bandwidth between a minimum and a maximum that depends on the number of table entries and the two decoupling table parameters. When choosing the value of these parameters some considerations must be made. Note that the objective for this methodology is to decrease the minimum bandwidth that can be assigned to a flow and to increase the maximum bandwidth assignable in order to be as flexible as possible. In order to be able to assign a small amount of bandwidth the k parameter must be high. However, the higher k is, the smaller the maximum bandwidth that can be assigned. And thus, the flexibility to assign the bandwidth decreases. We can solve this by increasing the value of w.

Table 1 shows two different scenarios, each one with a different pair of values for the w and k parameters: DTable4 ($k = 2$, $w = 4$) and DTable8 ($k = 4$, $w = 8$). Note that we refer the different DTable scenarios according to the w value used in each case. This table shows the minimum and maximum bandwidth that can be assigned to seven flows (referred to as Virtual Channels, VCs) with different number of table entries. This number of table entries correspond to 7 flows with different latency requirements, and thus, different distances between any pair of consecutive entries in the arbitration table. We have called these flows D2, D4, D8, D16, D32, D64, and D64', indicating the distance between any pair of consecutive table entries. As we can see, when we increase the k parameter, the minimum bandwidth decreases. However, to maintain the same maximum bandwidth in the two scenarios, we have had to increase the w parameter in the same proportion.

Table 1. Table configuration. $N = 64$, $MTU = 32$.

VC	#entries	%entries	DTable4 $k=2, w=4$		DTable8 $k=4, w=8$	
			$min\phi_i$	$max\phi_i$	$min\phi_i$	$max\phi_i$
D2	32	50	25	100	12.5	100
D4	16	25	12.5	50	6.25	50
D8	8	12.5	6.25	25	3.125	25
D16	4	6.25	3.125	12.5	1.5625	12.5
D32	2	3.125	1.5625	6.25	0.78125	6.25
D64	1	1.5625	0.78125	3.125	0.390625	3.125
D64'	1	1.5625	0.78125	3.125	0.390625	3.125
Total	64	100	50	200	25	200

However, increasing the value of the w parameter has two disadvantages. First of all, the memory resources to store each entry weight are going to be higher. Secondly, the latency of the flows is going to increase, because each entry is allowing more information to be transmitted, and thus, the maximum time between any consecutive pair of table entries is higher.

Our objective is to have a good flexibility when assigning the bandwidth to the flows but without increasing too much the w parameter. In order to achieve this we propose to use different MTUs for the different flows, instead of considering the general network MTU that the technology fixes for all the flows. Note that this means that some flows are going to have a specific MTU smaller than the general MTU. The use of different MTUs for different flows can be done at the communication library level.

The advantage of having a flow with a specific MTU smaller than the general MTU is that we can assign a table entry a minimum weight equal to the new

MTU. When we use the general MTU for all the flows we cannot do this. As stated before, in this case, the general MTU is the smallest value that ensures that there will never be necessary to cycle through the entire table several times in order to gather enough weight for the transmission of a single packet. Being able to assign the table entries of a flow with a weight smaller than the general MTU allows to decrease the minimum bandwidth that can be assigned to that flow. If the ith flow uses a specific MTU of size MTU_i, the maximum bandwidth that can be assigned to that flow is the same, but the minimum bandwidth depends on the proportion between the specific MTU and the general MTU:

$$min\phi_i = \frac{n_i \times MTU_i}{pool} = \frac{n_i \times MTU_i}{N \times MTU \times k} = \frac{n_i}{N} \times \frac{MTU_i}{MTU} \times \frac{1}{k}$$

Note that varying the w and k parameters affect the minimum and maximum bandwidth that can be assigned to all the flows. However, assigning a specific MTU to a flow only affects that flow minimum bandwidth.

Note that with this method we can achieve small minimum bandwidths with a low value for the k parameter. Note also that now k can be even lower than 1. This allows to use a small w and still getting big maximum bandwidths.

Table 2 shows two different scenarios, each one with a different pair of values for the w and k parameters: DTable1 ($k = 0.5$, $w = 1$) and DTable2 ($k = 1$, $w = 2$). If we compare these values with the values in Table 1, we can see that now we can assign a small amount of bandwidth to those flows with lots of entries with a small w parameter.

Table 2. Table configuration with different MTUs. $N = 64$, $MTU = 32$.

| | | | | DTable1 | | DTable2 | |
| | | | | $k = 0.5$, $w = 1$ | | $k = 1$, $w = 2$ | |
VC	#entries	%entries	MTU_i	$min\phi_i$	$max\phi_i$	$min\phi_i$	$max\phi_i$
D2	32	50	$MTU/32$	3.125	100	1.5625	100
D4	16	25	$MTU/32$	1.5625	50	0.78125	50
D8	8	12.5	$MTU/16$	1.5625	25	0.78125	25
D16	4	6.25	$MTU/8$	1.5625	12.5	0.78125	12.5
D32	2	3.125	$MTU/4$	1.5625	6.25	0.78125	6.25
D64	1	1.5625	$MTU/2$	1.5625	3.125	0.78125	3.125
D64'	1	1.5625	MTU	3.125	3.125	1.5625	3.125
Total	64	100		14.0625	200	7	200

When a message from a given flow arrives at the network interface, if its size is greater than its specific MTU, the message is splitted in several packets of a maximum size given by the specific MTU of the flow, as can be seen in Figure 1. A possible disadvantage of assigning specific MTUs smaller than the general MTU could be that the bandwidth and latency overhead of fragmenting the original message in several packets could probably affect performance of the flows. However, most restrictive latency flows (for example network control or voice traffic) usually present low bandwidth requirements, and small packet size. For example, in [14] several payload values for voice codec algorithms are shown. These values range from 20 bytes to 160 bytes. In that way, if we fix a small MTU for these flows, no fragmentation will be usually necessary because, in fact, the packets of those flows are already smaller than the new MTU. Therefore, the cornerstone of this proposal is to tune the specific MTU of each flow according to the specific characteristics of the flows.

Fig. 1. Process of message fragmentation into packets

In the performance evaluation section we are going to use the same kind of traffic (with the same average packet size) for all the flows in order to make a fair comparison. Moreover we are going to assign smaller specific MTUs to those flows with more table entries in order to decrease the minimum bandwidth that can be assigned. Therefore, results are going to show the negative effect of an excessive packetization.

4   Performance Evaluation

In this section, we evaluate the latency performance of the DTable scheduler. For this purpose, we have developed a detailed simulator that allows us to model the network at the register transfer level, following the Advanced Switching (AS) specification. Note, however, that our proposals can be applied to any interconnection network technology.

We compare the performance of the different scenarios with a different w parameter showed in the previous section (DTable1, DTable2, DTable4, and DTable8) and the SCFQ and DRR schedulers. We have chosen the SCFQ algorithm as an example of "sorted-priority" algorithm and the DRR algorithm because of its very small computational complexity. In order to simulate these algorithms we use the credit aware versions of both algorithms (SCFQ Credit Aware and DRR Credit Aware respectively) that we proposed in [8] for being used in networks with a link-level flow control network like AS.

4.1   Simulated Architecture

We have used a perfect-shuffle Bidirectional Multi-stage Interconnection Network (BMIN) with 64 end-points connected using 48 8-port switches (3 stages of 16 switches). In AS any topology is possible, but we have used a MIN because it is a common solution for interconnection in current high-performance environments. The switch model uses a combined input-output buffer architecture with a crossbar to connect the buffers. Virtual output queuing has been implemented to solve the head-of-line blocking problem at switch level.

In our tests, the link bandwidth is 2.5 Gb/s but, with the AS 8b/10b encoding scheme, the maximum effective bandwidth for data traffic is only 2 Gb/s. We are assuming some internal speed-up (x1.5) for the crossbar, as is usually the case in most commercial switches. AS gives us the freedom to use any algorithm to schedule the crossbar, so we have implemented a round-robin scheduler. The time that a packet header takes to cross the switch without any load is 145 ns, which is based on the unloaded cut-through latency of the AS StarGen's *Merlin* switch [12].

A credit-based flow control protocol ensures that packets are only transmitted when there is enough buffer space at the other end to store them, making sure that no packets are dropped when congestion appears. AS uses Virtual Channels (VCs) to aggregate flows with similar characteristics and the flow control and the arbitration is made at VC level. The MTU of an AS packet is 2176 bytes, but we are going to use 2048 bytes (a power of two) for simplicity but without loosing generality. The credit-based flow control unit is 64 bytes, and thus, the MTU corresponds to 32 credits.

The buffer capacity is 32768 bytes (16×MTU) per VC at the network interfaces and 16384 bytes (8×MTU) per VC both at the input and at the output ports of the switches. If an application tries to inject a packet into the network interface but the appropriate buffer is full, we suppose that the packet is stored in a queue of pending packets at the application layer. Regarding the latency statistics, a packet is considered injected when it is stored in the network interface.

4.2   Simulated Scenario and Scheduler Configuration

As stated before, we are going to compare the performance of the DTable scheduler using different values for the w parameter (DTable1, DTable2, DTable4, and DTable8) with the performance of the SCFQ and DRR algorithms. Note that all the scenarios have the same maximum bandwidth values, differing only in the minimum bandwidth values (see Tables 1 and 2). Table 3 shows the amount of bandwidth ϕ_i that we have actually assigned to each VC. This table also shows the configuration of the different DTable scenarios and the SCFQ and the DRR schedulers. Specifically, in the case of the DTable scheduler, this table shows the total weight (T. w.) that we have distributed among the table entries of each VC and the weight assigned to each table entry (E. w.) of each VC. For example, in the DTable1 case, the bandwidth pool is 1024 credits ($k = 0.5$), and thus, in order to assign 25% of bandwidth to this VC, 256 credits must be assigned to it. Therefore, 8 credits have been assigned to each one of its 32 table entries.

We are going to inject an increasing amount of traffic of all the VCs and study the throughput and latency performance of the different possibilities at different network load levels. The traffic load is composed of self-similar point-to-point flows of 1 Mb/s. The destination pattern is uniform in order to fully load the network. The packets' size is governed by a Pareto distribution, as recommended in [6]. In this way, many small-sized packets are generated, with an occasional packet of large size. The minimum payload size is 56 bytes, the maximum 2040 bytes, and the average 176 bytes, which represents enough packet size variability. The AS packet header size is 8 bytes. The periods between packets are modelled with a Poisson distribution.

Table 3. Bandwidth configuration of the DTable scheduler scenarios

VC	ϕ_i	DTable1		DTable2		DTable4		DTable8		SCFQ	DRR
		E. w.	T. w.	E. w.	T. w.	E. w.	T. w.	E. w.	T. w.	Weight	Quantum
D2	25	8	256	16	512	32	1024	64	2048	0.25	256
D4	25	16	256	32	512	64	1024	128	2048	0.25	256
D8	25	32	256	64	512	128	1024	256	2048	0.25	256
D16	12.5	32	128	64	256	128	512	256	1024	0.125	128
D32	6.25	32	64	64	128	128	256	256	512	0.625	64
D64	3.125	32	32	64	64	128	128	256	256	0.3125	32
D64'	3.125	32	32	64	64	128	128	256	256	0.3125	32
Total	100		1024		2048		4096		8196	1	1024

4.3   Simulation Results

The figures of this section show the average values and the confidence intervals at 90% confidence level of ten different simulations performed at a given input load. For each simulation we obtain the normalized average throughput, the average message injection latency, and the maximum message injection latency of each flow. Note that in the DTable1 and DTable2 scenarios we use specific MTUs for the VCs that are smaller than the general MTU. Therefore, in these cases, a message can be splitted in several packets. In the rest of cases (DTable4, DTable8, SCFQ, and DRR) a mesage is going to be transmitted in only one packet. Note that in the DTable1 and DTable2 scenarios we consider the latency of the message as a whole. This means that, in these cases, to calculate the latency of a message we consider the time since we inject the first packet of a message into the network interface up to the last packet of the message arrives at its destination. Note that this may suppose a certain overhead. No statistics on packet loss are given because, as has been said, we assume a credit-based flow control mechanism to avoid dropping packets. We obtain statistics per VC aggregating the throughput of all the flows of the same VC, obtaining the average value of the average latency, and the maximum latency of all the flows. Note that the maximum latency shows the behavior of the flow with the worst performance.

Figure 2 shows the normalized injection rate of the aggregated of flows associated with each VC and the normalized throughput results per VC of the DTable1 scenario. The rest of scenarios for the DTable scheduler and the DRR and SCFQ schedulers obtain similar throughput results. As we can see, when the load is low, all the VCs obtain the bandwidth they inject. However, when the load is high (around 95%) the VCs do not yield a corresponding result, obtaining a bandwidth proportional to their assigned bandwidth. Note that the VCs do not obtain all the bandwidth that they were supposed to have assigned because the network is not able to provide 100% throughput.

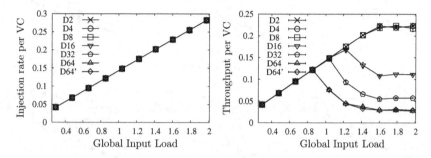

Fig. 2. Normalized injection rate and throughput per VC

Figure 3 shows the average latency performance. When the load is very low, all the VCs present a similar low latency. This is because at this load level there are few packets being transmitted through the network, and thus, there are few conflicts between them. However, when the load increases, the latency also increases because some packets must wait in the buffers until others have been transmitted. It is at this point when the scheduling algorithm assumes an important role and the VCs obtain a different latency depending on the scheduler configuration. However, when the load of the VC begins to outstrip its throughput, the latency of the scheduler starts to grow very fast. This is because the buffers used for that VC begin to be full. Finally, the buffers become completely full and the latency stabilizes at a given value which depends on the buffers' size and the bandwidth assigned to that VC.

Note that when using the SCFQ algorithm those VCs that have assigned the same bandwidth (in this case the D2, D4, and D8 VCs, and the D64 and D64' VCs) obtain the same latency performance. In the case of the DRR algorithm, all the VCs obtain a similar latency performance until a VC reaches the point when its load begins to outstrip its throughput. In that point, the latency of that VC grows very fast and obtains a different latency performance. This happens for all the VCs as load grows. When using the DTable scheduler, all the VCs, including those with the same bandwidth assignment, obtain a different latency performance depending on the separation between any consecutive pair of their table entries. The smaller the distance, the better latency performance they obtain.

These different latency performance behaviors are explained by the fact that the maximum time that a packet at the head of a VC queue is going to wait until being transmitted is different depending on the scheduler algorithm. In the case of the SCFQ algorithm, this time is proportional to the assigned bandwidth. In the case of the DTable scheduler, we can control this time by controlling the maximum separation between any consecutive pair of entries assigned to the same VC. In this way, we provide some VCs with a better latency performance

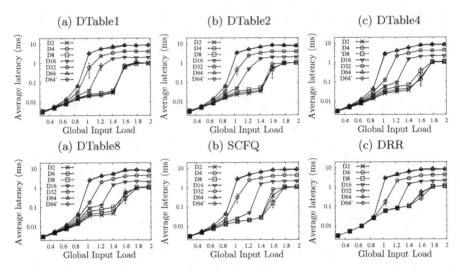

Fig. 3. Average latency per VC of the different scheduling scenarios

and other VCs with a worse latency performance. In the case of the DRR algorithm, the latency performance depends more on the frame length than on the quantum that each VC has been assigned. This is because when the quantum for a VC has been expended sending packets, all the frame must be cycled through before sending more packets of the same VC.

Finally, Figure 4 shows the percentage of improvement on average latency of the SCFQ algorithm over the four possibilities of the DTable scheduler and the DRR algorithm. Analyzing this figure we can compare the DTable performance comparing it not only with the SCFQ scheduler, but also with the DRR scheduler. Moreover, we can compare the difference between using the same general MTU for all the VCs or using specific MTUs for the VCs. This figure shows that in general, the SCFQ algorithm provides a better latency performance than the DTable scheduler in all the cases. However, this algorithm is the most complex. The DRR provides a worse performance for the most latency restrictive VCs and better for the less latency restrictive VCs than the DTable scheduler. This is because with the DTable scheduler we can provide a different level of latency performance to the VCs, priorizing those VCs with higher latency requirements. This is not possible with the DRR algorithm. Regarding the different scenarios of the DTable scheduler we can see that DTable1 provides a better latency performance than DTable2, and DTable4 than DTable8. This is because in general, the higher the value of the w parameter, the worse the latency performance. However, the effect of splitting the messages in several packets must also be taken into account.

Table 4 shows the bandwidth overhead per VC that is produced by using smaller specific MTUs than the general MTU. This packetization also has effect on the latency of the message. Note that each packet must be processed by the network elements (routing, scheduling, etc.). Moreover, if a table entry allows us

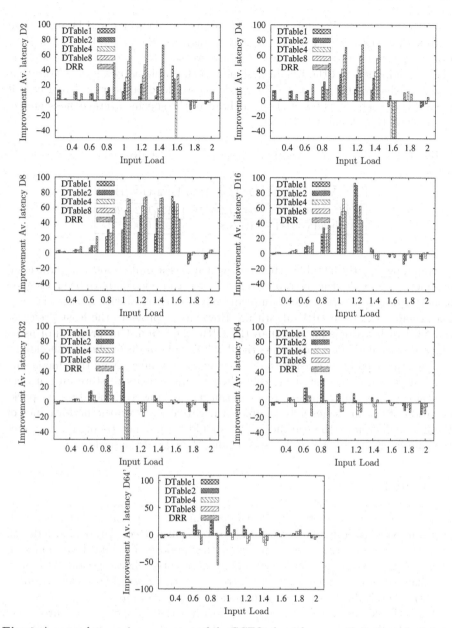

Fig. 4. Average latency improvement of the SCFQ algorithm over the other schedulers considered

to transmit a small number of packets of the new MTU size, it is possible that in order of transmitting all the packets belonging to the same message more than one table entry must be used, and thus, the latency increases. Figure 4 shows clearly the first effect when considering a low load for the D2 and D4 VCs. In this case, the latency of the DTable1 and Dtable2 scenarios is rather worse than for the others cases. We obtain a better latency for DTable1 and Dtable2 than DTable4 and Dtable8 when the latency is high for the D2, D4, and D8 VCs. However, for the rest of VCs we obtain a worse latency because the specific MTUs are higher and the weight assigned to the table entries lower. Note that this bad effect of the excessive packetization would disappear in a real case if the MTU of each VC is selected on the basis of the specific average message size of the flows that would use the VC.

Table 4. Packetization bandwidth overhead per VC with average packet size of 176 bytes

VC	D2	D4	D8	D16	D32	D64	D64'
MTU_i (bytes)	64	64	128	256	512	1024	2048
Overhead (%)	11.7	11.7	3.82	1	0.4	0.06	0

Summing up, the DTable scheduler, which has a quite good computational complexity, provides the most preferential VCs (those which have been assigned a shorter distance between any consecutive pair of entries) with a better latency performance than the DRR algorithm. However, it provides the least preferential VCs with a worse latency than the DRR algorithm. Our proposal of using different specific MTUs increments the flexibility of our decoupling methodology without the need of increasing the w parameter too much. Note that increasing this parameter would entail more hardware requirements to store and process the table entries and a worse latency performance. However, the excessive packetization of the messages may produce a negative effect on the performance of the flows. Therefore, the specific MTUs should be assigned taking into account the characteristics of the traffic, specifically, the size of the packets.

5   Conclusions

A key component for networks with QoS support is the output scheduling algorithm, which selects the next packet to be transmitted. An ideal scheduling algorithm should satisfy two main properties: good end-to-end delay and simplicity. Table-based schedulers try to address these two characteristics. However, they have several problems that we try to solve with a new table-based scheduler, the DTable scheduler, and several proposals to configure it.

The DTable scheduler is a simple algorithm that properly configured may provide the flows with different levels of latency performance. Moreover, given a flow or aggregated of flows with some latency requirements, we can assign a certain amount of traffic to that flow in a flexible way. The decoupling methodology that allows us to do this relies on two table configuration parameters. One of these parameters, the w parameter, determines the maximum weight that can be assigned to a single table entry, and thus, the maximum data that can be transmitted per table entry. The other, the k parameter, determines the maximum weight that can be distributed among all the table entries.

In this paper we have proposed a method to increase this flexibility. This method consists in using different MTUs for the different flows. This allows us to employ smaller values for the k parameter, and thus, for the w parameter. This is quite positive because a high value for the w parameter entails higher hardware requirements and worse latency performance. However, the specific flow MTU must be assigned taking into account the characteristics of the traffic flow. A too small specific MTU may decrease the latency of the flow.

In a real case the w parameter is probably going to be fixed by the network technology. The network manager should then choose an appropiate value for the k parameter and the specific MTUs. In order to do this, the characteristics of the traffic and the proportion of each kind of traffic must be taken into account. Note that different flows can exhibit very different message sizes. As future work we are focusing our attention on applying our proposals to a multimedia environment with different kind of traffics.

References

1. Advanced Switching Interconnect Special Interest Group. *Advanced Switching core architecture specification. Revision 1.0*, December 2003.
2. F. J. Alfaro, J. L. Sánchez, and J. Duato. QoS in InfiniBand subnetworks. *IEEE Transactions on Parallel and Distributed Systems*, 15(9):810–823, September 2004.
3. H. M. Chaskar and U. Madhow. Fair scheduling with tunable latency: A round-robin approach. *IEEE/ACM Transactions on Networking*, 11(4):592–601, 2003.
4. S. J. Golestani. A self-clocked fair queueing scheme for broadband applications. In *INFOCOM*, 1994.
5. InfiniBand Trade Association. *InfiniBand architecture specification volume 1. Release 1.0*, October 2000.
6. R. Jain. *The art of computer system performance analysis: Techniques for experimental design, measurement, simulation and modeling*. John Wiley and Sons, Inc., 1991.
7. R. Martínez, F.J. Alfaro, and J.L. Sánchez. Decoupling the bandwidth and latency bounding for table-based schedulers. *International Conference on Parallel Procesing (ICPP)*, August 2006.
8. R. Martínez, F.J. Alfaro, and J.L. Sánchez. Implementing the advanced switching minimum bandwidth egress link scheduler. *IEEE International Symposium on Network Computing and Applications (IEEE NCA06)*, July 2006.
9. R. Martínez, F.J. Alfaro, and J.L. Sánchez. Providing Quality of Service over Advanced Switching. *International Conference on Parallel and Distributed Systems (ICPADS)*, July 2006.
10. K. I Park. *QoS in Packet Networks*. Springer, 2005.
11. M. Shreedhar and G. Varghese. Efficient fair queueing using deficit round robin. In *SIGCOMM*, pages 231–242, 1995.
12. StarGen. *StarGen's Merlin switch*, 2004. http://www.stargen.com/products/merlin_switch.shtml.
13. D. Stiliadis and A. Varma. Latency-rate servers: a general model for analysis of traffic scheduling algorithms. *IEEE/ACM Transactions on Networking*, 1998.
14. A. Tyagi, J. K. Muppala, and H. de Meer. VoIP support on differentiated services using expedited forwarding. In *IEEE International Performance, Computing, and Communications Conference (IPCCC)*, February 2000.

A Cache-Pinning Strategy for Improving Generational Garbage Collection

Vimal K. Reddy, Richard K. Sawyer, and Edward F. Gehringer

Department of Electrical and Computer Engineering
Box 7911, North Carolina State University
Raleigh, NC 27695-7911, USA
{vkreddy, rksawyer, efg}@ncsu.edu

Abstract. In generational garbage collection, the youngest generation of the heap is frequently traversed during garbage collection. Due to randomness of the traversal, memory access patterns are unpredictable and cache performance becomes crucial to garbage-collection efficiency. Our proposal to improve cache performance of garbage collection is to "pin" the youngest generation (sometimes called the nursery) in the cache, converting all nursery accesses to cache hits. To make the nursery fit inside the cache, we reduce its size, and, to prevent its eviction from the cache, we configure the operating system's page-fault handler to disallow any page allocation that would cause cache conflicts to the nursery. We evaluated our scheme on a copying-style generational garbage collector using IBM VisualAge Smalltalk and Jikes research virtual machine. Our simulation results indicate that the increase in frequency of garbage collection due to a smaller nursery is overshadowed by gains of converting all nursery accesses to cache hits.

1 Introduction

Generational garbage collected systems [19] manage the heap as several subheaps, known as generations, based on age of the objects. An object is allocated in the youngest generation, sometimes called the nursery, and is promoted to an older generation if its lifetime exceeds the threshold of its current generation. Since most objects die, i.e. become garbage young, the nursery is the region most likely to produce garbage. Hence, garbage collection is mostly done on the nursery and extensive, global garbage collection is only resorted to if memory reclaimed from the nursery is insufficient.

When the nursery cannot satisfy a memory allocation request, garbage collection is invoked on it. The garbage collector begins with a root set of pointers, usually live program variables, and traverses each of the objects pointed to, scanning them recursively for more pointers until a closure of live objects is reached. Memory is either explicitly reclaimed, like in a mark-sweep garbage collector, or implicitly reclaimed by copying over live objects to a new heap region, like in copying garbage collectors. The main overhead of the garbage collector is in the transitive traversal of the live object set. Since the nature of objects pointing to other objects is random, the traversal

Y. Robert et al. (Eds.): HiPC 2006, LNCS 4297, pp. 98–110, 2006.
© Springer-Verlag Berlin Heidelberg 2006

leads to access patterns that are unpredictable and hard to capture with traditional cache prefetching algorithms.

Our proposal to reduce garbage-collection time is to satisfy all garbage-collection memory requests within the cache hierarchy. To do so, we reserve a portion of the L2 cache for the nursery and pin it in cache, so that cache lines belonging to the nursery are never evicted. To pin the nursery, we reduce its size to fit inside the cache and modify the page-placement policy of the operating system to only allow nursery pages to reside in page frames that map to the reserved cache region. This ensures there are no cache conflicts to the nursery.

In this paper, we focus on a copying-style garbage collector where the nursery is segregated into two generations, the new space and the old space. Allocation requests are satisfied by the new space until it is exhausted, at which time garbage collection is invoked. The garbage collector traverses the live object set in the new space and in the process, copies each live object over to the old space. The copied objects are recursively scanned for more pointers and the process continues until a closure of live objects is reached. As a result, by the end of a garbage-collection cycle all live objects are relocated from new space into old space, which now becomes a compacted new space where memory allocation continues.

Aside from poor cache performance due to the random, transitive traversal, there are additional overheads associated with copying collectors that motivate cache-pinning. Writing live objects to the temporally idle old space causes write misses and eviction of dirtied dead objects from the cache causes unnecessary write backs to main memory. Since garbage collection halts the user program, any perceived overhead affects overall program performance.

The nursery is usually much larger than contemporary L2 caches. It must be reduced in size to make it suitable for cache-pinning. Since a smaller nursery fills up faster, it increases the garbage-collection frequency. Despite this, the overall time of garbage collection *decreases*, since all accesses to the nursery are converted to cache hits. Moreover, since each garbage collection scans less memory, pause times for nursery collections are markedly reduced, leading to less annoyance for users and greater ability to meet real-time deadlines.

Our cache pinning strategy is a novel way to mitigate disruptions caused by garbage collection without increasing execution time. A small nursery would reduce the pause time for a single garbage collection, but the overall garbage collection time would increase due to the increase in garbage collection frequency. On the other hand, a large nursery would reduce the garbage collection frequency, but lead to a big pause time and cause noticeable annoyance to a user. Pinning a small nursery in the cache frees garbage collection from cache miss overheads and leads to a negligibly small pause time. The increase in garbage collection frequency is not perceived due to the small pause time, and for the same reason, the overall garbage collection time is also not increased. In fact, our results indicate that pinning can decrease overall garbage collection time, leading to less interfering and better performing garbage collection.

The remaining sections in the paper are organized as follows. Section 2 introduces our cache-pinning strategy on direct-mapped and set-associative caches. Section 3 discusses simulation methodology and benchmarks used in this study. Section 4 presents results from IBM VisualAge Smalltalk and the Jikes research virtual machine. Section 5 discusses related work and Section 6 concludes the paper.

2   Cache-Pinning Strategy

The idea is to "pin" the nursery in the L2 cache; that is, arrange the virtual-to-physical memory mapping so that there will be no conflict misses to lines from the nursery. We reduce the size of the nursery to a suitable size to fit inside the L2 cache. A portion of the cache is reserved for the nursery, and the page-fault handler is modified to ensure that only pages from nursery are placed in these "reserved" frames. This prevents conflicts to the nursery in the cache, so it stays pinned. The cache policies themselves remain unmodified. Next, we look at the mapping strategy for direct-mapped caches (Section 2.1) and set-associative caches (Section 2.2).

2.1   Direct-Mapped Caches

In direct-mapped caches, we reserve a portion of the cache that is just large enough to hold the nursery. For example, we might reserve the bottom (highest numbered) few lines of the cache. Next, we identify page frames whose memory maps to the reserved cache lines. The physical address space can be viewed as a set of regions, each of which is exactly as large as the cache. The bottom few page frames in each region map to the reserved region of the direct-mapped cache, as shown in Fig. 1. These page frames are said to form a "bucket." Each region contains its own bucket. The goal of our strategy is to place the nursery into page frames within these buckets. For a direct-mapped cache, each bucket's size is equal to the size of the nursery. Let us assume that one of these buckets is used to hold the nursery (this assumption makes it easy to see what is happening, but it is stronger than necessary, and will be relaxed later). Then the other buckets cannot be used by this process, though they certainly can be used by other processes. The page-placement logic must operate this way:

- A nursery page should be placed in a page frame belonging to one of the buckets.
- A non-nursery page should not be located in a page frame belonging to one of the buckets.

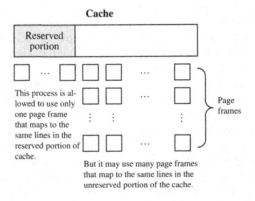

Fig. 1. "Pinning" a region of memory in the cache

Fig. 2 shows this strategy applied to a 1MB direct-mapped cache on a system with a physical address space of 4096 page frames, each 4KB large. A nursery of size 256 KB will require 64 pages for storage. We reserve the bottom 256 KB (8192 lines) of the cache for the nursery. A bucket consists of the last 64 page frames occurring in any 1MB (= size of the cache) region of physical address space. Only one of these buckets is chosen to contain the nursery, the choice being left to the page fault handler. The remaining buckets are unused to avoid conflicts. The list of buckets and the page frames they contain are passed to the page-fault handler to implement the replacement policy described above.

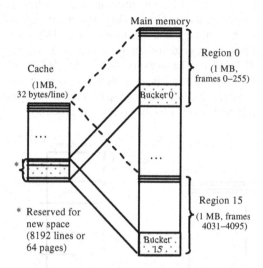

Fig. 2. Mapping strategy for a 1MB direct-mapped cache with 32-byte lines, to reserve a 256KB nursery in the cache

2.2 Set-Associative Caches

If the cache is set associative, we have two options for reserving space in it. The first is to reserve a certain number of "ways" of an n-way set-associative cache. This approach requires the cache hardware to be modified to ensure that a reserved cache line is filled only with a memory block from the nursery [16]. So, rather than reserve "ways" for the nursery, we reserve sets. This is consistent with our policy for direct-mapped caches, where we reserved lines, since each line in a direct-mapped cache is a set.

We choose to reserve the bottom few sets of the cache for the nursery. Again, we view the physical address space as divided into regions that are the same size as the cache. A bucket in each region contains lines that map to reserved sets in the cache. However, unlike the direct-mapped case, the size of a bucket is not equal to that of the nursery. Rather, a certain number of buckets, equal to the associativity of the cache, make up the nursery, as shown in Fig. 3. The other buckets remain unused by this process. The process of selecting a page frame from one of the buckets remains the same. The only difference is that the page-placement can use frames from more than one bucket for pages of the nursery. The page-placement logic must operate this way:

- A nursery page should be placed in a page frame belonging to one of k buckets, where k is the associativity of the cache.
- A non-nursery page should not be placed in a reserved page frame.

In reality, it is not necessary for all page frames used by the youngest generation to come from just k buckets. The page frames can be drawn from more than k buckets, as long as no more than k page frames map to the same sets in the cache. This will still prevent conflict misses for the nursery. Relaxing this restriction means that it is not necessary to select contiguous page frames to hold the nursery. This allows the memory manager and page-fault handler much greater freedom in where to place the nursery pages in memory. Similarly, for direct-mapped caches, page frames can be drawn from more than one bucket as long as they do not conflict in the cache.

Fig. 3 shows how this strategy can be applied to a 1MB, 4-way set-associative cache with 32-byte lines, for a system with a physical address space containing 4096 page frames, each 4KB large. The bottom 2048 sets (256KB) of the cache are reserved for the nursery. As shown in Fig. 3, a bucket consists of the bottom 256KB of a 1MB region, and contains 2048 lines, or 16 pages. Four such buckets make up the nursery, and the remaining buckets go unused (by this process).

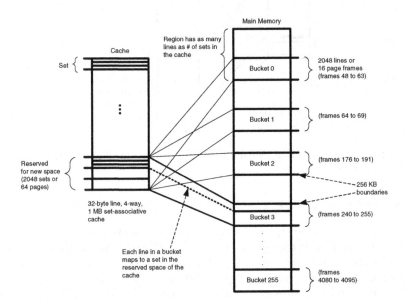

Fig. 3. Mapping strategy for a 32-byte line, 4-way, 1 MB set-associative cache to reserve a 256 KB large youngest generation in the cache

3 Simulation Methodology and Benchmarks

For Smalltalk experiments, we ran IBM VisualAge Smalltalk (VAST) [12] on top of Shade [6]. Shade simulates only the cache hierarchies and not the entire memory subsystem. Specifically, paging is not simulated. To achieve the effect of pinning the nursery in the L2 cache, we made modifications to Shade so that virtual addresses of the nursery were translated to another address space with the following properties:

- All nursery addresses map to the reserved portion of the cache.
- Non-nursery addresses *do not* map to the reserved portion of the cache.

For Java experiments, we used Dynamic SimpleScalar version 1.0.1 [10] that simulated a Jikes research virtual machine (RVM). The functional cache simulator (simcache) was used to calculate miss-rate improvements, and the cycle-based simulator (sim-outorder) was used to calculate speedup. A virtual-to-physical page mapping was added – virtual addresses are translated to physical addresses and the physical address is passed to the cache simulator. Also, sim-cache was modified to be a single-pass cache simulator [11].

For Smalltalk benchmarks, we used a collection of Smalltalk programs shown in Table 1, as there is no real benchmark suite available for Smalltalk. For Java benchmarks we used memory intensive benchmarks from the DaCapo [8] suite shown in Table 2.

A point to note is that our base runs, which do not use pinning, have the advantage of indexing the cache with virtual addresses. In real systems, it is possible for two adjacent virtual pages to be mapped to physical addresses that compete for the same cache lines. The assumption that physical address equals virtual address tends to underestimate the contention for L2 cache lines. This gives our base case, which does not use pinning, better performance than would be seen in a real system.

Table 1. Smalltalk Benchmarks

Benchmark	Description
Swim	Solves a system of shallow water equations using finite difference approximations.
Tomcatv	Vectorized mesh generation program.
Bench	Collection of programs performing basic tasks: sorting, matrix multiplication, etc.
Xml parser	A tree-based SAX (Simple API for XML) parser from IBM VAST.
ES compiler	Compiler for IBM VAST.
Richards	Simulates the kernel task dispatcher of an operating system.

Table 2. DaCapo Benchmarks

Benchmark	Description
antlr	Parses one or more grammar files and generates parser and lexical analyzers.
batik	Renders a number of SVG files.
bloat	Performs a number of optimizations and analysis on Java bytecode files.
chart	Uses JFreeChart to plot a number of complex line graphs and renders them as PDF.
Fop	Takes an XSL-FO file, parses it and formats it, generating a PDF file.
hsqldb	Executes a JDBC-like in-memory benchmark, executing a number of transactions against a model of a banking application.
jython	Interprets a series of Python programs.
pmd	Analyzes a set of Java classes for a range of source code problems.
ps	Reads and interprets a PostScript file.
xalan	Transforms XML documents into HTML.

4 Results

To put the cache pinning strategy into perspective, we first discuss the overhead of garbage collection by referring to the graphs in Fig. 4a and Fig. 4b. These statistics were gathered using a program analysis tool available in IBM VAST [12], which also uses a generation copying-garbage collector like Jikes RVM. The runs were carried out on a native Sun UltraSPARC system.

As seen in Fig. 4a, garbage collection of the nursery (shown as scavenge time) is a significant component of overall execution time. For small nursery sizes, the number of garbage collections is high, as shown in Fig. 4b, and the total collection time is high. For big nursery sizes, number of garbage collections is fewer and the total garbage collection time is smaller.

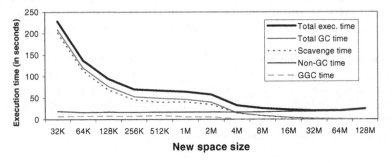

Fig. 4a. Components of execution time for Swim on IBM VAST virtual machine. Note that scavenge (i.e. garbage collection of the nursery) time tracks the total execution time. GGC refers to global garbage collection. Nursery size is twice the new space size.

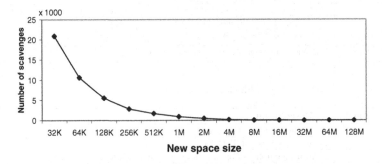

Fig. 4b. Number of scavenges for Swim on the IBM VAST virtual machine

Some interesting insights can be drawn from these graphs. Firstly, since the pinning strategy increases garbage collection frequency, a significant reduction in garbage collection pause time is needed due to pinning, so that overall performance is better than just having a large nursery. Secondly, the garbage collection pause time increases with the size of the nursery. For instance, a 512KB new space has almost the same overall garbage collection time as a 256KB new space, though number of garbage collections are far fewer for a 512KB new space. For real-time performance, small and deterministic pause times are more desirable.

In the next few experiments, we try to find out if our hypothesis will work: the cost of increasing the frequency of garbage collection due to a smaller nursery will be overcome by benefits of converting nursery accesses to cache hits and at the same time, pause times will be minimal and deterministic for real-time performance.

To find the optimum portion of a cache to reserve for the nursery, we ran experiments both on Jikes RVM and IBM VAST by setting aside 12.5%, 25% and 50% of the cache. Fig. 5 shows the global miss-rate (i.e., fraction of accesses that went to memory) improvements for {1, 2, 4, 8, 16}-way caches averaged across all benchmarks in the DaCapo suite. Overall, a reduction in the global miss rates is seen across all configurations.

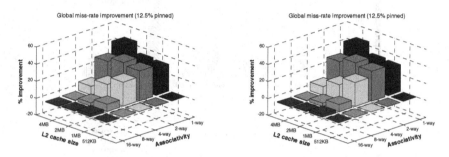

Fig. 5. DaCapo: Percentage improvement in global cache miss rate when 12.5% and 25% of the L2 cache is reserved for the nursery

Fig. 6 shows global miss rates for DaCapo on 4-way set-associative caches. In few cases, global miss rate increases due to pinning. This is due to increase in cache pressure on the user program due to pinning. The effect is prominent with 50% reservation in small caches, as seen in Fig. 6. Reserving 12.5% and 25% of the cache for the nursery seem to provide good benefits. For the remaining experiments, we choose 25% of L2 cache to be reserved for the nursery.

For Smalltalk benchmarks, the global cache miss rate for direct-mapped caches is shown in Fig. 7. There is a significant decrease in global miss-rate. The average improvement is between 35% and 40%. However, we noticed that returns diminish significantly with increased associativity. On 2-way associative L2 caches the average speedup was between 4% and 25% across configurations. For highly associative and large caches (2MB or more), speedups were smaller or none.

The Smalltalk programs used in this study are small and not memory intensive, thus invoking garbage collection fewer times. Hence, the gains due to pinning are not significant with set-associative caches where associativity itself captures most of the conflicts to the youngest generation. The diminishing trend is also visible in the DaCapo suite, but pinning is still able to achieve modest gains, as these benchmarks are more memory intensive than the Smalltalk programs.

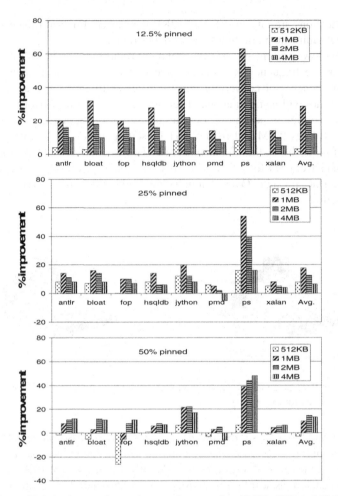

Fig. 6. DaCapo: Global miss-rate improvements for reserving 12.5%, 25% and 50% of a 4-way set-associative L2 cache

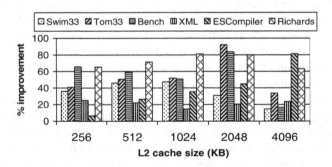

Fig. 7. Smalltalk: Global miss-rate improvement for reserving 25% of a direct-mapped L2 cache

Fig. 8 and Fig. 9 show the effect of pinning on garbage collection time for the DaCapo suite. As seen in Fig. 8, pinning significantly decreases pause times of the garbage collector. The garbage collection time reduces by a factor of 5 on average and up to a factor of 10 due to pinning. Fig. 9 shows that pinning makes pause times deterministic. This is useful in real-time programs where interference from the garbage collector is undesirable.

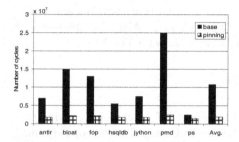

Fig. 8. DaCapo: Pause times for nursery collections with and without pinning

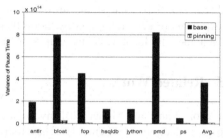

Fig. 9. DaCapo: Variance of nursery pause times with and without pinning

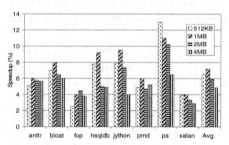

Fig. 10. DaCapo: Speedup for 2-way associative caches

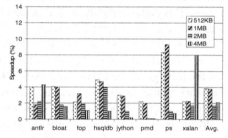

Fig. 11. DaCapo: Speedup for 4-way associative caches

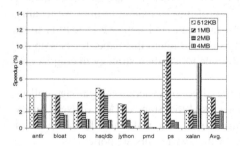

Fig. 12. DaCapo: Speedup for 8-way associative caches

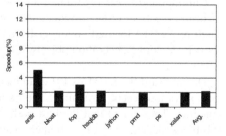

Fig. 13. DaCapo: Speedup for a future generation configuration: 4 MB, 32-way associative cache with memory latency of 400 cycles

We finally evaluate speedups due to pinning on a contemporary superscalar processor with memory latency of 200 cycles. We present results only for the DaCapo suite. We compare results with page coloring [15], which also improves cache performance by mapping consecutive virtual pages to consecutive physical page frames. Mapping in this manner takes advantage of spatial locality of memory accesses and can reduce conflict misses in the cache.

As seen in Fig. 10, Fig. 11 and Fig. 12, pinning achieves a speedup of 5%-7% on a 2-way associative cache and up to 4% on higher associative caches. We also evaluated a future generation configuration that used a 32-way, 4 MB L2 cache with 400 cycle memory latency. Gains of up to 5% are seen, as shown in Fig. 13.

5 Related Work

Wilson, Lam, and Moher [20] investigated the effects of generational copying garbage collection in Scheme 48. They found that direct-mapped caches outperformed set-associative caches when the cache size was near the size of the nursery. For caches sizes larger than the nursery, set-associative caches performed better. They found majority of misses in a cache larger than the young generation were conflict misses. This paper expands upon concepts in Wilson, Lam, and Moher by pinning the youngest generation in the L2 cache.

Boehm [4] also had the goal of reducing cache misses of garbage collected programs. Boehm reduced cache misses for a non-copying collector using two strategies: "prefetch-on-grey" and "lazy sweeping". The former strategy prefetches data during the mark phase to improve cache performance. The latter strategy postpones the sweep phase until allocation time, so that when reallocation occurs the allocated block is already in the cache because it was just accessed during the sweep.

Similar to Boehm, Reinhold [17] used a non-copying collector in his research. Reinhold found that caches perform best with infrequent garbage collection. However, his studies only included direct-mapped caches, which now are almost extinct.

Chilimbi and Larus [5] studied a real-time profiling technique to improve locality of heap objects in order to reduce the cache miss rate and improve execution time. Their researched focused on longer-lived objects. Although this paper focuses on the short-lived young generation objects, it is possible to use their strategy in conjunction with pinning.

Diwan, Tarditi, and Moss [9] examined the effect of generational garbage collection on various cache structures. Their studies used the Standard ML of New Jersey compiler. They found that the best performance was achieved when the cache had a sub-block placement (write-validate) coupled with a write-allocate policy.

The Standard ML of New Jersey compiler was also used by Cooper, Nettles, and Subramanian [7]. Their goal was to reduce page faults by modifying the page-fault handler. Ideally, pages that no longer contained useful information during garbage collection were not written back to secondary storage. The goal of this paper, however, is to reduce cache misses.

6 Conclusion

Garbage collection time can be diminished by reserving a portion of the L2 cache for the nursery in a generational garbage-collected system. Results from Java and Small-talk benchmarks show a decrease in global cache miss rates for direct-mapped and set-associative caches. This in turn translates to reduced garbage collection pause times and modest decreases in execution times even when using highly set-associative caches. More importantly, pinning leads to a deterministic garbage collection pause time which is desirable for real-time performance of garbage collected programs.

Acknowledgments

We thank Prof. Eric Rotenberg for discussions and comments during this research. We thank reviewers for their constructive feedback. We also thank Prof. Eric Sills for assistance in securing computing power from the NC State HPC grid. Finally, we would like to thank the IBM VisualAge Smalltalk group for assistance in setting up the Smalltalk environment used in this study.

References

[1] A.W. Appel. Simple generational garbage collection and fast allocation. Software Practice and Experience, 19(2), pp. 171-183, 1989.

[2] S.M. Blackburn and K.S. McKinley. Ulterior reference counting: fast garbage collection without a long wait. OOPSLA 2003, pp. 344-358.

[3] S.M. Blackburn, P. Cheng and K.S. McKinley. Oil and Water? High Performance Garbage Collection in Java with MMTk. ICSE 2004, pp.137-146.

[4] H. Boehm. Reducing garbage collector cache misses. ISMM 2000, pp 59-64.

[5] T.M. Chilimbi and J.R. Larus. Using generational garbage collection to implement cache-conscious data placement. ISMM 1998, pp. 37-48.

[6] R. Cmelik, D. Keppel. Shade: A fast instruction-set simulator for execution profiling. SIGMETRICS Performance Evaluation Review, vol. 22, issue 1, pp. 128–137, May 1994.

[7] E. Cooper., S. Nettles, and I. Subramanian. Improving the performance of SML garbage collection using application-specific virtual memory management. LFP 1992, pp. 43-52.

[8] DaCapo benchmarks version beta050224. http://ali-www.cs.umass.edu/DaCapo/

[9] A. Diwan, D. Tarditi, and E. Moss. Memory system performance of programs with intensive heap allocation. ACM Trans. Comput. Syst. 13, 3, pp. 244-273, August 1995.

[10] Dynamic SimpleScalar version 1.0.1. http://www-ali.cs.umass.edu/DSS/

[11] M.D. Hill and A. J. Smith. Evaluating Associativity in CPU Caches. IEEE Transactions on Computers, vol. 38, no. 12, December 1989.

[12] IBM VisualAge Smalltalk. www-3.ibm.com/software/awdtools/smalltalk

[13] Jikes Research Virtual Machine version 2.4.0. http://jikesrvm.sourceforge.net/

[14] R. Jones and R. Lins. Garbage Collection: Algorithms for automatic dynamic memory management. John Wiley & Sons, ISBN: 0471941484, 1996.

[15] R.E. Kessler and M.D. Hill. Page placement algorithms for large real-indexed caches. ACM Trans. Comput. Syst. 10, 4, pp. 338-359. November 1992.

[16] V. K. Reddy. Caching strategies for improving generational garbage collection in Small-talk. MS Thesis, North Carolina State University, August 2003.

[17] M.B. Reinhold. Cache performance of garbage-collected programs. PLDI 1994, pp. 206-217.

[18] R.K. Sawyer. Page Pinning Improves Performance of Generational Garbage Collection. MS Thesis, North Carolina State University, March 2006.

[19] D. Ungar. Generation Scavenging: A non-disruptive high performance storage reclamation algorithm. SESPSDE 1984. pp. 157-167.

[20] P.R. Wilson, M.S. Lam, and T.G. Moher. Caching considerations for generational garbage collection. LFP 1992, pp. 32-42.

A Realizable Distributed Ion-Trap Quantum Computer

Darshan D. Thaker[1], Tzvetan S. Metodi[1], and Frederic T. Chong[2]

[1] University Of California at Davis
{ddthaker, tsmetodiev}@ucdavis.edu
[2] University Of California at Santa Barbara
chong@cs.ucsb.edu

Abstract. Recent advances in trapped ion technology have rapidly accelerated efforts to construct a near-term, scalable quantum computer. Micro-machined electrodes in silicon are expected to trap hundreds of ions, each representing quantum bits, on a single chip. We find, however, that scalable systems must be composed of multiple chips and we explore inter-chip communication technologies. Specifically, we explore the parallelization of modular exponentiation, the substantially dominant portion of Shor's algorithm, on multi-chip ion-trap systems with photon-mediated communication between chips.

Shor's algorithm, which factors the product of two primes in polynomial time on quantum computers, has strong implications for public-key cryptography and has been the driving application behind much of the research in quantum computing. Parallelization of the algorithm is necessary to obtain tractable execution times on large problems. Our results indicate that a 1024-bit RSA key can be factored in 13 days given 4300 (each of area 10 by 10 centimeters) ion-trap chips in a multi-chip system.

1 Introduction

Although quantum computers may have an exponential advantage over classical computers, the challenges in manipulating and preserving quantum data require substantial performance and design optimization to allow large problems to remain tractable. One of the most celebrated large-scale quantum applications is Shor's algorithm for finding non-trivial factors of a large composite number in polynomial time [1], which leads to the ability to break the RSA public key cryptosystem. A classical computer may take millions of years for factoring numbers as large as 1024 bits, and even on a quantum computer such a calculation may take hundreds of years to complete. Consequently, this study focuses on effective parallelization of quantum computations on practical quantum architecture systems such as the Quantum Logic Array microarchitecture [2], which can improve large-scale execution times to useful levels. In particular, we find that area and reliability constraints require scalable systems to be constructed from multiple chips. Algorithms must then be designed to account for substantially different intra- and inter-chip communication bandwidth, much as in the classical multiprocessor domain.

The primary difficulty in the realization of quantum computation is that quantum data is inherently very unstable as it constantly interacts with the environment: a process called *decoherence*. To enable computation with such unreliable components, a

Y. Robert et al. (Eds.): HiPC 2006, LNCS 4297, pp. 111–122, 2006.
© Springer-Verlag Berlin Heidelberg 2006

rich theory of quantum fault-tolerance has been proposed [3, 4, 5]. Our previous work builds on these theories and the physical implementations of quantum computers, to design a scalable and reconfigurable Quantum Logic Array (QLA) architecture [2]. The QLA is a data-path based, reconfigurable architecture design that offers an efficient way of structuring circuits with the much advanced trapped-ion technology [6, 7]. The QLA design supports the execution of quantum algorithms such as Shor's factoring algorithm both flexibly and scalably through the complete overlap of computation and communication. We find, however, that the area requirements for building a single QLA chip to perform Shor's algorithm for numbers as large 1024 bits would require unrealistic technological milestones to be met. To relax these requirements we explore *multi-chip* solutions with the QLA architecture that make such large-scale problems feasible for discussion.

The focus of this paper is the evaluation of quantum modular exponentiation, which is by far the most computationally intensive part of Shor's factoring algorithm. Previous studies in this area have devised extensive quantum adder circuits that implement modular exponentiation [8, 9, 10, 11, 12]. However, this has been done with little regard to feasibility, or implementation on an existing quantum architecture models such as the QLA. We leverage the existing work on quantum adders to study and evaluate current and novel adder circuits on a multi-chip version of the QLA architecture. In particular, we show a circuit which through serialization requires the least possible area on a single QLA chip, and demonstrate that even maximal serialization yields area too great to physically overcome, in addition to unreasonable execution time. We introduce a *distributed quantum parallel prefix adder* based on the classical parallel prefix adder [13], and evaluate its communication and computation requirements on the multi-chip QLA architecture.

The paper is organized as follows: Section 2 provides a high-level description of the Quantum Logic Array architecture model, and a brief description of the logical and inter-chip interconnect technology. Sections 3 and 4 introduce Shor's algorithm and give a simple, spatially-optimized design. Our multi-chip solutions are introduced and compared in Section 5. Finally results and conclusions are discussed in Section 6 and Section 7 respectively.

2   Quantum Computation and the QLA Architecture

Quantum mechanical systems such as the different electronic spins of an atom, or the polarization states of a photon can be used to hold and manipulate binary information just as the old vacuum tubes or the modern semiconductors do. Unlike classical binary bits, however, the fundamental units of quantum information, *qubits*, always exist in a superposition of the binary 0 and 1 states with some probability of obtaining one or the other upon observation. Measuring a quantum state not only yields the measured result with some probability, but forces the state to be in the measured value. A collection of N qubits exists in a superposition of 2^N different states at any given time denoted by the binary bit-strings representing the numbers 0 through 2^{N-1}. Logic gates are implemented as 2^N dimensional, complex valued, unitary matrices which act on the system state vector.

Our prior work, the QLA architecture [2] is designed to support large-scale quantum computation through the efficient manipulation of arbitrarily large number of qubits. The major components of the QLA are: fault-tolerant logical qubits denoted with the letter Q, trapped atomic ions as the basic implementation technology at the bottom of the architecture and a teleportation based logical interconnect controlled.

The most important restriction on quantum circuits is the inability to form a sequence of gates that copies the state of a qubit, also known as the *no-cloning theorem* [14]. For a two-qubit operation the qubits must be brought together, and we cannot simply propagate the state of one to the other - we must physically transport each qubit. At inter-logical qubit distances ballistically moving the each ion-qubit holding valuable quantum data is too risky and will destroy the state.

2.1 Single-Chip Logical Interconnect

At small distances, the ballistic direct channels are enough for the communication requirements within logical qubits. However, the execution of any algorithm will depend of the efficient interaction among logical qubits. The concept of *quantum teleportation* [15] is utilized within each processor to provide high-speed, low-latency, fault-tolerant network interconnection between the logical qubits.

Quantum teleportation utilizes two qubits that have been previously prepared in an Einstein-Podolsky-Rosen (EPR) state [16]. The drawback is that these EPR qubits still have to be physically moved. However EPR pairs are replaceable and with enough resources we can establish entanglement between the source and the destination just in time for the communication to be completed.

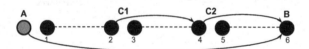

Fig. 1. A schematic of a single bandwidth channel between the logical qubits of our ion-trap processor. Rather than distributing an EPR pair of qubits directly from the source to the destination, the channel is divided into islands (here denoted as $C1$, and $C2$), that can be used to expand the entanglement over the entire channel.

In the intra-chip interconnect, our quantum processor solves the EPR transport problem by combining the concepts of *quantum repeaters* [17] and *entanglement purification* [18, 19, 17]. The quantum repeaters are islands that are strategically placed in the channels between the logical qubits to limit the distance traveled by each EPR pair. EPR pairs only travel to two neighboring islands, whose entanglement can then be efficiently purified using the purification protocols with some additional ancillary EPR pairs. Figure 1 the concept of quantum repeaters by showing a simple schematic of the channel between the logical qubits. The channel shown has a bandwidth of 1 physical bit at a time, and the ancillary EPR qubits are pipelined to the two opposing islands as they purify the data EPR pairs. EPR pairs $(1, 2)$, $(3, 4)$, and $(5, 6)$ can be distributed and purified in parallel over the channel separating the source (A), and the destination (B).

The arrows in Figure 1 show how teleporting each individual pair, entanglement can be distributed between the source and the destination for one final teleportation stage.

2.2 Multi-chip Interconnect

The above scheme however is not good enough when designing a multi-chip quantum network. The inter-chip distances, the difficulty in chip alignment, and the high failure rates at such conditions make it impossible to use the highly refined physical traps that make up the chips themselves. Instead of physically transporting ions encoded as an EPR pair between chips, two remote ions can establish entanglement through photon mediation [20, 21]. Photons emitted by the ions are collected with an optical fiber and sent to entangling stations to perform a collective measurement on the ions without knowing which ion sent which photon beam. The two ions are then projected into the two qubit entangled state The application of an X gate on the second qubit easily converts this state to the familiar EPR pair. The communication process is then completed by teleporting the source ion to the other side as described above.

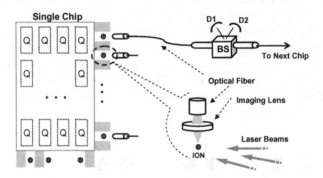

Fig. 2. Each chip has a connection pad on its sides with evenly distributed optical fiber locations. Remote trapped ions at each location on two opposing chip pads are entangled using photons emitted by the ions.

The ion-photon mapping method of creating entanglement between remote qubits is not possible inside the processors, because of the complex classical control required for quantum computation. However, outside of the processors it allows us to make an arbitrary quantum network by connecting adjacent processors as shown in Figure 2. Each chip has a connection pad on its sides with evenly distributed optical fiber locations. The bandwidth between two chips is defined by the number of optical fibers connecting them, which is limited by the chip dimensions. Due to the size of the optics, the fibers must be spaced at least $700\mu m$ apart [22], where a time of approximately $600\mu s$ for the entanglement operation in order to achieve near unit fidelity. The inherent entanglement purification of this channel [23] makes it much faster than the quantum repeater channel, thus we will not observe added latency when jumping from one chip to the next.

3   Shor's Algorithm

Shors' algorithm is widely studied due to its exponential advantage over conventional algorithms and due to its application to crytography. Shor's algorithm can be used to factor a product of two primes in polynomial time and comprises of the quantum fourier transform and quantum modular exponentiation. Although Shor's insight was in evaluating the quantum fourier transformation in polynomial time, it is the modular exponentiation that consumes many times greater computational resources than the rest of the algorithm. In other words, if an ion trap processor that was 1 meter long on each side, could be built, it would take about 45 days to perform a heavily parallelized version of modular exponentiation. Whereas the same processor would be able to solve the quantum fourier transform in 0.5 days [2].

Performing Modular Exponentiation: A considerable amount of work has been done in developing implementations of adders for modular exponentiation [8, 10, 9, 12]. Nonetheless, the actual technology for quantum computing and the issue of fault-tolerance have not been considered in depth. Our previous work [2] develops a detailed architecture for a fault-tolerant, ion-trap based quantum processor. We employed a fast quantum carry-lookahead adder developed by [12] to efficiently perform modular exponentiation. The drawback of this design was the size of the quantum processor. It would need to be about $0.9m^2$ to accommodate all the required qubits. The next two sections address this very issue of size.

Table 1. Increase in area (in cm^2) with the number of logical qubits

Number of Logical Qubits	Area of processor
100	$2.65\ cm^2$
200	$5.31\ cm^2$
300	$7.963\ cm^2$
350	$9.29\ cm^2$
400	$10.617 cm^2$

4   Single-Chip Solution

Our goal in this section is to design a circuit for modular exponentiation that is spatially optimized. The tradeoff is going to be extra time. From table 1, we determine that if we needed a processor that was less than $10cm^2$ it would restrict us to 350 logical qubits for the entire processor.

4.1   Serialized Adder

For performing addition in a small space, we choose the quantum ripple carry adder by Vedral et.al [8], which analogous to its classical counterpart, needs minimum extra qubits for an addition. Fig 3(a) shows a quantum circuit for determining sum and carry given inputs a and b and carry-in cin. The left-part of Fig 4 shows the original Vedral

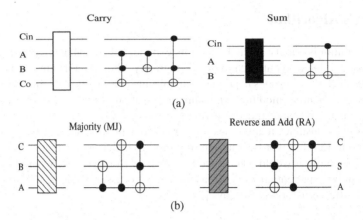

(a)

(b)

Fig. 3. (a) Standard method of computing carry and sum. (b) Using a majority (MJ) block and a reverse majority and add (RA) block to get the same result [11].

adder [8]. Note that the adder circuit has two carry blocks for the same inputs. The second block (on the right hand side of the circuit) performs the reverse computation, to erase the ancilla bits.

The Vedral adder can be improved if the reverse computations with the computations of the carry could be combined. This can be done using the *MJ* and *RA* blocks as shown in 3(b) as proposed by [11]. The *MJ* block computes the majority of three bits in place. The *RA* block adds two bits and reverses the majority. This allows us to save the n extra qubits that were needed for the Vedral ripple carry adder.

In Shor's algorithm, a can be as large as n^2. Thus we need $2n$ qubits to store a. Another n qubits are needed to store the result ($x^a \bmod n$). The circuits of Vedral et.al use addition to perform modular addition; modular addition to perform controlled modular multiplication and finally modular exponentiation from repeated controlled modular multiplication. Modular addition, multiplication and exponentiation each require n temporary qubits. Our modified Vedral adder needs 1 temporary qubit. This gives a total of $6n + 1$ qubits. We can reduce this requirement by observing that during addition, n qubits hold classical values only and thus can be eliminated. Also during modular addition, a set of n temporary qubits holds either 0 or $2^i \bmod n$. These n qubits can

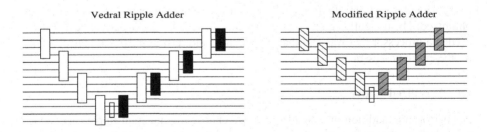

Fig. 4. 4-bit adder circuits using components from Figure 3(b) and Figure 3(a). The modified ripple carry adder requires $n - 1$ fewer qubits for adding two n qubit numbers.

be replaced by n classical bits and one qubit to keep track of entanglement. Our total required qubit count is now $4n + 2$. To this we add another 6 qubits which are required to perform a fault-tolerant toffoli gate; giving us a final count of $4n + 8$ qubits. The cost of keeping the required number of qubits low is that we have to perform all operations in serial order. They cannot be parallelized.

Table 2. Time and area for a serialized adder

N	Area in cm^2	ECC steps	Total Time
128	14.068	6.64 e9	3074 days
512	54.839	4.28 e11	542.87 years
1024	109.202	3.43 e12	4350.58 years

Table 2 shows the total area and time required for one complete modular exponentiation. In the table, ECC steps refers to the number error correction steps required. As can be seen, total area for the serialized version of the adder for the modular exponentiation of a 1024 bit number is quite large, $109.202cm^2$. At the other end, while the area for a 128 bit number is acceptable, the time it would take is approximately 8 years. The next section shows how we can reduce both the time and space requirements by using a parallel implementation of an adder.

5   Multi-chip Solutions

To exploit parallelism, we explore the performance of larger quantum systems composed of multiple interconnected chips, which we shall refer to as *Qchips*. Specifically, we evaluate distributed quantum adder designs and the cost of inter-chip communication on a multi-chip system.

5.1   Carry-Select Adder

Let $n = g * k$, where n is size of the number to be factored in bits, which is divided into g groups of k bits each [10]. Thus there are g Qchips each of which is responsible for adding k bits. We then use a carry-select adder spread out over these Qchips. Let the Qchips be $Qc_0, Qc_1, ..., Qc_g$. Each Qchip computes the sum and carry for its k bits in parallel. Since Qc_i does not have the carry out from Qc_{i-1}, it computes for both possibilities, the carry out being 0 and 1. When the carry from Qc_{i-1} is available, a multiplexer is used to select the correct value from the ones Qc_i calculated. In this manner, most of the computation can be done in parallel, except for the multiplexer. Two useful properties of such an adder are that the only communication between Qc_i and Qc_{i-1} is one bit, the carry from Qc_{i-1}, and that Qc_i communicates with Qc_{i-1} and Qc_{i+1}. Finally, we have to return the ancilla to their original state. This can be done by using an extra k qubits in each Qchip.

As outlined in [10], each Qchip will require about $6k$ qubits for the complete modular exponentiation. In order to perform parallel fault-tolerant toffoli's we conservatively increase it to $7k$ per Qchip. To calculate the time it would take for a complete, if

Table 3. Number of Qchips of fixed area required for a given problem size

Carry Select Adder				Parallel Prefix Adder			
N	$4\ cm^2$	$8\ cm^2$	$10\ cm^2$	N	$4\ cm^2$	$8\ cm^2$	$10\ cm^2$
128	5	2	2	128	11	5	4
512	23	11	9	512	51	25	20
1024	47	23	19	1024	109	54	43

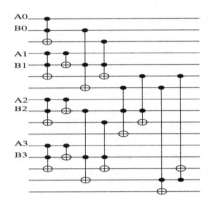

Fig. 5. Circuit of a parallel prefix adder for 4 bits. For reasons of clarity and space the reverse and sum computations are not shown.

inter-Qchip communication was free, we replace our parallel carry-select adder into the Vedral circuit. We also implement improvements suggested by van Meter and Itoh [10] namely using a lookup table to reduce the number of multiplications and their modulo adder block. Unlike their approach however, we only use one multiplier; this allows us to keep the area requirements within manageable levels.

For the carry-select adder, let $M1_i$ be a message from Qc_i to Qc_{i+1}. This is the carry out that Qc_i generated. When the ancilla have to be reset, Qc_i will receive the same qubit as message $M2_i$ from Qc_{i+1}. While all $M1$ messages cannot be parallelized, all $M2_j$ messages can be overlapped with $M1_{j+1}$ messages. Thus if we have g Qchips, the total communication cost we incur will only be g messages.

5.2 Parallel Prefix Adder

We now introduce our third adder which is a distributed parallel prefix adder. In a classical parallel prefix adder, partial information about the incoming carry is utilized to avoid ripple-carry computation. Let $C[i, j]$ denote the *carry status* on the interval $[i, j]$. It can have three possible values: k: kill, g:generate or p:propagate.

When adding a_i and b_i, we can determine something about the outgoing carry, c_{i+1}, without knowing c_i. If $a_i = b_i = 0$ then the $c_{i+1} = 0$ and $C[i, i + 1] = k$. Similarly, if $a_i = b_i = 1$, then carry is generated and $c_{i+1} = 1$. Also $C[i, i+1] = g$. If $a_i \neq b_i$ then

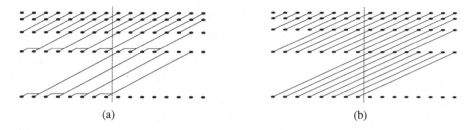

Fig. 6. Prefix graphs for Parallel-Prefix Adders [13]. Diagonal lines show communication between qubits at each stage. Each diagonal line carries two messages. Vertical line is Qchip boundary. The tradeoff lies between amount of communication and "fan-out". (a) The [2,2,1,1] scheme. (b) The [1,1,1,1] scheme.

we cannot determine c_{i+1} and the carry is said to have "propagated". $C[i, i + 1] = p$. The calculation of carry status is both associative and idempotent, which allows us to merge intervals. For any $i < k < j$,

$$p[i, j] = p[i, k] \cdot p[k, j] \tag{1}$$
$$g[i, j] = g[k, j] \oplus (g[i, k] \cdot p[k, j]) \tag{2}$$

The second expression is possible because $g[i, k]$ and $p[i, k]$ cannot both be 1 simultaneously.

Knowles [13] has shown that in the first stage of addition, each pair of qubit calculates its p and g values, the last stage calculates the sum while the intermediate stages all compute the *carry status*. There can be many approaches to communicating the p and g stages between qubits as shown by the prefix graphs in Figure 6. Each line represents two bits of data, p and g. Since we have to return the ancilla to their original state, these bits would also have to travel in the opposite direction at a later stage. Thus each lateral line in the graph, that crosses the Qchip boundary (represented by the dashed line), represents four messages. For our distributed adder, a scheme like $[8, 4, 2, 1]$ would reduce the amount of inter-chip communication while increasing "fan-out". Since it is not possible to do "fan-out" in a quantum processor [24], we would be forced to do those computations in a serial fashion. On the other hand, a $[1, 1, 1, 1]$ scheme (Figyre 6(b)) increases the amount of communication. For our adder we settle on a $[2, 2, 1, 1]$ scheme (Figure 6(a). A quantum circuit of the parallel prefix adder for 4-qubits, based

Table 4. Time (in years) for a complete Modular Exponentiation when per message time is 800 E-6 sec

N	$4\ cm^2$		$8\ cm^2$	
	Carry Select	Par. Prefix	Carry Select	Par. Prefix
128	0.10	0.22	0.19	0.29
512	2.99	0.34	2.99	0.43
1024	19.9	1.35	15.64	1.69

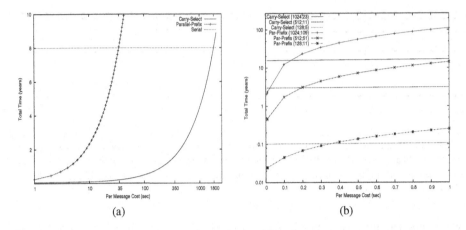

Fig. 7. (a) Shows when the multi-chips solutions outperform the serial version. (b) Performance of the Carry-Select and the Parallel-Prefix Adders.

on $[2, 1]$ scheme, is shown in Figure 5; reverse and sum computations are not included. Communication requirements of our parallel prefix adder will increase with the number of stages in the addition (This is primarily due to our choice of the $[2, 2, 1, 1]$ scheme). Messages sent, between two Qchips, at stage i can be given by $M_i = M_{i-1} + 2^{i-1}$

6   Results

Results for all our designs were obtained by hand calculations. Figure 7(a) shows the cost per message at which the multi-chip solutions outperform the single Qchip. Since the number of messages between Qchips increases rapidly for the parallel prefix adder (while staying constant for the carry select), the cross-over point is at 35 sec/message; and almost 1800 sec/message for the carry select. In Figure 7(b), we compare the multi-chip solutions with message costs of $\leq 1sec$. We can see that the parallel prefix will clearly be the best solution as message costs decrease. To arrive at Table 4, we employ ideas from Section 2.2 and let cost/message equal $800\mu sec$.

To optimize the parallel prefix solution further, we consider the effects of having larger bandwidth and a greater number of multipliers for modular exponentiation. Increasing the bandwidth involves using the interconnect ideas from Section 2.2. Since one Toffoli gate takes at least 0.8 sec [2], while sending an EPR pair between two Qchips takes only $600\mu sec$ [22], given enough bandwidth we can easily pre-communicate all EPR pairs while the computation is in progress. Additional multipliers allow us to increase concurrency in the algorithm while requiring more Qchips. Figure 8 shows that it would take almost 13 days to factor a 1024-bit number if we could employ 4300 Qchips, and send 10,000 messages simultaneously between two Qchips. Each of these 4300 Qchips is $10cm^2$ in area.

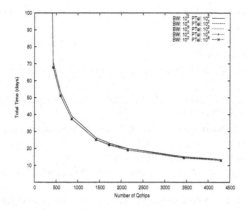

Fig. 8. Optimized Performance of the Parallel Prefix Adder with increasing number of Qchips

7 Conclusion

The technology for quantum computation has made extraordinary progress towards a practical quantum computer. The key to constructing a scalable system from these technologies, however, lies in classical principles of system design, parallelism, and communication. Our results indicate that a multi-chip implementation of ion-trap quantum computation is an attractive design point for the near future. Inter-chip communication through photon-mediated interaction allows low-latency and reasonable bandwidth. With careful partitioning of algorithms such as modular exponention, classically intractable problems can be solved by exploiting parallelism without overwhelming inter-chip bandwidth. Our hope is that this approach will facilitate practical implementation of future quantum algorithms as they arise.

References

1. P. Shor, "Polynomial-time algorithms for prime factorization and discrete logarithms on a quantum computer," *35'th Annual Symposium on Foundations of Computer Science* , pp. 124–134, 1994.
2. T. S. Metodi, D. D. Thaker, A. W. Cross, F. T. Chong, and I. L. Chuang, "A quantum logic array microarchitecture: Scalable quantum data movement and computation," *Proceedings of the 38th International Symposium on Microarchitecture* **MICRO-38**, 2005.
3. P. W. Shor, "Scheme for reducing decoherence in quantum computer memory," *Phys. Rev. A* **54**, p. 2493, 1995.
4. A. Steane, "Error correcting codes in quantum theory," *Phys. Rev. Lett* **77**, pp. 793–797, 1996.
5. D. Gottesman, "A class of quantum error-correcting codes saturating the quantum hamming bound," *Phys. Rev. A* **54**, p. 1862, 1996.
6. J. I. Cirac and P. Zoller, "Quantum computations with cold trapped ions," *Phys. Rev. Lett* **74**, pp. 4091–4094, 1995.
7. M. Riebe, H. Haffner, C. Roos, and et. al., "Deterministic quantum teleportation with atoms," *Nature* **429**(6993), pp. 734–737, 2004.

8. V. Vedral, A. Barenco, and A. Ekert, "Quantum networks for elementary arithmetic operations," *Phys. Rev.* **A54**, p. 147, 1996.
9. D. Beckman, A. Chari, S. Devabhaktuni, and J. Preskill, "Efficient networks for quantum factoring," *Unpublished* , 1996.
10. R. V. Meter and K. M. Itoh, "Fast quantum modular exponentiation," *E-Print: quant-ph/0408006* , 2004.
11. S. Cuccaro, T. Draper, S. Kutin, and D. Moulton, "A new quantum ripple-carry addition circuit," *Unpublished* , 2004.
12. T. Draper, S. Kutin, E. Rains, and K. Svore, "A logarithmic-depth quantum carry-lookahead adder," *E-Print: quant-ph/0406142* , 2004.
13. S. Knowles, "A family of fast adders," *Unpublished* , 1985.
14. W. Wootters and W. Zurek, "A single quantum cannot be cloned," *Nature* **299**, pp. 802–803, 1982.
15. C. H. Bennett and et. al., "Teleporting an unknown quantum state via dual classical and EPR channels," *Phys. Rev. Lett.* **70**, pp. 1895–1899, 1993.
16. J. S. Bell, "On the Einstein-Podolsky-Rosen paradox," *Physics* **1**, pp. 195–200, 1964.
17. W. Dur, H. J. Briegel, J. I. Cirac, and P. Zoller, "Quantum repeaters based on entanglement purification," *Phys. Rev.* **A59**, p. 169, 1999.
18. C. Bennett and et. al., "Purification of noisy entanglement and faithful teleportation via noisy channels," *Phys. Rev. Lett.* **76**, p. 722, 1996.
19. D. Deutsch, A. Ekert, R. Jozsa, C. Macchiavello, S. Popescu, and A. Sanpera, "Quantum privacy amplification and the security of quantum cryptography over noisy channels.," *Phys. Rev. Lett.* **77**, pp. 2818–2821, 1996.
20. C. Monroe, "Quantum information processing with atoms and photons," *Nature* **416**, p. 238, 2002.
21. C. Cabrillo, J. Cirac, P. Garca-Fernndez, and P. Zoller, "Creation of entangled states of distant atoms by interference," *Phys. Rev. A* **59**, pp. 1025–1033, 1999.
22. B. Blinov, D. Moehring, L. Duan, and C. Monroe, "Observation of entanglement between a single trapped atom and a single photon," *Nature* **428**, pp. 153–157, 2004.
23. L. Duan, M. Lukin, J. Cirac, and P. Zoller, "Long-distance quantum communication with atomic ensembles and linear optics," *Nature* **414**, p. 413, 2001.
24. M. A. Nielsen and I. L. Chuang, *Quantum Computation and Quantum Information*, Cambridge University Press, Cambridge, UK, 2000.

Segmented Bitline Cache: Exploiting Non-uniform Memory Access Patterns

Ravishankar Rao[1], Justin Wenck[1],
Diana Franklin[2], Rajeevan Amirtharajah[1], and Venkatesh Akella[1]

[1] University of California, Davis
[2] California Polytechnic State University, San Luis Obispo

Abstract. On chip caches in modern processors account for a sizable fraction of the dynamic and leakage power. Much of this power is wasted, required only because the memory cells farthest from the sense amplifiers in the cache must discharge a large capacitance on the bitlines. We reduce this capacitance by segmenting the memory cells along the bitlines, and turning off the segmenters to reduce the overall bitline capacitance.

The success of this cache relies on accessing segments near the sense-amps much more often than remote segments. We show that the access pattern to the first level data and instruction cache is extremely skewed. Only a small set of cache lines are accessed frequently. We exploit this non-uniform cache access pattern by mapping the frequently accessed cache lines closer to the sense amp. These lines are isolated by segmenting circuits on the bitlines and hence dissipate lesser power when accessed.

Modifications to the address decoder enable a dynamic re-mapping of cache lines to segments. In this paper, we explore the design-space of segmenting the level one data and instruction caches. Instruction and data caches show potential power savings of 10% and 6% respectively on the subset of benchmarks simulated.

1 Introduction

A fundamental shift is occurring in microprocessor design. The unending quest for performance has led to an unquenchable thirst for power. The focus is shifting from pure performance to power-efficient performance. As this shift occurs, architects take a new look at existing structures. We take another look at the design of a cache.

Caches have traditionally been designed such that each element in the cache is accessed in the same amount of time, requiring the same amount of power. In reality, all elements within the cache are not accessed equally, so a cache should be designed to exploit the patterns that emerge in an application. In essence, this is merely an extension of the original idea of caches. The basic hypothesis is that a subset of the memory has locality, and that a cache can exploit this locality. Temporal and spatial locality are exploited by storing recently used data items, and data in the successive memory locations, in the cache. Once an item is placed in the cache, however, it becomes equal to all other elements in

Y. Robert et al. (Eds.): HiPC 2006, LNCS 4297, pp. 123–134, 2006.
© Springer-Verlag Berlin Heidelberg 2006

the cache. This assumes there is no locality within the cache, an assumption that wastes an opportunity for optimization. We propose designing the cache such that the position within the cache determines the power consumption for accessing that set. This offers the opportunity to partition the cache at the circuit level into different power domains and explore micro-architectural techniques to remap the memory accesses into the appropriate power domain to save power. In order to implement a low-overhead segmented bitline cache, we exploit two key characteristics in the design of modern caches. We first observe that caches require more power on the bitlines because every element must drive the signal along the whole length of the bitline, irrespective of its physical location in the cache. We exploit this by placing segmenters on the bitline, requring an element to drive only the distance from its segment to the output, resulting in a lower power to attain the same performance.

In this paper, we present a segmented bit line cache implementation that minimizes the overhead of segmenting. We develop a power model of the segmented bitline cache using CACTI [8] and HSPICE. We then describe and evaluate two methods to dynamically map sets to bitline segments of the cache to take advantage of different power domains based on the access pattern. Finally, we evaluate these methods in the context of level one instruction cache and data caches on a subset of integer and floating point SPEC2000 benchmarks.

The rest of the paper is organized as follows. We begin by describing related work in power efficient cache design in Section 2. We provide empirical evidence for the non-uniform access patterns in Section 3, which provides motivation for the work. Section 4 describes the implementation of our segmented bit line cache, followed by the clustering and mapping algorithms in Section 5. Evaluation methodology and results are presented in Section 6. Finally, we outline our future work and conclusions in Section 7.

2   Related Work

In this section we discuss several related research projects. In particular, two important areas of research are related to this work: circuit level optimizations of the bitlines for power savings and architectural techniques for power efficient caches.

At the circuit level, bitline isolation has been extensively studied. Kamble and Ghose [4] propose a local bitline for segments which are then connected to a common line across isolating switches. Higher metal layers are used to implement the common bitline. Since these layers have a lower capacitance, it reduces the overall bitline capacitance and hence saves dynamic power.

Yang and Kim propose a low power SRAM using a hierarchical bitline and local sense amplifiers (HBLSA-SRAM) [9]. The conventional SRAM bitline is divided into a several sub-bitlines each with its local sense amplifier. The sub-bitlines are connect to the read/write/pre-charge circuits through a common bit line. The HBLSA-SRAM reduces the write power in bitlines without noise degradation by applying a low swing signal to the bit line and a full swing

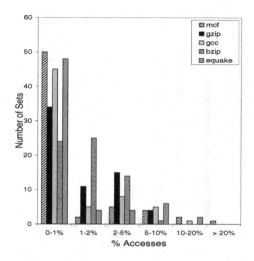

Fig. 1. Histogram: L1 Instruction Cache

signal to the sub-bitline. Yang and Falsafi [10] note that significant energy is wasted in statically pulling up the bitlines in all cache sub-arrays. They show the potential leakage power savings in not pre-charging the unaccessed sub-arrays. The authors study the energy and performance trade-offs of bitline isolation, and propose prediction techniques to exploit its full potential.

3 Motivation

In order for us to gain power savings from our segmented bitline cache, the frequently accessed sets must be mapped to the lower-powered segments. Non-uniform cache access patterns in caches are critical to the success of this design. This section provides preliminary data which demonstrate these patterns.

We simulated an out-of-order processor using Simplescalar [3]. The cache parameters were taken from the Itanium L1 cache [2]. The instruction and data caches are 4-way set associative, 16KB caches with a 64B set size. A subset of the SPEC2000 benchmark suite was run to completion to generate detailed access distribution data.

Figures 1 and 2 give histograms showing the number of sets and the frequency at which they were accessed for L1 instruction and data caches. There were a total of 64 sets. A uniform access pattern would have resulted in about 1.6% of access to each set.

We can see that for the L1 Instruction cache shown in Figure 1, *mcf*, *gcc*, and *equake* have 1-3 cache sets that account for at least 10% of the accesses each. For *mcf*, about 78% of the sets are accessed less than 1% of the time. These applications are good candidates for our approach. For some applications, like *bzip*, most cache sets are accessed at the mean access rate. They will be poor candidates.

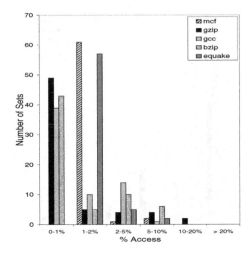

Fig. 2. Histogram: L1 Data Cache

In the data cache shown in Figure 2, the accesses are not quite as skewed. Only one application has sets taking more than 10% of the total accesses. In addition, two applications, *mcf* and *equake*, have the bulk of their accesses near the mean rate.

4   Segmented Bitlines

Data and tag arrays in caches are typically partitioned into banks and further divided into blocks and sub-blocks. A detailed evaluation of various configurations can be found in [1]. The different partitions give different power-performance trade-offs. In this paper, since we primarily explore level 1 caches, we use the lowest access time configuration as given by CACTI [8] We then add segmenters to the bitline and study two, four and eight segment bitline. Although this approach adds delay to the critical path of the bitline segment furthest from the sense amplifiers, it reduces the length of the bitline for the nearest cache rows thereby reducing the capacitance which has to be discharged, and hence the bitline power.

Cache Design and Operation. In a standard SRAM cache a physical bitline is one continuous wire segment that connects many SRAM cells to the sense amplifier (SA), precharge and write circuitry. Every time the bitline voltage levels need to be changed, either for a read or a write, the entire bitline must be charged or discharged. As CMOS processes continue to scale down, the parasitic interconnect capacitance begins to dominate overall cache energy. Bitlines could consume up to 50% of the total cache power [4]. When accessing an SRAM cell that is physically close to the read/write/precharge circuitry, energy is wasted by charging/discharging the other end of the bitline. To prevent this waste of

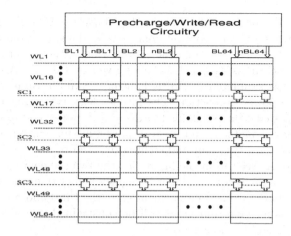

Fig. 3. Segmented Bitlines

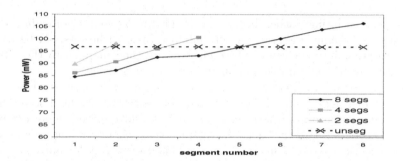

Fig. 4. Cache Power for accessing different segments

energy we use a segmented bitline shown in Figure 3. BL is the bitline and nBL is its complement. WL refers to a wordline. Each box has 16 SRAM cells connected between the bitline and its complement. The block diagram shows the bitline broken into 4 segments. We define a segment to be a portion of the bitline that can be isolated from any other bitline segment. The bitline can easily be divided into 2, 8, or any other combination of segments. Each segment is separated by a segmenter, which consists of a full CMOS transmission gate. Each transmission gate has a control signal (SC), which is operated each cycle according to the address of the data being accessed. When the clock is low all segmenters are turned on to enable the precharge of the entire bitline. Although it would take less power to pre-charge only part of the bitline up to the segment being accessed, in high performance caches, pre-charging overlaps address decoding and hence the segment address may not be decoded during pre-charge.

Power Model. The power model was developed using of CACTI and HSPICE. The only difference between a conventional cache and a segmented bitline cache are the additional segmenters along the bitline. Hence, the power consumption of the rest of the cache is assumed to be as given by CACTI. A column with 64 SRAM cells was simulated in HSPICE using TSMC $0.18\mu m$ technology to obtain the bitline power dissipation. The bitlines with two, four and eight segments were simulated to account for the delay and power consumption of the segmenters. Figure 4 plots the total cache power dissipation for accessing the different segments for a two, four and eight segmented bitline cache at 100MHz and 1.7V voltage. As seen from the figure, for the two segment case, the power dissipated to access the second segment is slightly higher than the unsegmented cache power. Similarly, in the four segment case, the third and fourth segments require higher power. Thus, our gains will be constrained by the percentage of accesses to the higher-power segments.

5   Clustering and Mapping

In a segmented bitline cache, the cache is divided into two, four or eight bitline segments. We define clustering as determining which cache sets can be partitioned together into a cluster. These clusters are then mapped to the bitline segments. Clustering and mapping introduce several challenges.

First, we do not have perfect knowledge of which sets will be accessed. Second, even if we had perfect knowledge, it would not necessarily be feasible to allow arbitrary permutation of sets to clusters. This would require a more complex logic to decode a set in the cache, increasing access times eroding some of our gains due to the shorter bitlines. We need a balance between flexibility in which cache lines may be clustered together and the speed of finding elements in the cache. Finally, access patterns change with time requiring re-mapping at regular intervals. Performing a re-mapping can be an extremely high-overhead operation. Either the sets that need to be re-mapped must be copied to their new locations, or those sets must be flushed from the cache. Swapping cache lines can be very expensive in terms of power consumption, and flushing the sets from the cache would result in an increased miss-rate. In this paper, we propose to limit the re-mapping to context switches. We assume that during a context switch, much of the state in the cache is lost, so invalidating the cache will incur a negligible performance penalty.

Clustering can be done either statically or dynamically. With static clustering, a single clustering is used across all applications and throughout the run-time of the application. Alternately, the clustering could be changed at run time, which is called dynamic clustering.

The simplest implementation of static clustering is to cluster the cache lines based on most significant address bits. With 64 cache lines, for instance, 6 bits are required to determine the row address. Now, a subset of these bits determine the segments and the others the line in the segment. For example, with 4 segments 2 MSBs determine the segment and 4 bits determine which line within the segment is being accessed. A configuration register stores the mapping of the clusters to the segments. The address bits control the mux (in this case, two 4:1 muxes are required.), which selects the new segment address. The only increase in the critical path of the decoder is the multiplexer delay.

Once we have a clustering scheme, we need to map the cluster into bitline segments. The basic idea is to map the most frequently accessed clusters into the lowest power segments. We associate a counter with every cluster and increment it each time the cluster is accessed. We study two versions of mapping: static mapping and dynamic re-mapping. In static mapping, a single mapping of clusters to segments, obtained by profiling, is used throughout the application.

The dynamic re-mapping uses the access counters to re-map at regular intervals. Two versions of dynamic re-mapping are explored in this paper. In the first version, the segment access information of only the previous interval is used for re-mapping. We refer to this case as 'dynamic counter flush' or **dcf** because the counters used for re-mapping are flushed each time a re-mapping occurs. In this second case called 'dynamic no counter flush' or **dncf**, the cumulative access counts to the cluster from the beginning of the application to the current interval is used to decide re-mapping.

6 Results

Simulations were run with Simplescalar. Counters were added to generate cache line access statistics and re-mapping was implemented on every 1 million cycles. The experiments were run on a subset of the SPEC2000 benchmarks, drawn from both the floating-point and integer suites, using a 16KB 4-way set associative cache for integer and data caches. We used SimPoint3.0 [7] to reduce the simulation time. SimPoint provides representative intervals and weights for each application Multiple standard simulation points each 100 million instructions long were simulated. For each benchmark, all the intervals were simulated, and the results were weighted using the interval weights given by SimPoint. The weighted percentage of access to each segment was then used as an activity factor to scale the power of the corresponding segment.

We perform a series of experiments to evaluate the benefits of these techniques. The graphs show the total cache power savings relative to an unsegmented cache. First, we compare static mapping based on a profile and dynamic re-mapping, with that of a dynamic clustering. We then vary the number of segments from 1 to 8 segments. The operation of a segmented bitline cache at higher frequency is discussed.

6.1   Static vs Dynamic Re-mapping

Figures 5 and 6 compare three choices of mapping:static, dynamic with cumulative counters, and dynamic with counters that reset each interval. We present the results for eight segments. and compare with that of a best case dynamic clustering labeled as oracle.

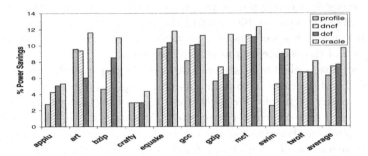

Fig. 5. Static vs. Dynamic: Instruction Cache

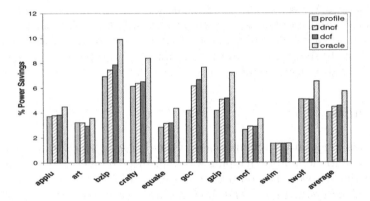

Fig. 6. Static vs. Dynamic: Data Cache

For the instruction cache, the average power savings from a static mapping is 6.3% while the two dynamic re-mapping techniques provide approximately 7.6% average power savings. The dynamic clustering indicates that there is a potential to save up to 10% power on average. For some applications like *applu*, the dynamic techniques achieve close to maximum power savings, while others like *gzip* indicate that the potential is much more than what is obtained by the current dynamic re-mapping techniques. Also, note that maximum power savings are obtained for applications like *mcf* and *gcc* for which access patterns are most-skewed. The data caches have less power savings. Static mappings provide about 4% savings, whereas dynamic re-mapping performs slightly better at 4.5% power savings. The dynamic clustering results indicate a potential power savings of about 5.7% on average. This is consistent with the more uniform access distribution we noted earlier for data caches. Again, as noted earlier, we see a direct relation between power savings and non-uniformity of the access distribution. Applications like *mcf* and *equake* give least power savings because most of their cache sets are accessed at about the mean rate.

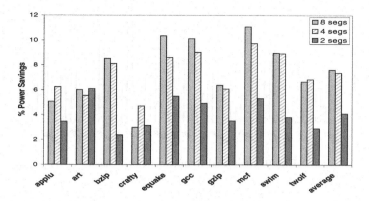

Fig. 7. Number of Segments: Instruction Cache

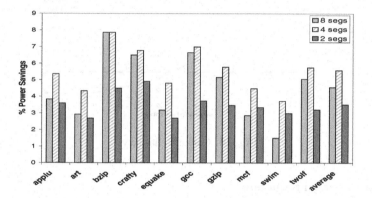

Fig. 8. Number of Segments: Data Cache

6.2 Number of Segments

The number of segments is an interesting design trade-off. With more segments, you can save more from the low-power segment, but you lose more from the high-power segment. We vary the number of segments from 2 through 8. Since the dynamic remapping with counter resetting performs better than static mapping and does not require prior knowledge, we present only the **dcf** results. Figures 7 and 8 show the results for the instruction cache and the data cache respectively.

For instruction caches, increasing the number of segments increases the power savings. While eight segments gives about 7.6% power savings on average, the savings reduce slightly to 7.4% and 4.1% for 4 and 2 segments respectively. This is because instruction caches showed greater non-uniformity in their access patterns, allowing them to greatly benefit from the extra segments. The data caches show a good inflection point giving best power savings at four segments. A more uniform access pattern of the data cache leads to more accesses to the high-power segments, decreasing power savings for eight segments.

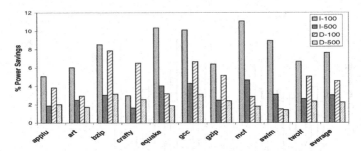

Fig. 9. Increased frequency: Both Caches

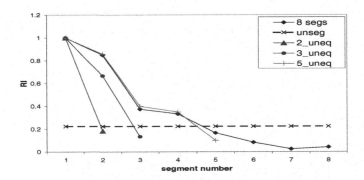

Fig. 10. Cache Reliability

6.3   Increasing Frequency

The results presented in the previous sections are at 100 MHz where the segmenter delay was small compared to the clock cycle. As the frequency is increased, the voltage to which the bitline (or its complement) discharges is reduced. Adding segmenters to this bitline, further increases the bitline delay, thereby reducing the discharge voltage of segments even more.

This reduces the power savings obtained by segmentation because the power dissipated on the bitlines is directly proportional to the voltage to which the bitline discharges. Its effect on the applications is shown in Figure 9 for 8 segments for the **dcf**.

The decrease in bitline discharge also potentially reduces the reliability of segments further away from the sense-amp circuitry since reduced swing increases the likelihood of sense-amp failure. We assume that reliability is inversely proportional to the discharge voltage of the bitline.

So using our remapping techniques, if most accesses are limited to these segments, an overall higher reliability is possible. Even so, in the eight segment case, half the cache has a lower reliability than an unsegmented bitline. To reduce the low reliable cache area we simulated unequal segments in our bitline. Three cases of unequal segments are considered in this paper: First, a two segment bitline,

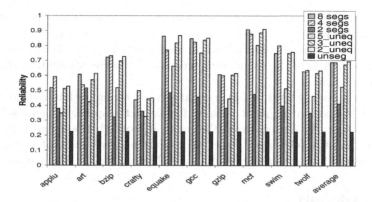

Fig. 11. Application Reliability

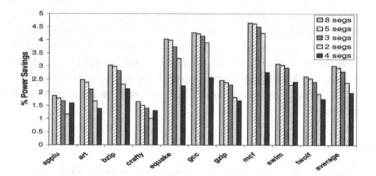

Fig. 12. Unequal Segments: Instruction Cache

with eight rows in the first segment and 56 in the second segment. Second, a three segment bitline, with eight rows in the first segment, 16 in the second and 40 rows in the third segment. Third, a five segment bitline, with eight rows each in the first four segments, and a fifth segment with the remaining 32 rows.

We quantify reliability in terms of 'Reliability Indicator' (RI) which is inversely proportional to the final voltage to which the bitline discharges. A typical sense-amp requires a differential voltage of 0.5V [6] between the bitline and its compliment.The reliability of a segment is thus inversely proportional to the lowest voltage on the bitline. For instance, if the bitline discharges to 1.2 V (with Vdd = 1.7V), the RI is 0, and if the bitline discharge is to 0V, the RI indicator is 1. Figure 10 compares the reliability of unequal segment configurations with that of an unsegmented cache and an eight segment bitline cache. First, we observe that an unsegmented cache itself has a reduced RI at higher frequency. The eight segment bitline has segments with very low RI. This is improved with unequal segments, where only one segment has a RI lower than an unsegmented bitline. We compute the reliability of an application as the sum of, product of segment RI and its percentage access for each configuration. The reliabilities thus obtained for all segment configurations is plotted in Figure 11 for the instruction cache with dcf mapping. It is clear from the figure that all cache configurations provide much better reliabilities than an unsegmented cache. Clearly, this is due to the effectiveness of the re-mapping techniques. Similar results were obtained for the data cache.

The power savings due to these unequal segment configurations is compared to that of eight and four segment bitlines, for the **dcf** mapping for an instruction cache, in Figure 12.As seen from the figure in both the data and instruction caches, for most applications, power savings are between those of an eight and a four segment bitline. Similar results were obtained for the data cache.

7   Conclusions

In this paper, we present architectural techniques that are combined with simple circuit modifications of a cache. We exploit the non-uniform access patterns of different applications to obtain dynamic power savings. We explore partitioning the cache lines into clusters and mapping them to segments dissipating different power. A re-mapping occurs on a context switch to limit the performance and power overhead. Instead, the re-mapping could be guided by the changing program phases. SimPoint has tools to detect phase changes in applications at runtime [5]. By integrating these tools clusters could be re-mapped only during the phase changes.

References

[1] B. S. Amrutur and M. A. Horowitz. Speed and power scaling of srams. *IEEE Journal of Solid-State Circuits*, 35(2):175–185, February 2000.
[2] D. Bradley, P. Mahoney, and B. Stackhouse. The 16kb single-cycle read acess cache on a next-generation 64b itanium microprocessor. In *International Solid State Cirtuits Conference*, 2002.
[3] D. C. Burger and T. M. Austin. The simplescalar tool set, version 2.0. Technical Report CS-TR-1997-1342, University of Wisconsin, Madison, June 1997.
[4] K. Ghose and M. B. Kamble. Reducing power in superscalar processor caches using subbanking, multiple line buffers and bit-line segmentation. In *International Symposium on Low Power Electronics and Design*, pages 70–75, 1999.
[5] J. Lau, S. Schoenmackers, and B. Calder. Transition phase classification and prediction. In *11th International Symposium on High Performance Computer Architecture*, Feb 2005.
[6] J. M. Rabaey. Digital integrated circuits: A design perspective. 1996.
[7] T. Sherwood, E. Perelman, G. Hamerly, and B. Calder. Automatically characterizing large scale program behavior. In *10th International Conference on Architectural Support for Programming Languages and Operating Systems*, Oct 2002.
[8] S. J. Wilton and N. P.Jouppi. Cacti: An enhanced cache access and cycle time model. In *IEEE Journal of Solid-State Circuits*, May 1996.
[9] B.-D. Yang and L.-S. Kim. A low-power sram using hierarchical bit line and local sense amplifiers. In *IEEE Journal of Solid-State Circuits*, June 2005.
[10] S.-H. Yang and B. Falsafi. Near-optimal precharging in high-performance nanoscale cmos caches. In *36th International Symposium on Microarchitecture*, 2003.

Trade-Offs in Transient Fault Recovery Schemes for Redundant Multithreaded Processors

Joseph Sharkey[1], Nayef Abu-Ghazaleh[1], Dmitry Ponomarev[1],
Kanad Ghose[1], and Aneesh Aggarwal[2]

[1] Department of Computer Science
[2] Department of Electrical Engineering
State University of New York at Binghamton
{jsharke, nayef, dima, ghose}@cs.binghamton.edu,
aneesh@binghamton.edu

Abstract. CMOS downscaling trends, manifested in the use of smaller transistor feature sizes and lower supply voltages, make microprocessors more and more vulnerable to transient errors with each new technology generation. One architectural approach to detecting and recovering from such errors is to execute two copies of the same program and then compare the results. While comparing only the store instructions is sufficient for error detection, register values also have to be compared to support fault recovery. In this paper, we propose novel checkpoint-assisted mechanisms for efficient fault recovery that dramatically reduce the number of register values to be compared for detecting soft errors and perform comprehensive investigation of these and other existing recovery schemes from the standpoint of performance, power and design complexity.

1 Introduction and Motivations

The continuous downscaling of CMOS technology leads to smaller transistor feature sizes and the use of lower skupply voltages with each new process generation, making the microprocessor chips more vulnerable to soft (or transient) errors. These transient errors, also known as "single event upsets", occur for various reasons, for example when cosmic alpha particles energize or discharge internal nodes of logic or SRAM bitcells, resulting in incorrect operation. It is projected that the rate at which the transient errors occur will grow exponentially [14] and will soon represent one of the most significant issues in the design of future generation high-performance microprocessors.

One popular approach to addressing these challenges is to execute two copies of the same program and then compare the sequence of results generated by each thread [1, 4, 5, 6, 7, 8, 10, 11]. Any discrepancy between the two result sequences indicates the occurrence of a transient error. Such redundant execution can be implemented in the framework of a superscalar processor. However, despite the well-known fact that the execution of just a single thread leaves the processor resources fairly underutilized, running two simultaneous copies while sharing all resources results in very significant performance degradations [1, 6, 9].

Y. Roberts et al. (Eds.): HiPC 2006, LNCS 4297, pp. 135 – 147, 2006.

Alternatively, a Simultaneous Multithreaded (SMT) processor naturally provides multiple contexts that can be used to execute two copies of the same program (which we call the *main thread* and the *verification thread*) with less impact on performance [4, 7, 8, 11]. Several solutions have been proposed in recent literature to employ SMT support for redundant multithreading, including the schemes that just detect the transient errors [7] as well as those that support recovery capabilities [11].

The key to avoiding performance loss in the redundant multithreaded environment is to use *staggered execution*, i.e. to delay the execution of the verification thread by a number of instructions (defined as *slack* in the rest of the paper) behind the main thread. With growing memory latencies, a larger amount of slack between the two threads can help in hiding the memory access delays experienced by the main thread. To take advantage of the staggered execution, the slack is built and maintained during the normal execution and it is consumed (the verification thread catches up with the main thread) on L2 cache misses. Another advantage of maintaining a sufficient amount of slack is that the actual branch outcomes supplied by the main thread can be used by the verification thread instead of branch predictions. This, in turn, eliminates the execution of the wrong-path instructions from the verification thread, further increasing the execution efficiency and reducing the contention for the use of shared datapath resources.

The basic scheme to provide the transient fault *detection* capabilities in an SMT processor, called SRT (Simultaneously and Redundantly Threaded processor) was introduced in [7]. In SRT, only the results (addresses and data) of the store instructions are compared, because any faults in the registers eventually propagate through the dependency chains to a store. However, if the capability to recover from such faults is also essential, then not only the values to be stored into the memory, but also all values written into the register file need to be verified. Otherwise, the recovery to a precise verified state following a transient error may be impossible, as such a state may never exist.

In this paper, we perform a comprehensive study of the trade-offs in the design of fault recovery schemes, encompassing the issues of performance, energy and design complexity. These schemes include the previously published methods, as well as the ones that are proposed here. We begin by describing the architecture of the baseline SRT machine used for fault detection.

2 Baseline Architecture

The baseline redundant multithreaded processor that we use for our evaluations is based on the SRT model of [7] for transient fault detection. We assume that both main and verification threads perform separate register allocations, so that the register file is also protected.

To introduce the slack between the execution of the two threads, we implemented the slack fetch mechanism described in [7]. The address and data of each store instruction are verified before the store is permitted to update the memory. To verify the address and data of store instructions, an ordered non-coalescing queue, called the *store buffer* (SB) is used, as in [7]. The SB is shared between the threads to synchronize and verify store values as they retire in program order. Data from the store buffer is forwarded to subsequent loads only when the store is retired in the thread issuing the load. The work of [7] proposes two alternatives for the input replication of load data. We implement the *load value queue* (LVQ) – which was shown to provide superior performance [7]. When a load commits from the main thread, it writes both its address and data into the LVQ. Subsequently, when the same load issues in the verification thread, the address is verified and the data is read from the LVQ (i.e., the verification thread does not access the D-cache). This increases performance because the verification thread does not experience cache misses and does not compete for the cache ports.

Finally, to eliminate the wrong-path instructions in the verification thread, we use the *branch outcome queue* (BOQ) [7]. This buffer delivers the committed branch outcomes from the main thread to the verification thread, effectively providing near oracle branch prediction for the verification thread (except in the case where a transient fault causes an incorrect branch resolution in the main thread).

3 Transient Fault Recovery Schemes for SMT

In this section, we describe several possible transient fault recovery schemes that provide recovery capabilities on top of SRT.

3.1 SRT+: Augmenting SRT with Full Register Checking

The first technique that we consider is a trivial augmentation to SRT to check all register values in addition to the data and the addresses of all store instructions. To reduce the pressure on the register file, this requires the addition of a queue (called Register Value Queue – RVQ), where the register results produced by the main thread are written after they are committed. These results are removed from the RVQ only after they are verified by the trailing thread. In this scheme, all register values are checked and an instruction that caused the transient fault can be identified precisely at the earliest possible opportunity. However, a large RVQ is needed to support sizable slack and significant energy is expended in the course of verifying all of the produced register values – those that have to be written to the RVQ, read from it, and compared.

3.2 SRTR and Dependence-Based Checking Elision (DBCE)

The next technique that we examine is called SRTR (SRT with Recovery) and it was introduced in [11]. In addition to checking the store instructions, the SRTR scheme also validates register values, but in contrast to SRT+ it does so selectively. To reduce the pressure on the RVQ and the number of verifications, the authors of [11] also proposed Dependence-Based Checking Elision (DBCE) – a mechanism to limit verifications to only the instructions at the end of short dependency chains, avoiding (or eliding) the verification of the other register values. As reported in [11], about 35% of all register checks are avoided (elided) on the average across SPEC 95 benchmarks using the DBCE scheme.

The original SRTR scheme requires that the result verification occurs prior to instruction commitment (using the writeback-to-commit time), thus putting a limit on the amount of slack that can be maintained. To accommodate a relatively short slack, the SRTR scheme uses the branch predictions (rather than the branch outcomes as in SRT) from the main thread to feed to the verification thread. As a result of the small slack and the use of branch prediction in the verification thread, the SRTR scheme has some performance overhead compared to the SRT design and also incurs some additional changes in the datapath mainly stemming from the need to support speculative instructions in the verification thread. The performance challenges faced by the SRTR scheme will only be exacerbated in the environments with lower branch prediction accuracies and/or D-cache hit rates as well as higher memory latencies.

We observe that it is possible to move the verification actions in the SRTR/DBCE scheme to the post-commit stages by committing the instructions from the main thread and establishing the RVQ entries at that time, just as in SRT+ scheme. The key is not to allow the commitment of any instruction from a dependency chain in the verification thread until the entire chain is verified. The state of the verification thread then can be used to restart the execution following the detection of a fault. This modification allows the DBCE scheme to be used with larger slack and use branch outcomes instead of branch predictions to avoid the execution of wrong-path instructions by the verification thread.

3.3 Checkpoint-Assisted Fault Recovery Schemes

In this paper, we propose novel schemes to further reduce the number of register values that need to be verified to guarantee recovery to a safe state compared to what is proposed by the DBCE scheme. The philosophy of the DBCE is to support a rollback to the latest checked and committed instruction following the detection of a fault and to begin the re-execution from that point. While such an approach completely avoids unnecessary re-executions of already verified instructions, the datapath complexities and performance overheads involved are non-negligible. In essence, from the standpoint of precise state reconstruction, the SRTR scheme treats transient faults like branch mispredictions or exceptions because it maintains the results of all unchecked instructions, just as the results of all speculative instructions are maintained for branch misprediction recovery or interrupt handling.

However, even in current and future technologies, the absolute rate at which transient faults will occur is very low, several orders of magnitude smaller than, for example, the rate of branch mispredictions or exceptions. Therefore, it is unnecessary to start the re-execution at the exact instruction that caused transient fault; even if the rollback occurs to a point which requires several tens of thousands of instructions to be re-executed, there is almost no impact on performance. The key question here is not how far to rollback and how many instructions to re-execute (within reasonable distance), but how to guarantee that a precise and completely verified register and memory state is always available and can be constructed at any point. In the rest of this section, we describe two checkpoint-based mechanisms to facilitate such a recovery. After the detection of an error, the processor state is rolled back to a complete and fully-verified register and memory state checkpoint and the execution restarts.

3.3.1 Lifetime-Based Checking Elision (LBCE)
It has been noticed by several researchers that most of the register instances in a datapath are short-lived [19]. A value produced by the instruction X is *short-lived*

(SL) if the architectural register allocated as a destination of X has been renamed again before the value generated by X is committed. In [20], it was shown that about 84% of all produced values are short lived. Using this notion of short-lived values, [20] proposes *lifetime-based checking elision* (LBCE) in which the verifications of *control-independent short-lived* (CISL) values are avoided. Only the non-CISL results are saved within the RVQ after the instruction commitment and are verified against similar values produced by the verification thread.

To support the capability to recover to a precise and completely verified state following a detection of a transient fault, LBCE relies on the creation of the periodic register and memory state checkpoints. To buffer a large number of store instructions between two consecutive checkpoints, we use the approach described in [21] and also used in a few others works. The memory updates received between two consecutive checkpoints are stored within the local cache hierarchy, but their propagation to the main memory is avoided until it is safe to do so. Each cache line updated in this manner is marked as volatile, using one extra bit for each cache line. When a processor needs to rollback to a checkpoint, all cache lines marked volatile are invalidated using a gang-invalidate signal. When the new checkpoint is created, all volatile bits set since the creation of previous checkpoint are cleared. A recent paper [23] also describes how to correctly incorporate caches with the volatile lines into a multiprocessor system.

Since transient faults are very infrequent events, we can create checkpoints at very large intervals. In fact, a checkpoint can be created on demand, when one of the sets within the cache has all its lines in Volatile status and a cache miss occurs that targets this set. At this point, the creation of a new checkpoint is initiated and, once the checkpoint is created, the volatile bits can be cleared. However, as the percentage of volatile lines in the cache increases, the victim selection algorithm becomes less flexible (the volatile lines cannot be replaced). In the worst case, this effectively transforms the cache into direct-mapped structure and degrades the cache hit rates. In order to avoid such performance degradations caused by the lower D-cache hit rates, we also force the checkpoint creation every 100000 instructions. Therefore, 100000 instructions are re-executed after transient fault detection in this scheme in the worst case. In the result section, we quantify the percentage of checkpoints created for these various reasons. We also show that the average number of instructions between two consecutive checkpoints is generally very large. A recent paper [24] also showed that in commercial workloads the I/O operations could occur more frequently, effectively requiring the creation of a checkpoint at that instant. To support these situations, in the results section we also evaluate the performance of the LBCE scheme with smaller checkpointing periods, as low as 500 instructions.

For more details of the LBCE technique, including the hardware implementation to detect the CISL values, we refer the reader to [20].

3.3.2 An RVQ-Free Recovery Scheme (RVQ_F)

We will now describe the checkpoint-assisted recovery scheme that completely eliminates the RVQ from the datapath. In this scheme, the decision to create a checkpoint can be triggered at the time of committing an arbitrary instruction from the main thread. At this point, the main thread is stalled and the verification thread is allowed to completely catch up (consume the slack). At that time, the contents of the architectural register state from both threads can be compared against each other, and if any mismatch occurs, then a transient fault is detected. Otherwise, new checkpoints of the register file and commit-time rename table can be created. Also, the volatile bit in the cache can be cleared.

Table 1. Comparison of the key features of the transient fault recovery schemes. Quantitative comparisons are provided in the results section.

	SRT+	SRTR	LBCE	RVQ_F
Checkpoints required	No	No	Yes	Yes
RVQ required	Large	Medium	Small	None
Additional Logic Needed	None	Track and form dependency chains	Detect short-lived values	None
Transient-Fault Detection Latency	Short	Short	Short to medium	Large
# of register verifications	All register values	~65% of register values	~ 30% of register values	Only on checkpoint creation
Useful work lost on every fault	None	None	Small to medium	High
Reasons for stalling the main thread	RVQ is full	RVQ is full	RVQ is full	Checkpoint creation

While this scheme simplifies the datapath compared to the LBCE technique from the previous section, it incurs some performance overhead. First, the main thread needs to stall during the checkpoint creation – that is not required by the LBCE. Second, the bulk-comparison of the architectural registers will require a number of cycles to be wasted: for example, for 64 architectural registers (as in the Alpha ISA), the comparisons will consume 16 cycles (if 4 comparisons can be performed per cycle). For small checkpointing periods, these overheads can be significant; we evaluate the sensitivity of these schemes to the checkpointing frequency in the results section. Finally, the RVQ_F scheme is likely to delay the detection of transient errors, as the detection can only occur during the checkpoint creation. In the next section, we compare all of the described techniques in terms of their performance, energy consumption, complexity and other metrics. Table 1 summarizes the key features of the four transient fault recovery schemes examined in this paper. A detailed quantitative comparison of the schemes follows later.

Table 2. Simulated processor configuration

Parameter	Configuration
Machine width	4-wide fetch, 4-wide issue, 4-wide commit
Window size	64 entry issue queue, 64 entry load/store queue, 128-entry ROB
Pipeline Depth	5 cycles fetch to dispatch, 3 cycles issue to execute
Function Units and Lat (total/issue)	4 Int Add (1/1), 2 Int Mult (3/1) / Div (20/19), 2 Load/Store (2/1), 4 FP Add (2), 2 FP Mult (4/1) / Div (12/12) / Sqrt (24/24)
Phys. Registers	300 combined integer and floating-point
L1 I-cache	64 KB, 4–way set–associative, 32 byte line
L1 D-cache	64 KB, 4–way set–associative, 32 byte line
L2 Cache unified	1 MB, 8–way set–associative, 128 byte line
Memory latency	100 cycles
TLB	64 entry (I), 128 entry (D), fully associative

4 Simulation Methodology

For estimating the performance impact of the schemes described in this paper, we used M-Sim [12] – a significantly modified version of the Simplescalar 3.0d simulator [1] that separately models pipeline structures such as the issue queue, re-order buffer, and physical register file, both for superscalar and SMT machines [5,6]. The SRT model described in Section 2 was implemented in this framework. The details of the studied processor configuration are shown in Table 2.

We simulated a total of 24 integer and floating point benchmarks from the SPEC 2000 suite [3], using the precompiled Alpha binaries available from the Simplescalar website [1]. Predictors and caches were warmed up for the first 1 billion instructions and the statistics were gathered for the next 100 million instructions.

5 Results and Discussions

Figure 1 compares the performance of the transient fault recovery schemes that rely on the RVQ. Results are presented in terms of harmonic means across all simulated SPEC 2K benchmarks. The first variation is the SRT scheme which only supports fault detection – this represents an upper bound on the performance, as there is no

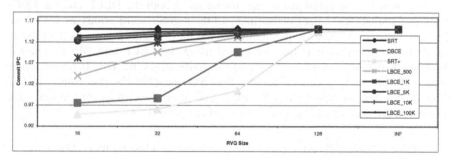

Fig. 1. Harmonic mean of commit IPC for various redundant multithreaded architectures for various sizes of the RVQ

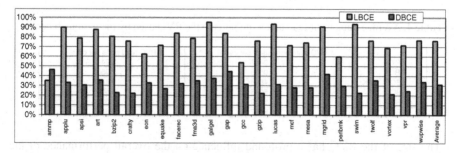

Fig. 2. Percentage of register verifications elided using the LBCE and DBCE schemes

recovery overhead. The other lines correspond to the SRT+, the DBCE scheme, and the LBCE scheme with various checkpointing periods. The number next to the LBCE label in the ledged signifies the checkpointing period (number of instructions) used for the corresponding configuration.

For these experiments, the target slack of 256 instructions (shown to be optimal in [7] and also confirmed by our experiments) was used. Because not all instructions are verified through the RVQ (loads, stores and branches are not), the performance saturates for all schemes at the RVQ size of 128 entries, with the saturation in the LBCE scheme occurring at much smaller RVQ sizes. The SRT+ scheme results in significant performance losses compared to simple SRT at smaller RVQ sizes. For example, the average performance losses are 21%, 19% and 15% for the RVQ sizes of 16, 32 and 64 entries respectively. The DBCE reduces the performance overhead of SRT+ and lowers the performance degradations to 18%, 16%, and 4.9% respectively for 16, 32 and 64-entry RVQ compared to the SRT+ design. Next, the LBCE scheme with the small checkpointing period of 500 lowers these percentages further to 10%, 5% and 1.6%. Finally, the LBCE scheme with a large checkpointing period of 100,000 instructions lowers these percentages to 1.4% 0.5% and 0.3%. Notice that the LBCE scheme with larger checkpointing periods provides better performance as the overhead of checkpoint creation is small. In summary, a 16-entry RVQ with the LBCE scheme provides almost the same performance as the SRT without any recovery overhead or as the SRT+ with 128-entry RVQ.

The reason for the performance improvements in both the DBCE and the LBCE schemes for small RVQ sizes is that many of the register verifications are elided and therefore fewer instructions require entries in the RVQ. Figure 2 presents the percentage of register verifications that are elided using the LBCE and DBCE schemes. While the LBCE scheme elides 76.1% of the verifications, the DBCE scheme elides about 32% of the verifications for the Spec2000 benchmarks (the results in [11] showed 35% for the Spec95 benchmarks). The larger percentage of register value checks that are elided by LBCE are manifested in higher IPCs.

Table 3. Number of cycles when the leading thread stalls because the RVQ is full

	16-entry RVQ	32-entry RVQ	64-entry RVQ	128-entry RVQ	256-entry RVQ
SRT+	81700029	79098756	72103164	32808131	281531
DBCE	75477795	72893922	53269057	8252604	38
LBCE_10K	36841417	24369222	6297368	315680	0

Table 4. Average number of read and write ports to the RVQ used by the various schemes

	# RVQ write ports used per cycle	# RVQ read ports per cycle
SRT+	2.9447	2.9447
DBCE	2.0505	2.0505
LBCE_100K	0.4898	0.4898
LBCE_10K	0.5045	0.5045
LBCE_5K	0.5219	0.5219
LBCE_1K	0.6417	0.6417

The size of the RVQ has a profound influence on the overall performance of the schemes that require a RVQ, as shown in Table 3. Whenever the RVQ is full, the main thread is stalled and the verification thread is run, preventing further progress of the main thread momentarily. As seen from Table 3, the LBCE scheme has a significant advantage over the others that use a RVQ, as it stores only the non-CISL values. At about an RVQ size of 256 entries, both DBCE and LBCE avoid any stalls of the main thread. In contrast, some stalls still occur for the SRT+ scheme at this RVQ size. Therefore, a smaller RVQ size is sufficient for the LBCE scheme to provide similar performance.

Next, we examine the impact on dynamic power dissipation within the RVQ of our technique. We compare two configurations that achieve the same performance, specifically a 32-entry RVQ with LBCE scheme and 128-entry RVQ with SRT+ scheme. The savings in dynamic power of LBCE scheme comes from two sources. First, much fewer access to the RVQ are performed because 76% of the checks are elided, and second the size of the RVQ is significantly smaller. Combined, these two artifacts result in 89.1% savings in dynamic power within the RVQ compared with the SRT+ design. Of course, additional power would be dissipated in the auxiliary datapath structures required by the LBCE scheme, which will somewhat lower these reported savings. However, if the point of comparison is the DBCE mechanism, then it also requires additional power to detect and form the dependency chains in both threads. A more detailed power related analysis of these mechanisms is beyond the scope of this paper.

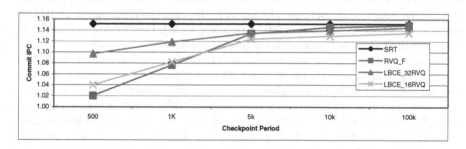

Fig. 3. Harmonic mean of commit IPC for various transient fault recovery schemes for various checkpointing intervals

Additionally, because many of the register verifications are elided, fewer reads and writes to the RVQ are performed each cycle with the LBCE scheme compared to the DBCE and SRT+ techniques. Table 4 presents the average number of read ports and write ports used per cycle to the RVQ for the various transient fault recovery schemes. The SRT+ technique, with allocates an RVQ entry for each and every register value, uses nearly 3 read ports and 3 write ports each cycle on average. Comparatively, the DBCE scheme uses only 2 read ports and 2 write ports on average each cycle and the LBCE technique uses less than one. This allows for a reduction in the number of ports to the RVQ with the LCBE scheme in addition to the reduction in RVQ size – which provides additional energy and power savings.

Now, we examine the checkpoint-based transient fault recovery solutions. Figure 3 presents the harmonic mean of commit IPC for the RVQ_F and LBCE schemes (the only two schemes that rely on checkpointing) for various checkpointing periods. The SRT scheme that does not provide recovery is also shown for comparison. For the small checkpointing periods, LBCE outperforms RVQ_F because the overhead of the frequent checkpoint creations offsets the advantages offered by the RVQ_F scheme. For example, the LBCE scheme with a 32-entry RVQ outperforms the RVQ_F scheme by 4% for a checkpointing period of 1000 instructions and 8% for the checkpointing period of 500 instructions. On the other hand, for large checkpointing periods, the RVQ_F scheme provides better performance. For the period of 100K instructions, the RVQ_F scheme outperforms the LBCE scheme by 1.5%.

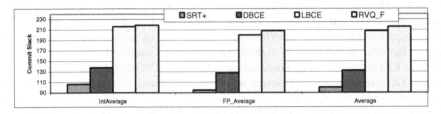

Fig. 4. Effective slack length measured in number of instructions at commitment

The RVQ_F is quite efficient for large checkpointing frequencies because it elides most of the register checks (other than the ones that are needed for checkpoint creation) by the nature of the scheme. For example, for checkpointing period of 100000 instructions, 99.8% of all register verifications are elided. For 500-instruction checkpointing period, the percentage of elided checks is about 80%.

The next metric that we examine is the effective slack length as measured at commit time. The results for the 64-entry RVQs are presented in Figure 4. For this configuration, the LBCE scheme achieves a slack of 207 instructions, on the average – more than twice that of the processor with the basic SRT+ which achieves a slack of only 101 instructions. The DBCE scheme achieves the slack of 135 instructions. These results show that the LBCE technique can maintain a large slack, and take advantage of it, with a small RVQ size. In fact, the amount of the effective slack in the LBCE scheme even with 16-entry RVQ is almost the same as the effective slack of the SRT+ scheme with infinite RVQ (again, the results of Figure 1 can be used to understand why that is the case). Finally, the RVQ_F scheme achieves an average slack of 210 instructions.

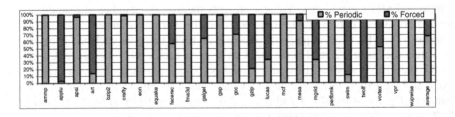

Fig. 5. Breakdown of the percentage of checkpoints created periodically versus the percentage of forced checkpoints due to cache behavior for the LBCE and RVQ_F schemes. Results are presented for the checkpointing period of 100000 instructions.

Next, we evaluate the impact of the checkpointing mechanism used by LBCE and RVQ_F in order to support recovery from transient faults. Recall that there are two triggers for the creation of checkpoints in these schemes. Checkpoints are created periodically, or when required due to the absence of non-volatile data in the cache set for victim selection. Figure 5 presents the data on the percentage of checkpoints created due to each of these triggers. As seen from the graph, 69% of the created checkpoints are induced periodically. The percentage of checkpoints that are created due to the absence of non-volatile lines in the accessed set of the cache is relatively small on the average, but can be quite high for the memory bound programs. For example, applu, art, swim, and twolf all experience high levels of memory traffic and therefore incur more such checkpoints.

It is conceivable that the use of volatile bits in the cache can somewhat degrade the cache hit rates because of the additional constraints imposed on the cache replacement policies. However, our results indicate that this impact is minimal. On the average, the L1 D-cache hit rates decreased from 94.6% to 94.5%, and the largest decrease was 2.3% observed on ammp.

6 Related Work

A popular approach for concurrent error detection and recovery is to execute two copies of the same program and then compare the results [1,4,5,6,7,8,10,11]. Ray, Hoe, and Falsafi [6] propose mechanisms for performing such redundant execution within a superscalar processor. Smolens et. al. [9] study the performance impact of redundant execution and identify the various bottlenecks that limit the performance in such environments. The DIVA design of [1] supplemented the out-of-order core with simple in-order checker logic.

The fault-tolerant architectures in [4,7,8,11] use the inherent hardware redundancy in SMT and CMP architectures for concurrent error detection. While the SRT scheme described in [7] only aims at detecting transient faults using the SMT support, the follow up study of [11] augments the work of [7] by adding the recovery capability. The resulting scheme, called SRTR (SRT with Recovery) is perhaps the closest in spirit to the proposal. We extensively discussed the SRTR scheme and contrasted it to techniques proposed here throughout the paper.

RMT explored the design space of using multithreading for fault detection [15], and was extended by CRTR [16] to provide fault recovery using CMPs.

The concept of partial soft error coverage was introduced in [5], where the redundant execution is only performed during the low-ILP phases of the main program, when the resources are sufficiently underutilized. In [2], the execution of the redundant thread only happens when the main thread experiences the L2 cache miss or the verification buffer is full.

Several industrial designs support fault tolerance. The Compaq NonStop Himalaya [12] employs off-the-shelf microprocessors in lock-step fashion and compares the outputs every cycle. The IBM S/390 [18] uses replicated, lock-stepped pipelines within the processor itself.

7 Summary and Concluding Remarks

The choice of the best transient fault recovery scheme is dictated by the checkpointing interval as well as datapath complexities that can be tolerated. We can expect aggressive modern out-of-order processors to use checkpoint-based recovery mechanisms. Some of the schemes studied in this paper assume the existence of such a facility. There is always a tradeoff between the performance, the complexity, and the energy consumption that guide the choice of the soft error detection and recovery scheme. The main conclusions of our study, in the light of such considerations, are as follows.

If large checkpointing intervals can be tolerated, then the RVQ_F scheme provides the best performance because of the least number of register values comparisons – only the architectural register values need to be compared at the time of checkpoint creation. Furthermore, RVQ_F scheme eliminates the need for an RVQ and all associated overhead. However, at smaller checkpointing intervals, the LBCE mechanism is more attractive because it achieves better performance for a smaller RVQ size relative to SRT+ and DBCE. Furthermore, the data on the usage of read and write ports shows that the LBCE technique can not just use a smaller RVQ compared to SRT+, but it can also use fewer register file ports, thereby reducing the overall power dissipation (and the overall complexity) of the verification logic. If a large RVQ can be supported, then schemes that do not rely on checkpointing, such as SRT+ and DBCE, are both reasonable choices.

References

1. T. Austin, "DIVA: a reliable substrate for deep submicron microarchitecture design," Proc. Micro-32, 1999.
2. Qureshi, M., et al, "Microarchitecture-Based Introspection: A Technique for Transient Fault Tolerance in Microprocessors", in DSN 2005.
3. J. G. Holm, and P. Banerjee, "Low cost concurrent error detection in a VLIW architecture using replicated instructions" Proc. ICPP-21, 1992.
4. M. Gomaa, et. al., "Transient-Fault Recovery for Chip Multiprocessors," Proc. ISCA-30, 2003.
5. M. Gomaa, T.N.Vijaykumar, "Opportunistic Transient Fault Detection", ISCA 2005
6. J. Ray, J. Hoe, and B. Falsafi, "Dual use of superscalar datapath for transient-fault detection and recovery," Proc. Micro-34, 2001.
7. S. Reinhardt, and S. Mukherjee, "Transient fault detection via simultaneous multithreading," Proc. ISCA-27, June 2000.
8. E. Rotenberg, "AR-SMT: A microarchitectural approach to fault tolerance in microprocessors," Proc. 29th Intl. Symp. On Fault-Tolerant Computing Systems, 1999.
9. J.Smolens, et. al., "Efficient Resource sharing in Concurrent error detecting Superscalar microarchitectures ," Proc. Micro- 37, 2004.
10. K. Sundaramoorthy, Z. Purser, and E. Rotenberg, "Slipstream processors: Improving both performance and fault tolerance," In Proc. Micro-33, December 2000.
11. T. Vijaykumar, I. Pomeranz, and K. Cheng, "Transient-fault recovery using simultaneous multithreading," Proc. ISCA-29, 2002.
12. J. Sharkey. "M-Sim: A Flexible, Multi-threaded Simulation Environment." Tech. Report CS-TR-05-DP1, Department of Computer Science, SUNY Binghamton, 2005.

13. D. Tullsen, et al. "Exploiting Choice: Instruction Fetch and Issue on an Implementable Simultaneous Multithreading Processor." in Proc International Symposium on Computer Architecture, 1996.
14. P. Shivakumar, et al. "Modeling the Effect of Technology Trends on the Soft Error Rate of Combinational Logic", in Proc DSN, 2002.
15. S. Mukherjee, et al. "Detailed Design and Evaluation of Redundant Multithreading Alternatives", in Proc ISCA 2002.
16. M. Gomaa, et al. "Transient-fault Recovery for Chip Multiprocessors" in Proc ISCA 2003.
17. Compaq zComputer Corporation, "Data Integrity for Compaq Non-Stop Himalaya Servers", 1999.
18. T. Slegel, et al. "IBM's S/390 G5 Microprocessor Design", IEEE Micro, 1999.
19. D. Ponomarev, et al., "Reducing Datapath Energy through the Isolation of Short-Lived Operands", Proc. PACT 2003.
20. N. Abu-Ghazaleh, et al., "Exploiting Short-Lived Values for Low-Overhead Transient Fault Recovery", Proc. ASGI 2006.
21. J. Martinez, et al., "Cherry: Checkpointed Early Resource Recycling in Out-of-Order Processors", Proc. MICRO 2002.
22. O.Ergin, et al., "Increasing Processor Performance through Early Register Release", Proc. ICCD 2004.
23. M. Kirman, et al., "Cherry-MP: Correctly Integrating Checkpointed Early Resource Recycling in Chip Multiprocessors", Proc. MICRO 2005.
24. J. Smolens, et al, "Fingerprinting: Bounding Soft-Error Detection Latency and Bandwidth", Proc. ASPLOS 2004.

Supporting Speculative Multithreading on Simultaneous Multithreaded Processors

Venkatesan Packirisamy, Shengyue Wang, Antonia Zhai, Wei-Chung Hsu, and Pen-Chung Yew

Department of Computer Science,
University of Minnesota, Minneapolis
{packve, shengyue, zhai, hsu, yew}@cs.umn.edu

Abstract. Speculative multithreading is a technique that has been used to improve single thread performance. Speculative multithreading architectures for Chip multiprocessors (CMPs) have been extensively studied. But there have been relatively few studies on the design of *speculative* multithreading for *simultaneous multithreading* (SMT) processors. The current SMT based designs - IMT [9] and DMT [2] use load/store queue (LSQ) to perform dependence checking. Since the size of the LSQ is limited, this design is suitable only for small threads. In this paper we present a novel cache-based architecture support for *speculative simultaneous multithreading* which can efficiently handle larger threads. In our architecture, the associativity in the cache is used to buffer speculative values. Our 4-thread architecture can achieve about 15% speedup when compared to the equivalent superscalar processors and about 3% speedup on the average over the LSQ-based architectures, however, with a less complex hardware. Also our scheme can perform 14% better than the LSQ-based scheme for larger threads.

1   Introduction

With increasing amount of resources available for the processor, architects are going for multithreading-based designs like CMPs and SMTs. At present, these architectures are mainly used to improve the processor throughput. Using these multithreaded architectures to improve single thread performance still poses a challenge. Speculative multithreading [5, 11] is one way to utilize the multiple threads to improve single thread performance. Here, threads are automatically extracted from a sequential program by a compiler and executed in parallel to improve its execution time. Architecture support is needed to detect any dependence violation, and also to buffer the results created by speculatively created threads.

Existing SMT based speculative multithreading approaches either use complex hardware [7] or use limited resources like LSQs [9, 2] to buffer speculative results, and to record load addresses to check for dependence violations. The advantage of LSQ-based method is that the LSQs are already available to the processor, so the technique does not need any major modifications to the

Y. Robert et al. (Eds.): HiPC 2006, LNCS 4297, pp. 148–158, 2006.

processor architecture as in the case of [7]. Also, in the LSQ, entries are created for each load and store operations, so the dependence checking granularity is at the byte level. At the byte level, there could be no false dependences. But the LSQ entries for speculative threads are not cleared till the thread commits. So the main disadvantage in using LSQs is their limited size since it is not cost effective (or power efficient) to have large LSQs. Due to this consideration, LSQ based architectures can support only small threads. But our research [14] shows that if we need to consider a more realistic overhead of forking a thread, it becomes more difficult to justify at small granularities. Hence, it is important to support larger threads.

In this paper, we propose a novel cache-based architecture to implement speculative multithreading in SMT processors that only requires a few extra bits to each cache line in existing L1 cache in SMT. Also our approach can handle large threads since now the entire cache can be used to buffer results and to check for dependences.

The rest of the paper is organized as follows: section 2 discusses related work, section 3 discusses our cache-based architecture to support speculative multithreading, section 4 discusses results and in section 5 we conclude the paper.

2   Related Work

Speculative multithreading architectures have been studied intensely during the past decade. Earlier architectures were based on special hardware structures for dependence checking like the *address resolution buffer (ARB)* in [4], and the *memory disambiguation table (MDT)* in [5]. These special hardware structures are of limited size and need extra cycles to access them. To avoid these limitations cache-based architectures like *speculative versioning cache (SVC)* [13] and *STAMPede*[10] were proposed.

When compared to speculative multithreading on *chip multiprocessors (CMPs)*, there are very few studies on supporting *speculative multithreading* for SMTs. In [7], private L1 cache for each context is used to buffer speculative values and do dependence checking. In DMT [2] and in IMT [9] an enhanced LSQ is used.

The main limitation of the LSQ-based approach is the limited size of the queue. To overcome this limitation we propose a cache-based scheme in this paper. We draw many ideas from the cache architectures proposed for CMPs. The difference is that the CMP-based architectures have private L1 cache for each core and is used to buffer results. The dependence checking hardware is also distributed among different L1 caches. In our approach, all the contexts in the SMT share the same cache.

Concurrent to our work, Stampede [11] has extended the cache protocol described in [10], to support shared cache architectures. Their technique was studied in the context of multi-core processors using shared cache. In [3], shared L2 cache based technique was used to speculatively parallelize database applications. Though they mention that it could be applied to SMT processors all their results and conclusions are for CMPs, while our scheme is specifically aimed at

SMT processors. Also we propose a novel two-thread scheme and our four-thread scheme uses fewer bits per cache line than their scheme. A detailed comparison of our technique with the STAMPede technique is beyond the scope of this paper.

3   Speculative Simultaneous Multithreading

3.1   Basic SMT Architecture

We consider a SMT architecture where many resources like fetch queue and issue queue are fully shared [12]. Fig. 1 gives a block diagram of the SMT architecture. To implement speculative multithreading, we need hardware support to buffer results from speculative threads, detect dependence violation between threads, and synchronize threads to communicate register values. The only modifications we need are the signal table and the modified L1 data cache. Our inter-thread register synchronization scheme is very similar to [15].

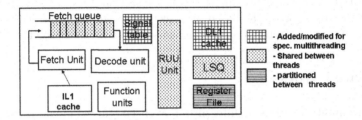

Fig. 1. SMT Block Diagram

We use a novel cache-based scheme to support buffering of speculative values and to enforce memory dependences. In section 3.2, we first present a simplified scheme that supports only one speculative thread, and in section 3.3 we extend this scheme to four (or more) threads.

3.2   Simplified Two-Thread Scheme

In this section, we consider a SMT processor with only two threads. Here, we only need to introduce two extra states to each cache line - *Speculative Valid (SV) and Speculative Dirty (SD)*. Each cache line also needs two extra bits - *Speculative Load (SL) and Speculative Modified (SM)* to support data dependence checking. In the proposed scheme, all of speculative data are kept only in the shared L1 cache, and all of the data stored in L2 cache are non-speculative. Fig. 2 presents the cache-line state transitions in this scheme. In fig. 2 the transitions are of the form 'Command from processor / Action taken'.

Speculative value buffering. When a speculative thread writes, the value is stored in the shared L1 data cache with the SM bit of the cache line set and the cache line transitions to the SD state. The value stays in the cache till the thread is committed or squashed. Thus, the L1 D-cache acts as a *store buffer* that buffers speculative updates.

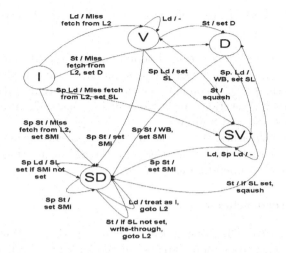

Fig. 2. Two Thread Scheme - Cache State Transitions

Dependence Violation Detection. When a speculative thread issues a load operation, it first checks if a speculative thread has already written the value. However, by having just one SM (speculative modified) bit for each cache line, we cannot be sure which word in a particular cache line was written by the speculative thread. To allow more precise dependence information, we could maintain one SM bit (SMi) for each word in the cache line. If the SMi bit is not set, the SL (speculative load) bit will be set and the cache line transitions to SV (speculative valid) state, as this load could cause a possible dependence violation, when a non-speculative write arrives later.

Here, when a non-speculative thread writes into a cache line, if the SL bit is already set, it indicates that the speculative thread has read a stale value. The speculative thread will be squashed and restarted.

Non-speculative thread execution. If the state of the cache line being written to is SD (speculatively dirty), the non-speculative thread writes the value directly to L2 cache. Also, it writes the portion of the data non-overlapped with the speculatively modified data (indicated by SMi bits) into the L1 cache. This merging is done, so that the speculative thread can get the most recent non-speculative value from L1 cache. Also this simplifies the commit operation. Reads by a non-speculative thread to a speculatively modified line (SD) are treated as a *cache miss* and the value is directly taken from the L2 cache.

Replacement policy. Speculatively modified cache lines or the lines with the SL bit set cannot be evicted from the cache. If evicted, we lose information which can lead to incorrect execution. When we have to replace a line, a line which has none of the SL and SM bits set is selected. If a non-speculative thread needs to replace a line and couldn't find a clean line, the speculative thread is squashed to relinquish its lines. This is done to avoid blocking the non-speculative thread and thus avoiding deadlock. In case of speculative thread, the thread is suspended till it becomes non-speculative.

Commit and Squash. When a thread commits, both the SL and SMi bits are cleared. Unlike other schemes where every speculative value needs to be written to the cache at the point of commit (which could potentially take hundreds of cycles), the commit operation can be done in just one cycle in our scheme by gang-clearing both SL and SMi bits.

When a thread squashes, the SL bit in all cache lines is cleared (gang-clear). The valid bit for a cache line is also cleared if the SM bit is set. This is like the conditional gang-clear operation used in Cherry[8].

3.3   Four-Thread Scheme

When executing more than one speculative thread, the L1 D-cache needs to buffer results from two or more threads, so the two-thread scheme cannot be directly applied. In this section we propose a scheme which can efficiently handle more than one speculative thread. The basic idea is to use the entire set in the cache to buffer different versions of the same line created by the different threads. We will use a 4-thread system to simplify our explanation.

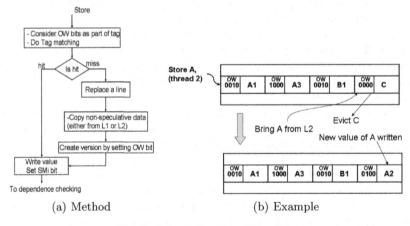

<table>
<tr><td>(a) Method</td><td>(b) Example</td></tr>
</table>

Fig. 3. Speculative Store Handling

Speculative Buffering. The L1 D-cache has to buffer results from multiple threads, so we introduce *Owner* bits (OW) (one for each thread) which keep the speculative thread id that wrote into the cache line. For a non-speculative cache line, the OW bits are cleared. Buffering of speculative values is explained in Fig 3(a). Fig 3(b) shows an example where thread 2 tries to write a new version of A to a set which already contain versions from thread 1 and thread 3.

Speculative Load Execution. A cache line can be read by any of the four threads, so a single SL bit is not sufficient to indicate which thread has caused dependence violation. We introduce a SL bit for each thread on each line of cache (4 bits for 4 threads). The execution of a *speculative load* instruction is explained in fig 4. We can see that the *speculative load* can either load from its own version (i.e., a hit), from predecessor thread's version (i.e., a partial hit) and from L2 cache (i.e., miss - fig. 4(b)). of the cache line.

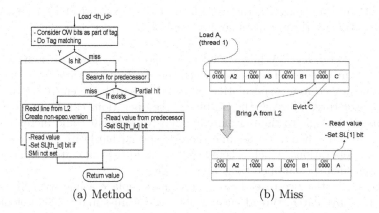

(a) Method   (b) Miss

Fig. 4. Speculative Load Handling Example

Dependence Detection. When a store executes, it checks whether the versions of the cache line belong to any of its own successor threads. If SL bit is set for any of the successor threads, the successor thread is squashed along with its successors. The oldest squashed thread is then restarted.

Non-Speculative Thread Execution. Execution of non-speculative load and store is very similar to the speculative thread execution. But the non-speculative thread does not set any SL, OW or SMi bits. Also, the non-speculative store writes the portion of the data non-overlapped with speculatively modified data (SMi bits) into all versions in the L1 cache. This merging is done so that the speculative threads will get the most recent non-speculative version.

Commit and Squash. To squash a thread, the SL[thread_id] is cleared for all of the lines in the cache. This can be done as a gang-clear operation. Also the line is invalidated if any of the SMi bit is set. This is accomplished by a conditional gang-clear operation as in two-thread scheme.

To commit a thread, the SL [thread_id] bit and the SMi bits of the thread are cleared. The commit must ensure two things. It should make sure that there is only one non-speculative version present in L1 cache. If a cache line to which the current thread wrote has another version which is earlier than that of the current thread, then that version needs to be written back and invalidated. Also, in our scheme, we require that only the non-speculative thread can send values to the successor threads. So once a thread becomes non-speculative, it has to send its speculatively modified values to all successor threads. To ensure these two conditions are met, we need to commit each cache line belonging to the current thread whose SMi bit is set. To commit a cache line, we write-back and invalidate any versions that belong to earlier threads, and we need to merge the modified data with the versions belonging to successor threads. Though this step might be time consuming, we can see that this is simple to implement, and we can potentially overlap this with the execution of the next thread. Our simulation shows that this overhead causes no potential performance degradation.

Implementation Issues. While executing a *speculative load*, we may have to search the entire set in the cache to get the predecessor thread's cache line. Also, while detecting mis-speculation, we need to search the entire set to find if any successor thread has set the SL bit. These operations can be implemented by adding more logic to the tag matching hardware but it could increase cache hit time. In our scheme, we assume there is special hardware that does these "whole-set" operations, which is kept separate from the tag matching hardware. We assume such special operations take 3 cycles.

As we see in the two-thread case, we cannot replace a line with SL or SM bit set. If a speculative thread encounters a cache miss and if it is not able to find a clean line to replace from the cache, it can either suspend and wait till it becomes non-speculative or it can squash the successor threads and consume its cache lines. A thread will be forced to wait if it has no successors to squash. While waiting, a thread occupies shared resources like fetch queue, RUU and LSQ. There may be a situation where all the resources are occupied by the suspended thread and the non-speculation thread is unable to proceed, thus, causing a deadlock. To avoid this scenario, the speculative thread will give up its resources when it is stalled.

4   Experimental Results

4.1   Experimental Methodology

In our experiments, we used a detailed superscalar simulator based on Simplescalar 3.0. We modified Simplescalar to a trace based simulator that accepts Itanium traces. The trace for the input program is generated by Pin [6]. Traces are collected for four threads at a time and the simulator is called to consume the traces. Table 1 details the processor parameters used.

To generate parallel threads, we use a compiler framework based on Intel's ORC compiler [1]. The compiler selects loops in each benchmark that are suitable for parallel execution, performs optimizations such as code scheduling to enhance overlap between threads. The compiler also generates synchronization instructions for frequently occurred cross-iteration data dependences. Our compiler framework and the loop selection methodology are described in detail in [14].

Table 1. Processor Parameters

Fetch Width	4 Bundles (3 instructions each)
Decode, issue and commit width	8, 4 and 4 instructions
Function Units	4 integer, 4 floating point, 4 memory ports
Latency	1 cycle for integer, 12 cycle for floating point
Register Update Unit(ROB)	256 entries
LSQ size	128 entries
Branch predictor	Bimod, 2K entries
L1D,I Cache	64K, 4 way associative, 32B blocksize, 1 cycle
Unified L2	1MB, 8 way associative, 64B blocksize, 18 cycles
Memory latency	120 cycles for 1st chunk, 18 cycles subsequent chunks
Branch mis-prediction penalty	6 cycles

4.2   Results

Table 2 shows the details of benchmarks (from SPEC2000) used to evaluate our scheme. Since our primary objective here is to evaluate our proposed cache scheme when the SMT processor is executing in parallel mode, we only focus our simulations on the parallel regions in each benchmark.

Table 2. Details of Benchmarks

Benchmark	No of loops selected	coverage of selected regions
Mcf	6	60%
Twolf	15	32%
Vpr (place)	3	65%
Equake	4	90%
Art	12	52%

We consider the following configurations:

Superscalar: This is an out-of-order superscalar processor with parameters described in Table 1.

SMT-2: This is an out-of-order SMT processor which can support two threads at a time using the two-thread scheme described in section 3.2. This configuration has the same number of functional units as in the superscalar.Each line of cache has 9 extra bits (8 SMi and 1 SL).

SMT-4: This SMT processor can support four threads using the four-thread scheme described in section 3.3. It also has the same number of functional units as in (1) and (2). Each line of cache has 16 extra bits (8 SMi, 4 SL and 4 OW bits).

SMT-LSQ: This SMT processor supports 4 threads and uses the LSQ-based mechanism as in [9][2]. It has the same number of functional units, but uses extra space for enhanced LSQs that support speculation. Each thread has 128 LSQ entries.

CMP: This is a multi-core based speculative processor using the Stampede protocol[10]. This uses four identical cores and each core is a superscalar processor described in Table 1.

Fig. 5 shows the relative speedups of the different configurations over the superscalar configuration. From Fig. 5 we can see that the two-thread SMT scheme achieves about 10% speedup over the superscalar version. The two-thread SMT achieved this with very simple modifications to cache. The four-thread version achieved 15% speedup over superscalar. This performs better than the two-thread version but needed more complex hardware.

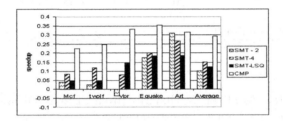

Fig. 5. Speedup of different configurations over the Superscalar configuration

Fig. 6 shows the execution time breakdown for the different configurations normalized to the execution time of the superscalar configuration. The explanation of the different categories are given in Table 3.

From Fig. 5, we can see that, usually the four thread configurations SMT-4 and SMT-LSQ perform better than SMT-2 configuration. This is because most loops selected by our compiler have good thread level parallelism and can benefit from more threads. However SMT-2 has the advantage of causing fewer squashes because it has only one speculative thread. Due to this SMT-2 performs better than SMT-4 and SMT-LSQ for the benchmark *Art*.

The CMP configuration performs about 15% better than the SMT-4 configuration. This is because the CMP uses four separate cores and, hence, has four

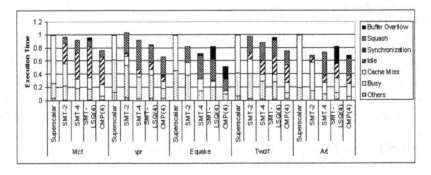

Fig. 6. Cycle breakdown normalized to the Superscalar configuration showing where the execution cycles are spent for different configuration

Table 3. Execution State Category

Category	Explanation
Cache Miss	Stall due to data cache miss
Idle	Lack of threads to execute
Synchronization	Thread is waiting for signal from predecessor
Squash	Thread is squashed due to dependence violation
Buffer Overflow	Thread made to wait due to lack space to buffer results
Others	Thread stalled due to instruction cache miss, branch mis-prediction, etc.

times more functional units and cache capacity. From these results, we can see that the performance of SMT-4 configuration is quite close to that of the CMP configuration even with much limited resources.

Fig. 6 shows that the SMT-LSQ is stalled for a significant amount of time in some benchmarks due to buffer overflow, thus making it slower than SMT-4 configuration. But in some loops, SMT-LSQ can perform better than SMT-4 configuration. This is because the SMT-LSQ has fewer squashes due to its fine grained dependence checking mechanism. This effect is observed in the benchmark *vpr*. Also, some loops have large number of squashes, in this case it is more beneficial to suspend the threads than to let them execute and later squash. This is because the squashed threads waste resources which could have been allocated to the non-speculative thread. In case of SMT-LSQ large threads are suspended and hence do not waste resources. This effect is observed in some of the loops in the benchmark *twolf*. The performance of SMT-4 can be improved if we have runtime feedback information, so that we can selectively turn off speculative threads on loops with frequent squashes.

In the SMT-LSQ configuration, we used an aggressive 128-entry per thread LSQ which might be unrealistic to implement in reality. Even with this configuration, it is still not able to support some of the loops without overflowing the queue. Fig. 6 shows that the SMT-4 scheme is able to achieve about 3% speedup over such SMT-LSQ configuration. But for loops with an average thread size of more than 150 dynamic instructions, the SMT-4 configuration performs about 14% better than SMT-LSQ configuration.

5   Conclusion

In this paper, we proposed a cache-based scheme to support speculative multithreading in SMT processors. Our two-thread scheme requires 9 bits to be added to each cache line and with this simple modification we can achieve about 10% speedup over the superscalar processors. Our four-thread scheme with slightly more complex hardware can perform about 15% better than superscalar processors. Also, we showed that this cache-based approach can outperform the LSQ based approach by 14% for large loops. From our paper it is clear that speculative threads can be easily supported in SMT processors with minimal changes in hardware.

158   V. Packirisamy et al.

Acknowledgements. This work was supported in part by the National Science Foundation under grants CCR-015571, CCR-0105574 and EIA-0220021, and grants from Intel.

References

[1] Open research compiler for itanium processor family. http://ipf-orc.sourceforge. net/.

[2] H. Akkary and M. Driscoll. A dynamic multithreading processor. In *MICRO-31*, December 1998.

[3] C. B. Colohan, A. Ailamaki, J. G. Steffan, and T. C. Mowry. Tolerating dependences between large speculative threads via sub-threads. In *Proceedings of the 33th ISCA*, Boston,MA.

[4] M. Franklin and G. S. Sohi. ARB: A hardware mechanism for dynamic reordering of memory references. *IEEE Transactions on Computers*, 45(5), May 1996.

[5] Venkata Krishnan and Josep Torrellas. A chip multiprocessor architecture with speculative multithreading. *IEEE Transactions on Computers, Special Issue on Multithreaded Architecture*, September 1999.

[6] C-K Luk, R. Cohn, R. Muth, H. Patil, A. Klauser, G. Lowney, S. Wallace, V.J. Reddi, and K. Hazelwood. Pin: building customized program analysis tools with dynamic instrumentation. In *Proc. ACM SIGPLAN 05 Conference on Programming Language Design and Implementation*, 2005.

[7] P. Marcuello and A. Gonzalez. Exploiting speculative thread-level parallelism on a smt processor. In *Proceedings of the 7th International Conference on High-Performance Computing and Networking*, April 1999.

[8] Jose F. Martinez, Jose Renau, Michael C. Huang, Milos Prvulovic, and Josep Torrellas. Cherry: checkpointed early resource recycling in out-of-order microprocessors. In *Proceedings of Micro-35*, Istanbul, Turkey, 2002.

[9] I. Park, B. Falsafi, and T.N. Vijaykumar. Implicitly-multithreaded processors. In *Proceedings of the 30th ISCA*, June 2003.

[10] J. G. Steffan, C. B. Colohan, A. Zhai, and T. C. Mowry. A scalable approach to thread-level speculation. In *Proceedings of the 27th ISCA*, June 2000.

[11] J. G. Steffan, C. B. Colohan, A. Zhai, and T. C. Mowry. The stampede approach to thread-level speculation. In *ACM Transactions on Computer Systems (TOCS)*, volume 23, August 2005.

[12] Dean Tullsen, Susan Eggers, Joel Emer, Henry Levy, Jack Lo, and Rebecca Stamm. Exploiting choice: Instruction fetch and issue on an implementable simultaneous multithreading processor. In *Proceedings of the 23rd ISCA*, May 1996.

[13] T.N. Vijaykumar, S. Gopal, J.E. Smith, and G. Sohi. Speculative versioning cache. In *IEEE Transactions on Parallel and Distributed Systems*, volume 12, pages 1305–1317, December 2001.

[14] S. Wang, X. Dai, K.S. Yellajyosula, A. Zhai, and P-C Yew. Loop selection for thread-level speculation. In *18th International Workshop on Languages and Compilers for Parallel Computing*, October 2005.

[15] A. Zhai, C. B. Colohan, J. G. Steffan, and T. C. Mowry. Compiler optimization of scalar value communication between speculative threads. In *Proceedings of the 10th ASPLOS*, Oct 2002.

Dynamic Internet Congestion with Bursts[*]

Stefan Schmid[1] and Roger Wattenhofer[2]

[1] Computer Engineering and Networks Laboratory (TIK), ETH Zurich, CH-8092 Zurich,
Switzerland
`schmiste@tik.ee.ethz.ch`
`http://dcg.ethz.ch/members/stefan.html`
[2] Computer Engineering and Networks Laboratory (TIK), ETH Zurich, CH-8092 Zurich,
Switzerland
`wattenhofer@tik.ee.ethz.ch`
`http://www.distcomp.ethz.ch/members/roger.html`

Abstract. This paper studies throughput maximization in networks with dynamically changing congestion. First, we give a new and simple analysis of an existing model where the bandwidth available to a flow varies multiplicatively over time. The main contribution however is the introduction of a novel model for dynamics based on concepts of network calculus. This model features a limited form of amortization: After quiet times where the available bandwidth was roughly constant, the congestion may change more abruptly. We present a competitive algorithm for this model and also derive a lower bound.

1 Introduction

The problem of avoiding congestion in the Internet has been studied with zeal for many years. The TCP congestion control mechanism of todays Internet successfully employs a window-based scheme to prevent the Internet from being overloaded. Thereby, the size of the so-called *TCP congestion window* is an approximation of the available network capacity. When TCP suffers a packet loss, it assumes that the network is congested and reduces the window's size. Consequently, the sending rate is cut down, and the Internet hosts collaboratively alleviate the load.

In the past, the transport layer and in particular the congestion problem was first studied empirically, and later embraced by the queuing theory and control theory communities. In order to analyze and compare protocols theoretically, a traffic model is needed. Queuing and control theory researchers have refined their early Poisson traffic models to an astonishing level of detail. However, probabilistic models are intricate to analyze. Probabilistic models that are simple enough to be analytically tractable might never model traffic accurately enough, as the nature of network traffic is self-similar and *bursty* [18].

In their seminal paper Karp, Koutsoupias, Papadimitriou, and Shenker [10] have proposed to study congestion control from a worst-case perspective instead. Karp et al. model congestion control as an online game between a flow and an adversarial network. In particular, the available bandwidth of the network changes over time and the

---

[*] Research supported by the Swiss National Science Foundation.

Y. Robert et al. (Eds.): HiPC 2006, LNCS 4297, pp. 159–170, 2006.

flow gets only a limited feedback—namely, whether packets have been lost or not—about the currently available bandwidth.

In this paper, we follow the algorithmic online approach proposed by Karp et al. [10] to broaden our understanding of congestion control. We build upon [10] by focusing on the dynamics of congestion. In particular, we integrate a notion of bursts happening in a worst-case manner. Although we do not claim that our models accurately represent what happens in the Internet, we believe that they are interesting and may ignite a further discussion on future variants of congestion control.

Concretely, after a new analysis of a model by Karp et al., we introduce the burst model: Instead of considering an adversary which always changes the bandwidth similarly each round, our adversary may accumulate some power in quiet rounds and then change the congestion more abruptly in later rounds. For this adversary, a lower as well as an upper bound are derived for the competitive ratio.

The paper is organized as follows. Section 2 reviews related work and also gives a short overview of the relevant network calculus concepts. After setting the stage in the model section (Section 3), we study the case of multiplicatively changing congestion in Section 4. In Section 5 our new model is presented in detail and analyzed. We state open problems in Section 6 and conclude the paper in Section 7.

2   Related Work

TCP lies at the heart of today's Internet, and many aspects of TCP are still subject to active research. For a reference on TCP, we refer the reader to [17]. TCP congestion control has been studied intensively, both from an empirical and from a theoretical perspective. Due to space constraints, we are bound to concentrate on the closest related work only.

In our work, we analyze congestion control from a *worst-case perspective* using competitive analysis. Generally, we think that a better algorithmic (worst-case) understanding of the transport layer is necessary. Whereas all other layers have received quite a lot of attention in the past (e.g., cf. [2] for the link layer, and [15] for the network layer), there has been comparatively little algorithmic networking research about the transport layer. Some notable exceptions are for instance adversarial queuing theory [6], the study of the TCP ACK problem [9], or mechanism design [7].

Our model is due to Karp et al. [10] who define several optimization problems related to congestion control. The authors investigate the issue of regulating the rate of a single unicast flow when the bandwidth available to it is unknown and changes over time. In our paper, we extend [10] in two respects: First, we provide a new analysis of a model where the bandwidth changes multiplicatively; our analysis is simpler and gives *strict* competitive bounds. Second, we enhance their model with *bursts*: Thereby, the congestion may change more after a time of quiescence.

The work by Karp et al. has already had an interesting follow-up by Arora and Brinkman [4] who study *randomized* algorithms for a dynamically changing congestion. In particular, they propose an asymptotically optimal randomized online algorithm against an adversary which may change the congestion by a constant factor in every round. Unfortunately, they assume a fairly weak oblivious adversary (see also the discussion in Section 6): Their algorithm uses randomization only in the first round, while the sending rate of all other rounds is computed deterministically. The adversary however is not allowed to be adaptive in these deterministic rounds.

The idea that an adversary may accumulate power over time has already appeared in the area of packet routing and is related to the adversarial queuing theory by Borodin et al. [6]. The problem considered there is as follows: Given a packet switched network and an adversary which continuously injects packets that have to be routed from a source to a destination node, how much buffer space is needed at the nodes, and what is the delivery time? In the paper by Aiello et al. [1], the adversary is allowed to inject any sequence of packets into the network, as long as in any w consecutive rounds, the total load created by the paths associated with the packets inserted in this time period is at most wr on any edge, for some $w \geq 1, r \leq 1$. The adversary studied by Andrews et al. [3] is similar to our adversary. Given two parameters $b \geq 1, r \leq 1$, for any $T \geq 1$ consecutive time steps, the adversary may inject as many packets as it wants, as long as the total load created by the paths associated with these packets is at most $Tr + b$ on any edge. These two adversary models have been compared by Rosén in [16]. A contribution of our paper is to introduce a modified version of the adversary in [3] on the *transport layer*.

Short Overview of Network Calculus. We now give a short introduction to those concepts of network calculus which are relevant to our work. Network calculus is a relatively new technique to analyze deterministic queuing systems found in communication networks. For a detailed introduction to network calculus, see [14].

In network calculus, there exists the notion of *arrival curves* which provide deterministic limitations to the network traffic sent by sources. Given that the data flows indeed correspond to these limitations, it is possible to make statements about the deterministic behavior of the network (maximal delays, maximal queue lengths, etc.).

Arrival curves are defined as follows. Let R be a data flow, and let $R(t)$ be the total number of bits R has sent until time t. Let α be an increasing function defined for all times $t \geq 0$. We say that R has an arrival curve α if and only if for all $s \leq t$:

$$R(t) - R(s) \leq \alpha(t - s)$$

In other words, the total number of bits sent until time t by flow R may never exceed the bits sent by R until some time s plus $\alpha(t - s)$. In this paper, we look at a so-called *leaky bucket arrival curve* defined as $\alpha(t) = c_1 t + c_2$ for some non-negative constants c_1, c_2. Figure 1 visualizes the constraints imposed upon a flow R by such an arrival curve: The total number of bits sent may increase by c_2 at once and with a rate c_1 over time, unless there is a conflict with a constraint from a previous round. Informally, the total number of bits must always be less or equal the minimum constraint that arises if the curve α is attached to all points of $R(t)$.

Note that such an arrival curve incorporates a limited form of amortization: If flow R only sends a few bits for several rounds, the constraints of earlier rounds get weaker and allow R to send up to c_2 bits at once in some later round.

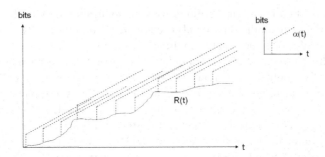

Fig. 1. Leaky bucket arrival curve: The number of bits sent by flow R may never exceed the constraints from earlier times (dashed lines), i.e., $\forall s \leq t : R(t) \leq R(s) + \alpha(t - s)$

3   Model

In the Internet, there is no central authority allocating bandwidth to hosts. On the contrary, individual hosts are responsible for setting their sending rate.[1] In this paper, we consider the problem of regulating the rate of a unicast flow from one host to another such that the throughput is maximized. The bandwidth available to the flow thereby fluctuates according to the varying requirements for bandwidth of other competing flows. A host is not provided direct information about the competing demands for bandwidth or the Internet topology, but does receive some limited information as to whether the flow is experiencing packet drops, and must determine its transmission rate solely on the basis of this information.

We assume that time is divided into infinitely many successive *rounds* and consider a *worst-case model* where in every round t, an adversary ADV selects the available bandwidth u_t. Thereby, u_t represents the maximum rate at which a host can transmit without experiencing packet drops. The host on the other hand runs an algorithm ALG which decides the sending rate x_t of round t, and receives immediate feedback as to whether packet drops have occurred, i.e., whether $x_t > u_t$. ALG can then choose the rate x_{t+1}.

We assume a *severe cost model* [10] where a host cannot transmit anything in round t if $x_t > u_t$, but can transmit at a rate x_t if $x_t \leq u_t$. Formally, the *gain* of ALG in round t is defined as follows:

$$gain_{ALG}(x_t, u_t) := \begin{cases} x_t & \text{, if } x_t \leq u_t \\ 0 & \text{, otherwise} \end{cases}$$

---

[1] Usually, this is done automatically by TCP. However, by using the User Datagram Protocol (UDP), selfish programs can try to maximize their own throughput and may have no incentive to reduce congestion collaboratively. Although it is generally believed that routers are configured to give priority to TCP packets [8]—with the consequence that UDP packets are dropped first if the Internet gets congested—at least in theory it is possible to design networking software from scratch that circumvents this restriction by sending UDP packets which look like TCP packets.

An optimal offline algorithm OPT knows the sequence $\{u_t\}$ in advance and achieves a gain of

$$gain_{OPT}(x_t, u_t) = u_t$$

in round t. These gains reflect two major issues: The online algorithm experiences an *opportunity cost* if its sending rate is smaller than the available bandwidth (case $x_t < u_t$), and a *retransmission overhead* if its packets are dropped due to congestion (case $x_t > u_t$).

We are in the realm of *competitive analysis* [5] and define the (strict) *competitive ratio* ρ achieved by ALG as the total amount of data (over all rounds) sent by OPT divided by the total amount of data sent by ALG (cf. Definition 3.1).

Definition 3.1. *[ρ-competitive] We say that an algorithm ALG is* (strictly) *ρ-competitive compared to an optimal offline algorithm OPT if for all input sequences I, it holds that*

$$gain_{OPT}(I) \leq \rho \cdot gain_{ALG}(I).$$

The goal of the online algorithm designer is to minimize ρ. Henceforth, we will assume that ALG knows the initial bandwidth, i.e., $x_0 = u_0$.

Observe that an unrestricted adversary could frustrate every online algorithm by always selecting $u_t := x_t - \varepsilon$ for some arbitrary small $\varepsilon > 0$. The natural way out proposed by Karp et al. [10] is to assume that the available bandwidth does not change too drastically over time. In this paper, we study different ways to restrict the adversary. In Section 4, we consider the multiplicative model proposed by Karp et al. In Section 5, we extend this model to allow for changes with *bursts*.

We will call rounds t where the online algorithm successfully transmits its packets without loss *good rounds*, and rounds t where $x_t > u_t$ *bad rounds*, cf. Definition 3.2.

Definition 3.2 (Good and Bad Rounds). *A round t where $x_t \leq u_t$ is called* good, *a round t where $x_t > u_t$ is called* bad.

We defer the description of the different adversaries to the corresponding sections. However, we now define the following class of online algorithms.

Definition 3.3 ($\mathcal{ALG}(G, B)$). *Let $\mathcal{ALG}(G, B)$ be the online algorithm which chooses*

$$x_{t+1} := \begin{cases} G \cdot x_t & , \text{if } x_t \leq u_t \\ B \cdot x_t & , \text{otherwise} \end{cases}$$

for some $G \geq 1$ and $B \leq 1$. That is, the algorithm $\mathcal{ALG}(G, B)$ increases the rate by a factor G after a good round, and decreases it by a factor B after a bad round.

The sending rate x_{t+1} of an algorithm $\mathcal{ALG}(G, B)$ depends solely on the binary feedback whether its probing rate x_t was larger than the available bandwidth u_t in the previous round or not.

4 Multiplicative Adversaries

In this section, we look at multiplicative changes of the available bandwidth. We first consider a model where the adversary can increase the bandwidth at most by a factor $\mu \geq 1$ per round and can decrease it arbitrarily (cf. Definition 4.1). Later, we will study a model where also the reduction is constrained multiplicatively (cf. Definition 4.2).

So let's look at the adversary $\mathcal{ADV}_{mult}$ (cf. Definition 4.1) proposed by Karp et al.

Definition 4.1 ($\mathcal{ADV}_{mult}$). *$\mathcal{ADV}_{mult}$ may choose the new bandwidth u_{t+1} in the interval $[0, u_t \cdot \mu]$, i.e.,*

$$\mathcal{ADV}_{mult} : u_{t+1} \in [0, u_t \cdot \mu],$$

for some given $\mu \geq 1$.

First, we restate the lower bound given in [10].

Theorem 4.1. *[10] Against $\mathcal{ADV}_{mult}$, no online algorithm can achieve a competitive ratio smaller than μ.*

Proof. Consider the following adversary ADV: In every round t, it chooses

$$u_t := \begin{cases} \mu & , \text{if } x_t \leq 1 \\ 1 & , \text{otherwise} \end{cases}$$

Thus, whenever an online algorithm ALG sends at a rate larger than one, all its packets are dropped because of congestion. On the other hand, if ALG transmits at a rate of 1 or less, the rate of OPT is at least a factor μ larger. Moreover, since ADV changes the available bandwidth at most by a factor of μ per round, it is indeed of type $\mathcal{ADV}_{mult}$. $\square$

In [10], it is shown that the algorithm $\mathcal{ALG}(\mu, \frac{\sqrt{\mu}}{\sqrt{\mu}+\sqrt{\mu-1}})$ yields a competitive ratio of

$$\rho = (\sqrt{\mu} + \sqrt{\mu - 1})^2$$

against $\mathcal{ADV}_{mult}$. However, [10] uses a different definition for the competitive ratio which allows for (possibly large) additive constants. By our strict definition (cf. Definition 3.1), the ratio can be much larger. To see this, assume an adversary which reduces the available bandwidth in every round by a factor slightly larger than $\frac{\sqrt{\mu}+\sqrt{\mu-1}}{\sqrt{\mu}}$. In this case, $\mathcal{ALG}(\mu, \frac{\sqrt{\mu}}{\sqrt{\mu}+\sqrt{\mu-1}})$ is only successful in the first round, and hence $gain_{ALG} = u_0$, while

$$gain_{OPT} \approx u_0 \cdot \sum_{i=0}^{\infty} \left(\frac{\sqrt{\mu}}{\sqrt{\mu}+\sqrt{\mu-1}}\right)^i.$$

Therefore, the (strict) competitive ratio is

$$\rho = \frac{gain_{OPT}}{gain_{ALG}} \approx \frac{\sqrt{\mu}+\sqrt{\mu-1}}{\sqrt{\mu-1}}.$$

For small μ, ρ can get very large (for instance $\rho > 100$ if $\mu = 1.0001$).

In the following, we give a simple proof that the algorithm $\mathcal{ALG}(\mu, 1/2)$ has a strict competitive ratio 4μ. According to Theorem 4.1, this is asymptotically optimal.

Theorem 4.2. $\mathcal{ALG}(\mu, 1/2)$ *is 4μ-competitive against $\mathcal{ADV}_{mult}$.*

Proof. First, we show by induction that in every good round t, $u_t \leq 2\mu x_t$. For $t = 0$, $u_0 = x_0$ and the claim holds. For the induction step, consider the round $t - 1$ before the good round t. There are two possibilities: either round $t-1$ has been bad ($x_{t-1} > u_{t-1}$) or good ($x_{t-1} \leq u_{t-1}$). If round $t - 1$ has been bad, we have $x_t = x_{t-1}/2$ and $u_t \leq u_{t-1}\mu < x_{t-1}\mu = 2\mu x_t$, hence $u_t/x_t < 2\mu$, and the claim holds. If on the other hand round $t - 1$ was good, the algorithm increases the bandwidth at least as much as the adversary. Together with the induction hypothesis, the claim also follows in this case.

Having studied the gain in good rounds, we now consider bad rounds. We show that in the bad rounds following a good round t, the adversary can increase its gain at most by $2\mu x_t$. So let t be the good round preceding a sequence of bad rounds, i.e., $x_t \leq u_t$, $x_{t+1} > u_{t+1}$, $x_{t+2} > u_{t+2}$, etc. We know that $x_{t+1} = \mu x_t$, so—because it is a bad round—u_{t+1} must be smaller than μx_t. Furthermore, we have $x_{t+2} = x_{t+1}/2 = \mu x_t/2$ and hence $u_{t+2} < \mu x_t/2$, $x_{t+3} = \mu x_t/4$ and hence $u_{t+3} < \mu x_t/8$, and so on. By a geometric series argument, the gain of the adversary in the bad rounds is upper bounded by $2\mu x_t$.

Therefore,

$$\rho = \frac{gain_{OPT}(good) + gain_{OPT}(bad)}{gain_{ALG}(good)}$$
$$< \frac{2\mu \cdot gain_{ALG}(good) + 2\mu \cdot gain_{ALG}(good)}{gain_{ALG}(good)}$$
$$< 4\mu.$$

$\square$

To conclude this section, we give another kind of proof to show that the algorithm $\mathcal{ALG}(\mu, 1/\mu^3)$ has a good competitive ratio for small μ. For our analysis, we assume a slightly more restricted adversary $\mathcal{ADV}^*_{mult}$ (cf. Definiton 4.2).

Definition 4.2 ($\mathcal{ADV}^*_{mult}$)**.** $\mathcal{ADV}^*_{mult}$ *chooses the new bandwidth u_{t+1} from the interval $[u_t/\mu, u_t \cdot \mu]$, i.e.,*

$$\mathcal{ADV}^*_{mult} : u_{t+1} \in [u_t/\mu, u_t \cdot \mu].$$

Theorem 4.3. $\mathcal{ALG}(\mu, 1/\mu^3)$ *is $(\mu^4 + \mu)$-competitive against $\mathcal{ADV}^*_{mult}$.*

Proof. The fact that ALG reduces its rate by a factor μ^3 after a bad round implies that the next round is always good: Assume, for the sake of contradiction, that round $t + 1$ is the first bad round following another *bad* round t, which—by the induction hypothesis—follows a good round $t - 1$. Hence, $x_{t-1} \leq u_{t-1}$. Moreover, observe that $u_{t+1} \geq u_t/\mu \geq u_{t-1}/\mu^2$, but on the other hand, $x_{t+1} = x_t/\mu^3 = \mu x_{t-1}/\mu^3 = x_{t-1}/\mu^2$. Therefore, $x_{t+1} \leq u_{t+1}$. Contradiction!

We now first analyze the gain of a good round t and show that $u_t < \mu^4 x_t$. There are two cases: Either round $t - 1$ has also been good, or not. If it has been a good round, then round t is at least as competitive as round $t - 1$ because $x_t = \mu x_{t-1}$. If on the other hand round $t - 1$ has not been good, we have $u_{t-1} < x_{t-1}$, $x_t = x_{t-1}/\mu^3$ and $u_t \leq \mu u_{t-1}$. Therefore, $x_t = x_{t-1}/\mu^3 > u_{t-1}/\mu^3 \geq u_t/\mu^4$, and the claim follows.

Next, we study the gains in a bad round t. In this case, it holds that $u_t < \mu x_{t-1}$: Since $x_{t-1} \leq u_{t-1}$, $x_t = \mu x_{t-1}$ and $u_t < x_t$, and hence $u_t < \mu x_{t-1}$.

Therefore,

$$\rho = \frac{gain_{OPT}(good) + gain_{OPT}(bad)}{gain_{ALG}(good)}$$

$$< \frac{\mu^4 \cdot gain_{ALG}(good) + \mu \cdot gain_{ALG}(good)}{gain_{ALG}(good)} = \mu^4 + \mu.$$

$\square$

Since $\mathcal{ADV}^*_{mult}$ is a special case of $\mathcal{ADV}_{mult}$, Theorem 4.2 also applies for $\mathcal{ADV}^*_{mult}$. Hence, it is possible to run $\mathcal{ALG}(\mu, 1/\mu^3)$ against $\mathcal{ADV}^*_{mult}$ if μ is small, and $\mathcal{ALG}(\mu, 1/2)$ otherwise, which yields the following corollary.

Corollary 4.4. *There is a deterministic online algorithm which is* $\min\{\mu^4 + \mu, 4\mu\}$-*competitive against* $\mathcal{ADV}^*_{mult}$.

5 Network Calculus Adversary

5.1 Description of $\mathcal{ADV}_{nc}$

In this section, we introduce the adversary $\mathcal{ADV}_{nc}$ which is based on *network calculus* [14] concepts. We will extend the model introduced in Section 4 by a form of limited amortization which allows for more drastic bandwidth changes after times of quiescence.

$\mathcal{ADV}_{nc}$ has two parameters: A *rate* $\mu \geq 1$ and *maximum burst factor* $B \geq 1$. In every round, the available bandwidth u_t varies according to these parameters in a multiplicative manner. More precisely, $\mathcal{ADV}_{nc}$ can select the new bandwidth u_{t+1} from the interval

$$\mathcal{ADV}_{nc} : u_{t+1} \in [\frac{u_t}{\beta_t \mu}, u_t \cdot \beta_t \cdot \mu],$$

that is, the available bandwidth can change by a factor of at most $\beta_t \mu$. Thereby, β_t is the *burst factor at time t*. This burst factor is explained next.

On average, the available bandwidth can change by a factor μ per round. However, there can be times of only small changes, but then the bandwidth changes by factors larger than μ in later rounds. This is modeled with the burst factor β_t: At the beginning, β_t equals B, i.e., $\beta_0 = B$. For $t > 0$, the burst factor β_t is computed depending on β_{t-1} and the actual bandwidth change c_{t-1} that has happened in round $t - 1$. More precisely,

$$\beta_t = \min\{B, \beta_{t-1} \frac{\mu}{c_{t-1}}\}$$

where

$$c_t := \begin{cases} \frac{u_{t+1}}{u_t} & \text{, if } u_{t+1} > u_t \\ \frac{u_t}{u_{t+1}} & \text{, otherwise} \end{cases}$$

This means that if the available bandwidth has changed by a factor less than μ in round t, i.e., $c_t < \mu$, the burst factor *increases* by a factor μ/c_t, and hence the bandwidth can change more in the next round—and vice versa if $c_t > \mu$.

In other words, the adversary can save adversarial power for forthcoming rounds. However, this amortization is limited as β_t never becomes larger than B for all rounds t. Also note that $\forall t : \beta_t \geq 1$, as $c_t \leq \mu\beta_t$ by the definition of $\mathcal{ADV}_{nc}$.

5.2   Analysis

At first sight, it seems that $\mathcal{ADV}_{nc}$ has roughly the same power as $\mathcal{ADV}^*_{mult}$: In order to change the bandwidth with a factor larger than μ, $\mathcal{ADV}_{nc}$ must have changed the bandwidth by a factor less than μ in previous rounds.[2] However, as we will see in the following, an online algorithm cannot exploit these quiet rounds sufficiently, and the competitive ratio does depend on B.

Theorem 5.1. *The competitive ratio is at least* $\Omega\left(\mu\sqrt{B}/\log B\right)$ *against* $\mathcal{ADV}_{nc}$.

Proof. Consider the following adversary ADV. ADV will select $u_t = 1$ whenever the burst factor β_t is not maximal in a round t, i.e., if $\beta_t < B$. If $\beta_t = B$, ADV continues choosing $u_t = 1$ until $x_t \leq 1$ for the first time. Then, if $x_t \leq 1$ and $\beta_t = B$, it selects $u_t = \mu\sqrt{B}$ but immediately sets the available bandwidth back to $u_{t+1} = 1$ in the next round. Therefore, no online algorithm can ever transmit at a rate larger than 1. Since ADV must be of type $\mathcal{ADV}_{nc}$, it can do this trick at most every $\lceil \log B/\log\mu \rceil$ rounds: After these two bursts (from 1 to $\mu\sqrt{B}$ and from $\mu\sqrt{B}$ back to 1), the burst factor becomes 1, and it takes $\lceil \log B/\log\mu \rceil$ rounds to increase it again to B: $\mu^i \geq B \Leftrightarrow i \geq \log B/\log\mu$.

Let us call the time period between two rounds where ADV raises the bandwidth from 1 to $\mu\sqrt{B}$ a *phase*. In every phase, ALG has a gain of at most

$$gain_{ALG} \leq 2 + \lceil \log B/\log\mu \rceil.$$

On the other hand, the optimal algorithm's gain is at least

$$gain_{OPT} \geq 1 + \lceil \log B/\log\mu \rceil + \mu\sqrt{B}.$$

Hence,

$$\rho = \frac{gain_{OPT}}{gain_{ALG}} \geq \frac{1 + \lceil \log B/\log\mu \rceil + \mu\sqrt{B}}{2 + \lceil \log B/\log\mu \rceil} \in \Omega\left(\mu\frac{\sqrt{B}}{\log B}\right).$$

$\square$

Note that the lower bound given in Theorem 5.1 even holds for online algorithms which get perfect (instead of only binary) feedback about the bandwidth of the previous round.

Although we were not able to find an algorithm which yields a tight upper bound, it can be shown that $\mathcal{ALG}(\mu\sqrt[3]{B}, 1/2)$ comes close to the bound of Theorem 5.1.

---

[2] Except for the first rounds of course, where a burst B comes "for free". However, as mentioned in Section 3, we consider infinite games only.

Theorem 5.2. *The competitive ratio of $\mathcal{ALG}(\mu\sqrt[3]{B}, 1/2)$ is $\mathcal{O}\left(\mu^{3/2}B^{2/3}\right)$ against* $\mathcal{ADV}_{nc}$.

Proof. We use again the proof technique of Section 4. First, we analyze the missed gain in bad rounds:

$$gain_{OPT}(bad) \leq \sum_{i=0}^{\infty}\left(\frac{1}{2}\right)^{i} \cdot \mu\sqrt[3]{B} \cdot gain_{ALG}(good)$$

$$\leq 2\mu\sqrt[3]{B} \cdot gain_{ALG}(good) \in \mathcal{O}\left(\mu\sqrt[3]{B}\right) \cdot gain_{ALG}(good)$$

Next, the good rounds are tackled. Let t be the last bad round before a good round $t+1$. Hence, $x_t > u_t$, $x_{t+1} = x_t/2 \leq u_{t+1}$, and $x_{t+2} = \mu\sqrt[3]{B}x_t/2$.

There are two cases: Either round $t+2$ is also good, or not. If round $t+2$ is good, $u_{t+2} \leq \mu^2 Bx_t$. We have

$$\rho \leq \frac{u_{t+1} + u_{t+2}}{x_{t+1} + x_{t+2}} \leq \frac{\mu B + \mu^2 B}{1/2 + \mu\sqrt[3]{B}/2} \in \mathcal{O}\left(\mu B^{2/3}\right)$$

More good rounds would reduce this ratio, because ALG grows faster than ADV.

If round $t+2$ is not good, it holds that $x_t > u_t$ and $x_{t+2} = \mu\sqrt[3]{B}x_t/2 > u_{t+2}$. Now observe that $u_{t+1} < \mu^{3/2}B^{2/3}x_t$. Assume, for the sake of contradiction, that $u_{t+1} \geq \mu^{3/2}B^{2/3}x_t$. Then the burst factor in round $t+1$ is at most $\beta_{t+1} \leq \sqrt[3]{B}/\sqrt{\mu}$, and thus

$$u_{t+2} \geq \frac{u_{t+1}}{\mu\beta_{t+1}} \geq \frac{\mu^{3/2}B^{2/3} \cdot \sqrt{\mu}}{\sqrt[3]{B} \cdot \mu}x_t = \mu\sqrt[3]{B}x_t > x_{t+2}.$$

Contradiction. Hence,

$$\rho \leq \frac{u_{t+1}}{x_{t+1}} \leq \frac{\mu^{3/2}B^{2/3}x_t}{x_t/2} \in \mathcal{O}\left(\mu^{3/2}B^{2/3}\right)$$

Thus, in conclusion,

$$\rho = \frac{gain_{OPT}(good) + gain_{OPT}(bad)}{gain_{ALG}(good)}$$

$$\leq \frac{\mathcal{O}\left(\mu^{3/2}B^{2/3}\right) \cdot gain_{ALG}(good) + \mathcal{O}\left(\mu\sqrt[3]{B}\right) \cdot gain_{ALG}(good)}{gain_{ALG}(good)}$$

$$\in \mathcal{O}\left(\mu^{3/2}B^{2/3}\right)$$

$\square$

6   Open Research Questions

Karp et al. have already pointed out several future research directions, for instance the study of different cost models. In this paper, we have extended their work by a novel model for the dynamics of the available bandwidth.

We believe that our network calculus model opens up many exciting questions. For example, the lower bound and the upper bound we presented are not tight. It would be interesting to know if there are asymptotically better online algorithms, or whether our lower bound is too pessimistic. Another challenge is the design of *randomized* online algorithms. In fact, Arora and Brinkman [4] have addressed this problem for the multiplicative adversary $\mathcal{ADV}_{mult}$ and presented an algorithm with competitive ratio $\mathcal{O}(\log \mu)$. By using Yao's minimax principle [5], it can be shown that this is asymptotically optimal. However, the authors assume a weak oblivious adversary: Their scheme uses randomization only in the first round, while all later rounds are deterministic. But the adversary is not allowed to be adaptive even in these deterministic rounds! The case of a stronger adversary is still an open problem. It is straight-forward to extend the algorithm by Arora and Brinkman for $\mathcal{ADV}_{nc}$: Over-pessimistically, we can assume that $\mathcal{ADV}_{nc}$ changes the bandwidth by a factor $B \cdot \mu$ in every round, which yields a competitive ratio of $\mathcal{O}(\log(B\mu))$. However, also here, it would be interesting to study a more powerful adversary which can be adaptive in deterministic rounds.

Finally, we believe that our network calculus adversary is an interesting model for dynamics in completely different fields of research.

7 Conclusion

This paper has studied online algorithms which aim at maximizing throughput in the presence of dynamic bandwidth changes. We have derived an asymptotically optimal algorithm for a multiplicative model. Moreover, a novel model for the congestion dynamics has been presented together with a lower and an upper bound for the competitive ratio. We hope that our models will give an impetus for future research. Generally, we believe that a better algorithmic (worst-case) understanding of the transport layer is necessary. Whereas all other layers have received quite a lot of attention in the past, the transport layer has always been a step-child of algorithmic networking research.

References

1. W. Aiello, E. Kushilevitz, R. Ostrovsky, and A. Rosén. Adaptive Packet Routing for Bursty Adversarial Traffic. In *Proceedings of the 13th Annual ACM Symposium on Theory of Computing (STOC)*, pages 359–368, New York, NY, USA, 1998.
2. D. Aldous. Ultimate Instability of Exponential Back-off Protocol for Acknowledgement Based Transmission Control of Random Access Communication Channels. *IEEE Transactions on Information Theory*, 1987.
3. M. Andrews, B. Awerbuch, A. Fernández, T. Leighton, Z. Liu, and J. Kleinberg. Universal-Stability Results and Performance Bounds for Greedy Contention-Resolution Protocols. *Jounal of the ACM*, 48(1):39–69, 2001.
4. S. Arora and B. Brinkman. A Randomized Online Algorithm for Bandwidth Utilization. In *Proceedings of the 13th Annual ACM Symposium on Discrete Algorithms (SODA)*, pages 535–539, Philadelphia, PA, USA, 2002.
5. A. Borodin and R. El-Yaniv. *Online Computation and Competitive Analysis*. Cambridge University Press, 1998.

6. A. Borodin, J. Kleinberg, P. Raghavan, M. Sudan, and D. P. Williamson. Adversarial Queuing Theory. In *Proceedings of the 28th Annual Symposium on Foundations of Computer Science (STOC)*, 1996.
7. J. Feigenbaum, C. H. Papadimitriou, and S. Shenker. Sharing the Cost of Multicast Transmissions. *Journal of Computer and System Sciences*, 63(1):21–41, 2001.
8. K. P. Gummadi, H. V. Madhyastha, S. D. Gribble, H. M. Levy, and D. Wetherall. Improving the Reliability of Internet Paths with One-hop Source Routing. In *Symposium on Operating Systems Design & Implementation (OSDI)*, 2004.
9. A. Karlin, C. Kenyon, and D. Randall. Dynamic TCP Acknowledgement and Other Stories about $e/(e-1)$. In *Proceedings of the 41st Annual Symposium on Foundations of Computer Science (STOC)*, 2001.
10. R. M. Karp, E. Koutsoupias, C. H. Papadimitriou, and S. Shenker. Optimization Problems in Congestion Control. In *Proceedings of Symposium on Foundations of Computer Science (FOCS)*, pages 66–74, 2000.
11. F. Kelly. Mathematical Modelling of the Internet. In *Bjorn Engquist and Wilfried Schmid: Mathematics Unlimited*. springer, 2001.
12. F. Kelly, A. Maulloo, and D. Tan. Rate Control in Communication Networks: Shadow Prices, Proportional Fairness and Stability. *Journal of the Operational Research Society*, 49, 1998.
13. T. V. Lakshman and U. Madhow. The Performance of TCP/IP for Networks with High Bandwidth-Delay Products and Random Loss. *IEEE/ACM Transactions on Networking*, 5(3):336–350, 1997.
14. J.-Y. Le Boudec and P. Thiran. *Network Calculus*. Springer LNCS 2050 Tutorial, 2001.
15. F. Leighton, B. Maggs, and S. Rao. Universal Packet Routing Algorithms. In *IEEE Symposium on Foundations of Computer Science (FOCS)*, pages 256–269, 1988.
16. A. Rosén. A Note on Models for Non-Probabilistic Analysis of Packet Switching Networks. *Inf. Process. Lett.*, 84(5):237–240, 2002.
17. R. Stevens and G. R. Wright. *TCP/IP Illustrated Vol. 2 (The Implementation)*. Addison-Wesley, 1995.
18. W. Willinger, W. Leland, M. Taqqu, and D. Wilson. On the Self-Similar Nature of Ethernet Traffic. *IEEE/ACM Transactions on Networking*, pages 1–15, 1994.

A Repair Mechanism for Fault-Tolerance for Tree-Structured Peer-to-Peer Systems

Eddy Caron[1], Frédéric Desprez[1], Charles Fourdrignier[2],
Franck Petit[2], and Cédric Tedeschi[1]

[1] LIP Laboratory
UMR CNRS-ENS Lyon-UCB Lyon-INRIA 5668
46 Allée d'Italie, 69364 Lyon Cedex 07, France
frederic.desprez@ens-lyon.fr
[2] LaRIA Laboratory
University of Picardie
5, rue du Moulin Neuf, 80000 Amiens, France
cedric.tedeschi@ens-lyon.fr

Abstract. Facing the limits of traditional tools of resource management within computational grids (related to scale, dynamicity, etc. of the platforms newly considered), new approaches, based on peer-to-peer technologies are emerging. The resource discovery and in particular the service discovery is concerned by this evolution. Among the solutions, a promising one is the indexing of resources using trie structures and more particularly prefix trees. The major advantages of trie-structured approaches is the capability to support search queries on ranges of values with a latency growing logarithmically in the number of nodes in the trie. Those techniques are easy to extend to multicriteria searches. One drawback of using tries is its inherent poor robustness in a dynamic environment, where nodes join and leave the network, leading to the split of the tree into a forest, which results in the impossibility to route requests. Within most recent approaches, the fault-tolerance is a prevention mechanism, often replication-based. The replication can be costly in term of resources required. In this paper, we propose a fault-tolerance protocol that reconnects subtrees *a posteriori*, after crashes, to have again a connected graph and then reorder the nodes to rebuild a consistent tree.

1 Introduction

These last few years have seen the development of large scale grids connecting distributed resources (computation resources, storage facilities, computation libraries, etc.) in a seamless way. This is now an efficient alternative to supercomputers to solve large problems such as high energy physics, simulation, bioinformatic, etc. However, existing middlewares used in grids require most of the time a stable and centralized infrastructure. They usually loose their performance on dynamic and large scale platforms without centralized management of resources.

Y. Robert et al. (Eds.): HiPC 2006, LNCS 4297, pp. 171–182, 2006.

To cope with the characteristics of these emerging kind of platforms, it has been suggested to use peer-to-peer technologies within computational grids [8].

Peer-to-peer technologies offer algorithms allowing the search and retrieval of objects over the net (data items, files, services, etc.). Among these technologies, Distributed Hash Tables (DHT) were initially designed for very large scale platforms, for example to share files over the Internet. However, DHTs have several major drawbacks. Among them, their discovery mechanism usually works on exact searches of a given key. Some work has then been done to allow complex requests to be submitted over DHTs or more generally in **structured** peer-to-peer systems, i.e. systems based on request routing. Some of these works are based on *tries* (also called *prefix trees*). A trie structure supports range queries in a logarithmic time in the number of nodes of the trie.

Fault-tolerance is a mandatory feature for peer-to-peer systems to avoid the loss of data stored on nodes and to allow a correct routing of messages. The crash of one or several nodes in a trie leads to the loss of objects references stored in the trie and to the split of the trie into several subtries, also called a *forest*. Fault-tolerance within structured peer-to-peer systems usually uses replication. Using such an approach, each node and each link of the trie would have to be duplicated k times, k being the replication factor. Keeping such structure up is costly, mainly in terms of resources used. Afterward, the purpose is to find for the value of k the right trade-off between the replication cost and the robustness of the system. In this paper, we study an alternative to the replication approach based on the reconnection of the subtries and the *a posteriori* reordering of a consistent trie. When the trie is disconnected, a first solution consists in rebuilding a trie adding nodes of remaining subtries one by one. This naive method can lead to a prohibitive cost when the number of remaining nodes is large (which is usually the case in peer-to-peer systems). For example, loosing one node can lead to a complete reconstruction of the trie. A second approach consists in reconnecting the subtries to get the original trie back at a minimum cost. This is this kind of algorithm we describe in this paper in a distributed and asynchronous environment. It can also be used to complete the replication process.

A brief history of peer-to-peer technologies is provided in Section 2, followed by the formal description of the particular trie structure we use (Section 3) and of the distributed system we place ourselves. We focus our study on fault-tolerance mecanisms related to them. Then, in Section 4 we present the repair algorithm we designed and give its proof before a conclusion and future work Section.

2   Related Work

With the spread of the peer-to-peer technologies going along with the file sharing over the Internet, purely decentralized search systems have emerged. Such tools first took the shape of *unstructured* mechanisms, *i.e.,* based on the flooding of search requests [10, 9]. These mechanisms resulted in overloading the network while providing non-exhaustive responses. Addressing both the scalability and the exhaustiveness issues within peer-to-peer systems, the distributed hash tables [13, 14, 18, 20], *a.k.a.,* the *structured* peer-to-peer group, are highly scalable in the sense that the number of logical hops required to route and the local state grows logarithmically with the number of nodes participating in the system. Moreover, DHTs prevent from loosing routing paths and objects' references by use of replication and periodic scans. Unfortunately, DHTs present several major drawbacks (homogeneous capacity assumptions, topology awareness, etc.). Among them, the rigidity of the requesting mechanism, *i.e.,* exact match on a given key hinders its use over real search systems.

A series of work gives the opportunity to allow flexible meanings of retrieval over structured peer-to-peer networks. First achievement in this way has been the ability to describe resources with semi-structured language, such XML, as described in [3]. [19] enhances DHTs with traditional database operations. Several approaches, based on space filling curves, such as Squid [15] or [17] support multi-dimensional range queries. [1] maps one-dimensional data space to d-dimensional Cartesian space by using the inverse Hilbert mapping. Built on top of multiple DHTs, SWORD [11] is an information service aiming at discovering computing resources on the grid by answering multi-attribute range queries.

We focus in this work on trie-structured retrieval solutions, also supporting range queries but outperforming previous approaches in the sense that logarithmic (or constant if we assume an upper bound on the depth of the trie) latency is achieved by parallelizing the resolution of the query in the several branches of the trie. Prefix Hash Tree (PHT) [12] builds a trie of the entire key-space on top of a DHT. The purpose of this architecture is to use the trie as a logical layer allowing complex searches on top of any DHT-like network. The architecture of PHT results in the multiplication of the complexities of the trie and of the underlying DHT.

The Skip Graphs structure proposed in [2] is similar to a trie but is built with the skip lists technology, allowing the use of their inherent fault-tolerance properties. But again, the complexity of the number of messages generated to process range queries is in $O(m \log(n))$, m being the number of nodes pertained by the range and n the total number of nodes in the graph.

Other approaches propose to rely on a trie for each purpose, *i.e.,* indexing the key-space, mapping the nodes of the trie on the network, and routing the requests. Among them, Nodewiz [4] assumes a set of static reliable nodes to host the trie, which is unfortunately hard to ensure on peer-to-peer platforms. P-Grid [7] builds a trie on the whole key-space (*i.e.,* the whole set of potential keys). Each leaf of this trie corresponds to a subset of the key-space. The fault-tolerance is achieved by probabilistic replication.

As a more general consideration, none of these approaches address the topology/physical locality awareness issue, *i.e.,* no information about the underlying network is taken into account to build the logical (overlay) network, what can raise a significant performance problem, physical locality being broken when the logical network is built. Moreover, the several fault-tolerance solutions are mostly replication-based, or DHT-based, also involving heavy replication mechanisms.

Initially designed for the purpose of service discovery over dynamic computational grids and attempting to solve the above drawbacks of existing approaches, we recently developed a novel architecture, based on a logical Greatest Common Prefix Tree formally described in Section 3, that is dynamically built as objects (services, but extensible to data items, files, etc.) are declared.

3   Preliminaries

Greatest Common Prefix Tree. Let an ordered alphabet A be a finite set of letters. Denote $\prec$ an order on A. A non empty *word* w over A is a finite sequence of letters $a_1, \ldots, a_i, \ldots, a_l$, $l > 0$. The *concatenation* of two words u and v, denoted $u \circ v$ or simply uv, is equal to the word $a_1, \ldots, a_i, \ldots, a_k, b_1, \ldots, b_j, \ldots, b_l$ such that $u = a_1, \ldots, a_i, \ldots, a_k$ and $v = b_1, \ldots, b_j, \ldots, b_l$. Let ϵ be the *empty word* such that for every word w, $w\epsilon = \epsilon w = w$. The *length* of a word w, denoted by $|w|$, is equal to the number of letters of w—$|\epsilon| = 0$.

A word u is a *prefix* (respectively, *proper prefix*) of a word v if there exists a word w such that $v = uw$ (resp., $v = uw$ and $u \neq v$). The *Greatest Common Prefix* (resp., *Proper Greatest Common Prefix*) of a collection of words $w_1, w_2, \ldots, w_i, \ldots$ $(i \geq 2)$, denoted $GCP(w_1, w_2, \ldots, w_i, \ldots)$ (resp. $PGCP(w_1, w_2, \ldots, w_i, \ldots)$), is the longest prefix u shared by all of them (resp., such that $\forall i \geq 1, u \neq w_i$). A *[Proper] Greatest Common Prefix Tree* ([P]GCP Tree, also a particular kind of *trie*) is a labeled rooted tree such that both following properties are true for every node of the tree:

1. The node label is a proper prefix of any label in its subtree;
2. The node label is the Proper Greatest Common Prefix of all its son labels.

In the following we use the word **trie** to designate our PGCP tree.

Distributed Lexicographic Placement Table. The *distributed system* considered in this paper consists of a set of asynchronous physical nodes organized in a *Distributed Hash Tables (DHT)*. Each physical node maintains one or more nodes of the logical PGCP Tree. Note that a DHT is used, but it can be replaced by any system, distributed or not, allowing the retrieval of any node from any other node. We also consider that the potential existing fault-tolerance mechanisms provided by this layer are not used within our architecture. We propose in this paper a fault-tolerance mechanism at the PGCP Tree level.

When one wants to insert an object labeled o into the trie, a message is generated containing o, according to which the message is routed within the trie until reaching the node labeled v such that v is the smallest label in the trie that shares with o the greatest common prefix of any node of the trie with o. More formally, if L denotes the whole set of label currently in the trie, the set $U = \{l \in L \mid GCP(l, o) = p\}$ where $p = max_{|m|}\{m = PGCP(l, o), l \in L\}$. The label of the target node is $t = min_{|w|}\{u \in U \mid u = pw\}$. Once found, the target node performs the insertion. If $t \neq o$, node(s) are created. If $o = tu$ $(u \neq \epsilon)$, a new node labeled o is created as a new son of the node labeled t. If $t = ou$

$(u \neq \epsilon)$, a new node is created as the father of the node labeled by t. Finally, if none of these conditions are satisfied, it means that o and t must be siblings but no node in the trie is labeled by their common prefix. Thus two nodes are created, a node labeled $GCP(o, t)$, father of the node labeled by t and also father of the other newly created node labeled by o. The distributed routing algorithm (that also performs the creation and the mapping of nodes) requires a number of hops bounded by twice the depth of the trie [5].

Physical nodes communicate by *message passing*. We assume two sending functions. The former, simply referred to *SEND*, is used by any physical node to send a message to another node asynchronously, i.e., without waiting any acknowledgement. The latter, called *SYNC-SEND*, waits for an acknowledgement for each message sent. We assume that each physical node may crash. So, when a physical node crashes, one or more logical nodes are lost.

4   Protocol

In this section, we give a detailed explanation of how the protocol works. We divide the algorithm code in two parts. The former shows the first phase developed with our technique during which a unique trie is recovered without considering any lexicographic property. During the second phase, the trie is reorganized to eventually form a distributed greatest common prefix tree.

4.1   Trie Recovery

After a node p detects the loss of its father $(p.father)$, it searches for a new father to link on. Making a traversal of the DHT, Node p collects in Variable PN all the addresses of each remaining physical node. Collecting the addresses in PN, p builds the set of logical nodes stored by the physical nodes in PN. Next, using a *PIF* (*Propagation of Information with Feedback*) Protocol [6, 16], p computes T, the set of logical nodes in its subtrie, which is made of its "real" descendants and its "temporary" relinked descendants. This first step of the recovery protocol ends when p chooses a temporary father $(p.tmpfather)$ in the subset $N \setminus T$. When, a node q is linked to a node p, then p considers q as a temporary son—stored in $p.tmpsons$. Note that Variable $p.tmpsons$ is required to compute T using a PIF in the subtrie of p. If $N \setminus T = \emptyset$ (i.e., there is no node for which p may link on), then p is considered as the root of the trie.

The above technique suffers of a drawback: Several nodes without father may make which could become a "bad" choice. In particular, they can choose as a temporary father a node belonging to the subtrie of another node being in the same situation. By doing this in parallel, cycles may appear. Our strategy is to detect and to break *a posteriori* such cycles as follows.

After the choice of its temporary father tf, a node p sends a message "HELLO" with its ID $(p.id)$ to tf. In the next step, tf transmits the message to its own father, and so on. Step by step, one of the two following situations eventually arises:

1. The "real" root of the trie receives the message "HELLO". In that case, the root notifies p that it is not involved in a cycle.
2. The message is received by a "false" root, i.e., a node having also lost its own father. the false root propagates the message to its temporary father.

Note that, in the above latter case, due to asynchrony of the network, it is possible that the false root receives the message "HELLO" sent by p before it executed its own recovery phase. In that case, the false root is still without a temporary father. The message "HELLO" is then delayed until the false root chooses its own temporary father.

Therefore, the message "HELLO" sent by p keeps circulating among its ancestors, carrying the list of false roots' IDs which were met during its traversal. Upon receipt of a message "HELLO", if the first item of the list carried by the message is equal to the ID of the receiver, then a cycle is detected. In that case, a leader election is computed among the IDs of the list—e.g., by choosing the smallest ID. The leader becomes the root of the subtrie, breaks its link which its father, and executes the recovery phase again. (The other "false" roots involved in the cycle remain connected to the subtrie rooted by the leader.) Note that a cycle may be created again. However, in the worst case, at each relaunching of the recovery phase, at least one subtrie becomes the subtrie of one false root. In other words, the number of cycles is periodically divided by at least 2. Therefore, the system eventually contains one (rooted) trie only.

4.2   Trie Reorganization

The trie reorganization is initiated once the trie recovery is done. Each node p having a temporary son q—i.e., q is a false root with its subtrie—initiates a routing mechanism closed to the original key insertion [5]. Let us consider the following cases:

1. The value $p.val$ is a prefix of the value of q—Figure 1, Case (i). In that case, q (and its subtrie) is placed in the subtrie of p following one of the four cases shown in Figure 1, Cases (a) to (d).
2. The value $p.val$ is not a prefix of the value of q. Then, p moves q to its father which now has the responsibility to place q.

Note that new services may keep inserting during the trie reconstruction. So, a new subtrie may have been created at the same place where the false root initially was. Thus, our method requires to take in account that any false root being placed in the trie can meet a node having the same value. In that case, the two tries must be merged. That is the aim of the merging protocol, initiated by the sending of a message "MERGE". Upon receipt of this message, a node p executes Procedure $Gluing(q)$, which moves the sons of q to p before withdrawing q from the trie (including the sons of q's father). Then, if necessary, p restarts recursively merging and placements among its sons, in order to merge both subtries eventually.

Algorithm 1 Recovery Protocol for each node p

```
1.01  upon receipt of <Disconnected from Father> do
1.02       PN := Physical Node Set in the DHT (collected by a DHT traversal);
1.03       N := Logical Node Set in PN (collected by polling the nodes in PN);
1.04       T := Logical Node Set in my subtrie (collected using a PIF wave)
1.05       using p.sons ∪ p.tmpsons;
1.06       if p.tmpfather ≠⊥ then send <DISCONNECT> to p.tmpfather;
1.07       if N \ T = ∅
1.08       then //I am the root
1.09            p.father :=⊥; p.tmpfather :=⊥;
1.10       else  p.tmpfather := random choice among N \ T;
1.11            send-sync <LINK> to p.tmpfather;
1.12            send <HELLO,p.id> to p.tmpfather;
1.13       endif
1.14  upon receipt of <HELLO,list> from q do
1.15       if First(list) = p.id
1.16       then //A cycle is detected
1.17            leader := LeaderElection(list);
1.18            if p = leader
1.19            then Executes "upon receipt of <Disconnect from Father> do",
1.20                      except PN and N;
1.21            endif
1.22       elseif p.Father ≠⊥
1.23       then send <HELLO,list> to p.father;
1.24       elseif p.tmpfather ≠⊥
1.25       then list := list + p.id;
1.26            send <HELLO,list> to p.tmpfather
1.27       elseif p.father =⊥
1.28       then // Both father and tmpfather are unknown, i.e.,
1.29            I am a false root which is still not linked
1.30            Executes "upon receipt of <Disconnect from Father> do"
1.31            if it is still not working;
1.32            if tmpfather ≠⊥
1.33            then  list := list + p.id;
1.34                  send <HELLO,list> to p.tmpfather;
1.35            else  send <NOCYCLE> to First(list);
1.36       else // I am the real root, so there is no cycle.
1.37            send <NOCYCLE> to First(list);
1.38       endif
1.39  upon receipt of <NOCYCLE> from q do
1.40       send <MOVE,p> to p.tmpfather;
1.41       send-sync <UNLINK> to p.tmpfather;
1.42       p.tmpfather :=⊥;
1.43  upon receipt of <LINK> from q do
1.44       tmpsons := tmpsons ∪ {q};
1.45  upon receipt of <UNLINK> from q do
1.46       tmpsons := tmpsons \ {q};
```

Algorithm 2 Reorganization Protocol for each node p

1.01 **upon** receipt **of** <MOVE,fs> **from** q **do**
1.02   **if** $fs.val = p.val$
1.03   **then** //I send to myself that a fusion is needed.
1.04   **send** <MERGE,fs> **to** p
1.05   **elseif** $p.val = prefix(fs.val)$
1.06   **then** **if** $\exists s \in p.sons|\ s.val = prefix(fs.val)$
1.07   **then** // fs is in the subtrie of s, Case (a) in Figure 1
1.08   **send** <MOVE,fs> **to** s;
1.09   **elseif** $\exists s \in p.sons|\ fs.val = prefix(s.val)$
1.10   **then** // s is in the subtrie of fs, Case (b) in Figure 1
1.11 $p.sons := p.sons \cup \{fs\}$; $p.sons := p.sons \setminus \{s\}$;
1.12   **send** <MOVE,s> **to** fs;
1.13   **elseif** $\exists s \in p.sons\ |\ p.val < PGCP(s.val, fs.val)$
1.14   **then** // fs and s have a PGCP which is greater than $p.val$
1.15   // Case (c) in Figure 1
1.16 $Newnode(PGCP(fs.val, s.val), s, fs)$; $p.sons := p.sons \setminus \{s\}$;
1.17   **else** // fs is one of my sons, Case (d) in Figure 1
1.18 $p.sons := p.sons \cup \{fs\}$;
1.19   **endif**
1.20   **else** **if** $p.father \neq \bot$
1.21   **then** **send** <MOVE,fs> **to** $p.father$
1.22   **else** **if** $fs.val = prefix(p.val)$
1.23   **then** // I am in the subtrie of fs
1.24   **send** <MOVE,p> **to** fs;
1.25   **else** // p and fs are brothers
1.26 $p.sons := p.sons \cup Newnode(PGCP(fs.val, f.val), fs, p)$;
1.27   **endif**
1.28   **endif**
1.29   **endif**

2.01 **upon** receipt **of** <MERGE,fs> **from** q **do**
2.02 $Gluing(q)$;
2.03   Sorting of $p.sons$ in the lexicographic order in Table t_s;
2.04   **for** $i = 0$ **to** $t_s.length()$ **do**
2.05   **if** $t_s[i].val = t_s[i+1].val$
2.06   **then** **send** <MERGE,$t_s[i+1]$> **to** $t_s[i]$;
2.07 $i := i+1$;
2.08   **elseif** $t_s[i].val = prefix(t_s[i+1].val)$
2.09   **then** **send** <MOVE,$t_s[i+1]$> **to** $t_s[i]$;
2.10 $p.sons := p.sons \setminus \{t_s[i+1]\}$;
2.11 $i := i+1$
2.12   **elseif** $p.val < PGCP(t_s[i].val, t_s[i+1].val)$
2.13   **then** $p.sons := p.sons \cup Newnode(PGCP(t_s[i].val, t_s[i+1].val),$
2.14 $t_s[i], t_s[i+1])$;
2.15 $p.sons; = p.sons \setminus \{t_s[i], t_s[i+1]\}$;
2.16 $i := i+1$;
2.17   **endif**
2.18   **done**

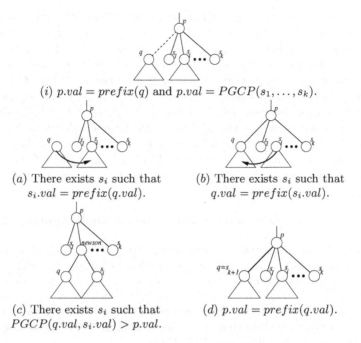

(i) $p.val = prefix(q)$ and $p.val = PGCP(s_1, \ldots, s_k)$.

(a) There exists s_i such that
$s_i.val = prefix(q.val)$.

(b) There exists s_i such that
$q.val = prefix(s_i.val)$.

(c) There exists s_i such that
$PGCP(q.val, s_i.val) > p.val$.

(d) $p.val = prefix(q.val)$.

Fig. 1. A false root q is linked to a node p such that $p.val = prefix(q.val)$

4.3   Correctness Proof

In this subsection, we discuss the correctness of our protocol. In order to do this, we first need to make the realistic assumption that under the considered context, the crash frequency is low enough to make the trie fully built sometime. (In the opposite way, the trie could never be built and unusable most of the time. More generally it is impossible to say anything about termination otherwise.) In other words, we fairly assume that no crash occurs after a crash until the trie is fully built, i.e., no two consecutive crashes interfere each other, at one given time.

Assuption 1. *If a node crashes at time t, then for every $t' > t$, no crash occurs.*

Lemma 1. *Under Assumption 1, the recovery protocol (Algorithm 1) termi-nates, and when this occurs, the system contains one trie only.*

Proof. The validation mainly consists in showing that the protocol terminates and that the reorganization of the trie is eventually initiated (by sending a message NOCYCLE).

Assume by contradiction that under Assumption 1, no node eventually sent a message NOCYCLE. So, neither Line 1.35 nor Line 1.37 in Algorithm 1 is executed. Note that in the first case (Line 1.35), the node becomes the "real" node after the crash of its father. So, in both cases, this means that NOCYCLE never reaches the "real" root of the trie. The height of the trie being finite, this means that every Message HELLO traverses cycles only. When a message HELLO is received by its initiator, the cycle is broken by the node which is elected among the false roots participating in the cycle—Lines 1.16 to 1.21. Therefore, cycles are created infinitely often. Let C be the number of created cycles. In the worst case, a cycle is made of at least two nodes. So, C is initially bounded by $F/2$, where F is the number of false root created by the crash. When a cycle is broken, at most one leader is elected. So, at most $C/2$ leaders are able to link another node again. In the next phase, the number of cycles is less than or equal to $C/2$. Since under Assumption 1, cycles may be created only when false roots are linked to other nodes (executing Lines 1.10 and 1.11), C never grows and is eventually equal to 0. This contradicts that cycles are created infinitely often.

We now consider the phase of trie reorganization shown in Algorithm 2.

Lemma 2. *Under Assumption 1 and assuming that the system contains one trie only, the reorganization protocol (Algorithm 2) terminates, and when this occurs, the trie is a PGCP tree.*

Proof. Clearly, each trie of the forest following the crash of a node is a *PGCP* tree. So, its remains to show that executing Algorithm 2, the whole trie eventually satisfies the condition to be a *PGCP* tree.

From the algorithm, it is easy to observe that, in the absence of merging, there are only two cases to consider depending on the value of Node p and its false son fs :

1. The value of p is a prefix of fs's value—Line 1.05. In that case, following the four cases described in Figure 1, fs is eventually placed at the right place in the subtrie of p—refer to Lines 1.06 to 1.19. The resulting trie is a *PGCP* tree.
2. The value of p is not a prefix of fs. Again, there are two cases to consider:
   (a) Node p has no father ($p.father = \perp$)—Line 1.22 to 1.28. In that case, if $fs.val$ is a prefix of p, then p (and its subtrie) becomes the node to be placed in fs—Line 1.24. Otherwise, p and fs become the two sons of a new root node q such that $q.val = PGCP(p, fs)$—Line 1.26. The trie is then clearly a *PGCP* tree.
   (b) Node p has a father. Then, fs is moved to the father of p—Line 1.21. By induction of the above discussion, either fs eventually moves on a node q such that $q.val = prefix(fs.val)$ or fs eventually reaches the root of the trie. The former case is equivalent to Case 1, the latter to Case 2a.

If p and fs merge, then there are four cases to consider after p and fs glued together into p:

1. There exists a pair of sons s_i, s_j of p such that $s_i.val$ is a prefix of $s_j.val$. Then, s_j is moved toward s_i—Lines 2.08 to 2.11. This case is similar to the above Case 1 (Cases (a) or (b) in Figure 1).

2. There exists a pair of sons s_i, s_j of p such that $PGCP(s_i, s_j) > p.val$. Then, s_i and s_j become the two sons of a new son q of p such that $q.val = PGCP(p, fs)$—Lines 2.12 to 2.16. This case is also similar to the above Case 1 (Case (c) in Figure 1).

3. There exists a pair of sons s_i, s_j of p such that $s_i.val = s_j.val$. This case is solved by initiating a recursive merging between s_i and s_j—Lines 2.05 to 2.07. This case is solved by induction on s_i and s_j.

4. There exists no pair of sons s_i, s_j of p satisfying either Case 1, 2, or 3. In that case, the subtrie of p clearly satisfies the properties of a $PGCP$ tree.

From Lemmas 1 and 2 follows:

Theorem 1. *Under Assumption 1, Algorithm 1 and Algorithm 2 provide a PGCP tree reconstruction after the crash of a physical node.*

5   Conclusion and Future Work

In this paper, we have presented a fault-tolerant protocol in case of node crashes in a Proper Common Greatest Prefix tree search system. This protocol can be coupled with a replication strategy to lower the costs related to high replication factors. This protocol allows the reconnection and repair of subtries after the crash of one or more nodes. This algorithm guarantees to recover a consistent PGCP tree after a finite time and thus to avoid partially replication.

Our future work will consist in connecting the two mechanisms (replication and repair) in order to minimize the cost of fault-tolerance on dynamic platforms. We will also develop and validate experimentally the mechanisms exposed in this paper on the Grid'5000 platform of the french ministry of research. The aim of such experimentation will be to see the performance of the repair algorithm and to see its capacity to answer clients' requests facing different levels of dynamicity. Moreover, we will be able to see starting from which level of dynamicity the repair mechanism is no more efficient alone, and then how we can progressively inject some replication as the dynamicity level increases.

References

1. A. Andrzejak and Z. Xu. Scalable, Efficient Range Queries for Grid Information Services. In *Peer-to-Peer Computing*, pages 33–40, 2002.
2. J. Aspnes and G. Shah. Skip Graphs. In *Fourteenth Annual ACM-SIAM Symposium on Discrete Algorithms*, pages 384–393, January 2003.
3. M. Balazinska, H. Balakrishnan, and D. Karger. INS/Twine: A Scalable Peer-to-Peer Architecture for Intentional Resource Discovery. In *Proceedings of Pervasive 2002*, 2002.
4. S. Basu, S. Banerjee, P. Sharma, and S. Lee. NodeWiz: Peer-to-Peer Resource Discovery for Grids. In *5th International Workshop on Global and Peer-to-Peer Computing (GP2PC) in conjunction with CCGrid, May 2005*, 2005.

5. E. Caron, F. Desprez, and C. Tedeschi. A dynamic prefix tree for the service discovery within large scale grids. In IEEE, editor, *The Sixth IEEE International Conference on Peer-to-Peer Computing, P2P2006*, Cambridge, UK., September 6-8 2006.

6. E.J.H. Chang. Echo Algorithms: Depth Parallel Operations on General Graphs. *IEEE Trans. on Software Engineering*, SE-8:391–401, 1982.

7. A. Datta, M. Hauswirth, R. John, R. Schmidt, and K. Aberer. Range Queries in Trie-Structured Overlays. In *The Fifth IEEE International Conference on Peer-to-Peer Computing*, 2005.

8. I. Foster and A. Iamnitchi. On Death, Taxes, and the Convergence of Peer-to-Peer and Grid Computing. In *IPTPS'03*, pages 118–128, 2003.

9. Gnutella. http://www.gnutella.com.

10. KaZaA 2005. The KaZaA Web Site. http://www.kazaa.com.

11. D. Oppenheimer, J. Albrecht, D. Patterson, and A. Vahdat. Distributed Resource Discovery on PlanetLab with SWORD. In *Proceedings of the ACM/USENIX Workshop on Real, Large Distributed Systems (WORLDS)*, December 2004.

12. S. Ramabhadran, S. Ratnasamy, J. M. Hellerstein, and S. Shenker. Prefix Hash Tree An indexing Data Structure over Distributed Hash Tables. In *Proceedings of the 23rd ACM Symposium on Principles of Distributed Computing*, St. John's, Newfoundland, Canada, July 2004.

13. S. Ratnasamy, P. Francis, M. Handley, R. Karp, and S. Shenker. A Scalable Content-Adressable Network. In *ACM SIGCOMM*, 2001.

14. A. Rowstron and P. Druschel. Pastry: Scalable, Distributed Object Location and Routing for Large-Scale Peer-To-Peer Systems. In *International Conference on Distributed Systems Platforms (Middleware)*, November 2001.

15. C. Schmidt and M. Parashar. Enabling Flexible Queries with Guarantees in P2P Systems. *IEEE Internet Computing*, 8(3):19–26, 2004.

16. A. Segall. Distributed Network Protocols. *IEEE Transactions on Information Theory*, IT-29:23–35, 1983.

17. Y. Shu, B.-C. Ooi, K.-L. Tan, and A. Zhou. Supporting Multi-Dimensional Range Queries in Peer-to-Peer Systems. In *Peer-to-Peer Computing*, pages 173–180, 2005.

18. I. Stoica, R. Morris, D. Karger, M. Kaashoek, and H. Balakrishnan. Chord: A Scalable Peer-to-Peer Lookup service for Internet Applications. In *ACM SIGCOMM*, pages 149–160, 2001.

19. P. Triantafillou and T. Pitoura. Towards a Unifying Framework for Complex Query Processing over Structured Peer-to-Peer Data Networks. In *DBISP2P*, 2003.

20. B. Y. Zhao, L. Huang, J. Stribling, S. C. Rhea, A. D. Joseph, and J. D. Kubiatowicz. Tapestry: A Resilient Global-scale Overlay for Service Deployment. *IEEE Journal on Selected Areas in Communications*, 22(1):41–53, January 2004.

Utility-Based Adaptation of RED Parameters

Rachel P. Villacorta[1] and Cedric Angelo M. Festin[2]

[1] Department of Electrical and Electronics Engineering, University of the Philippines,
Diliman, QC, 1101 Philippines
rachel.villacorta@up.edu.ph
[2] Department of Computer Science, University of the Philippines,
Diliman, QC, 1101 Philippines
cmfestin@up.edu.ph

Abstract. Random Early Detection (RED) is effective in decreasing losses on responsive flows but performs poorly with User Datagram Protocol (UDP) traffic since these do not react to congestion notification. However, it is important to address UDP flows' requirements since UDP traffic forms a considerable part of Internet traffic. This paper shows how Value-Based Utility (VBU) with RED manages UDP flows. VBU allows packet dropping and queueing decisions to be more sensitive to UDP traffic packet loss requirements. Three RED variations incorporating VBU were developed. Results indicate adoption of VBU with RED enhances management of UDP traffic. This results in the equitable distribution of loss based on the UDP flows' requirement.

Keywords: RED, buffer management, packet loss utility.

1 Introduction

Random Early Detection (RED) [1] is one of the most popular algorithms that is used to detect and avoid congestion. The idea behind RED is that it randomly notifies the connections of incipient congestion. RED allows occasional burst traffic by maintaining a regular, evenly spaced interval when marking packets. It avoids global synchronization by randomly dropping packets.

RED is mainly designed to be used in conjunction with Transport Control Protocol/Internet Protocol (TCP/IP) but it could still be used by other protocols in the transport layer. However, this claim is refuted by some studies showing that RED is ineffective on non-responsive flows [4,5]. Since these flows do not back off in response to congestion notification, they will experience drops imposed by RED's policies.

2 Utility-Based Enhancement of RED

Value-Based Utility (VBU) [6] is a framework for measuring the level of satisfaction (happiness) or dissatisfaction (unhappiness) of users. Its definition is given by:

$$U_{i,QoS,m,\Delta t}(p,b) = G/N - (N-G)/N * p/q \qquad (1)$$

Y. Robert et al. (Eds.): HiPC 2006, LNCS 4297, pp. 183–192, 2006.
© Springer-Verlag Berlin Heidelberg 2006

where U is the utility of a flow i for the required Quality of Service (QoS), taken at point m during the interval $\Delta t = (t_2 - t_1)$

p is the target percentage of packets meeting the required QoS

b is the target QoS bound

G is the number of packets that satisfy the requirements of a certain flow

N is the total number of packets of a flow

q is the allowed percentage of error and is equal to $1 - p$

VBU was used to manage a First-In-First-Out (FIFO) buffer based on utility thresholds [6] while in this research, VBU is used to enhance the performance of RED with the addition of enabling the 'gentle' [2,3] and 'wait' parameters which are used to make RED less aggressive in dropping packets. We implement three algorithms namely RED with VBU (R1), RED with VBU using three logical buffers (R2) and RED with VBU, logically partitioned buffers and adaptation of thresholds (R3).

2.1   R1: RED with VBU

In R1, we interrupt the original flow of RED's algorithm by adding the use of VBU. When the average queue size, avg is between the minimum threshold, min_{th} and twice the maximum threshold, max_{th}, the Utility of the flow is computed with an incremented N[1]. If the Utility measured is less than 0, then the packet is enqueued. Otherwise, the original RED algorithm is followed.

We use VBU instead of just using the dropping probability function to be able to differentiate the satisfied flows from the unhappy ones. Using VBU would aid in increasing the number of happy flows since flows are now treated according to their needs.

We have classified the flows into three categories[2] in order to cover a wide range of loss requirements. The flows are classified as High Expectation Flows (HEFS), Medium Expectation Flows (MEFS), and Low Expectation Flows (LEFS). This classification is similar to the study in [6] wherein HEFS are flows that have the strictest requirement with loss as the target QoS metric[3]; the MEFS with the less stricter loss requirement and the LEFS with the least strict loss requirement.

2.2   R2: RED with VBU Using Three Logical Buffers

We implemented two versions of R2, R2+ and R2-, to check if VBU will have an effect on a partitioned buffer. The buffer is now logically partitioned into three in order to differentiate the three different types of flows. The largest buffer is dedicated to HEFS, the next largest to MEFS and the smallest to LEFS. R2+ uses VBU while R2- does not.

After computing and comparing the average queue size of the entire buffer to the min_{th} and max_{th} of the entire buffer, the average queue size of the flow a

---

[1] N is the total number of transmitted packets.

[2] This classification will be used in all the schemes and experiments.

[3] QoS metrics could be other measurements aside from packet loss.

packet which arrived belongs to is computed. This is compared to the max_{th} of that type of flow's buffer. If avg is less than the max_{th} of that class' buffer, then the packet is a candidate for enqueueing. Otherwise, the packet is dropped.

2.3   R3: RED with VBU, Logically Partitioned Buffers and Adaptation of Thresholds

R3 improves on R2+ by dynamically changing the maximum threshold of the logical buffers. The total happiness[4] of each of the three types of flows are computed after some time t since the last change of the max_{th} of the buffers. These are then ranked in decreasing order. The maximum threshold of the flow that is last in rank is increased by getting resources from the flow that is first or second in rank. If there are no extra resources from either flow, then the maximum threshold of the flow which is last in rank will not be adjusted. Moreover, we adjust the maximum thresholds of the types of flows in an attempt to increase the utilities of the HEFS while minimizing the negative effect on the other types of flows. Thus, this adjustment favors HEFS, the MEFS and lastly the LEFS. However, the needs of each flow is still respected by ranking the total happiness of each type of flow. Eventhough the LEFS are our last priority, we still increase the maximum threshold of the LEFS if they are the last in rank in total happiness. The same is also done with the MEFS.

2.4   R3 and Some RED Variants

R3's design is different from previous implementations of RED. The Utilities from VBU are used to differentiate flows and dynamically tune flow's maximum thresholds. This is different from Adaptive RED (ARED) [7] where there is no flow differentiation. Moreover, max_p[5] is dynamically changed based on the fluctuations of the average queue size while in R3, the maximum thresholds of the logically partitioned buffers are dynamically tuned based on the obtained utilities. R3 is also different from RED with Dynamic Thresholds (RED-DT) [8] which aims to penalize unresponsive flows while R3 aims to treat these flows according to their needs. Adjusting parameters in RED-DT is a function of the number of active flows and each flow's buffer occupancy while in R3, adjusting of the thresholds is a function of the happiness of each type of flow. Lastly, R3 is different from RED with in/out bit (RIO) [9] which preferentially drops 'out'[6] packets while R3 drops packets primarily based on whether a flow is happy or not. R3 also offers a minimum level of protection to each type of flow by allocating buffer space to each one of them while in RIO 'out' packets are enqueued if there

---

[4] The Total Happiness of a type of flow is measured by dividing the number of the utilities which are greater than or equal to zero to the total number of flows of that type.

[5] max_p is a RED parameter that holds the maximum value for the initial packet-marking probability.

[6] 'out' packets are packets that do not meet a certain specified profile.

is still available space after enqueueing 'in'[7] packets. Lastly, R3 also allows flows to share resources depending on their level of happiness while in RIO, sharing of resources does not necessarily happen.

3   Experimental Environment

Experiments were performed using Network Simulator 2 (NS2) on a single hop model with the length of each simulation equal to 800 seconds. The observation window is from 200 seconds to 800 seconds to remove the effect of the start up phase and to acquire enough samples. There are initially 9 sources. This number was later increased to 36 sources to test the scalability of the system. Each source generates UDP traffic that uses the exponential distribution for the traffic's packet interarrival time. Each source connected to the router uses a 10Mbps link to ensure that no packet is lost in this link and that packet transmission delay is minimal. The propagation delay is set to 1msec which is negligible. The router has a buffer size equivalent to 50 packets. Values used for the bandwidth of the link connecting the router to the sink node are 750, 850 and 950 kbps for 9 sources and 3.2, 3.4 and 3.6Mbps for 36 sources. In the experiments, we changed the bandwidth to see the effect of having larger link capacities. Flows as mentioned earlier are classified into HEFS, MEFS and

Table 1. Expectation Levels

Class of Flows	p[8]
High Expectation Flows (HEFS)	0.99
Medium Expectation Flows (MEFS)	0.9
Low Expectation Flows (LEFS)	0.8

Table 2. Traffic Mix

Name	No. of Sources	No. of HEFS	No. of MEFS	No. of LEFS
mix 1	9	3	3	3
mix 2	36	12	12	12
mix 3	36	24	8	4
mix 4	36	8	24	4
mix 5	36	4	24	8
mix 6	36	8	4	24
mix 7	36	4	8	24

LEFS. These are shown in Table 1. Traffic mix is varied to see the effect of having more flows of a particular type over the other types of flows (Table 2).

---

[7] 'in' packets are packets that meet a certain specified profile.

[8] p is the target percentage of packets meeting the loss requirement of flows.

Utility measurements[9] are yet another form of measuring packet loss in this study. In the Utilities, packet losses are associated to the flow's requirements. We determine the Total Happiness Plot of each class of flows by getting the ratio of the number of utilities of each class which are greater than or equal to zero, to the total number of flows per class. As an example, at certain time T, for a total of X HEFS, if there are Y HEFS having utility greater than or equal to zero, then the measured and plotted utility at time T is Y/X.

4 Baseline Experiments on RED

Initial experiments on RED, like in FIFO, show that the packet loss requirement of flows is not addressed as shown in Fig. 1. The LEFS got the highest number of satisfied flows while the HEFS got the lowest number of satisfied flows. As already mentioned, this will be addressed by R3 scheme.

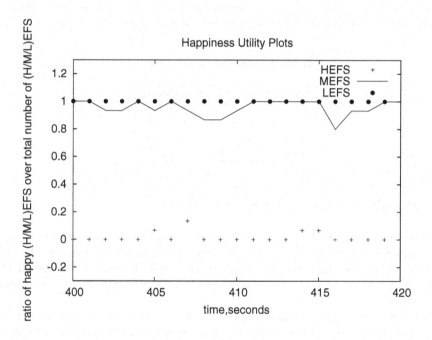

Fig. 1. For a RED queue, Total Happiness Plots of HEFS, MEFS and LEFS. This is the same scenario we got from FIFO. The LEFS have the highest Utility while the HEFS the lowest.

For the rest of the experiments, since RED has several parameters - max_{th}, min_{th}, max_p and w_q[10], preliminary experiments were performed to isolate the

---

[9] Utilities could also be measured on delay or jitter.

[10] w_q is a RED parameter for the queue weight.

cases that are significant. We will get the parameter combination that gave two of the highest Utilities, one of the lowest Utilities and a random case. The experimental values used are shown in Table 3. Some of these values are recommendations from [1] which we will use to cover a range of experiments.

Table 3. Experimental Values

RED Parameters	Values
max_p	0.02, 0.1, 0.5
w_q	0.002, 0.02, 0.2
max_{th}	10, 20, 30,40
min_{th}	5, 10, 15, 20, 25

Shown in Table 4 are cases used in this study. Cases 2 and 4 gave the highest Utility for HEFS while case 1 gave one of the lowest. Case 3 is used to examine conditions for Utilities of HEFS in between the extremes.

Table 4. RED Parameters' Setting

cases	max_p	w_q	max_{th}	min_{th}
case 1	0.1	0.002	40	25
case 2	0.02	0.002	30	20
case 3	0.02	0.2	30	10
case 4	0.02	0.002	40	25

5   Results on Using VBU with RED

When we compare RED and R1 for all cases and for all traffic mixes, RED's performance is enhanced[11] due to the additional information provided by the Utility. Shown in Fig. 2a is the graph of the Total Happiness Plots of HEFS taken from one of the best enhancement produced by R1. This is the result from mix 3 of case 3.

We can see from Fig. 2a that R1 improved the Utility of HEFS. Although there are cases when the Utilities of HEFS are not equal to 1, the Utilities produced by R1 are higher than those produced by RED. In some cases, differentiation of flows provided by VBU does not improve the Utility of HEFS. This may be due to limited bandwidth, buffer space or some traffic mixes that use more HEFS than the other types of flows. Moreover, improving the Utility of HEFS did not degrade the performance of the LEFS and MEFS except for some cases in MEFS. This is due to more HEFS being enqueued in the buffer using the flow differentiation offered by VBU.

---

[11] In this study, we base a queueing mechanism's performance primarily on the Utility of the HEFS.

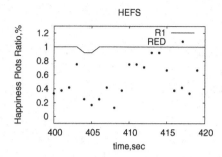

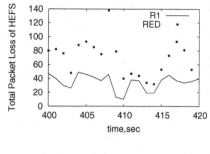

Fig. 2a. Total Happiness Plots of HEFS at case 3 mix 3. Almost all the time, all HEFS are happy in R1 while in RED, many HEFS are dissatisfied.

Fig. 2b. Total Packet Loss for HEFS at case 3 mix 3. Losses in R1 are kept low while losses in RED went up to a total of 200 packet drops.

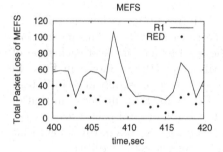

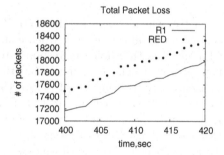

Fig. 2c. Total Packet Loss for MEFS at case 3 mix 3. Losses in RED are now lower than in R1.

Fig. 2d. Total Packet Loss at case 3 mix 3. Packet loss in RED are a little higher than in R1.

Looking at the aggregated losses of each type of flow in Figs. 2b to 2c, we can see that using VBU, less drops occur in HEFS but more drops are experienced by MEFS and LEFS. This shows that packet losses are redistributed using VBU making it possible for HEFS to be more satisfied. If we look closely at the total loss in Fig. 2d, we can see that the total losses in R1 is less than that of RED. This means that the big decrease in the losses of HEFS negated the effect of increased losses in MEFS and LEFS.

On the other hand, when we compare R1 and R2-, we saw that R2- is better than R1 except when there are more HEFS in the system. This means that a partitioned buffer is still better than using a single buffer except when there are more HEFS in the system. However, the disadvantage of buffer partitioning is an increase in the total losses since there are now three limited buffers as compared to a single limited buffer. Figs. 3a to 3b show the graphs from case 1 of mix 4.

When we compare R2+ and R2-, R2+ showed an improvement in the Utilities. This is another manifestation that the additional information provided by VBU improves the performance of an algorithm.

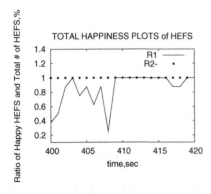 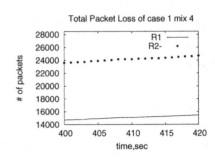

Fig. 3a. Total Happiness Plots of HEFS at case 1 mix 4. Almost all the time, HEFS in R2- are happy.

Fig. 3b. Total Packet Loss at case 1 mix 4

Lastly, when comparing R3 and R2+, Utilities of R3 either increased or showed no significant change when compared to those of R2+. This means that adjusting the thresholds in response to the level of happiness of each type of flow proved beneficial in increasing the Utilities of HEFS. Shown in Figs. 4a to 4b are graphs from case 4 mix 3. Using a 5% level of significance in T-test, we get $t = 28.848$ which means that HEFS' Utility for R3 is higher than that of R2+. On the other hand, when we look at the graph of total packet loss, we see that total packet loss of R2+ is higher than that of R3. Again, this shows that dynamically changing the thresholds aided in increasing the Utilities.

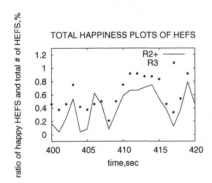 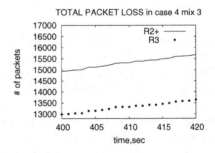

Fig. 4a. Total Happiness Plots of HEFS at case 4 mix 3 (R3 & R2+)

Fig. 4b. Total Packet Loss at case 4 mix 3 (R3 & R2+)

6   Summary of Results

The use of VBU in differentiating flows enhanced the performance of RED. RED is unable to differentiate flows according to their requirements. As a result, many HEFS were not satisfied since these flows have the strictest requirements compared to the other types of flows. With the use of VBU, packets are enqueued or dropped in response to the loss requirement of the flow it belongs to. If dropping of a newly arrived packet will make the flow it belongs to unhappy, then this packet will be enqueued. Thus, this scheme resulted to having more satisfied HEFS.

Allocating buffer space for each type of flow while using VBU proved effective in addressing the requirements of the flows. Pre-assigning of buffer space to each type of flow gave additional protection to each of them. Flows that have stricter requirements will now have greater chances of being appropriately addressed since each type of flow will not have to compete with the other types of flow for buffer space. As a result, there were more satisfied HEFS in the system. However, constraining each type of flow to a limited buffer will cause more drops when too much flows of the same type compete with each other for space. The impact of these packet drops is more pronounced particularly with HEFS since these have the strictest loss requirement than with MEFS and LEFS. This case was illustrated with the decrease of the Utilities of HEFS when the system load consists of a considerable number of HEFS compared to the other types of flows. In the case where there are more MEFS in the system, the decrease in the Utility of MEFS is smaller than with the decrease in HEFS. Also, when there are more LEFS, the Utility of LEFS was not affected but caused LEFS to lose more packets. This means that nothing much can be done in an environment wherein the resources are limited especially for HEFS whose requirement is harder to meet.

Lastly, adapting the maximum thresholds in response to the measured Utility gave further improvement in the Utility of HEFS while minimizing the negative effects on the other types of flows. The use of VBU in dynamically tuning the thresholds gave the necessary feedback for the scheme to know which type of flow can share resources. However, working in a limited environment proved that the effect of this adaptation is not highly significant since there are no more spare resources to share.

These solutions improved RED's algorithm. However, a necessary consequence of the schemes' effort to increase the Utilities of HEFS is the increase of the total packet loss. This is due to more flows of other types giving up more resources to make more HEFS satisfied.

7   Conclusions

The solutions developed could provide a new framework for congestion control and avoidance mechanisms wherein both the problems of congestion and meeting flows' requirements are addressed. Protection of non-TCP flows, which have their own requirements, is also offered by these solutions. The solutions developed could be used to properly address the current trend in the Internet where some flows are penalized (e.g. UDP flows) because of their non-responsiveness to congestion notifications. UDP flows and all the other types of flows could declare on the outset their requirements and base on this information, each class of flows will be dealt with accordingly.

In order to deliver a better than best-effort service, the developed solutions can be applied in a DiffServ environment. In this environment, aggregate of flows are provided different levels of prioritization depending on the contracted agreements between the user and the service provider. This service is an improvement on what the Internet can offer wherein guarantees of any sort is not to be expected. In using VBU with RED, we have shown that flows were treated according to their loss requirements. Flows were differentiated and higher priority is given to those flows which have higher expectation. VBU can then be used in providing different levels of service to each class of flows.

Finally, R3 can be implemented at the edge of the network instead of at the core to decrease the processing time and complexity inside the network. R3 could then be used as the traffic conditioner in a DiffServ environment at the boundary nodes. The Utility could be used to gauge the satisfaction of the incoming flows and based on this information, perform marking, shaping, dropping of packets or adjusting of the thresholds.

References

1. Floyd, S., Jacobson, V.: Random early detection gateways for congestion avoidance. IEEE/ACM Transactions on Networking.**1:4** (1993) 397-413
2. Floyd, S.: Recommendation on using the 'gentle' variant of RED. http://www.icir.org/floyd/red/gentle.html. (2000) (Accessed March 7, 2005)
3. Pentikousis, K.: Active Queue Management. ACM Crossroads Student Magazine (2001)
4. Parris, M., Jeffay, K.: A better-than-best-effort service for continuous media UDP flows. NOSSDAV '98. (1998)
5. Bonald, T., May, M., Bolot, J.: Analytic Evaluation of RED Performance. Proceedings of INFOCOM. (2000) 1415-1424
6. Festin, C., Sorensen, S.: Utility-Based Buffer Management for Networks. ICN 2005, LNCS 3420, (2005) 518-526
7. Feng, W. C., Kandlur, D., Saha, D., Shin, K.: A Self-Configuring RED Gateway. Proceedings of INFOCOM. (1999)
8. Vukadinovic, V., Trajkovic, L.: RED with Dynamic Thresholds for Improved Fairness. ACM Symposium on Applied Computing. (2004)
9. Clark, D., Fang, W.: Explicit Allocation of Best-effort Packet Delivery Service. IEEE/ACM Transactions on Networking.**6:4** (1998)

Capturing an Intruder in Product Networks

Navid Imani[1], Hamid Sarbazi-Azad[2,1], and Albert Zomaya[3]

[1] IPM School of Computer Science, Tehran, Iran
[2] Sharif University of Technology, Tehran, Iran
[3] University of Sydney, Sydney, Australia
{Imani, Azad}@ipm.ir, zomaya@it.usyd.edu.au

Abstract. In this paper, we envision a solution to the problem of capturing an intruder in a product network. This solution is derived based on the assumed existing algorithms for basic member graphs of a graph product. In this problem, a team of cleaner agents are responsible for capturing a hostile intruder in the network. While the agents can move in the network one hop at a time, the intruder is assumed to be arbitrarily fast in a way that it can traverse any number of nodes contiguously as far as no agents reside in those nodes. Here, we consider a version of the problem where each agent can replicate new agents. Hence, the algorithm start with a single agent and new agents are created on demand.

1 Introduction

One of the important problems concerning the distributed system is the network security. The reports show that a huge amount of money is spent annually to recover the information which is lost due to the attack of worms and viruses. Spywares and cookies are some other example of software agents which often enter the hosts without permission and gain unauthorized access to the information which may later be used for undesirable purposes.

Here we only assume a general kind of such threat namely a piece of software (e.g., a virus) which moves in the network from node to node; we will call such an element, the *intruder*. We assume we have a team of software agents that collaborate in order to protect the network and thus have the goal of neutralizing the intruder and cleaning the entire network. The intruder is also considered to be a software agent. We assume that, the intruder can only escape the cleaner agents and although it can be harmful to the hosts, it cannot damage or stop the cleaner agents. Furthermore, in order to consider the worst case problem, we assume that given a situation the intruder always chooses the best possible move. The purpose of our algorithm then is to move and replicate the cleaner agents in the network such that the intruder is finally captured by the former agents. Number of agents to be involved, number of moves the agents have to perform, and time are some of measures of efficiency for an intruder capturing algorithm. The problem of intrusion capturing is studied both in general and also on a number of well-known network topologies in [17, 15, 1, 2, 8, 4].

Our model takes the assumption that agents are able to replicate; i.e. they can create a copy of themselves whenever needed. Hence, without loss of generality, we

Y. Robert et al. (Eds.): HiPC 2006, LNCS 4297, pp. 193–204, 2006.
© Springer-Verlag Berlin Heidelberg 2006

assume that there exists only a single agent at the beginning and other agents are created gradually as the algorithm proceeds. All the agents are identical, hence we do not distinguish the agents by any names or labels, and when discussing the position of the agents we are only interested to know which nodes are occupied by the agents rather than being concerned with the cell where a particular agent resides.

A possible strategy is to start with a single agent from an arbitrary point (called *home*) and replicate new agents and move existing agents for checking the presence of the intruder, in such a way that no gap is created for the intruder to enter an already cleaned part of the network. In this way, the intruder will eventually be discovered.

Our algorithm does not discuss the position of the initial agent in the network since it only takes some constant steps (in terms of hops) for the agent to move to any other node of the network which is often negligible. This assumption does not threat the validity of our model because this constant in the worst case is in the order of network diameter while the total number of movements required by the algorithm is almost always in a higher order. Even for a network whose search number is equal by 1, i.e. the network can be searched using a single agent, the total number of required moves will be in the order of network diameter in the best case. As one may expect, in our model each agent can only move to a neighboring node and that can be done via the links connecting the two nodes.

Cartesian product networks encompass a large class of interconnection networks which are practically used nowadays. Many important topologies such as meshes, tori, k-ary n-cubes, hypercubes, generalized hypercubes, and hyper petersen networks are examples of such product networks. Hence, it comes as no surprise to see that product networks have been vastly studied in the literature as in [Bao98, 5, 6, 7, 9, Sabidussi59].

In the context of graph theory, the Cartesian product is considered as an effective method for constructing large graphs from several specified smaller graphs. The graph constructed in this way contains its basic graphs as subgraphs and can preserve many desirable properties of the basic graphs, such as regularity, vertex-transitivity, Hamiltonicity, etc. The study of product networks is interesting in the sense that many parameters of the network, such as node degree, diameter, network size, bi-section width, chromatic number [12, 13], domination number [10], fault diameter [16,3], edge disjoint spanning trees [14] can be easily calculated from the same parameters in their underlying basic graphs. This close relationship between the properties of the product graph and its basic graphs helps us to extend and generalize the same graph algorithms and problem solutions which were put forth for the basic graphs for their products as well.

In this paper, assuming intruder capturing algorithms for some graphs, we try to derive algorithm for intruder capturing in a graph product of the former graphs, accordingly. This work can be considered as an effort in generalizing the existing algorithms for more complex network topologies; in a way that the algorithm for the new structure (graph product) is performed in some high-level macro steps each consisting of some basic steps of lower level algorithms.

The rest of this paper is organized as follows. In section 2, we define the terminology and definitions that we will use later throughout the paper. In section 3, we probe into the problem of constructing spanning trees for graph products where we propose a novel method for obtaining a spanning tree in a product graph using a new product operator over graphs. Next, we introduce a spanning tree presentation for each intruder capturing algorithm using the notion of spanning search trees in section 4. Section 5 chiefly deals with our main problem, i.e. capturing an intruder in product networks and it is where we propose our two algorithms namely, *dimension order* and *dimension priority* for capturing an intruder in product networks. Finally, we conclude the paper in section 6.

2 Terminology and Definitions

The Cartesian product of two graphs G_1 and G_2, denoted by $G_1 \times G_2$, is a graph with vertex set $V(G_1) \times V(G_2)$. An edge joins a vertex $x = (x_1, x_2)$ to another $y = (y_1, y_2)$ ($x_j, y_j \in V(G_j)$, $j = 1, 2$) if and only if either $x_1 = y_1$ and $(x_2, y_2) \in E(G_2)$ or $x_2 = y_2$ and $(x_1, y_1) \in E(G_1)$. Hereafter, for the sake of simplicity, we refer to G_i graphs, $i = 1, 2$, as basic graphs and to any Cartesian product of some G_i's as a product graph.

Let $A = \{a_1, \ldots, a_k\}$ be the set of searcher agents employed to neutralize the network G. It is quite obvious that the total number of agents, denoted by $|A|$, is not constant over the run-time of the algorithm because we presumed that the searcher agents are created on demand. Throughout this paper we often refer to a "searcher agent" as "cleaner agent" or simply as "agent". Based on the premise that the intruder can only hide in the nodes, at any given instant an agent can only move to a neighboring node. The agents move in a way that the intruder cannot enter a node which has already neutralized, i.e. we are dealing with a contiguous node search problem. Let $C(t) \in V(G)$ denote the set of clean nodes at an instant t, then we can easily conclude that C is growing over the time; i.e. $\forall k > 0$, $C(t) \subseteq C(t+k)$.

Let us build a graph $H \subseteq G$ whose vertex set is C and each two c_i and c_j vertices of C are connected if and only if $<c_i, c_j> \in E(G)$. As the clean nodes do not get contaminated again and the agents can only move to their neighboring nodes, we may simply infer that H is a connected graph at any instant t. We call H the *clean territory* hereafter. $I = G \backslash H$ is the part of G that has not been cleaned yet, hence we call it *intruder territory*. It is also obvious that for some $v \in H$, v is incident to some vertex in I. Let $\partial(H)$ be the entire set of such vertices, we call this set the *battle front* of H. The battle front should always be occupied by the agents or the intruder will be able to expand its territory. We assume that each agent can only sense whether or not its neighboring nodes are clean and performs its actions according to this limited information provided about its neighboring nodes. A searcher agent in our model at any discrete time instant is capable of doing one the following two actions:

- Move: It can move to any of its neighboring nodes in order to extend the clean territory.
- Replicate: It can produce a new cleaner agent and inject it into a neighboring node.

Definition 1. Given a network G and a cleaner agent a in vertex v of G, the set of actions that a performs during the run-time of an algorithm while it is in v is called a *step* of the algorithm for agent a. It is easily seen that each step of the intrusion capturing algorithm for a, consists of some replication actions and a single optional movement. Let us denote by $r_a(v, u)$, the action corresponding to the replication of a new agent by agent a in v to a neighboring vertex u and by $m_a(v,u)$ the movement of a from v to a neighboring vertex u. Then, one *step* of algorithm for agent a in v can be represented by the set $st_G(a,v)=\{r(v,u_1),...,r(v,u_m),m(v,w)\}$, where $u_1,...,u_m$ and w are some arbitrary neighbors of v and $w \neq u_i$, $1 \leq i \leq m$.

The way that the deployment of the cleaner agents change in the network as dictated by the algorithm, is crucial to the validity and performance of the algorithm. Given a graph G, the way that the agents are distributed in G at an intermediate step of the algorithm is named the *configuration* of the agents in G. Each such configuration can be displayed by a binary configuration vector C of length $|V(G)|$. An element C_i of the vector is 1 iff v_i is occupied by an agent. By extending this notation to the case of product graphs, a binary $|V(G_1)| \times ... \times |V(G_n)|$ configuration matrix results.

3   Constructing Spanning Trees in a Product Network

The problem of constructing a spanning tree in a graph product has been studied in the literature. In [14], an algorithm for deriving edge disjoint spanning trees in graph products has been envisioned.

In this section, we propose a novel approach for constructing a spanning tree for a graph product based on the assumed spanning trees in its basic member graphs. We later use this approach of building spanning tree for generalizing our intruder capturing algorithms for product networks.

Definition 2. Given two graphs G and H, the *injective product* of G and H on the vertex $v \in V(G)$ denoted by $G \underset{v}{\times} H$, is a graph defined by

$$V(G \underset{v}{\times} H) = V(G \times H),$$

$$E(G \underset{v}{\times} H) = \{(u_1,v_1),(u_2,v_2) \mid [(u_1,u_2) \in E(G) \wedge (v_1 = v_2)] \vee [(v_1,v_2) \in E(H) \wedge (u_1 = u_2 = v)]\}$$

Clearly, $G \underset{v}{\times} H$ is a subgraph of $G \times H$; yet, the definition of $G \underset{v}{\times} H$ demands that two vertices in two different subgraphs homogeneous to G can be connected if and only if the first element in their vertex pair is a special vertex v of G. Hence, $E(G \underset{v}{\times} H)$ can be rewritten as $E(G \underset{v}{\times} H) = \{(u_1,v_1),(u_2,v_2) \in E(G \times H) \mid u_1 = u_2 \Rightarrow u_1 = u_2 = v\}$. As an example, the injective product of two cycle graphs of length 4 on an arbitrary vertex v, $C_4 \underset{v}{\times} C_4$, is depicted in figure 1.

Theorem 1. [11] Given two graphs G and H and a vertex v of G, the following properties are observable for $G \underset{v}{\times} H$:

  I.   If G and H are connected then $G \underset{v}{\times} H$ is connected for any given v in $V(G)$.

 II. $\left| E(G \underset{v}{\times} H) \right| = \left| E(G) \right|.\left| V(H) \right| + \left| E(H) \right|$.

III. $G \underset{v}{\times} H$ consists of one H component and $|V(H)|$ separate G components.

IV.   Each vertex u of $G \underset{v}{\times} H$ is contained in some subgraph homogenous to G, yet u may or may not be contained in a graph homogenous to H.

The definition of injective product can be further extended so that the graph product includes more than one H component.

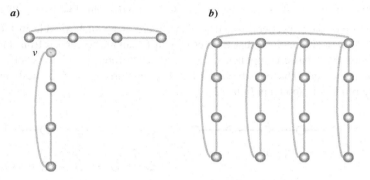

Fig. 1. The injective product of two cycles on v. *a*) The basic C_4 graphs. *b*) $C_4 \times_v C_4$.

Definition 3. Given two graphs G and H, the *injective product* of G and H on the vertex set $s \subseteq V(G)$, $G \underset{s}{\times} H$ is a graph defined as

$$V(G \underset{s}{\times} H) = V(G \times H),$$

$$E(G \underset{s}{\times} H) = \{(u_1, v_1), (u_2, v_2) \mid [(u_1, u_2) \in E(G) \wedge (v_1 = v_2)] \vee [(v_1, v_2) \in E(H) \wedge (u_1 = u_2) \wedge (u_1, u_2 \in s)]\}$$

Clearly, if $s = V(G)$, we can conclude $G \underset{s}{\times} H = G \times H$.

Theorem 2. Given a product graph $G = G_1 \times G_2 \times \cdots \times G_n$, let $ST_1, \ldots, ST_n$ and $r_1, r_2, \ldots, r_n$ be the spanning trees created on $G_1, \ldots, G_n$ and roots of the spanning trees, respectively. The spanning tree built on G denoted by ST is a tree with the root $r = (r_1, r_2, \ldots, r_n)$ and it can be constructed as $ST = ST_1 \underset{r_1}{\times} ST_2 \underset{r_2}{\times} \cdots \underset{r_{n-1}}{\times} ST_n$.

Proof: As ST_i's are connected, using the property I in theorem 1 and induction, we can state that ST is connected as well. On the other hand, as ST_i has the same number of vertices as G_i, by definition, the injective product of them would have $\prod_{i=1}^{n} |V(G_i)|$ vertices, which is equal to the number of vertices in G. To prove that ST is a spanning tree for G, it is enough to prove that there cannot be a cycle in ST, i.e. it is a tree. Let us assume that there exists some cycle C in ST. Two different cases are possible for the vertices of C. Either, they have a fixed vertex v in some i^{th} element of their vertex, i.e. they all belong to the same ST_i component, or the vertices of C are chosen from at least two different components ST_i and ST_j. The former case cannot happen since we assumed that ST_i's are trees and thus they cannot contain any cycles. In the latter case, we assume that the vertices in C belong to exactly two ST_i and ST_j components. Then, there should be at least two vertices in C such that at least two of the vertices in their n vertex set does not match, i.e. they belong to different ST_x basic components, $1 \leq x \leq n$. Let $u = <u_1, u_2, \ldots u_i, \ldots u_j, \ldots u_n>$ and $v = <u_1, u_2, \ldots u'_i, \ldots u'_j, \ldots u_n>$ be two such vertices which are contained in ST_i and ST_j components, respectively. As u and v are both in the cycle, there should be exactly two discriminated paths between them. It is while according to the definition of the injective product each such a path should pass from a node $<u_1,u_2,...,r_i,...r'_j,...u_n>$ which means that there cannot exists more than one paths between u and v. Thus, there is no such a cycle in ST. In the same way, we can conclude that the vertices in C cannot be chosen from more than two different ST_x basic components, $1\leq x \leq n$. Therefore, ST has no cycles and the theorem is proved. The construction of a spanning tree in a 5×5 mesh using the injective product is shown in figure 2.

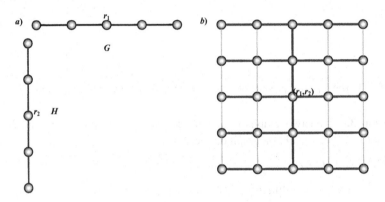

Fig. 2. Embedding an spanning tree in *a)* Basic graphs *b)* Graph product, Mesh 5×5

Definition 4. A given vertex v in ST is called an m-intersection of ST if it has neighboring vertices in m different G_i basic graphs, $1 \leq i \leq n$. It is obvious that most of the vertices in ST are 1- intersection nodes while the root node $r=(r_1,r_2,...,r_n)$ along with the vertices in the subgraph homogeneous to G_n are n- intersections.

4   Spanning Tree Presentation of the Algorithm

In this section, we envision a new presentation for the intruder capturing algorithm using the spanning trees. As we shall see later, this novel presentation will change our look to the problem, from an algorithmic viewpoint to a more tangible graph theoretical standpoint. Let A be a known contiguous algorithm for capturing an intruder in a graph G. As we saw before, A consists of some steps for each agent in some vertices of G. As we have the goal of cleaning the entire graph and knowing that the only way for a cleaner to capture an intruder is to be in the same vertex as the intruder, we can easily state that for each given vertex v of G, there exists some agent a such that a does some action in v during the run time of the algorithm. This is true since the cleaner should sweep the entire nodes in the network. Moreover, as A is contiguous, once a node is cleaned it need not to be rechecked. On the other hand, the path that each agent takes in the graph is a connected subgraph of G. From all the arguments stated above, we can conclude that the paths through which the agents move and replicate build a spanning tree for G.

Definition 5. Given a graph G and an algorithm A for capturing an intruder in G, *spanning search tree* for A denoted by SST is a spanning tree constructed from A in the following way:

- The root of SST is the home vertex in G.
- If during the runtime of the algorithm, some agent a performs a movement action $m_a(u,v)$, then we draw an edge between u and v and label it m.
- If during the runtime of the algorithm, some agent a performs a replication action $r_a(u,v)$, then we draw an edge between u and v and label it r.
- Each edge of the tree is also assigned a number which indicates the order of performing that particular action with respect to the other actions; e.g. an action which is performed first is labeled 1 and so on. Two edges with the same number correspond to concurrent steps of the algorithm.

  Numbers on the edges define the way of searching the spanning tree, with the exception that some edges may be traversed in parallel. Whenever the actions which are defined for an agent in each vertex v of SST are performed consecutively, we can simply assign the numbers to the vertices instead of labeling the outgoing edges with consecutive numbers. It is easily seen that for any given vertex v of the tree, only a single edge with label m can exist; i.e. all of the other exiting edges should be labeled r. This is true since each vertex in the battle front is occupied by a single agent and the additional required agents for the next stage of the algorithm should be created with replication operations corresponding to the edges which are labeled r. Figure 3 shows a spanning search tree corresponding to an algorithm for capturing an intruder in a 5×5 mesh.

5   Capturing an Intruder in Product Networks

In this section, we propose an algorithm for capturing an intruder in a product network based on the known algorithms for capturing an intruder in the basic member graphs of the graph product. Given $P=G_1 \times G_2 \times \ldots \times G_k$ as the graph product, we are interested in finding an algorithm A for capturing an intruder in P. We assume that there are known algorithms for capturing an intruder in G_i graphs. In particular, we

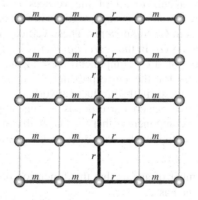

Fig. 3. Spanning search tree for $M_{5\times5}$

denote each such an algorithm for G_i by A_i and the corresponding spanning search tree by SST_i. We also presume that each A_i algorithm demands $N_a(G_i)$ cleaner agents and it can decontaminate G_i in $N_s(G_i)$ steps. In an attempt to obtain simple yet more scalable capturing algorithms, we try to utilize the inherent recursive structure of the graph products in as much as possible. That is, in order to obtain an algorithm for capturing an intruder in a graph product, we decompose the product graph into a set of subgraphs homogenous to a basic member graph G_i and perform the intruder capturing algorithm in each of those subgraphs independently as if it is the only graph we are dealing with. Hereafter, we call any such subgraph homogeneous to G_i, a G_i component of P.

Theorem 3. [11] Let $P=G_1{\times}G_2{\times}...{\times}G_k$ and v be a vertex in P, then v is contained in exactly k different subgraphs $G_1',G_2',...,G_k'$ of P homogeneous to $G_1,G_2,...,G_k$, respectively.

5.1 Dimension Order Algorithm

The first algorithm which we propose for capturing an intruder in the product network, replicate agents so that the root node of the spanning trees created over all of the G_1 components of the graph product be occupied by agents. These agents in turn will initiate the intruder capturing algorithm in their corresponding G_1 components. Cleaning all of the G_1 components of P then it would be tantamount to decontaminating the entire network. In particular, the dimension order algorithm consists of two phases: 1) Replication phase, and 2) Searching phase.

The *replication phase* starts with a single agent in the root of the constructed spanning tree on P. This agent replicates new agents and injects them to its neighboring n-intersections which in turn replicate new agents to their neighboring n-intersections until all the vertices in the one, and only G_n component of P be occupied by agents. In the next phase, the agents in the n-intersections which are the roots of G_{n-1} components, replicate new agents to their neighboring $n-1$-intersections which in turn replicate new agents to lower levels of the spanning trees. This process continues until all the vertices in all of the G_{n-1} components are occupied by the agents. These vertices are the root vertices of the spanning trees built on G_{n-2} components of P and hence can replicate new agents to all of the vertices in G_{n-2} components of P. Proceeding with this approach at some stage, vertices in G_3 components replicate new agents to all of the vertices in G_2 components. These vertices are the root nodes of the spanning trees constructed on G_1. In the *searching phase*, having one agent in the root node of each spanning tree on a G_1 component, these agents would run the intruder capturing algorithm on G_1 for the corresponding G_1 component and this would decontaminate the whole vertices in the product network.

The spanning search tree which corresponds to this algorithm is obtained by labeling all of the edges in components other than G_1 with r and labeling the spanning trees for each of the G_1 components according to the assumed intruder capturing algorithm on G_1.

Theorem 4. [11] The number of agents that the dimension order algorithm demands, denoted by $N_a(P)$, is obtained as

$$N_a(P) = N_a(G_1) \prod_{i=2}^{k} |V(G_i)| \cdot$$

Theorem 5. The number of steps it takes for the dimension order algorithm to complete on a product graph P, denoted by $N_s(P)$, is obtained as

$$N_s(P) = \sum_{i=1}^{k} N_s(G_i)$$

where $N_s(G_i)$ denotes the number of steps required for A_i to complete on G_i.

Proof: To obtain the total number of steps, we sum the number of steps which the algorithm takes for each phase. In the replication phase, it takes $N_s(G_i)$ steps for an $i+1$-intersection node to replicate the agents to i-intersections which are included in G_i components. Hence, the total number of steps of the replication phase is $\sum_{i=2}^{k} N_s(G_i)$. In the searching phase, all the agents can search their corresponding G_1 component in parallel and in $N_s(G_i)$ steps. The total number of steps of the algorithm is calculated as the sum of the number of steps for these two phases which is equal to the proposed equation.

5.2 Dimension Priority Algorithm

Although the dimension order algorithm is quite simple, the number of steps that is needed for the algorithm to complete might not be minimal. Our second algorithm, for each cleaner agent in a k'-intersection, vertex v of H performs the intruder capturing algorithm in k' different components of P which encompass v. As these k' subgraphs include all neighboring nodes of the agent in SST, this technique assures that no node of the network is missed by the algorithms. The only issue that may arise is that each of the k' different algorithms are acting on the agent simultaneously. Hence, we should guarantee that the actions that an agent performs according to an algorithm A_i does not conflict the actions that some other algorithm A_j advises for that agent.

As we discussed before, in our scenario we assume that the cleaner agents are capable of performing two kinds of actions, i.e. moving to a neighboring node or replicating a new cleaner agent and injecting it into a neighboring node. Obviously, each of the k' algorithms can do the replication in its own subgraph without having to be worried about the other subgraphs, as it only changes the configuration of the nodes which are exclusively in that subgraph. Hence, this does not make any conflicts between the k' different algorithms. Moving the agents, on the other hand, changes the configuration of the nodes which are mutually shared by some of the k' subgraphs. Thus, it changes the configuration of other subgraphs as well and this may result in a conflict. In particular, the problem which arises in the movement action is that if the algorithm for subgraph G_i moves the agent in v, there would not be any agents available in v for the other algorithms if they were to move the agent. In order to cope with this issue, we demand each agent to move in just one of its subgraphs. This can be safely done by letting one algorithm A_i out of k' algorithms run as ordinarily while replacing each movement action in all of the remaining $k'-1$ algorithms by an equivalent replication action. We call the m edge which is not affected by the replacements, the *free edge*.

Theorem 6. Given an agent a, let $sp_{G_i}(a,v) = \{r(v,u_{i,1}), r(v,u_{i,2}),...,r(v,u_{i,m_i}),m(v,w_i)\}$ be one step of the intruder capturing algorithm for G_i in an arbitrary k'-intersection v, $1 \le i \le k'$. We define $\widehat{sp}_{G_i}(a,v)$ to be the same step as sp_{G_i} but with the exception that each movement action $m(v,w_i)$ of a in $sp_{G_i}(a,v)$ is replaced by an equivalent replication action $r(v,w_i)$; Hence, $\widehat{sp}_{G_i}(a,v)$ can be derived as $\widehat{sp}_{G_i}(a,v) = \{r(v,u_{i,1}), r(v,u_{i,2}),...,r(v,u_{i,m_i}),r(v,w_i)\}$. One step of the algorithm for capturing an intruder in P, the graph product of G_is, for a in v is then obtained as

$$sp_p(a,v) = \left(\bigcup_{i=1,i\ne j}^{k'} \widehat{sp}_{G_i}(a,v) \right) \cup sp_{G_j}$$

for some arbitrary j such that $sp_{G_j}(a,v)$ has a movement action.

$sp_p(a,v)$ can equivalently be written as

$$sp_p(a,v) = \{r(v,u_{1,1}),...,r(v,u_{1,m_1}),r(v,w_1),...,r(v,u_{j,1}),...,r(v,u_{j,m_j}),m(v,w_j),...,r(v,u_{k',1}),...,r(v,u_{k',m_{k'}}),r(v,w_{k'})\}.$$

Proof: It is easy to see that the replacement of the movement actions with replication actions can not endanger the validity of the algorithms. Hence, after the replacement of the moves any algorithm A_i would still run as desired on any G_i graph, but perhaps it demands more agents. To see that the algorithms can run in parallel, it is to see that the equation for $sp_p(a,v)$ step of the algorithm consists of $k'-1$ steps without movements and a single step sp_{G_j} which could have at most one movement. Hence, each such $sp_p(a,v)$ step consists of at most one movement actions. As $sp_p(a,v)$ was chosen arbitrarily, we can conclude that the k algorithms can run in parallel.

The only problem which is still remained unsolved is in which order the algorithm should move/replicate the agents. In order to avoid any gap in the battle front, our algorithm lets the vertices in higher dimensions have priority over those in the lower dimensions. In particular, an i-intersection has priority in performing its actions over the $i-1$-intersection nodes and so on. Between the vertices which have the same priority in their dimensions, say i-intersections, a node which is further from the root node of SP_i has the highest priority. If this also comes to be equal between some nodes, the distance from the root node in the next lower dimensions are taken into account. If after these two rules, some nodes have the same priority, then these nodes can safely performs their actions in parallel. In more specific level, for each vertex v, the replication actions always have priority over the movement actions. The details of the algorithm for capturing an intruder in a product network are as follows.

Algorithm Dimension-Priority-Intruder-Capturing--DPIC()

a) Starting from the home node h in P where the one and only agent a initially occupies, perform step $sp_p(a,h)$ of the algorithm for capturing an intruder as described in theorem 6. This can be done by running steps $sp_{G_j}(a,h)$ of the algorithms in each of the k' subgraphs of P that contain h.

b) Let $\partial(H)$ be the battle front for P. At any intermediate stage of the algorithm, for each node v in $\partial(H)$ and agent a in v, perform step $sp_P(a,v)$ of the algorithm for capturing an intruder. If $sp_P(a,v)=\varnothing$ that is a does not have any actions to do (e.g v is leaf node), proceed with another node of $\partial(H)$.

c) Repeat step b until $\forall v \in \partial(H)$, we obtain $sp_P(a,v)=\varnothing$.

Corollary 1. Given a graph G, let A and SST be an algorithm for capturing an intruder in G and the corresponding spanning search tree, respectively. The total number of agents required by A denoted by $N_a(G)$ can be obtained as $N_a(G)=N_r(G)+1$, where $N_r(G)$ is the number of edges labeled r in SST.

Theorem 7. [11] The number of agents required by the algorithm for capturing an intruder in the product network can be obtained as

$$N_a(P) = N_r(G_1)\prod_{i=2}^{k}\left|V(G_i)\right| + \sum_{j=2}^{k}\left(N_r(G_j)+N_m(G_j)\right)\prod_{i=j+1}^{k}\left|V(G_i)\right|.$$

where $N_m(G_i)$ denotes the number of edges labeled m in SST_i.

Theorem 8. The number of actions that is required to be preformed by the agents according to A in order to clean the entire graph product is obtained as

$$N_{m+r}(P) = \prod_{i=1}^{k}\left|V(G_i)\right|-1.$$

Proof: The total number of the actions that is demanded by the A algorithm for decontaminating the network equals the number of edges in SST. As SST is a spanning tree, it consists of all vertices in the network and as it is a tree the number of edges is simply obtained as the number of nodes in the network minus one which is the proposed equation.

6 Conclusions

In this paper, we investigated the problem of capturing an intruder in a product network based on the assumed existing algorithms for intruder capturing in its basic member graphs. We proposed a novel approach for deriving a spanning tree in a product network using the newly defined injective product operation which is put forth as a general case of the well-known Cartesian product. Later, we modeled each intrusion capturing algorithm with a new tree structure, namely the spanning search tree. Finally, we reduced the problem of algorithms for capturing an intruder in a product network to finding the spanning search tree for the graph product using the existing spanning search trees for intruder capturing in its basic member graphs. We proposed two different algorithms based on the discussed method. The proposed algorithms and ideas presented in this paper are quite general and can be applied to intruder capturing in any product network, while the notion of the spanning search tree is considered as a simple yet general solution to the problem of intrusion capturing in any general network.

References

1. L. Barri`ere, P. Flocchini, P. Fraignaud and N. Santoro, *"Capture of an intruder by mobile agents"*, *Proc. 14-th ACM Symposium on Parallel Algorithms and Architectures (SPAA)*, Winnipeg, Manitoba, Canada, 200-209, 2002.
2. L. Barri`ere, P. Fraignaud, N. Santoro and D.M. Thilikos, *"Searching is not jumping"*, *Proc. 29th Workshop on Graph Theoretic Concepts in Computer Science (WG)*, Springer Verlag, LNCS 2880, 34-45, 2003.
3. K. Day, A. Al-Ayyoub, *"Minimal Fault Diameter of Highly resilient Product Networks"*, *IEEE Trans. On Parallel and Distributed Systems*, Vol. 11. No. 9. September 2000.
4. M. Demirbas, A. Arora, and M. Gouda, *"A pursuer-evader game for sensor networks"*, In *Proceedings of the Sixth Symposium on Self- Stabilizing Systems*, 2003, San Francisco, CA, USA, June 24-25, LNCS 2704, pp 1-16.
5. V.V. Dimakopoulos and N.J. Dimopolulos, *"A Theory for Total Exchange in Multidimensional Interconnection Networks,"* *IEEE Trans. Parallel and Distributed Systems*, vol. 9, no.7, pp. 639-649, 1998.
6. A. Ferna´ndez and K. Efe, *"Generalized Algorithm for Parallel Sorting on Product Networks,"* *IEEE Trans. Parallel and Distributed Systems*, vol. 8, no. 12, pp. 1211-1225, 1997.
7. A. Ferna´ndez and K. Efe, *"Efficient VLSI Layouts for Homogeneous Product Networks,"* *IEEE Trans. Computers*, vol. 46, no. 10, pp. 1070-1082, 1997.
8. P. Flocchini, M. J. Huang and F. L. Luccio, *"Contiguous search in the hypercube for capturing an intruder"*, *Proc. 19th IEEE International Parallel and Distributed Processing Symposium (IPDPS)*, Denver, Colorado 2005.
9. F. Harary, *"On the Group of the Composition of Two Graphs,"* *Duke Math. J.*, vol. 26, 1959.
10. M. A. Henning, R. Rall, *"On the Total Domination Number of Cartesian Products of Graphs"*, *Graphs and Combinatorics* Vol. 21, p. 63–69, 2005.
11. N. Imani, H. Sarbazi-Azad, A. Y. Zomaya, "Intrusion capturing in product networks", Technical Report TR-3-2006, IPM, School of Computer Science, Tehran, Iran March 2006.
12. S. Klavzar , H. Yeh, *"On the fractional chromatic number, the chromatic number, and graph products"*, *Discrete Mathematics*, v.247, n.1-3, p. 235-242, 28 March 2002.
13. S. Klavzar, *"Coloring graph products -- A survey"*, *Discrete Mathematics*, Vol. 155, Number 1, pp. 135-145(11), 1 August 1996.
14. S. Ku, B. Wang, T. Hung, *"Constructing Edge-Disjoint Spanning Trees in Product Networks"* *IEEE Transaction on Parallel And Distributed Systems*, Vol. 14, No. 3, March 2003.
15. S. Peng, M. Ko, C. Ho, T. Hsu, and C. Tang, *"Graph searching on Chordal graphs"*, *Algorithmica*, 27: 395-426, 2000.
16. Min Xu, Jun-Ming Xu, Xin-Min Hou, *"Fault diameter of Cartesian product graphs"*, *Information Processing Letters* 93 ,245–248 , 2005.
17. P. Yospanya, B. Laekhanukit, D. Nanongkai, and J. Fakcharoenphol, *"Detecting and cleaning intruders in sensor networks"*, *Proceedings of the National Comp. Sci. and Eng. Conf. (NCSEC'04)*.

Efficient In-Network Evaluation of Multiple Queries

Vinayaka Pandit[1] and Hui-bo Ji[2]

[1] IBM India Research Laboratory
pvinayak@in.ibm.com
[2] Australian National University
hui-bo.ji@anu.edu.au

Abstract. Recently, applications in which relational data is generated in a distributed and streaming manner have emerged from diverse domains. Processing queries on such data has become very important. In-network evaluation of a query is a technique in which the query is evaluated in the network without transferring all the data to a central location. So far, algorithms for in-network of evaluation of a single query have been proposed. They are not designed to exploit common computations across multiple queries. There is a need to develop techniques for efficient in-network evaluation of multiple queries. We consider the problem of in-network evaluation of multiple queries on relational data generated on a distributed network of machines. We present a novel algorithm based on an algorithm for dynamic regrouping of queries.

1 Introduction

We are witnessing the emergence of a new class of applications in which the data is generated by data sources distributed over a large distributed network (often of the scale of WANs) [2]. Applications querying such data have become popular in diverse domains such as sensor networks, publish-subscribe systems, and financial services [16, 7, 1]. There are many applications in which the generated data is in relational form [11]. In this paper, we consider continuous versions of SQL queries over such distributed, relational, streaming data.

Platforms which expose a simple programming model to build applications which work on distributed, streaming, relational data are becoming popular [7, 14]. The user writes continuous version of an SQL query. The platform compiles the query into an equivalent execution tree. Each node of the execution tree represents a logical relational operator. In order to carry out computation, it can also communicate with its parents and children which could be executing at different locations of the network. Such an execution tree is called as *operator tree*. Note that placing the entire operator tree at one location is a valid way of evaluating the query which transfers the entire data to that location. *In-network evaluation* aims to efficiently evaluate the output of the operator tree by placing the nodes of the operator tree at appropriate locations in the network so as to minimize communication.

Y. Robert et al. (Eds.): HiPC 2006, LNCS 4297, pp. 205–216, 2006.
© Springer-Verlag Berlin Heidelberg 2006

Note that, multiple queries can be represented by combining the individual operator trees corresponding to individual queries as one *global operator tree*. Such a representation of multiple queries does not exploit the possibility of sharing common computations. In this paper, we consider the problem of evaluating multiple queries on a distributed network of machines by sharing common computations at appropriate locations in the network. We observe that, in order to share computation, the global operator tree has to be reorganized into an equivalent tree in which common sub-expressions are exploited. The main difference between our problem and the problem considered by recent papers on in-network evaluation of a single query is that they do not consider the possibility of reorganizing the global operator tree to achieve efficiency.

Ahmad and Cetintemel [1] consider the problem of in-network evaluation of a single operator tree on a distributed network. They consider the problem of minimizing the end-to-end delay. Furthermore, they assume that all the output tuples are delivered at a single location called the *proxy*. In comparison, we do not impose the restriction of accessing the output of all the queries from a single location. They propose a heuristic which traverses the operator tree in the post-order and at each step, places the currently visited node at one of the following locations: (i) one of its children's locations, (ii) the proxy location, and (iii) the node and the subtree below it are placed at a common location. The heuristic picks one of the three choices greedily. To focus on the communication aspect of the problem, they assume the processing cost of each node to be zero. Srivastava et.al [16] consider the same problem as Ahmad and Cetintemel in which the processing costs of the nodes can be non-zero. They consider the special case when the machines form a tree. They give an approximation algorithm for the problem based on dynamic programming. They view each operator as a filter and introduce the novelty of modeling network links as special filters. Each step of the dynamic programming involves solving a pipelined set-cover problem [9].

Pandit et.al [10] consider different efficiency measures for in-network evaluation of an operator tree. They formulate the end-to-end delay minimization as a a facility location problem and propose (and evaluate) a local search heuristic. They also consider the problem of computing load balanced placement in which the total communication across the network is minimized. They propose an algorithm based on minimum cost multi-way balanced partitioning of a graph.

Although in-network evaluation of multiple queries has not been studied before, multi-query optimization is a well studied problem [13, 12, 15] in database literature. Early work on multi-query optimization [12, 15] focused on finding the optimal query plan for a very small number of queries using exhaustive search techniques. These approaches are very expensive for large number of queries typically expected in the domain of applications highlighted before. Roy [13] proposes important heuristics to reduce the cost of the exhaustive search algorithms and applies these techniques for materialized view selection and maintenance. All these algorithms were designed for centralized databases. They do not deal with the trade-offs introduced by the network delays in the optimization. Furthermore, applications like financial services would require in-network evaluation of a large number of queries, thus resulting in a very large search space. So, there is a need to develop new techniques to solve the problem considered in this paper.

We observe that the problem of in-network evaluation of multiple queries has to deal with reorganizing the global operator tree in order to share computations as well as place the transformed operator tree efficiently on the network of machines. These two aspects of the problem are closely inter-related. We develop a framework motivated by an algorithm for dynamically regrouping queries proposed by Chen and Dewitt [3]. Our algorithm is based on a bottom-up traversal of the operator tree. At each level, it considers the possibility of exploiting common sub-expressions depending on whether the resulting tree can be placed efficiently on the network of machines.

2   Problem Formulation

2.1   Motivation

Consider two queries $Q1 = (A \bowtie B) \bowtie C$ and $Q2 = B \bowtie C$ which are represented as a global operator tree as shown in Figure 2(a). The inputs A, B, and C are the streaming relations, also called as *base relations*. Each box computes the join of its two input relations and its output is treated as a streaming relation. The output of a box which does not represent any of the queries, for example, the output of the box computing $A \bowtie B$, is called as *intermediate relation*. Suppose the cost of $A \bowtie (B \bowtie C)$ and $(A \bowtie B) \bowtie C$ are equal, then, the optimal way of evaluating the two queries on a single machine is to evaluate $Q2$ as $B \bowtie C$ and to evaluate $Q1$ as $A \bowtie (B \bowtie C)$ as shown in Figure 2(b). Multi-query optimization techniques presented in [13] are designed to exploit such common sub-expressions to optimize the performance of a centralized database.

In case of distributed query processing, there are other considerations which influence the optimization. Firstly, a given base relation may be *available* only at the location where it is produced. If it is involved in a computation at a different location, it has to be transferred there. Similarly, the results of a query has to be *attached* to the location of the end-user. The cost of transferring results to the attached location has to be considered as well. In general, we consider the base relations and the final queries as *tied* to specific locations in the network. As observed in the previous papers on in-network query processing [1], the communication cost dominates the cost of local processing. We consider the following example to illustrate the trade-offs introduced by the distributed setting which is absent in centralized setting.

Figure 1 shows two scenarios of evaluating the queries $Q1 = (A \bowtie B) \bowtie C)$ and $Q2 = B \bowtie C$ on a network of two machines, M_1, and M_2. In both the scenarios, the set of base relations and final queries tied to each machines is specified as shown. Consider Figure 1(a). Any attempt to share the computation of $Q2$ results in increased communication cost and it is beneficial not to reorganize the global operator tree. Whereas, in Figure 1(b), it is beneficial to reorganize the global operator tree to share the computation of $Q2$. Thus, the optimizations carried out are not only dependent on the common sub-expressions in the queries, but also depend on the way the relations and views are attached in addition to the network delays in transferring intermediate results.

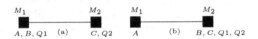

Fig. 1. Example Scenarios

2.2 Operator Tree and Topology Graph

The operator tree of a query is given by $Q = (J, R, I, O)$. R represents the set of base relations and O represents the different final views that the query computes. I represents the set of intermediate relations of Q. J is the set of operator nodes of the operator tree. Suppose $i_1 \in R \cup I$ and $i_2 \in R \cup I$ are the two inputs to a node $j \in J$, then the output of j is i_1 **Op** i_2 where **Op** is the relational operator corresponding to j. The incidence graph of the query is a tree. Suppose there are n user queries which have already been compiled into operator trees $Q_1, \ldots, Q_n$. Then, the main query Q is essentially the union of all the queries, $Q = \cup_{i=1}^{i=n} Q_i$. An equivalent query of Q is an operator tree $Q_F = (J_F, R, I_F, O)$ such that it computes the same set of final views as Q. The base relations in R are called as *producers* and the final views in O are called as *consumers*.

Chen et.al [4] showed that an effective heuristic for optimizing large number of continuous queries with similar join operator is to pull up the select operators over the join operators. As a first step towards efficient multi-query in-network evaluation we consider the problem of efficient evaluation of multiple queries consisting of only join operators [3], i.e, all the operator nodes in J are joins. It is a common practice in database literature to consider the operator tree to be *left-deep*. In our context, a tree is left-deep when at least one of the inputs of every node is a base relation. So, the inputs of a node at level i are a base relation, and the output of a node at level $(i-1)$. In our formulation, we assume $Q_1, \ldots Q_n$ to be left-deep while the main query Q itself may not be left-deep.

Suppose we represent a base relation by a unique symbol. Let us associate a string of symbols with the output of a join node as follows. If s_1 and s_2 represent the strings corresponding to the inputs of a join node $j \in J$, the output of j is denoted by $concat(s_1, s_2)$. Note that, in case of a left-deep tree, s_2 is a symbol corresponding to a base relation. So, a string of symbols can be used to unambiguously specify a left-deep tree. We shall use this way of conceptualizing an intermediate or a final view wherever it simplifies the exposition.

The distributed network of machines is given by $G = (V, E)$ where V is the set of machines and E is the set of communication links. Each link $e \in E$ has an attributed called *cost* which is an estimate of the delay across it. The cost of an edge between two nodes u, v is denoted by $c_{u,v}$ and by c_e when the edge in question is unambiguous. The producers and consumers are tied to the machines where they are generated and consumed respectively.

2.3   Cost Model

In traditional databases, statistics of the static tables are used for estimating cost. In our setting, we use the rate of arrival and rate of output as the main cost estimates. Suppose a node has two inputs with rates r_1, r_2 and a selectivity factor of s, then, the *processing cost* of the node is $r_1 \cdot r_2$ and its output rate is equal to $D(s \cdot r_1 \cdot r_2)$ where D is a function dependent on the scheduling policy at the node. Suppose a node $j_1 \in J$ with output rate r_1 is assigned to machine M_1 in G. Suppose the output of j_1 is input to a node $j_2 \in J$ which is assigned to machine M_2 in G. The *communication cost* of the data transfer between j_1 and j_2 is given by $r_1 \cdot l(M_1, M_2)$ where $l(M_1, M_2)$ denotes the length of the shortest path from M_1 to M_2. Given an assignment function $A : (J \in Q) \rightarrow (V \in G)$, the communication cost between nodes j_1 and j_2 is denoted by $CS_A(j_1, j_2)$.

Estimating the output rate of nodes is essential in order to compute costs. When the joins are *correlated*, i.e, the probability of two tuples joining at a node depends on the path they have taken, estimating the output rate is a hard problem [16]. Instead, we work with a simpler but, limited model. For every intermediate view in the global tree, we assume that its join selectivity with each of the base relations is given. The input has to specify $|R| \cdot |I|$ explicit selectivity factors corresponding to the I intermediate views. This *augmented selectivity information* is sufficient for the purpose of our algorithm.

2.4   Objective Function

In-network query evaluation is specified by the global operator tree $Q = (J, I, R, O)$ and the topology graph $G = (V, E)$. The goal is to compute a global operator tree $Q_F = (J_F, I, R_F, O)$ equivalent to Q and an assignment function $A : J_F \rightarrow V$. The cost of a solution (Q_F, A) is given by

$$C(Q_F, A) = \sum_{s \in R, t \in O} \sum_{(j_1, j_2) \in SP(s,t)} CS_A(j_1, j_2).$$

$SP(s, t)$ denotes the shortest path from s to t. If a pair $(s \in R, t \in O)$ is not connected, then the communication cost between them is assumed to be zero. We call this problem as *distributed, continuous, multi-query optimization* (DCMQOPT). The problems considered in [16,1,10] differ from our formulation as they assume that Q_F will be same as Q. In other words, they do not consider reorganizing the operator tree to save processing and communication costs.

3   Algorithm

Previous algorithms for in-network query evaluation did not consider the trade-offs discussed in Section 2.1. Chen and Dewitt [3] considered the problem of sharing computations across multiple dynamic queries. They showed the efficacy of their approach when the number of queries is large. As our algorithm is motivated by their ideas, we briefly summarize their work.

A query (or an intermediate view) x is said to be a *sub-query* of another query y if x is defined over a subset of producers over which y is defined. For example, BC is a sub-query of ABC. In other words, y *contains* x. In an operator tree Q, if the output of a join node j is input to a join node p, then, the output of the node p contains the output of j. The node j is called a *child* of the node p. The children of the same parent are called as *siblings*.

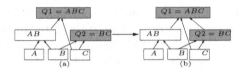

Fig. 2. Capturing sub-query relationships

Algorithm 1 briefly summarizes their approach. It is designed to exploit common sub-expressions in a top-down manner. The main component is the heuristic **Minimize_Graph**. It retains a minimum subset of views at level i which is sufficient to compute the set of views retained at level $i + 1$.

Algorithm 1. Chen and Dewitt Algorithm

Pass 1

1: **for** $i = 2$ to *num_level* **do**
2:     Reflect all the sub-query relations between the nodes at level i and level $i + 1$ as shown by the dotted edge from $Q2$ to $Q1$ in Figure 2.
3: **end for**

Pass 2

1: **for** $i = num\_level - 1$ to 2 **do**
2:     Use a heuristic called *Minimize_Graph* to retain a minimum subset of relations, say, I_i at level i such that every relation at level $i + 1$ has a sub-query in I_i.
3: **end for**

DCMQOPT poses a complex trade-off involving the processing cost and the communication cost. So, we consider an algorithmic approach which systematically decouples these two costs. Specifically, we are interested in traversing the operator tree in a specific order to identify common sub-expressions to save processing cost. At each stage, we share the identified common sub-expressions if only they can be computed in a communication-efficient manner. We first point out the difficulty in extending their algorithm in such a way.

Consider invoking a procedure to place the modified operator tree after step 2 in each iteration of Pass 2. The modified tree is accepted if the placement cost is lesser than before. But, when the set of views to be retained at level i is decided (so that all views at level $i + 1$ can be computed), the set of views to be retained at level $i - 1$ is not yet decided. So, there is uncertainty about the inputs (and hence, input rates) to the transforms at level i. So, estimating the processing cost and communication cost (See Section 2.3) of the transforms at level i is difficult. Thus, during a top-down traversal, it is difficult to estimate the impact of the decisions made at each step.

Alternatively, let us consider a bottom-up traversal of the operator tree. When a decision is made on the set of views to be retained at level i is made, the same decision has already been made regarding views at all the levels below i. Now, it is easy to observe that the input rates of the transforms at all the levels can be computed using the augmented selectivity information (See Section 2.3). Thus, we can traverse the tree in such a way that the impact of the decisions made at each step (regarding the shared sub-expressions) on the overall cost can be reasonably estimated. The details are presented in Algorithm 2.

Algorithm 2. Bottom-up Regrouping Placement(BRP)

1: $Q_C = Q$
2: **for** $i = 2$ to $num\_level - 1$ **do**
3:   $Cost_1 = $ Place_Single_Query(Q_C)
4:   $Q_N = Q_C$
5:   Add dotted edges between nodes in levels i and $i + 1$ of Q_N which capture sub-query relations (as shown in Figure 2).
6:   $F_i(Q_N) = F_i(Q_C)$
     {$F_i(Q)$ denotes the set of final views of Q at level i.}
7:   Let I_i^N be minimum cost subset of $I_i(Q_C)$ such that every node of Q_N at level $i + 1$ has a sub-query in $R_i(Q_N) = I_i^N \cup F_i(Q_N)$.
     {$R_i(Q)$ and $I_i(Q)$ are defined similar to $F_i(Q)$. In effect, the previous two steps are identifying common sub-expressions.}
8:   For every node of Q_N at level $i + 1$ add an edge to its cheapest sub-query in $R_i(Q_N)$.
9:   $Cost_2 = $ Place_Single_Query(Q_N)
10:   **if** $Cost_2 < Cost_1$ **then**
11:     $Q_C = Q_N$
12:   **end if**
13: **end for**

In step 5, we exploit the existing sub-query relations by creating new dotted edges which represent shareable sub-expressions. After this step, there may be many alternative ways of computing a view at level $i + 1$. However, the cost of taking each alternative can be estimated using the augmented selectivity information described in Section 2.3. In steps 7 through 9, we identify a set of beneficially shareable sub-expressions at level i.

Figure 3 shows the example considered in Section 2. For $Q = Q1 \cup Q2$, the two trees placed in steps 3 and step 9 are shown in the figure. If $(Cost_2 < Cost_1)$, then the tree is modified to share the computation for $Q2$. Otherwise, the operator tree is not modified.

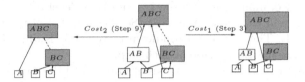

Fig. 3. Illustration of BRP Algorithm

Time Complexity. During an iteration, if there are x nodes at level i and y nodes at level $i + 1$, then, Step 5 takes $O(xy)$ time. Thus, over all levels, Step 5 takes $O(n^3)$ time where n is the number of nodes in Q. At Step 7, we select a subset of non-final nodes I_i at level i so that all the nodes at level $i + 1$ have a sub-query in $F_i \cup I_i$. The main trade-offs involved are (i) retaining a non-final node which is a sub-query of many nodes at $i + 1$ and (ii) retaining non-final nodes whose output rates are not too high. This can be formulated as a *set cover* problem which is NP-Complete [5]. The natural greedy heuristic is known to give the best approximation and we use it to compute I_i. Over all levels it runs in $O(n^2 \log n)$ time. At Steps 3 and 9, we use the placement algorithm proposed in [1]. Note that our framework is not tied to any particular placement and any of the algorithms proposed in [10,16] can also be used. Our algorithm runs in time $O(n^3 + n \cdot P(t))$ where $P(t)$ denotes the time taken by the placement module.

4   Experimental Evaluation

We now carry out empirical evaluation of our algorithm on a simulation framework developed by us. The framework can be used to generate operator trees which contain common sub-expressions so that they can be reorganized to share computation. It can be used to generate topology graphs with properties of large, distributed networks. We can tie streams and views to specific machines of the topology graph. Our algorithm is integrated with the framework. It also computes a normalized cost which is indicative of the average end-to-end delay.

On programming platforms like [7,14], it is possible to obtain real instances of queries written by different users which contain enough computational redundancy. One of the disadvantages of our framework is that it is currently not integrated with such a platform. So, it has to generate instances of global operator trees in which it is possible to reorganize the tree to share computation across multiple final views. We briefly discuss how to generate such instances.

An operator tree can be thought of as a set of paths from the streams to final views. Each edge in these paths satisfies the sub-query relationship between its end-points. Intuitively, common sub-expressions are those intermediate or final views through which many paths can pass through. Consider an operator tree $Q = (J, R, I, O)$. Let T_O be the tree defined by the data flow from the streams in R to the final views in O via intermediate views I. Let $n_O = |I|$. Let T_A be the augmented graph on the nodes in $R \cup I \cup O$ reflecting the data flow of the query and containing additional edges to reflect all the sub-query relations (refer to Section 3). Let $\mathcal{P}$ be a set of paths in T_A from the nodes in R to the set of final views in O such that they pass through as few intermediate nodes as possible. Let n_A be the number of intermediate nodes that paths in $\mathcal{P}$ pass through. An instance Q with lot of shareable computations satisfies the property that n_A is much smaller than n_O. We modify the algorithm by Melançon and Philippe [8] to generate random operator trees which contain sufficient computational redundancy.

We also generate random topologies for the network of machines on which the computed operator tree Q_F is placed. As mentioned in the introduction, the data sources are distributed over very large networks like WANs and the Internet. So, we are interested in generating random graphs which satisfies properties exhibited by such networks. We use an iterative algorithm similar to the Markov Chain simulation based generators proposed by Gkantsidis et. al [6] to generate the topology graphs with realistic edge costs for our experiments. It is possible to programmatically tie the streams and views of the operator tree to specific machines of the topology graph.

Our experimental set-up is as follows. We generate a random topology of machines and an operator tree which is known to be an instance that can be reorganized to share computations. The streams and the final views of the operator tree are tied randomly to different machines of the topology graph. We then compute placement by three algorithms: Random Placement(RP), Greedy Placement(GP) and our Bottom-up Regrouping Placement(BRP). In RP, every node of the operator tree is randomly assigned to one of the machines while the streams and final views are tied to their respective locations. In GP, the greedy placement algorithm proposed in [1] is used to compute the placement. BRP is our algorithm described in Section 3. A normalized integer score indicative of the end-to-end delay is computed for each placement. We present both experimental results and qualitative results. As the greedy heuristic was originally proposed for a single query with all the final views accessed from a single proxy, we discuss how we implement it for our purposes.

The main point to be emphasized about the algorithm in [1] is that it does not reorganize the operator tree. We modify the input so that it meets the input specification of their algorithm. In the operator tree, from every final view, we create a dummy operation whose output is marked as a final view. The original final views are now intermediate views tied to their respective machines. We add a new proxy machine into the topology graph. For an original final view v, let $m(v)$ denote the machine it is tied to. For every original final view v, we add an edge of cost zero from $m(v)$ to the proxy machine. This modified input meets the input specification of their algorithm.

Figure 4 shows a comparison of the three algorithms. The first column shows the details of the different operator trees considered for optimization. For each query Q, the number of nodes in the query, the number of base relations in the query, and the number of shareable common sub-expressions(CSE)s are mentioned. For each tree, the experimental results for topology graphs of different sizes are given in the second column. Specifically, the cost of the objective function under the assignment computed by the three algorithms are given.

Observations: The experimental results in Figure 4 indicate the efficacy of our approach. Let us emphasize some important observations. On a single machine, sharing computations is always beneficial. The results on Q_3, Q_4, Q_5 show that on a single machine, we do get improved performance. On queries Q_3, Q_4 which contain high degree of redundant computations, our algorithm computes very efficient placements. It shows that in realistic scenarios of evaluating large number of queries with highly redundant computations, our algorithm can improve the performance significantly. Currently, we manually count the number of maximum shareable sub-expressions. Although, we have tested our algorithm on larger instances, we have presented results on only those cases for which we could compute the maximum number of shareable sub-expressions.

In Section 2.1 we argued that, an optimal evaluation of multiple queries may chose not to share a sub-expression even though it can be shared. Qualitatively, it is important to verify that our algorithm does make such decisions on non-trivial

Operator Tree				Experimental Results			
Tree	#Nodes in Q	#BaseRels	#CSEs	#Nodes in G	RP	GP [1]	BRP
Q_1	11	5	1	5	50	43	35
				10	109	68	26
				20	103	98	54
Tree	#Nodes in Q	#BaseRels	#CSEs	#Nodes in G	RP	GP [1]	BRP
Q_2	16	6	3	5	102	68	64
				10	200	115	59
				20	226	197	138
				#Nodes in G	RP	GP [1]	BRP
				1	40	40	28
Tree	#Nodes in Q	#BaseRels	#CSEs	2	40	40	28
Q_3	26	7	3	5	280	213	78
				10	303	186	97
				20	745	439	259
				#Nodes in G	RP	GP [1]	BRP
				1	40	40	36
Tree	#Nodes in Q	#BaseRels	#CSEs	2	40	40	36
Q_4	31	10	8	5	283	164	109
				10	228	140	40
				20	995	521	278
				#Nodes in G	RP	GP [1]	BRP
				1	82	82	75
Tree	#Nodes in Q	#BaseRels	#CSEs	5	723	430	430
Q_5	41	16	4	10	1204	547	488
				20	4810	2803	2352

Fig. 4. Experimental evaluation of different algorithms

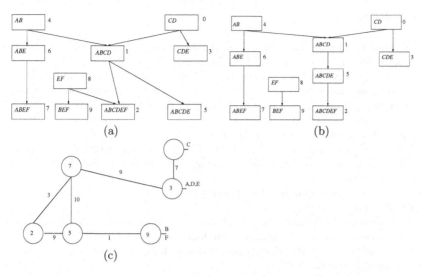

Fig. 5. An example where the algorithm chooses not to share computation based on the topology graph and how the streams and views are attached

examples. Figure 5(a) shows an example. The boxes represent transforms which compute views represented by the strings inside them. Views are indexed by the integers adjacent to the boxes. As shown in Figure 5(b), computation of the view $ABCDE$ can be reused to reduce the number of transforms by one. Figure 5(c) shows a network of six machines with edge costs and a specific attachment of streams and views of the example operator tree. In this case, any assignment which shares the computation of the view $ABCDE$ incurs a higher cost than an assignment which does not do so. Indeed, our algorithm chooses to keep the operator tree as it is. In this example, in addition to the augmented selectivity information, we input the selectivity of the join between AB and CD. On larger examples, our algorithm decides to share smaller number of sub-expressions than what is possible.

5   Future Work

We show that in-network evaluation of multiple queries involves dealing with issues which are not addressed by previous work on single query evaluation. We present a novel algorithm for in-network evaluation of multiple queries and provide empirical evidence of its efficacy. Our work can be extended in several directions. Our algorithm needs to be extended when other relational operators are present and the selectivities of operator nodes are correlated. It also has to be extended for the case when individual query trees can be bushy instead of being left-deep. In future, we intend to demonstrate the efficacy of our algorithm in practice by integrating it with distributed publish-subscribe systems like [7].

References

1. Y. Ahmad and U. Cetintemel. Network aware query processing for stream based applications. In *Proceedings of Very Large Data Bases (VLDB)*, 2004.
2. G. Banavar, T. Chandra, B. Mukherjee, J. Nagarajarao, R. Strom, and D. Sturman. An efficient multicast protocol for content based publish-subscribe systems. In *Proceedings of the international conference on distributed computing systems*, 1999.
3. J. Chen and D. Dewitt. Dynamic regrouping of continuous queries. Technical report, University of Wisconsin-Madison, 2002.
4. J. Chen, D. DeWitt, and J. Naughton. Design and evaluation of alternative selection placement strategies in optimizing continuous queries. In *ICDE*, pages 345–356, 2002.
5. U. Feige. A threshold of ln n for approximating set cover. *J. ACM*, 45(4):634–652, 1998.
6. C. Gkantsidis, M. Mihail, and E. Zegura. The markov chain simulation method for generating connected power law random graphs. In *ALENEX*, 2003.
7. Y. Jin and R. Strom. Relational subscription middleware for internet-scale publish-subscribe. In *DEBS*, 2003.
8. G. Melançon and F. Philippe. Generating connected acyclic digraphs uniformly at random. *Information Processing Letters*, 90(4):209–213, 2004.
9. K. Munagala, S. Basu, R. Motwani, and J. Widom. The pipelined set-cover problem. In *Proceedings of the International Conference on Database Theory*, 2005.
10. V. Pandit, R. Strom, G. Buttner, and R. Ginis. Performance modeling and placement of transforms for distributed stream processing. IBM Research Report, 2006.
11. B. Plale and K. Schwan. Dynamic querying of streaming data with the dquob system. *IEEE Transactions on Parallel and Distributed Databases*, 14(4), 2003.
12. A. Rosenthal and U. Chakravarthy. Anatomy of a modular multiple query optimizer. In *VLDB*, pages 230–239, 1988.
13. P. Roy. *Multiquery optimization and Applications*. PhD thesis, IIT Bombay, 2000.
14. R.Strom. Extending a content based publish-subscribe system with relational subscriptions. Technical report, IBM Research, 2003.
15. T. Sellis. Multiple query optimization. *ACM Transactions on database systems*, 10(3), 1986.
16. U. Srivastava, K. Munagala, and J. Widom. Operator placement for in-network stream query processing. In *Proceedings of ACM Symposium on Principles of Database Systems (PODS)*, 2005.

Island Model Parallel Genetic Algorithm for Optimization of Symmetric FRP Laminated Composites

Rahul[1], D. Chakraborty[1], and A. Dutta[2]

[1] Department of Mechanical Engineering, Indian Institute of Technology Guwahati,
Assam 781039 India
[2] Department of Civil Engineering, Indian Institute of Technology Guwahati,
Assam 781039 India
{rahul, chakra, adutta}@iitg.ernet.in

Abstract. This paper presents an approach to optimal design of composite structures using Island Model Parallel Genetic Algorithm (IMPGA) with a probabilistic migration strategy and in conjunction with 3D Finite Element Method (FEM). The subject problem is computationally intensive and consumes large amount of computer space-time; an attempt has been made to spawn a variable number of processes at each node of IMPGA for FEM analysis. Comparison shows that integrated IMPGA-FEM module outperforms SGA-FEM module with respect to convergence as well as computational time significantly. The present study observes that the speed-up obtained from IMPGA-FEM module is better than the theoretical speed-up. It has also been observed that the incorporation of a probabilistic migration strategy in the IMPGA lead to a much faster and improved converged solution. Results show that for optimization with IMPGA there exists a minimum size of sub-population on each processor below which the performance deteriorates.

Keywords: Island Model Parallel Genetic Algorithm, Optimization, FRP Composites.

1 Introduction

Though Genetic Algorithms (GAs) have demonstrated the potential to overcome many of the problems associated with gradient-based methods and are most effective when the design space is large, high computational time and storage requirement often forces us to work with a reduced design space. This some times limits the efficacy of GAs in achieving global optimal solutions. This dilemma could be solved by implementing GA in a parallel computing environment, where full advantage can be taken of the low communication requirements of GAs, and to use specialized models allowing sufficiently detailed representation without excessive computational requirements. Island model GA [1, 2] is ideally suited for parallel computing environment leading to IMPGA. Bryce Bockman [3] has mentioned two important algorithms for effective implementation of IMPGA. One is the Stepping Stone algorithm, which is the core of the migration routine, and another one is the

Y. Robert et al. (Eds.): HiPC 2006, LNCS 4297, pp. 217–228, 2006.
© Springer-Verlag Berlin Heidelberg 2006

development of precedence matrix to avoid any possible deadlock between the processes. Tanese [4] studied the advantages of distributed GA over conventional GA and the effect of migration strategies on the performance of distributed GA. It was observed that distributed GA consistently performed better than conventional GA. Mühlenbein et. al. [5] implemented the island model with rank-based selection to act as function optimizers and concluded that parallel search pays off only if the search space is large and complex. Norman [6] tested parallel GA using randomly communicating subpopulations with different migration strategies. Literature review reveals that parallel GA enjoys distinct advantages over SGA both in terms of reduced computational time as well as improved converged fitness. A preliminary attempt has been made by the present authors [7], which showed the efficacy of IMPGA in optimization of composite laminates and its superiority over SGA. However, detailed studies in terms efficiency, speedup, scalability, effectiveness, of different migration strategies, and combined effect of IMPGA and spawned set of processes, at each node of IMPGA, on computational time have not been reported.

By appropriate selection of stacking sequence and material of each lamina, the designer can impart directional strength in FRP laminates. This allows the designer to achieve a reduced weight/cost of the component while ensuring the required safety. Therefore, an important issue in the design of laminated FRP structures is the optimal selection of number of plies, fiber angle of each ply, material of each ply, thickness of each ply and thus the laminate thickness. From the manufacturing constraints, ply angle and ply thickness are to be elected from a set of discrete values and the optimum design of laminated FRP structure becomes a discrete optimization problem. GAs are non-deterministic, has the ability to work in a discrete search space and in the recent time have been successfully applied to the problem of composite design optimization [8, 9, 10, 11]. Qu et. al. [12] used deterministic and reliability based designs of composite laminates for cryogenic environments. Multi-objective optimization of hybrid composite laminates using SGA and FEM has also been reported for static and dynamic loading [13,14].

Since the optimization of laminated FRP composite structures involves large number of design variables and a complex search space, use of distributed genetic algorithm seems to be a promising tool for achieving faster and better convergence in search of optimal laminate. Therefore, in the present work, IMPGA has been implemented in a distributed memory platform for optimization of symmetric and balanced laminates in order to study its effectiveness by varying the number of processors as well as the population size in each processor. A 3D Finite Element Analysis (FEA) has been used for assessing the impact-induced failure [15, 16, 17] of the laminate along with the IMPGA. Also, each processor of IMPGA spawns number of processes to distribute the organism among spawned process for carrying out FEA on a sequential basis. Both deterministic as well as probabilistic migration strategies have been implemented in the parallel island model in order to study their influence in the converged fitness. Impact induced delamination and matrix cracking has been used as failure criteria for the optimization of laminate. A design problem for weight optimization of a transversely impacted Graphite/Epoxy(T300/5208)-Aramid/Epoxy(Kevlar 49) hybrid composite plate has been carried out.

2 The Parallel Genetic Algorithm

Optimization of laminated FRP composites using GA is computationally intensive, consuming large amounts of computer space and time. The need to speed up the computational process has guided to the implementation of Parallel GA. Among several paradigms for how populations are evolved in parallel GA, one common method is known as the Island Model approach, which ideally suits the problem under consideration.

The principle of IMPGA is based on the hypothesis that several competing sub-populations could be more search-effective than a wider one in which all the members were held together. Therefore, it's not surprising that a distributed model for GA exists, which is inspired by these biological observations. This model is called the multi-population large-grained GA, or just the Island Model. In the Island Model, the overall population of chromosomes is partitioned into a number of sub-populations. Each sub-population evolves independently for optimizing the same objective function. Some logical topology for how the populations are interconnected is defined and periodically each sub-population replaces its chosen chromosome, as per migration strategy, with the best of its neighbor's.

The migration of string between subpopulations is a key feature of the IMPGA. Since each processor starts with a different initial population, genetic drift will tend to drive these populations into different directions. By introducing migration the island model is able to exploit differences in the various subpopulations. This variation represents a source of genetic diversity. However, migration of large number of individuals too often may drive out any local differences between islands, thus destroying global diversity. On the other hand, if migration occurs not often enough, it may lead to premature convergence of the subpopulations.

3 Optimum Design of Laminated Composite

In the present work, weight minimization of Graphite/Epoxy-Aramid/Epoxy hybrid composite plate, while subjected to transverse impact, has been considered as the optimization problem. The design variables are, ply angle, ply material and ply thickness of each ply along with the total number of plies in the laminate and hence the thickness of laminate. The objective functions for weight minimization is:

$$Minimize\ f_{wt}(t, \rho_{ply}) = (L \times B) \sum_{i=1}^{N} t_i \rho_i \qquad (1)$$

Where t, ρ, L and B represent ply thickness, ply material density, length and breadth of the plate respectively. The combined effect of critical matrix cracking (e_M) and delamination (e_D) at interface, both proposed by Choi *et al* [16, 17], has been taken as the failure criterion. Whichever occurs first is taken as the cause of failure. The failure criterion is:

$$Failure\ Index: F.I = max\{e_D,\ e_M\};\ fails\ if\ either\ e_D\ or\ e_M \geq 1 \qquad (2)$$

Whenever a laminate satisfies above criterion (Eq.2), it is removed from the GA population by imposing heavy penalty in the fitness calculation, whereas laminates, which do not fail under above criterion, is being accepted with proportional bonus in the fitness calculation for the next generation.

4 Numerical Results and Discussions

In the present work, a computer code has been developed in 'C', which has two distinct modules viz. the FEM module and the parallel GA module. The FEM module consists of transient dynamic analysis and appropriate failure criteria for assessing the failure of laminated composites under impact loading [15]. The IMPGA module uses Message Passing Interface (MPI-1 and MPI-2) libraries as well as migration routines for optimization of laminated composite structures. The IMPGA module runs with a definite size of sub-population on different processors and on each processor FEA is carried out on a sequential basis. Also, each processor of IMPGA can spawn number of processes (2≤No. of Spawned Process ≤ Sub-Population) to distribute the organism among spawned process for carrying out FEA on sequential basis. The code has been run on a parallel platform, PARAM Padma, which is having one Teraflop peak computing power and having Power4 RISC processors.

A square laminated plate ($0.0762m\times0.0762m$) with arbitrary ply orientations and thicknesses, clamped along all the edges, subjected to transverse impact (of an aluminum spherical impactor [7, 15]) at the center has been considered for the analysis. The symmetric laminate may have any number of plies and each ply may be made of either Aramid/Epoxy or Graphite/Epoxy. The ply orientation of each ply could be between $-90°$ and $90°$ with increments of $15°$. The ply thickness of each ply can vary between $0.1mm$ to $0.5mm$ with increments of $0.1mm$. The laminated plate has been impacted by $0.0127m$ diameter aluminum sphere with initial impactor velocity of $9m/s$. Numerical experimentations have been done with different genetic parameters for studying the convergence of both sequential as well as parallel GAs and the tuned probabilities of the various operators are shown in Table 1.

Table 1. Probabilities of the GA Operators

Operators	Probability
Crossover	1.00
Ply Orientation Mutation (Gene Swap)	0.75
Ply Angle Mutation	0.05
Ply Thickness Mutation	0.10
Ply Material Mutation	0.05

In the present work, the GA employs the selection schemes of elitist and binary tournament selection. The GA implementation of the present problem uses a string of genes to represent one forth of a balanced symmetric laminated composite plate, mainly to reduce the search space and to expedite the convergence. The length of the gene string is kept fixed throughout the optimization process. Although the gene string length is fixed, having empty plies makes it possible to change the laminate thickness during the optimization process.

To accommodate two or more materials, three strings of genes have been introduced, namely ply orientation, ply material and ply thickness with provision for thickness alterations [7]. Genes in the second and third string will once again determine whether the ply location is empty or filled with a ply of prescribed material/thickness. Corresponding genes in the first string determine the ply orientation if the ply is present. By employing separate ply material gene strings, the number of materials that may be used in the stacking sequence may be changed easily by adjusting the size of the material gene string. The design variables i.e. the ply thickness, ply material, ply orientation and the number of plies have been initially chosen using random initialization. The initial population starts with laminates having randomly chosen number of plies and corresponding to each ply, ply thickness, ply material and fiber orientation are also chosen at random.

Single point crossover is the main genetic operator while mutation induces random changes in the genes and prevents the search from getting stuck in a local optimum. Gene swap is used to swap the positions of two genes on the hybrid chromosome. In all these, only one forth of the actual gene strings has been presented for carrying out the genetic operations by exploiting the fact that the search is for balanced as well as for symmetric laminates.

In addition to the existing GA operators, in the IMPGA implementation, migration routine has been used to facilitate the exchange of chromosomes among the subpopulations at different processors.

In the present study two migration strategies have been considered. In the first migration strategy (MS1), the fittest laminate in a subpopulation migrates to a neighbour and replaces the least fit laminate of the subpopulation. In second migration strategy (MS2), laminates are accepted when their fitness is better than the fittest laminate with probability 1.0. Laminates which are just as fit as the least fit laminate are accepted with probability $P_0 = 0.1$. Laminates with fitness between the best and least fit laminate are accepted with probability P_1, where P_1 is a linear interpolation between P_0 and 1.0. In all the cases, the migration frequency has been unity.

Table 2. Comparison of SGA and IMPGA CPU Time for different number of processor for 500 generations

Processors (n)	Time (t_p)	Speedup Actual $(s = t_s/t_p)$	Speedup Theoretical	Efficiency $(E = s/n)$
1 (Serial)	t_s= 715729.61s	-	-	-
4	161691.97s	4.43	4	1.10
5	119092.67s	6.01	5	1.20
8	64347.58s	11.12	8	1.39
10	47142.99s	15.18	10	1.51
16	31720.28s	22.56	16	1.41
20	42363.03s	16.90	20	0.84

Fig. 1 shows the convergence of the weight minimization problem with 10 processor IMPGA and 80-population size for two different migration strategies (MS1 and MS2). In comparing the two migration strategies, five numbers of runs has been given corresponding to each migration strategy. Fig. 1 shows the results corresponding to the best convergence in each case. It could be observed from Fig. 1 that MS2 gives faster convergence with lighter laminate compared to that in case of MS1. This is due to the fact that in the case of MS1, premature convergence takes place due to selective pressure. Therefore, in the present work, MS2 has been used in the IMPGA for optimization of laminated composites.

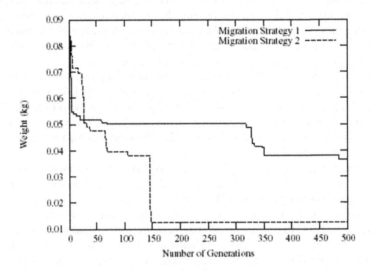

Fig. 1. Comparison of optimum laminate weight of MS1 with MS2 for weight minimization problem using 10 numbers of processors for a population size of 80

On one of the nodes of PARAM Padma platform, optimization of the laminated plate subjected to impact has also been carried out using SGA corresponding to same genetic parameters and population. To show the efficacy of parallelism, for the same weight minimization problem, speedup, s $(= t_s/t_p)$ which is the ratio of sequential run time t_s and parallel run time t_p has been compared with the theoretical time reduction $1/n$, where 'n' represents number of processors. It has been observed that actual time reduction outperforms the theoretical time reduction as evident from Table 2. Fig. 2 shows the theoretical time reduction and the actual time reduction (speed-up) with increasing number of processors. It is clearly visible form Fig. 2 that the ratio 's' increases almost linearly up to 10 processors and marginally declines in the increasing trend beyond 10 processors, which is due to the increase in communication time with increasing number of processors. The efficiency of parallelism $E = s/n$ with increasing number of processors has also been presented in Table 2 and is plotted in Fig. 3. It shows nearly linear increase of efficiency up to 10 processors. The decline in efficiency beyond 10 processors indicates increase in communication time with the increase of number of processors. It is important to note that in all the cases presented

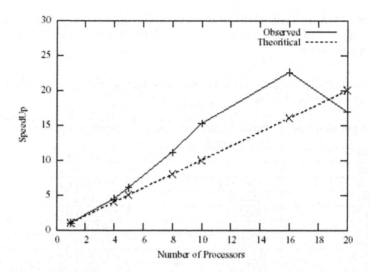

Fig. 2. Comparison of actual speedup with theoretical reduction in time for weight minimization problem using different number of processors for a population size of 80

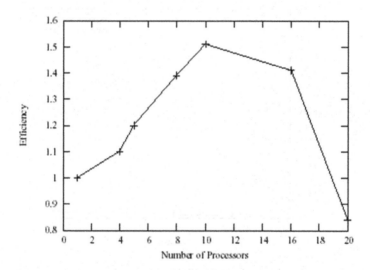

Fig. 3. Efficiency plot for weight minimization problem using different number of processors for a population size of 80

in Table 2, the efficiency is greater than 1.0, which shows the superior computational performance of IMPGA over SGA.

Further, to understand the behaviour of the algorithm with increasing number of population, keeping number of processors fixed, scalability analysis has been carried out and the same is compared with sequential algorithm. The CPU time taken by

IMPGA and SGA has been tabulated in Table 3. The study of the plot in Fig. 4 shows sharp increase in SGA CPU time relative to IMPGA CPU time with increase in number of population.

Table 3. Scalability Analysis; Comparison of SGA and IMPGA CPU Time for different number of populations for 200 generations

Population Size	Parallel Algorithm (*sec*); 10 Processors	Serial Algorithm (*sec*)
20	-	129165.99 *s*
40	-	267289.91 *s*
50	13193.79 *s*	350693.11 *s*
80	18449.28 *s*	571108.98 *s*
100	25024.53 *s*	≈715000.00 *s*
120	30446.63 *s*	≈868000.00 *s*
140	35632.34 *s*	≈1040000.00 *s*
160	38113.66 *s*	≈1205000.00 *s*
180	47213.07 *s*	≈1390000.00 *s*
200	53645.23 *s*	≈1583000.00 *s*

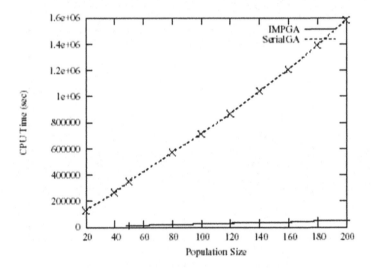

Fig. 4. Comparison of SGA CPU time with IMPGA CPU time for weight minimization problem using 10 numbers of processors for a population size of 80

In the weight minimization problem considered in present case, both SGA and IMPGA start with some maximum number of plies having arbitrary thickness and ply orientations. As the laminates converge towards optimum, the number of plies reduces in successive generation. Further, the rate of ply deletion in-turn is faster with IMPGA than SGA. The superior computational performance of IMPGA can be attributed to better mixing of genes, leading to faster convergence to optimal solution.

The reason of the above observation lies in the fact that thickness mutation operator evokes ply deletion which causes reduction in number of plies which in-turn reduces the finite element computational time for the laminate. The same is reflected in efficiency and scalability analysis which depicts the significant difference in computational time between integrated IMPGA-FEM module and SGA-FEM module. Since the optimization process using GA starts with a randomly generated initial population, it is possible that a solution or two may not attain satisfactory level. Hence, in the present work convergence has been tested with five different initial populations for each optimization problem and the best laminate obtained out of five such runs of the same problem has been presented as the converged near optimal solution in Table 4 along with the mean and standard deviation corresponding to each case.

Table 4. Comparison of Optimum laminates obtained by IMPGA and SGA for weight minimization

Processors	Fitness	Laminate Weight	Laminate Thickness
1	9.5529×10^{-1} $\bar{x} = 9.5461 \times 10^{-1}$ $\sigma = 4.3430 \times 10^{-4}$	$4.6916 \times 10^{-2}\ kg$ $\bar{x} = 4.8310 \times 10^{-2}\ kg$ $\sigma = 8.2690 \times 10^{-4}\ kg$	$0.0058\ m$ $\bar{x} = 0.0058\ m$ $\sigma = 0\ m$
4	9.652824×10^{-1} $\bar{x} = 9.652109 \times 10^{-1}$ $\sigma = 4.050670 \times 10^{-5}$	$3.632509 \times 10^{-2}\ kg$ $\bar{x} = 3.632509 \times 10^{-2}\ kg$ $\sigma = 0\ kg$	$0.004600\ m$ $\bar{x} = 0.004600\ m$ $\sigma = 0\ m$
8	9.878926×10^{-1} $\bar{x} = 9.708914 \times 10^{-1}$ $\sigma = 9.815632 \times 10^{-3}$	$1.263481 \times 10^{-2}\ kg$ $\bar{x} = 3.040252 \times 10^{-2}\ kg$ $\sigma = 1.025819 \times 10^{-2} kg$	$0.001600\ m$ $\bar{x} = 0.003850\ m$ $\sigma = 1.299038 \times 10^{-3}\ m$
10	9.878926×10^{-1} $\bar{x} = 9.709083 \times 10^{-1}$ $\sigma = 9.805880 \times 10^{-3}$	$1.263481 \times 10^{-2}\ kg$ $\bar{x} = 3.040252 \times 10^{-2}\ kg$ $\sigma = 1.025819 \times 10^{-2} kg$	$0.001600\ m$ $\bar{x} = 0.003850\ m$ $\sigma = 1.299038 \times 10^{-3}\ m$
16	9.652808×10^{-1} $\bar{x} = 9.649430 \times 10^{-1}$ $\sigma = 5.823054 \times 10^{-4}$	$3.632509 \times 10^{-2}\ kg$ $\bar{x} = 3.664096 \times 10^{-2}\ kg$ $\sigma = 6.317400 \times 10^{-4} kg$	$0.004600\ m$ $\bar{x} = 0.004640\ m$ $\sigma = 8.000000 \times 10^{-5}\ m$
20	9.652788×10^{-1} $\bar{x} = 9.649600 \times 10^{-1}$ $\sigma = 5.578570 \times 10^{-4}$	$3.632509 \times 10^{-2}\ kg$ $\bar{x} = 3.664096 \times 10^{-2}\ kg$ $\sigma = 6.317400 \times 10^{-4} kg$	$0.004600\ m$ $\bar{x} = 0.004640\ m$ $\sigma = 8.000000 \times 10^{-5}\ m$

To show the performance of the parallel algorithm with increasing number of processors, for a fixed population size of 80, convergence pattern for weight optimization has been studied and the results obtained are presented in Table 4. Fig. 5 shows the convergence of weight minimization problem with increasing number of processors.

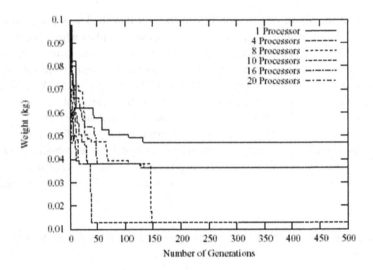

Fig. 5. Comparison of optimum laminate weight of SGA with IMPGA for weight minimization problem using different number of processors for a population size of 80

Table 4 shows the comparison of the laminates obtained using IMPGA on parallel platform and the SGA for the weight minimization problem discussed above. It could also be observed from the Table 4 that the efficacy of the IMPGA increases with increased number of processors up to 10 processors and the optimum laminate is much lighter and thinner compared to that obtained from SGA. But beyond 10 processors, when the same optimization has been carried out with 16 and 20 processors, the optimal laminates have been observed to be inferior though they are still lighter and thinner compared to that obtained from SGA. The reduction in fitness value or the increase in weight has been observed to be more pronounced with the increase in number of processors for a fixed population. This increase in weight/thickness beyond 10 processors in the present case shows that at each node, IMPGA requires a minimum size of subpopulation (critical sub-population size) below which the performance deteriorates. A laminate having higher value of failure index is given a penalty in the fitness calculation and the one with lower failure index is provided with bonus.

To observe the reduction in computational time for weight optimization, a fixed population size of 25 (5 at each node of IMPGA) has been run for 10 number of generations and observations are presented in Table 5.

Table 5 shows comparison of CPU time between IMPGA-FEM modules, one in which FEA is carried out sequentially at each node of IMPGA and other in which FEA is carried out at spawned processes (spawned at each node of IMPGA). From Table 5, nearly 4.76 times reduction in computational time has been observed which is quite close to the maximum theoretical reduction. The slight deviation from the maximum theoretical reduction is because of increased communication among processes.

Table 5. Comparison of IMPGA-Sequential FEA (at each node of IMPGA) CPU Time with IMPGA-Sequential FEA at spawned processes (at each node of IMPGA) CPU Time using 5 number of processors and population size of 25 for 10 generations

FEA at Each Node of IMPGA	Total CPU Time *(sec)*
Sequential FEA	2221.36 *sec*
Sequential FEA at spawned processes	466.61 *sec*

5 Conclusion

The integrated module comprising of IMPGA and FEM gives faster convergence along with lighter balanced symmetric laminate as compared with those obtained from integrated FEM and SGA module. Further, it has been observed that the speed up obtained from the present module is better than the possible theoretical speed up. A probabilistic migration strategy has been found to yield better convergence compared to the deterministic one. It has also been observed that for efficient working of IMPGA, a minimum size of sub population on each processor is a necessary requirement. Also, by spanning number of processes (at each node of IMPGA) for computation of FEA sequentially, it is possible to achieve near maximum theoretical computational time reduction. In summarizing, the IMPGA with an improved migration strategy in searching optimal laminates outperforms the SGA in terms of reduced computational time and better-converged solution.

References

1. Lin S.C., Punch W.F. and Goodman E.D. Coarse-grain parallel genetic algorithms: categorization and analysis, *IEEE Symposium on Parallel and Distributed Processing*, 1994; 27-36.
2. Pettey C.B., Leuze M.R. and Grefenstette J.J. A parallel genetic algorithm, *Proceedings of the 2nd International Conference on Genetic Algorithms and their application (ICGA)*, John J. Greffenstette (Ed.), Lawrence Erlbaum Associates Publishers, 1987
3. Bryce B. A Library for Island Model Parallel Genetic Algorithms, *Report, CSCE*, 2002;Department Pacific Lutheran University
4. Tanese R. Distributed genetic algorithms, *Proceedings of the 3rd International Conference on Genetic Algorithms and their application (ICGA)*, J.D. Schaffer (Ed.), 1989; 434-439, Morgan Kaufmann, San Mateo CA
5. Mühlenbein H, Schomisch M. and Born J. The parallel genetic algorithm as function optimizer, *Proceedings of the 4th International Conference on Genetic Algorithms and their application (ICGA)*, 1991;271-278, R-K. Belew, L.B. Booker (Eds.), San Diego, CA
6. Norman M.G. A genetic approach to topology optimization for multiprocessor architectures, *Tech. Report ECSP-TR-7*, 1988;*Univ. of Edinburgh, Dept. of Physics*
7. Rahul, Chakraborty D. and Dutta A. Optimization of FRP composites against impact induced failure island model parallel genetic algorithm, *Composites Science and Technology*, 2005;65: 2003-2013
8. Sadagopan D. and Pitchumani R. Application of genetic algorithms to the optimal tailoring of composite materials, *Composites Science and Technology*, 1998;58: 571-589.

9. Walker M.and Smith R. A technique for the multiobjective optimization of laminated composite structures using genetic algorithms and finite element analysis, *Composite Structures*, 2003;62:123-128.

10. Soremekun G., Gurdal Z., Haftka R. and Watson L. Composite laminate design optimization by genetic algorithm with generalized elitist selection, *Computers and Structures,* 2001;79: 131-143.

11. Sivakumar K. Iyengar N. and Deb K. Optimum design of laminated composite plates with cut-outs using a genetic algorithm, *Composite Structures,* 1998;42: 265-279.

12. Qu S., Venkataraman S.and Haftka R. Deterministic and reliability based optimization of composite laminates for cryogenic environments, *AIAA* 2000; 2000-4760

13. Deka D. Sandeep G. Chakraborty D and Dutta A. Multi-objective Optimization of Laminated Composites using Finite Element Method and Genetic Algorithm, *Journal of Reinforced Plastics and Composites*, 2005;23, (3): 273,286

14. Rahul, Sandeep G., Chakraborty D. and, Dutta A. Multi-objective optimization of hybrid laminates subjected to transverse impact, *Composite Structures*, 2006;73:360–369

15. Wu Hsi-Yung T. and Chang F.K. Transient dynamic analysis of laminated composite plates subjected to transverse impact, *Comput. Struct.* 1989;31 (3) : 453-466

16. Choi H.Y., Wu His-Yung T.and Chang Fu-Kuo. A new approach towards understanding damage mechanism and mechanics of laminated composites due to low velocity impact: Part II – analysis, *J. of Compo. Mat.*, 1991;25 :1012-1038 .

17. Choi H. Y. and Chang F K. A model for predicting damage in Graphite/Epoxy laminated composite resulting from low velocity point impact, *J. of Compo. Mat*, 1992;26: 2134-2169.

Experiments with Wide Area Data Coupling Using the Seine Coupling Framework*

Li Zhang[1], Manish Parashar[1], and Scott Klasky[2]

[1] TASSL, Rutgers University, 94 Brett Rd. Piscataway, NJ 08854, USA
[2] Oak Ridge National Laboratory, P.O. Box 2008, Oak Ridge, TN, 37831

Abstract. Emerging scientific and engineering simulations often require the coupling of multiple physics models and associated parallel codes that execute independently and in a distributed manner. Realizing these simulations in distributed environments presents several challenges. This paper describes experiences with wide-area coupling for a coupled fusion simulation using the Seine coupling framework. Seine provides a dynamic geometry-based virtual shared space abstraction and supports flexible, efficient and scalable coupling, data redistribution and data streaming. The design and implementation of the coupled fusion simulation using Seine, and an evaluation of its performance and overheads in a wide-area environment are presented.

1 Introduction

Scientific and engineering simulations are becoming increasingly sophisticated as they strive to achieve more accurate solutions to realistic models of complex phenomena. A key aspect of these emerging simulations is the modeling of multiple interacting physical processes that comprise the phenomena being modeled. This leads to challenging requirements for coupling between multiple physical models and associated parallel codes that execute independently and in a distributed manner. Coupled systems provide the individual models with a more realistic simulation environment, allowing them to be interdependent on and interact with other physics models in the coupled system and to react to dynamically changing boundary conditions. For example, in plasma science, an integrated predictive plasma edge simulation couples an edge turbulence code with a core turbulence code through common grids at the spatial interface [11].

However, achieving efficient, flexible and scalable coupling of physics models and parallel application codes presents significant algorithmic, numerical and computational challenges. From the computational point of view, the coupled simulations, each typically running on a distinct parallel system or set of processors with independent (and possibly dynamic) distributions, need to periodically exchange information. This requires that: (1) communication schedules between individual processors executing each of the coupled simulations are computed efficiently, locally, and on-the-fly, without

---

* The research presented in this paper is supported in part by National Science Foundation via grants numbers ACI 9984357, EIA 0103674, EIA 0120934, ANI 0335244, CNS 0305495, CNS 0426354 and IIS 0430826, and by Department of Energy via the grant number DE-FG02-06ER54857.

Y. Robert et al. (Eds.): HiPC 2006, LNCS 4297, pp. 229–241, 2006.

requiring synchronization or gathering global information, and without incurring significant overheads on the simulations; and (2) data transfers are efficient and happen directly between the individual processors of each simulation. Furthermore, specifying these coupling behaviors between the simulation codes using popular message-passing abstractions can be cumbersome and often inefficient, as they require matching sends and receives to be explicitly defined for each interaction. As the individual simulations become larger, more dynamic and heterogeneous and their couplings more complex, implementations using message passing abstractions can quickly become unmanageable. Clearly, realizing coupled simulations requires an efficient, flexible and scalable coupling framework and simple high-level programming abstractions.

This paper presents experiences with wide-area coupling for a coupled fusion simulation using the Seine [6] coupling framework. The objective of this paper is to evaluate the ability of Seine to support the coupling requirements of the recent CPES [1] DoE SciDAC Fusion Simulation Project. Seine provides a semantically specialized virtual shared space coupling abstraction and efficient, flexible and scalable mechanisms for data coupling, redistribution and transfer [6]. The Seine shared space abstraction is derived from the tuple space model. It presents an abstraction of a transient interaction space that is semantically specialized to the application domain. The specialization is based on the observation that interactions in the target applications can be specified on an abstract spatial domain that is shared by the interacting entities, such as a multidimensional geometric discretizations of the problem domain (e.g., grid or mesh). Further, the interactions are local in this domain (e.g., intersecting or adjacent regions). The shared spaces provided by Seine are localized to these regions of interaction, which are sub-regions of the overall abstract domain. This allows the Seine abstraction to be efficiently and scalably implemented and allows interactions to be decoupled at the application level [6].

The Seine coupling framework differs from existing approaches in several ways. First, it provides a simple but powerful abstraction for interaction and coupling in the form of a virtual semantically-specialized shared space. This may be the geometric discretization of the application domain or an abstract multi-dimensional domain defined exclusively for coupling purposes. Processes register regions of interest, and associatively read and write data associated with the registered region from/to the space in a decoupled manner. Registering processes do not need to know of, or explicitly synchronize with, other processes during registration and computation of communication schedules. Second, it supports efficient local computation of communication schedules using lookups into a directory, which is implemented as a distributed hash table. Finally, it supports efficient and low-overhead processor-to-processor socket-based data streaming and adaptive buffer management. The Seine model and the Seine-based coupling framework complement existing parallel programming models and can work in tandem with systems such as MPI, PVM and OpenMP.

This paper presents the coupling, data redistribution and data transfer requirements of the coupled fusion simulations, and describes a prototype implementation of these simulations using Seine. The paper then describes experiments with wide-area coupling and demonstrates that the Seine-based implementation can potentially meet the

---

[1] Center for Plasma Edge Simulation.

data coupling requirements of the project. The experiments investigate the behavior and performance of Seine-based coupling between simulations running at Oak Ridge National Laboratory (ORNL) in TN, and Rutgers University (RU) in NJ, and measure the time required for data redistribution and streaming as well as throughputs achieved for different distribution patterns and data sizes. These experiments are intended to be a proof-of-concept to demonstrate the feasibility of using the Seine coupling framework to support data coupling in real Fusion simulations.

The rest of the paper is organized as follows. Section 2 presents related work. Section 3 presents an overview of the Seine coupling framework. Section 4 describes the coupled fusion simulations. Section 5 presents the Seine-based prototype implementation and experimental evaluation of the simulations. Section 6 presents conclusions and outlines future directions.

2   Background and Related Work

Parallel data redistribution (also termed as the MxN problem) is a key aspect of the coupling problem. It addresses the problem of transferring data from a parallel program running on M processors to another parallel program running on N processors. Different aspects of this problem have been addressed by recent projects such as MCT [4], InterComm [3], PAWS [2], CUMULVS [1], DCA [5], DDB [9] etc., with different foci and approaches. These systems differ in the approaches they use to compute communication schedules, the data redistribution patterns that they support, and the abstractions they provide to the application developer. Most of these existing systems gather distribution information from all the coupled models at each processor and then locally compute data redistribution schedules. This implies a collective communication and possible global synchronization across the coupled systems, which can be expensive and limit scalability. Further, abstractions provided by these systems are based on message passing, which requires explicit definition of matching of sends and receives and synchronized data transfers. Moreover, expressing very general redistribution patterns using message passing type abstractions can be quite cumbersome.

The Seine geometry-based coupling framework provides a simple but powerful high-level abstraction, based on a virtual associative shared space, to the application developer. Communication schedules are computed locally and in a decentralized manner using a distributed directory layer. All interactions are completely decoupled and data transfer is socket-based and processor-to-processor, and can be synchronous or asynchronous. The Seine framework is introduced below.

3   The Seine Geometry-Based Coupling Framework

Seine is a dynamic geometry-based interaction/coupling framework for parallel scientific and engineering applications. Note that the geometry may be based on the geometric discretization of the application domain or an abstract multi-dimensional domain defined exclusively for coupling purposes. Seine spaces can be dynamically created and destroyed. They complement existing parallel programming models and can co-exist with them during program execution.

3.1 The Seine Geometry-Based Coupling Model

Conceptually, the Seine coupling/interaction model is based on the tuple space model where entities interact with each other by sharing objects in a logically shared space. However there are key differences between the Seine model and the general tuple space model. In the general tuple space model, the tuple space is global, spans the entire application domain, can be accessed by all the nodes in computing environments, and supports a very generic tuple-matching scheme. These characteristics of the general tuple model have presented several implementation challenges. In contrast, Seine defines a virtual dynamic shared space that spans a specific geometric region, which is a subset of the entire application domain, and is only accessible by the dynamic subset of nodes to which the geometric region is mapped. Further, objects in the Seine space are geometry-based, i.e. each object has a geometric descriptor that specifies the region in the application domain that the object is associated with. Applications use these geometric descriptors to associatively *put* and *get* objects to/from a Seine space. These interactions are naturally decoupled.

The Seine API consists of a small set of simple primitives as listed in Table 1. The *register* operation allows a process to dynamically register a region of interest, which causes it to join an appropriate existing space or create a new space if one does not exist. The *put* operator is used to write an object into the space, while the *get* operator

Table 1. Primitives provided by the Seine framework

Primitives	Description
init(bootstrap-server-IP)	Uses a bootstrap mechanism to initialize the Seine runtime system.
register(object-geometric-descriptor)	Registers a region with Seine.
put(object-geometric-descriptor, object)	Inserts a geometric object into Seine.
get(object-geometric-descriptor, object)	Retrieves and removes a geometric object from Seine. This call will block until a matching object is *put*.
deregister(object-geometric-descriptor)	De-registers a region from Seine.

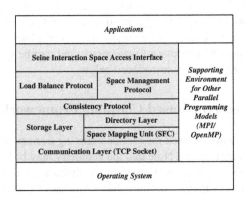

Fig. 1. Architectural overview of the Seine framework

retrieves a matching object from the space, if one exists. If no matching object exists, it will block until a matching object is *put* into the space. The *deregister* operation allows a processor to de-register a previously registered region.

3.2 Design of the Seine Geometry-Based Coupling Framework

A schematic overview of the Seine architecture is presented in Figure 1. The framework consists of three key components: a directory layer, a storage layer, and a communication layer. The distributed directory layer enables the registration of spaces and the efficient lookup of objects using their geometric descriptors. It detects geometric relationships between shared geometry-based objects and manages the creation of shared spaces based on the geometric relationship detected, the lifetime of shared spaces including merging or splitting, and the destruction of shared spaces. The storage layer consists of the local storage associated with registered shared spaces. The storage for a shared space is distributed across the processors that have registered the space. The communication layer provides efficient data transfer between processors. Since coupling and parallel data redistribution for scientific application typically involves communicating relatively large amounts of data, efficient communication and buffer management are critical. Further, this communication has to be directly between the individual processors. Currently Seine maintains the communication buffers at each processors as a queue, and multiple sends are overlapped to better utilize available bandwidth. Adaptive buffer management strategies are being integrated.

To share an object in the space, the geometric region of the object first needs to be *registered* with Seine. During registration, the Seine runtime system first maps the region defined in the n-dimensional application space to a set of intervals in a 1-dimensional index space using the Hilbert Space Filling Curve (SFC) [8]. The index intervals are then used to index into the Seine directory to locate the processor(s) to which the region is mapped. Note that the mapping is efficient and only requires local computation.

A new registration request is compared with existing spaces. If overlapping regions exist, a union of these regions is computed and the existing shared spaces are updated to cover the union. Note that this might cause previously separate spaces to be merged. If no overlapping regions exist, a new space is created. After registration, objects can be *put/get* to/from the shared space. When an object is *put* into the space, the update has to be reflected to all processors with objects whose geometric regions overlap with that of the object being inserted. This is achieved by propagating the object or possibly corresponding parts of the object (if the data associated with the region is decomposable based on sub-regions, such as multi-dimensional arrays) to the processors that have registered overlapping geometric regions. As each shared space only spans a local communication region, it typically maps to a small number of processors and as a result update propagation does not result in significant overheads. Further, unique tags are used to enable multiple distinct objects to be associated with the same geometric region. Note that Seine does not impose any restrictions on the type of application data structures used. However, the current implementation is optimized for multi-dimensional arrays. The *get* operation is simply a local memory copy from Seine's buffer to the application's buffer. Further details of Seine can be found in [6,7].

3.3 Coupling Parallel Scientific Applications Using Seine

Developing coupled simulations using the Seine abstraction consists of the following steps. First, the coupled simulations register their regions of interests, either in the geometric discretization of the application domain or in an abstract n-dimensional domain defined exclusively for coupling purposes. The registration phase detects geometric relationships between registered regions and results in the dynamic creation of a virtual shared space localized to the region and the derivation of associated communication schedules. Coupling between the simulations consists of one simulation writing data into the space and the other simulation independently reading data from the space. The actual data transfer is point-to-point between the corresponding source and destination processors of the respective applications.

4 Coupling Requirements in the CPES SciDAC Fusion Simulation Project

4.1 An Overview of the CPES Fusion Simulation Project

The CPES DoE SciDAC Fusion project is developing a new integrated predictive plasma edge simulation code package that is applicable to the plasma edge region relevant to both existing magnetic fusion facilities and next-generation burning plasma experiments, such as the International Thermonuclear Experimental Reactor (ITER) [10]. The plasma edge includes the region from the top of the pedestal to the scape-off layer and the divertor region bounded by a material wall. A multitude of non-equilibrium physical processes on different spatio-temporal scales present in the edge region demand a large scale integrated simulation. The low collisionality of the pedestal plasma, magnetic X-point geometry, spatially sensitive velocity-hole boundary, non-Maxwellian nature of the particle distribution function, and particle source from neutrals, combine to require the development of a special, massively parallel kinetic transport code for kinetic transport physics using a particle-in-cell (PIC) [12] approach. However, a fluid code is more efficient in terms of computing time, for studying the large scale MHD phenomena, such as Edge Localized Modes (ELMs) [12]. Furthermore, such an event is separable since its time scale is much shorter than that of the transport. The kinetic and MHD codes must however be integrated together for a self-consistent simulation as a whole. Consequently, the edge turbulence PIC code (i.e., XGC [13]) will be connected with the microscopic MHD code (i.e., M3D) using common grids at the spatial interface to study the dynamical pedestal-ELM cycle.

4.2 Data Coupling in the CPES Fusion Simulation Project

The coupled parallel simulation codes, XGC and M3D, will be run on different numbers of processors on different platforms. The overall workflow illustrating the coupling between XGC and M3D code is shown in Figure 2. The coupling begins with the generation of a common spatial grid. XGC then calculates two dimensional density,

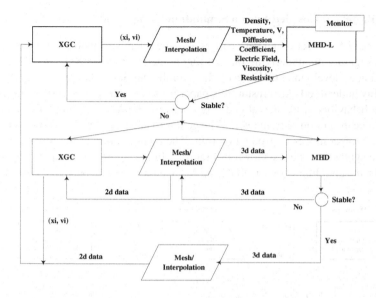

Fig. 2. Workflow illustrating the coupling between XGC and M3D

temperature, bootstrap current, and viscosity profiles in accordance with neoclassical and turbulent transport, and sends these to M3D. The input pressure tensor and current information are used by M3D to evolve the equilibrium magnetic field configuration, which it then sends back to XGC to enable it to update its magnetic equilibrium and to check for stability. During and after the ELM crash, the pressure, density, magnetic field and current will be toroidally averaged and sent to XGC. During the ELM calculation, XGC will evaluate the kinetic closure information and kinetic E_r evolution and send them to M3D for a more consistent simulation of ELM dynamics. The XGC and MHD codes [12] use different formulations and domain configurations and decompositions. As a result, a mesh interpolation module (referred to as MI) is needed to translate between the mesh/data used in the two codes.

Challenges and Requirements. In the CPES project, XGC will be running on a large number of processors while M3D will typically run on 128 or fewer processors. As a result, coupling these codes will require data redistribution. Note that in this case, the redistribution is actually MxPxN, where the XGC code runs on M processors, the interpolation module (MI) runs on P processors, and the M3D code runs on N processors.

The fusion simulation application imposes strict constraints on the performance and overheads of data redistribution and transfer between the codes. Since the integrated system is constructed so as to overlap the execution of XGC with stability check by M3D, it is essential that the result of the stability check is available by the time it is needed by XGC, otherwise the large number (1000s) of processors running XGC will remain idle offsetting any benefit of a coupled simulation. Another constraint is the overhead that the coupling and data transfer imposes on the simulations.

4.3   A Prototype Coupled Fusion Simulation Using Seine

Since the CPES project is at a very early stage, the scientists involved in the project are still investigating the underlying physics and numerics, and the XGC and M3D codes are still under development. However, the overall coupling behaviors of the codes are reasonably understood. As a result, this paper uses synthetic codes, which emulate the coupling behaviors of the actual codes but perform dummy computations, to develop and evaluate the coupling framework. The goal is to have the coupling framework ready when the project moves to production runs. The configuration of the mock simulation using the synthetic codes is shown in Figure 3. In the figure, the coupling consists of two parts, the coupling between XGC and MI and the coupling between MI and M3D.

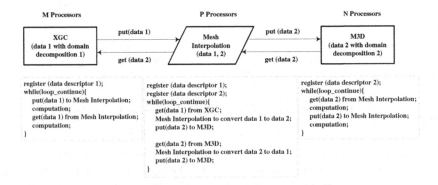

Fig. 3. Configuration of the mock coupled fusion simulation

Domain decompositions: The entire problem domain in the coupled fusion simulation is a 3D toroidal ring. The 3D toroidal ring is then sliced to get a number of 2D poloidal planes as the computation domains. Each plane contains a large number of particles, each of which is described by its physical location using coordinates and a set of physics variables. Each 2D poloidal plane is assigned to and replicated on a group of processors. Since XGC and M3D use different domain decompositions, the numbers of planes in the two codes are different, and MI is used to map the XGC domain decomposition to the M3D domain decomposition.

Coupled fusion simulations using Seine Shared Spaces: Recall that coupling in Seine is based on a spatial domain that is shared between the entities that are coupled. This may be the geometric discretization of the application domain or may be an abstract multi-dimensional domain defined exclusively for coupling purposes. The prototype described here uses the latter.

Given that the first phase of coupling between XGC and M3D is essentially based on the 2D poloidal plane, a 3D abstract domain can be constructed as follows: The X-axis represents particles on a plane and is the dimension that is distributed across the processors. The Y-axis represents the plane id. Each processor has exactly one plane and may have some or all the particles in that plane. The Z-axis represents application variables associated with each particle. Each processor has all the variables associated with each particle that is mapped to it. Using this abstract domain, Seine-based coupling is achieved as follows. Each XGC processor registers a region in the 3D abstract domain based on the 2D poloidal plane and the particles assigned to it, and the variables associated with each particle. The registered region is specified as a 6-field tuple and represents a 2D plane in the 3D abstract domain, since each processor is assigned particles on only one poloidal plane. Each processor running MI similarly registers its corresponding region in the 3D abstract domain. Note that since MI acts as the "coupler" between XGC and M3D, these processors register regions twice - once corresponding to the XGC domain decomposition and the second time corresponding to M3D domain decomposition. Once the registration is complete, the simulations can use the operators provided by Seine, i.e., *put* and *get*, to achieve coupling.

5   Prototype Implementation and Performance Evaluation

The schematic in Figure 4 illustrates a prototype implementation of a Seine-based coupled simulation. Note that, while the figure illustrates a MxN coupled simulation, the configuration for a coupled MxPxN simulation is similar. The Seine implementation requires a Seine-proxy, which is a local daemon process that resides on each processor using Seine. The Seine distributed directory layer deterministically maps the shared abstract domain onto the Seine infrastructure processors. The Seine distributed directory runs on X processors, which may or may not overlap with the M, P and N processors running XGC, MI and M3D respectively. The Seine-proxy at each processor is initialized by the *init* call within the application code. Once the Seine-proxy is initialized, it handles all the processor interaction with Seine including *register*, *put* and *get* operations.

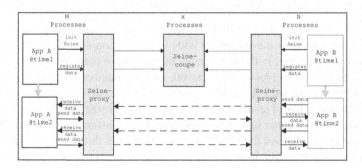

Fig. 4. A prototype schematic of coupling and data redistribution using the Seine framework

5.1   Experiments with Wide-Area Coupling Using the Prototype Seine-Based Fusion Simulation

The experiments presented in this section were conducted between two sites: a 80 nodes cluster with 2 processors per node at Oak Ridge National Laboratory (ORNL) in TN, and 64 node cluster at the CAIP Center at Rutgers University in NJ. The synthetic XGC code ran on the ORNL cluster and the MI module and the synthetic M3D code ran on the CAIP cluster. That is, in the MxPxN coupling, site M was at ORNL and sites P and N were at CAIP. The two clusters had different processors, memory and interconnects. Due to security restrictions at ORNL, these experiments were only able to evaluate the performance of data transfers from ORNL to CAIP, i.e., XGC pushing data to the MI module, which then pushes the data to M3D.

In the experiments below, the XGC domain was decomposed into 8 2D poloidal planes, while the M3D problem domain was decomposed into 6 2D poloidal planes. The number of particles in each plane was varied in the different experiments. Each particle is associated with 9 variables. Since the *get* operation in Seine is local and does not involve data communication, the evaluations presented below focus on the *put* operation, which pushes data over the network. The experiments evaluate the operation cost and throughput achieved by the *put* operation.

Cost of the put *operation:* In this experiment, 7,200 particles were used in each poloidal plane resulting in an abstract domain of size 7,200x8x9 between XGC and MI and 7,200x6x9 between MI and M3D. The number of processors at site M, which ran the XGC code, was varied. As the number of processors at site M increased, the absolute time for the *register* and *put* operations decreased since operation costs are directly affected by the size of the region. The decrease in absolute time cost is because the size of the entire abstract domain is fixed and as the number of processor increases, each processor registers/puts a smaller portion of this domain, resulting in a decrease in the absolute operation cost. Since the size of the region varies in the above metric, a normalized cost for the operations is calculated by dividing the absolute cost of an operation by the size of region involved. The normalized cost increases as the system size increases. Several factors contribute to this increase, including blocked-waiting time within a *register* operation, and message and data transfer costs associated with a *register* or *put* operation. A detailed analysis of this behavior can be found in [7].

Throughput achieved: The goal of this experiment is to measure the per processor throughput that can be achieved during wide-area data coupling for different system and abstract domain sizes. In the experiment, the number of particles per poloidal plane was varied to be 7,200, 14,400, and 28,800, and the number of processors running XGC at site M were varied to be 8, 16, 32, 64 and 128. Throughput per processor in this experiment was calculated as the ratio of the average data size used by a *put* operation to the average cost of a *put* operation. Note that data transfers from the processors at site M occur in parallel and the effective application level throughput is much higher. The per processor throughput at site M is plotted in Figure 5(a), and the estimated effective system throughputs computed assuming different levels of concurrency for the data transfer are plotted in Figure 5(b) and (c). Two observations can be made from Figure 5(a). First, the per processor throughput at site M decreases with the number of processors used at site M for all the abstract domain sizes tested. This is because the wide-area link is shared and when the number of processors increases the bandwidth available to each processor decreases, resulting in a lower throughput on each processor. Second, for the same number of processors at site M, in most cases, the per processor throughput for smaller abstract domain sizes is higher than the throughput for larger abstract domain sizes. This is because, for larger abstract domain sizes, the size of data to be redistributed is correspondingly larger, resulting in a more congested network.

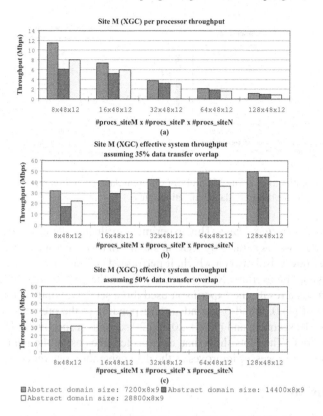

Fig. 5. (a) Per-processor throughput for XGC at site M; (b) Estimated effective system throughput assuming a data transfer overlap of 35%; (c) Estimated effective system throughput assuming a data transfer overlap of 50%

Further, the processors at site P are connected to the processors at both site M and site N. Consequently, site M processors have to compete with site N for connections with site P, which further causes the throughput to decrease for larger abstract domain sizes.

In the CPES Fusion project, site M (running XGC) throughput is a key requirement that must be met by the coupling framework. An initial estimate for the transfer rate from XGC to MI is 120Mbps. The estimated effective system throughput, based on the per processor bandwidth measured above and assuming 35% and 50% overlap in the per processor data transfer respectively, are plotted in Figure 5(b) and (c). Assuming that the system running XGC has 32 IO nodes, as seen from these plots, the estimated effective system throughputs are 34 - 42Mbps assuming a 35% overlap and 50 - 60Mbps assuming a 50% overlap. While these figures are still not close to the Fusion throughput requirement, we believe that these are conservative numbers and that Seine can support the required throughput when used in a real production scenario. This is because (1) these experiments assumed an extreme case where data was continuously generated by XGC, which is not realistic, and (2) these experiments use the Internet for the wide-area data-transfers while a real production run would use a dedicated and customized high-speed interconnect.

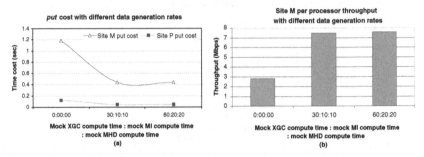

Fig. 6. (a) *put* operation cost at site M and P for different data generation rates; (b) Per processor throughput at site M for different data generation rates

Effect of data generation rates: This experiment evaluated the effect of varying the rate at which data was generated by XGC at site M. In this experiment, XGC generated data at regular intervals, between which, it performed computations. It is estimated by the physicists that on average, XGC requires 3 times the computation as compared to MI and M3D. As a result, the experiment used three sets of computes times for XGC, MI and M3D of (1) 0, 0 and 0 seconds (corresponding to the previous experiments), (2) 30, 10 and 10 seconds, and (3) 60, 20 and 20 seconds respectively. The results are plotted in Figure 6. The plots show that, as expected, the cost of the *put* operation and the throughput per processor improves as the data generation rate reduces.

6   Conclusion

The paper presented experiments and experiences with wide area coupling for a fusion simulation using the Seine coupling framework. The goal of these experiments is to evaluate the ability of Seine to support the coupling requirements of the ongoing CPES DoE SciDAC Fusion Simulation Project. Seine presents a high-level semantically specialized shared space abstraction to application and provides efficient and scalable data coupling, data redistribution and data transfer services. The experimental results using a prototype coupled fusion simulation scenario demonstrate the performance, throughput and low simulation overheads achieved by Seine.

Note that the experiments presented here are a proof-of-concept and demonstrate feasibility. As the project progresses and real codes and more detailed requirements become available, the Seine framework and abstractions will have to be tuned to ensure that it can support true production runs.

References

1. J.A. Kohl, G.A. Geist. Monitoring and steering of large-scale distributed simulations, IASTED Intl Conf on Appl. Modl. and Sim., Cairns, Queensland, Aus., Sept. 1999.
2. K. Keahey, P. Fasel, S. Mniszewski. PAWS: Collective interactions and data transfers. High Perf. Dist. Comp. Conf, San Francisco, CA, Aug. 2001.
3. J.Y. Lee and A. Sussman. High performance communication between parallel programs. 2005 Joint Wksp on High-Performance Grid Computing and High Level parallel Programming Models. IEEE Computer Society Press. Apr. 2005.

4. J.W. Larson, R.L. Jacob, I.T. Foster, and J. Guo. The Model Coupling Toolkit. Intl Conf on Comp. Sci. 2001, vol. 2073 of Lec. Nt. in Comp. Sci., pg 185-194, Berlin, 2001. S.-V.
5. F. Bertrand and R. Bramley. DCA: A distributed CCA framework based on MPI. The 9th Intl Wksp on High-Level Par. Prog. Mdls and Sup. Env., Santa Fe, NM, Apr. 2004. IEEE Press.
6. L. Zhang and M. Parashar. Enabling Efficient and Flexible Coupling of Parallel Scientific Applications. The 20th IEEE Intl Par. and Dist. Proc. Symp. Rhodes Island, Greece. IEEE Comp. Soc. Press.
7. L. Zhang and M. Parashar. Seine: A Dynamic Geometry-based Shared Space Interaction Framework for Parallel Scientific Applications. Con. and Comp.: Prac. and Exp. John Wiley and Sons. 2006.
8. T. Bially. A class of dimension changing mapping and its application to bandwidth compression. PhD thesis, Poly. Inst. of Brklyn., Jun. 1967.
9. L.A. Drummond, J. Demmel, C.R. Mechoso, H. Robinson, K. Sklower, and J.A. Spahr. A data broker for distributed computing environments. Intl Conf on Comp. Sc., pg 31-40, 2001.
10. ITER. http://www.iter.org.
11. I. Manuilskiy and W.W. Lee. Phys. Plasmas 7, 1381 (2000).
12. W.W. Lee, S. Ethier, W.X. Wang, W.M. Tang, and S. Klasky. Gyrokinetic Particle Simulation of Fusion Plasmas: Path to Petascale Computing. SciDAC 2006. June 25-29. Denver.
13. C. Chang, S. Ku, and H. Weitzner, Numerical Study of Neoclassical Plasma Pedestal in a Tokamak Geometry. Phys. Plasmas 11, 2649 2667 (2004).

HeteroMPI+ScaLAPACK: Towards a ScaLAPACK (Dense Linear Solvers) on Heterogeneous Networks of Computers

Ravi Reddy[1] and Alexey Lastovetsky[2]

[1] GS Laboratory Private Limited, Pune, India
ravi@gs-lab.com
[2] School of Computer Science and Informatics, University College Dublin,
Belfield, Dublin 4, Ireland
Alexey.Lastovetsky@ucd.ie

Abstract. The paper presents a tool that ports ScaLAPACK programs designed to run on massively parallel processors to Heterogeneous Networks of Computers. The tool converts ScaLAPACK programs to HeteroMPI programs. The resulting HeteroMPI programs do not aim to extract the maximum performance from a Heterogeneous Networks of Computers but provide an easy and simple way to execute the ScaLAPACK programs on such networks with good performance improvements. We demonstrate the efficiency of the resulting HeteroMPI programs by performing experiments with a matrix multiplication application on a local network of heterogeneous computers.

1 Introduction

In this paper, we present a tool, which ports conventional parallel programs that are designed to run on *massively parallel processors* (MPP) such as Scalable Linear Algebra Package (ScaLAPACK) programs [1] to Heterogeneous Message Passing Interface (HeteroMPI) programs [2] for Heterogeneous Networks of Computers (HNOCs). The resulting HeteroMPI programs do not aim to extract the maximum performance from a heterogeneous network but provide an easy and simple way to execute the conventional parallel programs on HNOCs with good performance improvements. Before we describe the details of the porting procedure, we present briefly the ScaLAPACK and HeteroMPI packages.

ScaLAPACK is a well-known standard package of high-performance linear algebra routines for distributed-memory message passing MIMD computers and networks of workstations supporting PVM [3] and/or MPI [4]. It is a continuation of the LAPACK project [5], which designed and produced analogous software for workstations, vector supercomputers, and shared-memory parallel computers. Both libraries contain routines for solving systems of linear equations, least squares problems, and eigenvalue problems.

HeteroMPI is an extension of MPI for programming high-performance computations on heterogeneous networks of computers. The main idea of HeteroMPI is to automate the process of selection of a group of processes, which would execute the heterogeneous parallel algorithm faster than any other group.

Y. Robert et al. (Eds.): HiPC 2006, LNCS 4297, pp. 242–253, 2006.

The first step in this process of automation is the specification of the performance model of the heterogeneous parallel algorithm in a performance model definition language. Performance model is a tool supplied to the programmer to specify his or her high-level knowledge of the application in a generic form. This knowledge is used by the HeteroMPI runtime system to find the most efficient implementation of the heterogeneous parallel algorithm on HNOCs.

The second step involves the writing of a HeteroMPI application. A typical HeteroMPI application consists of the following steps:

1. Accurate determination of the platform parameters using HeteroMPI characterization API;
2. Optimal data partitioning using HeteroMPI data partitioning API. This step of heterogeneous decomposition is parameterized by the platform parameters determined in the first step;
3. Determination of the optimal algorithmic parameters using HeteroMPI estimation API;
4. Efficient mapping of processes to the computers of the executing heterogeneous network. HeteroMPI group management operations automate this step.
5. Finally the execution of the HeteroMPI program using the HeteroMPI's command line interface.

The tool that we present in this paper mainly assists scientists trying to port their homogeneous parallel algorithms to HNOCs. It is usually a difficult design task to come up with a practical and efficient heterogeneous counterpart of a homogeneous parallel algorithm on HNOCs. The problem of optimal heterogeneous data distribution has proved to be NP-complete even for such a simple linear algebra kernel as matrix multiplication on HNOCs [6]. Once the heterogeneous parallel algorithm is designed, its portable and efficient implementation on heterogeneous platforms requires writing of a lot of complex code to automate several tedious and error-prone tasks [7]. The scientists can use this tool for porting their homogeneous parallel algorithms for HNOCs without any rewriting or redesigning. *It can be seen as a first step towards the realization of a ScaLAPACK for HNOCs.*

The tool takes two inputs. The first input is a ScaLAPACK program containing the homogeneous parallel algorithm that solves the problem on MPPs. The other input is the performance model of the homogeneous parallel algorithm employed in the ScaLAPACK program described in HeteroMPI's performance model definition language. It generates a HeteroMPI program, which uses a multiprocessing algorithm consisting of the following steps:

- The whole computation is partitioned into a large number of equal chunks;
- Each chunk is performed by a separate process;
- More than one process is allowed to be run on each processor. During the creation of a HeteroMPI group of processes, the mapping of the parallel processes in the group is performed such that the number of processes running on each processor is as proportional to its relative speed as possible.

In other words, while distributed evenly across parallel processes, data and computations are distributed unevenly over processors of the heterogeneous network, and this way each processor performs the volume of computations as proportional to its speed as possible. At the same time during the creation of a HeteroMPI group of processes, the mapping algorithm invoked tries to arrange the processors along a 2D grid so as to optimally load balance the work of the processors.

We start with literature survey on the multiprocessing approaches to solving parallel problems and proposals for heterogeneous ScaLAPACK. Then we describe the details of the porting procedure of the ScaLAPACK programs to HeteroMPI programs. This is followed by experimental results with a matrix multiplication application on a local network of heterogeneous computers demonstrating the efficiency of the resulting HeteroMPI programs. We conclude the paper by outlining our future research goals.

2 Literature Survey

The section surveys related papers from the literature. The papers surveyed are mainly: papers presenting proposals for heterogeneous ScaLAPACK and papers presenting multiprocessing approaches to solve parallel problems on HNOCs.

Beaumont *et al.* [8] discuss data allocation strategies to implement matrix products and dense linear system solvers on heterogeneous computing platforms as a basis for a successful extension of the ScaLAPACK library to heterogeneous platforms. They show that extending the standard ScaLAPACK block-cyclic distribution to heterogeneous 2D grids is difficult. In most cases, a perfect balancing of the load between all processors cannot be achieved and deciding how to arrange the processors along the 2D grid is a challenging NP-complete problem. They formally state the optimization problem to be solved and present both an exact solution (with exponential cost) and a heuristic solution.

Kalinov and Lastovetsky [9] analyze two strategies:

- HeHo - heterogeneous distribution of processes over processors and homogeneous block distribution of data over the processes;
- HoHe - homogeneous distribution of processes over processors with each process running on a separate processor and heterogeneous block cyclic distribution of data over the processes.

Both strategies were implemented in the mpC language [10, 11]. The first strategy is implemented using calls to ScaLAPACK; the second strategy is implemented with calls to LAPACK and BLAS [12]. They compare the strategies using Cholesky factorization on a network of workstations. They show that for heterogeneous parallel environments both the strategies HeHo and HoHe are more efficient that the traditional homogeneous strategy HoHo (homogeneous distribution of processes over processors and homogeneous distribution of data over the processes as implemented in ScaLAPACK). The main disadvantage of the HoHe strategy is non-Cartesian nature of the data distribution. This leads to additional communications that can be essential in the case of large networks. The HeHo strategy is easy to accomplish. It allows the reuse of high-quality software, such as ScaLAPACK, developed for homogeneous distributed memory systems in heterogeneous environments and to obtain a good speedup with minimal expenses.

Kishimoto and Ichikawa [13] adopt a multiprocessing approach to estimate the best processing element (PE) configuration and process allocation based on an execution-time model of the application. The execution time is modeled from the measurement results of various configurations. Then, a derived model is used to estimate the optimal PE configuration and process allocation. Kalinov and Klimov [14] investigate the HeHo strategy where the performance of the processor is given as a function of the number of processes running on the processor and the amount of data distributed to the processor. They present an algorithm that computes optimal number of processes and their distribution over processors minimizing the execution time of the application.

3 Porting a Legacy ScaLAPACK Program

This section is divided into three sub-sections. We start with the legacy ScaLAPACK program that is to be ported. This is followed by description of the homogeneous parallel algorithm used in the ScaLAPACK program in HeteroMPI's performance model definition language. In the second sub-section, we explain the structure of the HeteroMPI program output by the porting procedure. Finally we explain the issues involved in the porting procedure and how they are resolved.

3.1 Inputs

There are two inputs provided to the tool. The first input is the ScaLAPACK program computing matrix multiplication using the routine **PDGEMM**. There are four basic steps involved in calling a ScaLAPACK routine. The reader is directed to the ScaLAPACK users' guide [15] for more details.

The second input is the performance model definition **pdgemm** of the matrix multiplication routine **PDGEMM**. HeteroMPI allows application programmers to describe a performance model of their implemented homogeneous algorithm. This model allows specification of all the main features of the underlying parallel algorithm that have an essential impact on application execution performance on HNOCs. These features are:

- The total number of processes executing the algorithm.
- The total volume of computations to be performed by each of the processes in the group during the execution of the algorithm,
- The total volume of data to be transferred between each pair of processes in the group during the execution of the algorithm, and
- The order of execution of the computations and communications by the involved parallel processes in the group, that is, how exactly the processes interact during the execution of the algorithm.

HeteroMPI provides a small and dedicated model definition language for specifying this performance model. This language uses most of the features in the specification of network types of the mpC language. A compiler compiles the description of this

```
/* 1 */ algorithm pdgemm(int n, int b, int t, int p, int q)
/* 2 */ {
/* 3 */    coord I=p, J=q;
/* 4 */    node {I>=0 && J>=0: bench*((n/(b*p))*(n/(b*q))*(n/t));};
/* 5 */    link (K=p, L=q)
/* 6 */    {
/* 7 */       I>=0 && J>=0 && I!=K :
/* 8 */          length*((n/(b*p))*(n/(b*q))*(b*b)*sizeof(double))
/* 9 */             [I, J]->[K, J];
/* 10 */      I>=0 && J>=0 && J!=L:
/* 11 */         length*((n/(b*p))*(n/(b*q))*(b*b)*sizeof(double))
/* 12 */            [I, J]->[I, L];
/* 13 */   };
/* 14 */   parent[0,0];
/* 15 */   scheme
/* 16 */   {
/* 17 */      int i, j, k;
/* 18 */      for(k = 0; k < n; k+=b)
/* 19 */      {
/* 20 */         par(i = 0; i < p; i++)
/* 21 */            par(j = 0; j < q; j++)
/* 22 */               if (j != ((k/b)%q))
/* 23 */                  (100.0/(n/(b*q))) %% [i,((k/b)%q)]->[i,j];
/* 24 */         par(i = 0; i < p; i++)
/* 25 */            par(j = 0; j < q; j++)
/* 26 */               if (i != ((k/b)%p))
/* 27 */                  (100.0/(n/(b*p))) %% [((k/b)%p),j]->[i,j];
/* 28 */         par(i = 0; i < p; i++)
/* 29 */            par(j = 0; j < q; j++)
/* 30 */               ((100.0×b)/n) %% [i,j];
/* 31 */      }
/* 32 */   };
/* 33 */ };
```

Fig. 1. Specification of the performance model of the homogeneous algorithm employed by **PDGEMM** in the HeteroMPI's performance definition language

performance model to generate a set of functions. The functions make up an algorithm-specific part of the HeteroMPI runtime system.

The tool takes as input the performance model definition **pdgemm** shown in Figure 1. This performance model definition describes the simplest scenario performed by the **pdgemm** routine in ScaLAPACK, which uses outer-product algorithm using the *logical LCM hybrid algorithmic blocking* strategy [16]. The performance model definition describes the parallel matrix-matrix multiplication of two dense square matrices A and B of size **n×n**. The distribution blocking factor **b** used in the matrix-matrix multiplication is assumed to be equal to the algorithmic blocking factor. The performance model definition also assumes that the matrices are divided into whole number of blocks of size equal to distribution blocking factor, that is, (**n%(b×p)**) and (**n%(b×q)**) (see explanation of variables below) are both equal to zero.

The reader is referred to [11,17] for explanation of the main constructs, namely **coord**, **parent**, **node**, **link**, and **scheme**, used in a description of a performance

```
int main(int argc, char **argv) {
    static int p, q, n, t, input_p, output_p;
    int* mdlparams;
    HMPI_Group gid;
    HMPI_Init(&argc, &argv);
    // Estimation of speeds of the processors
    if (HMPI_Is_member(HMPI_PROC_WORLD_GROUP)
        HMPI_Recon(&dgemm, &input_p, 2, &output_p);
    // Model parameter initialization
    if (HMPI_Is_host())
        mdl_params[0] = n; mdl_params[1] = 64; mdl_params[2] = t;
    // HMPI Group creation
    if (HMPI_Is_host())
        HMPI_Group_heuristic_auto_create(&gid, &HMPI_Model_pdgemm,
                                    &hfunc, mdl_params);
    if (HMPI_Is_free())
        HMPI_Group__heuristic_auto_create(&gid, &HMPI_Model_pdgemm,
                                    NULL, NULL);
    // Execution of the algorithm
    if (HMPI_Is_member(&gid)) {
        MPI_Comm algocomm = *(MPI_Comm*)HMPI_Get_comm(&gid);
        HMPI_Group_topology(&gid, &nd, &dp);
        p = dp[0]; q = dp[1]; // optimal process grid arrangement
        ictxt = Csys2blacs_handle(algocomm);
        //Legacy ScaLAPACK program pdgemm code using ictxt
    }
    // HMPI Group Destruction
    if (HMPI_Is_member(&gid))
        HMPI_Group_free(&gid);
    HMPI_Finalize(0);
}
```

Fig. 2. The most relevant fragments of generated HeteroMPI code computing matrix-matrix multiplication using **PDGEMM** on heterogeneous networks

model. Briefly, Line 1 is a header of the performance model declaration. It introduces the name of the performance model **pdgemm** parameterized with the scalar integer parameters **n, b, t, p,** and **q**. Parameter **n** is the size of square matrices *A, B,* and *C*. It is assumed that the benchmark code multiplies two **b×t** and **t×b** matrices. Parameter **b** is the size of the distribution blocking factor. Parameters **p** and **q** are output parameters representing the number of processes along the row and the column in the process grid arrangement. Line 3 is a *coordinate declaration* declaring the coordinate system to which the processor nodes of the network are related. Line 4 is a *node declaration*. It relates the virtual processors to the coordinate system declared and specifies the (absolute) volume of computations to be performed by each of the processors. Lines 5-13 are a *link declaration*. This specifies the links between the virtual processors, the pattern of communication among the abstract processors, and the total volume of data to be transferred between each pair of virtual processors during the execution of the algorithm. Line 14 is a *parent declaration*. It specifies the coordinates of the parent processor node in a given coordinate system. Line 15 introduces the *scheme declaration*. The **scheme** block describes how exactly virtual processors interact during the execution of the algorithm.

3.2 Target HeteroMPI Program

The HeteroMPI program shown in Figure 2 resulting from the porting procedure performs typically the following steps:

1. The initialization of HeteroMPI runtime using the function **HMPI_Init**;
2. This is followed by dynamic refreshment of the estimation of the processor speeds using the characterization function **HMPI_Recon**. The benchmark code used in the call to **HMPI_Recon** is a serial BLAS version of the parallel ScaLAPACK routine. In this case, the BLAS routine **dgemm** multiplying two dense matrices is used to dynamically refresh the processor speeds. The benchmark code allocates, multiplies, and frees two **b×t** and **t×b** matrices where **b** is the distribution blocking factor and **t** is is equal to the size of the matrix used in the parallel application divided by the square root of the total number of processes that are available for computation. This is a heuristic used because some of the processes may not be chosen by the mapping algorithm employed by the HeteroMPI group constructor function (presented subsequently) to participate in the execution the parallel application.
3. Creation of a HeteroMPI group of processes using the group management function **HMPI_Group_auto_create** to obtain a handle to the HeteroMPI group of MPI processes. This function detects the optimal number of processes that can execute the parallel application, that is, finds the optimal arrangement of processes in a grid. During the creation of a HeteroMPI group of processes, the mapping of the parallel processes in the group is performed such that the number of processes running on each processor is proportional to its speed. At the same time, the processors are arranged along the 2D grid **p×q** so as to optimally load balance the work of the processors. The mapping algorithm is explained in detail in [11]. Since the number of 2D process grid arrangements is large, the HeteroMPI program uses the HeteroMPI function **HMPI_Group_heuristic_auto_create** instead of the HeteroMPI function **HMPI_Group_auto_create**, which evaluates all the possible 2D process grid arrangements. The function **HMPI_Group_heuristic_auto_create** uses heuristics to reduce the number of process arrangements to evaluate. The design and implementation of the HeteroMPI group constructor functions are explained in detail in [17];
4. The function **HMPI_Group_heuristic_auto_create** returns an HeteroMPI handle to the group of MPI processes in **gid**. The second parameter **HMPI_Model_pdgemm** is a handle that encapsulates all the features of the performance model. These features are in the form of a set of functions generated by the compiler from the description of the performance model. The third parameter **hfunc** is a heuristic function used to reduce the number of 2D process arrangements to evaluate. The fourth parameter **mdl_params** is an input parameter to the performance model, which consists of problem size to be solved, the algorithmic blocking factor used (which is equal to the distribution blocking factor) and the size of matrix used in the benchmark code. The only input provided by the application programmer is the problem size to be solved;
5. Conversion of the handle to the HeteroMPI group of MPI processes obtained previously to an MPI communicator using the function call **HMPI_Get_comm**;
6. Conversion of the MPI communicator to an integer BLACS handle, which can be passed into grid creation routine. This is done using the interim BLACS routine **Csys2blacs_handle**;

7. Creation of the BLACS context using the integer BLACS handle. This is done using the interim BLACS routine **Cblacs_gridinit**;

8. The legacy ScaLAPACK code is then executed using the BLACS context obtained;

9. This is followed by freeing the group using operation **HMPI_Group_free** and the finalization of HeteroMPI runtime system using operation **HMPI_Finalize**.

It can be seen that the HeteroMPI program automates the most tedious and error-prone tasks that are involved in porting a homogeneous parallel application.

3.3 Porting Issues

There are three important issues to be considered in the porting procedure.

1. The total number of processes to be allocated to each participating computer when the user starts up the application. Some basic rules to choose the number of processes to allocate per each processor can be followed:

- First of all, the number of processes running on each computer should not be less than the number of processors of the computer just to be able to exploit all the available processor resources. So the lower bound on the number of processes to be run on a computer is given by the number of processors on the computer.
- The upper bound on the number of processes executed on each processor is roughly equal to the ratio of speed of the fastest processor to speed of the slowest processor on the executing network of computers.

2. The blocking factor used to distribute the rows and the columns of the matrices involved in the computation. It is observed that for a process arrangement, execution times are the same no matter what algorithmic blocking factor is used. However to ensure efficient data distribution, ScaLAPACK [15] recommends that any blocking factor between 32 to 64 be used to distribute the rows and the columns of the matrices involved in the computation of the linear algebra kernel. The tool uses a value of 64;

3. The optimal arrangement of processes in the grid. This is determined by the HeteroMPI group constructor functions **HMPI_Group_auto_create** or **HMPI_Group_heuristic_auto_create**.

4 Experimental Results

A local network of 15 different heterogeneous Linux workstations hcl01 to hcl15 is used in the experiments. The computers used in the experiments are connected to communication network, which is based on 2 Gbit Ethernet with a switch enabling parallel communications between the computers. The experimental results are obtained by averaging the execution times over a number of experiments. Figure 3 shows the experimental results using the routine **pdgemm** performing parallel matrix-matrix multiplication on this heterogeneous network. The speedup calculated is the ratio of the execution time of the ScaLAPACK program over the execution time of the HeteroMPI+ScaLAPACK program. The reader is referred to [17] for details on the execution of the HeteroMPI program using HeteroMPI's command line interface.

Table 1. Optimal process grid arrangements (**p,q**) detected by the HeteroMPI group constructor function **HMPI_Group_heuristic_auto_create**. **n** is the size of the matrix. The third column gives the time taken to refresh the speeds of the processors at runtime. The fourth column gives the time taken to evaluate the process arrangements during the creation of the HeteroMPI group of processes that would execute the parallel application. The last column gives the execution time of the parallel application.

N	(p,q)	Processor speed update time (sec)	HeteroMPI Group creation time (sec)	Execution time (sec)
1024	(4,2)	0.09	1.29	17
2048	(8,2)	0.10	2.59	20
3072	(6,3)	0.21	1.38	21
4096	(8,2)	0.26	1.32	26
5120	(10,2)	0.31	2.70	30
6144	(6,3)	0.37	7.02	41
7168	(7,2)	0.44	1.76	53
8192	(8,2)	0.51	2.83	69
9216	(9,2)	0.58	5.33	100
10240	(10,2)	0.67	7.85	138
11264	(11,2)	0.76	6.36	215
12288	(12,2)	0.88	48.19	266
13312	(13,2)	1.18	10.73	312
14336	(14,2)	3.41	23.64	354
15360	(15,2)	8.97	54.65	405
16384	(16,2)	11.78	34.83	513
17408	(17,2)	14.28	23.99	772
18432	(18,2)	24.15	100.94	956
19456	(19,2)	30.49	32.45	1323
20480	(8,4)	33.59	41.97	2063

The absolute speeds of the processors are obtained based on serial version **dgemm** of the corresponding parallel routine **pdgemm**. The absolute speeds in million floating point operations per second (MFlop/s) is obtained by multiplication of two dense 1536×1536 matrices for the processors. The absolute speeds are {2171, 2099, 1761, 1787, 1735, 1653, 1879, 1635, 3004, 2194, 4580, 1762, 4934, 4096, 2697}. It can be seen that the fastest processor is *hcl13* and the slowest processor is *hcl08*. It should be noted that a process is run per processor to obtain these measurements. The ratio of absolute speed of the fastest processor to the absolute speed of the slowest processor is 4934/1635 = 3. This is the number of processes run on each processor in the network during the execution of the parallel application. So the total number of processes available to the HeteroMPI+ScaLAPACK program for computation is 25×3 = 75 since there are 25 processors in the network. The HeteroMPI+ScaLAPACK program detects the optimal process grid arrangement from the set of all possible 2D process grid arrangements of 75 processes in a reasonable amount of time as presented in Table 1. The number of possible 2D process arrangements can be calculated to be 338 (using the formula m×(1+1/2+1/3+...+1/m) where m=75). The ScaLAPACK program uses a 5×5 grid of processes (using one process per processor configuration).

Table 1 shows the optimal process grid arrangements determined by the HeteroMPI group constructor functions for the problem sizes experimented. The second column gives the optimal process grid arrangements for the problem sizes shown in first column. The third column gives the time taken to refresh the speeds of the processors at runtime during the **HMPI_Recon** function call. The fourth column gives the time taken to evaluate the process arrangements during the creation of the HeteroMPI group of processes using the **HMPI_Group_heuristic_auto_create** function call. This time varies due to different number of process arrangements evaluated for a given values of **n** and **b**. The last column gives the execution time of the parallel application. It includes the processor speed update time and the group creation time. It can be seen that the processor speed refreshment time and the group creation time are much less than the actual execution time of the parallel application. The function **HMPI_Group_heuristic_auto_create** uses heuristics to reduce the number of 2D process grid arrangements (**p,q**) to evaluate. One such heuristic used is that one-dimensional process arrangements where either **p** or **q** or both is equal to 1 are not evaluated.

The speedups of the HeteroMPI+ScaLAPACK program over ScaLAPACK program for these problem sizes are shown in Figure 3. As can be seen from the results, the resulting HeteroMPI programs deliver good performance improvements on HNOCs for problem sizes beyond 12288. There are two reasons for such good speedups observed. First reason is the better load balance achieved through proper allocation of processes involved in the execution of the algorithm to the processors. During the creation of a HeteroMPI group of processes, the mapping of the parallel processes in the group is performed such that the number of processes running on each processor is as proportional to its speed as possible. In other words, while distributed evenly across parallel processes, data and computations are distributed unevenly over processors of the heterogeneous network, and this way each processor performs the volume of computations as proportional to its speed as possible. It can be seen that for problem sizes larger than 12288, more than 25 processes must be involved in the execution to achieve good load balance. Since only 25 processes are involved in the execution of the ScaLAPACK program, good load balance is not achieved. However just running more than 25 processes in the execution of the ScaLAPACK program would not resolve the problem. This is because in such a case the optimal process arrangement and the efficient mapping of the process arrangement to the executing computers of the underlying network must also be determined. This is a complex task automated by HeteroMPI.

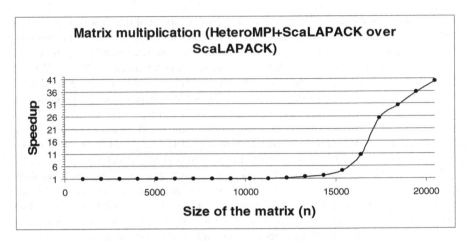

Fig. 3. Speedup of the HeteroMPI+ScaLAPACK program over the ScaLAPACK program employing matrix-matrix multiplication using the routine **pdgemm**

The second reason is the optimal 2D grid arrangement of processes. During the creation of a HeteroMPI group of processes, the function **HMPI_Group_heuristic_auto_create** estimates the time of execution of the algorithm for each process arrangement evaluated. For each such estimation, it invokes mapping algorithm, which tries to arrange the processors along a 2D grid so as to optimally load balance the work of the processors. It returns the process arrangement that results in the least estimated time of execution of the algorithm.

5 Conclusions and Future Work

In this paper, we have presented a tool that ports ScaLAPACK programs to heterogeneous platforms. The tool converts the ScaLAPACK programs to HeteroMPI programs. These HeteroMPI programs do not aim to extract the maximum performance from a heterogeneous network but provide an easy and simple way to execute the conventional parallel programs on HNOCs with good performance improvements. We have taken the first step towards the realization of a heterogeneous ScaLAPACK for HNOCs. Our future work will involve the development of a Heterogeneous ScaLAPACK library, which will include dense linear solvers of ScaLAPACK redesigned for HNOCs. The design and implementation of this library will include: (a) Design of performance models for each of the level-1, level-2, and level-3 PBLAS routines; (b) Design of performance models for each of the dense linear solvers of ScaLAPACK routines.

References

[1] Blackford, L., Choi, J., Cleary, A., Demmel, J., Dhillon, I., Dongarra, J., Hammarling, S., Henry, G., Petitet, A., Stanley, K., Walker, D., Whaley, R.: ScaLAPACK: A Portable Linear Algebra Library for Distributed Memory Computers – Design Issues and Performance. In: Proceedings of the 1996 ACM/IEEE Supercomputing Conference. IEEE Computer Society, CD-ROM/Abstracts Proceedings, Pittsburgh PA USA (1996)

[2] Lastovetsky, A., Reddy. R: HeteroMPI: Towards a Message-Passing Library for Heterogeneous Network of Computers. Journal of Parallel and Distributed Computing 66 (2006) 197-220

[3] Geist, A., Beguelin, A., Dongarra, J., Jiang, W., Manchek, R., Sunderam, V. S.: PVM: Parallel Virtual Machine, Users' Guide and Tutorial for Networked Parallel Computing. The MIT Press: Cambridge, MA (1994)

[4] Dongarra, J., Huss-Ledermann, S., Otto, S., Snir, M., Walker, D.: MPI: The Complete Reference. The MIT Press (1996)

[5] Anderson, E., Bai, Z., Bischof, C., Demmel, J., Dongarra, J., Du Croz, J., Greenbaum, A., Hmmarling, S., McKinney, A., Ostrouchov, S., Sorenson, D. LAPACK Users' Guide. Release 1.0, SIAM, Philadelphia (1992)

[6] Beaumont, O., Boudet, V., Rastello, F., Robert, Y.: Matrix Multiplication on Heterogeneous Platforms. IEEE Transactions on Parallel and Distributed Systems 12 (2001) 1033-1051

[7] Lastovetsky, L.: Scientific Programming for Heterogeneous Systems - Bridging the Gap between Algorithms and Applications. In: Proceedings of the 5th International Symposium on Parallel Computing in Electrical Engineering (PARELEC 2006), IEEE Computer Society Press (2006)

[8] Beaumont, O., Boudet, V., Petitet, A., Rastello, F., Robert, Y.: A Proposal for a Heterogeneous Cluster ScaLAPACK (Dense Linear Solvers). IEEE Transactions on Computers 50 (2001) 1052-1070

[9] Kalinov, A., Lastovetsky, A.: Heterogeneous Distribution of Computations Solving Linear Algebra Problems on Networks of Heterogeneous Computers. Journal of Parallel and Distributed Computing 61 (2001) 520-535

[10] Lastovetsky, A., Arapov, A. Kalinov, A., Ledovskih, I.: A Parallel Language and Its Programming System for Heterogeneous Networks. Concurrency: Practice and Experience 12 (2000) 1317-1343

[11] Lastovetsky, A.: Adaptive parallel computing on heterogeneous networks with mpC. Parallel Computing 28 (2002) 1369-1407

[12] Dongarra, J., Croz, J.D., Duff, I.S., Hammarling, S.: A set of level-3 basic linear algebra subprograms. ACM Transactions on Mathematical Software 16 (1990) 1-17

[13] Kishimoto, Y., Ichikawa, I.: An Execution-Time Estimation Model for Heterogeneous Clusters. In: 13th Heterogeneous Computing Workshop (HCW 2004), Proceedings of 18th International Parallel and Distributed Processing Symposium (IPDPS'04). IEEE Computer Society (2004)

[14] Kalinov, A., Klimov, S.: Optimal mapping of a parallel application processes onto heterogeneous platform. In: 4th Heterogeneous Computing Workshop (HCW 2005), Proceedings of 19th International Parallel and Distributed Processing Symposium (IPDPS'05), IEEE Computer Society (2005)

[15] Blackford, L., Choi, J., Cleary, A., D'Azevedo, E., Demmel, J., Dhillon, I., Dongarra, J., Hammarling, S., Henry, G., Petitet, A., Stanley, K., Walker, D., Whaley, R.: ScaLAPACK User's Guide. SIAM (1997)

[16] Petitet, A., Dongarra, J.: Algorithmic Redistribution Methods for Block-Cyclic Decompositions. IEEE Transactions on Parallel and Distributed Systems 10 (1999) 1201-1216

[17] Reddy, R.: HeteroMPI: A Message Passing Library for Heterogeneous Networks of Computers. PhD Dissertation, University College Dublin, Dublin, Ireland (2005)

Parallel Implementation of a Spline Based Computational Approach for Singular Perturbation Problems

Rajesh K. Bawa[1] and S. Natesan[2]

[1] Department of Computer Science, Punjabi University,
Patiala - 147 002, India
[2] Department of Mathematics, Indian Institute of Technology,
Guwahati - 781 039, India

Abstract. In this paper, a parallelizable computational technique for singularly perturbed reaction-diffusion problems is analyzed and implemented on parallel computer. In this technique, the domain is decomposed into non-overlapping subdomains, and boundary value problems are posed on each subdomain with suitable boundary conditions. Then, each problem is solved by the adaptive spline based difference scheme on each subinterval on parallel computer. Detailed theoretical analysis is provided to prove the convergence of the technique. To check the validity of the method, parallel implementation is performed on a numerical example and results are presented.

1 Introduction

We consider the following singularly perturbed self–adjoint boundary–value problem (BVP):

$$Lu(x) \equiv -\varepsilon u''(x) + b(x)u(x) = f(x), \quad x \in D = (0,1) \tag{1}$$

$$u(0) = A, \quad u(1) = B, \tag{2}$$

where $0 < \varepsilon \ll 1$ is a small parameter, b and f are sufficiently smooth functions, such that $b(x) \geq \beta > 0$ on $\overline{D} = [0,1]$. Under these assumptions, the BVP (1)-(2) possesses a unique solution $u(x) \in C^2(D) \cap C(\overline{D})$. In general, the solution $u(x)$ may exhibit two boundary layers of exponential type at both end points $x = 0$, and $x = 1$.

Singular perturbation problems (SPPs) model convection - diffusion and reaction - diffusion processes in engineering applications that arise in diverse areas, including linearized Navier-Stokes equation at high Reynolds number and the drift-diffusion equation of semiconductor device modelling. For $\varepsilon \ll 1$ the solution of the BVP (1)-(2) has a boundary layer of thickness $O(\sqrt{\varepsilon})$ near the boundaries $x = 0, 1$, and it is clear that a classical finite difference/element scheme on uniform mesh will not give a satisfactory numerical solution in this case. Therefore, a separate treatment is necessary to deal with these types of problems.

Y. Robert et al. (Eds.): HiPC 2006, LNCS 4297, pp. 254–262, 2006.

To solve the singularly perturbed reaction–diffusion BVP (1)-(2), various computational techniques exist in the literature, for example one can refer the books [FHM1, MOS1, RST1]. Most of them are of sequential type, only very few parallel computational methods exist. To cite a few, Paprzycki and Gladwell [PG1] proposed a mesh chopping algorithm for solving singular perturbation problems. Boglaev [Bo1] proposed a domain decomposition iterative algorithm and applied to singularly perturbed elliptic and parabolic problems. A numerical method is proposed by Natesan et al. in [NVR1] for a class of singularly perturbed convection–diffusion problems. Vigo-aguiar and Natesan [VN1] developed a parallel boundary–value technique for singular perturbation problems, which will accommodate maximum number of processors for computation. Bawa and Natesan [BN1] proposed a quintic spline based computational method for such problems on sequential computer, which is very well suitable for parallelization. Using a similar idea of domain decomposition and exploiting the layer resolving nature of adaptive spline, Bawa [Ba1], proposed a simple and efficient computational domain decomposition based spline technique in which stretching of fast moving component of solution required in using quintic spline can be avoided and also maximum absolute error can be reduced to much extend in comparison to applying it on whole domain. In this paper, we have analysed this technique theoretically in detail and implemented on a parallel machine by solving a numerical example.

More precisely, we divide $\overline{D}$ (the domain of definition of the BVP (1)-(2)) into three non-overlapping subdomains, and then pose suitable BVPs on each subdomain, by taking the given differential equation with suitable boundary conditions. To obtain the boundary values, we use the asymptotic expansion solution of the BVP (1)-(2). Now, each BVP can be solved by the adaptive spline difference scheme proposed by [St1]. Since all the BVPs are independent, it is very well possible to use parallel processors to solve each BVP, as a consequence one can save lot of CPU time. Speedup, and scalability of the present method are discussed.

2   Computational Technique

Adaptive Spline Difference Scheme

Here, we present the adaptive spline scheme for the BVP (1)-(2).

Consider the whole domain $\overline{D} = [0,1]$ with equally spaced knots $x_i = i/N$, $i = 0, \cdots, N$. Using the notation u_i for approximation of $u(x)$ at mesh point x_i and $S(x_{i-1}, q_{i-1}) = u_{i-1}$, $S(x_i, q_i) = u_i$ as interpolatory constraints and following [St1], the adaptive spline function $S(x, q)$ can be defined as solution of the differential equation:

$$-\varepsilon S''(x,q) + b_i S(x,q) = \frac{(x - x_{i-1})}{h}(-\varepsilon M_i + b_i u_i) + \frac{(x_i - x)}{h}(-\varepsilon M_{i-1} + b_{i-1} u_{i-1}),$$

$$(3)$$

where $x_{i-1} \le x \le x_i$, $S''(x_i, q) = M_i$ and $q = \sqrt{b/\varepsilon}h$.

Solving this for $S(x, q)$ and using the continuity of its first derivative at mesh point x_i, we get the following equation:

$$u_{i-1} - 2u_i + u_{i+1} = h^2 \left[q_{i-1}^{-2} M_{i-1} \left(1 - \frac{q_{i-1}}{\sinh q_{i-1}} \right) + \right.$$
$$\left. + 2q_i^{-2} M_i \left(1 - \frac{q_i}{\tanh q_i} \right) + q_{i+1}^{-2} M_{i+1} \left(1 - \frac{q_{i+1}}{\sinh q_{i+1}} \right) \right] \tag{4}$$

Substituting $M_i = (b_i u_i - f_i)/\varepsilon$ in (4), we get the following tridiagonal finite difference scheme, for $i = 1, \cdots, N-1$:

$$
\begin{cases}
\left[1 - \frac{h^2 b_{i-1}}{q_{i-1}^2 \varepsilon} \left(1 - \frac{q_{i-1}}{\sinh q_{i-1}} \right) \right] u_{i-1} + \left[-2 + 2\frac{h^2 b_i}{q_i^2 \varepsilon} (1 - \frac{q_i}{\tanh q_i}) \right] u_i + \\
+ \left[1 - \frac{h^2 b_{i+1}}{q_{i+1}^2 \varepsilon} \left(1 - \frac{q_{i+1}}{\sinh q_{i+1}} \right) \right] u_{i+1} = -\frac{h^2}{q_{i-1}^2 \varepsilon} \left(1 - \frac{q_{i-1}}{\sinh q_{i-1}} \right) f_{i-1} + \\
+ \frac{2h^2}{q_i^2 \varepsilon} \left(1 - \frac{q_i}{\tanh q_i} \right) f_i - \frac{h^2}{q_{i+1}^2 \varepsilon} \left(1 - \frac{q_{i+1}}{\sinh q_{i+1}} \right) f_{i+1}.
\end{cases} \tag{5}
$$

For the purpose of theoretical analysis, this system can be expressed in the following compact matrix form:

$$\left(J - h^2 Q \widehat{B} \right) U = \widehat{C} + h^2 Q F, \tag{6}$$

where

$$
J = \begin{bmatrix}
-2 & 1 & & & & \\
1 & -2 & 1 & & & \\
& & \cdot & \cdot & \cdot & \\
& & \cdot & \cdot & \cdot & \\
& & & \cdot & \cdot & \cdot \\
& & & 1 & -2 & 1 \\
& & & & 1 & -2
\end{bmatrix},
$$

$$
Q = \begin{bmatrix}
\frac{-2}{q_1^2 \varepsilon}(1 - \frac{q_1}{\tanh q_1}) & \frac{1}{q_2^2 \varepsilon}(1 - \frac{q_2}{\sinh q_2}) & & & \\
\frac{1}{q_1^2 \varepsilon}(1 - \frac{q_1}{\sinh q_1}) & \frac{-2}{q_2^2 \varepsilon}(1 - \frac{q_2}{\tanh q_2}) & \frac{1}{q_3^2 \varepsilon}(1 - \frac{q_3}{\sinh q_3}) & & \\
& \vdots & & \vdots & \vdots \\
& & \frac{1}{q_{N-3}^2 \varepsilon}(1 - \frac{q_{N-3}}{\sinh q_{N-3}}) & \frac{-2}{q_{N-2}^2 \varepsilon}(1 - \frac{q_{N-2}}{\tanh q_{N-2}}) & \frac{1}{q_{N-1}^2 \varepsilon}(1 - \frac{q_{N-1}}{\sinh q_{N-1}}) \\
& & & \frac{1}{q_{N-2}^2 \varepsilon}(1 - \frac{q_{N-2}}{\sinh q_{N-2}}) & \frac{-2}{q_{N-1}^2 \varepsilon}(1 - \frac{q_{N-1}}{\tanh q_{N-1}})
\end{bmatrix},
$$

$$\widehat{C} = \begin{bmatrix} -A\left[1 - \frac{h^2 b_0}{q_0^2 \varepsilon}\left(1 - \frac{q_0}{\sinh q_0}\right)\right] \\ 0 \\ \cdot \\ \cdot \\ \cdot \\ 0 \\ -B\left[1 - \frac{h^2 b_N}{q_N^2 \varepsilon}\left(1 - \frac{q_N}{\sinh q_N}\right)\right] \end{bmatrix}, \quad \widehat{B} = \begin{bmatrix} b_1 \\ b_2 \\ \cdot \\ \cdot \\ \cdot \\ b_{N-2} \\ b_{N-1} \end{bmatrix}, \quad \text{and} \quad F = \begin{bmatrix} f_1 \\ f_2 \\ \cdot \\ \cdot \\ \cdot \\ f_{N-2} \\ f_{N-1} \end{bmatrix}.$$

Non-overlapping Decomposition of the Domain $\overline{D} = [0, 1]$

We decompose the computational domain $\overline{D} = [0, 1]$ into three non–overlapping subdomains as $\overline{D} = D_1 \cup D_2 \cup D_3$, where $D_1 = [0, k\sqrt{\varepsilon}]$, $D_2 = [k\sqrt{\varepsilon}, 1 - k\sqrt{\varepsilon}]$, and $D_3 = [1 - k\sqrt{\varepsilon}, 1]$, $k = \sigma \ln(N)$, and $\sigma > 0$ is a parameter. Here N is the number total number of intervals. Then, we solve the differential equation (1), subject to suitable boundary conditions in each subdomain.

The BVPs corresponding to the three subdomains are given by

$$-\varepsilon u''(x) + b(x)u(x) = f(x), \quad x \in D_1 \tag{7}$$
$$u(0) = A, \quad u(k\sqrt{\varepsilon}) = \overline{A}, \tag{8}$$

$$-\varepsilon u''(x) + b(x)u(x) = f(x), \quad x \in D_2 \tag{9}$$
$$u(k\sqrt{\varepsilon}) = \overline{A}, \quad u(1 - k\sqrt{\varepsilon}) = \overline{B}, \tag{10}$$

and

$$-\varepsilon u''(x) + b(x)u(x) = f(x), \quad x \in D_3 \tag{11}$$
$$u(1 - k\sqrt{\varepsilon}) = \overline{B}, \quad u(1) = B. \tag{12}$$

To determine the boundary conditions at the interfaces (transition points), we take the zeroth–order asymptotic approximation of the BVP (1)-(2) given by

$$\widetilde{u}(x) = u_0(x) + v_0(x) + w_0(x),$$

where u_0, v_0, and w_0 are given by

$$\begin{cases} u_0(x) = f(x)/b(x), \\ v_0(x) = [A - u_0(0)]\exp[-\sqrt{b(0)/\varepsilon}x], \\ w_0(x) = [B - u_0(1)]\exp[-\sqrt{b(1)/\varepsilon}(1 - x)]. \end{cases}$$

The values of $\overline{A}$, $\overline{B}$ are given by

$$\overline{A} = \widetilde{u}(k\sqrt{\varepsilon}), \quad \text{and} \quad \overline{B} = \widetilde{u}(1 - k\sqrt{\varepsilon}).$$

The BVPs (7)-(8), (9)-(10), and (11)-(12) can be solved by the adaptive spline difference scheme (5).

Parallel Computation

The BVPs (7)-(8), (9)-(10), and (11)-(12) are completely independent of each other, because they have been defined on non–overlapping subdomains with suitable boundary conditions. These BVPs can be solved in a parallel computer, one can assign an individual processor for each BVP, which reduce the CPU time adequately, and one has the possibility to use more number of nodes on each subdomain.

3   Convergence Analysis

In this section, we provide some theoretical results, such as the truncation error, stability and convergence results of the adaptive spline difference scheme (5).

First, we shall present the maximum principle, and the stability estimate.

Lemma 1. *Let v be a smooth function satisfying $v(0) \geq 0$, $v(1) \geq 0$ and $f(x) \leq 0, \forall x \in D$. Then $v(x) \geq 0$, $\forall x \in \overline{D}$. Further, we have the following uniform stability estimate:*

$$|v(x)| \leq C[|v(0)| + |v(1)| + \max_{y \in \overline{D}} |f(y)|], \quad \forall x \in \overline{D}.$$

hereafter C denotes a positive constant independent of the parameter ε, the mesh points x_i, and the step size h.

Proof. One can prove this result following the method given in [FHM1, MOS1].
♣

Theorem 1. *Let u be the solution of the BVP (1)-(2), and u_i be the numerical solution of the difference scheme (5). Then, we have*

$$|u(x_i) - u_i| \leq Ch^2.$$

Proof. Replacing the the approximate solution $U = (u_1, \cdots, u_{N-1})^t$ by the exact solution $\widetilde{U} = (u(x_1), \cdots, u(x_{N-1}))^t$ in (6), we obtain

$$\left(J - h^2 Q \widehat{B}\right) \widetilde{U} = \widehat{C} + h^2 Q F + \widetilde{T}(h), \tag{13}$$

where $\widetilde{T}(h) = (t_1(h), \cdots, t_{N-1}(h))^t$ is the truncation error generated from this replacement and is given by

$$\|\widetilde{T}(h)\| = O\left(\frac{h^4}{\varepsilon}\right), \tag{14}$$

where $\| \cdot \|$ is the matrix maximum norm.

Subtracting (6) from (13) and letting $E = \widetilde{U} - U$, we have

$$\left(J - h^2 Q \widehat{B}\right) E = O\left(\frac{h^4}{\varepsilon}\right). \tag{15}$$

Following [Va1], it can be shown that,

$$\left\| \left(J - h^2 Q \widehat{B} \right)^{-1} \right\| \leq C \left(\frac{\varepsilon}{h^2} \right). \tag{16}$$

So, from (15), we have

$$\|E\| \leq \left\| \left(J - h^2 Q \widehat{B} \right)^{-1} \right\| \, \|\widetilde{T}(h)\|. \tag{17}$$

Thus, one can obtain

$$|u(x_i) - u_i| \leq C h^2. \quad \clubsuit \tag{18}$$

Lemma 2. *Let us consider the BVP:*

$$-\varepsilon u''(x) + b(x)u(x) = f(x), \quad x \in (c,d) \tag{19}$$

$$u(c) = \alpha, \quad u(d) = \beta, \tag{20}$$

and the same differential equation with a perturbation in the left and right hand side boundary conditions, that is, to say $u(c) = \alpha + O(\varepsilon)$, and $u(d) = \beta + O(\varepsilon)$. We refer the second problem as a Perturbed BVP (PBVP). Let u_1 and u_2 be respectively the solutions of these problems. Then we have the following estimate:

$$|u_1(x) - u_2(x)| \leq C\varepsilon, \quad \forall x \in [c,d],$$

Proof. Let $w(x) = u_1(x) - u_2(x)$. Then $w(x)$ satisfies the following BVP

$$Lw(x) = 0, \quad x \in (c,d)$$

$$w(c) = O(\varepsilon), \quad w(d) = O(\varepsilon).$$

Applying Lemma 1, to the previous BVP, we get $|w(x)| \leq C\varepsilon$. $\quad \clubsuit$

The following theorem is the main result of this article, which conveys the relation between the numerical solution using the transition boundary conditions and the exact solution.

Theorem 2. *Let u be the solution of the BVP (19)-(20) and u_i be the numerical solution of the respective PBVP by applying the difference scheme as given in (5). Then,*

$$|u(x_i) - u_i| \leq C(\varepsilon + h^2), \quad \forall x \in [c,d].$$

Proof. We have

$$|u(x_i) - u_i| \leq |u(x_i) - u_2(x_i)| + |u_2(x_i) - u_i|,$$

where $u_2(x)$ is the solution of the Perturbed BVP.

Applying Theorem 1 to the second part in the right hand side of the above inequality, we get $|u_2(x_i) - u_i| \leq C h^2$. Combining this with the result of Lemma 2, we obtain the required estimate. $\quad \clubsuit$

By observing the facts that $\overline{A} = u(k\varepsilon) + O(\varepsilon)$, and $\overline{B} = u(1 - k\varepsilon) + O(\varepsilon)$, where u is the solution of the BVP (1)-(2), we obtain the following result.

Theorem 3. *Let u be the solutions of the BVP (1)-(2), and u_i be the numerical solution of one of the subdomain problems in the boundary layers or in the regular layer obtained by the difference scheme given in (5). Then,*

$$|u(x_i) - u_i| \leq C(\varepsilon + h^2), \quad \forall x_i \in [0, 1].$$

4   Parallel Implementation

We have implemented the present method in parallel environment using Sun Blade 150 servers on Solaris9 OS, 1GB memory using MPI (Message Passing Interface) library optimized for this particular setup. An example has been taken, which is generally used as a test example for these types of problems. From the numerical tables, one can easily see the accuracy, and performance of the method over other methods.

Example 1. Consider the singularly perturbed self–adjoint BVP:

$$-\varepsilon u''(x) + u(x) = -\cos^2(\pi x) - 2\varepsilon \pi^2 \cos(2\pi x), \quad x \in (0, 1)$$
$$u(0) = 0, \quad u(1) = 0.$$

The exact solution is given by

$$u(x) = \frac{[\exp(-x/\sqrt{\varepsilon}) + \exp(-(1-x)/\sqrt{\varepsilon})]}{1 + \exp(-1/\sqrt{\varepsilon})} - \cos^2(\pi x).$$

The transition boundary condition is given by

$$\widetilde{u}(x) = -\cos^2(\pi x) + \exp(-x/\sqrt{\varepsilon}) + \exp(-(1-x)/\sqrt{\varepsilon}).$$

As mentioned earlier, we implement the computations with three processors. For this, we divide the interval $[0, 1]$ into three non–overlapping subdomains as $\overline{D} = D_1 \cup D_2 \cup D_3$, where $D_1 = [0, k\sqrt{\varepsilon}]$, $D_2 = [k\sqrt{\varepsilon}, 1 - k\sqrt{\varepsilon}]$, and $D_3 = [1 - k\sqrt{\varepsilon}, 1]$. Here, we took $k = 2\ln(N)$, in order to have distribution of mesh points analogous to the well-known piece-wise uniform mesh namely, Shishkin's mesh and divide the job into the three processors using Master-Slave approach. We repeat the procedure for different values of ε and N.

It should be noted that the length of the interval is D_2 is bigger than D_1 and D_3, as a consequence the workload is not properly divided into the processors. This can be overcome by using more than three processors. In that case, the interval D_2 can be divided into more subintervals to attain a monotonic speedup. This idea has been introduced in the paper of Vigo-Aguiar and Natesan [VN1].

The numerical results are given in terms of maximum point-wise errors in the tables for various values of ε, and N. Table 1 shows the point-wise error for three values of ε.

For comparison we took the numerical results by applying the spline scheme in the whole domain and results are given in Table 2. It is clear that the proposed scheme improves the results significantly.

The CPU time of both sequential and parallel computers have been given in Figure 1 (a), which reveals the fact that the parallel computer reduces adequately the CPU time. Figure 1 (b) shows the speedup of the present method.

Table 1. Maximum error by applying spline scheme in whole domain $\overline{D}$ for Example 1

	Domain	N=32	N=64	N=128	N=256
$\varepsilon = 10^{-06}$	[0,1]	2.8793e-04	1.3446e-04	5.7463e-05	2.0412e-05
$\varepsilon = 10^{-08}$	[0,1]	3.0570e-05	1.5223e-05	7.5177e-06	3.6606e-06
$\varepsilon = 10^{-10}$	[0,1]	3.0748e-06	1.5401e-06	7.6955e-07	3.8384e-07

Table 2. Maximum error of proposed technique for Example 1

	Domain	N=32	N=64	N=128	N=256
	D_1	9.5512e-07	9.8008e-07	9.9223e-07	9.9273e-07
$\varepsilon = 10^{-06}$	D_2	0.0	0.0	0.0	0.0
	D_3	9.5512e-07	9.8008e-07	9.9223e-07	9.9273e-07
	D_1	9.5497e-09	9.7999e-09	9.928e-09	9.9718e-09
$\varepsilon = 10^{-08}$	D_2	0.0	0.0	0.0	0.0
	D_3	9.5497e-09	9.7999e-09	9.928e-09	9.9718e-09
	D_1	9.5498e-11	9.8003e-11	9.9230e-11	9.9711e-11
$\varepsilon = 10^{-10}$	D_2	0.0	0.0	0.0	0.0
	D_3	9.5498e-11	9.8003e-11	9.9230e-11	9.9711e-11

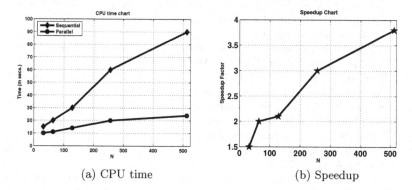

(a) CPU time   (b) Speedup

Fig. 1. CPU time and Speedup in the parallel computer

5   Conclusions

A parallel computational technique for singularly perturbed BVPs of the form (1-2) has been analysed and implemented in this article. Firstly, a non-overlapping domain decomposition of the computational domain is devised, and then independent BVPs on each subdomain are proposed. Suitable boundary conditions have been supplied from the asymptotic approximate solution, which are not difficult to calculate either theoretically, or numerically. Stability, and truncation error have been discussed, and error estimate has been obtained. It is observed

262   R.K. Bawa and S. Natesan

that the implementation of proposed scheme reduces the maximum absolute error and the CPU time to much extend.

Acknowledgements. The authors wish to thank Ms. G. Verma for initial help in verifying their theoretical findings. Also the authors express their sincere thanks to the unknown referees for their valuable suggestions and comments.

References

[Ba1]   Bawa, R.K.: Parallelizable Computational Technique for Singularly Perturbed Boundary Value Problems Using Spline. Lecture Notes in Computer Science **3980** (2006) 1177–1182

[BN1]   Bawa, R.K., Natesan, S.: A computational method for self-adjoint singular perturbation problems using quintic splines. Comput. Math. Applic. **50** (2005) 1371–1382

[Bo1]   Boglaev, I.: Domain decomposition in Boundary layer for Singular Perturbation problems. Appl. Numer. Math. **35** (2000) 145–156

[FHM1] Farrell, P.A., Hegarty, A.F., Miller, J.J.H., O'Riordan, E., Shishkin, G.I.: Robust Computational Techniques for Boundary Layers. Chapman & Hall/CRC Press (2000)

[MOS1] Miller, J.J.H., O'Riordan, E., Shishkin, G.I.: Fitted Numerical Methods for Singular Perturbation Problems. World Scientific (1996)

[NVR1] Natesan, S., Vigo-Aguiar, J., Ramanujam, N.: A numerical algorithm for singular perturbation problems exhibiting weak boundary layers. Comput. Math. Applic. **45** (2003) 469–479

[PG1]   Paprzycki, M., Gladwell, I.: A parallel chopping method for ODE boundary value problems. Parallel Comput. **19** (1993) 651–666

[RST1]   Roos, H.-G., Stynes, M., Tobiska, L.: Numerical Methods for Singularly Perturbed Differential Equations. Springer (1996)

[St1]   Stojanović, M.: Numerical solution of initial and singularly perturbed two-point boundary value problems using adaptive spline function approximation. Publications de L'institut Mathematique **43** (1988) 155–163

[Va1]   Varah, J.M.: A lower bound for the smallest singular value of a matrix. Linear Alg. Appl. **11** (1975) 3–5

[VN1]   Vigo-Aguiar, J., Natesan, S.: A parallel boundary value technique for singularly perturbed two-point boundary value problems. The Journal of Supercomputing **27** (2004) 195–206

Collaborative Grid Process Creation Support in an Engineering Domain

Thomas Friese[1], Matthew Smith[1], Bernd Freisleben[1],
Julian Reichwald[2], Thomas Barth[2], and Manfred Grauer[2]

[1] Dept. of Mathematics and Computer Science, University of Marburg
Hans-Meerwein-Str., D-35032 Marburg, Germany
{friese, matthew, freisleb}@informatik.uni-marburg.de
[2] Information Systems Institute, Faculty of Economics, University of Siegen
Hölderlinstr. 3, D-57068 Siegen, Germany
{reichwald, barth, grauer}@fb5.uni-siegen.de

Abstract. The software development process for many real-world applications requires many experts from different domains to cooperate. This is especially true for applications which are to be deployed in a distributed and heterogeneous environment such as the service-oriented Grid. In this paper, we present a collaborative synchronized Grid process creation environment for the service-oriented Grid, in which experts from different engineering and IT domains can interactively work on a single application process model, to quickly and efficiently design applications spanning multiple problem domains. An example application from the engineering domain of metal forming is presented and the enabling technologies are discussed.

1 Introduction

Service-oriented Grid computing has gained tremendous interest in various application domains. Many of those applications stem from an academic environment and have traditionally been designed as monolithic solutions that are hard to adapt, even to slight changes in the application requirements. Required adaptations must be implemented by programmers specialized both in Grid middleware and the applications. The paradigm shift to service-orientation in Grid middleware opens the possibility to use a far more flexible software development approach, namely to compose applications from standard components, promising easier development and modification of Grid applications. Even though, Grid technology has only seen a slow adoption in commercial application domains such as engineering. We see two main reasons for this slow adoption: On the one hand, the inherent complexity of current service-oriented Grid middleware systems is still prohibitive for everyday use by an application domain expert who has no background in middleware development, Grid computing or even computer science. On the other hand, an engineering solution to a concrete problem is often a team effort undertaken by a number of involved engineers, and other

Y. Robert et al. (Eds.): HiPC 2006, LNCS 4297, pp. 263–276, 2006.

non-IT personnel. Current support for collaborative software development is often limited to the use of CVS, email and conference calls. Such methods offer only limited support to ease the entry of engineers not trained in formal software development processes into the Grid.

The Business Process Execution Language for Web Services (BPEL) [16] has gained much attention and broad adoption for composition of component based business applications. The focus of the BPEL language is to enable the composition of basic web services into more complex applications. Its popularity in the business application domain makes BPEL very promising and interesting for process creation in the Grid domain, since many process execution, management and creation tools are expected to be developed in the future or are even currently under development.

In this paper, we present an engineering application from the domain of metal forming and discuss how a novel collaborative Grid process creation environment is used to support engineers in their Grid application development process. The resulting process representation of the engineering Grid application is executed by a Grid enabled BPEL process execution engine. The visualization of the process in our distributed process editor helps to bridge the gap between Grid application developers and domain experts by allowing them to interactively refine the target application from a high level perspective down to the actual executable code. A Grid extension to the BPEL language is proposed and implemented to facilitate easy parallel execution of BPEL tasks in the Grid.

The term *process* is used in four different connotations throughout this paper. First, the actual collaboration of engineers to solve a given engineering problem is referred to as the engineering process. This paper describes a tool to support the software development process that leads to the creation of a Grid application. This Grid application is based on a process description, referred to as a Grid process. The Grid process describes the orchestration of basic component services into a more complex application. Finally, the actual activities during the production of a cast metal part are referred to as casting process.

The paper is organized as follows. After an overview of related work in section 2, a sample scenario from an engineering domain (casting as a sub-domain of metal forming) is introduced in 3. The design and implementation of a collaborative Grid process editor that supports the software development process for Grid experts and domain experts is presented in section 4. Section 5 describes the collaborative development of the Grid application for the metal casting sample scenario. Section 6 concludes the paper and outlines areas for future work.

2   Related Work

Supporting business processes with software systems and especially service-oriented architectures realized with web services have received considerable attention in both academia and industry. Several other research projects try to cope with similar subjects in related fields.

The Geodise project [24, 20] focuses on optimization, design and fluid dynamics, especially in aerodynamics. Its main goal is to provide a distributed problem solving environment (PSE) for engineers working in the mentioned fields by utilizing e.g. MATLAB and adding Grid functionality to it. Although first Geodise implementations were based on the Globus Toolkit version 2, the core Geodise Toolbox is now part of the managed program of the *Open Middleware Infrastructure Institute (OMII)* [17].

A Grid-enabled problem solving environment for engineering design where distributed parties are able to collaborate has been introduced by Goodyer et al. [11]. The system makes use of the gViz Library [4] which allows collaborative visualization on the Grid and provides the user to start Grid jobs on Globus Toolkit based hosts. The main focus is put on collaborative application steering and result visualization of given simulation problems.

The P-GRADE Portal [18] aims to be a workflow-oriented computational Grid portal, where multiple clients can collaboratively participate in design, development and execution of a workflow, and multiple Grids may be incorporated in the workflow execution. The P-GRADE Portal is based on the Globus Toolkit version 2 for file transfer operations and job execution, the workflow execution is done by a proprietary implementation. P-GRADE neither uses Grid service and business process standards such as BPEL, nor does the proposed collaborative editing approach support real time collaboration on a process in an on-line meeting style.

The GridNexus Project [5] is a GUI for workflow creation based on Ptolemy II and JXPL, a XML scripting language based on Lisp, which it uses as a Grid foundation. It allows a very fine grained algorithm design (right down to the arithmetic operations) to be integrated with Grid service interactions, raising the complexity of the resulting workflow. GridNexus does not allow the collaborative creation of workflows nor does it support business standards such as BPEL.

The mentioned software systems are examples for the large variety of problem solving environments, collaborative Grid application systems and collaborative workflow development systems. However, none of the mentioned systems provides both a problem solving environment for engineering problems as well as sophisticated support for the collaborative software development process for Grid applications and their execution in a service-oriented Grid environment. Collaboration support often relies on out-of-band collaboration and synchronization techniques such as exchanging e-mail or CVS like server based communication.

3 Sample Application

In this section, a simplified view on a sample application from an engineering domain is presented to motivate the need for support in the distributed software development process of a Grid software system for engineering applications. The concrete use case comes from casting, a sub-domain of metal forming. Only those parts relevant to the Grid are briefly sketched; they do not reflect the entire complex field of metal forming. For more information regarding the complexity involved in collaborative engineering particularly in the field of metal forming and casting, the reader is referred to e.g. [15, 23].

In the metal casting industry, customers' quality requirements, e.g. allowed tolerances in a casting product's geometry compared to the specification, are constantly increasing. Therefore, the use of numerical simulation and simulation-based optimization is gaining importance, since the creation of prototypes is prohibitively expensive and time consuming. The benefit of simulated prototyping is constrained by the accuracy of the simulation environment. Both the creation and use of the simulation application require great expertise in the metal casting domain. Furthermore, applying numerical simulation for this purpose introduces an extremely high demand for computational capacity since a single - sufficiently precise - simulation run typically lasts several hours up to days. Since many small and medium sized engineering enterprises are not capable of acquiring and maintaining high performance computing resources, outsourcing of computational demanding tasks is necessary. Grid computing promises to offer the infrastructural components to realize this outsourcing activity as easy as plugging into the electrical power Grid. However, currently the implementation of a Grid application still requires these firms to involve Grid specialists to adapt and maintain their applications in a Grid environment.

To sumarize, the utilization of numerical simulation in the casting industry requires a variety of competencies:

- knowledge about the physical properties of casting in industrial practice (casting engineer)
- modeling a casting engineering process for simulation (casting engineers together with IT specialists)
- adapting existing simulation software to the Grid (Grid specialists consulting the casting engineers)
- setting up and maintaining a simulation and/or optimization environment for the engineers' customers (Grid specialists, casting engineers and their customers)
- interpreting a simulation's result (casting engineer and customer).

These requirements lead to a software platform which enables the integration of the aforementioned competencies and resources during the software design process. Since most of the possible users of simulation in the casting industry are small to medium enterprises (SME), lacking at least one of the requirements, the Grid software platform must be able to facilitate both renting computational resources on demand as well as the collaborative involvement of Grid experts, casting engineers and their customers.

As a concrete sample scenario, we introduce the engineering process of collaborative development of a metal casting model. From a software development point of view, the Grid relevant development cycle starts with a problem definition, expressed by the casting engineer and progresses through some iterations of model definition, simulation and refinement. The given problem definition is then modeled as an initial casting process model by a numerical simulation expert. Usually, this expert is located in another company due to the already mentioned lack of personnel or know-how in small and medium engineering enterprises. The numerical simulation expert periodically discusses the evolution of the initial model with the casting engineer during the design phase. Both experts have to combine their expertise to successfully define an accurate model for the casting process. To verify the accuracy of the resulting model, it typically is numerically simulated. The results must be reviewed and compared to knowledge about real casting processes held by the casting engineer. If this first simulation run does not match reality, the model needs to be calibrated and further model variants are created by the simulation expert and the casting engineer.

During this model calibration phase, an optimization expert is also involved in creating model variants. When a single model is calibrated, the optimization of the model begins by automatically generating a number of n new models by varying the parameters in the casting process model. They can be evaluated in parallel, and the results from the simulation runs flow back to the optimization algorithm. This procedure iterates until the optimized casting process meets the requirements set by the casting engineer. The simulation software, which runs n instances in parallel, requires distributed computing resources and therefore suggests the application of Grid technology.

In the next section, we will introduce a tool to support the different actors in collaboration during the software development process for the final Grid application allowing them to perform their engineering process for a concrete casting process model. This development process brings together domain experts from different engineering domains and IT experts from the Grid domain.

4   Collaborative Grid Process Creation Support

The main goal for a tool that supports collaborative Grid process creation is to enable developers with only limited expertise in distributed application development or programming for the Grid to create applications. Ideally, it should enable domain experts without a computer science background to construct applications as solutions to their domain specific problems that utilize the available Grid resources. In our example descibed above, it is the Grid application supporting the engineering process by providing distributed simulation and optimization functionality. In the service-oriented Grid, the application consists of Grid services as the basic components and a process description defining their orchestration to provide functionality for numerical simulation and optimization. In the sample application, the casting engineer, the numerical simulation expert and the Grid expert are the parties who collaborate to create the required Grid process. The two engineering domain experts define and discuss a high level application flow before involving the Grid expert who identifies and implements the necessary Grid components and adapts the high level process definition to the pecularities of the Grid.

4.1   Design of the Grid Process Editor

Two main considerations drive the design of our proposed Grid process editor as a tool supporting fast development of Grid application from a high level perspective: It should provide the ability to adapt to the needs of different groups of developers, allowing Grid middleware experts to inspect and manipulate fine details of a Grid process (high-fidelity editing) while hiding complicated details from application domain experts (low-fidelity editing). Furthermore, a Grid process editor should foster collaboration among experts in a distributed environment. Ideally, it should support synchronized collaborative work (i.e. same-time, different-place collaboration [8]) on the process under development, building on the underlying communication infrastructure.

The overall design of our Grid process editor is based on a model driven approach. Figure 1(a) shows a conceptual overview of the core components of the collaborative Grid process editor. A core model that can be shared among networked nodes is used to represent a concrete process that in turn is presented to the user through a view component with a corresponding controller, allowing for editing operations. A target system mapping defines the transition into actually executable code for a concrete process execution environment. The model sharing component propagates changes of the model to other editors currently acting on the same shared model. This model sharing component must also be able to lock the process model to prevent local modification when a distributed coordination protocol requires this locking operation.

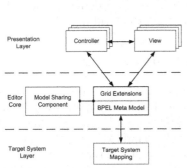 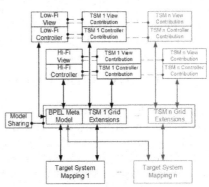

(a) Design overview showing the component and layer separation.

(b) Internal dependencies between target system mapping (TSM), core model and presentation contributions.

Fig. 1. Overall design of the collaborative process editor

The field of process execution for the Grid is a rapidly evolving field. Currently, there is no standard for the expressive power of the core process models that everyone agrees upon. This lack of a common standard and the greater flexibility in implementing new constructs lead to the choice of the internal design shown in figure 1(b). We chose the BPEL language for the core process standard since there is currently no Grid process execution standard but BPEL is the de-facto standard in the business process execution environment. Implementations of a process execution target system may extend the core model and provide their contributions to the model and view components corresponding to their newly introduced constructs.

4.2 Implementation of the Grid Process Editor

The previously introduced overall design has been used to implement a graphical process editor for the Globus Toolkit 4 based MAGE [21] platform. The implementation of this process editor is based on the Eclipse Graphical Editing Framework (GEF) [10] for the presentation layer and the Eclipse Communication Framework (ECF) [9] for the model sharing component.

Core Model. There are three basic types of elements in the core process model. The base class of every model element is the class **Element**. Two direct descendants of this base class are **ContainerElement** and **Connection**. Containers may directly contain other model elements. An example for such a container is the **Sequence** class that represents a BPEL sequence of activities. Sub-classes of the connection class are used to represent links between activities in the process, such as the links between activities in a BPEL Flow (a flow is not ordered like a sequence but a collection of activities that are executed based on transition conditions). A globally unique ID [13] is assigned to every element in the process model, allowing to uniquely identify the elements even across different nodes sharing a single process model through the model sharing component. Conceptually, every element of the process model represents a collection of attributes. The **Element** class implements the **IPropertySource** interface allowing the Eclipse platform to directly display the attribute values of a selected model element in the standard property view. The root element of every proces model based on this meta-model implementation is an instance of the **Process** class.

Every model element implements the **IPropertySource** interface. It defines operations to retrieve a list of **IPropertyDescriptors** and access methods to set the value of a particular property or to get a property value from the model element. As a means of selectively displaying certain element properties, the core process model allows to select a set of filter rules on the model instance. Before returning the result list in the implementation of the **getPropertyDescriptors** operation, the list is filtered by a **PropertyVisibilityFilter** filter that removes every property descriptor defined in the exclusion list defined for the element. The property access methods of the element still provide access to filtered attributes, allowing wizards that automate tasks for the user full access to every element and property of the process model.

The model needs to be serialized and deserialized for storage and transmission. Serialization of the model elements is handled by proxy classes that implement the serialization capability for the core model elements. The proxy classes handle storage of the model elements, therefore, they are referred to as *storage proxies*. Every storage proxy instance holds a reference to an associated model object. A storage proxy instance for a concrete instance of a model element can be obtained from a factory that receives the model element object in the proxy creation request and constructs or returns the associated proxy instance. Storage proxies for child objects of a container model element are constructed recursively. The default implementation of the storage proxy elements implement the methods toDOMElement and fromDOMElement in the IStorageProxy interface. These methods return or interpret a tree of XML elements representing the elements and connections in the process model. The structure of the resulting XML elements and attributes is governed by the concrete implementation of the IStorageProxy instances returned by the proxy factory.

Similar to the regular serialization implementation for simple storage of the process model, a concrete target system mapping implementation may perform serialization of the model into the required process representation by providing another proxy factory. The basic serialization implementations queries all elements for their unfiltered properties and serializes them into attributes and child elements, enabling storage and retrieval of the complete model information.

Target System Mapping. A target system mapping for the MAGE process execution environment has been implemented as a second set of storage proxies with a corresponding factory. This MAGE process execution engine is based on the ActiveBPEL [1] engine and introduces additional Grid specific concepts such as a Grid-For-Each (GFE) construct. This construct supports the very common case of carrying out a number of calculations on a separation of the input parameter domain. It encapsulates discovery of computational resources, splitting of input parameters, collecting and merging of execution results. Contributions to the core BPEL model, view and controller components of the process editor have been implemented to represent the Grid specific extensions supported by the MAGE process execution engine.

The storage proxy implementation for the process model element in the MAGE target system mapping handles creation of all necessary process descriptions, deployment descriptors and WSDL artifacts needed for the process execution engine. The GFE enables non-Grid experts to easily model the parallel execution of tasks in the Grid, since it hides the complexity of discovering nodes, deploying the services, splitting the parameter range and executing the services. The engineers only need to configure the GFE to a certain parameter range and define the single node process, the rest is done by the MAGE process execution environment. Providing such high level constructs representing complex yet common tasks in the Grid is a vital step towards enabling the use of the Grid for non-Grid experts.

Presentation Layer. The process editor contributes an implementation of a graphical editor as its presentation layer implementation. Its purpose is to provide a graphical representation of the process model to the user that visualizes the process structure and lets the user add, remove and rearrange components, connect them and edit the properties associated with the process activities and other elements.

Visualization of the process model elements is handled by corresponding view elements, editing of the underlying model is performed by edit parts associated with the model elements. The Eclipse platform abstracts the source of an operation as a request, actual modification of the model elements is then performed by a `Command` retrieved from the target edit part. During an editing operation, the edit part also displays feedback (such as drag handles for move and resize operations) through the view elements to the user. Edit parts do not handle editing directly, rather they delegate editing to edit policies associated with the part. Before returning a command or signaling approval to an edit request, the core model is queried for any edit locks set on the model by the model sharing component.

An Eclipse workbench window shows a so called *perspective* that is a collection of multiple views on the workspace of the user or currently selected elements. The process editor uses the properties view in addition to the graphical representation of the process model. The properties of a particular model element are accessible to the user through a tree that allows editing of the individual properties. The set of visible properties is determined by the previously described view filters.

Every view on a model element is updated if a property change occurs due to a user interaction or a change received from the model sharing component. This mechanism automatically integrates with the property view provided by the Eclipse platform. Additionally, a property visibility filtering mechanism has been implemented for the editor. This allows to adjust the level of detail in the view presented to different groups of users.

Model Sharing. The model sharing implementation of the process editor uses facilities provided by the Eclipse Communication Framework (ECF) [9]. In order to collaborate, the users of the process editor join a collaboration channel (access protection to the channel may be set up by the collaboration initiator). The node of the collaboration initiator also acts as a coordinator to the collaboration. After joining the collaboration, a new editing partner requests the model from the channel. The initiating partner then serializes the model and transmits it to the newly joining party. The model is then deserialized and used as input for the graphical editor of the new collaboration partner.

The underlying communication channel implements a protocol that ensures reliable message transmission to a selected partner or to all communication partners in the channel. Actual update of the distributed model happens by relay of the edit commands upon their execution.

For this purpose, every command is derived from a `BaseCommand` class that triggers serialization of the command and transmission to other connected editors in its `execute` method, if the editor is connected to a channel. Every command implementation is required to call its super classes `execute` method to ensure transmission of the command. Every command implements the `IAdaptable` interface and returns a serializable and transmissible version of the command to the model sharing component that transmits the command to other editors in the channel. References to model elements are encoded using the globally unique identifier of the element.

This model sharing enables the different actors in the application design, development and usage phases to collaboratively work on the same process model, each editing the model part which belongs to their area of expertise, while at the same time being able to look at the big picture and get instant feedback from colleagues.

5 Collaborative Development of the Sample Application

As a first step towards the optimization process, the casting engineer and numerical simulation expert join a collaborative process design session using the distributed process editor described above. The distributed process editor allows users to join the collaboration and create, delete, connect and move basic activities concurrently. This interaction may be augmented by audio or video-conferencing giving the participants an on-line meeting room environment for their collaborative work. Figure 2(a) shows a snapshot of the distributed process editor with the minimal core process for the Grid application being worked on. The ability to concurrently edit the process model is a critical element in the overall development process since communication difficulties between the different experts can be quickly identified and solved. As stated before, they rely on cooperation to apply their combined expertise to design an application that can support the engineering process in a satisfactory manner. After finishing an initial sketch of the ideal application workflow from their domain perspective, they involve a Grid expert to help them adapt their process to the Grid environment and identify the necessary component services for their Grid process. The Grid expert will introduce infrastructural requirements such as service discovery and infrastructure management into the purely application oriented workflow of the domain experts.

As a result of this second step, a process for the Grid application and the specification of the required component services can be used by the Grid expert to implement or select the basic components. A basic skeleton implementation of the required services can be automatically generated using a model driven service generator [19, 22] leading to fast availability of the component services. For the metal casting sample application, the following two services were identified and implemented:

The Distributed Polytop Service. This service is an implementation of the distributed polytop optimization algorithm [2] (DPA) which belongs to the class of direct search methods. It has its roots in the Complex algorithm [3], a predecessor of the Nelder-Mead Simplex [14]. The DPA was designed regarding efficiency and scalability in distributed systems.

During its runtime, it requires an a priori unknown number of evaluations of both an objective function and corresponding constraint functions, in this case calculated by the metal casting simulation software CASTS [12]. The service has to save its state each time an evaluation request occurs, and it passes the data set which is to be evaluated to the process execution engine instead of directly invoking the simulation service. Considering these conditions, the service was implemented by utilizing the Web Service Resource Framework (WSRF [1]), which allows the creation of stateful web service resources. Apart from a service operation which allows a client to set necessary parameters needed by the polytop algorithm, the only Grid service operation `iterate(IterateRequest)` takes care of starting and restarting the algorithm at the appropriate position - according

---

[1] http://www.globus.org/wsrf/

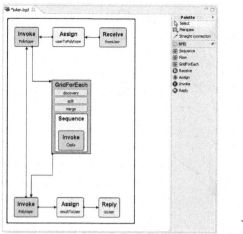
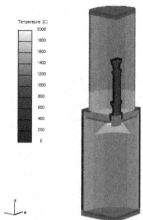

(a) Screenshot of the graphical distributed process editor window displaying the metal casting process.

(b) Visualization window showing a simplified model of a casted gas turbine blade.

Fig. 2. Development and application environment: distributed process editor and engineering application

to its internal state and according to the input data inside the `IterateRequest` data structure. A resulting data set is returned immediately after invoking the operation, telling the process execution engine if further evaluations are needed or if the polytop algorithm reached a predefined stop condition.

The Casts Service. The main purpose of this service is to wrap the metal casting legacy software CASTS as a Grid service. However, the *Casts Service* does not only provide a service-wrapped version of CASTS, but it also takes care of the following operations: It is capable of modifying the input model of the casting process according to a set of parameters passed to the service. This parameter set is the input received from the distributed polytop algorithm. The service executes the CASTS legacy application on a number of different execution platforms. In this case, a 128 node cluster computer with two 64Bit AMD Opteron CPUs and 2GB main memory per node was utilized, leading the execution subsystem to incorporate the local resource manager Torque [7] and the scheduling system Maui [6]. The execution state of a cluster job is monitored and exposed by the Casts Service. The execution subsystem is highly modularized so that the service also works on single workstations without local queuing/scheduling. The service also provides functionality to evaluate the simulation result (which is done by CritCASTS, a legacy software system bundled with CASTS) and determining the objective function value as well as the constraint function values.

Utilization of WS-GRAM [2] for running CASTS jobs in a Grid service wrapped command line was inappropriate due to the complexity of the internal tasks of

---

[2] http://www.globus.org/toolkit/docs/4.0/execution/wsgram/

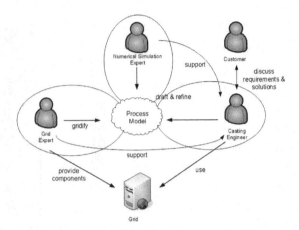

Fig. 3. Overview of the collaborative process creation scenario for the metal casting application

the Casts Service. The chosen approach of a custom wrapper service exposes the necessary information and results much cleaner to the process execution engine. Using WS-GRAM would have required to create a sophisticated shell script as a CASTS wrapper and the inclusion of logic in the Grid application process for parsing the output of WS-GRAM, which truly belongs into the Casts Service.

The concurrent execution of many simulations has been modeled using the Grid-For-Each construct, a Grid specific extension of the BPEL language. The GFE construct neatly encapsulates details of the concurrent execution in the simulation tasks, keeping middleware complexity from the domain experts.

An overall view of this collaborative and distributed development scenario is shown in figure 3. The grey zones mark the network domains of the different experts, they are geographically distributed, and their collaboration takes place via the shared and synchronized process model. The distributed collaborative process editor allows them to synchronously edit the model and directly see the operations of other connected partners.

6   Conclusions

In this paper, we introduced a collaborative process creation environment in which experts from different engineering and IT domains can interactively work on a single synchronized Grid process model, to quickly and efficiently design complex applications spanning multiple problem domains. We demonstrated the feasibility of our approach using an engineering application for metal casting. A new BPEL construct the Grid-For-Each (GFE) was used to facilitate the adoption of Grid computing in the engineering community. The GFE can be used to transparently split an input value set into distinct parameter ranges, distribute processing jobs over multiple Grid nodes and then collect and merge the results. The application process is executed by a Grid enabled BPEL process execution engine, which takes care of most of the Grid specific operations required to execute an application in the Grid. The visualization of the application process in our distributed process editor helps to bridge the gap between Grid application developers and domain experts by allowing them to interactively refine the target application from a high level perspective down to the actual executable code. The resulting process model is very flexible allowing a standard application core to be easily adopted to new problems, which is a common requirement in the engineering field.

Future work includes the extension of the collaborative process creation environment by further BPEL constructs to ease the distributed application development and include utilities and wizards to further speed up the time to market for legacy applications. Furthermore, the integration of a video conferencing software into the platform would also increase the productivity of the development environment without having to rely on external video conferencing solutions.

Acknowledgements

This work is partially supported by the German Ministry of Education and Research (BMBF) (D-Grid Initiative, In-Grid Project), Siemens AG (Corporate Technology, München) and IBM (Eclipse Innovation Grant). We would like to thank Jürgen Jakumeit of ACCESS Materials + Processes for his involvement in the design process of the Grid Casts application.

References

1. ActiveBPEL, LLC. ActiveBPEL - BPEL Execution Engine. http://www.activebpel.org.
2. T. Barth. *Verteilte Lösungsansätze für simulations-basierte Optimierungsprobleme.* Wissenschaftlicher Verlag Berlin, 2001.
3. M. Box. A New Method of Constrained Optimization and a Comparison with Other Methods. *Computer Journal,* 8:42–52, 1965.
4. K. Brodlie, D. Duce, J. Gallop, M. Sagar, J. Walton, and J. Wood. Visualization in Grid Computing Environments. *Proceedings of IEEE Visualization,* pages 155–162, 2004.
5. J. L. Brown, C. S. Ferner, T. C. Hudson, A. E. Stapleton, R. J. Vetter, T. Carland, A. Martin, J. Martin, A. Rawls, W. J. Shipman, and M. Wood. GridNexus: A Grid Services Scientific Workflow System. *International Journal of Computer Information Science,* 6:72–82, 2005.
6. clusterresources.com. Maui Cluster Scheduler. http://www.clusterresources.com/pages/products/maui-cluster-scheduler.php.
7. clusterresources.com. Torque Resource Manager. http://www.clusterresources.com/pages/products/torque-resource-manager.php.
8. F. Cosquer, P. Veríssimo, S. Krakowiak, and L. Decloedt. Support for Distributed CSCW Applications. In *Distributed Systems,* volume 1752, pages 295–326. Springer, 2000.
9. eclipse.org. Eclipse Communication Framework (ECF). http://www.eclipse.org/ecf.

10. eclipse.org. Graphical Editing Framework. `http://www.eclipse.org/gef`.
11. C. Goodyer, M. Berzins, P. Jimack, and L. Scales. A Grid-Enabled Problem Solving Environment for Parallel Computational Engineering Design . *Advances in Engineering Software*, 37:439–449, 2006.
12. G. Laschet, J. Neises, and I. Steinbach. Micro-Macrosimulation of Casting Processes. 4^{ieme} *école d'été de Modélisation numérique en thermique*, 1998.
13. P. Leach, M. Mealling, and R. Salz. A Universally Unique IDentifier (UUID) URN Namespace. IETF RFC 4122, 2005. `http://www.ietf.org/rfc/rfc4122.txt`.
14. J. Nelder and R. Mead. A Simplex Method for Function Minimization. *Computer Journal*, 7:308–311, 1965.
15. T. Nguyen and V. Selmin. Collaborative Multidisciplinary Design in Virtual Environments. In *Proceedings of the 10th International Conference on CSCW in Design.*, pages 420–425, Nanjing, China, 2006.
16. Oasis WSBPEL TC. Web Services Business Process Execution Language Version 2.0 - draft, December 2005.
17. omii.ac.uk. Open Middleware Infrastructure Institute (OMII). `http://www.omii.ac.uk/`.
18. G. Sipos and P. Kacsuk. Collaborative Workflow Editing in the P-GRADE. *Proceedings of the MicroCAD*, 2005.
19. M. Smith, T. Friese, and B. Freisleben. Model Driven Development of Service-Oriented Grid Applications. In *Proc. of the International Conference on Internet and Web Applications and Services*, pages 139–146, Guadeloupe, 2006. IEEE Press.
20. W. Song, Y.-S. Ong, H.-K. Ng, A. Keane, S. Cox, and B. Lee. A Service-Oriented Approach for Aerodynamic Shape Optimization Across Institutional Boundaries. *Proceedings of ICARCV*, 2004.
21. The Distributed Systems Group, Univ. of Marburg, Germany. The Marburg Ad Hoc Grid Envrionment - MAGE. `http://ds.informatik.uni-marburg.de/MAGE`.
22. The Distributed Systems Group, University of Marburg, Germany. Grid Development Tools for Eclipse. `http://ds.informatik.uni-marburg.de/MAGE/gdt`.
23. S. Woyak, H. Kim, J. Mullins, and J. Sobieszczanski-Sobieski. A Web Centric Architecture for Deploying Multi-Disciplinary Engineering Design Processes. In *Proceedings of the 10th AIAA/ISSMO Multidisciplinary Analysis and Optimization Conference*, Albany, New York, USA, 2004.
24. G. Xue, W. Song, S. Cox, and A. Keane. Numerical Optimization as Grid Services for Engineering Design. *Journal of Grid Computing*, 2(3):223–238, 2004.

A Multi-attribute Data Structure with Parallel Bloom Filters for Network Services[*]

Yu Hua[1,2] and Bin Xiao[1]

[1] Department of Computing
Hong Kong Polytechnic University, Kowloon, Hong Kong
{csyhua, csbxiao}@comp.polyu.edu.hk
[2] School of Computer Science and Technology
Huazhong University of Science and Technology, Wuhan, China

Abstract. A Bloom filter has been widely utilized to represent a set of items because it is a simple space-efficient randomized data structure. In this paper, we propose a new structure to support the representation of items with *multiple* attributes based on Bloom filters. The structure is composed of Parallel Bloom Filters (PBF) and a hash table to support the accurate and efficient representation and query of items. The PBF is a counter-based matrix and consists of multiple submatrixes. Each submatrix can store one attribute of an item. The hash table as an auxiliary structure captures a verification value of an item, which can reflect the inherent dependency of all attributes for the item. Because the correct query of an item with multiple attributes becomes complicated, we use a two-step verification process to ensure the presence of a particular item to reduce false positive probability.

1 Introduction

A standard Bloom filter can represent a set of items as a bit array using several independent hash functions and support the query of items [1]. Using a Bloom filter to represent a set, one can query whether an item is a member of the set according to the Bloom filter, instead of the set. This compact representation is the tradeoff for allowing a small probability of false positive in the membership query. However, the space savings often outweigh this drawback when the false positive probability is rather low. Bloom filters can be widely used in practice when space resource is at a premium.

From the standard Bloom filters, many other forms of Bloom filters are proposed for various purposes, such as counting Bloom filters [2], compressed Bloom filters [3], hierarchical Bloom filters [4], space-code Bloom filters [5] and spectral Bloom filters [6]. Counting Bloom filters replace an array of bits with counters in order to count the number of items hashed to that location. It is very useful

---

[*] This work is partially supported by HK RGC CERG B-Q827 and POLYU A-PA2F, and by the National Basic Research 973 Program of China under Grant 2004CB318201.

Y. Robert et al. (Eds.): HiPC 2006, LNCS 4297, pp. 277–288, 2006.

to apply counting Bloom filters to support the deletion operation and handle a set that is changing over time.

With the booming development of network services, the query based on *multiple* attributes of an item becomes more attractive. However, not much work has been done in this aspect. Previous work mainly focused on the representation of a set of items with a single attribute, and they could not be used to represent items with multiple attributes accurately. Because one item has multiple attributes, the inherent dependency among multiple attributes could be lost if we only store attributes in different places by computing hash functions independently. There are no functional units to record the multiple attributes dependency by the simple data structure expansion on the standard Bloom filters and the query operations could often receive wrong answers. The lost of dependency information among multiple attributes of an item greatly increases the false probability. Thus, we need to develop a new structure to the representation of items with multiple attributes.

In this paper, we make the following main contributions. First, we propose a new Bloom filter structure that can support the representation of items with multiple attributes and allow the false positive probability of the membership queries at a very low level. The new structure is composed of Parallel Bloom Filters (PBF) and a hash table to support the accurate and efficient representation and query of items. The PBF is a counter-based matrix and consists of multiple submatrixes. Each submatrix can store one attribute of an item. The hash table captures a verification value of an item, which can reflect the inherent dependency of all attributes for one item. We generate the verification values by an attenuated method, which tremendously reduces the items collision probability. Second, we present a two-step verification process to justify the presence of a particular item. Because the multiple attributes of an item make the correct query become complicated, the verification in the PBF alone is insufficient to distinguish attributes from one item to another. The verification in the hash table can complement the verification process and lead to accurate query results. Third, the new data structure in the PBF explores a counter in each entry such that it can support comprehensive data operations of adding, querying and removing items and these operations remain computational complexity $O(1)$ using the novel structure. We also study the false positive probability and algebra operations through mathematic analysis and experiments. Finally, we show that the new Bloom filter structure and proposed algorithms of data operations are efficient and accurate to realize the representation of an item with multiple attributes while they yield sufficiently small false positive probability through theoretical analysis and simulations.

The rest of the paper is organized as follows. Section 2 introduces the related work. Section 3 presents the new Bloom filter structure, which is composed of the PBF and hash table. Section 4 illustrates the operations of adding, querying and removing items. In Section 5, we present the corresponding algebra operations. Section 6 provides the performance evaluation and Section 7 concludes our paper.

2 Related Work

A Bloom filter can be used to support membership queries [7], [8] because of its simple space-efficient data structure to represent a set and Bloom filters have been broadly applied to network-related applications. Bloom filters are used to find heavy flows for stochastic fair blue queue management scheme [9] and summarize contents to help the global collaboration [10]. Bloom filters provide a useful tool to assist the network routing, such as route lookup [11], packet classification [12], per-flow state management and the longest prefix matching [13].

There is a great deal of room to develop variants or extensions of Bloom filters for specific applications. When space is an issue, a Bloom filter can be an excellent alternative to keeping an explicit list. In [14], authors designed a data structure called an exponentially decaying bloom filter (EDBF) that encoded such probabilistic routing tables in a highly compressed manner and allowed for efficient aggregation and propagation.

In addition, network applications emphasize a strong need to engineer hash-based data structure, which can achieve faster lookup speeds with better worst-case performance in practice. From the engineering perspective, authors in [15] extended the multiple-hashing Bloom filter by using a small amount of multi-port on-chip memory, which can support better throughput for router applications based on hash tables.

Due to the essential role in network services, the structure expansion of Bloom filters is a well-researched topic. While some approaches exist in the literature, most work emphasizes the improvements on the Bloom filters themselves. Authors in [16] suggested the multi-dimension dynamic bloom filters (MDDBF) to support representation and membership queries based on the multi-attribute dimension. Their basic idea was to represent a dynamic set A with a dynamic $s \times m$ bit matrix that consists of s standard Bloom filters. However, the MDDBF lacks a verification process of the inherent dependency of multiple attributes of an item, which may increase the false positive probability.

3 Analytical Model

In this section, we will introduce a novel structure, which is composed of PBF and a hash table, to represent items of p attributes. The hash table stores the verification values of items and we provide an improved method for generating the verification values.

3.1 Proposed Structure

Figure 1 shows the proposed structure based on the counting Bloom filters. The whole structure includes two parts: PBF and a hash table. PBF and the hash table are used to store multiple attributes and the verification values of items, respectively. PBF uses the counting Bloom filters [2] to support the deletion operation and can be viewed as a *matrix*, which consists of p parallel *submatrixes*

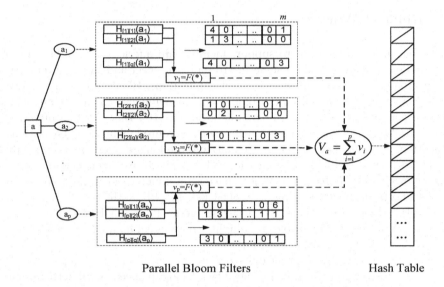

Parallel Bloom Filters   Hash Table

Fig. 1. The proposed structure based on counting Bloom filters

in order to represent p attributes. A submatrix is composed of q parallel *arrays* and can be used to represent one attribute. An array consists of m counters and is related to one hash function. q arrays in parallel are corresponding to q hash functions. Assume that a_i is the ith attribute of item a. We use $H_{[i][j]}(a_i)(1 \le i \le p, 1 \le j \le q)$ to represent the hash value computed by the jth hash function for the ith attribute of item a. Thus, each submatrix has $q \times m$ counters and PBF composed of p submatrixes utilizes $p \times q \times m$ counters to store the items with p attributes.

The hash table contains the verification values, which can be used to verify the inherent dependency among different attributes from one item. We measure the verification values as a function of the hash values. Let $v_i = F(H_{[i][j]}(a_i))$ be the verification value of the ith attribute of item a. The verification value of item a can be computed by $V_a = \sum_{i=1}^{p} v_i$, which can be inserted into the hash table for future dependency tests.

3.2   Role of Hash Table

The fundamental role of the hash table is to verify the inherent dependency of all attributes for an item and avoid the query collision. The main reason for the query collision in terms of multiple attributes is that the dependency among multiple attributes is lost after we insert p attributes into p independent submatrixes, respectively. Then, the PBF *only* knows the existence of attributes and cannot determine whether those attributes belong to one item. Meanwhile, the verification based on PBF itself is not enough to distinguish attributes from ne item to another. Therefore, the hash table can be used to confirm whether the queried multiple attributes belong to one item.

Thus, if a query receives answer *True*, the two-step verification process must be conducted. First, we need to check whether queried attributes exist in PBF. Second, we need to verify whether the multiple attributes belong to a single item based on the verification value in the hash table.

3.3 Verification Value

Traditionally, the hash values computed by hash functions are only used to update the location counters in the counting Bloom filters. In the proposed structure, we utilize the hash values to generate the verification values, which can stand for existing items.

The basic method of generating the verification value is to add all the hash values and store their sum in the hash table. For example, the value of variable v_i is $v_i = F(H_{[i][j]}(a_i)) = \sum_{j=1}^{q} H_{[i][j]}(a_i)$ for the ith attribute of item a. In this case, the function F is a sum operation. Then, the verification value of item a is $V_a = \sum_{i=1}^{p} \sum_{j=1}^{q} H_{[i][j]}(a_i)$. Thus, V_a can be inserted into the hash table and stands for an existing item a. However, in the basic method, the values computed by different hash functions are possible to be the same and their sums might be the same, too. Thus, different items might hold the same verification values in the hash table and this will lead to the verification collision.

The improved method utilizes the *sequential information* of hash functions to distinguish the verification values of different items. We allocate different weights to sequential hash functions in order to reflect the difference among hash functions. As for the ith attribute of item a, the value from the jth hash function in the ith submatrix is defined as $\frac{H_{[i][j]}(a_i)}{2^j}$, which is similar to the idea of the *Attenuate Bloom Filters* [17]. In attenuate Bloom filters, higher filter levels are attenuated with respect to earlier filter level and it is a lossy distributed index. Therefore, as for the item a, the verification value of the ith attribute is defined as $v_i = F(H_{[i][j]}(a_i)) = \sum_{j=1}^{q} \frac{H_{[i][j]}(a_i)}{2^j}$. The verification value of item a is $V_a = \sum_{i=1}^{p} \sum_{j=1}^{q} \frac{H_{[i][j]}(a_i)}{2^j}$. This verification value of item a can be inserted into the hash table.

4 Operations on Data Structure

Given a certain item a, it has p attributes and each attribute can be represented using q hash functions as shown in Figure 1. We denote its verification value by V_a, which is initialized to zero. Meanwhile, we can implement the corresponding operations, such as adding, querying and removing items, with a complexity of $O(1)$ in the parallel Bloom filters and the hash table.

4.1 Adding Items

Figure 2 presents the algorithm of adding items in the proposed structure. We need to compute the hash values of multiple attributes by hash functions and then generate the verification values based on the improved method. Meanwhile, the values of corresponding location counters in PBF are incremented and corresponding operations are denoted by $PBF[H_{[i][j]}(a_i)] + +$. Finally, we insert the verification value of item a into the hash table.

Add_Item (Input: Item a)

Initialize $V_a = 0$
for $(i = 1; i \leq p; i + +)$ **do**
  for $(j = 1; j \leq q; j + +)$ **do**
    Compute $H_{[i][j]}(a_i)$
    $V_a = V_a + \frac{H_{[i][j]}(a_i)}{2^j}$
    $PBF[H_{[i][j]}(a_i)] + +$
  end for
end for
Insert V_a into the hash table

Fig. 2. The algorithm of adding an item with multiple attributes

4.2 Querying Items

Figure 3 shows the multi-attribute query algorithm, which realizes the two-step verification process. After computing the hash values of multiple attributes, we need to check whether the attributes exist in PBF. If any $PBF[H_{[i][j]}(a_i)]$ is 0 for item a, the query returns answer *False* in order to show that the queried item a does not exist. Otherwise, the hash values are added to generate the verification value V_a. If the value V_a is also in the hash table, we can determine that item a exists.

4.3 Removing Items

The operation of removing items needs to remove both the attributes in PBF and the verification values in the hash table. Figure 4 shows the algorithm for removing an item. As for an item a, we compute the hash values of its attributes and subtract 1 from the values of corresponding location counters in order to remove multiple attributes in PBF. Afterwards, the verification value of item a, V_a, is also removed from the hash table.

5 Algebra Operations

The algebra operations of Bloom filters are helpful to implement the representation and membership query of items from different sets. The operations, such as union and intersection, are still applicative in the PBF structure. We first introduce the union and intersection operations of standard Bloom filters and then describe the corresponding operations of PBF and hash table. We illustrate

---

Membership_Query_Item (Input: Item a)

Initialize $V_a = 0$
for $(i = 1; i \leq p; i++)$ **do**
  for $(j = 1; j \leq q; j++)$ **do**
    Compute $H_{[i][j]}(a_i)$
    if $PBF[H_{[i][j]}(a_i)]==0$ **then**
      Return *False*
    end if
    $V_a = V_a + \frac{H_{[i][j]}(a_i)}{2^j}$
  end for
end for
if V_a is in the hash table **then**
  Return *True*
end if
Return *False*

Fig. 3. The algorithm for querying an item with multiple attributes

---

Remove_Item (Input: Item a)

Initialize $V_a = 0$
for $(i = 1; i \leq p; i++)$ **do**
  for $(j = 1; j \leq q; j++)$ **do**
    Compute $H_{[i][j]}(a_i)$
    $V_a = V_a + \frac{H_{[i][j]}(a_i)}{2^j}$
    $PBF[H_{[i][j]}(a_i)] --$
  end for
end for
Remove V_a from the hash table

Fig. 4. The algorithm of removing an item with multiple attributes

these operations in an example. Finally, we compare the false positive probability of the standard Bloom filter and our proposed structure with respect to union and intersection operations.

5.1   Standard Bloom Filter

A set S can be represented as a Bloom filter using a mapping relation: $S \rightarrow BF(S)$. We use two Bloom filters $BF(A)$ and $BF(B)$ to represent sets A and B with the same number of bits and hash functions.

Definition 1. *The union of two Bloom filters, $BF(A)$ and $BF(B)$, can be represented as $BF(A \cup B)$ by logical OR operation of their bit vectors.*

Theorem 1. *The false positive probability of $BF(A \cup B)$ is larger than that of $BF(A)$ or $BF(B)$.*

Proof. We use $|A|$, $|B|$ and $|A \cup B|$ to represent the numbers of the sets A, B and $A \cup B$. Thus, we have $|A \cup B| \geq max\{|A|, |B|\}$. The false positive probability of set $A \cup B$ is $(1 - (1 - \frac{1}{m})^{k|A \cup B|})^k$, which is larger than the false positive probability of sets A or B, $(1 - (1 - \frac{1}{m})^{k|A|})^k$ or $(1 - (1 - \frac{1}{m})^{k|B|})^k$.

Definition 2. *The intersection of two Bloom filters, $BF(A)$ and $BF(B)$, can be represented as $BF(A \cap B)$ by logical AND operation of their bit vectors.*

Theorem 2. *The false positive probability of $BF(A \cap B)$ is smaller than that of $BF(A) \cap BF(B)$ with probability*

$$(1 - (1 - \frac{1}{m})^{k|A - (A \cap B)|})(1 - (1 - \frac{1}{m})^{k|B - (A \cap B)|})$$

Proof. Intuitively, a bit is set in both filters if it is set by items in $A \cap B$, or in $A - (A \cap B)$ and $B - (A \cap B)$ [7]. In fact, we have

$$BF(A) \cap BF(B) = BF(A \cap B) \cup BF(A - (A \cap B)) \cap BF(B - (A \cap B))$$

Meanwhile, the items in $A \cap B$ produce the same bits for filters $BF(A \cap B)$ and $BF(A) \cap BF(B)$. Thus, we can conclude that $BF(A \cap B)$ is smaller than that of $BF(A) \cap BF(B)$ when $BF(A - (A \cap B)) \cap BF(B - (A \cap B)) = 1$.

Given a standard Bloom filter and from the conclusion in [7], we know $P(BF(A - (A \cap B)) = 1) = 1 - (1 - \frac{1}{m})^{k|A - (A \cap B)|}$, and $P(BF(B - (A \cap B)) = 1) = 1 - (1 - \frac{1}{m})^{k|B - (A \cap B)|}$. Thus, the event that the false positive probability of $BF(A \cap B)$ is smaller than that of $BF(A) \cap BF(B)$ occurs with probability

$$(1 - (1 - \frac{1}{m})^{k|A - (A \cap B)|})(1 - (1 - \frac{1}{m})^{k|B - (A \cap B)|})$$

5.2   Practical Operations for PBF

Although the union and intersection operations of PBF are similar to those of standard Bloom filters, they are different because PBF is counter-based filters. The counter-based Bloom filters utilize the one-way hashed computation. Because we cannot accurately know the actual relationship between two data sets represented by two Bloom filters, the union operation result is possible not to exhibit the actual effects very accurately. Hence, we consider the conservative viewpoint as the policy of our union operation in order to statistically display the approximate result. The union operation in PBF obtains the bigger counter values from two arrays in the corresponding positions. On the contrary, the intersection operation in PBF obtains the smaller counter values.

Given the new structure to represent items with multiple attributes, the union and intersection operations will get an updated hash table to store verification values. The union operation integrates the verification values of two hash tables into a new one. The intersection operation maintains the verification values, which appear at both hash tables, in the new hash table. Figure 5(a) and (b) show an example to realize the union and intersection operations of PBF and hash tables respectively.

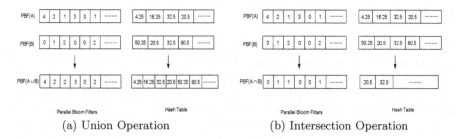

(a) Union Operation   (b) Intersection Operation

Fig. 5. The union and intersection operations of PBF for multiple attributes

5.3 Comparisons of False Positive

We compare the false positive probability applying the union and intersection operations in both the standard Bloom filter and the newly proposed structure. We can compute the false positive probability of union and intersection operations for multiple attributes, which are shown in Figure 6 (Note that PBF refers to the whole structure including PBF and the auxiliary hash table). We carry out the comparison by the false positive probability of $BF(A) \cup BF(B)$ minus that of $PBF(A) \cup PBF(B)$ in Figure 6(a) and $BF(A) \cap BF(B)$ minus $PBF(A) \cap PBF(B)$ in Figure 6(b) respectively. We set the parameters as $m = 1280$ and $k = 6$.

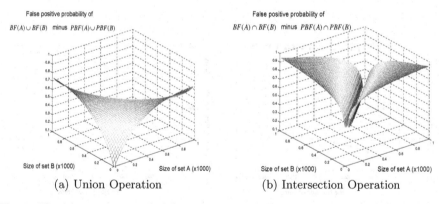

(a) Union Operation   (b) Intersection Operation

Fig. 6. The false positive probability of union and intersection operations for multiple attributes

Figure 6(a) displays that the false positive probability of $PBF(A) \cup PBF(B)$ is less than that of $BF(A) \cup BF(B)$, especially when the set size becomes larger. Figure 6(b) displays that the false positive probability of $PBF(A) \cap PBF(B)$ is also much lower than that of $BF(A) \cap BF(B)$. The PBF structure fully supports algebra operations and maintains low errors. For example, we can realize the intersection operation based on PBF in order to know the common items of two data sets. Similarly, we use the union operation to get the total information of two data sets.

6   Performance Evaluation

We simulate the basic and improved methods of generating verification values and compare the false positive probability for Standard Bloom Filter (SBF) and PBF in terms of increasing number of items. In order to make the multi-attribute operations feasible in the SBF, we can concatenate multiple attributes into one parameter as the input to MD5 hash functions. Thus, the SBF in this paper uses the concatenated multiple attributes as an input of hash functions and the approach is an extension to the Bloom filters in [7] for items with multiple attributes.

6.1   Verification Values

We compare the false positive probability of *Basic Method (BM)* with that of *Improved Method (IM)* based on the same hash functions and available space sizes. Each item has four attributes and each attribute is computed by six hash functions, *i.e.*, $p = 4$ and $q = 6$, respectively. Figure 7 illustrates the simulation results. Compared with the basic method, the improved method can obtain the smaller false positive probability under different space sizes. As a result of considering the sequential information of hash functions, the improved method can distinguish the hash values of attributes of items very well. Thus, the average false positive probability of *IM* can be bounded by 0.002, which is much smaller than *BM*.

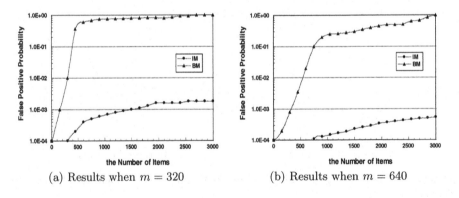

(a) Results when $m = 320$ (b) Results when $m = 640$

Fig. 7. The false positive probability of Basic and Improved Methods

6.2   Parallel Bloom Filters

In this simulation, we use the MD5 as the hash function for its well-known properties and relatively fast implementation. The value of an attribute can be hashed into 128 bits by calculating the MD5 signature. Then, we divide the 128 bits into four 32-bit values and utilize the modulus of each 32-bit value by the filter size m.

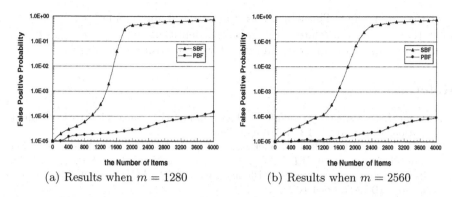

(a) Results when $m = 1280$ (b) Results when $m = 2560$

Fig. 8. The results of comparisons between SBF and PBF

Figure 8 shows the false positive probability of SBF and PBF in terms of different space sizes. Note that PBF in this figure stands for the new data architecture, which consists of PBF and the hash table. We set three attributes for each item and each attribute is computed by seven hash functions, *i.e.*, $p = 3$ and $q = 7$. Meanwhile, the space sizes available are m=1280, 2560 counters, respectively. It can be seen that given a certain number of items, the bounds on the PBF are always smaller than the bounds on the SBF. The upper probability of PBF is much smaller than that of SBF. Meanwhile, the variation trends of PBF are smooth in terms of the increasing number of items. The main reason is that the verification step based on the hash table can enhance the accuracy and efficiency of PBF, which can support the operations with multiple attributes. Therefore, although we use the simple method of attributes concatenation to realize the attributes-based operations in SBF, the PBF shows that its false positive probability is much lower than that of SBF.

7   Conclusion

Bloom filter is a kind of space-efficient data structure and can be widely used for information representation and membership query in current network environments. The space efficiency is achieved with certain false positive probability in membership query. The standard Bloom filter cannot efficiently support the representation and query of *multiple* attributes for the burgeoning and higher-level network services.

In this paper, we have presented a novel structure and practical algorithms which outperform the conventional standard Bloom filters algorithms by using the two-step verification process. Our proposed architecture extends the standard Bloom filters to efficiently support the membership query with multiple attributes. By using the verification values in the hash table, we illustrate how the false positive probability of the proposed structure can be reduced significantly. Meanwhile, the operations on both Bloom filters and the hash table have the complexity of $O(1)$. Hence, the total complexity of our proposed structure is of the order $O(1)$. Through theoretical analysis and simulations, we further show that the novel architecture can be efficiently applied in network services for its small space requirement and very low false positive probability.

References

1. Bloom B.: Space/time Trade-offs in Hash Coding with Allowable Errors. Communications of the ACM, **13** (1970) 422–426
2. Fan L., Cao P., Almeida J., Broder Z.A.: Summary cache: a scalable wide area web cache sharing protocol. IEEE/ACM Transaction on Networking, **8** (2000) 281–293
3. Mitzenmacher M.: Compressed Bloom filters. IEEE/ACM Transaction on Networking, **10** (2002) 604–612
4. Zhu Y.F., Jiang H., Wang J.: Hierarchical Bloom Filter Arrays (HBA): A Novel, Scalable Metadata Management System for Large Cluster-based Storage. Proceedings of the 5th IEEE International Conference on Cluster Computing (Cluster), (2004) 165–174
5. Kumar A., Xu J., Wang J., Spatschek O., Li L.: Space-Code Bloom filter for efficient per-flow traffic measurement. Proceedings of the IEEE INFOCOM, **3** (2004) 1762–1773
6. Saar C., Yossi M.: Spectral Bloom filters. Proceedings of the ACM SIGMOD, (2003) 241–252
7. Broder A., Mitzenmacher M.: Network applications of Bloom filters: a survey. Internet Mathematics, **1** (2005) 485–509
8. Xiao B., Chen W., He Y.X., Sha E.H.M.: An active detecting method against SYN flooding attack. Proceedings of the 11th International Conference on Parallel and Distributed Systems (ICPADS), **1** (2005) 709–715
9. Feng W.C., Kandlur D.D., Saha D., Shin K.G.: Stochastic Fair Blue: A Queue Management Algorithm for Enforcing Fairness. Proceedings of the IEEE INFOCOM, **3** (2001) 1520–1529
10. Cuenca-Acuna F.M., Peery C., Martin R.P., Nguyen T.D.: PlantP:Using gossiping to build content addressable peer-to-peer information sharing communities. Proceedings of the 12th IEEE High Performance Distributed Computing, (2003) 236–246
11. Broder A., Mitzenmacher M.: Using multiple hash functions to improve IP lookups. Proceedings of the IEEE INFOCOM, **3** (2001) 1454–1463
12. Baboescu F., Varghese G.: Scalable packet classification. Proceedings of the ACM SIGCOMM, (2001) 199–210
13. Dharmapurikar S., Krishnamurthy P., Taylor D.E.: Longest Prefix Matching Using Bloom Filters. Proceedings of the ACM SIGCOMM, (2003) 201–212
14. Kumar A., Xu J., Zegura E.W.: Efficient and scalable query routing for unstructured peer-to-peer networks. Proceedings of the IEEE INFOCOM, **2** (2005) 1162–1173
15. Song H.Y., Dharmapurikar S., Turner J., Lockwood J.: Fast Hash Table Lookup Using Extended Bloom Filter: An Aid to Network Processing. Proceedings of the ACM SIGCOMM, (2005) 181–192
16. Guo D.K., Wu J., Chen H.H., Luo X.J.: Theory and Network Application of Dynamic Bloom Filters. Proceedings of the IEEE INFOCOM, (2006)
17. Rhea S.C., Kubiatowicz J.: Probabilistic location and routing. Proceedings of the IEEE INFOCOM, **3** (2002) 1248–1257

Receive Side Coalescing for Accelerating TCP/IP Processing

Srihari Makineni, Ravi Iyer, Partha Sarangam, Donald Newell, Li Zhao,
Ramesh Illikkal, and Jaideep Moses

Intel Corporation, 2111 NE 25th Ave.,
Hillsboro 97124, USA
{srihari.makineni, ravi.iyer, parthasarathy.sarangam,
donald.newell, li.zhao, ramesh.illikkal,
jaideep.moses}@intel.com

Abstract. With rapid advancements in Ethernet technology, Ethernet speeds
have increased by 10 fold, from 1 to 10Gbps, in a period of 2-3 years. This sudden increase in speeds has outpaced the rate at which processor and memory
speeds have been increasing, raising concerns that TCP/IP processing will not
scale to these levels. As a result, applications running on commercial servers
will not be able to take advantage of the increased Ethernet bandwidth. This has
led to a flurry of activity in the industry and academia focused on finding ways
to scale up TCP/IP processing to 10Gbps and beyond. In this paper, we propose
a novel technique called "Receive Side Coalescing" (RSC) that increases
TCP/IP processing efficiencies significantly. RSC allows NICs to identify
packets that belong to same TCP/IP flow and coalesce them into a single large
packet. As a result, TCP/IP stack has to process fewer packets reducing per
packet processing costs. NIC can do this coalescing of packets during interrupt
moderation time hence packet latency is not affected. We have collected packet
traces and analyzed those to find out how much coalescing is possible in different scenarios. Our analysis shows that about 50% reduction in number of packets is possible. We have prototyped RSC on Windows and Linux to understand
the benefits, and the results show that 2-7% of savings in CPU utilization is
possible at 1Gbps speeds. Projection models developed to estimate processing
costs at 10Gbps show that RSC can save up to 20% of the CPU.

Keywords: Receive Side Coalescing, RSC, TOE, TCP/IP acceleration, defragmentation, receive offload.

1 Introduction

TCP/IP is the most commonly used protocol to process data both in enterprise data
centers and on the Internet. However, TCP/IP processing is inherently very expensive
and will not be able to scale to 10Gbps rates process packets unless a significant
number of overheads are eliminated. We have measured TCP/IP performance on
Intel® XeonTM processor (2.6GHz) based server platform running windows OS.
Results shown in figure 1 highlight the fact that the transmit side can achieve higher

Y. Robert et al. (Eds.): HiPC 2006, LNCS 4297, pp. 289–300, 2006.
© Springer-Verlag Berlin Heidelberg 2006

throughputs at lower CPU utilizations compared to that of the receive side processing. CPU utilizations shown in the graph are for two logical CPUs. At 65536 byte payload size, receive side throughput maxed out at 2Gbps with average CPU utilization across the two processors is around 65%. This highlights the fact that one CPU can barely achieve 2Gbps of receive throughput, which means multiple CPUs need to be expensed to achieve 10Gbps throughput.

While the TCP/IP protocol portion [4] has been proven to be not much compute intensive, there are significant number of other overheads that come into play when the entire processing (application to NIC) that happens is considered. Some recently published papers shed more light [6], [21], [25], [14] on these overheads by analyzing their impact.

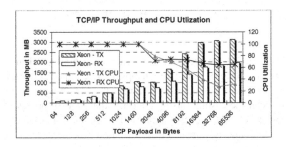

Fig. 1. TCP/IP Performance

Within TCP/IP processing, transmit side processing costs have been reduced significantly over the years with the development of techniques such as zero copy transmit and Large Segment Offload (LSO). With the help of these optimizations, today's CPUs can achieve throughputs of 7-8Gbps at close to 100% utilization. In addition, the transmit side processing scales well with the CPU speeds because it is not very memory intensive. On the other hand, receive side processing is much more difficult to optimize. Today's server platforms have to spare an entire CPU to achieve just 2-3Gbps of throughput even when receiving large packets (> 1KB). CPU cycles required to receive and process a single byte of data, when transferring 8KB payloads, is in the order of 6 cycles. Reasons for such high cost of receive side processing are: descriptor and TCP/IP header processing, data copy from kernel to application buffer generate compulsory cache misses, per packet processing costs, OS overheads, etc. Since receive side processing is memory intensive and memory speeds increase rather slowly (as compared to CPU speeds), receive side performance does not scale as well with increases in CPU speeds. To address receive side processing challenge, several solutions have been proposed. These solutions range from techniques that target specific overheads [7] involved in receive side processing to new protocols like RDMA [12] to complete offload solutions like TOE [1]. These techniques are reviewed in section 3 of this paper.

Our contribution in this paper is that we propose and evaluate a new technique, called Receive Side Coalescing (RSC), to accelerate TCP/IP receive side processing. RSC combines incoming packets of the same TCP connection into larger packets. This technique reduces number of packets that software has to process thus reducing overheads involved in processing packets. Our analysis showed that up to 50% reduction in number of packets to be processed can be achieved.

The rest of the paper is organized as follows: In section 2, we provide an overview of TCP/IP receive side processing and highlight various overheads involved. In section 3, we review various techniques/solutions that have been proposed to solve receive side processing challenge. In section 4, we show data from our analysis on how much coalescing is possible in the network. In section 5, we describe how RSC works, provide details on RSC SW prototypes and show the results from these prototypes. We conclude the paper with summary and future work.

2   TCP/IP Receive Processing

In this section we provide a high level overview of the processing that takes place from the time a NIC receives an Ethernet frame till the incoming data is handed over to the intended application. It is not our intention to provide a detailed description of this processing, but to provide sufficient context for readers while highlighting some major overheads involved in this processing.

Receive-side processing begins when NIC hardware receives an Ethernet frame from network. NIC extracts Ethernet frame delineation bits and CRC value and validates the frame. Today's NICs also perform checksum computations for the TCP and IP portions of the packet and compare those with checksum values TCP and IP headers. In order to notify the software stack about incoming packets and their placement in the memory, NIC uses descriptors that are arranged in a circular ring fashion. Descriptor data structure is typically 16bytes and contains among other things, address of a memory buffer (NIC buffer) to store the incoming packet data. NIC copies the incoming data at the memory location specified in the descriptor using onboard DMA engine. Once the packet is placed in memory, NIC updates a status field inside the descriptor to indicate to the driver that this descriptor holds a valid packet and generates an interrupt. This kicks off the SW processing of the received packet.

Figure 2 shows the overall flow for receive side processing. The Ethernet device driver reads the descriptor and makes sure that NIC has indicated that this is a valid packet. Driver then classifies the packet as either IP packet or some other. If it is an IP packet then it forwards it to the TCP/IP stack for further processing. Since this descriptor was updated by NIC earlier, it results in invalidating processor's copy (if found in cache). So the processor would have to fetch the descriptor from the main memory. If the descriptor size is 16 bytes then each cache line (64 bytes) can accommodate up to 4 descriptors. Similarly, accessing packet headers by TCP/IP stack also results in cache misses as this data was just placed in the memory by the NIC. TCP/IP headers combined, without any option fields, is 40 bytes long. So each packet header will result in one compulsory cache miss. The next step in processing is to identify the connection to which this packet belongs. TCP/IP software stores state information of each open connection in a data structure, called the TCP/IP Control Block (TCB). Since there can potentially be several thousand open connections, hence many TCBs, TCP/IP software uses a well known search mechanism called hashing for fast lookup of the right TCB. The hash value is calculated by using the IP address and port number of both the source and destination machines. Several fields (sequence numbers for received/acknowledged bytes, application's pre-posted buffers, etc.) inside the TCB are updated whenever a new packet is received. TCP/IP stack then needs to figure out where to copy the packet payload (data portion). It checks to see if the target application has already posted a buffer to receive incoming data. If a buffer is available, stack copies the data from the NIC buffer into that buffer. Otherwise, it will wait for the application to provide a buffer. TCP/IP stack may be forced to copy the data into a temporary buffer if application didn't provide one. When the incoming data is first copied, source buffer results in compulsory cache misses as the data has to be read from main memory.

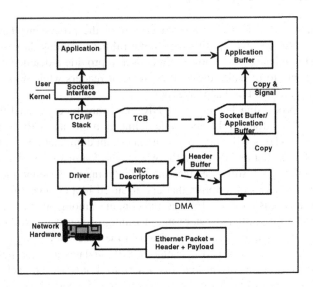

Fig. 2. Data Flow in Receive-Side processing

3   Existing Techniques to Accelerate TCP/IP Receive Processing

Over the years several solutions have been proposed to improve the performance of TCP/IP receive processing. Some of these are implemented in majority of the NICs that ship today.

Interrupt Moderation
Network interrupt processing is an expensive operation even on today's machines. Interrupt moderation technique was developed to reduce the number of interrupts NIC generates. NIC instead of generating one interrupt for every incoming packet, it generates an interrupt after some number of packets are received. This interrupt moderation is typically exposed as a configurable parameter on today NICs. This is a commonly available feature in today's NICs.

Jumbo Frames

Jumbo frames technology was developed to reduce number of network packets required to transfer larger payloads. Most of today's NICs support multiple size jumbo frames. Jumbo frames not only reduce numbers of packets on the network but can also significantly reduce TCP/IP stack processing time hence CPU utilization. However, jumbo frames are not widely used because enabling jumbo frames on end machines is not sufficient - all the intermediate routers and switches need to enable jumbo frames as well.

Receive Side Scaling

Receive side scaling is relatively a new technique developed by Microsoft to allow NICs to distribute interrupts across multiple CPUs in a system. Without this feature, all the interrupts would go to one processor (typically CPU0). As a result, maximum receive throughput a machine can achieve depends on what a single CPU can achieve.

TCP Offload Engines (TOE)

TOE offloads entire TCP/IP processing from the main CPU. TOEs are typically implemented on the NIC. TOEs are expensive, require changes to operating systems and face other problems that are described well in a recent paper [17]. Given these issues, it is not known yet how this idea will succeed in the market.

Remote Direct Memory Access (RDMA)

RDMA is a set of specifications developed by the RDMA consortium to solve the problem of directly placing incoming network data into user application buffers without having to go through intermediary copies of that data. RDMA provides the ability of one computer to directly place information in another computer's memory with minimal demands on memory bus bandwidth and CPU processing overhead, while preserving memory protection semantics. RDMA over TCP/IP defines the interoperable protocols to support RDMA operations over standard TCP/IP networks.

Direct Cache Access (DCA)

DCA [7] allows NIC to place incoming packet data directly into a processor's cache. This can potentially eliminate compulsory cache misses that occur during packet processing. It also speeds up data copies as the incoming payload will be in processor's cache instead of main memory.

4 Receive Side Coalescing (RSC)

RSC is a stateless and software transparent offload mechanism. RSC coalesces on the NIC packets that are for the same TCP/IP connection. Thus RSC reduces number of packets that a TCP/IP stack needs to process. This will reduce per packet processing costs significantly. RSC, in concept, is exactly opposite of Large Segment Offload (LSO) [14] that happens on the transmit side. Figure 2 illustrates the effect of RSC. When there is no RSC, NIC sends all the packets it receives to the stack for processing. With RSC, NIC sends fewer but larger packets. RSC requires NIC driver to allocate large buffers (4KB or higher) so that it can place payload from more than one packet in each buffer. Coalescing of packets also leads to reduction in descriptor usage as well. NIC performs coalescing of incoming packets while waiting for the

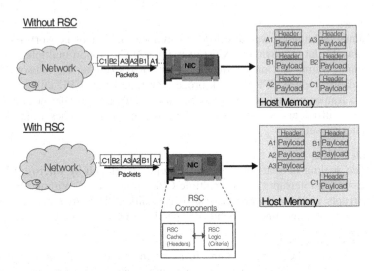

Fig. 3. Illustration of RSC concept

interrupt moderation timer to expire. As a result, RSC does not add any extra delay in delivering the packets to the processor.

Coalescing TCP/IP packets is possible because TCP is a byte stream protocol and applications can't make any assumptions about boundaries of a message. For example, an application can receive tail end of a message and the beginning of the next message in the same 'receive' call.

In order to perform coalescing, RSC needs to save some information (IP, TCP header info, descriptor number, number of packets and bytes coalesced, current checksum values, etc.) about each connection for which coalescing is in progress. This information is saved in a cache memory, called RSC cache on the NIC. Each coalesced packet takes up about 100 bytes in the cache.

RSC logic on NIC extracts TCP/IP header fields from incoming packets and does a series of tests to determine whether to coalesce this packet or to stop existing coalescing. We are not listing the coalescing criteria that RSC uses here due to space constraints.

If an incoming packet is the first packet for a TCP/IP connection, and RSC logic decides to start coalescing, then the packet's TCP/IP header is removed and relevant information from the header is saved in the RSC cache. Packet's payload is then copied (DMA) into a buffer provided by the NIC driver. RSC does not hold onto the payload while coalescing is in progress so it does not need any additional memory on NIC. When a second packet on the same connection comes and if it meets coalescing criteria, then the entries in the RSC cache are updated (how many bytes received, starting offset in the buffer for next packet's payload, etc). TCP/IP headers are stripped from the packet and payload is copied next to the previous one in the same buffer. When the RSC logic decides to stop coalescing for a connection because an incoming packet does not meet coalescing criteria (out of order packet, payload does not fit in the remaining space in the buffer, PSH flag in TCP header is set, etc.), then modified header in the RSC cache for that connection is written back to the memory at the location specified in the descriptor.

At the time of interrupt all the headers from the RSC cache are written back. RSC can start coalescing again after the interrupt. As a result, RSC requires only a small amount of RSC cache (1-2 KB) to store information about 3-4 connections at a time.

5 How Much Coalescing Is Possible?

It is well known in the networking community that TCP/IP traffic is bursty in nature. Several studies have been conducted and multiple papers [2], [9], [22] have been published on this subject. RSC is mainly dependent on the existence of these bursts. RSC with a cache size of 2 or more entries does not require packets on a connection to come back to back to be able to coalesce those packets. It is sufficient if they come in close succession. In order to find out how much coalescing is really possible and how it varies with number of simultaneous flows (TCP/IP connections), RSC cache size, buffer size and interrupt moderation window, we have collected several sets of network packet traces and analyzed those. Results from this analysis are presented in this section.

For all the experiments described below, we have used the following setup. We have set up a dual processor server machine as the receiver. This receiver machine is connected to a number of client machines (senders) via two 1Gbps Ethernet switches. All the client and server machines used 1Gbps NICs and ran Chariot [26] (commercially available network performance analyzer) scripts to transfer data. Each client machine sets up 1 connection with the receiver and starts transmitting data. We have used an open source program, called ethereal [5], to collect network packet traces on the receiving machine. We have then analyzed these packet traces using internally developed scripts to find out how much coalescing is possible in each scenario based on certain coalescing criteria.

5.1 Simultaneous Connections

The main purpose of this experiment is to find out how much coalescing is possible when there are multiple simultaneous TCP/IP connections. We have assumed 8KB NIC buffer and a RSC cache with 4 entries. 8KB buffer allows us to coalesce up to 5 full sized TCP/IP packets (1460 bytes each). Four entry deep RSC cache allows us to keep track of up to 4 different connections. After every 20 packets, we have stopped coalescing and flushed RSC cache and started all over again. This is same as simulating a 20 packet interrupt moderation window. Results from this experiment are shown in the graph in figure 4. On the x-axis, we are showing various payload sizes for which we have collected and analyzed packet traces. On the y-axis, we are showing number of packets that RSC is able to eliminate out of a total of ten thousand packets that were captured. The results show that about 50 to 80% coalescing is possible, and there is not much variation in achievable coalescing between 4, 8 and 16 connections. One exception though is at 8KB point where the number of coalesced packets with sixteen connections is slightly smaller than that of 4 and 8 connections. This is because with the 20 packet hard stop we have for stopping the coalescing activity, we found ourselves often not able to coalesce the last packet (~890 bytes) of 8KB payload. But in case of 4 and 8 connections, more often we were able to coalesce all the packets (5 full + 1 partial) into one.

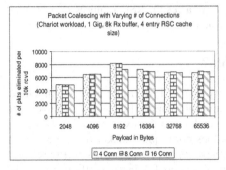
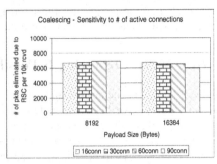

Fig. 4. Amount of coalescing possible with 2, 4 and 16 simultaneous connections

Fig. 5. Data Flow in Receive-Side processing

Next, we ran another experiment with 16, 30, 60 and 90 clients, with as many connections, sending data to a single receiver on a 1Gbps NIC. Results are shown in the graph in figure 5. These results confirm further that we can still achieve more than 50% coalescing (50% reduction in number of packets) even when there are large number of connections on which packets are received.

5.2 Sensitivity Studies

Next we wanted to study how various key parameters such as RSC cache size, buffer size and interrupt moderation window can impact how much coalescing is possible.

The graph in figure 6 shows sensitivity study results where we have varied number of entries in the RSC cache size. RSC cache size defines how many simultaneous connections for which coalescing can be in progress. The x-axis of the graph shows different cache sizes we have studied while the y-axis shows how much coalescing was possible. We have shown data for various payload sizes. These results point out RSC cache size of 3 entries allows us to achieve close to maximum coalescing.

Next, we study sensitivity to receive buffer size; size of the buffer dictates how many individual packets we can coalesce. The analysis of captured packets is done assuming RSC cache size of 4 entries and 20 packet interrupt window. There were 16 active connections from 16 different clients. It is clear from the results shown in the graph in figure 7 that 8KB buffer yields more coalescing for payloads larger than 4KB.

The last sensitivity study that we have done is to figure out the impact of varying interrupt moderation window. On today's 1G NICs, it is around 250us, which is equal to roughly 25 full size packets (this is the default setting on Gigabit NICs from Intel Corp.). We found out from the data that we can achieve more than 50% coalescing even with a 16 packet window. Increasing the interrupt moderation window from 16 packets to 32 packets shows 5-10% improvement in coalescing, and no significant gain was observed beyond 32 packet window.

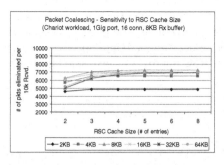
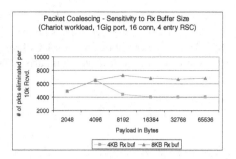

Fig. 6. Data Flow in Receive-Side processing   **Fig. 7.** Data Flow in Receive-Side processing

These sensitivity studies prove that significant coalescing is possible on machines that receive significant amount of TCP/IP data. In the next section we show how this coalescing translates into CPU utilization reduction.

6   RSC Performance

Next step in RSC evaluation is to find out how much CPU savings that we can achieve with RSC. In order to find this out, we have implemented a software version of RSC in Ethernet driver code.

To allow coalescing of multiple packets, we have forced the Ethernet driver to allocate larger buffers (4KB and 8KB). NIC does its normal processing of packets and sends them to the processor. In the Ethernet driver, we execute RSC code and decide which packets can be coalesced. We have varied the degree of coalescing to figure out how the benefits vary. We have used 3 different levels of coalescing and compared the results to a base case (denoted as "Regular" in the graphs) that does not have any coalescing. The three levels of coalescing used in our experiments are indicated in the graphs as "RSC2" (or "RSC_2"), "RSC3" (or "RSC_3"), and "RSC5" (or "RSC_5") to mean at most two, three and five packet coalescing respectively.

In all these tests, we have a used a fast client to transmit data to the receiver machine that is under the test. The client machine has enough capacity in terms of compute power as well as network bandwidth so that it does not become bottleneck in any way.

6.1   Results

We have first implemented RSC functionality in Linux OS and took measurements with and without RSC. Results are show in the graph in figure 8. The x-axis in the graph shows different payload sizes (application buffers) that we have taken measurements for. The y-axis shows measured CPU utilization and the secondary y-axis shows efficiency of processing in terms of cycles/byte. We have used ttcp [23] program on both the client and server machines to generate TCP/IP traffic between them. Across all the payload sizes and configurations the system was able to achieve maximum possible throughput (~950Mbps) on a 1Gbps NIC.

Across all the payload sizes, RSC shows a benefit over the base case. Two packet coalescing (RSC_2) offers 2-4% savings in CPU utilization while five packet coalescing (RSC_5) saves about 5-7% of CPU when processing at 1Gbps speeds. You can also see from the cycles/byte comparison how RSC improves receive side processing efficiency.

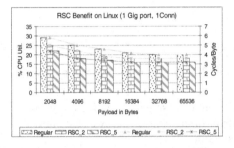

Fig. 8. RSC Performance on Linux at 1Gbps

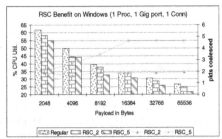

Fig. 9. RSC Performance on Windows at 1Gbps

Next, we have repeated the same experiment on a server machine running windows. We have modified the Gigabit Ethernet driver code and added RSC functionality to it. We have used a program, called ntttcp [23], to generate network traffic. Results are shown in the graph in figure 9. In this graph, we are showing on the secondary y-axis, the average number of packets that we were able to coalesce in each case. Results show that RSC offers benefit with windows stack too. If we consider the 8KB payload data point, RSC_2 brings down CPU utilization from 39.5% to 37.8% and RSC_5 brings this further down to 32.6%. We have also noticed that for 16KB payloads, RSC_2 CPU utilization turned out to be (0.1%) higher than the Regular case. This is due to the artifacts of running RSC in software (resulted in more cache pollution).

We wanted to find out what the benefit would be when processing at multi gigabit per second rates. So we have experimented with a current generation 10Gbps NIC on a windows machine. We have modified the 10G Ethernet driver and added RSC support. Even though we have used 10G NIC, the DP server system under test could only achieve 1.6Gbps throughput at 4KB payload size and 3.1Gbps at 65KB payload size with 100% CPU utilization. This test was conducted with 1 TCP/IP stream and 1 CPU. RSC2 benefit now ranges from 4 to 12%. As expected, the benefits are higher (4-18%) with RSC3 and RSC5 configurations. From these results, it is clear that when receiving packets at true 10Gbps speeds, even two packet coalescing (RSC2) can bring down the CPU utilization by more than 20% points.

7 Summary and Conclusions

Scaling of receive side TCP/IP processing beyond the current 1-2 Gbps has become important as the bandwidth demands in enterprise data centers have grown significantly in recent years. This requires that several overheads involved in receive side processing need to be reduced drastically. In this paper, we have proposed a novel technique, called Receive Side Coalescing (RSC) that addresses per packet processing overheads by reducing the number of packets that need to be processed by the stack. We have shown collected network packet traces and analyzed those to prove that more 50% coalescing is possible with a 4 entry RSC cache, 20 packet (~200us) interrupt moderation window and 8KB NIC buffers. In essence RSC offers similar benefits like that of 9KB Jumbo frames but without all the problems associated enabling the jumbo frames in the data centers. Next, we have shown results from our software implementation of RSC. These results showed that CPU utilization can be reduced by 1-5% with RSC2 (2 packet coalescing) and 3-7% with RSC5 (5 packet coalescing) even when processing at 1Gbps speeds. We have also shown RSC benefits increasing significantly when processing at 2-3Gbps speeds. We believe that RSC will be a key technique that allows TCP/IP processing to scale better. Going forward, we would like to implement RSC in hardware and use it in real environments so we can measure the benefits it offers for some key real world applications. This also allows us to find out any changes that may be required to the TCP/IP stacks to fully support RSC.

References

1. "Alacritech SLIC: A Data Path TCP Offload methodology", http://www.alacritech.com/html/techreview.html
2. E. Blanton and M. Allman, "On the Impact of Bursting on TCP Performance," Proceedings of the Workshop for Passive and Active Measurement, Mar 2005.
3. J. Chase et. al., "End System Optimizations for High-Speed TCP", IEEE Communications, Special Issue on High-Speed TCP, 2000.
4. D.D. Clark, J. Romkey, and H. Salwen, "An analysis of TCP processing overhead," IEEE Communications, vol. 27, no. 6, pp. 23–29, June 1989.
5. Ethereal Network Protocol Analyzer, http://www.ethereal.com/
6. A. P. Foong, T. R. Huff, H. H. Hum, J. P. Patwardhan, and G. J. Regnier, "TCP Performance re-visited," in Proc. IEEE Int. Symp. on Performance of Systems and Software, pp. 70–79, Austin, Mar. 2003.
7. R. Huggahalli, R. Iyer and S. Tetrick, Direct Cache Access for High Bandwidth Network I/O," 32nd Annual International Symposium on Computer Architecture (ISCA 2005), June 2005.
8. iSCSI, IP Storage Working Group, Internet Draft, available at http://www.ietf.org/internet-drafts/draft-ietf-ips-iscsi-20.txt
9. R. Jain and S. A. Routhier, "Packet Trains - Measurements and a NewModel for Computer Network Traffic," IEEE Journal on Selected Areas in Communications, vol. 4, pp. 986–994, Sept. 1986.
10. H. Jiang and C. Dovrolis, "Source-Level Packet Bursts: Causes and Effects," in Proceedings Internet Measurement Conference (IMC), Oct. 2003.
11. J. Kay and J. Pasquale, "The importance of non-data touching processing overheads in TCP/IP," in Proc. ACM SIGCOMM, pp. 259–268, San Francisco, Oct. 1993.
12. RDMA Consortium [Online]. Available at http://www.rdmaconsortium.org/.
13. C. Kurmann, M. Müller, F. Rauch, and T. M. Stricker, "Speculative defragmentation— A technique to improve the communication software efficiency for gigabit Ethernet," in Proc. 9th IEEE Symp. High Performance Distr. Comp, Pittsburgh, Aug. 2000.

14. S. Makineni and R. Iyer, "Architectural Characterization of TCP/IP Packet Processing on the Pentium M microprocessor," Int'l Conf. on High Performance Computer Architecture (HPCA-10), Feb 2004.
15. D. McConnell, "IP Storage: A Technology Overview", http://www.dell.com/us/en/biz/topics/vectors_2001-ipstorage.htm
16. J. C. Mogul. Observing TCP Dynamics in Real Networks. In ACM SIGCOMM, pages 305{317, 1992.
17. J. Mogul, "TCP offload is a dumb idea whose time has come," A Symposium on Hot Operating Systems (HOT OS), 2003.
18. J. Postel, Ed., "Internet Protocol - DARPA Internet program protocol specification," RFC 791, Sep. 1981.
19. J. B. Postel, "Transmission Control Protocol", RFC 793, Information Sciences Institute, Sept. 1981.
20. M. Rangarajan et al., "TCP Servers: Offloading TCP/IP Processing in Internet Servers. Design, Implementation, and Performance," Rutgers University, Technical Report, DCS-TR-481, March 2002.
21. G. Regnier, S. Makineni, R. Illikkal, R. Iyer, et al., "TCP onloading for data center servers," IEEE Computer, vol. 37, no. 11, pp. 48–58, Nov. 2004.
22. S. Sinha, S. Katula and D. Katabi, "Harnessing TCP's Burstiness with Flowlet Switching," 3rd ACM SIGCOMM Workshop on Hot Topics in Networks (HotNets III), San Diego, CA, Nov 2004
23. "The TTTCP Benchmark", http://ftp.arl.mil/~mike/ttcp.html
24. D. Yates., "Connection-Level Parallelism for Network Protocols on Shared-Memory Multiprocessor Servers," Ph.D. Dissertation, University of Massachusetts, Amherst, 1997
25. L. Zhao, S. Makineni, Ramesh Illikkal, D. Newell and L. Bhuyan, "TCP/IP Cache Characterization in Commercial Server Workloads, Seventh Workshop on Computer Architecture Evaluation using Commercial Workloads (CAECW-7), along with HPCA-10, February 2004
26. Chariot – More information available at http://www.ixiacom.com/solutions/display?skey=ixchariot

Minimizing Metadata Access Latency in Wide Area Networked File Systems

Jian Liang[1,*], Aniruddha Bohra[2], Hui Zhang[2], Samrat Ganguly[2], and Rauf Izmailov[2]

[1] Polytechnic University, Brooklyn, NY 11201
jliang@cis.poly.edu
[2] NEC Laboratories America, Princeton, NJ, 08540
{bohra, huizhang, samrat, rauf}@nec-labs.com

Abstract. Traditional network file systems, like NFS, do not extend to wide-area due to low bandwidth and high network latency. We present WireFS, a Wide Area File System, which enables delegation of metadata management to nodes at client sites (*homes*). The home of a file stores the most recent copy of the file, serializes all updates, and streams updates to the central file server. WireFS uses access history to *migrate* the home of a file to the client site which accesses the file most frequently.

We formulate the home migration problem as an integer programming problem, and present two algorithms: a dynamic programming approach to find the optimal solution, and a non-optimal but more efficient greedy algorithm. We show through extensive simulations that even in the *WAN* setting, access latency over WireFS is comparable to NFS performance in the *LAN* setting; the migration overhead is also marginal.

1 Introduction

With economic globalization, more and more enterprises have multiple satellite offices around the world. In such scenarios, network file systems provide a familiar interface for data access and are used extensively. Traditionally, network file systems have been designed for the local area networks, where bandwidth is ample and latencies are low. Common networked file systems like NFS [1] and CIFS [2] transfer large amounts of data frequently. All writes are transmitted to the server and require synchronous updates to the files there. Apart from wasting bandwidth, typical networked file systems require multiple round trips to complete a single file operation. The high network latency and the *chatty* nature of the protocols make file access over WAN slow and unreliable.

To improve network bandwidth utilization and to hide wide area latencies, Wide Area File Systems (WAFS) have been developed [3, 4]. These file systems reduce bandwidth utilization by (*i*) aggregating file system operations to reduce bandwidth requirements, and (*ii*) using content based persistent caching to eliminate duplicate block transfers. Unfortunately, current Wide Area File Systems ignore the file system access patterns, and are oblivious to the characteristics of the underlying network. An enterprise that

---

* This work was done when the author was at NEC Labs, America. J. Liang is supported in part by NFS grant CNS-0412029.

Y. Robert et al. (Eds.): HiPC 2006, LNCS 4297, pp. 301–312, 2006.

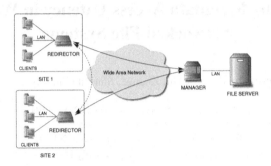

Fig. 1. WireFS Architecture

deploys existing WAFS has no way to take advantage of *temporal locality* across sites, e.g., different timezones and access patterns, or *network diversity*, which arises due to distinct network paths between sites and data centers.

In this paper, we present WireFS, a wide area file system, that takes an *organization-centric* view, enables data and meta-data sharing across multiple client sites, and minimizes metadata access latency in this system. WireFS takes advantage of temporal locality in file access, and allows data and metadata sharing across client sites. Figure 1 shows the WireFS architecture. WireFS uses Redirectors (WFSRs), that act as file servers for all clients that belong a site (island). These redirectors act as WAFS clients and communicate with a server side Manager (WFSM), which acts as a WAFS server. WFSM appears as the only client to the central file server which is the final authority on file system contents. In WireFS, Redirectors communicate not only with the Manager, but also with other Redirectors to allow data sharing, and cooperative metadata management.

WireFS uses a *home* based approach to minimize metadata access latency. Each file is assigned a home server, WFSR or WFSM, which controls access and serializes updates to the file. The most recent copy of a file is cached at its home server. The home maintains a single serialization point for all updates, therefore provides semantics and consistency guarantees similar to a centralized file server. Fault tolerance is achieved by maintaining a primary and a secondary home which maintain identical state. The home is not statically assigned and can be *migrated* closer to the clients accessing the file most frequently.

In this paper, we address the problem of home migration based on file system access history. Intuitively, a file that is accessed frequently by a client is moved closer to it. This is achieved by assigning the home of the file to the WFSR at the client site. Since the number of files in a modern file system is large, assigning homes to individual files is both inefficient and infeasible due to the overwhelming maintenance and lookup overhead. Instead, we decompose the file system namespace into a number of sub-trees and assign homes to these sub-trees. We formulate the problem of tree decomposition and home assignment to redirectors as an integer programming problem. We propose a dynamic programming algorithm to find the optimal solution in polynomial time. We also present a greedy algorithm as a heuristic which works much faster than the dynamic programming algorithm.

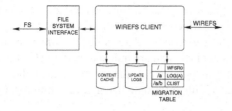

Fig. 2. WireFS Redirector (WFSR) architecture

The rest of the paper is organized as follows. Section 2 presents the WireFS architecture. The problem definition and algorithms for home migration are described in Sections 3. Section 4 describes the WireFS implementation and Section 5 presents the experimental setup and evaluation results. Section 6 describes the related work and Section 7 concludes the paper.

2 WireFS

WireFS is a wide area file system which enables delegation of metadata management, and uses content caching and duplicate elimination to reduce redundant data block transfers. The WireFS architecture has two logical components which capture the typical behavior of network file systems: *(i)* the Data Access Layer (DAL), that enables fast transfer of data across the WAN, and *(ii)* the Metadata Layer (MDL), that handles the synchronous metadata operations, e.g. lookup. In this paper, we focus on the design of algorithms for MDL to minimize the latency of metadata operations.

The MDL is composed of a set of WireFS redirectors, that serve all meta-data requests including file and directory lookup, creation and deletion of files, and updates to the file or directory metadata e.g. access time updates. The primary goal of MDL is to reduce the *latency* of the above operations in WireFS. In the following, we describe the architectural components of WireFS and the design of the meta data layer in detail.

2.1 WireFS Redirector

A WireFS redirector is deployed at each client site and has three main functions, *(i)* to export a file system interface to the clients at the site, *(ii)* to maintain a content addressable cache and communicate with other WFSRs or the WFSM to perform data transfers, and *(iii)* to maintain operation logs, perform serialization of updates, and handle metadata queries for files it is the designated *home*. Figure 2 shows the architecture of a WireFS redirector.

The file system interface (FSI) exported by the WFSR enables clients to communicate using an unmodified file system protocol. This interface translates client requests to WFS requests. On receiving the corresponding WFS reply, the FSI constructs the response to the original client request and sends it to the client. A pending request map is maintained by the FSI to match the responses to the corresponding requests. WireFS can support multiple file system protocols by defining the appropriate FSI.

Each WFSR maintains a large persistent *content-cache* that stores files as chunks indexed by content hashes, which can be used across files. Chunks are non-overlapping segments of file data whose extents are determined by content boundaries (*breakpoints*) using a fingerprinting technique, and are indexed by the SHA-1 collision resistant hash of the contents. WireFS associates a sequence of chunk indices in the file metadata which augments the default file information, e.g. access times, permissions, access control lists, etc.

2.2 WireFS Manager

The WireFS manager is deployed at the server site has a specialized role in the WireFS protocol. It communicates directly with the server and maintains a global view of the file system namespace. It also assigns and maintains the WireFS specific attributes of files like the home node, ownership information, generation numbers etc. The WFSM is the home node for all files until it delegates this responsibility to a WFSR. The WFSM is also responsible for the coordinated dissemination of commonly accessed files to multiple WFSRs to warm up the WFSR caches. Finally, the WireFS manager periodically reorganizes the file system namespace by reassigning homes of files according to the access history statistics.

2.3 WireFS Home

Each file in the file system namespace is assigned a *home*. The home is responsible for maintaining file consistency, provides a serialization point for all updates, and performs aggregation on file system operations to reduce network round trips. Homes maintain not only the update logs and serialization, but also maintain the latest version of the file metadata including access times, permissions, size, chunk indices, etc.

Each WFSR and the WFSM maintain a *migration table*, which contains a view of the file system namespace, statistics and access history, and per-file WireFS metadata. An entry in the migration table is indexed by the file identifier, and contains either the home node identifier, or the WireFS metadata for the file. WireFS metadata contains attributes defined in the file system, and a list of chunk indices, update logs, etc. The migration table is updated locally, on each operation to maintain statistics and access history, and remotely, by the WFSM.

On receiving a client request for metadata, for example file lookup, or data, for example read or write, the WFSR identifies the home of the file using the migration table and forwards the request to the home. The home provides the information and maintains a timestamped record of all updates as *update logs*. The home node aggregates updates, eliminates duplicate or redundant updates, and streams the update logs to the file server.

3 WireFS Meta-data Layer

3.1 Home Migration

We use a virtual namespace tree rooted at the directory "/" to model the file organization in NFS. Our solution is based on the following observation: if most of the accesses into

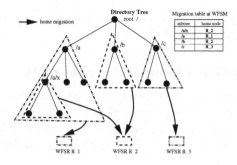

Fig. 3. The home migration of a directory tree and the corresponding migration table

a subtree in the directory tree come from one site (through a WFSR), we will assign the administration privilege of this subtree onto that site (WFSR). We call this task delegation as a *home migration*, and that WFSR the *home node* of this subtree. Notice home migrations can occur recursively in that a subtree migrated to one WFSR may have its own subtree migrated to another WFSR node. Therefore, the directory tree is decomposed into multiple sub-trees based on access statistics, and we want to design the assignment scheme for home migrations so that the total access latency is minimized. In addition, to allow fast (one-hop) resolution of home nodes, we will maintain a migration table at WFSM, the central server side, which keeps one pointer (the address of the home node) for each distinct migrated sub-tree. Figure 3 shows one example for home migration.

Formally, we label the WFSM as R_0, the n WFSRs as $R_1, R_2, \ldots R_n$, and the network latency (RTT) between R_i and R_j as $L_{R_i R_j}$. When a file lookup from R_i traverses a directory node D_x ($1 \leq x \leq m$, where m is the number of directory nodes), we call it one access of R_i on D_x. For each node D_x in the directory tree, a stack of n registers $\{C_{D_x R_i}, i \in [0, n]\}$ record the expected access times of each WFSR on D_x during the next time period T [1].

Now we formulate access latency optimization as an integer programming problem:

$$\min \sum_{x=1}^{m} \sum_{i=0}^{n} I_{D_x R_i} \left(\sum_{j=0}^{n} C_{D_x R_j} L_{R_j R_i} + M_{D_x R_i} \right) \tag{1}$$

subject to $I_{D_x R_i} \in 0, 1$

$$\sum_{i=0}^{n} I_{D_x R_i} = 1$$

Where $I_{D_x R_i} = 1$ if the subtree rooted at D_x will be migrated to R_i, 0 otherwise. $I_{D_x R_i}(\sum_{j=0}^{n} C_{D_x R_j} L_{R_j R_i})$ were the total access cost to the directory node R_i if we migrated the subtree rooted at it to the home node R_i. $M_{D_x R_i}$ is the transfer cost of migrating D_x from its current home node to R_i.

When there is no migration table size constraint, the optimal solution can be found by deciding the best home node for each directory node individually. Next, we present

---

[1] In the experiments we use an exponential weighted moving average (EWMA) counter to approximate the access register based on past historical information.

the algorithm to compute the optimal solution of the optimization problem when we have migration table size constraint.

3.2   Optimal Solution Under Constrained Migration Table

Let P_{max} ($<$ the directory size) be the maximal number of pointers that the migration table can contain. Deciding the P_{max} distinct subtrees is similar to many cache or filter placement problems in the literature [5, 6]. To find the optimal solution in a bounded-degree directory tree, we can solve the following problem using dynamic programming.

(i.) Let $access(D_x, k, H_p(D_x))$ be the optimal access cost for the directory (sub)tree rooted at D_x given that there are k pointers left for this subtree and the home node for the parent node of D_x is $H_p(D_x)$. We start with $access("/", P_{max}, R_0)$ on the root node and enumerate the rest of the nodes following breadth first search.

(ii.) At each directory node D_x, the optimal assignment is decided as
   - If $k = 0$, all nodes in the subtree will be assigned to $H_p(D_x)$ and
     $access(D_x, k, H_p(D_x)) = \sum_{z: nodes in subtree} S$, where $S = \sum_{j=0}^{n}(C_{D_z R_j} L_{R_j R_{H_p(D_x)}} + W_{D_z R_{H_p(D_x)}})$.
   - Otherwise, $access(D_x, k, H_p(D_x)) =$
     min { min [for all allocation schemes (z, A_z) of k-1 pointers on children of D_x
     $\sum_{j=0}^{n}(C_{D_x R_j} L_{R_j R_y} + W_{D_x R_y}) + \sum_{z: \text{ child of } D_x} access(z, A_z, y) \forall y \neq H_p(D_x)]$,
     min [for all allocation schemes (z, A_z) of k pointers on children of D_x
     $\sum_{j=0}^{n}(C_{D_x R_j} L_{R_j R_{H_p(D_x)}} + W_{D_x R_{H_p(D_x)}}) + \sum_{z: \text{ child of } x} access(z, A_z, H_p(D_x))] \}$

Next we present the analysis result on the dynamic programming algorithm.

Theorem 1. *The dynamic programming algorithm finds the optimal solution in* $O(P_{max}^D m^2 n)$ *time, where D is the maximal degree in the directory tree.*

Proof. The analysis is similar to the one for the k-median problems on trees [7] and is skipped in the paper.

3.3   A Greedy Algorithm Under Constrained Migration Table

While we can find the optimal solution in polynomial time, the large directory tree size m and large degree bound D makes it desirable to find a solution good enough and as quickly as possible. We observe that on the file directory tree, the nodes close to the root receive more lookup requests than the nodes close to the leaf nodes do. Therefore, when deciding home migration we can take the top-down order and start from the nodes at the top of the directory tree. For a set of candidate nodes, we will firstly pick the node whose subtree has the most access requests (from all users) for the home migration process. The cost of all nodes is assigned using a Depth First Search and the nodes are maintained in an ordered list. We move from the highest cost node in the list and assign the home for the subtree rooted at the node. This process continues until all k pointers in the migration table are filled. The remaining nodes' home is assigned to that of the closest ancestor.

Theorem 2. *The greedy algorithm finds an assignment scheme in* $O(m \log(m) + P_{max} m)$ *time.*

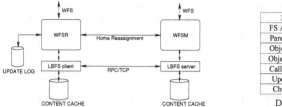

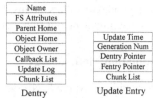

| Name |
| FS Attributes |
| Parent Home |
| Object Home |
| Object Owner |
| Callback List |
| Update Log |
| Chunk List |

| Update Time |
| Generation Num |
| Dentry Pointer |
| Fentry Pointer |
| Chunk List |

Dentry   Update Entry

Fig. 4. WireFS implementation

Fig. 5. Directory and update log entries in WFSR

Proof. The algorithm performs one tree traversal using the Depth First Search in $O(m)$ time. Operations on ordered list take $O(mlog(m))$ time, and for each node to added to the migration table, checking ancestors for m nodes takes $O(P_{max}m)$ time.

4 Implementation

WireFS is implemented by extending the Low Bandwidth File System (LBFS) [3]. LBFS provides content hashing, file system indexing, and chunk storage and retrieval. WireFS extends the LBFS implementation by including the WFSR update logs. Unlike the default LBFS, WireFS uses a modified NFS implementation which sends the file system requests to the LBFS client at each WFSR. At the WFSM, the LBFS server sits in front of the NFS server and is unmodified.

Figure 4 shows the WireFS implementation. In addition to the default LBFS, WireFS includes additional functionality for home migration and maintaining update logs. These are implemented as extensions to LBFS and use the SFS toolkit [8] to provide the asynchronous programming interface. Finally, the interaction between the WFSRs is independent of the LBFS protocol. WireFS receives all NFS requests from the clients, and uses the WFS protocol to identify the home node. The requests are passed on to LBFS client at the home node which in-turn uses the content cache and the LBFS server to service the requests.

WireFS associates additional metadata with each file system object. It is important to note that this information is not visible to either the server or the clients, but is generated and maintained by the WireFS redirectors transparently. The additional attributes enable WireFS specific optimizations over the wide-area-network. As shown in Figure 5, for each file, WireFS maintains a directory entry (dentry) which contains four additional attributes, a chunk list, callback list, *Home* information for the parent and the file itself, and *Owner* information. In addition to the extended attributes, update logs are maintained for any updates in queue for the server. Finally, each WFSR maintains a translation table which maps the file handles provided by the server at mount time to the path name of the file on the server.

5 Evaluation

In this section, we present an evaluation of WireFS home migration using trace driven simulation. We first describe our simulation methodology. We then show the behavior of the home based WireFS metadata access protocol and compare it against existing network and wide area file systems. Finally, we show the benefits of home migration in WireFSwhile comparing our two algorithms for reassignment.

5.1 Simulation Methodology

We use the publicly available NFSv3 traces from Harvard SOS project [9]. The Harvard traces include up to three month real campus NFSv3 traffic in different deployment scenarios. We choose the most diverse workload which is a mix of research, email and web workload. In our simulation, traffic traces of two weeks are extracted to evaluate WireFS performance under different configurations. The traces feature workload and operation diversity where 993 thousand distinct files with 64 thousand directory files are monitored. During the studied two week period, 384 million NFS RPC call/response pairs are recorded. 75 distinct host IP addresses are identified from the traces and are used for creating user groups.

To emulate an enterprise environment with branch offices, we partition the total 75 hosts into 10 groups (sites) with the access pattern following uniform or Zipf distribution. The site geographic distribution is emulated based on the Ping project traces [10]: we randomly picked 10 PlanetLab nodes scattered around the world, and emulated the wide-area network latency between them by extracting the round-trip time (RTT) information between them from the Ping project traces. The *RTT* between two sites varies from 2.4ms to 358ms with the average value of 157ms. The time zone for each site is included in our experiments by adding time offset to the trace data originating from that site.

We compare four network file systems in our simulated WAN setting. The first file system is a wide-area deployment of the NFSv3 system, called *WAN-NFS* in the rest of the paper. In WAN-NFS, all client groups access files from the remote central NFS server via NFS RPC procedures. The second file system, called *DHT* file system, utilizes the DHT based data management scheme (like SHARK [4]) that randomly distributes file objects among the participating sites. For simplicity, in the simulations we assume a file lookup takes only one-hop searching for remote file object access. The third file system is called *WireFS-node*, where home assignment is done on individual files based on their access statistics. The fourth system is called *WireFS-tree*, where home assignment is done based on the greedy algorithm described in Section 3.3.

In both WireFS-node and WireFS-tree, home migration decision is recomputed every T minutes, and the number of accesses to a file f from a site x at the end of the i-th period is calculated with an EWMA counter: $C_f^x(i) = \alpha \times C_f^x(i-1) + (1-\alpha) \times n_f^x(i)$, where $n_f^x(i)$ is the total access number of x on f during the i-th period and $C_f^x(0) = 0$. Unless explicitly stated, $T = 60$, $\alpha = 0.5$, and a migration table with the size $k = 50000$ are used in the following.

5.2 Results

Figure 6 shows the average meta-data lookup latency distribution in the four file systems where host grouping was based on Zipf distribution and time zone effect is considered.

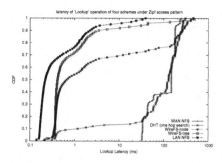 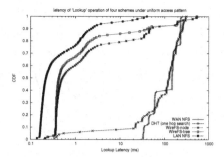

Fig. 6. The CDFs of average lookup latency for different systems with zipf-based grouping and time zone offset

Fig. 7. The CDFs of average lookup latency for different systems with uniform-based grouping and time zone offset

The NFS lookup latency performance in local area network (*LAN NFS*) is also included as the baseline for all these schemes.

We observe that WireFS-tree performs close to NFS-LAN and outperforms the other three schemes. The latency of more than 96% of the lookups in WireFS-tree is comparable to that in NFS-LAN; 92% of the lookups in WireFS-tree take less than 10ms, compared with 75% of WireFS-node, less than 15% of DHT system, and 0 of WAN NFS as all other sites are more than 10ms away from the central file server; only 2% of the operations in WireFS-tree underperformed the other schemes due to its worst case scenario with two-hop lookups. We repeat the above simulations with host grouping based on uniform distribution, and the result (as shown in Figure 7) was similar to that of Zipf distribution.

Figure 8 compares the performance of WFS-tree and WFS-node in terms of the distribution of local hit ratios (computed every T minutes) throughout the 2 weeks. We observe that WFS-tree has a hit ratio over 95% most of the time, while WFS-node experiences hit ratio oscillation during the experiment with average value less than 90%.

The performance difference between WireFS-tree and WireFS-node is caused by the *prefetching* nature of the subtree-based migration and the *caching* nature of the node-based migration. If file accesses from a site have a locality pattern within the directory tree hierarchy, *prefetching* avoids "cold" misses, due to first-time accesses; our experiment results clearly validated that assumption.

Figure 9 shows the time evolution of local hit ratios in WFS-tree. The aperiodic deterioration of hit ratios is explained by the spikes of remote first-time-access file ratios [2], which are also shown in Figure 9.

Figure 10 presents the time evolution of average lookup latency in WireFS-tree over the two-week time period. The first-time-file access ratio is shown in Figure 10. We observe that the latency spikes are consistent with the spikes of the first-time-access file ratios.

The effect of home migration is demonstrated by the immediate decline after each latency spike in Figure 10. The drop in the access latency shows that home migration

---

[2] Remote first-time-access file ratio is defined as the percentage of the files accessed by a remote group for the first time out of all files accessed during a time period.

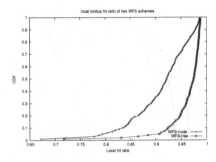

Fig. 8. CDF of local hit ratio for the two WireFS systems

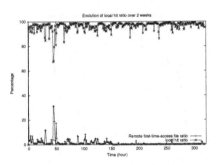

Fig. 9. WireFS-tree: local access hit ratio vs. remote first-time-access file ratio

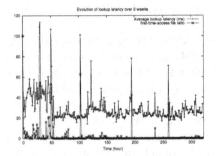

Fig. 10. WireFS-tree: average access latency vs. first-time-access file ratio

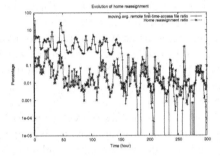

Fig. 11. WireFS-tree: home reassignment ratio vs. moving average remote first-time-access file ratio

reduces the wide-area accesses adaptively and quickly. Over the first 50 hours, most of the files are accessed for the first time by remote sites, which makes the average lookup latency oscillate dramatically. After this time, the latency stabilized until another first-time-access spike changed the pattern.

Figure 11 presents the time evolution of home reassignment ratio in WFS-tree system. Home reassignment ratio is defined as the percentage of the meta-data files whose home nodes change. This ratio is used as a metric for home migration overhead as each reassignment requires computation. The maximum 2.5% and average 0.1% home reassignment ratio demonstrates the marginal overhead home migration incurred after the initial few delegations. Moving average (3 runs) remote first-time-access file ratios [3] are also shown in Figure 11 to illustrate the main cause of home reassignments.

---

[3] Moving average (m rounds) remote first-time-access file ratio is the average value of the remote first-time-access file ratio in the consecutive m rounds. As we use an EWMA counter to record the historical access information, remote accesses in the current (one) round might not immediately affect the home assignment decision. Therefore, we pick $m = 3$ in Figure 11 to better reflect the reason behind the home reassignment evolution.

6   Related Work

Network file systems have been studied in the local area with stateless[1] and stateful servers [11, 12, 13, 14]. Satyanarayanan presents an overview of several traditional distributed file systems [15]. Recently, there has been significant research activity in providing data access (object or file system based) over the WAN. Multiple peer-to-peer architectures for decentralized data management have been proposed [16, 17, 18, 19]. However, the goal of such systems is to store large quantities of data, dispersed and replicated across multiple clients to improve fault resilience and reduce management overheads. In contrast, WireFS tries to improve performance of existing network file systems for interactive workloads. While WireFS is capable of storing large data, replication, and disconnected operation, such characteristics are not the primary concern.

Shark [4] uses geographically distributed cooperative caching proxies which enable fast parallel downloads in addition to difference elimination and content caching. Unlike WireFS, Shark is designed for environments like PlanetLab [20] where multiple clients are interested in the same large file concurrently e.g. for an experiment. Therefore, no attempt to improve the metadata access performance as well as inter-site read-write sharing is explored, which is the primary goal of WireFS.

7   Conclusions

In this paper, we presented *home migration*, a technique to minimize meta-data access latency in wide-area file systems. We first described the design of WireFS a wide-area networked file system. Next, a set of algorithms for home assignment and migration were proposed in the context of WireFS to improve performance of metadata accesses. Through trace driven simulations, we demonstrated that our technique improves the latency of metadata operations with low management and network overheads.

Acknowledgements

We thank the Harvard University SOS project team, especially Daniel Ellard, Jonathan Ledlie, and Christopher Stein for providing the NFS traces and answering several questions.

References

1. Callaghan, B., Pawlowski, B., Staubach., P.: NFS Version 3 Protocol Specification, RFC 1813. IETF, Network Working Group. (1995)
2. Microsoft Corporation: Cifs: Common internet file system. http://msdn.microsoft.com/library/default.asp?url=/library/en-us/cifs/protocol/portalcifs.asp (2006)
3. Muthitacharoen, A., Chen, B., Mazières, D.: A low-bandwidth network file system. In: Proc. of SOSP '01. (2001)
4. Annapureddy, S., Freedman, M.J., Mazières, D.: Shark: Scaling File Servers via Cooperative Caching. In: Proc. of NSDI'05. (2005)

5. Krishnan, P., Raz, D., Shavitt, Y.: The cache location problem. IEEE/ACM Trans. on Networking **8**(5) (2000) 568–582
6. Shah, R., et al.: Efficient dissemination of personalized information using content-based multicast. IEEE Trans. on Mobile Computing **3**(4) (2004) 394–408
7. Tamir, A.: An $o(pn^2)$ algorithm for the p-median and related problems on tree graphs. Operations Research Letters **19** (1996) 59–64
8. Mazieres, D.: A toolkit for user-level file systems. In: Proc. Usenix'01, Boston, MA (2001)
9. Ellard, D., Seltzer, M.: New NFS Tracing Tools and Techniques for System Analysis. In: Proc. of LISA '03. (2003) 73–86
10. Stribling, J.: All-Pairs-Pings for PlanetLab. http://pdos.csail.mit.edu/ strib/pl_app/. (2005)
11. Kistler, J., Satyanarayanan, M.: Disconnected Operation in the Coda File System. ACM Trans. on Computer Systems **10**(1) (1992) 3–25
12. Birrell, A.D., et al.: The Echo Distributed File System. Technical Report 111, DEC SRC (1993)
13. Nelson, M.N., Welch, B.B., Ousterhout, J.K.: Caching in the sprite network file system. ACM Trans. Comput. Syst. **6**(1) (1988) 134–154
14. Hartman, J.H., Ousterhout, J.K.: The Zebra Striped Network File System. ACM Trans. on Computer Systems. **13**(3) (1995) 274–310
15. Satyanarayanan, M.: A survey of distributed file systems. Technical Report CMU-CS-89-116, Carnegie Mellon University, Pittsburgh, Pennsylvania (1989)
16. Dabek, F., et al.: Wide-area cooperative storage with CFS. In: Proc. of SOSP'01. (2001)
17. Rowstron, A., Druschel, P.: Storage management and caching in PAST, a large-scale, persistent peer-to-peer storage utility. In: Proc. of SOSP'01. (2001)
18. Kubiatowicz, J., et al.: OceanStore: an Architecture for Global-Scale Persistent Storage. In: Proc. of ASPLOS'00. (2000) 190–201
19. Muthitacharoen, A., et al.: Ivy: A read/write peer-to-peer file system. In: Proc. of OSDI '02. (2002)
20. Peterson, L., et al.: A Blueprint for Introducing Disruptive Technology into the Internet. In: Proc. of HotNets'02. (2002)

Connecting Pervasive Frameworks Through Mediation

Florence T. Balagtas and Cedric Angelo M. Festin

Department of Computer Science
University of the Philippines
Diliman, Quezon City, Philippines
florence.balagtas@up.edu.ph, cmfestin@up.edu.ph
http://engg.upd.edu.ph/cs/index.html

Abstract. Context information helps an application decide on what to do in order to adapt to its user's needs. To easily develop ubiquitous applications, there has been increased research in the design and development of frameworks called pervasive computing frameworks. Although these frameworks help application developers create ubiquitous applications easily, interoperability has been a problem because of the different representation of context information and protocols used. This research attempts to solve this problem by creating a Context Information Mediator (CIM) which will serve as a translation gateway between different applications created using different frameworks. To test our system, we developed two versions of an inventory system application that keeps track of items inside a building. The idea here is to let these applications communicate with each other and share information through the CIM.

1 Introduction

Pervasive computing is a computing paradigm that aims to make digital environments composed of ubiquitous applications that are sensitive, responsive and adaptive to human needs without humans actually knowing what happens in the background [1]. Creating ubiquitous applications is quite difficult since different types of devices and different forms of data are to be processed and should be able to work seamlessly. To simplify the creation of ubiquitous applications, several researches in the area of pervasive computing are focused on the creation of pervasive computing frameworks such as the Aura Framework [2], Context-Toolkit framework [5] and One.world framework [6]. These frameworks aim to collect raw data from diverse devices and process the collected data into context information. These context information are then disseminated to diverse applications that run on different devices with the concern for security to avoid unauthorized use of these information [7]. The problem now lies in the representation of context information in different frameworks. Different frameworks have different formatting of context information which prevents them from sharing information. The need for sharing of information among different frameworks is

Y. Robert et al. (Eds.): HiPC 2006, LNCS 4297, pp. 313–325, 2006.

important to provide interoperability which is one of the major goals of pervasive computing.

The problem of interoperability has been present in several areas of computing and different types of mediator systems have been developed in order to address this problem. The *Context Interchange Architecture of Database systems (COIN)* [8] architecture has tried to detect and reconcile semantic conflicts among different database systems. The *P2P gateway* [9] has aimed to facilitate information sharing among different peer-to-peer file sharing systems that uses different protocols. The *Internet architecture* has been designed to facilitate the sharing of computer resources present in different networks through the use of the gateway.

This research aims to create a *Context Information Mediator* (CIM) which is used to get information from different servers that uses different frameworks, and convert these information into data which can be understood by the other frameworks. The CIM framework is developed by using the Java 2 Platform API and uses XML to represent the data in the system. The principles of the Internet architecture were applied when designing the protocols of the CIM.

2   Design and Implementation

This section discusses the steps to achieve the goal of creating a Context Information Mediator.

2.1   Inventory Application Implementation

The application that we have developed is an inventory system for an office area which similar to the Smart Toolbox and Smart Tool Inventory of [11]. What our inventory system application does is that, it keeps track of where a certain item in the inventory is located inside a building. This is done by having sensors monitoring certain areas of the building in order to know where certain items are located. For the purpose of this research, we will simulate the environment that contains the items and the sensors through a graphical user interface (GUI) made using Java Swing. The GUI will show a building that contains three rooms, wherein each room contains several inventory items. Dragging the items in our simulator simulates movement of an item from one location to the next.

The inventory system is composed of two major components. The *pervasive inventory system* which gets the location of the inventory items and stores this information, and the *application interface* which represents the client applications that subscribes to the pervasive inventory system in order to get information about the items. We have created two versions of this sample inventory application. The first version was created using the Aura Contextual Service Interface version 2.3 of the Aura framework. The second version was created using the Context-toolkit framework 2003 release. The idea here is to let the two applications that are written using the two different frameworks communicate with each other and share information through the Context Information Mediator (CIM).

2.2   Context Information Mediator Implementation

The Context Information Mediator architecture consist of two general communicating components, the client and the CIM server. Figure 1 shows a general view of the CIM architecture. Take note that although the figure shows only two clients connected to CIM, the CIM server can handle many clients. We have chosen to design our system base on the client-server model of computer networks since it is already a well established model and is mostly used in networking today. The client-server model is a design in computer networks in which client machines request and receive service by querying the server. The server then sends the needed information to its clients. This model is especially effective when clients and server have their own special tasks that they routinely perform.

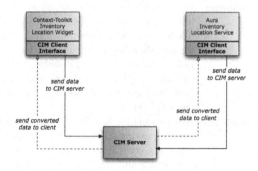

Fig. 1. General view of the CIM architecture

CIM Client Interface. The *CIM Client Interface* sits below the client[1] system and is responsible for forwarding and receiving information to and from the CIM server. In this part of the CIM architecture, we used the *layering strategy* that is used in computer networks.

CIM Server. The *CIM Server* is responsible for gathering all the context information and sends the converted information to its clients. Table 1 shows the subcomponents of the CIM server and their functions.

CIM Data Packets. The CIM data packet contains the information sent between the clients and the CIM server. The CIM packet is divided into two main sections, the *header* and the *data* section. The header section is further subdivided into two which is the *packet type* and the *client type*. The *packet type* identifies what type of packet is sent while the *client type* identifies what type of client has sent the packet. The *data section* contains the main data sent by

---

[1] In this section, the term *client* refers to the different servers implemented in different frameworks that requests and sends context information to the CIM server.

the sender of the packet. Table 2 shows the different packet types and their descriptions.

Table 3 shows the different types of senders supported by CIM. This helps CIM distinguish from which type of sender the packet was from and helps it decide on how to convert the data contained in the packet.

Table 1. CIM Server Subcomponents

SubComponent Name	Function
Client Request Listener	Responsible for listening to client requests to connect to CIM. It grants the client's request and transfers the request to the Client Registration Manager.
Client Registration Manager	Responsible for asking the client for registration information such as client type (Aura or Context-Toolkit) and is also responsible for assigning a unique identifier to the client. Once the registration process has been done, it assigns a Client Thread Manager to that particular client and the client can now send and receive context information from the CIM server.
CIM Client Thread Managers	Responsible for handling context information received from a particular client assigned to it and stores the data to a central repository. It is also responsible for converting and sending context information that is not present in the client. Aside from that it also monitors if a client is still active by sending AYA (Are You Alive) messages to the client. Once it has detected the client has been disconnected, it informs the main CIM server that which then terminates that client thread manager.

The data section of the CIM packet contains data that is formatted in XML. Although both Aura and Context-toolkit applications convert their data to XML, they still have different representations of the information modeled in XML.

Shown in Figure 2 is a snippet of the XML file in the Aura framework that contains the item *FAX* with item ID number 10004 located at room MH 215 together with the time it was moved.

Shown in Figure 3 is a snippet of the XML file for the Context-Toolkit framework that contains the item *iMac* with item ID number 10001 located at room MH 215 together with the time it was moved.

CIM Translation. This section will describe how the data is translated from the client to the CIM server and vice versa.

Table 2. CIM Packet Types

Packet Type	Description
REGREQ	Sent by the CIM Server to request for registration information from the client
REGOK	Sent by the CIM Server to the the client if registration is successful
REGFAILED	Sent by the CIM Server to the client if registration failed
REGDETAILS	Sent by the client as reply to the REGREQ packet. This contains information regarding the client
GOODBYE	Sent by client/server to signify termination
NEWCONTEXTINFO	Sent by client/server that contains the new context information
AYA	Sent by client/server to ask if a client/server is still alive
IAA	Sent by client/server as a reply to the AYA packet

Table 3. CIM Sender Types

Sender Type	Description
AURACLIENT	Sender is from the Aura Framework.
CTKCLIENT	Sender is from the Context-Toolkit Framework.
CIMSERVER	Sender is the CIM Server.

```
<QueryResultItems>
    <ITEM>
        <itemName>FAX</itemName>
        <itemID>10004</itemID>
        <itemLocation>mh 215</itemLocation>
        <updateTime>1140073743718</updateTime>
    </ITEM>
```

Fig. 2. XML format in Aura

```
<attribute attributeType="struct">Item#0
    <attributes>
        <attributeNameValue>
            <attributeName>ItemName</attributeName>
            <attributeValue>iMac</attributeValue>
        </attributeNameValue>
        <attributeNameValue>
            <attributeName>ItemID</attributeName>
            <attributeValue>10001</attributeValue>
        </attributeNameValue>
        <attributeNameValue>
            <attributeName>ItemLocation</attributeName>
            <attributeValue>mh 215</attributeValue>
        </attributeNameValue>
        <attributeNameValue>
            <attributeName>Timestamp</attributeName>
            <attributeValue>1138173984310</attributeValue>
        </attributeNameValue>
    </attributes>
</attribute>
```

Fig. 3. XML format in Context-Toolkit

CIM Client Interface Sends Data to CIM Server. One of the responsibilities of the CIM Client interface is to send new context information obtained by the client system to the CIM server. Before it sends the new data to the CIM server, it first extracts the DataObject (for Context-toolkit clients) or QueryResult object (for Aura clients) that contains the new information from the client system. It then starts to create a CIM packet that will be sent to the CIM server. The CIM packet is created by first appending the necessary headers. The following headers will be added to the packet: NEWCONTEXTINFO for the packet type and AURACLIENT or CTKCLIENT header for the client type. Finally, the XML form of the object extracted from the client is then appended to the data section of the packet. The XML form of the DataObject of a Context-toolkit client is created by using the XMLEncoder class of the Context-toolkit API. It creates an XML form of the DataObject that contains the necessary tags and inventory item information. For the XML form of the QueryResult object of an Aura client, it is created by using the CimXMLEncoder class of the CIM API. The CimXMLEncoder is adapted from the XMLEncoder of the Context-toolkit API. After the packet has been created, it is then sent to the CIM server.

CIM Server Translation. After the CIM Server has received a packet of type NEWCONTEXTINFO from its clients, it first determines the client type of the packet. It then forwards the data section of the packet to the translator that handles the translation of data for that certain client type. After the inventory item information has been extracted from the packet, the inventory database that contains all the inventory data gathered from the different clients are then updated. After the inventory database has been updated, the CIM server will create packets that contains the latest inventory information and will send it to the other clients. In this case, since there are two types of clients currently supported by the CIM framework, the CIM server will create two types of packets, one for the Aura clients and one for the Context-toolkit clients.

CIM Client Interface Receives Data from CIM Server. When the CIM Client receives new context information from the CIM Server, it then extracts the data from the packet and creates it into a DataObject (for Context-toolkit clients) or QueryResult object (for Aura clients). Client applications have a choice of calling several methods from the CimClient class to get the new data. Table 4 shows the methods and their description.

CIM Protocols. The CIM architecture has several protocols for establishing connection between client and server, terminating a connection and the exchange of data between client and server.

Establishing a Connection with the CIM Server. To establish a connection between the CIM Server and the CIM client, the client connects to the CIM Client Request Listener, which then forwards the request to the CIM Client Registration Manager. The CIM Client Registration Manager asks the client for registration information. If the client has successfully sent all the requirements, the

Table 4. CIM Client class data retrieval methods

METHOD	DESCRIPTION
getNewItemListFromCim()	This method returns a java.util.Vector object that contains the list of items in the inventory. This can be used by either an Aura client or a Context-toolkit client.
getNewItemListFromCimForAura()	This method returns a QueryResult object that contains the list of items in the inventory. This is for Aura clients only.
getNewItemListFromCimForCtk()	This method returns a DataObject object that contains the list of items in the inventory. This is for Context-toolkit clients only.

CIM Client Registration Manager sends it a REGOK message that signifies its successful connection to the CIM server. It then assigns a CIM Client Thread manager that is responsible for communicating with the connected client. In cases wherein a client was unable to satisfy the requirements, the CIM Client Registration Manager sends the client a REGFAILED message and disconnects the client.

Terminating a Connection with the CIM Server. In cases wherein the client application leaves or if the CIM Server terminates, each sends a GOODBYE message in order to signal the other that it is leaving. These scenarios show cases wherein there is a clean termination of both client and server. However, there are cases in which either of the two crashes and will not be able to send a GOODBYE message. To resolve this problem, we have created the AYA (Are you alive) packet that is constantly sent by both server and client to each other in order to monitor if the other still exists. The client/server that receives this type of message should reply with a IAA (I am alive) message in order to signify that it is still alive. In cases wherein the client/server is unable to reply with an IAA message, the sender of the AYA packet will terminate its connection with the dead client/server.

Context-Information Exchange. The CIM client sends the CIM server new context-information about its system. It does this by sending a packet of type NEWCONTEXTINFO which contains the new information. When the CIM server receives this type of information, it then updates its database and creates packets containing the newly updated information for both Aura and Context-toolkit clients. It then sends the packets to the other clients connected to CIM.

3   Performance Evaluation and Results

The experiments that we have conducted runs a single CIM server and one or more CIM clients. The first batch of experiments have been done by using an

Apple Powerbook, with a 1.5 Ghz PowerPC G4 processor and 512 MB DDR
SDRAM. All applications both client and server have been run in this single
computer. For the second batch of experiments, we run the CIM server in an
Intel PC, with a 2.66 GHz Pentium 4 processor and 512 MB RAM, while all the
clients run on the Apple Powerbook. The clients and server are connected via
a local area network (LAN). To start the experiments, we run the CIM server
which waits for clients to connect to it. We then run the environment simulator
that shows a graphical user interface of a building with three rooms, and the
eight inventory items. Each inventory item will be dragged from one room of the
building to the next to simulate movement of items. For the experiments that
we have done, we have focused on the following parameters that are relevant to
the analysis of CIM's performance. These parameters are data size, number of
clients connected to the CIM Server and the variety of clients connected to the
CIM Server.

Data Sizes. To get the data sizes of the packets, we have written the data that is
sent by a client to the server to a file in order to get the number of bytes a single
packet contains. The packet size varies based on the type of client that creates
the packet (Aura or Context-toolkit), and it also varies based on the number of
inventory item information the client has stored in the packet. In order to test
how the packet gets bigger as the number of items increases, we have created
packets containing 0 items up to 8 items in the inventory. We have done this three
times in order to get the average data sizes of the packets sent. Figure 4 shows
the relationship of data sizes (in bytes) to the number of items that a packet
contains. The number of items here pertains to the number of inventory items
that a certain client has information about. The Y-axis describes the average
data sizes of packets in bytes. The X-axis shows the number of items contained
in a packet. The graph shows the data sizes in bytes for both Context-toolkit and
Aura client. Observe that as the number of inventory item information increases,
the data size of the packets also increases. Also observe that the Context-toolkit
data packet is bigger compared to the Aura data packet. This is because the
data of the Context-toolkit client which represents the XML file being sent has
more tags. The average data size of a single Context-toolkit packet is 528 bytes
per item, while for a single Aura packet, we have an average of 154 bytes per
item. There is a slight variation on the data size of each packet since there is no
standard size on the information inside a packet.

Translation Time. To get the translation time of the CIM server, we have done
a variety of experiments by changing the values of the data sizes sent, changing
the number of clients connected and changing the variety of clients connected.
To test the effect of data size to the translation time of the CIM server, we
iteratively moved items to a particular room monitored by a client one by one
every n-milliseconds. As this happens, the client sends its updated inventory
information for the room that it monitors to the CIM server. At this point,
we measure how long the CIM server can convert these information base on
the adding and removing of items. For this experiment, we got the average

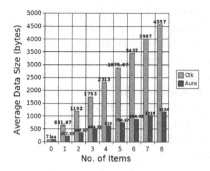

Fig. 4. Average data sizes of packets

translation time by running this test 5 times. We did this for both Aura clients and Context-toolkit clients.

Translation Time vs. Data Sizes. The graph in Figure 5 shows the relationship of the data size versus the time it takes to translate a certain packet for both Context-toolkit and Aura clients. As seen in the graph for both clients, as the data size gets bigger, the translation time also increases. The average translation time per data item for a Context-toolkit client is about 52.59 msecs/item, while for the Aura client we have 52.08 msecs/item. We can see that it takes more time to translate data for a Context-toolkit client as compared to an Aura client, however, the difference is very minimal.

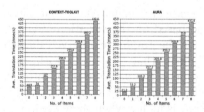

Fig. 5. Average Translation Time vs. Data Sizes

Translation Time vs. Number of Clients connected. To test the effect of number of clients to the translation time of the CIM server, we run several clients that monitors different rooms and continuously sends information about the rooms they monitor. The items are moved from one location to the next every n-milliseconds. To get the total translation time, we add up all the translation times for moving from the first room to the next and vice versa for all clients. We then get the average translation time by doing this experiment 5 times. We increase the number of clients as we did this experiment. We did this for 1, 2, 4, 8, 12 and 16 clients. We did the same experiments to test whether the variety of clients has some effect on how CIM translates data. However, we also varied the types of clients connected to the CIM server and the number of clients per variation. In Figure 6 we can see in the graphs, as the number of clients increases from 1, 2, 4, 8, 12 to 16, the average translation time of the CIM server also increases. The average translation time for one Context-Toolkit client is 211.5 msecs/client, while for the Aura client we have 208.63 msecs/client. The average translation time of the Context-toolkit clients is higher as compared to the Aura clients. This means that it takes longer time to translate the data given by the Context-toolkit clients. This is due to the fact that the data of the Context-toolkit client is more complicated as shown in section 2.2 as compared to the data of the Aura clients.

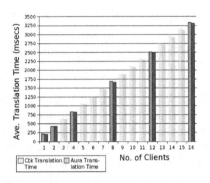

Fig. 6. Comparing Average Translation Time of both Context-toolkit and Aura Clients

Translation Time vs. Variety of Clients. The graph shown in Figure 7 shows the average translation time based on the variation of clients that the CIM server has to deal with. Based on the graph, we observed that for the experiments that has the same types of clients that CIM deals with, the Aura group of clients has a lower translation time as compared to the Context-toolkit group of clients. Comparing the results in which CIM has both a Context-toolkit and an Aura client, the translation time is higher when there are more Context-toolkit clients connected. We can see here that the variation of clients does not really affect the translation time of CIM, however, it is greatly affected by the type of clients that it has to deal with.

Scalability and Extensibility. The scalability of the CIM architecture is measured in terms of the number of clients that the CIM server can accommodate. To test the scalability of CIM, we have tried running 1, 2, 4, 8 and 16 clients connected to CIM and measured CIM's translation time given the number of clients. Please refer to Figure 8 to see the average translation time of clients. As we have seen in the graph, the translation time is doubled as the number of clients double which means that there is an increase in translation time. However, since the increase in translation time is minimal, the performance of the CIM server is not greatly affected. For this experiment, we only tried up to 16 clients, however,

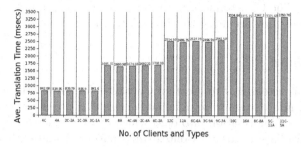

Fig. 7. Average Translation Time vs. Variety of Clients

based on our observation on the increase of translation time, the CIM server can accommodate more, especially if the CIM server is placed in a more powerful computer.

The extensibility of the CIM architecture was measured in terms of how it can be extended to support other frameworks. We have written two different applications using the Context-toolkit framework and the Aura framework which can interoperate using the CIM framework. CIM can support other types of clients that are implemented using the Java platform.

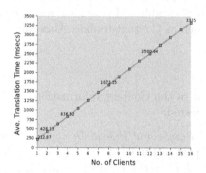

Fig. 8. Average translation time is doubled as the number of clients double

Summary of Experiment Results. Based on the results of our experiments we have the following observations: (1) The data size (in bytes) of packets increases as the number of inventory item information increases. Packets coming from Context-toolkit clients have a bigger data size than that of Aura clients. (2) The translation time increases as the data size of packets increases. (3) The translation time increases as the number of clients connected to the CIM server increases. (4) The translation time of Context-toolkit clients is higher as compared to the Aura clients. This means that it takes more time to translate data coming from a Context-toolkit client as compared to Aura clients. (5) The variety of clients connected to CIM has minimal impact on the translation time of CIM. (6) The CIM framework is scalable in terms of the number of clients it deals with. (7) The CIM framework can be extended to support other types of clients as long as they are using the Java platform.

4   Conclusion

The goal of this research is to create a Context-information Mediator which is used to obtain information from different servers that use different frameworks, and convert these information into data which can be understood by the other frameworks. We have created a Context-information Mediator that serves as a gateway between different frameworks in order for them to share information. Even though CIM was designed to be a gateway for different frameworks, it can still act as a server for clients of the same framework. This can serve as a server for those clients so that they do not need to create their own servers in order to pass information.

The *CIM client interface* serves as the link of a client to the CIM server which abstracts the different protocols needed by the client in order to connect to CIM. The *CIM Server* serves as the translator of different context-information that is provided to the clients connected to it.

We applied the principles of the Internet architecture when designing the protocols of the CIM. We used XML format for the representation of information from the different clients. The choice for choosing XML to represent our data is that it is already a well established standard and is mostly used by different pervasive frameworks such as Context-toolkit, Aura and One.world frameworks.

5   Future Work

For further development of the Context-Information Mediator architecture, the following ideas are suggested:

1. *Translation at End Points and Hybrid Design.* The current implementation for the translation part of the CIM framework is by having the clients send their own format of data to the CIM server and have the CIM server translate and process the data. Another option for the implementation of the translation part of our system is by having the end-points (clients) translate their data into a standard format understood by the CIM server. When data arrives at the CIM server, it does not have to do any translation. The advantage for this design is having the burden of processing distributed among the clients. However, for pervasive computing, we are not assured if the clients have enough processing power since a client system can reside in any type of device. Another option is to have a hybrid design in which a client can choose if it wants to do its own processing or if wants the CIM server to do the processing.

2. *Support for other pervasive computing frameworks.* Currently, the CIM framework supports clients created using the Aura framework and the Context-toolkit framework. CIM can be extended so that it will support other pervasive computing frameworks as well.

3. *A network of CIM servers*. Create several CIM servers that are connected to each other and allow the sharing of information across different CIM servers. This can help in distributing the processing of information for the different clients connected to the different CIM servers. These CIM servers can also act as data filterers wherein a client can choose not to share all of its information to the other clients in connected to the system.

References

1. Saha, D., Mukherjee, A.: Pervasive computing: A paradigm for the 21st century. Pervasive Computing, IEEE **36**(3) (2003) 25–31
2. Sousa, J.P., Garlan, D.: The aura software architecture: an infrastructure for ubiquitous computing. Technical Report CMU-CS-03-183, School of Computer Science, Carnegie Mellon University (2003)
3. Sousa, J.P., Garlan, D.: Aura: An architectural framework for user mobility in ubiquitous computing environments. In: Proceedings of 3rd IEEE/IFIP Conference on Software Architecture, Montreal (2002). (2002)
4. D., G., D.P., S., A., S., P., S.: Project aura: toward distraction-free pervasive computing. Pervasive Computing, IEEE **1** (2002) 22–31
5. Salber, D., Dey, A.K., Abowd, G.D.: The context toolkit: Aiding the development of context-enabled applications. In: Proceedings of CHI'99, ACM Press (1999) 434–441
6. Grimm, R., Davis, J., Lemar, E., Macbeth, A., Swanson, S., Anderson, T., Bershad, B., Borriello, G., Gribble, S., Wetherall, D.: System support for pervasive applications. ACM Trans. Comput. Syst. **22**(4) (2004) 421–486
7. Chen, G., Kotz, D.: Solar: A pervasive computing infrastructure for context-aware mobile applications. Technical Report TR2002-421, Dept. of Computer Science, Dartmouth College (2002)
8. Goh, C.H., Bressan, S., Madnick, S., Siegel, M.: Context interchange: new features and formalisms for the intelligent integration of information. ACM Trans. Inf. Syst. **17**(3) (1999) 270–293
9. Lui, S.M., Kwok, S.H.: Interoperability of peer-to-peer file sharing protocols. SIGecom Exch. **3**(3) (2002) 25–33
10. Clark, D.: The design philosophy of the darpa internet protocols. In: SIGCOMM '88: Symposium proceedings on Communications architectures and protocols, New York, NY, USA, ACM Press (1988) 106–114
11. Lampe, M., Strassner, M., Fleisch, E.: A ubiquitous computing environment for aircraft maintenance. In: SAC '04: Proceedings of the 2004 ACM symposium on Applied computing, New York, NY, USA, ACM Press (2004) 1586–1592

Error Resilient Video Streaming for Heterogeneous Networks

Divyashikha Sethia and Huzur Saran

Indian Institute of Technology (IIT), New Delhi, India
sethia@gmail.com, saran@cse.iitd.ernet.in

Abstract. We consider the problem of video streaming for a critical private web cast, for a medium sized audience with heterogeneous nodes having different bandwidths and reliabilities. The nodes can distribute video in a peer-to-peer manner by forming a multicast tree at application level. A majority of the nodes in the network have low bandwidths and low reliability and can only receive the video stream. A simulation model has been implemented to compare single video streaming scheme with error resilience schemes with stream replication and Multiple Description Coding (MDC) [6][7]. Results indicate that MDC error resilience scheme provides lower average outage, better video quality and network utilization as packet loss percentage and node failure probability increases. We discuss the significance of path diversity in multiple multicast trees for error resilience and the number of multicast trees.

Keywords: Video Streaming, Error Resilience, Multiple Description Coding (MDC), Path Diversity.

1 Introduction

Video streaming is used to web cast a live or recorded event using the Internet public infrastructure, for subscribers across the globe. In this work we look into the problem of web cast of a live critical private group event for a medium-sized group of heterogeneous nodes (numbering in hundreds or even thousands). The continuity of the video is important to the viewers. The event could be a private community event or a corporate lecture session with the viewers spread across the globe as shown in Fig 1.

Internet TV like P2P TV [1] provides channels for public viewing across the globe. An application of this type cannot be used to web cast a private group event. Video conferencing is used for real time two-way communication between people at different locations via video. A web cast is one-way communication of video to groups of viewers over the Internet. We cannot broadcast stream from video server to all the nodes, since it will make the server heavily loaded. The recent trend is to organize the nodes in a cascade or tree structure. The stream is forwarded in a peer-to-peer manner. Each peer can receive as well as forward the stream to other peers.

Y. Robert et al. (Eds.): HiPC 2006, LNCS 4297, pp. 326–337, 2006.

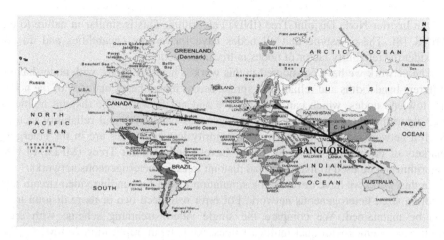

Fig. 1. Video streaming session of a private critical corporate lecture session from head quarters in Banglore, India to other global parts of the corporation

IP multicast cannot be used to form a multicast tree, since it has deployment problems from ISPs across the globe. Instead the nodes can form an application level multicast tree, which is an overlay network. Various application level protocols have evolved like ALMI (Application Level Multicast Infrastructure) [2], CoopNet [3], SplitStream [4] and Narada [5].

Application Level multicast can be used to deliver a private critical group event for which it is important to retain the continuity of video for the entire session. However, the network is prone to error conditions caused due to node and link reliability issues and bursty packet losses. A low reliability node or link can fail due to power or ISP connection failure. In a peer-to-peer video stream distribution using a single multicast tree organization, if a node or link goes down, its descendents in the tree will stop getting the video. They can rejoin the tree, but reconstructing the tree will be time consuming and will result in large outage.

Error resilience can be achieved by providing alternate paths and data to each peer. Instead of a single multicast tree, two or more trees can be maintained. The original stream can be replicated on each tree. There is another encoding technique Multiple Description Coding (MDC) [6][7], which encodes stream into independent descriptions, each of which can be sent across independent paths. Each node receives descriptions from different paths and the quality of decoded video depends on the number of descriptions received by the node. CoopNet and SplitStream provide error resilience by maintaining multiple disjoint multicast trees. It uses MDC to send independent streams across different trees.

Another important issue is tree management. The tree should be short in height, utilizing the outgoing bandwidths of all peers as much as possible. This will reduce the number of points of failures i.e. number of nodes in the path from the source to the leaf nodes. Multiple Multicast trees can be used to provide redundant paths for error resilience, and should be as diverse and disjoint as possible. CoopNet and SplitStream use an Interior Node Disjoint Tree (INDT) algorithm. This is similar in nature to the MINK [8]. These assume that all nodes have forwarding capabilities and do not consider the reliability factor of nodes in tree construction.

The network we have considered in this work is heterogeneous, with nodes having different bandwidths and reliabilities spread across the globe. A majority of the nodes are low-end nodes, which can only receive and not forward. The disjoint multicast tree algorithm should consider the heterogeneous bandwidth and reliability factors. Such a network would exist where the majority low-end nodes are in the emerging markets of developing countries. Multi Level Dual Disjoint Trees (MLDDT) [18] algorithm can be used to maintain dual disjoint trees for heterogeneous networks.

In this work, we have developed a simulation model for sending video stream to a medium sized heterogeneous network. For error resilience two or more disjoint trees can be maintained. We compare the single video streaming scheme with error resilience schemes with dual multicast trees, with stream replication and MDC. The performance is compared in terms of video quality and network utilization. We discuss the significance of path diversity and tree construction algorithm for maintaining multiple disjoint trees for a heterogeneous network.

The results indicate that for a critical, private web cast of a live event, the nodes can form an overlay network using an application multicast protocol. Dual multicast trees with MDC encoding provides better video quality in terms of lower outage, higher number of frames received and higher frame rate as error conditions get worse in a network. It has good network utilization. The MLDDT algorithm can further improve MDC scheme. We extend it to maintain multiple disjoint multicast trees for MDC. Average outage decreases, as there is increase in the MDC descriptions with each description sent across a unique multicast tree.

Future plans are to further improve the tree construction algorithm for mixed set of reliabilities and bandwidths for larger networks. The deployment of an application for a private web cast for a medium sized heterogeneous network is also an area of interest.

The rest of the paper is organized as follows. In section 2, the error resilience schemes have been discussed. Section 3 describes the path diversity issue for multiple multicast trees. Section 4 discusses the simulation model. In section 5 we give the results and finally conclude in section 6.

2   Error Resilience

Video stream can be sent across an application level multicast tree kind of topology. This is the traditional single description coding (SDC) stream case. Each node receives stream from a single parent along a single path in the multicast tree. Whenever there is a link or a node failure, its descendents in the multicast tree will also suffer. Tree reorganization can be time consuming resulting in outage or discontinuity of video on many nodes.

Error resilience can be obtained by using alternate paths and data. Instead of a single multicast tree, two or more multicast trees can be maintained. Each node receives video streams from two parent nodes along different paths. If one path fails, then the other path can provide the stream. With multiple paths each node will get to see an average network behavior and the probability of discontinuity in video will reduce. In our simulation we use dual multicast trees for comparing various video streaming schemes. The error resilience schemes are: Dual Trees with Stream Replication and Dual Trees with MDC.

2.1 Dual Trees with Stream Replication

The original single description video stream is replicated on both dual trees. Each node gets two identical streams from the two different paths (primary and secondary paths). Only one stream is active at a time, and is used to decode video. The other serves as a backup. Whenever the active stream is interrupted, the node experiences outage and switches to the backup stream. The two streams should be synchronized, so that video is displayed uninterruptedly when switching takes place, without duplicating or loosing frames. We refer to this scheme as TWO SDC in our simulation.

2.2 Dual Trees with MDC

The video stream is encoded into two independent sub streams using MDC. Each sub stream is sent across a different multicast tree. The objective of MDC is to encode a source into two bit streams such that a high-quality reconstruction is decodable from the two bit streams together, while a lower, but still acceptable, quality reconstruction is achievable if only one stream is received. MDC with overlay networks has been used in Peer-to-Peer Based Television [1]. It uses MDC with SplitStream Application Multicast system and Bit Torrent bartering [9]. We refer this scheme as MDC in our simulation.

3 Path Diversity

Error resilience requires streams to be sent across redundant paths. Paths are prone to network error conditions like node and link failure, and packet loss. If the paths are diverse, the redundant paths will be independent. A network error will affect the paths independently. The node will get to see average network error losses, and the chances of at least one of the paths being error free will be high. A study of path diversity with MDC has been done in [8] [10] [11].

This paper considers a heterogeneous network, with nodes with different bandwidths and reliability. There are three types of nodes High-end nodes, Medium-end nodes and Low-end nodes. High-end nodes (T1, E2, Cable) have high bandwidth and high reliability. Medium-end nodes (DSL) have medium bandwidth and reliability. Majority nodes are Low-end nodes (wireless) with low bandwidth and low reliability. Low-end nodes can only receive and cannot forward the stream further. These nodes will always be at the leaf level in all multicast trees. The High-end and Medium-end nodes can receive as well as forward stream. These are called the Server nodes.

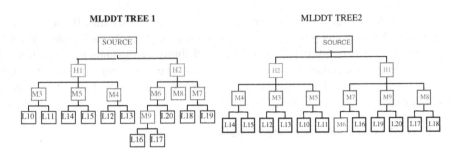

Fig. 2. Sample dual multicast trees based on MLDDT algorithm for network size of 20

CoopNet [3] and SplitStream [4] maintain multiple disjoint trees, using Interior Node Disjoint Tree (INDT) algorithm. A node is placed as an interior node in one tree, and as exterior node in rest of the trees. This property makes the trees inherently disjoint. A node failure will affect descendents in a single tree only. The nodes can have heterogeneous bandwidths, but each node should have the capability to receive as well as forward the stream further.

Traditional INDT algorithm does not consider the heterogeneous node reliability factor and cannot be used for our problem. First, the Low-end nodes can never forward stream to other nodes. Hence they cannot be interior nodes in any tree. They can be present only at the leaf level in all the trees. Second, the algorithm does not consider the reliability factor of a node. In INDT algorithm, a High-end node will be placed at a higher level as interior node in one tree, and at low-level as exterior or leaf node in rest of the trees. The goodness of its high reliability is used in only one tree. Hence it is not an optimal algorithm. In [11] an optimal disjoint tree algorithm has been suggested in which the good paths should be allowed to be joint in all the multiple disjoint trees. This optimal disjoint tree algorithm also assumes that all nodes have forwarding capabilities.

Multi Level Dual Disjoint Tree (MLDDT) [18] algorithm is an extension of INDT algorithm. It considers the significance of a nodes bandwidth and reliability factor in all disjoint trees. The nodes are placed level by level as per their bandwidth and reliability factor. We maintain dual multicast trees for comparing the error resilient video streaming schemes.

In this work we have further investigated the impact of the increasing number of disjoint trees for error resilience with MDC video streaming. MLDDT algorithm maintains dual disjoint trees. We extend it further to support multiple disjoint multicast trees, as Multi Level Multiple Disjoint Trees (MLMDT). Each MDC description can be sent across a unique disjoint tree. The number of trees can be increased keeping in view the bandwidth available.

4   Simulation Model

A simulation for a heterogeneous network has been performed to compare the different video streaming schemes and tree construction algorithms. It has been implemented using C and executed on a Linux machine. The video streaming schemes considered are: 1.Single Description Coding (SDC) 2. Dual multicast trees with stream replication (TWO SDC) 3. Dual multicast trees with MDC (MDC). The performance of the three streams is compared in terms of average outage, video quality and network utilization.

A central controller as in ALMI maintains the tree rooted at the source. It maintains a single tree for SDC scheme and dual disjoint trees for TWO SDC and MDC. The source node sends the stream (dummy video frames) across the multicast tree organization of nodes. The duration of the simulation run is 10000 time units.

Experiments have been performed for medium sized network model as mentioned in 4.1, by varying the reliabilities of the Server nodes, packet loss percentage and disjoint tree algorithm. The video packet-encoding scheme is discussed in section 4.2. The Error Model for the links is discussed in 4.3.

4.1 Network Model

The network model considered for simulation is heterogeneous, with 10% High-end nodes, 40% Medium-end nodes and majority 50% Low-end nodes. We consider the High-end nodes to have Cable connectivity, Medium-end nodes with DSL connectivity and low-end nodes being wireless. The nodes are spread across different geographic parts of the world. We consider a network size of 100 nodes for simulation. The video stream source sends out stream to first two nodes of the tree. The bandwidths and reliabilities considered for the three types of nodes in simulation are in Table 1. With video stream rate as 64 kbps and packet size as 500 bytes, the Cable nodes can have maximum out degree of 6, DSL out degree of 4 and wireless nodes with out degree 0.

Table 1. Network Model for heterogeneous network

Node	Failure Probability %	Download	Upload
Cable	High	500kbps	500 kbps
DSL	Medium	250 kbps	250 kbps
Wireless	Low	140 kbps	0

4.2 Video Packet-Encoding

The video packet-encoding scheme is based on the MDC implementation in [12]. Video stream bandwidth considered is 64 kbps, 8 frames/sec, and with RTP packet size around 500 bytes. For single stream encoding, each frame is encoded into two packets. If one of the packets is lost, the other can be approximated using error concealment techniques. For MDC encoding, each frame is encoded into two descriptions, with one packet per description. Each description is sent across a different path. The MDC coder in [12] gives the video quality in terms of Peak Signal to Noise Ratio (PSNR), for SDC and MDC schemes, with and with out packet loss, for different packet loss percentage as in Table 2. These results have been used in our simulation.

Table 2. Average frame PSNR values of MDC coder using temporal prediction considered for simulation

Packet loss %	MD no loss	SD no loss	MD w/ loss	SD w/loss
3	29.70	32.01	28.85	25.11
4	29.61	31.92	27.00	23.47
10	29.35	31.67	26.66	22.24
20	28.94	31.2	24.93	21.31

For multiple MDC descriptions we extend this scheme further and assume that each description consist of a single packet and is sent across a distinct tree. Each node gets each of the independent description from a distinct node.

For the error resilience schemes comparison dual trees are used. The two streams should be synchronized. They may be out of sync due to different link delays on the two paths. The stream, which has frame ahead of the other stream, is the Ahead stream and the other is the Lag stream. For synchronization a circular video buffer is maintained to buffer each stream. The Ahead stream is buffered until the Lag stream is available. This will help in video being synchronized when node switches from active stream to another stream in TWO SDC. For MDC this will help in getting best video quality by decoding frames with both descriptions. When there is packet loss or outage, it is possible only one of the descriptions is available. Then the frame will be decoded with a single description at a lower quality. If both description packets are lost then the entire frame will be lost.

4.3 Error Model

The node and link, both are prone to error losses. The node can go down, due to power cut. The link has packet losses and can also go down if there is problem with the ISP. The effect of a node or a link loss is the same, i.e. discontinuity of video on its descendents. In this simulation, node loss and link packet loss error conditions have been considered.

The node loss model considered is as an independent probability model. Each node has a failure probability % according to its type. The link packet loss model considered is two-state Gilbert–Elliot Model. The link can be in two states Good or Bad. The packet is lost when link is in Bad state. P (G, G) is the conditional probability of transition from Good to Good state and P (B, B) is the conditional probability from Bad to Bad state. Packet loss percentages 3, 4, 10 and 20 have been considered in our simulation. The corresponding two -state transition matrix is used to simulate the average bursty packet loss.

5 Results

The experiments are based on test cases by varying the packet loss percentage and the node reliabilities for different schemes and algorithms. The video streaming schemes are evaluated on the basis of average outage, network utilization and video quality [13] in terms of average frame PSNR, average frame rate and the number of frames received. *PSNR (Peak Signal To Noise Ratio)* is a widely used method to calculate the distortion introduced by the transformation of an input signal to an output.

5.1 Effect of Packet Loss Percentage

The impact of increasing packet loss percentage is observed for the three video streaming schemes as shown in Graph 1 (a), (b), and (c). The Server nodes failure probability is kept 0% ad the low-end node failure probability as 0.1%

Overall the outage remains zero, since all the forwarding server nodes are up. The leaf nodes keep going down. But do not affect other nodes. Results indicate that average frame rate, average frame PSNR and network utilization decrease with increase in packet loss percentage.

As compared to SDC, TWO SDC metric values are lower due to synchronization efforts between the two streams. At times there is a lag between the active and backup stream, when the node switching takes place. In that case the node has to wait for the backup stream to deliver frames starting from the last frame displayed from the first stream.

MDC fairs better than SDC as packet loss increases. Each node gets the independent stream from different paths. It gets to see an average packet loss. The number of frames received and the frame rate is higher even for higher percentage packet loss. The difference between the PSNR values for SDC and MDC schemes decreases as packet loss percentage increases. The network utilization is low for TWO SDC scheme since only one of the streams is used actively. The network load is high since there are two trees maintained, but the usage is low resulting in lower network utilization.

5.2 Effect of Medium-End Node Failure Probability

The network model considers that High-end node to be more reliable compared to Medium-end nodes. We study the impact of increasing Medium-end node failure probability on the three video streaming schemes with 3% packet loss and 0.05% High-end node failure probability. Average outage of MDC and TWO SDC is lower compared to SDC, since there are alternate paths and streams available as shown in Graph 2 (a). MDC is better than TWO SDC, since both the streams contribute to the decoded video. If one description is unavailable, it can decode a lower quality video from one of them. Number of nodes outaged in MDC is also low comparatively. When outage is low then the average frame rate will also be high. A higher frame rate indicates higher number of the frames received as shown in Graphs 2(b).

MDC experiments are based on two types of tree algorithms INDT and MLDDT. Since MLDDT algorithm places the Medium-end nodes below the High-end nodes in all the multicast trees, the impact of increase in failure probability of Medium-end nodes is less on MLDDT based experiments. It provides lesser outage and hence higher average frame rate as compared to INDT algorithm.

5.3 Effect of Number of Trees

We study the impact increasing the number of trees for MDC video streaming scheme. We assume that the number trees are equal to the number of MDC descriptions.

Each MDC description is sent across a unique tree. The experiments are based on 1,2,3 and 4 trees.

Graph 3. (a), (b) show the impact of medium-end node reliability for different number of trees. Packet loss is 3 % and High-end node failure probability low as 0.01%. The average outage decreases with increase in the number of trees. Each MDC description traverses a distinct path. Hence the probability that all descriptions are lost is very less. As outage decreases the average frame rate and number of frames received increases. Graph 3. (c), (d) show the impact of packet loss for different number of trees with MDC scheme. The Server-end node failure probability is set a 0%. The average frame PSNR and the average frame rate increases with the increase in the number of trees. A network error affects the disjoint paths independently. The average network error reduces with the increase in the number of trees.

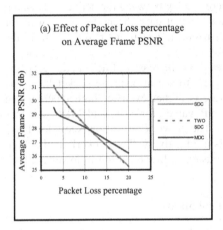

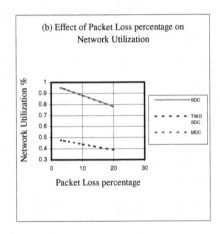

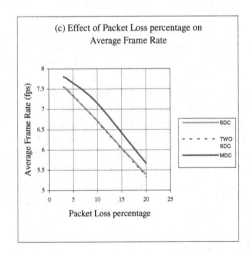

Graph 1. Effect of Packet loss percentage for different video streaming schemes

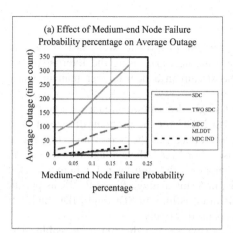

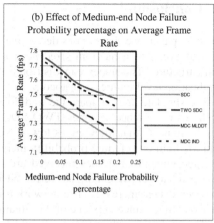

Graph 2. Effect of Medium-end Node Failure Probability percentage for different video streaming schemes

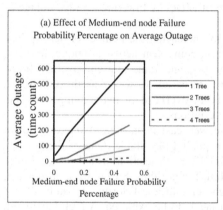

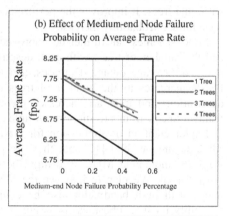

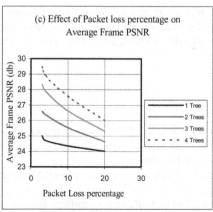

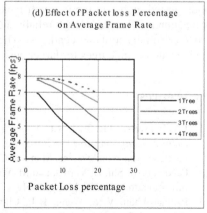

Graph 3. Effect of *Medium-end* node failure probability percentage and Packet loss percentage for different number of multicast trees with MDC scheme

6 Conclusion

In this paper we have considered the problem of private web cast of a live critical event, to a medium sized heterogeneous network, using the Internet public infrastructure. The nodes can distribute the stream in a peer-to-peer manner using application level multicast protocol.

We simulate and compare the traditional single description stream scheme (SDC) with error resilience schemes TWO SDC and MDC for the problem discussed. The results indicate that MDC gives lower average outage, as the Server node (High-end and Medium-end) failure probability increases. For MDC, the average frame rate and average number of frames received are comparatively high even for higher packet loss percentage. MDC fairs as well as SDC, in terms average frame PSNR as packet loss percentage increases. The network utilization is high in SDC and MDC and half in TWO SDC, since only one of the streams is used actively.

The multiple multicast trees should be as diverse and disjoint as possible. The factor of bandwidth and reliability of nodes has great impact on disjoint tree organization. MLDDT algorithm is compared with INDT algorithm for MDC scheme with dual trees. It improves the performance of MDC by lowering average outage as node failure probability increases.

MDC schemes can further be improved by increasing the number of descriptions and the number of trees, with each description being sent across a unique multicast tree. The average outage decreases, and the average frame rate and average frame PSNR increases as the number of trees are increased. However as the number of descriptions and trees are increased there is an overhead of maintaining trees and bandwidth constraints to support multiple descriptions.

Implementation of application for private web cast is an area of interest. Lot of MDC techniques exit [17], but they are not publicly available. One of the feasible ways to implement MDC is using Unequal Error Protection with Erasure Codes [14][15][16]. We plan to improve the MMDDT algorithm. So far we consider that nodes with high bandwidth also have high reliabilities. Priority of bandwidth or reliability for tree robustness is an area that needs to be explored. A high bandwidth node placed higher in the tree will make the tree short and reduce points of failures from source to leaves. A high reliable node placed higher in the tree will improve the video continuity to its descendants. The tree algorithm also needs to be extended for supporting dynamic trees for larger networks.

References

1. J.A. Pouwelse, J.R. Taal, R.L. Lagendijk, D.H.J. Epema, and H.J. Sips, *Real-time Video Delivery using Peer-to-Peer Bartering Networks and Multiple Description Coding*, IEEE Int'l Conference on Systems, Man and Cybernetics, October 2004.
2. Pendarkis, S. Shi, D. Verma, and M. Waldvogel. "ALMI: An Application Level Multicast Infrastructure", *USITS*, March 2001.
3. Padmanabhan, V.N., Wang, H.J., Chou, P.A., 2003. Resilient Peer-to-Peer Streaming. *Proc. Int'l Conf. Network Protocols.* Atlanta, GA.

4. Castro, M Druschel, P Kermarec, A M Nandi, A Rowstron and Singh, A. SplitStream: High bandwidth Content Distribution in a Cooperative Environment. In *Proc. SOSP* (Oct. 2003).

5. Chu, Y.H., Rao, S., Zhang, H., 2000. A Case for End System Multicast. *Proc. ACM Sigmetrics*. Santa Clara, CA.

6. V. K. Goyal. Multiple description coding: Compression meets the network. *IEEE Signal Processing Magazine*, pages 74–93, September 2001.

7. Y. Wang, M. T. Orchard, V. A. Vaishampayan, and A. R. Reibman, "Mulitple description coding using pairwise correlating transforms", *IEEE Trans. Image Processing*, vol. 10, no. 3, pp. 351–366, Mar. 2001

8. Matulya Bansal, Avideh Zakhor, Path Diversity Based Techniques for Resilient Overlay Multimedia.

9. J.A. Pouwelse, J.R. Taal, R.L. Lagendijk, D.H.J. Epema, and H.J. Sips, *Real-time Video Delivery using Peer-to-Peer Bartering Networks and Multiple Description Coding*, IEEE Int'l Conference on Systems, Man and Cybernetics, October 2004.

10. J.G. Apostolopoulos, "Reliable video communication over lossy packet networks using multiple state encoding and path diversity," *Visual Communications and Image Processing (VCIP)*, January 2001.

11. Begen, A.C., Altunbasak, Y., Eregun, O., Ammar, M.H., 2005. Multi-path selection for multiple description video streaming over overlay networks. *EURASIP Signal Processing: Image Communication*, **20**(1):39-60.

12. Amy R. Reibman, Hamid Jafarkhani, Yao Wang, Michael T. Orchard, Rohit Puri: Multiple-description video coding using motion-compensated temporal prediction. *IEEE Trans. Circuits Sust. Video Techn. 2002*

13. Video quality Studio http://www.visumalchemia.com/

14. R. Puri and K. Ramchandran, ``Multiple Description Source Coding Through Forward Error Correction Codes", 33rd Asilomar Conference on Signals, Systems and Computers, Pacific Grove, CA, October 1999.

15. A. E. Mohr, E. A. Riskin and R. E. Ladner, "Unequal Loss Protection: Graceful Degradation of Image Quality over Packet Erasure Channels through Forward Error Correction", Proceedings of the DCC, Snowbird, UT, March 1999".

16. A. Albanese, J. Blomer, J. Edmonds, M. Luby and M. Sudan, "Priority Encoded Transmission", IEEE Transactions on Information Theory, November 1996

17. Yao Wang, Shunan Lin: Error-resilient video coding using multiple description motion compensation. IEEE Trans. Circuits Syst. Video Techn. 12(6): 438-452 (2002)

18. Divyashikha Sethia, Huzur Saran, "Resilient Video Streaming with Path Diversity For Heterogeneous Networks", *Workshop on Recent advances in Peer-to-Peer Streaming, IEEE QShine, Waterloo, Canada, 2006*.

Exploring Thread and Memory Placement on NUMA Architectures: Solaris and Linux, UltraSPARC/FirePlane and Opteron/HyperTransport

Joseph Antony, Pete P. Janes, and Alistair P. Rendell

Department of Computer Science
Australian National University
Canberra, Australia
alistair.rendell@anu.edu.au

Abstract. Modern shared memory multiprocessor systems commonly have non-uniform memory access (NUMA) with asymmetric memory bandwidth and latency characteristics. Operating systems now provide application programmer interfaces allowing the user to perform specific thread and memory placement. To date, however, there have been relatively few detailed assessments of the importance of memory/thread placement for complex applications.

This paper outlines a framework for performing memory and thread placement experiments on Solaris and Linux. Thread binding and location specific memory allocation and its verification is discussed and contrasted.

Using the framework, the performance characteristics of serial versions of lmbench, Stream and various BLAS libraries (ATLAS, GOTO, ACML on Opteron/Linux and Sunperf on Opteron, UltraSPARC/Solaris) are measured on two different hardware platforms (UltraSPARC/FirePlane and Opteron/HyperTransport). A simple model describing performance as a function of memory distribution is proposed and assessed for both the Opteron and UltraSPARC.

1 Introduction

Creation of scalable shared memory multiprocessor systems has been made possible by cc-NUMA (cache-coherent Non-Uniform Memory Access) hardware. This approach uses a basic building block comprising one or more processors with local memory and an interlinking cache coherent interconnect [5]. Unlike UMA (Uniform Memory Access) systems which comprise processors with identical cache and memory latency characteristics, NUMA systems exhibit asymmetric memory latency and possibly asymmetric bandwidths between its building blocks. On such platforms the operating system should consider physical processor and memory locations when allocating resources (i.e. memory allocation and CPU scheduling) to processes.

Y. Robert et al. (Eds.): HiPC 2006, LNCS 4297, pp. 338–352, 2006.
© Springer-Verlag Berlin Heidelberg 2006

Pthreads [6] and OpenMP [17] are two widely used programming models which target shared memory parallel computers. Both, however, were developed for UMA platforms, making no assumptions about the physical location of memory or where a thread is executing. Although there has been debate about the merit of adding NUMA extensions to these programming models, this issue is yet to be resolved [7]. In part the aim of the work presented here is to develop the tools and protocols required for performing memory and thread placement experiments that can be used to address this issue.

Linux and Solaris are two examples of operating systems that claim to be "NUMA aware". Exactly what this implies is not always well defined, but suffice it to say that both Solaris and Linux provide application program interfaces (API) that give the user some level of control over where threads are executed and memory is allocated [11,14] and perform some form of default NUMA aware placement of threads and data. While this is useful, for the programmer wishing to explore NUMA issues it is also useful to have functions that will identify the CPU currently being used by that thread, and the physical location that corresponds to an arbitrary (but valid) virtual address within an executing process.

In this paper we compare the support provided for thread and memory placement by Solaris and Linux, and also outline how a user can interrogate these runtime environments to determine actual thread and memory placements. Using this infrastructure the performance characteristics of two contemporary NUMA architectures – the UltraSPARC [20] using the FirePlane interconnect [4] and the Opteron [12] using HyperTransport [9] - are explored through a series of latency, bandwidth and basic linear algebra (BLAS) experiments.

A novel placement distribution model (PDM) which describes performance as a function of bandwidth and latency is presented and used to analyse performance results. The PDM uses directed graphs representing processor, memory and interconnect layout to aid in the enumeration of contention classes. The distribution of these contention classes permit qualitative analysis of performance data from NUMA platform experiments.

The paper is structured into the following sections – thread and memory placement on Solaris and Linux is discussed in section 2. The experimental hardware and software platforms used are described in section 3 while section 4 outlines the latency, bandwidth experiments and the placement distribution model. Section 5 covers related work and section 6 presents our conclusions.

2   Thread and Memory Placement

Conceptually, both Solaris and Linux are similar in their approach to abstracting underlying groupings of processors and memory based on latency. Yet, the mechanics of using the two NUMA APIs are quite different. Below we provide a brief review of Solaris thread and memory placement APIs, before contrasting this with the Linux NUMA support. We then consider placement verification for both Solaris and Linux.

2.1   Solaris NUMA Support

Solaris represents processor and memory resources as locality groups [11]. A locality group (lgrp) is a hierarchical DAG (Directed Acyclic Graph) representing processor-like and memory-like devices, which are separated from each other by some access-latency upper bound. A node in this graph contains at least one processor and its associated local memory. All the lgrps in the system are enumerated with respect to the root node of the DAG, which is called the root lgrp. Two modes of memory placement are available, *next-touch*[1] and *random*[2]. The former is the default for thread private data, while the latter is useful for shared memory regions accessed by multiple threads as it can reduce contention.

A collection of APIs for user applications wanting to use lgrp information or provide memory management hints to the operating system is available through liblgrp [18]. Memory placement is achieved using madvise(), which provides advice to the kernel's virtual memory manager. The meminfo() call provides virtual to physical memory mapping information. We also note that memory management hints are acted upon by Solaris subject to resources and system load at runtime.

Threads have three levels of binding or affinity – *strong*, *weak* or *none* which are set or obtained using lgrp_affinity_set() or lgrp_affinity_get() respectively. Solaris' memory placement is determined firstly by the allocation policy and then with respect to threads accessing it. Thus there is no direct API for allocating memory to a specific lgrp, rather a first touch memory policy must be in place and then memory allocated by a thread that is bound to that specific lgrp. Within an lgrp it is possible to bind a specific thread to a specific processor by using the processor_bind() system call.

2.2   Linux NUMA Support

NUMA scheduling and memory management became part of the mainstream Linux kernel as of version 2.6. Linux assigns NUMA policies in its scheduling and memory management subsystems. Memory management policies include *strict*[3] allocation to a node, *round-robin*[4] memory allocations, and *non-strict preferred* binding to a node (meaning that allocation is to be *preferred* on the specified node, but should fall back to a default policy if this proves to be impossible). In contrast, Solaris specifies policies for shared and thread local data.

The default NUMA policy is to map pages on to the physical node which faulted them in, which in many cases maximises data locality. A number of system calls are also available to implement different NUMA policies. These system calls modify scheduling (struct task_struct) and virtual memory (struct vm_area_struct) related variables structures within the kernel.

---

[1] The next thread which touches a specific block of memory will possibly have access to it locally i.e. if remote memory is accessed it will possibly be migrated.

[2] Memory is placed randomly amongst the lgrps.

[3] Memory allocation is to occur at a given node. It will fail if there is not enough memory on the node.

[4] Memory is dispersed equally amongst the nodes.

Relevant system calls include mbind(), which sets the NUMA policy for a specific memory area, set_mempolicy(), which sets the NUMA policy for a specific process, and the sched_setaffinity(), which sets a process' CPU affinity. Several arguments for these system calls are supplied in the form of bit masks, and macros, which makes them relatively difficult to use. For the application programmer a more attractive alternative is provided by the libnuma API. Alternatively, numactl is a command line utility that allows the user to control the NUMA policy and CPU placement of a entire executable[5].

Within libnuma useful functions include the numa_run_on_node() call to bind the calling process to a given node and numa_alloc_onnode() to allocate memory on a specific node. Similar calls are also available to allocate interleaved memory, or memory local to the caller's CPU. In contrast to the Solaris memory allocation procedure, numa_alloc modifies variables within the process' struct vm_area_struct and the physical location of the thread that performs the memory allocation is irrelevant. The libnuma API can also be used to obtain the current NUMA policy and CPU affinity. To identify NUMA related characteristics libnuma accesses entries in /proc and /sys/devices. This makes applications using libnuma more portable those that use the lower level system calls directly.

2.3   Placement Verification

Solaris provides a variety of tools[6] to monitor process and thread lgroup mappings – lgrpinfo, pmadvise, plgrp and pmap. The lgrpinfo tool displays the lgroup hierarchy for a given machine. The pmadvise tool can be used to apply memory advice to a running process. The plgrp tool can observe and affect a running thread's lgroup; it can also give a diagrammatic representation of the lgroup affinities. The pmap tool permits display of lgroups and physical memory mapping for all virtual address associated with a running process.

Although libnuma provides a means for controlling memory and process placement on Linux systems, it does not provide a means for determining where a given area of memory is physically located. A kernel patch that attempts to addresses this issue is provided by Per Ekman [15]. The patched kernel creates per-PID /proc entries that include, among other things, information about which node a process is running on, and a breakdown of the locations of each virtual memory region belonging to that process. While we found that this package was generally sufficient as a verification tool it involved having to check quickly the /proc entries while the program was running. We also found that under some circumstance the modified kernel failed to free memory after a process had terminated.

Based on the work of Ekman [15] we designed an alternative kernel patch that provides a system call and user level function to return the memory locations for each page in a given virtual memory range. This utility proved considerably more convenient as it could be called from within a running application.

---

[5] It can also be used to display NUMA related hardware configuration and configuration status.

[6] http://opensolaris.org/os/community/performance/numa/observability

3   Experimental Platforms

Two NUMA platforms were used in this work: a twelve processor Sun Ultra-
SPARC V1280 [21] and a four processor AMD848 Opteron system based on the
Celestica A8448 [3] motherboard. We now briefly outline the NUMA character-
istics of each platform.

3.1   UltraSPARC/FirePlane

The V1280 has twelve 900MHz UltraSPARC III Cu processors each with a 32
Kb L1 instruction cache, 64 Kb L1 data cache and 8192 Kb L2 cache which
is off-chip. The system contains three boards which hold four processors each
and are linked using the FirePlane interconnect [4]. The system contains 8GB
of memory per board giving a total of 24GB for the entire system. The three
boards form a combined snooping based coherency domain. For larger systems,
i.e. > 24 processors, a directory based protocol is used at the point-to-point level.

A pair of processors and their associated memories are all linked using a
Dual CPU Data Switch (DCDS), i.e. there are four separate data paths each
running at 2.4 GB/s from processors or memories to the DCDS. The DCDSs
can sustain 4.8 GB/s to the board data switch. Since memory on the boards is
16-way interleaved across a board, a peak of 6.4 GB/s per board is achieved. The
point-to-point links among boards have a bi-directional bandwidth of 4.8 GB/s
per board, approaching a peak of 9.6 GB/s for the whole system. Since the four
processors on a board have similar memory access latencies, we will refer to it
as one node. A schematic illustration of the V1280 is given in Figure 1 (a).

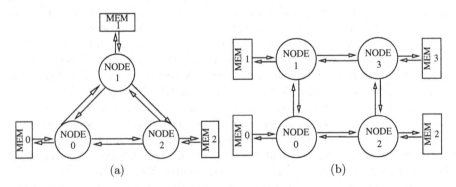

Fig. 1. (a) Schematic diagram of the V1280 UltraSPARC platform and (b) Celestica
Opteron platform

3.2   Opteron/HyperTransport

The Opteron system contains four 2.2Ghz AMD848 processors each with a 64
Kb L1 data and instruction cache and a 1024 Kb L2 cache. The Celestica A8440
motherboard is configured with 2GB of memory per processor giving a total
of 8GB for the entire system. The AMD848 Opterons have an on-chip memory

controller and uses coherent HyperTransport to link processor coherency traffic. The Opteron has two coherent HyperTransport links [9], each operating at 6.4 GB/s bi-directionally. The processors are arranged in a ring topology resulting in processors having at most two hops to reach the most distant processor. A schematic illustration of the Opteron system is given in Figure 1 (b).

3.3 Software Platform

While Solaris 10 was used on the V1280 system, the Opteron platform was capable of dual booting into either Solaris 10 or OpenSuSE 10. The Sun Studio 11 compilers were used on both Solaris platforms, while version 6.0 of the Portland Group compilers were used under Linux. Compiler flags for the highest optimisation levels were used on both compilers. To obtain accurate performance data the PAPI library [2] was used to access hardware performance counters under Linux whereas the libcpc [19] infrastructure was used under Solaris. The numeric libraries used under Linux are the ACML (version 3.0) from AMD, ATLAS (version 3.6) [23], GOTO BLAS [8] (version 1.00) while Sunperf (Sun Studio 11) was used under Solaris.

4 Results

This section discusses observed memory latency, serial memory bandwidth and parallel memory bandwidth for the two NUMA platforms.

4.1 Latency Characterisation

To determine the memory latency characteristics of the two platforms the lm bench [13] memory latency benchmark was modified to accept memory and thread placement parameters. Latencies to get data from level-one cache (L1) on the Opteron and UltraSPARC were measured as 3 and 2 cycles respectively, while accessing level-two cache (L2) took 20 cycles on both platforms. The latencies recorded for a thread bound to a particular node accessing memory at a specific location are given in Table 1. From these, the NUMA ratio[7] of the Opteron system is found to be 1.11 for one hop and 1.53 for two hops from any given processor, while on the V1280 there is only one NUMA level with a ratio of 1.2. While these are NUMA machines with low NUMA ratios, the emphasis of this paper is the sketching out and testing of the memory and thread placement framework with the view of extending it to NUMA systems with higher NUMA ratios.

4.2 Bandwidth Characterisation

To determine the memory bandwidth characteristics of the two platforms the Stream [10] benchmark was modified to accept memory and thread placement parameters. This benchmark performs four different vector operations, corresponding to vector copy, scale, add, and triad. On the Opteron system there are four

---

[7] NUMA ratio $= \frac{RemoteLatency}{LocalLatency}$

Table 1. Main Memory latencies (Cycles). lmbench uses a pointer chasing benchmark to determine memory latencies. Results were obtained for the Opteron and V1280 platforms by pinning a thread on a given node and placing memory on different nodes.

Thread	Memory Location						
	Opteron				V1280		
Location	0	1	2	3	0	1	2
0	225	250	250	345	220	265	265
1	250	225	345	250	265	220	265
2	250	345	225	250	265	265	220
3	345	250	250	225	–	–	–

nodes and four physically distinct memory banks, while on the UltraSPARC system there are three nodes and three memory banks. For a single thread it might be expected that the "best" possible Stream performance would be obtained when a thread is accessing vectors that are stored entirely in local memory. Conversely the "worst" possible performance would correspond to a thread accessing data stored in memory located as far away as possible.

Results for these two scenarios are given in Table 2. For the Opteron system running Solaris we find performance differences between best and worst memory placement that vary from a factor of 1.4 to 1.6. For Linux on the same platform we find a some what larger variation with factors between 1.09 to 2.35. On the V1280 system the effect is considerably less indicating relatively mild NUMA characteristics. (We note that the superior performance of the copy operation on the Opteron using Linux reflects the use of specialised instructions by the PGI compiler to perform the memory moves).

Table 2. Serial Stream bandwidths (GB/s) for the Opteron and V1280 systems. A single thread was pinned to a given node and had its memory placed on different nodes. Best and Worst refer to thread and memory placements which are expected to give the best and worst possible performance (See text for details).

	Opteron				V1280	
	Solaris		Linux		Solaris	
Test	Best	Worst	Best	Worst	Best	Worst
Copy	2.17	1.99	4.68	3.14	0.72	0.71
Scale	2.50	1.58	2.35	1.47	0.79	0.74
Add	2.75	1.17	2.55	1.54	0.83	0.81
Triad	2.24	1.51	2.44	1.52	0.85	0.79

The Stream benchmark was modified to create multiple threads, that concurrently ran separate instances of the original Stream benchmark. Results for this are presented in Table 3. In this case the worst case scenario on the Opteron would correspond to node 0's executing the Stream benchmark with all the data being serviced from memory 3, while the opposite happens on node 3, and nodes

1 and 2 are similarly exchanging data. Not surprisingly on both the Opteron and V1280 system the difference between good and bad memory placement has increased significantly over that observed for the serial benchmark.

Table 3. Parallel Stream bandwidths (GB/s). Threads were pinned to various nodes and had its memory placed locally ("Best") or remotely ("worst"). Four threads were run concurrently for the Opteron which twelve threads were run concurrently for the V1280 system.

| | Opteron | | | | V1280 | |
| | Solaris | | Linux | | Solaris | |
Test	Best	Worst	Best	Worst	Best	Worst
Copy	8.98	2.53	16.55	4.22	4.89	3.56
Scale	9.98	2.67	9.60	2.67	4.91	3.46
Add	10.85	2.85	10.33	2.94	5.22	3.57
Triad	9.17	2.68	9.87	2.96	5.14	3.71

4.3   Placement Distribution Model

In the above we considered "best" and "worst" case scenarios for the various Stream benchmarks. In the general case as well as on the Opteron system each vector or data quantity used in a Stream benchmark could be located in the memory associated with any one of the four available nodes. For the parallel add and triad benchmarks, on the Opteron system, this means that there are a total of $4^{12}*4!$ possible thread/memory combinations[8] while $4^8*4!$ copy and scale benchmarks are possible (add and triad benchmarks use 3 data quantities while copy and scale use 2 data quantities). Obviously evaluating the performance characteristics of each of these cases quickly becomes impossible for large NUMA systems. Thus, it is useful to develop a simple performance model which gives the *probability* of a given memory and thread placement experiments.

A placement distibution model (PDM) is developed to categorize the occurrence and type of possible placements. A directed graph of the NUMA platform is given to the model along with the data quantities used per thread. Figures 1 (a), (b) can be interpreted as graphs where links entering and exiting nodes are arcs. Traffic associated with each link can be modeled as weights along the links between nodes. Nodes are assumed to route traffic to their local memory controller or to other nodes along the most direct path. The model also assumes concurrent execution of all defined threads accessing its data quantities in tandem with other threads over the interconnect. This model can be used to characterise the communication requirements for any given memory placement experiment.

---

[8] A given data quantity could reside in 4 possible memory locations and each thread could run on 4 possible processors i.e. there are a total of 4^3 experiments for one thread and three data quantities. For all the 4 threads in the system there are $4^3 * 4 * 4^3 * 3 * 4^3 * 2 * 4^3 * 1 = 4^{12} * 4!$ possible combinations.

4.4 Placement Distribution Algorithm

An algorithm for the placement distribution model is presented in Algorithm 1. The PDM requires a graph $\mathbb{G}$, which represents the layout of memory $\mathbb{M}$, processor nodes $\mathbb{N}$ and a set $\mathbb{I}$ of ordered processor to memory or memory to processor data movements for a set of data quantities $\mathbb{D}$. These inputs are used to traverse over all possible configurations per thread of both thread and memory placement

---

Algorithm 1. The Placement Distribution Model

1: $\mathbb{N} \leftarrow \{node_1, node_2, \dots, node_i\}$ *The set of all processor nodes*
2: $\mathbb{M} \leftarrow \{mem_1, mem_2, \dots, mem_j\}$ *The set of memory nodes*
3: $\mathbb{L} \leftarrow \{link_1, link_2, \dots, link_k\}$ *The set of all links between nodes*
4: $\mathbb{T} \leftarrow \{data_1, data_2, \dots, data_l\}$ *The set of data quantities*
5: $\mathbb{E} \leftarrow \mathbb{N} \times \mathbb{M}$ *Cartesian product denoting data movement*
6: $\mathbb{G} \leftarrow <\mathbb{E}, \mathbb{L}>$ *Graph $\mathbb{G}$ representing memory and processor layout*
7: $\mathbb{D} \leftarrow \{<x,y> \mid x \in \mathbb{T}, y \in \mathbb{M}\}$ *A data quantity x resides in memory location y*
8: $\mathbb{I} \leftarrow \mathbb{E} \times \mathbb{D}$ *Set of inputs for thread, memory placement*
9: $\mathbb{I} \equiv \{<e, f> \mid e = <n,m> \in \mathbb{E}, f = <x,y> \in \mathbb{D}\}$
10: $W(l) \mid l \in \mathbb{L}$ *Weight matrix W*
11: $C(x,y)$ *Cost matrix C*

Require: $<n,m> \in \mathbb{E}$
12: **procedure** OPTPATH($<n,m>$)   *Optimal path from n to m where $n, m \in \mathbb{E}$*
13:         *Use appropriate alogrithm or heuristic*
14:     **return** $\{<x,y> \mid x,y \in \mathbb{L}\}$ *to get path between $<n,m>$*
15: **end procedure**

Require: $x \in \mathbb{D} \ \forall x \in \mathbb{Q}$
Require: $<x,y> \in \mathbb{L} \ \forall \ <x,y> \in \mathbb{P}$
16: **procedure** FLOWSIZE($\mathbb{Q}, \mathbb{P}$)   *Compute cost of moving data items across link $\mathbb{P}$*
17:     cost $\leftarrow 0$
18:     **for all** (link $\in \mathbb{P}$) **do**
19:        **for all** (qty $\in \mathbb{Q}$) **do**
20:          cost $\leftarrow$ cost $+ \mid$ qty $\mid * W(link)$
21:        **end for**
22:     **end for**
23:     **return** cost
24: **end procedure**

25: **procedure** COMPUTEDISTRIBUTION
26:     $\mathbb{Q}' \leftarrow \{x \mid x \in \mathbb{D}\}$ *Set of data quantities of interest*
27:     **for all** ($i \in \mathbb{I}$) **do**   *Loop over input $\mathbb{I}$ ($i \equiv <e,f>$)*
28:        links $\leftarrow OptPath(e)$ where $e \in i$ *Get the optimal path for a given e*
29:        **for all** (($j \leftarrow$ links) $\wedge$ ($f \in i$)) **do**   *Loop over links and use $f \in <e,f>$*
30:          $C(i,j) = C(i,j) + FlowSize(\mathbb{Q}', j)$
31:        **end for**
32:     **end for**
33: **end procedure**

---

for each data quantitity. A traversal implies data quantities are moved over a link and this entails a cost $W(l)$ per link l. Each traversal contributes to a cumulative cost entry in cost matrix C. Three procedures are defined in Algorithm 1 namely *OptPath*, *FlowSize* and *ComputeDistribution*. Procedure *OptPath* returns the optimal path, a set of ordered pairs of $<x,y>$ between two end points $<n,m>$ while procedure *FlowSize* computes a cost associated with moving data quantities contained in set $\mathbb{Q}$ over links contained in set $\mathbb{P}$ and procedure *ComputeDistribution* uses a set of data quantities as used per thread for all threads in set $\mathbb{Q}'$ and computes the cost for these data quantites for an ordered set of inputs $\mathbb{I}$.

A state machine was coded to perform walks along the links of graph $\mathbb{G}$, for all possible thread and memory placements given a specific processor/memory topology and data quantities. These walks model link traffic moving from a source node to a target node, traffic moving from a node to its local memory and traffic moving from one memory bank to another. In the event that there are two paths to the required destination of equal length, the traffic is split equally along each path. This assumption is made as a simplification to avoid complex specification of the underlying interconnect protocol. For example, if a placement dictates that Node 1 will be continuously accessing memory from Node 0, we increment variables belonging to each link along the route to record the quantity of the data movement. This results in a tuple holding values for link contention and node contention.

Using the PDM, for a given processor and memory layout, yields costs for thread and memory placement which are distributed in ranges which we term as *link contention classes*. The range of a link contention class gives the degree of contention at a node. For each contention class obtained from the PDM, 20 random configurations were generated i.e. thread and memory placement for all threads and data quantities which yields a link contention that lies in the range of all observed link contention classes. These placement configurations are subsequently used to perform copy and scale `Stream` measurements. In effect this process permits for a tractable analysis of possible performance characteristics for the benchmarks without resorting to running all experiments for all possible thread and memory placements.

Table 4 characterises the copy and scale `Stream` benchmarks according to the maximum level of contention on any given link. This table shows, for example, that on the Opteron system 51.9% of all possible memory placement configurations have link contentions greater or equal to 3 but less than 4, while 0.1% have a link contention of between 7 and 8. The ranges 3-4 and 7-8 are the *link contention classes*. The results show that on the Opteron system given random vector placement the probability of landing in a 3-4 link contention class is the highest, and within this class you might expect to see a performance degradation of about 20%. On the V1280 the effect is much less.

4.5 BLAS Experiments

Using the memory placement framework developed above, experiments were conducted for level 2 (DGEMV – matrix vector) and level 3 (DGEMM – matrix

Table 4. Copy and Scale (GB/s) Stream benchmark results for the placement distribution model. Contention classes denote the ranges of link contention for all the nodes in the system. %Freq gives the frequency of occurance of a given class in percent.

| Contention | | _Solaris _ | | _Linux _ | |
Class	%Freq	Copy	Scale	Copy	Scale
Opteron					
2-3	2.6	5.7	6.0	7.4	5.6
3-4	51.9	5.0	5.1	6.7	4.7
4-5	34.6	4.5	4.8	6.4	4.2
5-6	9.2	3.9	4.6	5.5	3.6
6-7	1.5	3.3	3.4	4.4	3.0
7-8	0.1	3.0	3.0	3.3	2.7
V1280					
08-12	10.3	4.0	4.0		
12-16	59.7	3.8	3.9		
16-20	24.8	3.7	3.8		
20-24	5.0	3.6	3.6		

multiply) BLAS operations. Results obtained for square matrices of dimension 1600 using ACML on the Opteron and sunperf on the V1280 are given in Table 5. In addition we also include results obtained from the triad Stream benchmark as these are representative of level 1 BLAS operations. The results show greatest NUMA effects on the Opteron system, where, as expected the variation is largest for triad, less for level 2 BLAS and almost unnoticeable for level 3 BLAS. This

Table 5. BLAS Stream Triad, Level 2 BLAS, Level 3 BLAS (GigaFlops) results for the placement distribution model. Results are averages for twenty random generated configurations per contention class. Tr = Triad.

| Contention | | _Solaris _ | | | _Linux _ | | |
Class	%Freq	Tr	L2	L3	Tr	L2	L3
Opteron							
3-4	1.9	0.5	1.6	15.3	0.5	1.5	15.6
4-5	38.1	0.4	1.4	15.2	0.4	1.4	15.6
5-6	38.2	0.4	1.5	15.2	0.4	1.4	15.6
6-7	16.0	0.4	1.4	15.2	0.4	1.3	15.6
7-8	5.0	0.3	1.3	15.2	0.3	1.3	15.6
8-12	3.4	0.3	1.1	14.9	0.3	0.8	15.6
V1280							
12-16	8.3	0.4	1.0	17.4			
16-20	48.3	0.3	1.0	15.8			
20-24	30.7	0.3	1.0	16.2			
24-28	10.2	0.3	1.0	17.4			
28-40	2.3	0.3	1.0	17.5			

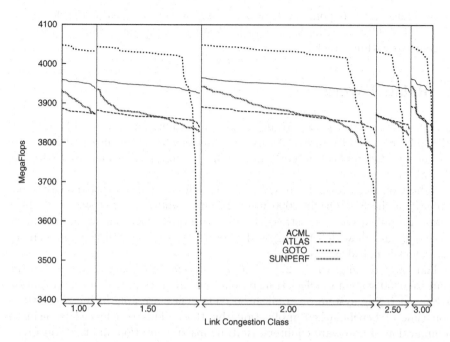

Fig. 2. Serial Opteron DGEMM ($C = A * B$) performance (MegaFlops) for square matrices of dimension 1600 using ACML, ATLAS, GOTO and Sunperf libraries. A total of 256 experiments were run for all possible thread and memory placements for one thread and three data quantities A, B and C.

reflects the fact that a well written DGEMM will spend most of its time working on data that is resident in the level 2 cache, but this is not possible for level 1 or level 2 BLAS where data must be streamed from memory to processor.

In Figure 2 we present floating point performance for serial DGEMM on the Opteron system using four different BLAS libraries: i) ACML, ii) ATLAS , iii) GOTO and iv) Sunperf. With three matrices and four different nodes on the Opteron system there are a total of 256 different thread and memory placement permutations. These have been ordered according to maximum contention on a given link and then sorted by performance observed within that group. As this is a serial matrix multiply, the link contention ranges from 1 to 3. Memory contention of 1 occurs when the three matrices are located on adjacent nodes with the compute thread is bound to the central node, i.e. the contention on the network is actually reduced compared to the case when all three matrices are local to the node accessing them (contention value of 3). Interestingly in some cases the best performance is obtained for a link contention of 2 indicating that on an idle machine non-local placement of some data quantities may be advantageous if it leads to enhanced overall memory bandwidth. The performance also shows considerable fine structure, especially for GOTO BLAS [8] which for most of the time exhibits the best performance, but in some cases also shows the worst performance. At this point we believe the reason for these sudden performance drops (of ≈16%) is cache line conflicts arising from slightly different memory placements within a node.

5   Related Work

Brecht [1] evaluates the importance of placement decisions on NUMA machines with different NUMA ratios. Application placement which mirrored hardware is beneficial for application performance and its importance increased with the NUMA ratio.

Robertson and Rendell [16] quantify the effects of memory bandwidth and latency on the SGI Origin 3000 using lmbench and stream. Using a 2D heat diffusion application, they stress the importance of good thread and memory placement and show that relying on the operating system for thread and memory placement is not always optimal.

Tikir and Hollingsworth [22] use link counters and a bus analyzer, on the SunFire 6800 system to effect transparent page migration, without modification to the operating system or application code. They are able to improve execution time of benchmarked applications by 16%. This is achieved by using a combination of hardware counters, runtime instrumentation and madvise().

6   Conclusions

The support for thread binding and memory placement provided by Solaris and Linux has been outlined and contrasted. For Linux, the kernel was modified in order to provide a user API that could be used to verify binding and determine physical memory placement from a user supplied virtual address. Using the various thread and memory placement APIs, a framework was outlined for performing NUMA performance experiments. Detailed measurements of the latency, bandwidth and BLAS performance characteristics of two different hardware platforms were undertaken. These showed the Opteron system to be "more NUMA" than the Sun system, despite the fact that it had only 4 processors. To assist in the analysis of the performance data, a simple placement distribution model of the NUMA characteristics for the two platforms was outlined. The PDM uses directed graphs to represent processor, memory and interconnect layout. It was found that if multiple level 1 or level 2 BLAS operations are run in parallel on the Opteron system performance differences of up to a factor of two were observed depending on memory and thread placement. For level 3 BLAS, differences are much smaller as there is much better re-use of data from level 2 cache.

The use of the PDM shows node local allocation of memory is not always the best strategy for the DGEMM kernel. The best peak results were obtained for a link contention of 2 i.e. non-local placement of data. This highlights the benefits of user-level discovery, at runtime, of processor and memory topologies and the use of this knowledge within the application to effect thread and memory placement specific to its needs.

Acknowledgments

This work was possible due to funding from the Australian Research Council, Gaussian Inc. and Sun Microsystems Inc. under ARC Linkage Grant LP0347178. We also wish to thank Peter Strazdins for his helpful suggestions and comments; Alexander Technology for access to the Opteron system.

References

1. T. Brecht. On the Importance of Parallel Application Placement in NUMA Multiprocessors. In *Proceedings of the Fourth Symposium on Experiences with Distributed and Multiprocessor Systems (SEDMS IV)*, pages 1–18, 1993.
2. S. Browne, J. Dongarra, N. Garner, G. Ho, and P. Mucci. A Portable Programming Interface for Performance Evaluation on Modern Processors. *The International Journal of High Performance Computing Applications*, 14(3):189–204, 2000.
3. Celestica Inc. AMD A8440 4U 4 Processor SCSI System. http://www.celestica.com/products/A8440.asp.
4. A. Charlesworth. The Sun Fireplane System Interconnect. In *Supercomputing '01: Proceedings of the 2001 ACM/IEEE conference on Supercomputing (CDROM)*. ACM Press, New York, New York, USA, November 2001.
5. D. E. Culler, A. Gupta, and J. P. Singh. *Parallel Computer Architecture: A Hardware/Software Approach*. Morgan Kaufmann Publishers, Inc., San Francisco, California, USA, 1999.
6. David R. Butenhof. *Programming with POSIX Threads*. Addison-Wesley Professional, 1997.
7. Dimitrios Nikolopoulos and Theodore Papatheodorou et. al. Leveraging Transparent Data Distribution in OpenMP via User-Level Dynamic Page Migration. In M. Valero, K. Joe, M. Kitsuregawa, and H. Tanaka, editors, *ISHPC*, volume 1940 of *Lecture Notes in Computer Science*, pages 415–427. Springer, 2000.
8. K. Goto and R. A. van de Geijn. Anatomy of High-Performance Matrix Multiplication. *In submission to ACM Transactions on Mathematical Software*, 2006.
9. Jay Trodden and Don Anderson. *HyperTransport System Architecture*. Addison-Wesley Professional, 2003.
10. John McCalpin. Stream: Sustainable memory bandwidth in high performance computers. http://www.cs.virginia.edu/stream.
11. Jonathan Chew. Memory Placement Optimisation. http://www.opensolaris.org/os/community/performance/mpo_overview.pdf.
12. C. N. Keltcher, K. J. McGrath, A. Ahmed, and P. Conway. The AMD Opteron Processor for Multiprocessor Servers. *IEEE Micro*, 23(2):66–76, 2003.
13. L. W. McVoy and C. Staelin. lmbench: Portable tools for performance analysis. In *USENIX Annual Technical Conference*, pages 279–294, 1996.
14. Novell. A NUMA API for Linux. http://www.novell.com/collateral/4621437/4621437.pdf.
15. Per Ekman. Linux kernel memory-to-node mappings. http://www.pdc.kth.se/~pek/linux/NUMA/.

16. N. Robertson and A. P. Rendell. OpenMP and NUMA Architectures I: Investigating Memory Placement on the SGI Origin 3000. In P. M. A. Sloot, editor, *International Conference on Computational Science*, volume 2660 of *Lecture Notes in Computer Science*, pages 648–656. Springer, 2003.

17. Rohit Chandra and Ramesh Menon et. al. *Parallel Programming in OpenMP*. Morgan Kaufmann, 2000.

18. Sun Microsystems. Solaris 10 : Extended Library Functions. `http://docs.sun.com/app/docs/doc/817-0679`.

19. Sun Microsystems. Solaris 10 : Programming Interfaces Guide. `http://docs.sun.com/app/docs/doc/817-4415`.

20. Sun Microsystems. *UltraSPARC III Cu User's Manual*. Sun Microsystems, Santa Clara, California, USA, January 2004. Version 2.2.1.

21. Sun Microsystems Inc. The Sun Fire V1280 Server Architecture. `http://www.sun.com/servers/midrange`, November 2002.

22. M. M. Tikir and J. K. Hollingsworth. Using Hardware Counters to Automatically Improve Memory Performance. In *SC*, page 46. IEEE Computer Society, 2004.

23. R. C. Whaley, A. Petitet, and J. Dongarra. ATLAS. *Parallel Computing*, 27 (1-2):3–35, 2001.

Low Power Scheduling of DAGs to Minimize Finish Times*

Sanjeev Baskiyar[1] and Kiran Kumar Palli[2]

[1] Computer Science and Software Engineering
[2] Electrical and Computer Engineering
Auburn University, AL 36849
baskiyar@eng.auburn.edu, palliki@auburn.edu

Abstract. We propose a schedule named Low Power Heterogeneous Makespan (LPHM) that attempts to minimize makespan as well as power consumption in the execution of any directed acyclic task graph on heterogeneous processors. We combine the techniques of Heterogeneous Earliest Finish Time (HEFT) [9] and voltage scaling [4]. The processors used for execution are considered to be continuously voltage scalable within the range of operation. After initial scheduling for minimum makespan, the processors are voltage scaled down to reduce power consumption whenever there is an idle time. This voltage scaling is performed without violating the precedence relationships among tasks. The simulation results show power savings of 22% over HEFT with no increase in makespan.

Keywords: DAG, Power, Scheduling, Makespan, Heterogeneous, Voltage Scaling.

1 Introduction

Heterogeneous computing uses diverse resources connected over high speed networks to support computationally intensive applications. These tasks may require to be scheduled in an optimum way so that the computation is done in minimum time consuming minimum resources. Task scheduling finds extensive use in ubiquitous computing where resources are constrained.

Scheduling tasks for minimum finish time using static, dynamic or hybrid scheduling algorithms has been well researched. Real-time environments have constraints such as speed, power and memory. Hence, it is important to address such issues in scheduling. There is little research which addresses both makespan and power consumption. Most of the work done to reduce the power consumption in a distributed system neglects the finish time. The power saving is obtained at the cost of longer execution time.

The classification of low power research as described in [8] is as follows:

---

* This work was supported in part by a grant from the National Science Foundation CCF 0411540.

Y. Roberts et al. (Eds.): HiPC 2006, LNCS 4297, pp. 353–362, 2006.

1. Power Estimation Techniques

- Entails power management at instruction, architecture, and gate level

2. Power Optimization Techniques

- *Hardware Optimization*

  - Behavioral level: Entails transformation, scheduling, and resource allocation
  - Architectural level: Use of low power flip flops, adders, etc.
  - Circuit level: Use of low power circuitry

- *Software Optimization*

  - Instruction-level: Low power compilation, Low power instructions, and scheduling
  - System-level: Dynamic power management, Low power memory management

Power consumption is a limiting factor for the functionality of devices operating on batteries with rapidly increasing computing and communication costs. There are many applications which require both low finish time and power consumption. Power consumption is a major issue in many real-time distributed systems such as real-time communication in satellites, as most applications running on power-limited systems inherently impose temporal constraints on finish time. A new area of interest is multi-hop radio networks used for sensor data traffic. New wireless communication systems are expected to evolve using this system. These networks are distributed networks operating on power constraints also called as power-aware distributed systems (PADS). Hence there is need for scheduling algorithms which would effectively reduce the overall power consumption and yet attain the best possible makespan.

LPHM addresses [10] both issues of minimizing makespan and reducing the power consumption. After the initial scheduling has been performed using HEFT for minimum finish time, it performs voltage scaling to reduce power consumption without increasing finish time. Also care is taken to meet precedence relationships.

The remainder of this paper is organized as follows. In the next section, we describe the necessary background on scheduling and related terminology. Different types of scheduling on homogeneous and heterogeneous processors have been discussed. Section 3 introduces the concepts in computing power. Section 4 discusses how voltage scaling can be used to reduce the power consumption of a processor. Section 5 details LPHM. Results have been discussed in Section 6. Section 7 presents conclusions and future work.

2 Background

In a distributed environment, an application can be decomposed into a set of computational tasks. These tasks may have data dependencies among them, thereby creating a partial order of precedence in which the tasks may be executed. DAGs are an important class of graphs having many applications such as those involving precedence among events. An example of a task DAG is shown in Figure 1.

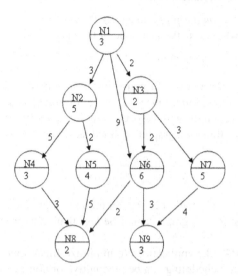

Fig. 1. A sample task DAG graph

In this paper, a DAG is represented by the tuple $G=(V,E,P,T,C,W)$ similar to [1], where V is the set of v nodes, E is the set of e edges between the nodes, and P is a set of p processors. $E(v_i, v_j)$ is an edge between nodes v_i and v_j. T is the set of costs $T(v_i, p_j)$, which represent the computation times of tasks v_i on processor p_j. C is the set of costs $C(v_i, v_j)$, which represent the communication cost associated with the edges $E(v_i, v_j)$. Since intra-processor communication is insignificant compared to inter-processor communication, $C(v_i, v_j)$ is considered to be zero if v_i and v_j execute on the same processor. W is the set of costs $W(v_i, v_j)$, which represent the power consumption costs of tasks v_i on processor p_j. The length of a path is defined as the sum of node and edge weights in that path.

Node v_p is a predecessor of node v_i if there is a directed edge originating from v_p and ending at v_i. Likewise, node v_s is a successor of node v_i if there is a directed edge originating from v_i and ending at v_s. We further define $pred(v_i)$ as the set of all predecessors of v_i and $succ(v_i)$ as the set of all successors of v_i. An ancestor of node v_i is any node v_p that is contained in $pred(v_i)$, or any node v_a that is also an ancestor of any node v_p contained in $pred(v_p)$.

The earliest start time of node v_i on processor p_j is represented as $EST(v_i, p_j)$. Likewise, the earliest finish time of v_i on processor p_j is represented by $EFT(v_i, p_j)$. $EST(v_i)$ and $EFT(v_i)$ represent the earliest start and finish times on any processor respectively. $T\_avail [v_i, p_j]$ is defined as the earliest time that processor p_j will be available to begin executing task v_i. Hence,

$$EST(v_i, p_j) = max\{ T\_avail [v_i, p_j], \max_{v_p \in pred(v_i)} (EFT(v_p, p_k)+C(v_p, v_i)) \} \qquad (1)$$

$$EFT(v_i, p_j) = T(v_i, p_j) + EST(v_i, p_j) \qquad (2)$$

The goal of LPHM is to minimize both makespan and total power consumption (makepower) which are defined as follows:

$$makespan = \max\{ EFT(v_i)\} \qquad (3)$$
where v_i is the exit node of the graph.

$$makepower = \sum W(vi, pj) \qquad (4)$$

Schedule length (makespan) is a major metric to measure the performance of a scheduling algorithm on a graph. Since a large set of graphs is used, it is necessary to normalize the schedule length with respect to its lower bound. Thus schedule length ratio (SLR) of an algorithm on a graph is defined as:

$$SLR = \frac{makespan}{\sum_{n_i \in CP_{MIN}} \min_{p_j \in Q}\{w_{i,j}\}} \qquad (5)$$

The denominator is the summation of the minimum computation costs of tasks on the critical path. Thus, SLR of a graph cannot be less than one since the denominator is the lower bound.

Scheduling of DAGs like applications to minimize finish time is a well researched NP-complete problem. Scheduling can be preemptive or non-preemptive. In this work we deal with non-preemptive scheduling.

In a real-time distributed environment, the availability of computing resources varies which results in both temporal as well as spatial heterogenity. Heuristic-based and guided-random-search based algorithms are two principal approaches of scheduling DAGs. Heuristic algorithms are classified as list scheduling, cluster scheduling and task duplication based scheduling. Guided random search based algorithms can be classified as Genetic Algorithms, Simulated Annealing and Local Search Technique. List scheduling tries to minimize a predefined cost function by first making a list of all tasks based on their priorities and then selecting the appropriate processor based on the heuristic. List scheduling algorithms have usually low complexity and give good finish times. Examples of list scheduling algorithms are HEFT, Modified Critical Path (MCP) [3], Dynamic Level Scheduling [2] and Mapping Heuristic [5].

Task duplication can minimize interprocessor communication and hence results in shorter finish times. HNPD [1] was shown to give better makespan than HEFT. It combines the techniques of HEFT and Scalable Task Duplication Scheduling (STDS) [7]. It uses task duplication to minimize finish time, however this approach increases power consumption. We therefore chose to perform voltage scaling on HEFT.

HEFT is an insertion-based algorithm, i.e, it tries to schedule a task between two already scheduled tasks. Tasks are ordered in the order of their upward rank, which for any task n_i is recursively defined as follows:

$$rank_u(n_i) = \overline{w_i} + \max_{n_j \in succ(n_i)} (\overline{c_{i,j}} + rank_u(n_j)) \qquad (6)$$

where $succ(n_i)$ is the set of immediate successors of task n_i, $\overline{c_{i,j}}$ is the average communication cost of edge(i, j) over all processor pairs, and $\overline{w_i}$ is the average of the computation cost of task n_i over all processors. Since the rank is computed recursively by traversing the graph upward, starting from the exit task, it is called upward rank. For the exit task n_{exit}, the upward rank is:

$$rank_u(n_{exit}) = \overline{w_{exit}} \ . \tag{7}$$

Basically, $rank_u(n_i)$ is the length of the critical path from task n_i to the exit task, excluding the cost of task n_i.

In HEFT the priority of each task is set to its upward rank. Next, the task list is sorted in decreasing order of upward rank. Next, it tries to insert tasks in idle slots such that the insertion of any task does not violate any precedence relationship. It has an $O(e \ x \ q)$ time complexity for e edges and q processors. For a dense graph, the number of edges is proportional to v^2 (v is the number of tasks), the time complexity is $O(v^2 \ x \ p)$.

3 Power Computation

We know that the processor clock frequency, f, can be expressed in terms of the supply voltage, V_{dd}, and the threshold voltage, V_t, as follows:

$$f = k(V_{dd} - V_t)^2 / V_{dd} \ . \tag{8}$$

Where k is a constant. From eqn. (8), we derive V_{dd} as a function of f, $F(f)$ as:

$$V_{dd} = F(f) = (V_t + \frac{f}{2k}) + \sqrt{(V_t + \frac{f}{2k})^2 - V_t^2} \ . \tag{9}$$

The processor power, p, can be expressed in terms of the frequency, f, switching capacitance, N, and supply voltage, V_{dd} as:

$$p = \frac{1}{2} fNV_{dd}^2 = \frac{1}{2} fNF(f)^2 \tag{10}$$

Given the number of clock cycles, η_i, for executing task i, its energy consumption, E_i, under supply voltage V_i and clock frequency, f_i, is given by:

$$E_i = (\eta_i / f_i) * p(f_i) \tag{11}$$

4 Voltage Scaling

Voltage scaling is technique in which the core power supply voltage of a system is varied depending on the processing load, to decrease the total power consumption. But reducing the power supply voltage also reduces the speed of execution. In some instances it is observed that with the reduction in supply voltage from $5.0V$ to $3.3V$, there is about 56% reduction in power consumption [6].

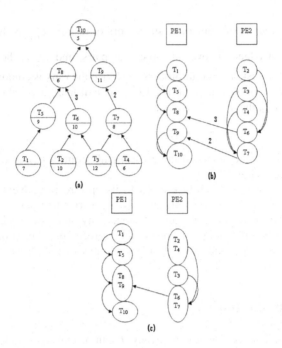

Fig. 2. Voltage Scaling

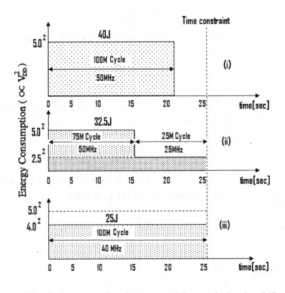

Fig. 3. An example of a power-delay optimization [4]

Consider the example in Figure 2, where T_2 and T_3 must precede T_6. Assume T_2 and T_3 take *20* and *15* seconds on their respective processors. So, the processor executing T_2 has an idle time of *5* seconds which means it can operate at a lower speed, i.e., at a lower supply voltage. Suppose T_4 takes *15* seconds or less to complete. Now, the processor executing task T_4 has some idle time and hence can operate at low power. Next, tasks T_2 & T_4 can be merged into a group (if they don't have an off chip data transfer) so that they both can be dynamically voltage scaled at the same time as shown in Figure 2(c). Such merging decreases the number of times one must dynamically switch the voltage. Thus given a voltage scaled schedule, tasks are merged wherever possible and then merged tasks which can operate at lower frequencies are determined.

Consider Figure 3 which shows energy consumption vs. execution time for different voltages and frequencies [4]. It shows the advantage of voltage and frequency scaling. When a task is executed at *5V* and *50MHz*, it consumes *40J* of energy. But if the power supply in the last quarter of the execution is scaled down to *2.5V* and frequency to *25MHz*, it meets the deadline and consumes only *32.5J* of energy. If the power supply is scaled down *4V* and frequency to *40MHz* from the start, it still meets the deadline requirements and consumes only *25J* of energy.

5 Scheduling Algorithm

In Figure 4 we detail our new algorithm, LPHM.

Procedure LPHM

1. Assign computation costs of tasks and edges with mean values over all processors. Assign power consumption of tasks on all processors.
2. Compute upper rank of all tasks by traversing graph upward, starting from the exit task.
3. Sort tasks in non-increasing order of upper rank.
4. **while** there are unscheduled tasks **do**
    (i)   Select the first task n_i from the above sorted list
    (ii) for all processors P_j compute $EFT(n_i, P_j)$ using insertion based scheduling
    (iii) Schedule task n_i on processor P_j that provides the minimum $EFT (n_i)$
5. **endwhile**
6. **for** all k, m, i, j
    (i) **If** $EFT (n_k, P_i) + C (P_i, P_j) > EFT (n_m, P_j)$ **then**
       $EFT (n_m, P_j) = EFT(n_k, P_j) + C (P_i, P_j)$
       // where n_k is a node which is scheduled immediately before n_m on
       //P_j
7. **endfor**
8. Sum power consumptions on all tasks using eqn. 10.

End LPHM

Fig. 4. Procedure LPHM

Using LPHM on the graph of Figure 5, we observe that if $EFT(N_1, P_1) + C(P_1, P_2) > EFT(N_2, P_2)$ we can increase $EFT(N_2, P_2)$ because P_2 has to anyway wait upon N_1 before executing N_4. Idle time on P_2, would be available if $EFT(N_2, P_2) < EST(N_4, P_2)$. Checking the above is equivalent to checking for any available idle time on all processors. Therefore, we replace lines 6 and 6.(i) in the above algorithm by the following which is less computationally intensive:

$$\textbf{if } EFT(n_k, P_j) < EST(n_m, P_j) \textbf{ then}$$
$$EFT(n_k, P_j) = EST(n_m, P_j)$$

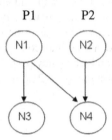

Fig. 5. Example graph

6 Simulation and Results

For purposes of simulations, random weighted directed-acyclic-graphs were generated as described in [1]. The following parameters were used:

- Number of nodes in the graph, v.
- Shape parameter (SP) of the graph, α.
- Mean out degree of a node, (*out_degree*)
- Communication to computation ratio, (CCR)
- Computation range, β

In each experiment, the above parameters are varied over the sets of values given below:

- SET (v)   = {10, 20, 40, 60, 80}
- SET(α)   = {0.5, 1.0, 2.0}
- SET(*out_degree*) = {1, 2, 3, 4, 5}
- SET(CCR)   = { 0.1, 0.5, 1.0, 5.0, 10.0}
- SET (β)   = {0.1, 0.25, 0.5, 0.75, 1.0}

The above combinations give 1,875 different DAGs. Assigning a number of input parameters and selecting each input parameter from a set of values generates diverse DAGs with a wide variety of characteristics. Since we generated 25 random graphs in each study, the total number of graphs is 46,875. Using these graphs, the performance of LPHM was compared with respect to various graph characteristics.

Figure 6 shows the percentage reduction in power consumption with respect to CCR. The power reduction increases with CCR. This is because, as CCR increases, the communication cost of tasks increase and thus more tasks are scheduled on same processor rather than choosing a new processor. This in turn, increases idle time be-tween tasks which can be used for power reduction.

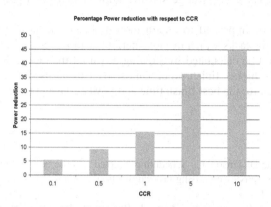

Fig. 6. Power reduction vs. CCR

Figure 7 shows power reduction with respect to computation range. As computa-tion range increases, heterogeneity among processors increases. In other words, dif-ferent processors execute similar tasks at different speeds. Hence we can find more idle time between tasks. Hence, we see a slight increase in power reduction as compu-tation range increases.

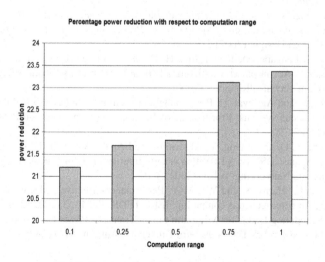

Fig. 7. Power reduction vs. computation range

7 Conclusions

This work reports a new scheduling algorithm which tries to minimize both make-span and power consumption. LPHM has been seen to obtain the same makespan as HEFT but with an average power savings of about 22%. The results are based on an experimental study using a large set of randomly generated graphs with various characteristics.

This work can be improved to obtain more reduction in power consumption by slowing down all the tasks which do not interfere with the critical path. Also, power consumption should be calculated by slowing down all the tasks which comes at the expense of slightly higher finish time. This study can be used to find the best frequency of operation for the best finish time and power consumption.

References

1. Baskiyar, S. and Dickinson, C.: Scheduling directed a-cyclic task graphs on a bounded set of heterogeneous processors, using task duplication. J. Parallel Distrib. Comput. Elsevier. 8 (2005) 911-921.
2. Chandrakasan, M, Potkonjak, R., Mehra, J., Rabaey, R., Brodersen, W. Optimizing power using transformations: IEEE Transactions on Computer-Aided Design of Integrated Circuits and Systems.1 (1995) 12-31.
3. Huizer, C. M. Power dissipation analysis of CMOS VLSI circuits by means of switch level simulation: IEEE European Solid State Circuits Conference, Grenoble, France. (1990) 61-64.
4. Ishihara, T. and Yasuura, H. Voltage scheduling problem for dynamically variable voltage processors: Proceedings of 1998 International Symposium on Low Power Electronics and Design. (1998) 197-202.
5. Luo, J. and Jha, N. Static and dynamic variable voltage scheduling algorithms for real-time heterogeneous distributed embedded systems: Proceedings of the 15th International Conference on VLSI Design. (2002) 719-726.
6. Pouwelse, J., Langandoen, K and Sips, H. Dynamic voltage scaling on a low-power microprocessor: At http://www.ubicom.tudelft.nl/docs/UbiCom-TechnicalReport_2000_3. PDF, UbiCom TechnicalReport. 3 (2000).
7. Ranaweera, S. and Agrawal, D. P. A scalable task duplication based scheduling algorithm for heterogeneous systems: Proceedings of International Conference on Parallel Processing. (2000) 383-390.
8. Sha, E., Zhuge, Q. and Zhang, Y. Algorithms and Analysis of Scheduling for Low-Power High-Performance DSP on VLIW Processors: International Journal of High Performance Computing and Networking. vol. 1. (2004) 3-16.
9. Topcuoglu, H., Hariri, S. and Wu, M-Y. Performance-effective and low-complexity task scheduling for heterogeneous computing: IEEE TPDS. Vol 13, 3 (2002) 260-274.
10. Palli, K. Scheduling DAGs for Minimum Finish Time and Power Consumption on Heterogeneous Processors: MS Thesis, Auburn University, Auburn, AL. (2005).

GPU-ClustalW: Using Graphics Hardware to Accelerate Multiple Sequence Alignment

Weiguo Liu, Bertil Schmidt, Gerrit Voss, and Wolfgang Müller-Wittig

School of Computer Engineering, Nanyang Technological University
Centre for Advanced Media Technology
{liuweiguo, asbschmidt, asgerrit, askwmwittig}@ntu.edu.sg

Abstract. Molecular Biologists frequently compute multiple sequence alignments (MSAs) to identify similar regions in protein families. However, aligning hundreds of sequences by popular MSA tools such as ClustalW requires several hours on sequential computers. Due to the rapid growth of biological sequence databases biologists have to compute MSAs in a far shorter time. In this paper we present a new approach to reduce this runtime using graphics processing units (GPUs). To derive an efficient mapping onto this type of architecture, we have reformulated the computationally most expensive part of ClustalW in terms of computer graphics primitives. This results in a high-speed implementation with significant runtime savings on a commodity graphics card.

1 Introduction

Dynamic programming (DP) is often used to compute the optimal local alignment of a pair of sequences [1]. However the extension of the DP method for simultaneous alignment of multiple sequences is impractical as the time and space complexities are in the order of the product of the lengths of the sequences. Thus, many heuristics to compute multiple sequence alignments (MSAs) in reasonable time have been developed.

Progressive alignment is a widely used heuristic [2]. Examples of popular tools which are using this approach include ClustalW [3], PRALINE [4], MUSCLE [5], and T-Coffee [6]. Typically, progressive alignment methods consist of three steps. Firstly, a distance value between each pair of sequences is computed. Secondly, a phylogenetic tree is calculated based on this distance matrix. Finally, pairwise alignment of various profiles is done following the branching order in the phylogenetic tree to form the final MSA. Unfortunately, progressive alignment programs suffer from a high computational complexity, for instance the alignment of a few hundred protein sequences using ClustalW requires several hours on a state-of-the-art workstation.

A popular technique to speedup this time consuming task is to use parallel processing. The runtime of progressive alignment programs is typically dominated by the first step (computation of pairwise sequence distances). There are two basic approaches of parallelizing this step: one is based on the systolisation of the pairwise distance computation algorithm (fine-grained); the other is based

Y. Robert et al. (Eds.): HiPC 2006, LNCS 4297, pp. 363–374, 2006.
© Springer-Verlag Berlin Heidelberg 2006

on the distribution of the computation of pairwise distances (coarse-grained). Several coarse-grained parallel implementations of ClustalW have been developed. The solutions presented by [7] and by [8] use message-passing on a PC cluster. Parallel ClustalW implementations have also been designed for more expensive shared memory machines [9]. Another fine-grained parallel solution based on reconfigurable hardware (FPGAs) has recently been presented in [10].

In this paper, we investigate how graphics processing units (GPUs) can be used as a computational platform to accelerate MSAs with ClustalW. The main advantage of GPUs compared to other accelerator architectures such as FP-GAs is that they are commodity components. In particular, most users already have access to PCs with modern graphics cards. For these users this direction provides a zero-cost solution. Even if a graphics card has to be bought, the installation of such a card is trivial (plug and play). However, there are still a number of challenges to be solved in order to enable scientists other than computer graphics specialists to facilitate efficient usage of these resources within their research area. The biggest challenge in order to solve a specific problem using GPUs is reformulating the proposed algorithms and data structures using computer graphics primitives (e.g. triangles, textures, vertices, fragments). Furthermore, restrictions of the underlying streaming architecture have to be taken into account, e.g. random access writes to memory is not supported and no cross fragment data or persistent state is possible (e.g. all the internal registers are flushed before a new fragment is processed). In this paper we show how MSA based on ClustalW can benefit from this type of computing power.

The rest of this paper is organized as follows. Section 2 provides a brief description of progressive alignment using ClustalW. Section 3 describes the algorithm for pairwise sequence distance computation. Important features of the GPU architecture are described in Section 4. Section 5 presents our mapping of the algorithm onto the GPU architecture. The performance is evaluated in Section 6. Section 7 concludes the paper.

2   Progressive Multiple Sequence Alignment

In this Section, we briefly describe the three stages involved in progressive alignment using ClustalW [3] (see Figure 1).

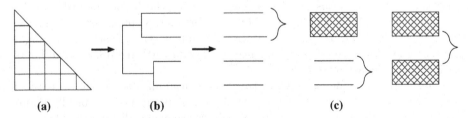

Fig. 1. The three stages of progressive multiple sequence alignment (a) distance matrix, (b) guided tree, (c) progressive alignment along the tree

(a) *Distance matrix*: A distance value between each pair of sequences is computed using the Smith-Waterman algorithm. These values are stored in a so-called distance matrix.

(b) *Guided tree*: This step uses the distance matrix obtained from the first step and forms a guided-tree using the neighbor-joining method [11]. The leaves of the tree contain the various sequences. The topology of the tree is totally dependent upon the sequences that are taken, i.e. closely related sequences are placed together and share a common branch in the guided-tree and divergent sequences are widely spaced in the tree. The guided-tree is used to find out closely related sequences or a group of sequences that are aligned progressively in the last step to form the final MSAs.

(c) *Progressive Alignment*: First closely related sequences or group of sequences are aligned and at the end most divergent sequences are aligned to get the final MSAs.

Profiling of the three stages of ClustalW for different numbers of globin sequences reveals that more than 93% of the overall runtime is spent on the first stage (see Table 1). Hence, we have decided to map only this stage onto a GPU.

Table 1. Profiling of the three stages of ClustalW using a different number of globin sequences on a Pentium4 3GHz (based on the code from [7])

Number of Sequences	Distance matrix	Guided tree	Progressive alignment
200	94.4%	0.03%	5.6%
400	93.4%	0.09%	6.5%
600	93.3%	0.20%	6.4%
800	94.0%	0.20%	5.8%
1000	93.6%	0.30%	6.1%

3   Pairwise Sequence Distance Computation

Given is a set of n sequences $S = S_1, \ldots, S_n$. For two sequences $S_i, S_j \in S$, their distance $d(S_i, S_j)$ is defined as follows:

$$d(S_i, S_j) = 1 - \frac{nid(S_i, S_j)}{min\{l_i, l_j\}} \qquad (1)$$

where $nid(S_i, S_j)$ denotes the number of exact matches in the optimal local alignment of S_i and S_j (with respect to the parameters α, β and sbt, which are explained below) and l_i (l_j) denotes the length of S_i (S_j).

The optimal local alignment of two sequences can be computed using the Smith-Waterman algorithm [1]. The algorithm compares two sequences by computing a distance that represents the minimal cost of transforming one segment into another. Two elementary operations are used: substitution and insertion/deletion (also called a gap operation). Through series of such elementary operations, any segments can be transformed into any other segment. The smallest number of operations required to change one segment into another can be taken into as the measure of the distance between the segments. Consider two strings S_1 and S_2 of length l_1 and l_2. To identify common subsequences, the Smith-Waterman algorithm computes the similarity $H(i, j)$ of two sequences ending at position i and j of the two sequences S_1 and S_2. The computation of $H(i, j)$, for $1 \leq i \leq l_1$, $1 \leq j \leq l_2$, is given by the following recurrences:

$$
\begin{aligned}
H(i, j) &= max\{0, E(i, j), F(i, j), H(i - 1, j - 1) + sbt(S_1[i], S_2[j])\} \\
E(i, j) &= max\{H(i, j - 1) - \alpha, E(i, j - 1) - \beta\} \\
F(i, j) &= max\{H(i - 1, j) - \alpha, F(i - 1, j) - \beta\}
\end{aligned} \tag{2}
$$

where sbt is a character substitution cost table. Initialization of these values are given by $H(i, 0) = E(i, 0) = H(0, j) = F(0, j) = 0$ for $0 \leq i \leq l_1$, $0 \leq j \leq l_2$. Multiple gap costs are taken into account as follows: α is the cost of the first gap; β is the cost of the following gaps. This type of gap cost is known as *affine gap penalty*. Some applications also use a *linear gap penalty*, i.e. $\alpha = \beta$. For linear gap penalties the above recurrence relations can be simplified to:

$$
H(i, j) = max\{0, H(i, j - 1) - \alpha, H(i - 1, j) - \alpha, H(i - 1, j - 1) + sbt(S_1[i], S_2[j])\} \tag{3}
$$

Each cell of the matrix H is a similarity value. The two segments of S_1 and S_2 producing this value can be determined by a trace-back procedure. Example 1 illustrates this procedure for the two sequences ATCTCGTATGATG and GTCTATCAC. The value $nid(S_1, S_2)$ of the two sequences S_1 and S_2 can then be computed by counting the number of exact character matches during the traceback procedure of the Smith-Waterman algorithm. For instance the nid-value for the sequences in Figure 2 is six. Unfortunately, this procedure is not very suitable for a parallel implementation on streaming architectures such as GPUs. Therefore, we have formulated a new recurrence relation for the nid-value computation that is more suitable for the GPU implementation. It facilitates nid-calculation without the computation of actual alignment. In the rest of this section, we first explain our idea for linear gap penalties and then generalize it for affine gap penalties. Its efficient implementation on a GPU will be described in Section 4.

Given are the two sequences S_1 and S_2, the linear gap penalty α and the substitution table sbt. The computation of $N(i, j)$ is given by the following recurrence relations:

$$
N(i, j) = \begin{cases} 0, & \text{if } H(i, j) = 0 \\ N(i - 1, j - 1) + m(i, j), & \text{if } H(i, j) = H(i - 1, j - 1) \\ & \qquad + sbt(S_1[i], S_2[j]) \\ N(i, j - 1), & \text{if } H(i, j) = H(i, j - 1) - \alpha \\ N(i - 1, j); & \text{if } H(i, j) = H(i - 1, j) - \alpha \end{cases} \tag{4}
$$

	\	A	T	C	T	C	G	T	A	T	G	A	T
\	0	0	0	0	0	0	0	0	0	0	0	0	0
G	0	0	0	0	0	0	2	1	0	0	2	1	0
T	0	0	2	1	2	1	1	4	3	2	1	1	3
C	0	0	1	4	3	4←3	3	3	3	2	1	0	2
T	0	0	2	3	6	5	4	5	4	5	4	3	2
A	0	2	2	2	5	5	4	4	7	6	5	6	5
T	0	1	4	3	4	4	4	6	5	9	8	7	8
C	0	0	3	6	5	6	5	5	5	8	8	7	7
A	0	2	2	5	5	5	5	4	7	7	7	10	9
C	0	1	1	4	4	7	6	5	6	6	6	9	9

Fig. 2. Example of the Smith-Waterman algorithm to compute the local alignment between two DNA sequences ATCTCGTATGAT and GTCTATCAC. The matrix $H(i,j)$ is shown for the linear gap cost $\alpha = 1$, and a substitution cost of $+2$ if the characters are identical and -1 otherwise. From the highest score ($+10$ in the example), a traceback procedure delivers the corresponding alignment, the two subsequences TCGTATGA and TCTATCA.

where

$$m(i,j) = \begin{cases} 1, & \text{if} S_1[i] = S_2[j] \\ 0; & otherwise \end{cases}$$

The value $nid(S_1, S_2)$ is then equal to $N(i_{max}, j_{max})$ where (i_{max}, j_{max}) are the coordinates of the maximum in matrix H. For affine gap penalties the recurrence relation for $N(i,j)$ is extended as follows:

$$N(i,j) = \begin{cases} 0, & \text{if } H(i,j) = 0 \\ N(i-1,j-1) + m(i,j), & \text{if } H(i,j) = H(i-1,j-1) \\ & \qquad + sbt(S_1[i], S_2[j]) \\ N(i,j - NE(i,j)), & \text{if } H(i,j) = E(i,j) \\ N(i - FE(i,j), j); & \text{if } H(i,j) = F(i,j) \end{cases} \quad (5)$$

where

$$m(i,j) = \begin{cases} 1, & \text{if } S_1[i] = S_2[j] \\ 0; & otherwise \end{cases}$$

$$NE(i,j) = \begin{cases} 1, & \text{if } E(i,j) = H(i,j-1) - \alpha \text{ or } j = 1 \\ NE(i,j-1) + 1; & \text{if } E(i,j) = E(i,j-1) - \beta \end{cases}$$

$$FE(i,j) = \begin{cases} 1, & \text{if } F(i,j) = H(i-1,j) - \alpha \text{ or } i = 1 \\ FE(i-1,j) + 1; & \text{if } F(i,j) = F(i-1,j) - \beta \end{cases}$$

4 GPU Architecture

The fast increasing power of the GPU and its streaming architecture opens up a range of new possibilities for a variety of applications. With the enhanced programmability of commodity GPUs, these chips are now capable of performing more than the specific graphics computations they were originally designed for. Recent work on GPGPU (General-Purpose computation on GPUs) shows the design and implementation of algorithms for non-graphics applications. Examples include scientific computing [12], image processing [13], and bioinformatics [14, 15]. Currently, the peak performance of high-end GPUs such as the GeForce 7900 GTX 512 is approximately ten times faster than that of comparable CPUs. Further, the GPU performance has been increasing from two to two-and-a-half times a year. This growth rate is faster than Moore's law as it applies to CPUs, which corresponds to about one-and-a-half times a year [16]. Consequently, GPUs will become an even more attractive alternative for high performance computing in the near future.

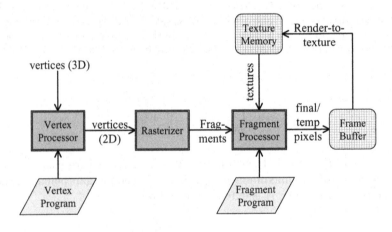

Fig. 3. Illustration of the GPU graphics pipeline

Computation on a GPU follows a fixed order of processing stages, called the graphics pipeline (see Figure 3). The pipeline consists of three stages: vertex processing, rasterization and fragment processing. The vertex processing stage transforms three-dimensional vertex world coordinates into two-dimensional vertex screen coordinates. The rasterizer then converts the geometric vertex representation into an image fragment representation. Finally, the fragment processor forms a color for each pixel by reading texels (texture pixels) from the texture memory. Modern GPUs support programmability of the vertex and fragment processor. Fragment programs for instance can be used to implement any mathematical operation on one or more input vectors (textures or fragments) to compute the color of a pixel.

In order to meet the ever increasing performance requirements set by the gaming industry, modern GPUs use two types of parallelism. Firstly, multiple processors work on the vertex and fragment processing stage, i.e. they operate on different vertices and fragments in parallel. For example, a typical mid-range graphics card such as the nVidia GeForce 7800 GTX has 8 vertex processors and 24 fragment processors. Secondly, operations on 4-dimensional vectors (the four channels Red/Green/Blue/Alpha (RGBA)) are natively supported without performance loss.

Several authors have described modern GPUs as *streaming processors*, e.g [17]. Streaming processors read an input stream, apply kernels (filters) to the stream and write the results into an output stream. In case of several kernels, the output stream of the leading kernel is the input stream for the following kernel. The vast majority of general-purpose GPU applications use only fragment programs for their computation. In this case textures are considered as input streams and the texture buffers are output streams. Because fragment processors are SIMD architectures, only one program can be loaded at a time. Applying several kernels thus means to do several passes.

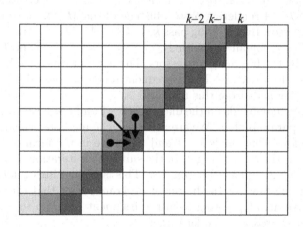

Fig. 4. Data dependency relationship in the Smith-Waterman DP matrix: each cell (i, j) depends on its left neighbor $(i, j - 1)$, upper neighbor $(i - 1, j)$ and upper left neighbor $(i - 1, j - 1)$. Therefore all cells along anti-diagonal k can be computed in parallel from the anti-diagonals $k - 1$ and $k - 2$.

5   Mapping onto the GPU Architecture

In this section we describe how to map our algorithm onto a GPU efficiently. We take advantage of the fact that all elements in the same anti-diagonal of the DP matrix can be computed independent of each other in parallel (see Figure 4). Thus, the basic idea is to compute the DP matrix in anti-diagonal order.

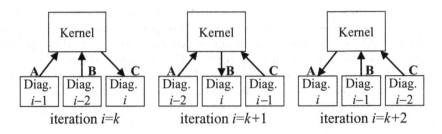

Fig. 5. Cyclic change of the functions of buffers A, B, and C for computation of anti-diagonals in the DP matrix

The anti-diagonals are stored as textures in the texture memory. The fragment program is then used to implement the arithmetic operations specified by the recurrence relations in E.q. (2) and (5).

Assume we are aligning two sequences of length M and K with affine gap penalties on a GPU. As a preprocessing step both sequences and the substitution matrix are loaded into the texture memory. We are then computing the DP matrix in $M + K - 1$ rendering passes, instead of the $M \times K$ steps required on a sequential processor. In rendering pass k, $1 \leq k \leq M+K-1$, the values $H(i,j)$, $E(i,j)$, and $F(i,j)$ for all i, j with $1 \leq i \leq M$, $1 \leq j \leq K$ and $k = i + j - 1$ are computed by the fragment processors. The new anti-diagonal is stored in the texture memory as a texture. The subsequent rendering pass then reads the two previous anti-diagonals from this memory.

Since diagonal k depends on the diagonals $k-1$ and $k-2$, we store these three diagonals as separate buffers. We are using a cyclic method to change the buffer function as follows: Diagonals $k-1$ and $k-2$ are in the form of texture input and diagonal k is the render target. In the subsequent iteration, k becomes $k-1$, $k-1$ becomes $k-2$, and $k-2$ becomes k. This is further illustrated in Figure 5. An arrow pointing towards the fragment program means that the buffer is used as a texture. An arrow pointing from the fragment program to a buffer means that the buffer is used as a render target.

Considering a set of n sequences. We first sort it according to the sequence lengths. Because in MSA each sequence in a set has to be compared to every other sequence, there are totally $n \times (n-1)/2$ pairwise comparisons. In order to take full advantage of the inherently parallelism and high memory bandwidth of GPUs, we pack the query and subject sequences into 2D textures, thus multiple pairwise comparisons can be done at the same time.

In our GPU application the computation of all the items in E.q. (2) and (5) is incorporated as follows: The value $max\{H(i,j), H(i-1,j), H(i,j-1)\}$ and $nid(S_1, S_2)$ are calculated for each cell and stored in the A-channel of an RGBA-color pixel in two render targets separately. In the first render target, the R-, G-, and B-channels are used for the computation of $H(i,j), E(i,j)$ and $F(i,j)$ (see Figure 6a). In the second target, R-, G-, and B-channels are used for the computation of $N(i,j), NE(i,j)$ and $FE(i,j)$ respectively (see Figure 6b).

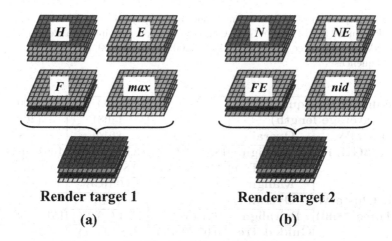

Fig. 6. Using the RGBA channels of two-dimensional texture buffers for the computation of H, E, F, max, N, NE, FE and nid

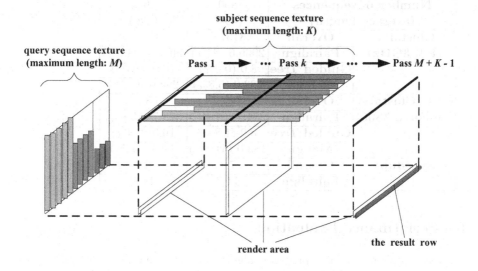

Fig. 7. The rendering process of one passes loop

Figure 7 illustrates the rendering process for one passes loop. Fragment processors write computation results of each pass to the render targets. During each pass, the dimension of the render area on the render targets will change according to the current pass number, the maximum length of query sequences and the maximum length of subject sequences. The final results (nid) will be found at the bottom row of the render targets at the last pass.

Table 2. Comparison of runtimes (in seconds) and speedups of ClustalW running on a single Pentium4 3GHz to our GPU-ClustalW version running on a Pentium4 3GHz with an NVIDIA GeForce 7800 GTX 512 for 200, 400, 600, 800, and 1000 input globin protein sequences

Number of sequences (average length)		200 (412)	400 (408)	600 (462)
ClustalW (P4, 3GHz)	Overall	194.9	891.9	1818.1
	Pairalign	183.8 (94.4%)	833.1 (93.4%)	1697 (93.3%)
	Guided Tree	0.07 (0.03%)	0.8 (0.09%)	4.1 (0.2%)
	Malign	11.0 (5.6%)	58.0 (6.5%)	117.0 (6.4%)
GPU-ClustalW (GeForce 7800)	Overall	27.2	134.1	272.4
	Pairalign	16.1 (59.2%)	75.3 (56.2%)	151.3 (55.5%)
	Guided Tree	0.07 (0.3%)	0.8 (0.6%)	4.1 (1.5%)
	Malign	11.0 (40.4%)	58.0 (43.3%)	117.0 (43%)
Speedups	Overall	7.2	6.7	6.7
	Pairalign	11.4	11.1	11.2

Number of sequences (average length)		800 (454)	1000 (446)
ClustalW (P4, 3GHz)	Overall	3157.6	4711.6
	Pairalign	2966.6 (94%)	4409.6 (93.6%)
	Guided Tree	8.0 (0.2%)	16.0 (0.3%)
	Malign	183.0 (5.8%)	286.0 (6.1%)
GPU-ClustalW (GeForce 7800)	Overall	445.2	680.7
	Pairalign	254.2 (57.1%)	378.7 (55.6%)
	Guided Tree	8.0 (1.8%)	16.0 (2.4%)
	Malign	183.0 (41.1%)	286.0 (42%)
Speedups	Overall	7.1	6.9
	Pairalign	11.7	11.6

6 Performance Evaluation

We have implemented the proposed algorithm using the high-level GPU programming language *GLSL* (OpenGL Shading Language) [18] and evaluated it on the following graphics card:

- *nVidia GeForce 7800 GTX*: 627 MHz engine clock speed, 1.83 GHz memory clock speed, 8 vertex processors, 24 fragment processors, 512 MB memory.

Tests have been conducted with this card installed in a PC with an Intel Pentium4 3.0GHz, 1GB RAM running Windows XP.

A set of performance evaluation tests have been conducted using different numbers of globin protein sequences, to evaluate the processing time of the GPU implementation versus that of the original ClustalW pairwise alignment stage on the PC. The ClustalW application is benchmarked on an Intel Pentium4 3GHz processor with 1GB RAM. We have used the ClustalW code from Li ([7], available online at http://web.bii.a-star.edu.sg/ kuobin/clustalw-mpi/index.html) for our evaluation. The results for this are shown in Table 2. As can be seen, our GPU implementation achieves speedups of almost ten compared to first stage of ClustalW and six compared to the overall runtime.

7 Conclusion and Future Work

In this paper we have demonstrated that the streaming architecture of GPUs can be efficiently used for MSA. To derive an efficient mapping onto this type of architecture, we have reformulated the computationally expensive first stage of the ClustalW algorithm in terms of computer graphics primitives. Our design is based on a new recurrence relation for calculating the number of exact matches in the optimal local alignment of two sequences. The evaluation of our implementation on a high-end graphics card shows a speedup of up to ten compared to a Pentium IV 3GHz. At least the same number of PCs connected by a fast switch is required to achieve a similar speedup using the ClustalW-MPI code from Li [7]. A comparison of these two parallelization approaches shows that graphics hardware acceleration is superior in terms of price/performance. Our solution is also easily scalable to several GPUs (within the same PC or across a network) by simply partitioning the individual pairwise sequence comparisons.

Table 2 also shows that the progressive alignment stage of ClustalW (malign) would dominate the runtime of ClustalW if the first stage would be accelerated by a factor of more than 16. The malign stage consists of computing several profile-profile alignments based on dynamic programming. It will be interesting to investigate how this can be efficiently mapped onto graphics hardware.

The first stage of several other MSA tools such as T-Coffee [6] and MUSCLE [5] is also based on the computation of pairwise distances. Hence, these tools could see a similar speedup form the accelerator presented in this paper.

Acknowledgment

The work was supported by the A*Star BMRC research grant No. 04/1/22/19/375.

References

1. Smith, T., Waterman, M.: Identification of common molecular subsequences. Journal of Molecular Biology **147** (1981) 195–197
2. Feng, D., Doolittle, R.: Progressive sequence alignment as a prerequisite to a correct phylogenetic trees. Journal of Molecular Evolution **25** (1987) 351–360
3. Thompson, J., Higgins, D., Gibson, T.: Clustalw: improving the sensitivity of progressive multiple sequence alignment through sequence weighting position specific gap penalties and weight matrix choice. Nucleic Acids Res. **22** (1994) 4673–4680

4. Heringa, J.: Two strategies for sequence comparison: Profile-preprocessed and secondary structure-induced multiple alignment. Comput. Chem. **23** (1999) 341–364
5. Edgar, R.: Muscle: multiple sequence alignment with high accuracy and high throughput. Nucleic Acids Res. **32** (2004) 1792–1797
6. Notredame, C., Higgins, D., Heringa, J.: T-coffee: A novel method for fast and accurate multiple sequence alignment. Journal of Molecular Biology **302** (2000) 205–217
7. Li, K.: Clustalw-mpi: Clustalw analysis using parallel and distributed computing. Bioinformatics **19** (2003) 1585–1586
8. Ebedes, J., Datta, A.: Multiple sequence alignment in parallel on a workstation cluster. Bioinformatics **20** (2004) 1193–1195
9. Mikhailov, D., Cofer, H., Gomperts, R.: Performance optimization of clustalw: parallel clustalw, ht clustal, and multiclustal. White papers, Silicon Graphics (2001)
10. Oliver, T., Schmidt, B., Nathan, D., Clemens, R., Maskell, D.: Using reconfigurable hardware to accelerate multiple sequence alignment with clustalw. Bioinformatics **21** (2005) 3431–3432
11. Saitou, N., Nei, M.: The neighbor-joining method: a new method for reconstructing phylogenetic trees. Mol. Biol. Evol. **4** (1987) 406–425
12. Krüger, J., Westermann, R.: Linear algebra operators for gpu implementation of numerical algorithms. ACM Trans. Graph **22** (2003) 908–916
13. Xu, F., Mueller, K.: Ultra-fast 3d filtered backprojection on commodity graphics hardware. In: IEEE International Symposium on Biomedical Imaging'04. (2004)
14. Horn, D., Houston, M., Hanrahan, P.: Clawhmmer: a streaming hmmer-search implementation. In: Proceedings of Supercomputing 2005. (2005)
15. Liu, W., Schmidt, B., Voss, G., Schröder, A., Müller-Wittig, W.: Bio-sequence database scanning on a gpu. In: Proceedings of 20th IEEE International Parallel & Distributed Processing Symposium (HICOMB Workshop). (2006)
16. Manocha, D.: General-purpose computations using graphics processors. Computer **38**(8) (2005) 85–88
17. Owens, J., Luebke, D., Govindaraju, N., Harris, M., Kruger, J., Lefohn, A., Purcell, T.: A survey of general-purpose computation on graphics hardware. In: Eurographics 2005. (2005) 21–51
18. Kessenich, J., Baldwin, D., Rost, R.: The opengl shading language, document revision 59. Technical report, http://www.opengl.org/documentation/oglsl.html (2005)

Load Balanced Block Lanczos Algorithm
over GF(2) for Factorization of Large Keys

Wontae Hwang and Dongseung Kim

Department of Electrical Engineering
Korea University, Seoul, Korea (Rep. of)
dkim@classic.korea.ac.kr

Abstract. Researchers use NFS (Number Field Sieve) method with Lanczos algorithm to analyze big-sized RSA keys. NFS method includes the integer factorization process and nullspace computation of huge sparse matrices. Parallel processing is indispensible since sequential computation requires weeks (even months) of CPU time with supercomputers even for 150-digit RSA keys. This paper presents details of improved block Lanczos algorithm based on previous implementation[4,10]. It includes a new load balancing scheme by partitioning the matrix such that the numbers of nonzero components in the submatrices become equal. Experimentally, a speedup up to 6 and the maximum of efficiency of 0.74 have been achieved using an 8-node cluster with Myrinet interconnection.

Keywords: parallel/cluster computing, cryptology, RSA key, load balancing, sparse matrix.

1 Introduction

Eigenvalue problems often used in mechanical structure engineering and quantum mechanical engineering are very computation intensive. Sequential computing is never suitable since it takes days even weeks of running time. Lanczos algorithm [3,7] to solve eigenproblems $Ax = \lambda x$ is widely used when A is a large, sparse, and symmetric matrix. Parallelization of the algorithm has been drawn special attention of many people. Especially, researchers in cryptology apply Lanczos algorithm to factorize long integers of RSA keys [15,16] often requiring thousands of MIPS year to compute by known algorithms. For example, a 140-digit RSA key requires 4,671,181 x 4,704,451 B matrix, which demands 585Mbytes (= 4,671,181 * 32.86 * 4bytes) if there are 32.86 nonzeros per column [1]. Thus, only a good data structure can store and keep track of nonzero elements, instead of storing the two dimensional matrix as it is. In addition, it was reported that the computation took five days of CPU time on Cray C916 with 100 CPUs and 810 Mbyte main memory [1]. Thus, parallel computing is necessary to reduce the computation time and to accommodate huge matrices in the main memory.

Y. Robert et al. (Eds.): HiPC 2006, LNCS 4297, pp. 375–386, 2006.
© Springer-Verlag Berlin Heidelberg 2006

This paper presents a parallel block Lanczos algorithm over GF(2) for the factorization by NFS(number field sieve) method [1]. The algorithm is to achieve efficient parallelization and load balanced data partitioning. Naive partitioning by allocating an equal number of rows/columns of the matrices to each processor results in uneven work load among processors, due to uneven allocation of nonzeros in the sparse matrix. By rearranging rows, we distribute even work load to all processors, and obtain improved performance. In addition, we develop a parallel program with proper group communication scheduling for minimizing communication overhead on distributed- memory parallel machine like a cluster computer.

2   Factorization of a Large Integer by NFS Algorithm

Recent RSA challenge [2,16-18] shows how difficult and complex the factoring problems are, such as a 120-digit key solved in 1993 and a 193-digit key in 2005. The analytical CPU times to solve the keys are impractically long even with today's very high performance computer. Many of the successful results are know to adopt NFS algorithm [8,9].

To factorize a number N, NFS algorithm finds (X,Y) integer pairs satisfying the following relationship:

$$X^2 \equiv 1 \ (\text{mod} \ N), \ Y^2 \equiv 1 \ (\text{mod} \ N), \text{and} \ \gcd(XY,N) = 1 \tag{1}$$

where gcd represents the greatest common divisor. The reason of finding such (X,Y) pairs comes from the fact that $\gcd(X - Y, N)$ is a factor of N [10]. In finding all pairs of X and Y, the algorithm starts to find integer pairs of (a,b) that satisfies $a \equiv b \ (\text{mod} N)$, where a and b are either squares or a square times of a smooth number. A number M is called smooth with respect to a bound δ if its prime factors are under the bound δ.

From the set S consisting of (a_i, b_i), $1 \leq i \leq k$, we need to find nonempty subsets S' such that for some integer K

$$\prod_{(a,b) \in S'} a \prod_{(a,b) \in S'} b = K^2$$

Then, the product of the a's and b's can be written in quadratic form like (1) above.

Let B be a matrix of size n_1 x n_2 where the elements are equal to the exponent *modulo 2* of the prime factors. n_1 is the number of prime factors of a and b, and n_2 is the total number of pairs found during the sieving phase.

Now, observe that finding pairs (X,Y) to satisfy (1) is equivalent to choosing a set of columns from the matrix B. Whether or not selecting a column is represented by a vector of size n_2 in which 1 is marked at the corresponding component if the column is selected, 0 otherwise. Thus, the subsets S' are obtained by finding the nullvectors x satisfying the relation $Bx \equiv 0 \ (\text{mod} 2)$.

3 Lanczos Algorithm of Finding Nullspace of $Bx = 0$

NFS finds solutions of the equation $Bx = 0$ where B and x are an $n_1 \times n_2$ matrix and an $n_2 \times 1$ vector, respectively. In the factoring application, many elements of B are zeros, thus B is a sparse matrix. Numerous methods are devised to solve it including Gaussian elimination. Conjugate Gradient(CG) and Lanczos algorithm are known suitable for such a sparse B matrix [3,5]. Since Lanczos algorithm requires a symmetric matrix, a new symmetric matrix $A = B^T B$ is generated. All the solutions of $Bx = 0$ also satisfy the relation of $Ax = 0$ in this case. Extended version of the algorithm is called block Lanczos algorithm, which finds N vectors at a time. Practically, N is chosen as the word size of a computer used. Detailed derivation and theoretical background can be found in [3,10].

The block Lanczos algorithm [4,10] has been developed for the factorization. Our implementation was based on the algorithm in [4], shown in Figure 1. Especially, for the solution of NFS, the number space is limited to $Z_2 = \{0,1\}$ of GF(2), thus, the matrices A and B consist of 0s and 1s, and numerical computation can be converted to logical operations.

Time complexity of the algorithm is introduced now. Let d be the average count of nonzeros per column in C. We assume that $n = n_1 \approx n_2$ for simplicity. Since we are dealing with matrices of big columns and rows, it is not unusual that $n_1, n_2 >> N$ where N is the word size of the computer for parallel bitwise computation to be explained later. Let m denote the number of iterations of the block Lanczos algorithm, which is approximately $m = n / N$ [10]. The overall time complexity is obtained as [10]

$$T_{Block\ Lanczos} \approx O(m(nd)) + O(m(nN))$$

Since $n >> N$ and $m = n / N$, the complexity of the algorithm is

$$T_{BlockLanczos} = O(\frac{n^2 d}{N}) + O(n^2). \qquad (2)$$

4 Parallel Block Lanczos Algorithm

Parallel block Lanczos algorithm is established by including bit-parallel operations of the matrix computation, data partitioning with proper communication, and work load balancing to achieve greatest utilization of the computing resources.

4.1 Bit-Parallel Operations by Logical Operations

If the values in the matrix computation are restricted to $\{0,1\}$ under GF(2), the matrix computation can be done by simple logical operations such as AND and XOR(exclusive-or). Computation under GF(2) of multiplication and addition/subtraction can be performed by logical operations AND(&) and XOR($\wedge$). Multiple logical values can be computed simultaneously by packing them into one word and such bitwise operations speeds up by N times with better use of the main memory assuming an N -bit vector is stored in one integer word.

The matrix of $E = CD$ can be found by computing their *inner products*. In the parallel Lanczos algorithm, *outer products* instead of inner products are computed and added to the corresponding E-components, until the complete results of the multiplication are found as shown below.

$$E = \sum_{k=1}^{m} c_k \cdot d_k$$

where c_k and d_k represent vectors composed of the k th column of C and the k th row of D, respectively. While the cost is the same as the inner product method, the computation of outer product in each stage can be done *in parallel* as summarized in Figure 2 [4].

4.2 Data Partitioning and Interprocessor Communication

Parallel algorithm shown in Figure 1 has been developed that consists of a few types of parallelization with associated group communication, depending on the matrix and vector involved as described below. After the partitioning and allocation of matrix to processors, data residing in separate processors are frequently needed. The broadcast, reduce, scatter, and so on in the algorithm are functions for the group communication.

Input: $B_{part} : (n_1 / nprocs) \times n_2 , Y : n_2 \times N$

Output: X and V_m

Initialize:

$$W_{-2}^{inv} \; {}^{N\times N} = W_{-1}^{inv} = 0 ; (V_{-2}^{n2\times N})_{part} = (V_{-1})_{part} = 0$$
$$BV_{-1\,part}^{(n1/nprocs)\times N} = 0 ; SS_{-1}^{T} \; {}_{N\times N} = I_N ; X_{part}^{(n2/nprocs)\times N} = 0$$

Broadcast(Y)

$$V_{0some} = (B_{part})^T * (B_{part} * Y) ;$$
$$V_0 = Reduce(V_{0some}); V_{0part} = Scatter(V_0)$$

Broadcast(V_0)

$$(BV_0)_{part} = B_{part} * V_0$$
$$(Cond_0^{N\times N})_{some} = (BV_{0\,part})^T * (BV_0)_{part}$$
$$(Cond_0) = Reduce((Cond_0)_{some})$$
$$i = 0$$

While $Cond_i \neq 0$ do
  if(node==0)
    $[W_i^{inv}, SS_i^T]$ =generateMatrixWandS($Cond_i, SS_{i-1}^T, N, i$)

Fig. 1. The pseudo code of parallel Block Lanczos algorithm

$$Reduce((V_{ipart})^T * V_{0part}))$$

$$Broadcast(W_i^{inv} * (V_i^T * V_0))$$

$$X_{part} = X_{part} + V_{ipart} * (W_i^{inv} * (V_i^T * V_0))$$

$$(B^T BV_i)_{some} = (B_{part})^T * BV_{ipart}$$

$$B^T BV_i = Reduce((B^T BV_i)_{some})$$

$$Broadcast(B^T BV_i)$$

$$C^{N \times N} = Reduce((BV_{ipart})^T * (B_{part} * (B^T BV_i)))$$

$$if(node==0) \{ K_i^{N \times N} = C^{N \times N} * SS_i^T + Cond_i$$

$$D_{i+1}^{N \times N} = I_N - W_i^{inv}(K_i)$$

$$E_{i+1}^{N \times N} = W_{i-1}^{inv}(Cond_i * SS_i^T)$$

$$F_{i+1}^{N \times N} = -W_{i-1}^{inv}(I_N - Cond_{i-1} * W_{i-1}^{inv})(K_{i-1})SS_i^T \}$$

$$Broadcast(SS_i^T); Broadcast(D_{i+1})$$

$$Broadcast(E_{i+1}); Broadcast(F_{i+1})$$

$$(V_{i+1})_{part} = (B^T BV_i)_{part} * SS_i^T + (V_i)_{part} * D_{i+1}$$
$$+ (V_{i+1})_{part} * E_{i+1} + (V_{i-2})_{part} * F_{i+1}$$

$$V_{i+1} = Gather((V_{i+1})_{part})$$

$$Broadcast(V_{i+1})$$

$$(BV_{i+1})_{part} = B_{part} * V_{i+1}$$

$$(Cond_i)_{some} = ((BV_{i+1})_{part})^T * (BV_{i+1})_{part}$$

$$Reduce((Cond_i)_{some})$$

$$i = i + 1$$

$$if(node==0) \ V_m = V_i$$

$$X = Gather(X_{part})$$

Return X and V_m

Fig. 1. (*Continued*)

As the first and most fundamental way of parallelization, matrix B is divided into P $(n_1/P) \times n_2$ submatrices (each is denoted by B_{part}) and allocated to P processors. $B_{part}Y$ is computed in each processor with a common (unpartitioned) matrix Y, where Y is supplied to every processor by a broadcast. Figures 3 & 4 depict the details for such matrix multiplication, where only two processors are used for simplicity. Another type of parallelization is devised for the computation of $B^T \cdot BY$. Since B_{part} is stored in each processor, the partial computation of $B^T \cdot BY$ can be done in each processor. Now matching part of BY in each processor is used and the partial product is evaluated. Then, since each processor retains an incomplete product, the complete matrix is obtained by combining them together using the collective communication *Reduce*.

Description of other types of parallelization [6] is omitted here.

Input: C and D of sizes $p \times q$ and $q \times N$, respectively Output: $E = CD$ where E is a $p \times N$ vector.

Initialize as $E = \mathbf{0}$.
for $i = 1, \cdots, p$ **do**
  for $k = 1, \cdots, q$ **do**
    if $c_{ik} = 1$ **then**
      $e_k = e_i \wedge d_k$

Fig. 2. Parallel matrix multiplication by outer product computation

4.3 Load Balancing Strategy

Since the outer product computation in $E = CD$ omits XOR operation if the corresponding component of C is zero as shown in Figure 2, the actual work load in the sparse matrix multiplication relies on the number of nonzero elements in the matrix. Thus, naive partitioning that allocates even rows/columns of B to individual processors does not guarantee even distribution of work load. In our algorithm, an equal number of rows to each node is assigned, and then rows are later interchanged among nodes in such a way that each partition contains approximately the same number of nonzeros. The swapping does not alter the nullspace of $Bx = 0$. Figure 5 illustrates the scheme where an 8X6 matrix B and a 6X2 matrix Y are employed. Only nonzero components of B are shown, moved to the left of the corresponding rows for illustration. Straightforward method (called *static partitioning*) of B shown in Figure 5a, which only divides the matrix with an equal number of contiguous rows (four rows in this case), leaves 18 and 13 nonzeros to Node1 and Node 2, respectively. However, *load balanced partitioning* shown in Figure 5b allows 15 and 16 nonzero elements, thus, the computational load of two nodes is nearly equal. In the implementation, for both simplicity and load balancing in the remaining computation, an equal number of rows to each node is to be assigned at first. Then, rows are later interchanged among them as depicted in Figure 6. The swapping does not alter the nullspace of $Bx = 0$.

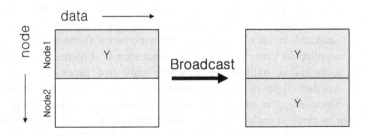

Fig. 3. Broadcast of matrix/vector

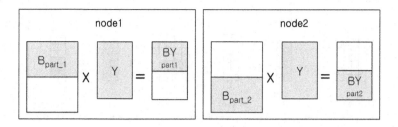

Fig. 4. Parallel computation of Type-B

We know there remains more to be done for perfect balancing. B is known at the start of the execution and never changes, but other matrices and vectors in the algorithm change while the computation is in progress; thus, it is not possible to partition with even load (nonzeros) in advance. Partitioning them equally during the iterations seems too complicated, thus, only B is equally partitioned.

4.4 Time Complexity of the Algorithm

Let α, β, and γ be the setup time, the inverse of bandwidth of the network, and a unit computation cost of one REDUCTION operation, respectively. The complexities of various group (collective) communications involving words under P processors are modeled respectively as

$$T_{scatter}(P,n) = \lceil \log(P) \rceil \alpha + \frac{P-1}{P} n\beta$$

$$T_{gather}(P,n) = \lceil \log(P) \rceil \alpha + \frac{P-1}{P} n\beta$$

$$T_{broadcast}(P,n) = \lceil \log(P) \rceil (\alpha + n\beta)$$
$$T_{reduce}(P,n) = \lceil \log(P) \rceil (\alpha + l\beta + n\gamma)$$

Overall communication cost per iteration of the algorithm in Figure 1 is summed to

$$T_{Lanczos-comm} =$$

$$2(\lceil \log(P) \rceil (\alpha + n\beta)) + 5(\lceil \log(P) \rceil (\alpha + N\beta)) +$$

$$3(\lceil \log(p) \rceil (\alpha + N\beta + N\gamma)) + (\lceil \log(P) \rceil (\alpha + n\beta + n\gamma)) +$$

$$m * (\lceil \log(P) \rceil \alpha + \frac{P-1}{P}(\frac{n}{P})\beta)$$

$$\approx O(\lceil \log(P) \rceil (\alpha + n\beta + n\gamma))$$

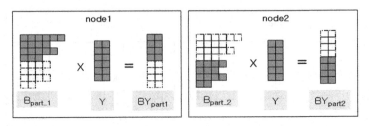

(a) Straightforward/naive partitioning

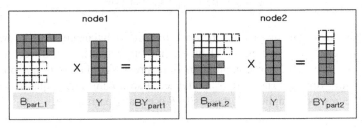

(b) Partition with even load distribution

Fig. 5. Two data partitioning strategies

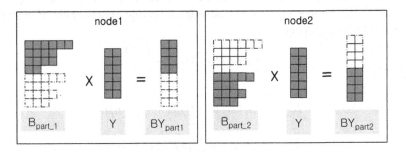

Fig. 6. New partitioning that takes into account both the number of rows and nonzero counts

Under the assumption of perfectly load balanced condition, pure computation cost can be estimated by dividing the sequential complexity by P. Thus, the execution time is found as [4,5]

$$T_{Lanczos-comp} = O(m(\frac{nd}{P})) + O(m(\frac{nN}{P})) .$$

The total running time is estimated as

$$T_{ParrelLanczos} = T_{Lanczos-comp} + T_{Lanczos-comm} =$$

$$O(m(\frac{nd}{P})) + O(m(\frac{nN}{P})) + O(\lceil \log P \rceil (\alpha + n\beta + n\gamma)m) .$$

Applying the relationship of $m = n/M$ results in the total complexity as follows:

$$T_{ParrelLanczos} = O(\frac{n^2 d}{NP}) + O(\frac{n^2}{P}) + O(\lceil \log P \rceil \frac{\alpha n + n^2 \beta + n^2 \gamma}{N}) \qquad (3)$$

Due to the term log P, the speedup may not improve linearly with the increase of parallel computing processors.

5 Experimental Results with Discussion

The experiments have been performed on two 8-node clusters; one with Gigabit ethernet interconnection, and another with Myrinet. Processors are 1.6GHz Intel Pentium 4s with 256Mbyte main memory for Gigabit cluster. Myrinet cluster consists of 1.83GHz AMD Athlon XP 2500+ CPUs with 1Gbyte memory. We install MPICH 1.2.6.13 and MPICH_gm 1.2.5.12 [13,14].

Input matrices Bs are synthetically generated by random placement of 1s with the restriction of d nonzeros per column. "t400k" is the largest matrix to fit to the main memory of a node in the cluster. Matrices have a row slightly larger than the column to avoid singularity like 100000 x 100100, 200000 x 200100, 400000 x 400100 for t100k, t200k, and t400k, respectively.

Previous results [4] were obtained on a shared memory parallel computer of SGI Origin 3800 with smaller inputs, thus, they can not be directly compared with ours. We instead compare our results of the load balanced scheme to those of straightforward partitioning algorithm.

Three sets of inputs are applied to the parallel computation with both static partitioning and load balanced allocation. The load balanced method always delivers better results with at least 6.30% up to 14.50% reduction in the execution time as shown in Table 1. The percentage values do not show any tendency as the matrix size grows. One of the reasons could be the fact that partial products computed by individual nodes during the matrix-matrix and matrix-vector computation have uneven number of 0s and 1s, thus, individual processors may not consume the same amount of computing time. In other words, only B matrix is partitioned with equal work load in the beginning, and all other matrices used in the algorithm are partitioned to have same number of rows/columns without the knowledge of nonzero count.

Table 1. Comparison of CPU times of static partition and load-balanced partition for t200k on the cluster with Myrinet (unit: hour)

No. of nodes	2	4	8
Static partition	3.09	1.84	1.10
Balanced partition	2.90	1.61	0.99
enhancement(%)	**6.30**	**14.50**	**11.03**

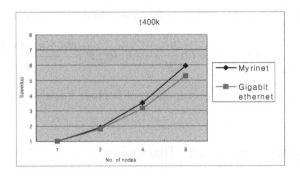

Fig. 7. Speedup for t400k matrix

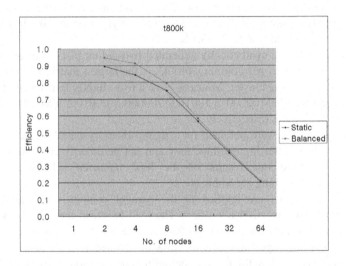

Fig. 8. Efficiency of the balanced method

The performance is surely affected by network speed, thus, faster networked Myrinet cluster always gives better results than Gigabit ethernet as observed in Figure 7. They also tell that the speedup increases as the increase of processors. However, the efficiency degrades as described below.

A big matrix of 800,000 x 800,100 (t800k, d =35) is employed on a larger cluster in KISTI supercomputer center [12] to observe the performance in wide range of parallel execution. Cluster in KISTI consists of 2.8GHz Intel Xeon DP processors with 3GB main memory interconnected with Myrinet2000. From Figure 8 we can observe 2.37% to 8.01% improvement in the load balanced method. Although the speedup grows by including more processors, the efficiency degrades, because the communication overhead grows faster than the rate of computing time reduction. Thus, if we want to achieve highest efficiency (rather than speedup), either the network should be sufficiently fast, or only a limited number of processors must be used to avoid the performance loss due to the increase in communication overhead.

6 Conclusions

For the factorization of recent RSA keys of over 150 decimal digits with Number Field Sieve, Lanczos algorithm is used. To reduce the computing time and overcome the limit of main memory capacity of a single processor, parallel computation is necessary. This paper presents an efficient method to maximize the performance of parallel Lanczos computation by allocating even work load to every processor taking into account nonzero counts of the matrix B, and by properly arranging group communication among processors to reduce the communication overhead. Research of further improvement of the algorithm on highly parallel machines is under progress.

Acknowledgement. This research was supported in part by KOSEF grant (R01-2006-000-11167-0) and in part by ETRI.

References

1. Cavallar, S., Dodson, B., Lenstra, A. K., Leyland, P. C.,Lioen, W. M., Montgomery, P. L., Murphy, B., te Riele, H., Zimmermann, P.: Factorization of RSA-140 Using the NumberField Sieve. In ASIACRYPT (1999), pp. 195-207.
2. Cavallar, S., Dodson, B., Lenstra, A. K., Lioen, W. M.,Montgomery, P. L., Murphy, B., te Riele, H., Aardal, K.,Gilchrist, J., Guillerm, G., Leyland, P. C., Marchand, J.,Morain, F., Muffett, A., Putnam, C., Zim-mermann, P.: Factorization of a 512-bit RSA Modulus. In *Theory and Application of Cryptographic Techniques* (2000), pp. 1-18.
3. Cullum, J. K., and Willoughby, R. A.: *Lanczos Algorithms for Large Symmetric Eigenvalue Computations Vol.I Theory*. Birkhauser, Boston, Basel, Stuttgart, 1985.
4. Flesch, I., Bisseling, R. H.: A New Parallel Approach to the Block Lanczos Algorithm for Finding Nullspaces over GF(2). Master's thesis, Department of Mathematics, Utrecht University, Utrecht, the Netherlands, November 2002.
5. Horn, R. A., Johnson C. R.: *Matrix Analysis*, Cambridge University Press, 1985.
6. Hwang, W.: *Improved Parallel Block Lanczos Algorithm over GF(2) by Load Balancing*, Master Thesis, Korea University, Dec. 2005.
7. Lanczos, C.: An Iteration Method for the Solution of the Eigenvalue Problem of Linear Differential and Integral Operators. *Journal of Re-search of the National Bureau of Standards 45*, 4 (Oct. 1950), pp. 255-282.
8. Lenstra, A., Lenstra, H. J., Manasse, M., Pollard, J.: The number field sieve. In *22nd Annual ACM Symposium on the Theory of Computation* (1990), pp. 564-572.
9. Montgomery, P. L.: Square roots of products of algebraic numbers. In *Mathematics of Computation 1943-1993: a Half-Century of Computational Mathematics* (May 1994), Walter Gautschi, Ed., Proceedings of Symposia in Applied Mathematics, American Mathematical Society, pp. 567-571.
10. Montgomery, P. L.: A Block Lanczos Algorithm for Finding Dependencies over GF(2). In *EUROCRYPT '95* (1995), vol. 921, pp. 106-120.
11. Kim, D., Kim, D.: Fast Broadcast by the Divide-and-Conquer Algorithm, 2004 *IEEE International Conference on Cluster Computing*, San Diego, USA, page 487-488, September. 2004.

12. KISTI Supercomputing Cener: http://www.ksc.re.kr
13. MPI: A Message-Passing Interface Standard: MPI Forum, 1995.
14. MPICH - A portable implementation of MPI.http://www.mcs.anl.gov/ mpi/mpich
15. Rivest, R., Shamir, A., Adleman, L.: A method for obtaining digital signatures and public key cryptosystems, *Comm. ACM* 21 (Feb. 1978), pp. 120-126.
16. RSA Laboratories: RSA Security Home page - http://www.rsasecurity.com Dec. 2005.
17. RSA Laboratory: Security Estimates for 512-bit RSA, RSA Data Security Inc.
18. RSA Lab.: http://www.rsasecurity.com/rsalabs/challenges/factoring/index.html, 1995.

Parallel Support Graph Preconditioners*

Meiqiu Wang and Vivek Sarin

Texas A&M University, College Station, TX 77840, USA
{mqwang, sarin}@cs.tamu.edu

Abstract. Support graph preconditioning is a relatively new technique that has gained attention in recent years. Unlike incomplete factorization-based preconditioning, this is a robust technique whose performance is not affected significantly by domain characteristics such as anisotropy and inhomogeneity. A major limitation of this technique is that it is applicable to symmetric diagonally dominant M-matrices only. In this paper, we outline an extension of the technique to symmetric positive definite matrices arising from finite element discretization of elliptic problems. An added advantage of our approach is the inherent parallelism that can be exploited to develop efficient parallel preconditioners. Our method allows trade-off between the preconditioner's parallelism and the rate of convergence of the iterative solver. In contrast, efforts to parallelize incomplete factorization-based preconditioners often result in much slower convergence. Numerical results show that our preconditioner achieves good parallel speedup on distributed memory multiprocessors such as Beowulf workstation clusters.

Keywords: Iterative methods, preconditioning, support graphs, parallel computing.

1 Introduction

Conjugate gradient (CG) method is a widely used iterative method for solving large sparse linear systems of the form $Ax = b$, where A is a symmetric positive definite (SPD) matrix. The rate of convergence of the algorithm is proportional to the square root of the spectral condition number of A, denoted by $\kappa(A)$. To increase the rate of convergence, one can use the preconditioned CG method (PCG) in which a preconditioner M is used to approximate A. PCG converges rapidly when $\kappa(M^{-1}A)$ is small (see, e.g., [10] for details). The main challenge is to construct a preconditioner such that the linear system $My = d$ can be solved efficiently in every iteration and also the condition number of the preconditioned system is reduced considerably.

Support graph preconditioners were first proposed by Vaidya [11] for symmetric, positive semidefinite, diagonally dominant M-matrices. The preconditioner M is constructed by dropping off-diagonal entries from the original matrix A.

---

* This research has been supported by NSF grants CCR 9984400, CCR 0431068, CCR 0427014, and DMS 0216275.

Y. Robert et al. (Eds.): HiPC 2006, LNCS 4297, pp. 387–398, 2006.

Since M is factored exactly, it is important to drop entries that lower the *fill-in* in the factors without compromising the preconditioner's quality. An effective preconditioner must make a trade-off between the number of entries dropped and the condition number of the resulting preconditioned system. Vaidya's approach is limited to symmetric, diagonally dominant matrices with non-positive off-diagonals. In addition, the preconditioners are not designed for parallelization.

Variants of this approach include extensions to a larger class of matrices: maximum weight bases preconditioner has been proposed for symmetric diagonally dominant matrices [2]; extensions have been proposed recently for simple problem instances from finite element method [3]. A compilation of algebraic tools and various results on support graph preconditioners can be found in [8, 9, 1]. Numerical experiments in [4, 5] show that for a model two-dimensional Poisson problem, support graph preconditioners are often superior to the incomplete Cholesky factorization preconditioners. In particular, the former require $O(n^{1.2})$ work whereas the latter require $O(n^{1.25})$ work. Furthermore, support graph preconditioners appear to be robust in the presence of varying boundary conditions and domain characteristics such as anisotropy and inhomogeneity. At present, however, numerical results for support graph preconditioners are limited to regular grids because of the limitation imposed on the matrix type.

Parallelizable variants of support graph preconditioners have also been proposed [8, 9] in which virtual nodes and edges are added to the original system to create a very sparse, parallelizable preconditioner for an expanded system. The increased system size is a major drawback of this approach. Furthermore, condition number bounds are available only for matrices derived from regular grid discretization with constant coefficients.

Our work is an ongoing effort to extend support graph preconditioners to matrices arising in the finite element method. Our approach uses a coordinate transformation at the element level to approximate the coefficient matrix by a symmetric diagonally dominant M-matrix. The quality of this approximation depends only on the mesh topology and not on the problem size. A support graph preconditioner is constructed for this approximation, and used as a preconditioner for the original coefficient matrix. We also propose a novel preconditioner which can be parallelized efficiently without compromising other desirable features such as robustness and effectiveness on anisotropic and inhomogeneous problems.

The rest of the paper is organized as follows: Sect. 2 outlines the technique to extend support graph preconditioners to the finite element method; Sect. 3 describes the proposed domain partitioned support graph (DPSG) preconditioner and its parallel implementation; Sect. 4 outlines strategies for anisotropic and inhomogeneous domains; Sect. 5 presents results of numerical experiments to highlight the characteristics of the parallel preconditioner; Sect. 6 concludes this paper.

2   Support Graph Preconditioners for Finite Element Matrices

In this section, we outline a technique to convert finite element matrices to symmetric diagonally dominant M-matrices that can be preconditioned via support graphs. For the sake of illustration, we consider the two-dimensional Poisson problem. It should be noted, however, that the technique is applicable to general three-dimensional elliptic problems.

Consider the model Poisson problem on a two-dimensional domain Ω with Dirichlet boundary conditions on the boundary Γ:

$$- \bigtriangledown \cdot \bigtriangledown u = f(\mathbf{x}) \text{ in } \Omega \ , \quad u(\mathbf{x}) = g \text{ on } \Gamma \ . \tag{1}$$

A standard Galerkin finite element discretization scheme can be used with a triangulation $\mathcal{T}$ to obtain the linear system

$$Au = b \ , \tag{2}$$

which is solved for u. A piecewise linear approximation of the solution on element e is given as

$$u_e = \phi^e \cdot \mathbf{u}^e = u_1^e \phi_1^e + u_2^e \phi_2^e + u_3^e \phi_3^e \ .$$

The global stiffness matrix A and the load vector b are assembled, respectively, from element stiffness matrices K_e and element load vectors b_e, as shown below:

$$A = \sum_{e \in \mathcal{T}} K_e \ , \quad b = \sum_{e \in \mathcal{T}} b_e \ ,$$

where K_e is a 3×3 matrix given by $K_e(i,j) = \int_e \bigtriangledown \phi_i^e \cdot \bigtriangledown \phi_j^e d\mathbf{x}$ and $b_e(i) = \int_e f \phi_i^e d\mathbf{x}$.

It is easy to show that A is SPD but not diagonally dominant, which prevents construction of a support graph preconditioner. Next, we describe a transformation that converts A into a symmetric diagonally dominant M-matrix. Consider the element gradient matrix B_e:

$$B_e = \begin{bmatrix} \phi_{1,x}^e & \phi_{2,x}^e & \phi_{3,x}^e \\ \phi_{1,y}^e & \phi_{2,y}^e & \phi_{3,y}^e \end{bmatrix} \ , \tag{3}$$

where $\phi_{j,x}^e$ and $\phi_{j,y}^e$ denote the partial derivatives along x and y, respectively. The partial derivative along two adjacent edges of an element can be expressed as:

$$\begin{bmatrix} \phi_{j,\tau}^e \\ \phi_{j,\eta}^e \end{bmatrix} = \begin{bmatrix} x_\tau & y_\tau \\ x_\eta & y_\eta \end{bmatrix} \begin{bmatrix} \phi_{j,x}^e \\ \phi_{j,y}^e \end{bmatrix} = G_e \begin{bmatrix} \phi_{j,x}^e \\ \phi_{j,y}^e \end{bmatrix} \ ,$$

where τ and η denote the edges. The element gradient matrix along (τ, η), denoted by B_e', is given below:

$$B_e' = G_e B_e = \begin{bmatrix} \phi_{1,\tau}^e & \phi_{2,\tau}^e & \phi_{3,\tau}^e \\ \phi_{1,\eta}^e & \phi_{2,\eta}^e & \phi_{3,\eta}^e \end{bmatrix} = \begin{bmatrix} l_{12} & 0 \\ 0 & l_{13} \end{bmatrix}^{-1} \begin{bmatrix} -1 & 1 & 0 \\ -1 & 0 & 1 \end{bmatrix} \ ,$$

where l_{ij} denotes the length of the edge (i, j). Thus,

$$A = \sum_{e \in T} B_e^T \Delta_e B_e = \sum_{e \in T} B_e'^T G_e^{-T} \Delta_e G_e^{-1} B_e' ,$$

where Δ_e is the area of element e. Furthermore, A can be approximated by the following matrix

$$A' = \sum_{e \in T} B_e'^T \Delta_e B_e' ,$$

which is a symmetric diagonally dominant M-matrix.

It can be shown that the condition number of $A'^{-1}A$ is bounded:

$$\kappa(A'^{-1}A) \leq \max_{e \in T} \kappa(G_e^T G_e) = \max_{e \in T} \frac{1 + |\cos(\theta_e)|}{1 - |\cos(\theta_e)|} ,$$

where θ_e is the angle between the edges along τ and η in element e. This indicates that the quality of the approximation can be improved by selecting pairs of edges for which $\cos(\theta_e)$ is as small as possible. In good quality meshes, usually $\pi/8 \leq \theta_e \leq 7\pi/8$ (see, e.g., [12]), which implies that $\kappa(G_e^T G_e) \leq 26$. In practice, it is observed that $\pi/4 \leq \theta_e \leq 3\pi/4$ and $\kappa(G_e^T G_e) \leq 6$. Our experiments shown in Table 1 have confirmed that the quality of the preconditioner A' is independent of the mesh width.

Table 1. The number of iterations of PCG using preconditioner A' for the Poisson problem on different meshes (n denotes the number of unknowns, *Iter* denotes the number of iterations)

n	229	874	3381	13440	53454	212854
Iter	9	9	10	10	10	11

Support graph preconditioners can be constructed for A' and used as preconditioners for A. Now we introduce some concepts for support graph. Let A be an $n \times n$ symmetric M-matrix. The graph of A is the weighted undirected graph $G_A = (V_A, E_A)$ defined on a vertex set $V_A = \{1, 2, \ldots, n\}$ with edges $E_A = \{(i, j) : i \neq j, A_{ij} \neq 0\}$ and edge weights $w_{ij} = |A_{ij}|$. Suppose $G_M = (V_A, E_M)$ is a graph defined on V_A whose edges are a subset of E_A. The *support path* of an edge $e \in G_A$ is a path $p \in G_M$ whose end nodes are the end nodes of e. G_M is a *support graph* of G_A if there exists a support path for every edge $e \in G_A$. For a support path p, the *support path weight* w_p is defined as the weight of the edge e which is supported by p. For an edge $e \in G_A$, *edge dilation* δ_e is the number of edges in its support path. The *dilation* δ of G_A is the maximum edge dilation over all edges in G_A. For an edge $e \in G_M$, *edge congestion* c_e is the number of support paths through e. In weighted graphs, c_e is the sum of weights of the support paths through e divided by w_e. The *congestion* c of G_M is the maximum congestion over all edges in G_M.

Lemma 1 (Congestion-Dilation Lemma). *Given two symmetric M-matrices A and M such that G_M is a support graph of G_A,*

$$\kappa(M^{-1}A) \leq \delta \cdot c \ . \tag{4}$$

For finite element matrices, there are two sources of approximations – the first approximation involves transformation of A to a symmetric M-matrix A' and the second approximation involves preconditioning A' by the support graph matrix M. As a result,

$$\kappa(M^{-1}A) \leq \kappa(M^{-1}A') \cdot \kappa(A'^{-1}A) \leq (\delta \cdot c) \max_{e \in \mathcal{T}} \left[\frac{1 + |\cos(\theta_e)|}{1 - |\cos(\theta_e)|} \right] \ .$$

3   DPSG: Domain Partitioned Support Graphs

The graph of the matrix arising from finite element method is the same as the mesh used to discretize the domain. The edges in the support graph must be a subset of the mesh edges. To construct an effective and parallelizable preconditioner, we divide the domain into subdomains of nearly equal size. Each triangular element belongs to a subdomain. In contrast, nodes and edges may be in the *interior* of a subdomain or at the *interface* between two adjacent subdomains. All the interface edges are included in the support graph. A maximum weight spanning forest is generated within each subdomain to identify the interior edges that should be included in the support graph (see [6] for efficient spanning tree algorithms). The root nodes of these spanning components must be interface nodes or Dirichlet boundary nodes. Figure 1 illustrates the scheme for a two-dimensional square domain. The requirement to include all interface edges may force selection of these edges during transformation of the corresponding element gradient matrices; however, this does not appear to degrade the quality of the preconditioner.

To estimate the quality of the preconditioner for the two-dimensional unit square Poisson problem, we assume that the domain is partitioned into $k \times k$ subdomains, each containing approximately n/k^2 nodes distributed uniformly within the subdomain. Note that a support path for any interior edge in a subdomain can be constructed using the spanning forest edges and the boundary edges of that subdomain itself. It is easy to see that dilation is $O(\sqrt{n}/k)$. Taking planarity into consideration, it can be shown that congestion is $O(\sqrt{n}/k)$ as well. Using Lemma 1 we obtain the bound $\kappa(M^{-1}A') = O(n/k^2)$. Table 2 indicates that the asymptotic number of iterations required by PCG remains unchanged when the mesh is refined to quadruple the number of nodes as long as the number of subdomains is quadrupled as well.

Reordering of nodes is necessary to reduce the *fill-in* during sparse Cholesky factorization of the preconditioner M. A minimum degree ordering guarantees zero *fill-in* when eliminating interior nodes since they are part of spanning trees. The interface nodes common to exactly two subdomains can be eliminated next, with *fill-in* proportional to the number of interface nodes. The remaining $O(k^2)$

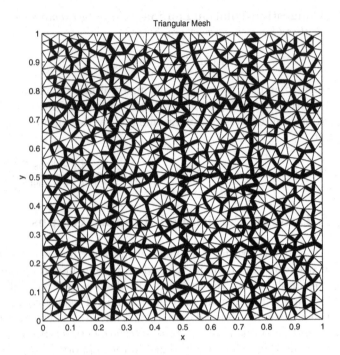

Fig. 1. A domain partitioned support graph constructed using 16 subdomains (support graphedges are shown in bold)

Table 2. Number of iterations of PCG for DPSG (n: number of unknowns, $k \times k$: number of subdomains, $Iter$: number of iterations)

n	$k \times k$	$Iter$
874	4×4	68
3381	8×8	103
13440	16×16	117
53454	32×32	123
212854	64×64	128

interface nodes form a planar graph with a topology similar to the original mesh. Algorithms such as nested dissection and reverse CutHill-McKee [7] can be used to order these nodes to minimize the factorization cost. It can be verified that the cost to eliminate interior nodes is $O(n)$ and the cost to eliminate interface nodes between two subdomains is $O(\sqrt{n}k)$. The remaining $O(k^2)$ interface nodes require $O(k^3)$ operations when nested dissection is used. Thus, the overall cost of factorization is $O(n+\sqrt{n}k+k^3)$. The additional memory required is $O(\sqrt{n}k+k^2 \log k)$.

The cost of each iteration of the PCG algorithm is $O(\text{nnz}(A) + \text{nnz}(R))$ where R is the Cholesky factor of M. Since the number of iterations is proportional to $\sqrt{\kappa(M^{-1}A)}$, the cost of the iterative solver is $O((n + \sqrt{n}k + k^2 \log k)\sqrt{n}/k)$, which simplifies to $O(n^{1.5}/k + n + \sqrt{n}k \log k)$. On a uniprocessor, k should be chosen to minimize the overal computational time and memory. For example, if we take $k = n^{3/8}$, then both the iterative cost and factorization cost are $O(n^{9/8})$.

3.1 Parallel Formulation

Parallelism can be exploited in several ways due to the partitioning imposed by DPSG preconditioner. For the two-dimensional Poisson problem, we assume a virtual two-dimensional grid of $\sqrt{P} \times \sqrt{P}$ processors where each processor *owns* a contiguous region of $k/\sqrt{P} \times k/\sqrt{P}$ subdomains. Processors can compute the maximum weight spanning forest for their subdomains concurrently, and complete the first phase of factorization involving elimination of interior nodes independently. Next, they cooperate with neighboring processors that own adjacent regions to eliminate interface nodes that are shared by exactly two processors. Elimination of the remaining nodes that are shared by more than two processors can also be parallelized; however, the gains are visible only when the number of processors is large.

There are three types of arithmetic operations in the CG algorithm: matrix-vector multiplication, vector inner products, and vector updates. These operations can be parallelized at the subdomain level with minimal communication overhead. The preconditioning step in PCG involves triangular solves with the factors of M which can be a bottleneck for parallelization. When using the DPSG preconditioner, we exploit the concurrency available in the factorization phase to parallelize the triangular solves. The lower triangular solve proceeds in three steps: the first step is a fully parallelizable step involving interior nodes only; the second step involves interface nodes that are shared by two processors only; the third step requires a global triangular solve involving the remaining nodes and is the least parallelizable. The steps are reversed for the upper triangular solve. Since the amount of work in the first two steps is considerably larger for large problems, reasonable speedup can be expected in the preconditioning step.

4   Preconditioning Variable Coefficients Problems

The preconditioning approach described earlier can be applied to a wider class of elliptic problems. Consider the d-dimensional problem

$$- \nabla \cdot (a(\mathbf{x}) \nabla u) = f(\mathbf{x}) \text{ in } \Omega , \quad u(\mathbf{x}) = g \text{ on } \Gamma . \tag{5}$$

where $a(\mathbf{x})$ is a $d \times d$ symmetric positive definite matrix that represents domain characteristics at the point $\mathbf{x}$. Appropriate choices of $a(\mathbf{x})$ can be used to model a wide variety of domain properties including anisotropy and inhomogeneity. The coefficient matrix arising from a piecewise linear approximation on a triangular mesh is given as

$$A = \sum_{e \in \mathcal{T}} B_e^T \Delta_e D_e B_e \; ,$$

where D_e is obtained by evaluating $a(\mathbf{x})$ inside element e. Using the transformation described in Section 2, we obtain an alternate expression for A

$$A = \sum_{e \in \mathcal{T}} B_e'^T G_e^{-T} \Delta_e D_e G_e^{-1} B_e' \; ,$$

which is used to derive the following approximation of A

$$A' = \sum_{e \in \mathcal{T}} B_e'^T \Delta_e D_e' B_e' \; ,$$

where D_e' is a positive diagonal matrix that is chosen to minimize the approximation error. It is easy to see that A' is a symmetric M-matrix which can be approximated by a support graph preconditioner. The quality of the approximation is determined by the following bound

$$\kappa(A'^{-1} A) \leq \max_{e \in \mathcal{T}} \kappa(D'^{1/2} G_e D_e^{-1} G_e^T D'^{1/2}) \; ,$$

which should be minimized by appropriate choice of D_e' in each element.

Inhomogeneous Domains. For problems involving different isotropic materials, the discontinuity is restricted to the interface between the materials. One can use a mesh that conforms to these interfaces such that elements do not straddle different materials. In this case, $D_e = d_e I$, where d_e is a scalar and I is the identity matrix. By choosing $D_e' = D_e$, we guarantee that the quality of the preconditioner is no different from that of an isotropic problem.

Anisotropic Domains. In anisotropic domains, $a(\mathbf{x})$ varies continuously over the domain. The eigenvectors and eigenvalues of $a(\mathbf{x})$ give orthogonal directions and corresponding magnitude of anisotropy, respectively, at the point $\mathbf{x}$. For the element transformation, one should choose the pair of edges closely aligned with these eigenvectors. Consider the case when every element e has a pair of edges aligned with the eigenvectors of D_e. In this case, the transformation matrix G_e^T is identical to the eigenvector matrix of D_e, leading to the following simplification

$$A = \sum_{e \in \mathcal{T}} B_e'^T G_e^{-T} \Delta_e (G_e^T \Lambda_e G_e) G_e^{-1} B_e' = \sum_{e \in \mathcal{T}} B_e'^T \Delta_e \Lambda_e B_e' \; ,$$

which indicates that A is a symmetric M-matrix. When the edges in element e do not coincide with the eigenvectors, one must choose a pair of edges that is most closely aligned with the eigenvectors. In this case, a good choice of D_e' is the diagonal of $G_e^{-T} D_e G_e^{-1}$.

5   Numerical Experiments

Numerical results show that our scheme performs well on both shared memory multiprocessors and distributed memory work stations. Due to space limitation, we report the results of numerical experiments only on a distributed-memory Beowulf cluster (1.4GHz 64-bit AMD Opterons, running SuSE-Linux, connected via Gigabit Ethernet). The code was implemented in C++ and parallelized using the MPI library. Two types of parallel preconditioners are used with PCG: diagonal preconditioner (DIAG) and DPSG. PCG iterations were terminated when the relative residual norm was reduced below 10^{-6}.

First we give the experimental results for the unit square Poisson Problem with Dirichlet boundary conditions. Table 3 shows that the parallel implementation of DPSG is able to exploit inherent coarse grained parallelism to achieve near linear speedup. The following notation has been used: P: number of processors, $Iter$: number of PCG iterations, T_{part}: partitioning time, T_{asm}: matrix assembly time, T_{mst}: time to compute maximum weight spanning forest, T_{fac}: factorization time, T_{solve}: time spent in PCG, T_{total}: total time, and E: efficiency. Time is reported in seconds. Although different components of the code parallelize to different degree, the overall speedup is very high.

Table 3. Performance of DPSG on Beowulf Cluster ($n = 424867$, $k = 64$)

P	$Iter$	T_{part}	T_{asm}	T_{mst}	T_{fac}	T_{solve}	T_{total}	E
1	194	4.6	8.6	7.3	8.1	75.8	104.3	1.00
2	193	2.2	4.2	3.5	3.1	35.4	48.3	1.08
4	193	1.1	2.0	1.6	1.0	16.2	21.9	1.19
8	194	0.5	1.0	0.8	0.8	7.7	10.8	1.21
16	193	0.2	0.5	0.4	1.1	4.7	6.9	0.95

For the inhomogeneous and anisotropic problems, we consider the problem domains as shown in Fig. 2. The inhomogeneous domain consists of two equal parts with $a(\mathbf{x})$ set to unity in the left part, and a constant μ in the right part, which is varied between 10^{-5} and 10^{-1} to obtain various instances of the problem. For the anisotropic problem, the anisotropic axes at a point $\mathbf{x}$ are along the tangent and normal to the circle passing through $\mathbf{x}$ centered at the origin. The anisotropy coefficient along the tangent is set to 1. The coefficient along the normal direction, ν, again is varied from 10^{-5} to 10^{-1}.

In Table 4, we increase the problem size as we decrease μ and ν, respectively. The data show that DPSG is a very effective preconditioner for both problems. Note that as ν decreases, the anisotropic problem becomes a one-dimensional problem for which the maximum weight spanning forests constructed by DPSG turn out to be very effective approximations.

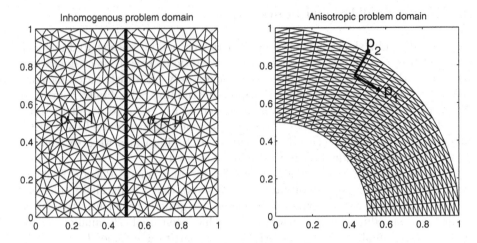

Fig. 2. Inhomogeneous and anisotropic problem domains

Table 4. Number of iterations of PCG for problems with varying coefficients

Inhomogeneous Problem				Anisotropic Problem			
μ	n	DIAG	DPSG	ν	n	DIAG	DPSG
10^{-1}	867	76	85	10^{-1}	561	58	50
10^{-2}	3426	152	128	10^{-2}	2145	145	50
10^{-3}	13403	301	157	10^{-3}	8385	297	39
10^{-4}	53335	548	183	10^{-4}	33153	574	30
10^{-5}	212891	1081	207	10^{-5}	131841	1089	20

Table 5. Number of iterations of PCG for cases with fixed parameters

Inhomogeneous Problem: $\mu = 10^{-3}$			Anisotropic Problem: $\nu = 10^{-3}$		
n	DIAG	DPSG	n	DIAG	DPSG
867	74	90	561	67	12
3426	150	131	2145	143	21
13403	301	157	8385	297	39
53335	599	176	33153	610	76
212891	1217	192	131841	1249	150

When the parameters μ and ν are fixed and we increase the problem size, Table 5 shows that the number of iterations required by DPSG grows slowly with the problem size compared to DIAG. The parallel performance results are shown in Table 6. For both cases DPSG achieves very high parallel efficiency, and proves to be a very effective parallel preconditioner. Low efficiency on large processors can be traced to two sources: one is the communication overhead, e.g. collective call such as inner product, becomes high; the other is the ratio of the communication and the computation becomes large. By increasing the problem size, the latter effect can be decreased and the efficiency can be improved.

Table 6. Performance comparison for problems with varying coefficients

Inhomogeneous Problem ($n = 425050$, $\mu = 10^{-3}$)

	DPSG (64×64 subdomains)							DIAG		
P	$Iter$	T_{part}	T_{asm}	T_{mst}	T_{fac}	T_{solve}	T_{total}	E	$Iter$	T_{total}	E
1	291	4.7	8.6	7.1	8.5	109.7	138.5	1.00	1537	218.2	1.00
2	289	2.3	4.2	3.4	3.0	52.3	65.3	1.06	1537	121.4	0.90
4	306	1.1	2.0	1.7	1.1	27.1	33.1	1.05	1537	63.0	0.87
8	294	0.5	1.0	0.8	0.7	13.8	16.7	1.03	1537	37.8	0.72
16	300	0.3	0.5	0.4	1.0	8.4	10.6	0.82	1536	24.2	0.56

Anisotropic Problem ($n = 118503$, $\nu = 10^{-3}$)

	DPSG (16×2 subdomains)							DIAG		
P	$Iter$	T_{part}	T_{asm}	T_{mst}	T_{fac}	T_{solve}	T_{total}	E	$Iter$	T_{total}	E
1	124	0.7	2.1	1.3	0.1	7.6	11.8	1.00	1123	36.8	1.00
2	124	0.3	1.0	0.6	0.1	4.0	6.1	0.98	1123	20.1	0.92
4	124	0.2	0.5	0.3	0.1	2.2	3.2	0.92	1123	11.5	0.80
8	124	0.1	0.3	0.1	0.1	1.3	1.8	0.84	1123	7.1	0.65
16	123	0.0	0.1	0.1	0.2	1.0	1.3	0.57	1123	5.1	0.46

6   Conclusions

In this paper, we have outlined a novel approach to transform SPD matrices arising from the finite element discretization of a class of elliptic problems into symmetric diagonally dominant M-matrices that can be approximated by support graph preconditioners. We have also presented an approach based on domain partitioning to construct support graph preconditioners that are robust and effective in addition to being parallelizable. Preliminary numerical experiments conducted on up to 16 processors of AMD-Opteron based Beowulf cluster indicate that our preconditioner can achieve high parallel efficiency without sacrificing robustness and effectiveness. The preconditioner should retain these advantages on larger multiprocessors as well.

References

1. E. G. Boman and B. Hendrickson. Support theory for preconditioning. *SIAM J. Matrix Anal. Appl.*, 25, no.3:694–717, 2004.
2. Erik G. Boman, Doron Chen, Bruce Hendrickson, and Sivan Toledo. Maximum-weight-basis preconditioners. *Lin. Alg. Appl.*, 11:695–721, 2004.
3. Erik G. Boman, Bruce Hendrickson, and Stephen Vavasis. Solving elliptic finite element systems in near-linear time with support preconditioners, Submitted 2004.
4. D. Chen and S. Toledo. Vaidya's preconditioners: Implementation and experimental study. *Electronic Transactions on Numerical Analysis*, 16:30–49, 2003.
5. Doron Chen. Analysis, Implemetation and Evaluation of Vaidya's Preconditioners. Master's thesis, School of Computer Science, Tel-Aviv University, Israel, Feb 2001.
6. T. Cormen, C. Leiserson, and R. Rivest. *Introduction to Algorihtms*. McGraw-Hill, second edition, 2001.
7. Alan George and Joseph W-H. Liu. *Computer solution of large sparse positive definite systems*. Englewood Cliffs, N.J. : Prentice-Hall, 1981.
8. K. D. Gremban, G. L. Miller, and M. Zagha. Performance evaluation of a parallel preconditioner. In *9th International Parallel Processing Symposium*, pages 65–69, Santa Barbara, Apr 1995. IEEE.
9. Keith D. Gremban. *Combinatorial Preconditioners for Sparse, Symmetric, Diagonally Dominant Linear Systems*. PhD thesis, School of Computer Science, Carnegie Mellon University, Oct 1996.
10. Yousef Saad. *Iterative Methods for Sparse Linear Systems*. SIAM, second edition, 2003.
11. P. Vaidya. Solving linear equations with symmetric diagonally dominant matrices by constructing good preconditioners. Unpublished manuscript. A talk based on the manuscript was presented at the IMA Workshop on Graph Theory and Sparse Matrix Computation, Oct 1991.
12. Xunlei Wu, Michael S. Downes, Tolga Goktekin, and Frank Tendick. Adaptive nonlinear finite elements for deformable body simulation using dynamic progressive meshes. In A. Chalmers and T.-M. Rhyne, editors, *EG 2001 Proceedings*, volume 20(3), pages 349–358. Blackwell Publishing, 2001.

Group Based Routing in
Disconnected Ad Hoc Networks

Markose Thomas[1], Arobinda Gupta[2], and Srinivasan Keshav[3]

[1] Google, Inc., Bangalore, India
[2] Dept. of Computer Science & Engineering
Indian Institute of Technology, Kharagpur
[3] School of Computer Science
University of Waterloo

Abstract. In this paper, we propose a routing protocol for disconnected ad hoc networks where most nodes tend to move about in groups. To the best of our knowledge, no routing protocol for disconnected ad hoc networks has been designed earlier keeping in mind possible group patterns formed by the movement of nodes. Our protocol works by identifying groups using an efficient distributed group membership protocol, and then routing at the group level, rather than at the node level. The protocol is designed so that existing concepts of routing in disconnected ad hoc networks can be extended to work at the group level. Initial simulations across a broad spectrum of parameters suggest that our protocol performs better in terms of delivery ratio and latency over traditional approaches like AODV [1], and also over disconnected routing approaches like the 2-Hop routing protocol [2].

1 Introduction

The existence of an end-to-end path is not guaranteed in many kinds of ad hoc networks because of various reasons, such as nodes switching off their radios or reducing their transmission range to conserve energy, node mobility, and application specific deployment. In such situations, traditional routing protocols for ad hoc networks such as [1, 3, 4] will fail in sending packets to destination nodes to which no path exists. An approach to solve this problem is to exploit the buffering capacities and mobility of the nodes participating in the network. If the nodes have sufficient mobility, then instead of waiting for a path to the destination, messages can be forwarded to intermediate nodes, which in turn would buffer these packets for some period of time and then forward them to other nodes. This process can be continued until some intermediate node eventually comes in contact with the destination node and delivers the message to it. Since such routing intrinsically relies on waiting for intermittently available paths, there are high latencies involved in message delivery. Applications which are able to tolerate such levels of latencies are referred to as *delay tolerant*.

Many of the delay tolerant routing protocols for disconnected networks do not assume any specific movement and location patterns of the nodes, and hence are unable to exploit any opportunity that these patterns may present. Several routing protocols have been proposed for disconnected ad hoc networks. The epidemic protocol [5] blindly

Y. Robert et al. (Eds.): HiPC 2006, LNCS 4297, pp. 399–410, 2006.
© Springer-Verlag Berlin Heidelberg 2006

floods each message to as many nodes as possible until the message reaches its destination. It is therefore very resource hungry with respect to energy consumption and buffer capacity. The 2-Hop protocol [2] showed that mobility can be used to keep throughput independent of the size of the network. However, it provides poor performance in terms of delivery ratios within practical time limits. Sushant Jain et. al. [6] formulate the general problem of routing in delay tolerant networks, and consider different levels of information availability in choosing a protocol. There has recently been research into protocols which use aggregated past information to predict future behavior ([7, 8, 9]), and thereby hope to make better routing decisions. The message ferrying approaches in [10] consider situations where dedicated *ferry nodes* are available to move around fixed routes, collect and relay packets.

An interesting class of applications within the delay tolerant disconnected ad hoc networking framework has to do with those in which nodes tend to move about in groups. This clustering of nodes leads to the formation of multiple groups of nodes (see [11, 12, 13]). Nodes are free to move about in their own groups, and also to occasionally relocate to another group. The nodes within each group will usually form a connected subnetwork, but the groups themselves would usually be in and out of communication range of each other. Further, the groups also have the ability to move about in the network, often mixing and merging with the other groups in the network and splitting into smaller groups. Some applications which work in such group environments include sensor networks deployed in wildlife [14], vehicular networks [15], and relief and military networks. None of the traditional routing protocols for disconnected environments mentioned earlier exploit the underlying group structure present in such applications.

By recognizing the presence of formation of such groups in the network, it becomes possible to combine the best ideas of both traditional ad hoc routing protocols as well as the ideas of routing in disconnected environments. For example, if members of a group are connected and if each node maintains routes to other members, it becomes possible to use traditional ad hoc routing to deliver messages and exchange routing information with nodes belonging to the same group quickly. Such a scheme would not be possible in any of the routing protocols proposed for disconnected networks. Moreover, if we treat each group as an entity in itself, much like an individual node in a normal disconnected network, then it becomes possible to apply concepts of routing for disconnected ad hoc networks to transfer messages across groups.

Our main contribution in this paper is a routing protocol for the above type of intermittently connected ad hoc networks, which is able to exploit the underlying group structure formed by node locations to provide better delivery ratios at lower latencies.

The rest of this paper is organized as follows. Section 2 gives details of the proposed routing protocol. Section 3 discusses some implementation details. In Section 4 we demonstrate, via initial simulations across many parameters, that the proposed group based protocol works better than existing routing protocols in many real life scenarios.

2   Protocol Overview

We define a group to be a set of connected nodes which maintain their connectivity for a sufficiently large period of time. The proposed protocol attempts to recognize and exploit such group structures found in many disconnected ad hoc networks, in order to provide better delivery ratios and lower latencies. The main design goals of the protocol can be classified as follows:

1. Recognize groups and maintain group identities in a decentralized and efficient manner.
2. Leverage principles of existing ad hoc routing for intra-group communication.
3. Adapt concepts of routing in disconnected environments for inter-group routing.
4. Reduce possible complexities in inter-group communication by using group leaders.

The first part of our design goal is achieved by a distributed group membership protocol. Our focus is to develop an efficient protocol, and hence we do not place the requirement that all group members have consistent views of each other at all times. However, the algorithm is expected to converge to a consistent view for all group members given sufficient amount of time. In Section 2.1 we discuss the first two of our design goals. The third and fourth design goals are described in Section 2.2.

2.1 Group Membership

Let *id* denote a globally unique and comparable identifier of a node. Nodes belonging to the same group are tagged with a common identifier called the *gid*. Groups are created and maintained so that the *gid* of a group is equal to the value of the lowest *id* of any member in the group. Initially, each node belongs to its own group. As time progresses, nodes that remain together perform merge operations and form larger groups.

Within each group, a proactive routing algorithm is run. This algorithm not only maintains up to date paths to other group members, but it also helps in timely detection of unannounced disconnections. In this paper, we use the DSDV [3] protocol, although any other proactive routing protocol can be used. Each update message (hereafter referred to as an *Update* packet) that is sent by DSDV now contains one additional field - *gid*. It's value is set to the *gid* of the node sending the *Update* packet. This helps nodes in discarding *Update* packets sent by members of other groups. Specifically, a node accepts a DSDV *Update* packet from its neighbor only if either

1. the *gid* in the *Update* packet has the same value as the node's own *gid*, or
2. the neighbor is already present in the routing table of the receiving node, and the *gid* in the *Update* packet has a lower value than the node's own *gid*.

If the first condition is satisfied, then the node has received an *Update* packet from a neighbor belonging to its group. The second condition signifies that the node's group is currently taking part in a merge operation. We explain the reason for these conditions in the sections below. Since nodes maintain paths only to other group members, each node can easily find out who the other group members are by looking at its routing table.

As time progresses, nodes of other groups may come within the communication range of one or more members of a group, or some existing members may move out of range from the group. The protocol should be able to adapt to these changing scenarios quickly. Our protocol handles the first case by an explicit merge operation. It handles the second case by implicitly removing the disconnected members from the original group, and by an explicit split operation on the nodes that have left the original group. The following two subsections describe how the proposed protocol handles these two scenarios (joining and leaving).

Group Merge. A group merge is triggered explicitly by a node if it detects that one or more nodes belonging to another group have come within its communication range for a sufficiently long period of time, called the minimum group merge wait time T_M. This minimum wait time ensures that groups are not merged accidentally when they come in contact only for a very short time in the course of their movements. Each node periodically sends beacon packets advertising its *gid*. A node receiving a beacon packet from another group registers a hit corresponding to the group from which the beacon was received. If a node stops receiving beacons from the other group for more than some time duration T_G, it resets the hit counter for that group. Once a node receives beacons from another group for a time period greater than T_M, a *MergeRequest* packet is sent to the node which sent the last beacon. The merge request contains the *id*, *gid*, and the DSDV routing table of the node. On receiving the merge request, a node takes a local decision to accept or reject the request. If it accepts the request, then it updates its *gid* to the lower value of both the *gid*'s, and also adds the nodes in the routing table of the other group to its own routing table. It then propagates this information in the form of a new DSDV *Update* packet to the other members of its group. It is easy to verify that the other members of the group will accept this *Update* packet because of the conditions stated at the start of Section 2.1. The other members of the group, on receiving this *Update* packet from a group member, lower their *gid* if the new *gid* has a smaller value, and update their routing tables. They also propagate a new DSDV *Update* packet immediately since new nodes have been added to the group (as a result of the merge). This process continues till all the nodes of the group have added the new nodes to their routing tables. The node receiving the merge request also sends a *MergeReply* packet to the sender of the request. The reply packet contains the *id*, *gid*, *status* (accept or reject), and the routing table of the node. Similar actions as above happen on receipt of the reply packet to update the *gid* if needed and the routing table of nodes in the other group. At the end of the merge operation, a new group with a common *gid* is formed.

It is possible that multiple nodes of the same group may decide to merge with different nodes of another group at about the same time. In such a case, pairs of nodes (one node from each group) merge initially. The *gid* of each node in the pair will finally be the lower of the *gid* of the two groups. Each node in a pair will also initiate a fresh round of DSDV updates within its original group. Thus, the number of fresh DSDV updates that will take place due to a merge will be limited by the number of pairs of nodes that started the merge operation. Since each such DSDV update will have the same *gid* across all pairs, the routing tables of all nodes will eventually be updated consistently.

Group Split. There are two main things to be done when members leave a group. Nodes still belonging to the group should realize quickly that such members have left the group. Also, since the *gid* of a group should reflect the value of the lowest *id* member in the group, the *gid* of the group which does not contain the lowest *id* node any more should be changed. However, nodes disconnect from a group in an unannounced fashion, with no prior information as to when they will leave the group. When a node leaves a group, the protocol depends on the underlying DSDV routing protocol to inform other members of its departure. If a node leaves a group, then eventually its neighbor would detect that it is no longer reachable and will remove it from its routing table. It will then send a DSDV *Update* packet to inform the other group members of this. Thus, within a short time, the node which moved out from the group will be dropped from the routing tables of all remaining members of the group. Also, the part of the group which splits and does not contain the lowest *id* node, will eventually be classified as belonging to a separate group using the following mechanism. Among the nodes which belong to the above split part, the node having the lowest *id* will eventually discover that its *id* is not equal to its *gid* and it does not have a path to any member having *id* lower than it. This node then forms a new group with itself as the leader by sending an announcement to all other nodes to which it has a path in its routing table to change their *gid* to its *id*. On receiving the announcement, all nodes which no longer have paths to their current *gid* (because they split from the old group) and which has the node sending the announcement as the lowest *id* member in its routing table, will change their *gid* to the *id* in the announcement.

It may happen that two groups which have just split still linger close to each other temporarily. This may cause nodes of the two groups to exchange *Update* packets in a way that leads to corruption of routing tables. In order to prevent this, we impose the condition that all nodes belonging to the new group (i.e. the one having the larger *gid*), clear their existing routing tables and rebuild them from scratch. The two conditions for accepting an *Update* packet, stated at the beginning of Section 2.1, will ensure that the new group will no longer accept *Update* packets from members of the old group (because the clearing of their routing tables will remove the old members from their tables, and their *gid* is now different). The conditions will also ensure that members of the old group will also no longer accept *Update* packets from the new group, because the second condition prohibits them from accepting packets coming from nodes in their routing tables with a higher *gid* than their current *gid*. Note that this may lead to an occasional loss of *Update* packets (for example, when a node lowers its *gid* due to a merge operation, and then it immediately receives an *Update* packet from a neighbor which is yet to lower its *gid*), but this loss of information is not expected to be serious, as it will be corrected in the next round of *Update* packets sent.

2.2 Routing

Once the groups are identified, any standard routing algorithm for disconnected networks can be used to route between groups. Here, on the assumption that groups often move in predictable ways, we show how to adapt the PRoPHET routing protocol [7] to work in our group scenario. Like PRoPHET, we maintain probability measures of the *delivery predictability*; but instead of them being measures of successful delivery to nodes, they indicate the chances of successful delivery to other groups. These measures are stored at each node in a vector called *probTable*, which contains an entry for each group. In order for this scheme to work, it is also necessary to have up to date information regarding the group to which a destination node belongs. The protocol gathers this information in a proactive manner and stores it in a vector called the *nodeInfo* vector, which contains an entry for each node. For the purpose of inter-group routing, the node with the lowest *id* is chosen as the group leader. The main responsibility of the leader is to consistently update and decay the *probTable* vector of a group, and to disseminate the *probTable* and *nodeInfo* vector to the rest of the group.

Updating and Propagating Delivery Predictability and Group Information. The method of updating the *probTable* vector is the same as the approach followed by PRoPHET [7]. Let $P_{(A,B)}$ be the probability of group A being able to deliver a message to group B. Each group maintains, in its *probTable* vector, the values $P_{(gid,B)}$, for all groups B. The group leader is responsible for carrying out the update of the *probTable* vector. It does so when a group member informs it that a group has become newly adjacent to its group (this information is sent to the leader in the form of a *GroupUpdate* packet). On receiving information that group B has become adjacent to its group, the leader of group A will update its group's delivery predictability value to group B as follows ($P_{init} \in (0, 1]$ is an initialization constant).

$$P_{(A,B)} = P_{(A,B)_{old}} + (1 - P_{(A,B)_{old}}) \times P_{init} \tag{1}$$

Also, a group A can deliver a message to group C indirectly through another group B. To take care of this, the delivery predictability value can also be updated transitively using the following equation ($\beta \in [0, 1]$ is a scaling constant, deciding how large a role transitivity should play).

$$P_{(A,C)} = P_{(A,C)_{old}} + (1 - P_{(A,C)_{old}}) \times P_{(A,B)} \times P_{(B,C)} \times \beta \tag{2}$$

If two groups do not meet each other for a while, they become less likely of being able to exchange messages in the future. Hence, it is necessary for the delivery predictability values to age or decay. This is done according to the following equation, where $\gamma \in (0, 1)$ is an aging constant and k is the time elapsed since the last decay operation was carried out.

$$P_{(A,B)} = P_{(A,B)_{old}} \times \gamma^k \tag{3}$$

The group leader periodically sends out the updated values of the *probTable* and *nodeInfo* vectors to the entire group in a *LeaderUpdate* packet. The *LeaderUpdate* packet serves two purposes. First, each node belonging to the group of the leader updates its own *probTable* and *nodeInfo* vectors using the values present in the *LeaderUpdate* packet. Second, these *LeaderUpdate* packets will also be received by nodes of other groups which lie within communcation range of some member of the group of the leader. These nodes, on receiving a new *LeaderUpdate* packet from an adjacent group, forward it to their own group leader using a *GroupUpdate* packet, so that their group leader can update the *probTable* vector using equations 1 and 2. Further, the group leader, on receiving a *GroupUpdate* packet also updates the values in its *nodeInfo* vector, using information from the *nodeInfo* vector contained in the *GroupUpdate* packet. Specifically, an entry in its *nodeInfo* vector is updated, if either the corresponding entry in the received *nodeInfo* vector has a higher sequence number (*seqNo*), or it has the same sequence number, but has a lower *gid* value.

The changes in the delivery predictability values when groups merge or split depend on the specific application scenario and the movement pattern of the new merged or split group. For example, when two groups merge, we may use the weighted average (according to size) of the *probTable* values of the two groups, or we may take the minimum of the two values, etc. For simplicity, we use the *probTable* values of the group having the lower gid. In the case of a split, the nodes which form a new group will start with an empty *probTable* vector, whereas no change occurs for the remaining nodes.

Forwarding Strategy. The *probTable* and *nodeInfo* vectors at each node are used to decide whether a message should be forwarded to an adjacent node belonging to another group. The protocol uses the following simple greedy strategy for forwarding a packet. A node uses the information present in its *nodeInfo* variable to learn the group to which the destination node belongs. We call this group as the *destination group*. A node forwards a message to an adjacent node belonging to another group, only if the other group has a higher delivery predictability value to the destination group than the group to which the sending node belongs. Thus messages are transferred from group to group, until it reaches the destination group. Once the message reaches the destination group, the message is sent to the destination node using the DSDV routing table.

3 Implementation Overview

The essential data structures maintained by each node is shown in Table 1. Some of these members (like *id,gid,seqNo,leaderSeqNo*) are simple data types, while the rest are multi-valued compound data structures. Not shown in the table are additional fields (like settling time etc.) in *rTable* (the routing table) that will be necessary for DSDV. The data structure *groupInfo* is maintained at each node to keep track of groups which are currently adjacent to it. Each entry has a field called *active*, which indicates whether the group corresponding to that entry is currently adjacent to the node. If an entry remains active for more than T_M time, then the node initiates a merge request (as explained in Section 2.1) with the group corresponding to that entry. The field *leaderSeqNo* in the *groupInfo* structure is used to discard old *LeaderUpdate* packets received from an adjacent group. In a similar fashion, the field *processedSeqNo* is used by a group leader to discard *GroupUpdate* packets that it has already processed.

All packet types that are used by our protocol are listed in Table 3. The *Update* packet corresponds to the DSDV update packet. The *LeaderUpdate* packet is sent periodically by each group leader to inform group members of updated *probTable* and *nodeInfo* values.

Since it is sent periodically, it also acts as a beacon packet, advertising the presence of its group to other groups. A node, on receiving a new *LeaderUpdate* packet from another group, sends out its contents in a *GroupUpdate* packet (which has the same fields as a *LeaderUpdate* packet) to its own leader and carries out the actions that were outlined in Section 2.1. The *MergeRequest* and *MergeReply* packets are used for sending merge requests and replying to them. The *rTable* member shown in the *Update*,

Table 1. Essential Node Structures

Name	Fields (if any)
id	-
gid	-
seqNo	-
leaderSeqNo	-
rTable	dst,nextHop,seqNo
	hopCount,nInfoTimer
groupInfo	gid,active,gInfoTimer,
	leaderSeqNo,processedSeqNo
nodeInfo	id,gid,seqNo
probTable	gid,deliveryPredictability

Table 2. Constants Used

Name	Description	Value
P_{init}	Initial delivery predictability value	0.5
β	Transitivity constant	0.5
γ	Aging constant	0.98
T_M	Merge wait period	50 sec
T_U	Interval for sending *Updates*	15 sec
T_L	Interval for sending *LeaderUpdate*	10 sec
T_N	Timeout period for *rTable* entry	45 sec
T_G	Timeout period for *groupInfo* entry	15 sec
T_{MR}	Min span between two merge request's	5 sec

Table 3. Type of Packets

Name	Members
Update	id,gid,rTable
LeaderUpdate	id,leaderSeqNo,probTable,nodeInfo
GroupUpdate	gid,leaderSeqNo,probTable,nodeInfo
MergeRequest	id,gid,gidOther,rTable
MergeReply	id,gid,status,rTable

MergeRequest, and *MergeReply* packets is essentially the routing table of the sending node, with only the following fields - *dst, seqNo, hopCount*. Table 2 lists some constants that are used in the protocol. These values were chosen based on estimates from initial simulation results.

It is possible that, during a merge operation, the *gid* of the node receiving the merge request may change even before it processes the merge request (due to another merge operation initiated elsewhere). In that case, the merge reply sent back would confuse the node which initiated the merge request, since the *gid* of the replying group is different from what it is expecting. To circumvent this, the *MergeRequest* packet contains a field called *gidOther*, which is the *gid* value of the adjacent group at the time the merge request is being sent. The node which sends the *MergeReply* copies the *gidOther* from the *MergeRequest* packet into the *gid* field of the *MergeReply* packet. The node receiving the *MergeReply* is then able to easily find out the group it had originally sent the *MergeRequest* to. The fact that the other group possibly has a different *gid* even after the merge operation is not significant, because as soon as the next round of DSDV updates are triggered, all nodes of the merged group will have a common *gid*.

4   Simulation and Results

We verify the effectiveness of our protocol through simulations across a wide range of scenarios on the NS-2 simulator [16]. The input parameters varied included mobility pattern, communication range, and the size of the network. We compare the performance of our protocol with that of the 2-Hop protocol and AODV. Results for AODV are included in order to highlight the ineffectiveness of standard routing protocols in disconnected environments. The performance metrics used for the comparison are delivery ratio and delay in receiving messages. We first discuss the simulation setup and then present results of our simulations.

Three different mobility patterns are chosen in the simulations - *Random Waypoint Model* [11], *Inplace Model* [12], and *Community Model* [7]. In *Inplace Model*, the topology is divided into different grids. Each group is assigned a specific grid, which it never moves out of. Groups are able to interact when they meet along grid boundaries. In *Community Model*, the topology is again divided into grids. There are some designated grids called the *gathering grids*. Each group is assigned a *home grid* and also a *gathering grid*. The movement of groups is such that they tend to travel to and fro between the home and gathering grids, even though there is a slight chance that they may move to some other grids temporarily.

For the Community model, each grid has dimensions 200m×200m. Each scenario has 3 randomly chosen gathering grids. If a group is in its home grid, it moves to its gathering grid with a probability of 0.90, while if it is in its gathering grid, it moves to its home grid with a probability of 0.95. For the Inplace model, each grid has dimensions 250m×250m. Node assignments to groups are generated using the model in [13]. About 3% of the nodes are not placed in any group at the beginning of the simulation. In [13], nodes have to choose a new destination after reaching their current goal. The new destination can be within its group, within another group, or can also be outside of all groups. The probability of choosing these 3 types of destinations are set to 0.98, 0.01 and 0.01 respectively. For all scenarios, nodes belonging to a group move in a bounding box of 100m×100m. Further, each group moves to its destination with a uniform random speed of (0,10] m/s. Likewise, each node within a group moved to its destination within the group with a uniform random speed (relative) of (0,5] m/s. The pause time for node movement within a group is set to 5 seconds, while the pause time for the group movement is set to 25 seconds.

We tested 3 scenarios having different network sizes. The first scenario has 50 nodes, 7 groups, a 1000m×1000m topology, 30 randomly established communicating pairs and a simulation time of 1000 seconds. The second scenario has 100 nodes, 13 groups, a 2000m×2000m topology, 50 randomly established communicating pairs and a simulation time of 2000 seconds, while the third scenario has 200 nodes, 25 groups, a 3000m×3000m topology, 70 randomly established communicating pairs and a simulation time of 3000 seconds. In all scenarios, each node has a buffer capacity of 1000 messages. All messages are generated as CBR traffic over a UDP connection. The underlying MAC protocol that used in our simulations is the IEEE 802.11 protocol. No measurements are taken during the first 50 seconds of the simulation to allow the protocols to stabilize.

We present results of simulation for scenario 2 (100 node case) in Figures 1 and 2. The results for the other two scenarios look similar, and due to lack of space, we do not include them here. Figure 1 shows the delivery ratio v/s communication range for different mobility models. Each point on the plot is the average value taken over 5 simulation runs. The plots also show best fit bezier curves and 95% confidence intervals. Figure 2 gives the CDF of the message delivery delays, when the communication range is 100m.

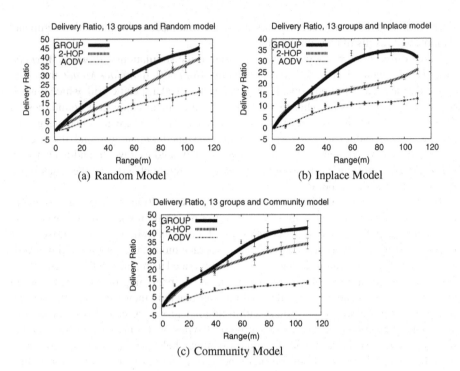

Fig. 1. Delivery Ratio % v/s Communication Range for 100 nodes, 13 groups and various mobility models

From Figure 1, we see that our protocol provides much better delivery ratios than the other two protocols for the Inplace model. Moreover, from Figure 2, we see that the delay our protocol incurs is significantly less than the 2-Hop protocol for the last set of delivered messages. This is expected , as in the Inplace model, groups move in more or less predictable ways, and our protocol is able to exploit this property effectively.

For the case of the Community model also, it is seen that our protocol performs better than the 2-Hop protocol, although the improvement is not as drastic as in the case of the Inplace model. One reason for this is that the movement of the groups is less predictable in the Community model than in the case of the Inplace model. For example, this can happen due to groups moving to grid locations which are neither their home grid nor their gathering grid. Another reason for this is that in the Community model, groups have more freedom to overlap and mingle with each other. This causes merging of some groups, even when they are not actually going to stay as a merged group in the near future. Due to this, there would be instances where groups merge, followed by a split very soon. This in turn, would cause otherwise unnecessary delays in rebuilding routing tables, and also a temporary loss of information regarding delivery predictability values.

For the case of the Random Waypoint model, we see that our protocol performs quite well in comparison to the 2-Hop protocol. However, in general, it is observed that the delivery ratio performance of our protocol varies and lies close to the delivery

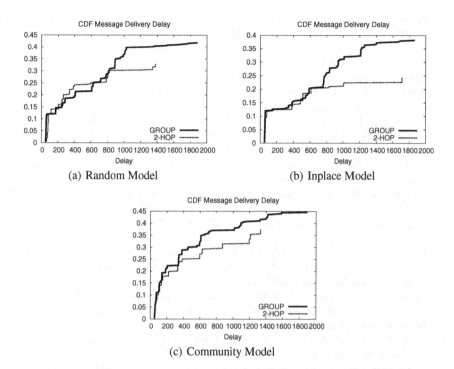

Fig. 2. CDF of message delivery delay for 100 nodes, 13 groups, various mobility patterns and 100m communication range

ratio curves for the 2-Hop protocol. This reasonably good performance of our protocol in the Random Waypoint model is slightly unexpected. This was also reported in the PRoPHET protocol simulations [7], and can be due to the fact that even in random motion, two groups that have met each other may not have moved far away from each other. In such cases, the delivery predictability values will still be useful.

5   Conclusion

In this paper, we proposed a routing protocol for disconnected networks where nodes tend to move in groups following particular mobility patterns. Initial simulations have shown that for the Inplace and Community model, our group protocol is able to deliver more messages than the 2-Hop protocol at comparable or better latencies.

References

1. Perkins, C.E., Belding-Royer, E.M.: Ad-hoc on-demand distance vector routing. In: WM-CSA. (1999) 90–100
2. Grossglauser, M., Tse, D.N.C.: Mobility increases the capacity of ad hoc wireless networks. IEEE/ACM Trans. Netw. **10** (2002) 477–486

3. Perkins, C., Bhagwat, P.: Highly dynamic destination-sequenced distance-vector routing (DSDV) for mobile computers. In: ACM SIGCOMM'94 Conference on Communications Architectures, Protocols and Applications. (1994) 234–244
4. Johnson, D.B., Maltz, D.A.: Dynamic source routing in ad hoc wireless networks. In: Mobile Computing. Volume 353. Kluwer Academic Publishers (1996)
5. Vahdat, A., Becker, D.: Epidemic routing for partially connected ad hoc networks. Technical Report CS-200006, Duke University (2000)
6. Jain, S., Fall, K., Patra, R.: Routing in a delay tolerant network. In: SIGCOMM '04: Proceedings of the 2004 conference on Applications, technologies, architectures, and protocols for computer communications, New York, NY, USA, ACM Press (2004) 145–158
7. Lindgren, A., Doria, A., Schelen, O.: Probabilistic routing in intermittently connected networks. In: SAPIR. (2004) 239–254
8. Musolesi, M., Hailes, S., Mascolo, C.: Adaptive routing for intermittently connected mobile ad hoc networks. In: Proceedings of the IEEE 6th International Symposium on a World of Wireless, Mobile, and Multimedia Networks (WoWMoM 2005). Taormina, Italy., IEEE press (2005)
9. Jones, E.P.C., Li, L., Ward, P.A.S.: Practical routing in delay-tolerant networks. In: WDTN '05: Proceeding of the 2005 ACM SIGCOMM workshop on Delay-tolerant networking, New York, NY, USA, ACM Press (2005) 237–243
10. Shah, R.C., Roy, S., Jain, S., Brunette, W.: Data mules: modeling a three-tier architecture for sparse sensor networks. In: Proceedings of the First IEEE International Workshop on Sensor Network Protocols and Applications. (2003) 30–41
11. Camp, T., Boleng, J., Davies, V.: A survey of mobility models for ad hoc network research. Wireless Communications and Mobile Computing (WCMC): Special issue on Mobile Ad Hoc Networking: Research, Trends and Applications 2 (2002) 483–502
12. Hong, X., Gerla, M., Pei, G., Chiang, C.: A group mobility model for ad hoc wireless networks. In: Proceedings of ACM/IEEE MSWiM'99, Seattle, WA. (1999) 53–60
13. Musolesi, M., Hailes, S., Mascolo, C.: An ad hoc mobility model founded on social network theory. In: Proceedings of ACM/IEEE MSWiM '04, Venice, Italy, New York, NY, USA, ACM Press (2004) 20–24
14. Juang, P., Oki, H., Wang, Y., Martonosi, M., Peh, L., Rubenstein, D.: Energy-efficient computing for wildlife tracking: Design tradeoffs and early experiences with zebranet. In: ASPLOS, San Jose, CA. (2002)
15. Franz, W., Eberhardt, E., Luchenbach, T.: Fleetnet - Internet on the road. In: Proceedings of 8th World Congress on Intelligent Transport Systems. (2001)
16. McCanne, S., Floyd, S.: ns network simulator. http://www.isi.edu/nsnam/ns (2005)

A Hybrid Routing Scheme for Mobile Ad Hoc Networks with Mobile Backbones

Ashish Pandey, Md. Nasir Ahmed, Nilesh Kumar, and P. Gupta

Department of Computer Science and Engineering
Indian Institute of Technology Kanpur
Kanpur – 208016, India
{pandey.ashish, nileshkb, nasir.ahmed}@gmail.com,
pg@cse.iitk.ac.in

Abstract. A flat mobile ad hoc network has an inherent scalability limitation. When the network size increases, per node throughput of an ad hoc network rapidly decreases. This is due to the act that in large scale networks, flat structure of networks results in long hop paths which are prone to breaks. These long hop paths can be avoided by building a physically hierarchical backbone network. These networks have some specific backbone capable nodes that have powerful radios and are functionally more capable than ordinary nodes.

In this paper, a hybrid routing protocol for large scale networks with mobile backbones has been proposed. This routing protocol uses different types of routing schemes in different layers of hierarchical network which makes it easily extendable to support QoS as well. Along with hierarchical structure, a low-overhead clustering scheme to elect backbone nodes has been proposed and works with our routing protocol without causing any extra overhead.

Keywords: MANET, clustering, hierarchical routing, quality of service.

1 Introduction

Mobile Ad-Hoc Networks (MANETs) are self-organizing, rapidly deployable, and require no fixed infrastructure. MANETs are multi-hop networks where nodes co-operate with each other in forming routes and forwarding data (i.e, nodes behave as routers) to respective source and destination. Nodes in such networks are highly mobile, or stationary, and may vary widely in terms of their capabilities and uses. MANETs may operate autonomously or may be used to expand the present Internet. Collaborative computing and communications in smaller areas (building organizations, conferences, etc.) can be set up using MANETS. Communications in battle-fields and disaster recovery areas are further examples of application environments.

With the evolution of Multimedia Technology, Quality of Service in MANETs has become an area of great interest. Besides the problems that exist for QoS in wire-based networks, MANETS impose new constraints because of the dynamic behavior and the limited resources of such networks. Since MANET is a highly dynamic multi-hop wireless network, the effective working of routing protocols plays the most

Y. Robert et al. (Eds.): HiPC 2006, LNCS 4297, pp. 411 – 423, 2006.
© Springer-Verlag Berlin Heidelberg 2006

important role in making such network useful. The main aim of this work to provide a routing protocol which works for a large scale network. The proposed protocol is easily extendible to support quality of service (QoS) requirements of applications. The target network may comprise hundreds or even thousands of nodes. Providing efficient routing protocol in such a network raises the need of physical hierarchical network namely, mobile backbone network. These networks have some specific backbone capable nodes (BCNs) having powerful radios and functionally more capable than regular nodes. For hierarchical network, there is a need to form clusters of nodes using these BCNs as cluster-head. But existing clustering schemes like lowest ID results in instability of clusters under the high mobility condition. Further, applications demanding quality of service in such a large scale network raises the need of protocols this supports QoS inherently and can easily be extended to achieve QoS if needed. In this paper a scalable routing protocol for large scale networks which is easily extendible to support quality of service has been proposed.

Section 2 presents a review of significant contribution in the area of hierarchical and non-hierarchical routing for Ad Hoc networks and their limitations. In Section 3, the design philosophy and the Methodology of our routing scheme have been presented. Section 4 discusses the clustering algorithm to evaluate cluster-head as well as formation of clusters. Simulation results and performance evaluation are given in section 5. Conclusions are given in the last section.

2 Related Works

It is well known fact that proactive protocols incur too much overhead in maintaining topology information and perform very badly in large networks (not scalable). On the other hand reactive protocols are good candidate for scalability but because of mobility their performance degrades very heavily, which is the main reason why this type of protocols are not suitable for scalability in high mobility conditions. To target the scalability of the network many hierarchical and non-hierarchical protocols have been purposed.

The Zone Routing Protocol [13] is a non-hierarchical protocol designed for large networks. In this protocol each node has a zone that is defined as all nodes that are within hops of the node. Routing within a zone uses a protocol such as Dynamic Source Routing (DSR) [14]. For sending a packet to a node outside the zone, the source sends a route request packet to nodes at the periphery of its zone. These nodes query their zones or their peripheral nodes, etc.

Another non-hierarchical protocol can be found in [1]. This protocol uses hybrid approach of proactive and reactive protocols. When the destination is within K-hop distance of the source node it uses predictive location based routing (PLB) [25], a proactive routing protocol. Otherwise, it uses ad hoc on-demand routing (AODV) [22], a reactive routing protocol. The idea of using proactive protocol in localized network and reactive protocol for the rest of the network is good as it results in the low overhead, but still, when the network size is large, flat structure of routing protocol results in frequent route breaks due to node mobility.

Many hierarchical routing protocols have been proposed, which rely on the construction and maintenance of a hierarchy in the ad hoc network. One set of protocols use a clustering algorithm at the lowest level. Communication between nodes from two different clusters takes place via the cluster- heads (CBRP [19] Landmark [18]).The problem with clustering algorithm is that it results in too much control overheads to maintain the cluster structure (cluster-head election, node leaving, node joining, cluster-head failures etc). Another problem is that the cluster-heads become the center of all outgoing routing traffic which results in increase in delay and dropping of packets because of overflow of buffer.

3 Proposed Routing Protocol

The objective of the proposed routing protocol is to design a scalable routing protocol for large scale networks and which is easily extendible to support quality of service parameters.

3.1 Mobile Backbone Networks (MBNs)

A mobile backbone network (MBN) is a hierarchical network consisting of a large area with hundreds or thousands of nodes [28]. There are two types of nodes in these networks: backbone capable nodes (BCNs) and regular nodes (RNs). The BCNs have an additional powerful radio and are functionally more capable than the RNs. Some of the BCNs are elected to function as backbone nodes (BNs) which form a network among themselves using long range radios. This higher level network is called backbone network. Since the BNs are also mobile and keep joining and leaving the backbone network in an ad hoc manner, the backbone network is actually a MANET. Thus, there are multiple MANETs in a multi-level MBN. All nodes in a network operate in the same channel but these networks operate in different channels to minimize the interference across levels. Thus, in two levels, two independent ad hoc networks run simultaneously at different channels. In underlying network, all nodes including BNs use low power radios while in backbone network, all nodes (i.e. BNs) use high power radios. The RNs, which are in underlying network, are generally limited in their battery, storage, and processing power. They use just a single low power radio. In order to access the complete network, every node associates itself to a BN. The backbone nodes (BNs) are part of both networks and use two different radios, low power in underlying network and high power in backbone network, for communication in different networks simultaneously.

3.2 Proposed Scheme

In two-level MBNs, there are two independent ad hoc networks running at different levels. There exists some work on routing protocol for providing scalability using MBNs which uses extended LANMAR routing protocol [28]. But LANMAR uses proactive protocols in both intra and inter cluster routing, and hence suffers from drawbacks of proactive protocols. It is known that proactive protocols result in too much overhead in maintaining topology information and hence perform very badly in large networks. On the other hand, a reactive protocol is a better candidate for scalability but in the long hop paths due to mobility of nodes; paths are prone to break quite frequently and lead to degrade the performance. The proposed scheme is a hybrid approach of reactive and proactive protocol, which exploits the advantages of each at the appropriate place. Two-level MBN is used to make the protocol work in large scale network. A new stable clustering scheme has been designed for backbone nodes election. This clustering scheme elects BNs from available BCNs using K-hop clustering, i.e. a BN will be at least at K-hop distance from another BN. These BNs form the backbone network using high power radios.

In underlying network Predictive Location Based Routing (PLB), a proactive protocol is used while Ad Hoc On-Demand Distance Vector Routing (AODV), a reactive protocol, is used in backbone network.

The algorithm to send a packet to a destination is given below:

1. When a node has to send a packet, it checks if there is an entry in the update table for that destination.

2. If there is an entry then it calculates the route to that destination using PLB, sends the packet to it and it is done.

3. If there is no such entry that means that destination is not in the K-hop distance of this node; the node sends that packet using PLB routing to the BN which is within K-hop of it.

4. This BN stores the packet and uses AODV routing protocol to establish and to maintain the route to the destination. For this, it sends a route request (RREQ) for the destination node in backbone network.

5. When a node in the backbone network receives this RREQ it checks if the destination node is within its K-hop. If not, then it forwards the RREQ. Else it sends RREP.

6. When the BN which initiated the query receives this RREP, an AODV path to destination is established. Then, the BN sends all stored packets to destination using AODV routing.

7. When a node in the backbone network receives this data packet, it checks if the destination node is within its K-hop. If not, then it forwards the packet, otherwise it sends the packet to destination node using PLB routing.

Apart from the capable of working in large scale networks, the proposed routing protocol is very effective in high mobility conditions and can be easily extended to support QoS. Firstly, it uses PLB an update based predictive location routing protocol, in localized network. Hence, every node has not only latest information about the locations of other nodes in its K-hop but it can predict their locations when the packet reaches to them. Secondly, a reactive protocol like AODV is vulnerable to mobility but proposed protocol uses that in high power links. The transmission range of these high power links is chosen such that it allows the mobility of BNs without causing the existing routes in AODV to break. Apart from the advantages of PLB and high power mobile backbones, the proposed routing algorithm also supports mobility to some extent using the step 7 of the mentioned algorithm. If a destination node moves out of the K-hops of its BN and enters in the K-hop of another BN which is in the route to this destination, then data packets are delivered to the destination using the new BN without causing the existing route to break.

Further, the proposed protocol is easily extendible to support QoS because of PLB and MBN. PLB, used in the localized network, assists QoS routing inherently. Hence, for a destination lying within K-hop of a node, PLB works efficiently to support QoS. For a destination lying outside the local network of a node, that node uses MBN to route packets. This MBN usually has large bandwidth and high power mobile backbone. This mobile backbone results in very low delay and high bandwidth which are the most important parameters for applications requiring QoS.

4 Proposed Clustering Scheme

There are three most desired properties of a clustering algorithm: simplicity, stability, and low overhead. Clustering algorithms available in [2] [17] [11] [16]. The Lowest ID (LID) [17] and Highest Degree (HID) [11] are the most widely used algorithms due to their low overhead and simplicity. For hierarchical structure, where clustering is often used to support routing, stability of the clustering algorithm is the most crucial property. Any overhead incurred by clustering algorithm may add delay and consume the bandwidth of network which is a very scarce resource in MANETs. This paper presents a new clustering scheme that provides stability and works with proposed routing protocol without causing any extra overhead. The scheme is basically a modified version of Lowest ID algorithm with enhanced cluster maintenance for greater stability and low overhead. The Lowest ID algorithm is extended to support multi-hop clustering as well.

4.1 Enhanced Lowest ID Clustering (ELIDC)

The proposed enhanced lowest ID clustering (ELIDC) scheme is an enhanced version of existing lowest ID clustering algorithm (LID) [17]. In this scheme, firstly, it is not necessary for nodes to know the IDs of its K-hop neighbors beforehand. They wait for a certain amount of time for getting the message from K-hop distanced nodes before participating in the cluster-head election process. Even if they don't get a message from their all K-hop neighbors, proposed scheme works fine. Secondly, if a message sent by a node does not reach to a particular destination node in certain period of time then destination stops considering that node as its K-hop neighbor for cluster-head election process. Thirdly, there is no restriction on network topology to be static during the execution of the algorithm.

These timers are set considering the K-hop propagation delay and certain number of retries. In the proposed routing protocol, nodes send PLB update packets to their k-hop neighbors. All messages required for our clustering scheme are merged in the update packets. Hence, the clustering scheme combined with the routing scheme results in almost no extra overhead in the network. In the mobile backbone network, only BCNs participate in the clustering algorithm. The ELIDC algorithm works as follows:

1. At the time of system initialization, all nodes defer by a random time to avoid collisions after which they start sending their cluster election packets up to K-hop and keep sending that until the cluster-head election process is finished.

2. Before starting cluster-head election, all nodes wait for a certain time (WAIT_TIME) so that they can receive cluster election packets from K-hop distanced nodes.

3. Upon receiving a cluster election message from some node, the node join that cluster form which it has received first cluster election packet.

4. After WAIT_TIME, every node starts cluster-head election algorithm using the set formed in previous step and associates a timer with each entry in the set. This timer is reset upon receiving a cluster election packet from the node corresponding to the entry.

5. When timer for any node in the set goes off, then that node is removed from the set, and is no more considered for cluster-head election.

6. Whenever a node becomes the smallest in the set, it makes itself a cluster-head, terminates the algorithm and starts sending cluster-head claims (beacons) to its K-hop neighbors continually.

7. Upon receiving cluster-head claim packet, a node makes the source node of the packet its cluster-head, informs its K-hop neighbors and terminates the algorithm. These neighbors upon receiving this message remove this node from their cluster-head election set.

The algorithm terminates when a node becomes smallest in its set or gets a cluster-head claim packet. We use this clustering scheme in our protocol only for selecting some nodes from available BCNs to act as BNs but there is as such no boundary or membership concept of clusters. When a regular node receives cluster-head claim (actually a field saying that source of this packet is a BN) in the PLB update packets then it makes the source of that packet its BN and uses it when the destination is not within its k-hop scope.

5 Simulations and Performance Evaluation

The proposed routing protocol has been tested on network simulator NS-2. In this section, the simulation environment, extension made in NS-2 to simulate proposed protocol, and methodology used for getting the results are explained and then evaluated the performance of the proposed protocol.

5.1 Enhancement in NS-2 Simulator

The current implementation of NS-2 simulator supports only single network interface in a mobile node. This limitation restricts the NS-2 to simulate the scenario where mobile nodes possess multiple network interfaces, e.g. mobile backbone networks (MBNs). Multiple interface support has been implemented in NS-2 since it is necessary for realistic simulation of the proposed routing protocol which is based on

MBNs. Basically the CMU's Monarch group's mobile node implementation [9] has been extended. This scheme does not force all nodes in the network to have same number of interfaces. Hence, the scenarios where only some nodes possess multi-interface and others are single interfaced mobile nodes can be simulated. Apart from helping in simulating the desired scenario, this extension adds a significant feature in NS-2.

5.2 Simulation Methodology

This section discusses the simulation methodology used to compare the performance of our proposed protocol with other existing protocols. It comprises the topography, traffic model, protocol implementation, and the performance metrics used for evaluating the results obtained through the simulation.

5.2.1 Simulation Topography

Extensive simulation experiments have been performed to measure the performance of the protocols. The experiment has been divided in two parts, one for different mobility conditions and the other for scalability of the network for different network sizes. For all simulations, 25% of the mobile nodes are backbone capable. Every backbone capable node had two IEEE 802.11 radios. One was the low power radio same to the ordinary node with transmission range 200m. The other is the high power radio with transmission range 800m. All nodes in the network use same mobility model, random waypoint model. For mobility, the performance of proposed protocol has been evaluated in a large scale network under different mobility speeds. Network size is kept fixed as 400 nodes. These nodes are randomly distributed in a rectangular area of 4000m by 2500m. Simulation has been done for different mobility speeds as high as {1, 5, 10, 15 and 20} with two different movement patterns for 600sec. Each node in the simulation has different pause time randomly distributed between 0 to 600 seconds. The scalability of the proposed protocol has been evaluated with different network size. For each network size, the node density to 40 nodes/km^2 has been kept. The mobility speed of the nodes was fixed to as high as 20m/s.

5.2.2 Traffic Model

For all simulation, the traffic sources used are 30 CBR (Constant Bit rate). Data packet size is fixed as 64 bytes and packet sending rate was 4packets/second. The clustering scheme for backbone nodes election has been used. The K in K-hop clustering was chosen as 2. The implementation of the AODV has been used as provided with the NS-2 simulator and implementation of PLB-AODV has been used as provided by Ahmed et al [1] on NS-2. Multiple interface support in NS-2 has been implemented since it is necessary for realistic simulation of the proposed routing protocol which is based on MBNs. Basically the CMU's Monarch group's mobile node implementation [9] has been extended.

5.2.3 Performance Metrics

For evaluating the performance of the proposed protocol against other protocols, a set of metrics has been chosen. The following three metrics capture the most basic overall performance of the protocol:

Packet delivery ratio: The ratio of the total number of data packets received by all nodes to the total number of packets originated by all nodes. The packet delivery ratio is similar to network throughput.

Average End-to-End delay: It is the delay experienced by the successfully delivered packets in reaching their destinations. This is a good metric for comparing protocols supporting real time application and QoS. This metric is a measure of how efficient the underlying routing algorithm is, because primarily the delay depends upon optimality of path chosen, the delay experienced at the interface queues and delay caused by the retransmissions at the physical layer due to collisions. This is a direct measure, which shows how well a routing protocol could handle the real time traffic.

Routing Overhead: Routing Overhead is the number of routing packets sent by the routing protocol to deliver the data packets to destinations.

5.3 Results and Analysis

5.3.1 Packet Delivery Ratio
Mobility: Figure 1 shows the graph obtained for packet delivery ratio under different mobility conditions. For 1m/s mobility, all three protocols have packet delivery ratio almost equal to 1. But as the mobility increases, the packet delivery ratio of our proposed protocol, PLB-AODV with MBN, clearly outperforms AODV and PLB-AODV. This happens because when the mobility increases, routing information at nodes tend to become obsolete very rapidly especially for AODV which is a reactive protocol. But in our protocol, topology of backbone network, which uses AODV, does not suffer from mobility due to high transmission range radio used in it.

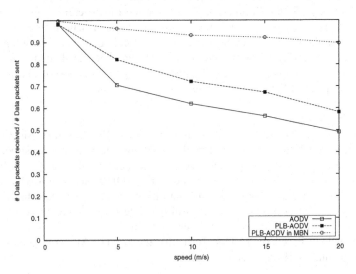

Fig. 1. Packet delivery ratio vs. mobility

Scalability with Network size: Figure 2 shows the graph obtained for packet delivery ratio against different network sizes. It proves the inability of non-hierarchical protocols such as AODV, PLB-AODV to be scaled to work in the large ad hoc network. When the size of network and hence topology (to maintain the same node density) increase, along with it average number of hops in a route also increases. This results in long hops paths between nodes which are inherently prone to break under mobility condition. This justifies the decrease in packet delivery fraction of AODV and PLB-AODV with the increase in network size. But the proposed protocol does not suffer from this since firstly, the number does not increase much and secondly, as seen in Figure 2, this protocol does not suffer from mobility.

5.3.2 Average End-to-End Delay (EED)

Mobility: Figure 3 shows the graph for average end-to-end delay against different mobility conditions. As shown in Figure 3, even in the low mobility there are significant differences in the average end-to-end delays of AODV, PLB-AODV and of this protocol. Since end-to-end delay is directly proportional to number of hops in the path, non-hierarchical protocols, e.g. AODV, PLB-AODV result in greater end-to-end delay. Since, this protocol is physical hierarchical protocol, number of hops and hence end-to-end delays experienced by packets are far less than other protocols. With increase in mobility, performance of AODV decrease badly as mobility causes routes breakage and re-computation and thereby, increases the end-to-end delay of packets waiting in the router's queue.

Scalability with network size: Figure 4 shows the graphs obtained for average end-to-end delay against different network sizes. The graph shows that with the increase in network size, average end-to-end delay incurred by all protocols increases. However, the protocol results in far less delay and hence proves its superiority over other protocols for real time applications in large scale network.

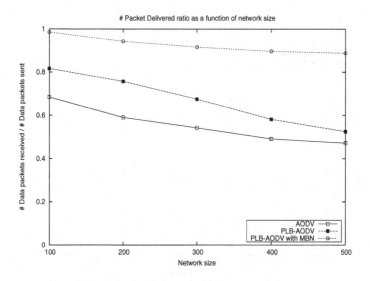

Fig. 2. Packet delivery ratio vs. network size

5.3.3 Routing Overhead

Figure 5 shows the graph for number of routing packets generated by the protocols to route data packets against the network sizes. Figure shows that AODV produces least routing overhead. This is because of its reactive nature which generates the any other packet than data packet only when it is needed. Where as PLB-AODV and our protocol are combinations of proactive and reactive protocols and hence, generates more number of packets, most of which are update packets generated by the PLB protocol.

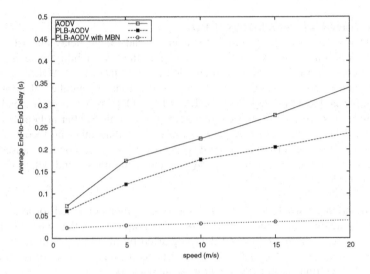

Fig. 3. Average end-to-end delay vs. Mobility

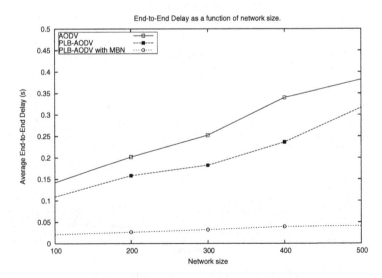

Fig. 4. Average end-to-end delay vs. Network Size

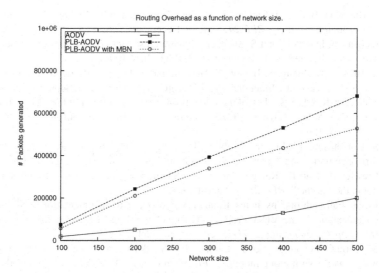

Fig. 5. Routing overhead vs. Network size

6 Conclusions

This paper has proposed a new routing protocol for large scale networks using mobile backbones, devised a new stable and low overhead clustering scheme, and extended mobile node implementation of NS-2 simulator to support multi-interface. In order to prove the correctness and efficiency, the proposed routing protocol has been implemented and simulated on NS-2 simulator. The performance of the proposed protocol was compared with the performances of some existing protocols under different mobility conditions and different network sizes. The metrics used for performance comparison are packet delivery ratio, average end-to-end delay, and routing overhead. The results obtained through rigorous simulations have been shown superiority of the proposed protocol in packet delivery ratio and end-to-end delay over other protocols but in routing overheads, AODV outperformed the proposed protocol.

The issue of providing quality of service (QoS) in large scale networks has been rarely addressed. The proposed routing protocol uses PLB and AODV which are capable of supporting QoS efficiently. Hence, this scheme can easily be extended to support quality of service in large scale networks. Further, most of the routing overheads are result of update packets generated by PLB. It is observed that most of them are redundant. Some techniques to reduce these redundant broadcasts can be applied which may improve the performance of the proposed scheme.

References

[1] N. Ahmed and P. Gupta, "Predictive location based qos aware routing scheme for highly dynamic mobile ad hoc networks". Master's Thesis Report, July 2004.

[2] S. Banerjee and S. Khuller. "A clustering scheme for hierarchical control in multi-hop wireless networks". *Proceedings of IEEE Infocom,* April 2001.

[3] S. Blake, D. Black, M. Carlson, E. Davies, Z. Wang, and W. Weiss. "An architecture for differentiated services". *RFC 2475, IETF*, December 1998.

[4] R. Braden, D. Clark, and S. Shenker. "Integrated services in the internet architecture: an overview". Request for Comments: 1633, IETF, June 1994.

[5] R. Braden, L. Zhang, S. Berson, S. Herzog, and S. Jamin., "Resource reservation protocol (RSVP)-version 1 functional specification". *IETF RFC 2205*, September 1997.

[6] Nahrstedt K. Chen S. "Distributed quality-of-service routing in ad hoc networks". *IEEE Journal on Selected Areas in Communication Special Issue on Ad hoc Networks*, 17(8): 1488 1505, 1999.

[7] S. Corson and J. Macker. "Mobile ad hoc networking (manet): Routing protocol performance issues and evaluation considerations". *RFC 2501, IETF*, January 1999.

[8] Crawley E., Nair R., Rajagopalan B., and Sandrick H. "A framework for qos based routing in the internet". *RFC 2386*, August 1998.

[9] Kevin Fall, editors. ns notes Kannan Varadhan, and documentation. "NS notes and documentation", the VINT Project, UC Brekeley, lbl, usc/isi, and xerox parc. Available from *http://www-mash.cs.brekeley.edu/ns/*, November 1997.

[10] M. Gerla G. Pei and X. Hong. "Lanmar: Landmark routing for large scale wireless and ad hoc networks with group mobility". *Proceedings of ACM Mobihoc*, August 2000.

[11] M. Gerla and J. T. Tsai. "Multicluster, mobile, multimedia radio network". *ACM- Baltzer Journal of Wireless Networks*, 1, (3) :255-265, 1995.

[12] P. Gupta and P. R. Kumar. "The capacity of wireless networks". *IEEE Transactions on Information Theory*, 46(2), March 2000.

[13] Z. J. Hass, Mark R. Pearlman, and P. Samar. "The zone routing protocol (zrp) for ad hoc networks". *Internet Draft, draft-ietf-manet-zone-zrp-04.txt, IETF*, July 2002.

[14] David B Johnson and David A Maltz. "Dynamic source routing in ad hoc wireless networks". In Imielinski and Korth, editors, *Mobile Computing*, volume 353, pages 152-81. *Kluwer Academic Publishers*, 1996.

[15] X. Hong K. Xu and M. Gerla. "An ad hoc network with mobile backbones". *Proceedings of IEEE ICC'02*, April 2002.

[16] P. Krishna, N. H. Vaidya, M. Chatterjee, and D. K. Pradhan. "A cluster-based approach for routing in dynamic networks". *Proceedings of ACM SIGCOMM Computer Communication Review*, pages 372-378, 1997.

[17] A. Pandey, S. Kushwaha, P. Gupta and C. J. Hwang, "A Weighted Probabilistic Algorithm for Cluster based Mobile Ad Hoc Networks", *Proceedings of IASTED International conference on Wireless Networks & Emerging Technologies (WNET 2003)* held at Banff, Canada, July 2003.

[18] X. Hong M. Gerla and G. Pei. "Landmark routing for large as hoc wireless networks". *Proceeding of IEEE GLOBECOM 2000*, November 2000.

[19] J. Li M. Jian and Y.C. Tay. "Cluster based routing protocol (cbrp) functional specification". *Internet Draft, draft-ietf-manet-cbrp-spec-00.txt*.

[20] S. Murthy and J. J. Garcia-Luna-Aceves. "An efficient routing protocol for wireless networks". ACM Mobile Networks and App. J., *Special Issue on Routing in Mobile Communication Networks*, pages 183-97, October 1996.

[21] V. D. Park and M. S. Corson. "A highly adaptive distributed routing algorithm for mobile wireless networks". *Proceedings of INFOCOM*, pages 1405-1413, April 1997.

[22] C. E. Perkins and E. M. Royer. "Ad-hoc on-demand distance vector routing". *Proceedings of 2nd IEEE Workshop of Mobile Comp. Systems and Applications*, pages 90-100, February 1999.

[23] Charles Perkins and Pravin Bhagwat. "Highly dynamic destination-sequenced distance-vector routing (DSDV) for mobile computers". *Proceedings of ACM SIGCOMM'94 Conference on Communications Architectures, Protocols and Applications*, pages 234-244, 1994.

[24] Das SR. Perkins C, Royer E. "Quality of service for ad hoc on-demand distance vector (aodv) routing". *Internet-Draft, July 2000.*

[25] Samarh H. Shah and Klara Nahrstedt. "Predictive location-based qos routing in mobile ad hoc networks". *Technical Report UIUCDCS-R-2001-2242/ UILU- ENG-2001-1749, Department of CS, UIUC,* September 2001.

[26] Md. Nasir Ahmed, Ashish Pandey, Nilesh Kumar & P. Gupta "A Hybrid Routing Protocol for Large Scale Mobile Ad Hoc Networks with Mobile Backbones", *Proceedings of ADCOM 2004*, Ahmedabad, India, December 2004.

[27] Kaixin Xu, Xiaoyan Hong, and Mario Gerla. "An ad hoc network with mobile backbones". *Proceedings of IEEE International Conference on Communications (ICC'02),* New York, NY, April 2002.

K-Tree: A Multiple Tree Video Multicast Protocol for Ad Hoc Wireless Networks[*]

B. Anirudh, T. Bheemarjuna Reddy, and C. Siva Ram Murthy

Department of Computer Science and Engineering
Indian Institute of Technology Madras, India 600036
badam@cse.iitm.ernet.in, arjun@cs.iitm.ernet.in, murthy@iitm.ac.in

Abstract. In this paper, we address the problem of video multicast over ad hoc wireless networks. Video multicasting demands high quality of service with a continuous reachability to receivers. However, the existing multicast solutions do not guarantee this because they are not resilient to mobility of the nodes. Uninterrupted video transmission requires continuous reachability to receivers which emphasizes the usage of path-diversity. Hence, we propose a multiple tree multicast protocol (K-Tree) which maintains maximally node-disjoint multicast trees in the network to attain robustness against path breaks. We further enhance the robustness by using the error resilient Multiple Description Coding (MDC) for video encoding. Through simulations we show how the protocol improves the video quality as we use two or three trees instead of a single tree.

1 Introduction

Ad hoc wireless networks are defined as the category of wireless networks that utilize multi-hop radio relaying and are capable of operating without any fixed infrastructure, involving unrestricted mobility of the nodes. The absence of any central coordinator or base station and also the dynamic nature of the network makes routing a complex one compared to infrastructure based networks. The routing model in ad hoc wireless networks is to route data through other nodes in the network, that is nodes act as routers also. A packet that needs to reach from a source to a destination has to pass through different nodes in the network. The topology of the network keeps changing continuously and routing information has to be updated. Hence unicasting data packet can cause many problems on ad hoc wireless networks.

Video multicast is the problem of multicasting video data, that is to transmit the video stream to more than one receiver simultaneously. Many cutting edge applications like digital classrooms require robust video multicasting solutions. Video multicast problem has been extensively studied and successfully solved for infrastructure based wired and wireless networks. The solutions are centered around utilizing central coordinators in multicasting, which are difficult

---

[*] This work was supported by the iNautix Technologies India Private Limited, Chennai, India and the Department of Science and Technology, New Delhi, India.

Y. Robert et al. (Eds.): HiPC 2006, LNCS 4297, pp. 424–435, 2006.

to extend to the peculiar ad hoc networks. Multicasting using trees and meshes involves path breaking and making due to the mobility of nodes in the tree or mesh. Reliable delivery of video traffic is important as there is time constraint which discourages automatic repeat request (ARQ) in case of a packet loss or packet error. This paper addresses the issue of reliable video multicast through multiple trees.

Uninterrupted transmission is not guaranteed by the existing multicast solutions, for ad hoc networks, as they have been predominantly designed for data multicast which is not sensitive to delay. Single tree based multicast is not well suited for video multicast in ad hoc networks due to the movement of nodes leading to link breaks in the tree and hence not ensuring continuous reachability. Mesh based multicast has some amount of robustness but the overhead is more and also mesh based protocols transmit the same data on different paths to a receiver and also one cannot assure more than one path to a receiver. The reception quality of video at receivers depends on the path to the source. If the path has breakups and makeups like the transient state of not receiving the video packets leads to interruptions. Hence to sustain path breaks we need some redundancy. By redundancy we mean multiple paths to receivers. The source has to multicast the video data using these redundant paths and minimize the number of interruptions there by improving the video quality. Hence we look to multiple tree multicast.

Multiple tree protocols not only assure more than one path to the receivers but also provide control on what to send on each of the trees. Having the control on what to send on each of the trees is of paramount importance to video encoding techniques. Conventional Single Description Coding (SDC) does not utilize the path-diversity in an efficient way. It provides redundant data at receiver, receiving the stream along different paths does not improve the quality. Error-resilient video coding techniques have been proposed to alleviate the problems of packet loss in a network. Examples of error-resilient video coding are Layered Coding (LC) and Multiple Description Coding (MDC). The LC encodes the video into layers of different importance. Base layer is the most important layer without which video cannot be decoded, while there are other layers called the enhancement layers which are not a must for video decoding. Also, enhancement layers, which enhance the quality of the video, can be used for decoding only if the base layer is available. Hence, base layer, which requires more protection, can be allocated to more trees while few trees might be given to enhancement layers. MDC generates multiple equally important, and independent complimentary sub-streams, also called descriptions. Each description can be independently decoded and is of equal importance in terms of quality. The quality of the decoded frame improves with the number of descriptions that are correctly received. This enhances the robustness of the multicast, because based on the bandwidth availability receivers can be allocated multiple number of descriptions. Many advances in MDC have made it more widely accepted than LC for video transmission in wireless networks. Apart from continuous reachability, continuous delivery of decodable video data is important. MDC can trivially use the continuous reachability to provide the continuous delivery of decodable video data. On the other hand in LC, we have to ensure continuous delivery of the base layer and only then can we use the enhancement layers. Hence use of MDC supplements the robustness provided by the path-diversity in the multiple trees in the simplest fashion. A more comprehensive comparative study on MDC and LC can be found in [1].

To the best of our knowledge, multiple tree multicast of video over ad hoc networks has not been studied extensively. Hence, we here propose multiple tree video multicast using MDC for robustness against path breaks. The rest of the paper is organized as follows. In section 2 we present related work. In section 3 we describe an online heuristic solution to finding multiple maximally node-disjoint trees. In section 4 we propose the protocol and elucidate it with graphic examples. In section 5 we present the simulation scenario and the simulation results. Finally, in section 6 we conclude with a discussion on future work possible.

2   Related Work

Multicasting in ad hoc wireless networks has been studied extensively in the literature. Different approaches include constructing structures like trees [2] or meshes [3] for multicasting. But none of these solutions are robust enough for multicasting video traffic, because they do not guarantee continuous reachability required for an uninterrupted transmission. Some attempts were made in the video unicasting to reduce the number of interruptions by exploiting path diversity [4][5][6]. The approach they used was to analyze how to use multiple paths for robust transmission of video. In [4] only base layer was transmitted with ARQ. The authors of [5][6] describe how MDC can use multiple paths to distribute descriptions over different paths. The emphasis in all this work has been to use multiple paths for the video unicast.

One-to-one multipath solutions cannot be trivially extended to one-to-many multipath transmissions without tremendous increase in the control overhead. Robust one-to-many transmission need structures like multiple trees. But not much work has been done in this direction. The Independent-Tree Ad Hoc Multicast Routing (ITAMAR) [7] creates multiple multicast trees simultaneously based on different metrics in a centralized fashion. One main problem of ITA-MAR is that routing overhead might be very large to get enough information of the network to build multiple trees, and the authors only show how ITAMAR works based on perfect network information. The Multiple Disjoint Trees Multicast Routing (MDTMR) [8] protocol creates two node disjoint trees one after the other in a serial fashion. However, the authors of [8] themselves point out that unless the network is dense all receivers might not participate in the trees leading to a drastic increase in the packet loss rate. Also the frequent flood of the network with *JoinReq* control packets leads to a tremendously high overhead. Our protocol is aimed at finding multiple maximal node disjoint multicast trees with lesser overhead in a distributed fashion ensuring connectivity to all the receivers.

3   Online Heuristic

In this section we look into an online heuristic solution to finding a given number of node-disjoint multicast trees. We seek to find out maximally node-disjoint multicast trees (allowing some amount of overlap) in the network and maintain them for our use. This overlap leads to reduction in robustness but finding node-disjoint trees might bring in the problems of feasibility. Moreover it is well known that maintaining network topology in ad hoc networks involves very high overhead as the topology keeps continuously changing due to unrestricted mobility of the nodes. Hence we need a distributed approach to finding the trees. Also in a realistic multicast scenario receivers keep joining and leaving the multicast group as time progresses. Hence a centralized heuristic approach would not work efficiently as the heuristic has to be applied again and again to account for receivers joining and leaving. The heuristic has to be online where requests are to join and to leave. These two requests are sufficient to model such a network where there is no mobility. Mobility is a characteristic feature of ad hoc networks and hence the topology of the network keeps changing. Hence to model this kind of system, we need to also include a request called the movement request. This request changes the network, in a certain way. The change might be reflected in the multiple multicast tree system, that is, it might lead to tree break or it might create possibilities for a better system (lesser overlap).

Heuristic approaches can minimize the overlap by minimizing the number of common nodes among different trees. Hence we propose a distributed online heuristic solution to minimize the number of intersections, while serving the requests online. Let us henceforth call such a system as K-Tree system where we are trying to maintain K maximally node-disjoint multicast trees. A join request would involve passing some control messages in the network and getting the new K-Tree system. A leave request would involve percolating up the trees a delete message until a node with degree more than two in that tree is met.

A movement request would involve changing a node position according to the request. For simplicity let us assume movements occur in bursts. The reason for this assumption is that it would be difficult to model a continually moving node theoretically. Hence we assume that a node moves in bursts of distance and these movement bursts are instantaneous to account for the fact that even mobile nodes participate in tree formations and repairs.

It is already easy to see that, that online algorithm is the best one which serves the request with minimum message passing, adding the least possible number of forwarding nodes, and making the least number of increase in the number of intersections among the trees in the system. But some thought would directly prompt us to the fact that there is mobility. Minimization of forwarding nodes can lead to a problem. When a node moves from one place to another, in the system, it might lead to a substantial number of receivers dependent on this to go orphan for the time until the system takes care of them. At the same time it would not be appropriate to include many forwarding nodes. This is because a tree system with more forwarding nodes leads to more movement requests leading to maintenance requests. This observation comes from the fact that not all movement requests can lead to maintenance requests. But as number of nodes increases, the fraction of movement requests leading to maintenance requests also increases.

Hence the problem now boils down to finding an algorithm which serves the requests by minimizing the control overhead and also the number of intersections among the trees. We try to solve this problem by a weight (a positive quantity used to judge a node) model.

Weight Model: To attain node disjointedness with a weight model we incorporate a penalty system. We penalize a node for participating in more than one tree. We associate a weight with each node. The penalty increases the weight of the node as the number of trees it participates in increases. The system, for each request, has to reduce the total increase in weight of the nodes in the network.

In assigning weights to the nodes it must be noted that a single node serving 2 trees must weigh more than 2 nodes, which are each serving a tree. In a similar way, it has to be noted that choosing a single node for l trees should increase the weight more than the case when l different nodes are chosen for the l trees. Hence the minimum weight of a node participating in l trees should be at least l times the weight of a node participating in $l - 1$ trees. The following cost model satisfies these conditions.

- The weight of a node in the graph is zero if it participates in no trees.
- The weight of a node in the graph which participates in one tree is x, for some $x > 0$.
- The weight of a node in the graph which participates in l trees, w_l, $l \leq k$ is $l * w_{l-1} + y$ for some $y > 0$.
- The cost of a path is the sum of the weights of the nodes in the path.
- The cost of an operation on the graph is the positive change in the total weight of the graph.

The term y can be used to quantify the number of intersections among the trees while node participations can be captured by the term x. If x is large compared to y then, in the total weight of a path, the contribution made by node participations will dominate the contribution made by the number of intersections and vice-versa.

4   The K-Tree Protocol

In this section we define the protocol which is modeled as an online algorithm based on the weight model mentioned in the previous section. The protocol defines how each of the above mentioned requests is handled while trying to minimize the total weight of the nodes. Minimizing the weight is the heuristic idea behind reducing the number of shared nodes among the trees. A scenario for the protocol would be to maintain maximal node-disjoint multicast trees according to a request sequence having the requests mentioned in the previous section. Each request in the sequence triggers an operation on the network which determines the changes that need to be done to the network in order to serve the request and maintain the multiple tree system. The changes are usually accomplished by passing messages in the network. These messages measure the overhead of the protocol. In order to reduce this overhead we relax our condition that each receiver should be a leaf in the subtree that it wants to participate in. By allowing a receiver to be a forwarder overhead can be reduced to an extent in realistic scenarios. For example, a forwarder who wants to later become a receiver can directly become so without any message passing.

We elucidate the protocol by partitioning it into two phases. The first phase describes how a receiver joins the multiple tree system. The next phase describes how a maintenance of the tree system is done.

4.1   Multiple Tree Initialization Phase

The tree initialization phase is initiated by the receivers. Each node in the network can participate in the K-Tree system either as a receiver or a forwarder. Hence each node represents its participation in the K-Tree system using a K-bit vector kvp. If the jth bit in kvp is set then the node is either a receiver or a forwarder for the jth tree, X_j. When some node wishes to become a receiver, it uses a K-bit vector to represent the trees that it wishes to join, kvj. If the jth bit is set in kvj then the node wishes to be a receiver in tree X_j. A node which wishes to become a receiver in X_j can trivially become so, if it is already a forwarder in X_j. Thus the receiver needs to join only those trees where it is not currently participating as a forwarder. Hence it forms a K-bit vector, kvs, which represents the trees for which it is not a forwarder but it wants to join. It then floods the network with a *Join* control packet expressing its wish to join in the trees represented by kvs.

The flooded *Join* control packet reaches different nodes in the network. A node when receives a *Join* control packet has to check if the packet has been already processed by the node before. To make this identification, each *Join* control packet carries the full path that it has traversed. If the node is in the path traversed by the packet, then the packet is dropped. Otherwise the node has to check if it can reply to the sender of this *Join* control packet. A reply can be sent only in the case when the node receiving the packet participates in any of the trees sought by the *Join* control packet. The receiving node has to check to find out to how many trees it can send a reply. The node then replies using a K-bit vector, kvr to the sender of *Join* control packet using the *Reply* control packet. If j^{th} bit in kvr is set then the node is replying for the tree X_j, meaning the node participates in X_j and also that the *Join* control packet sender wants to join X_j. The *Reply* control packet contains the complete path information to the source, that is, the nodes and their participation vectors in the path to the source for each of the trees represented by kvr.

As long as there are trees for which the node cannot reply (as it may not be participating in them at the moment), the *Join* control packet has to be forwarded to other nodes. That is, even if there is one tree for which a reply has not been sent by this node then the node needs to forward the *Join* control packet with the vector kvf. If the jth bit in kvf is set then it means the node cannot reply to the *Join* control packet sender for tree X_j as it does not participate in it.

The receiver when receives a *Reply* packet has to collect the reply and store all the paths that are obtained due to this reply. For each tree X_j, for which a *Join* control packet was sent, the receiver stores all paths that are obtained. A receiver that has a *Join* control packet which has been sent whose timer has not expired yet, maintains a cache called ReplyBuf which stores a certain number of replies per tree.

After the timer initiated while sending the *Join* packet expires, the receiver has to choose those paths to each of the trees that it sought, which add the least amount of weight to the multiple tree system, as this would be intuitively minimizing the number of intersections among the trees. The receiver collects a set of paths to each of the trees and now it has to choose one path to each of the trees. The heuristic we apply is to choose those paths to trees that add the minimum number of intersections that is the minimum amount of weight to the system. For doing this, it evaluates all possible path combinations that can be chosen to reach each of the trees. Then it chooses those paths which add the least amount of weight to the system. The weights are calculated using the weight model described in the section 3. Any combination of paths has a certain number of new overlaps created among the trees. These new overlaps increase the weights according to the weight model. Evaluating possible combinations gives the combination with the least increase in weight which intuitively minimizes the added number of intersections.

The receiver now finally has to send the *Ack* control packets to acknowledge the nodes in the paths chosen. It simply unicasts *Ack* control packets to its immediate parents in each of the trees and they in turn percolate it up. The nodes receiving *Ack* packets establish forwarding states and initialize timers for tree maintenance and tree tear down. Tree maintenance is triggered when a *KeepAlive* control packet does not arrive from the parent in time *KEEP_ALIVE_TIME*. And a forwarder discards the forwarding state when it does not receive passive *KeepAlive* control packets from the downstream nodes for *TREE_TEAR_DOWN_TIME*. Source starts sending *KeepAlive* control packets, passively in data packets or exclusively for every *SEND_KEEP_ALIVE* time units. Forwarders in turn keep forwarding it down the trees. The *KeepAlive* packet also has a K-bit vector representing the trees it wishes to refresh, it carries along with it the path information for these trees, that is, the nodes and their participation vectors. *KeepAlive* control packets are also used to update the path information by the nodes in the trees. It must be noted that for the protocol to work *KEEP_ALIVE_TIME* must be more than *SEND_KEEP_ALIVE*.

An example of tree initialization is shown in Figure 1(a). It shows a multicast session in which 4 receivers (nodes 3, 8, 12, and 18) have already joined along the 2 trees. It also shows weights of the nodes, determined based on nodes' current participation vectors. Node 5 is a new receiver who wants to join into the session. It broadcasts a combined join request which is replied by node 3 for

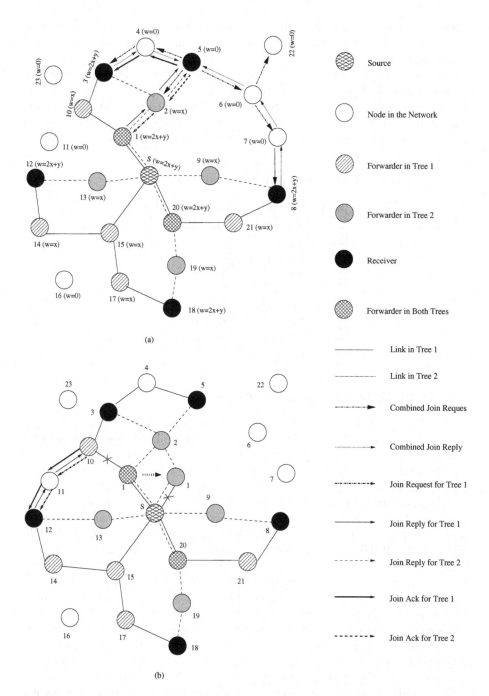

Fig. 1. The *K*-Tree protocol, *K*=2, (a) depicts tree initialization phase and (b) depicts tree maintenance phase

both trees and by node 2 for tree 2 and by node 1 for tree 1 and by node 8 for both trees. Each reply packet carries the complete path information to the source S. Let paths retrieved from reply packets be $p_{11}, p_{31}, p_{81}, p_{22}, p_{32}, p_{82}$ where p_{ij} represents the path to the source S via node i for tree j. Hence now the new receiver (node 5) has 9 combinations (i.e., 3 replies for tree 1 multiplied by 3 replies for tree 2) of paths to choose from. These possible combinations are evaluated (refer Table 1) and the set of paths p_{31} and p_{22} is chosen as the one adding the least weight (according to the weight model given in section 3).

Table 1. Comparison of Path Combinations

S. No.	Combination	Added Weight
1	p_{11} and p_{22}	$3x + 2y$
2	p_{11} and p_{32}	$4x + 2y$
3	p_{11} and p_{82}	$5x + 2y$
4	p_{31} and p_{22}	$3x + y$
5	p_{31} and p_{32}	$4x + 2y$
6	p_{31} and p_{82}	$5x + y$
7	p_{81} and p_{22}	$4x + y$
8	p_{81} and p_{32}	$5x + y$
9	p_{81} and p_{82}	$6x + 3y$

4.2 Multiple Tree Maintenance Phase

The multiple tree maintenance is done in a soft state manner. Whenever a receiver or a forwarder gets a data packet in a tree then it refreshes its timers for that tree, that is data packets are passive *KeepAlive* packets. When the timer expires, that is if there is no data packet or a *KeepAlive* control packet for a *KEEP_ALIVE_TIME* time then the node initiates this process.

The node initiating this process sends out a join request for those trees alone which are broken. It may happen that a few trees are broken simultaneously. Hence whenever there is a tree where there is no packet since *KEEP_ALIVE_TIME* time, all the other trees are tested for timeouts, and all the trees that are broken are found out and a join request for those trees is sent. This is done using the *Join* packet by setting only those bits in the K-bit vector. Because it is a maintenance operation, the tree might just be nearby, so the time out for starting the processing of the received *Reply* packets is maintained much lesser than the one corresponding to the tree initialization process. Further the node detecting path break sends a dummy *KeepAlive* packet to the subtree(s) under it. The dummy *KeepAlive* packet notifies the downstream nodes that a tree maintenance is in progress and that they should not initiate their own tree maintenance.

An example of tree maintenance is shown in Figure 1(b). When the node 1 moves away, the link in tree 1 between node 10 and node 1 is broken. The downstream node 10 detects the breakup and triggers the maintenance process. It sends out a short timed *Join* control packet. This control packet is replied by node 12. Then an exchange of *Reply* and *Ack* packets reconnects the node 10 to tree 1.

5   Simulation Scenario

We use the simulation model based on NS-2 simulator [9]. The IEEE 802.11 DCF is used as the underlying MAC layer protocol. The channel capacity is 11 Mbps, and a radio range of 250 meters is used. The random waypoint model is used to model mobility of the nodes. Each node moves with some constant speed (i.e., minimum speed is equal to maximum speed) with zero pause time. In each run, we simulate a 65 node ad hoc network within a 1350m X 1200m area. Each simulation is 900 seconds long and the results are averaged over 30 runs. We randomly choose one sender and 10 receivers in each simulation who join and leave K-Tree system randomly.

5.1   Metrics

We use the following metrics to evaluate the performance of the protocol.

1. **Ratio of Bad Frames (RBF):** It is the ratio of number of bad frames (totaled for all receivers) versus the number of frames that have to be received in all, by all the receivers. We define a frame as bad if none of the multiple descriptions of a frame are received at the receiver before the play back deadline for the frame.
2. **Normalized Packet Overhead (NPO):** It is the ratio of the total number of control and data packets exchanged in the network over the total number of data packets received by the receivers. This is used to illustrate the forwarding efficiency and also maintenance ability.

5.2   Simulation Results

In our simulations we have chosen the following values for the parameters of K-Tree protocol. $x = 1$, $y = 2$, $MAX\_REPS = 4$, $MAX\_REPLIES\_TIMER = 200$ ms, $SEND\_KEEP\_ALIVE = 125$ ms, and $KEEP\_ALIVE\_TIME = 250$ ms.

Effect of increasing K: Here we show how the protocol seamlessly scales from one tree to two trees and three trees without doubling and tripling the overhead and at the same time improving the video quality at the receivers. We compare the results for $K = 1$ against $K = 2$. In this case we use a two description coding of video. The total video transmission rate is fixed at 48 Kbps with 8 frames per second. MDC coder generates 2 descriptions with 24 Kbps in each description. We have kept packet rate as 8 packets per second with each packet having a size of 3 Kb in each description. When $K = 1$ we send both the descriptions on the same tree and when $K = 2$ we use one tree per description. Similarly we compare $K = 1$ and $K = 3$. Here we split the video into three descriptions each of 8 packets per second with packet size 2 Kb. In all our simulations, the play back deadline for a frame is 120 ms after it was created.

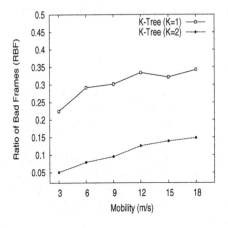 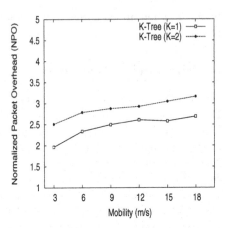

Fig. 2. Variation of RBF vs. Mobility for K-Tree ($K = 1$) and K-Tree ($K = 2$)

Fig. 3. Variation of NPO vs. Mobility for K-Tree ($K = 1$) and K-Tree ($K = 2$)

- Comparison of K-Tree ($K = 1$) and K-Tree ($K = 2$) : Figure 2 shows the expected decrease in the RBF when K is increased for different mobility values. RBF has almost fallen down by 60% when moving from $K = 1$ to $K = 2$. As expected RBF values for both the cases decrease with increasing mobility. Figure 3 shows the expected increase in the overhead due to increase in the number of forwarders. But it has to be noted that the NPO has not doubled, yet a significant improvement has been achieved in RBF. This directly follows from the fact that the protocol, K-Tree uses bit vectors in common join and maintenance control packets to reduce overhead.
- Comparison of K-Tree ($K = 1$) and K-Tree ($K = 3$) : Figure 4 shows the expected decrease in the RBF when K is increased. RBF has almost fallen

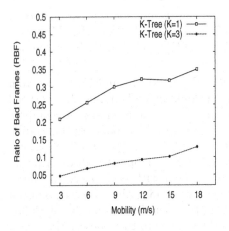 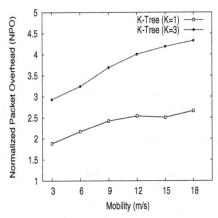

Fig. 4. Variation of RBF vs. Mobility for K-Tree ($K = 1$) and K-Tree ($K = 3$)

Fig. 5. Variation of NPO vs. Mobility for K-Tree ($K = 1$) and K-Tree ($K = 3$)

by 65% when moving from $K = 1$ to $K = 3$. Figure 5 shows the expected increase in the overhead due to increase in the number of forwarders. NPO has not tripled, yet there is a substantial decrease in RBF. This again shows how important the bit vectors in control packets are in reducing the overhead.

6 Conclusion and Future Work

In this paper we designed a protocol to obtain and maintain path-diversity for robustness for video multicast over ad hoc wireless networks. We have shown that the protocol scales to two or three trees without doubling or tripling the overhead, respectively. Further work may involve taking weights as functions of battery power and bandwidth available at nodes. This would help bringing in energy and bandwidth awareness into the protocol. Further work may also involve finding optimal allocations of trees and partitioning of receivers according to some good video encodings. We also would like to compare our protocol with the existing two-tree video multicast protocols (e.g., MDTMR) to show that the our protocol efficiently, in terms of overhead, provides high quality video as compared to them.

References

1. J. Chakareski, S. Han, and B. Girod, "Layered Coding vs. Multiple Descriptions for Video Streaming over Multiple Paths", *in Proc. of ACM Multimedia*, pp. 422-431, November 2003.
2. Sung Ju Lee, William Su, and Mario Gerla, "On-Demand Multicast Routing Protocol in Multihop Wireless Mobile Networks", *Mobile Networks and Applications*, vol. 7, no. 6, pp. 441-453, December 2002.
3. J. J. Garcia-Luna-Aceves and E. L. Madruga, "The Core-Assisted Mesh Protocol", *IEEE Journal on Selected Areas in Communications*, vol. 17, no. 8, pp. 1380-1994, August 1999.
4. S. Mao, S. Lin, S. Panwar, and Y.Wang, "Reliable Transmission of Video over Ad-hoc Networks using Automatic Repeat Request and Multi-path Transport", *in Proc. of IEEE VTC 2001*, pp. 615-619, October 2001.
5. S. Mao, S. Lin, S. Panwar, Y. Wang, and E. Celebi, "Video Transport over Ad hoc Networks: Multistream Coding with Multipath Transport", *IEEE Journal on Selected Areas in Communications*, vol. 21, no. 10, pp. 1721-1737, December 2003.
6. J. G. Apostolopoulos, "Reliable Video Communication over Lossy Packet Networks using Multiple State Encoding and Path Diversity", *in Proc. of Visual Communication and Image Processing 2001*, pp. 392-409, January 2001.
7. Sajama and Zygmunt J. Haas, "Independent-tree Ad hoc Multicast Routing (ITAMAR)", *Mobile Networks and Applications*, vol. 8, no. 5, pp. 551-566, October 2003.
8. W. Wei and A. Zakhor, "Multipath Unicast and Multicast Video Communication over Wireless Ad hoc Networks", *in Proc. of BROADNETS 2004*, pp. 496-505, October 2004.
9. NS-2: network simulator, *http://www.isi.edu/nsnam/ns/*

Towards Estimating Lifetime of Ad Hoc Wireless Networks

S. Jayashree[1] and C. Siva Ram Murthy[2,*]

[1] Computer Science and Artificial Intelligence Laboratory,
Massachusetts Institute of Technology, MA, USA
[2] Department of Computer Science and Engineering,
Indian Institute of Technology, Madras, India
jaya@mit.edu, murthy@iitm.ac.in

Abstract. Ad hoc wireless networks possess highly constrained energy resources. Even the energy aware protocols for ad hoc networks do not consider all the characteristics of the underlying batteries. Hence, they fail to efficiently utilize the available energy. Thus, a mechanism is required to measure the efficiency of the protocols of ad hoc networks, in terms of the network lifetime. To the best of our knowledge, there has been no reported work till date for analyzing the lifetime of the ad hoc networks for various protocols. This paper primarily provides a novel generalized analytical model for estimating lifetime of ad hoc networks, in the presence of the following two kinds of MAC protocols: (i) reservation-based TDMA protocols and (ii) a specific class of CSMA protocols that try to follow a pattern, such as a *round-robin* scheduling, for packet transmission. We prove through analytical and simulation studies that energy awareness is crucial in deciding the performance of the MAC protocols.

1 Introduction and Related Work

The nodes of an ad hoc wireless network, a group of uncoordinated nodes which self organize themselves to form a network, have constrained battery resources. For example, in search-and-rescue operations, battle-fields, and in other places where setting up of a network is difficult, it becomes difficult to replace or recharge the batteries of the dead nodes. On the other hand, ad hoc wireless networks, with characteristics such as the lack of a central coordinator and unrestricted mobility of the nodes, as in the case of battle-field networks, require nodes with a very high energy reserve. In such scenarios, there exists a need for battery/energy aware protocols at all the layers of the protocol stack. Several protocols have been proposed, in order to improve the lifetime of the ad hoc networks. However, not all the protocols, proposed for ad hoc networks, are energy efficient. In addition, there also exists a few energy unaware protocols, for ad hoc networks, which unknowingly provide a better lifetime for the nodes. Hence, there exists a need for analyzing the performance of all these protocols, in terms of lifetime. The lifetime of ad hoc networks can be defined as the time between the start of the network, when the network becomes operational to the death of the first node [1]. In this paper,

---

* This work was supported by the Department of Science and Technology, New Delhi, India.

Y. Robert et al. (Eds.): HiPC 2006, LNCS 4297, pp. 436–447, 2006.

we focus only on the MAC layer protocols [1], [2]. This study can easily be extended to study the effect of other higher layer protocols on the lifetime of the network.

The authors of [3] and [4] provide an analytical model that captures the behavior of the batteries in the presence of binary and pulsed discharge. We use the battery recovery model provided by Chiasserini and Rao in [3] for our analysis. The authors of [5] extend the basic model for batteries provided in [4] for ad hoc wireless networks that possess nodes powered by a dual battery setup. However, these works model a single node's battery discharge based on the traffic existing at that node or based on the discharge properties of the batteries. To the best of our knowledge, there has been no reported work till date for analyzing the lifetime of the ad hoc networks for various protocols. We found that many of the existing MAC protocols fail drastically, to provide a higher lifetime, in the presence of a limited energy source. Even many of the existing energy aware protocols, such as [6] and [7] do not take into consideration all the properties of the batteries, such as the recovery capacity effect [8]. Though they perform better than the non-power aware protocols, in the process of energy saving, they do not provide idle time for nodes and do not encourage battery recovery. Thus they do not exploit the underlying battery properties efficiently. This paper proposes a novel general analytical model for estimating the lifetime of all TDMA-based MAC protocols for ad hoc networks. The model can also be used for calculating the lifetime of ad hoc networks that use CSMA based MAC protocols, which follow a regular pattern (a combination of packet transmissions and idleness of the nodes) for packet scheduling, such as a round-robin scheduling. In the discussions that follow, we use the term *pattern-based CSMA* to denote these special CSMA protocols.

The rest of the paper is organized as follows. Section 2 describes the proposed analytical model in detail, followed by Section 3, which discusses two MAC protocols, BAMAC(k) [1] and RTMAC [2], to verify the proposed model. Section 4 discusses the factors affecting the accuracy of our analytical model and analyzes the possibilities for future work and Section 5 concludes the paper.

2   Analytical Model for Estimating Lifetime of Ad Hoc Networks

We now propose a novel theoretical model to estimate the lifetime of *homogeneous ad hoc wireless networks*, in which all the nodes possess batteries with similar characteristics. Our model uses discrete-time Markov chains with probabilistic recovery to capture the behavior of the batteries of the nodes. In this section, we provide a mechanism to model batteries, taking into consideration the underlying MAC protocol. The basics of batteries can be found in [8]. We assume the existence of a Smart Battery System (SBS) [1], which provides the state of a node's battery and hence, enables the control of its behavior such as charging and discharging.

2.1   Description of the Model

The model used for analyzing the batteries of the nodes in a homogeneous ad hoc wireless network assumes all the nodes to have similar battery and traffic characteristics. The behavior of the batteries of the nodes, using TDMA or pattern-based CSMA-based

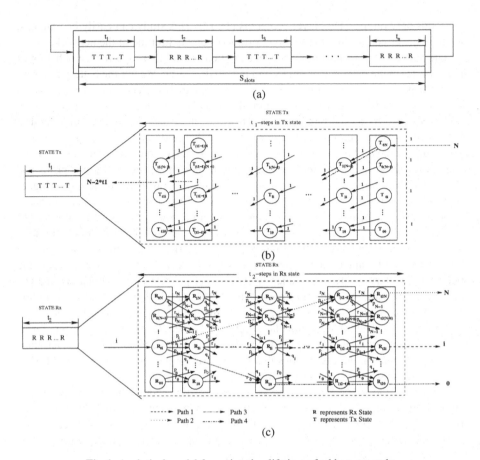

Fig. 1. Analytical model for estimating lifetime of ad hoc networks

protocols, is represented using a Markov model as shown in Figure 1(a). A pattern may have combinations of packet transmissions and idleness of the nodes. This combination then repeats itself perfectly in TDMA and with some probability in CSMA protocols. A generalized combination of transmission and idle (recovery) state of the nodes for all kinds of traffic is shown in Figure 1(a). Here, we assume the recovery state R to be either receiving or the idle state of the node. Here, a set of t_1 transmissions is followed by a set of t_2 recoveries, which is then followed again by a set of t_3 transmissions and so on, where t_i, $1 \le t_i \le S_{slots}$ and $\sum_{i=1}^{i=n} t_i < S_{slots}$ represents the number of transmissions or recoveries in the ith occurrence of Tx (transmission) or Rx (reception) burst, respectively and S_{slots} is the maximum number of slots in the pattern. In TDMA protocols, S_{slots} represents the number of slots available in the superframe. The state of the battery in the Markov model represents the remaining nominal capacity of the battery. Hence, the battery can be in any of the states from 0 to N, where N is the nominal capacity of the battery and state 0 represents its dead state. The battery model assumes that, in any Δt time interval, the battery can remain in any one of the two main states – transmission state (Tx) or the reception state (Rx). In Tx state, the node transmits its packets and discharge its battery. Whenever the node does not have packets to transmit, it enters into the Rx state. In Rx state, the node either receive packets and discharge or remain idle and recover charges with some probability. Here Δt, the basic time interval, is the time taken for one packet transmission. We define a term called cycle time, which is the time between two successive entries in to the first occurrence of the Rx or Tx states of the successive patterns.

If the state of the battery is denoted by the tuple $< Ni, Ti >$ and the initial state is given by the tuple $< N, T >$, in one time interval Δt, a battery which is in state $< N_i, T_i >$, goes to state $< N_{i-1}, T_{i-1} >$ if it discharges. If the battery remains idle in recovery state, it reaches $< N_{i+1}, T_i >$ or $< N_i, T_i >$ with probabilities R_{N_i,T_i} and I_{N_i,T_i} respectively, where the probability to remain in the same state on being idle is given by $I_{N_i,T_i} = 1 - R_{N_i,T_i}$. Hence, the battery can be modeled differently in each of these two states and the battery flip-flops between these two states. The generalized stochastic model representing the battery behavior in the network which operates using TDMA or pattern-based CSMA-based MAC protocols is shown in Figure 1(a). In each time interval Δt, if the node remains in Tx state, it transmits a packet and the battery discharges two units of its charge or, if the node remains in the Rx state, the neighbor nodes transmit and if the node does not receive any packets, the battery recovers one unit of the charge with probability R_{N_i,T_i}, where R_{N_i,T_i} is given by $R_{N_i,T_i} = e^{-g \times (N-N_i) - \phi(T_i)}, if\ 1 \leq N_i \leq N,\ 1 \leq T_i \leq T$ and 0 otherwise, where g is a constant value and $\phi(T_i)$ is a piecewise constant function of number of charge units delivered which are specific to the battery's chemical properties [1]. This value affects the battery recovery drastically. If a node receives a packet in the Rx-state, its battery discharges one unit of its charge. In the model shown above, R_{ij} (T_{ij}) represents the battery in the Rx (Tx) state at time unit i and j represents the remaining nominal capacity of the battery. R_{I0} and T_{I0} represent the battery in its dead (absorbing) state with nominal capacity 0 at any time unit I. If R and T represent the unit transition probability matrices for a single discharge and recovery state of the battery of a node in one basic time unit, the unit transition probability matrix for states Tx and Rx can be calculated from Figures 1(b) and (c), respectively. While the battery resides in Rx state, in each time unit Δt, it recovers one charge unit with probability $r_{\Delta t} = R_{N_{\Delta t}, T_{\Delta t}}(1 - q_{\Delta t})$ and enters into higher state, or remains in the same state with probability $p_{\Delta t} = I_{N_{\Delta t}, T_{\Delta t}}(1 - q_{\Delta t})$. Here, $q_{\Delta t}$ is the probability that the node receives a packet in Δt time unit. The transition probability matrices for this model can be calculated as follows.

$$
Tx = T^{t_1} = \begin{bmatrix} 1 & 0 & 0 & \ldots & 0 & 0 \\ 1 & 0 & 0 & \ldots & 0 & 0 \\ 1 & 0 & 0 & \ldots & 0 & 0 \\ 0 & 1 & 0 & \ldots & 0 & 0 \\ \vdots & \vdots & \vdots & & \vdots & \vdots \\ 0 & \ldots & 0 & 1 & 0 & 0 \end{bmatrix}^{t_1} \qquad Rx = R^{t_2} = \begin{bmatrix} 1 & 0 & 0 & 0 & \ldots & 0 & 0 \\ q_1 & r_1 & p_1 & 0 & \ldots & 0 & 0 \\ 0 & q_2 & r_2 & p_2 & 0 & \ldots & 0 \\ \vdots & \vdots & \vdots & & & \vdots & \vdots \\ & & & & \ldots & & \\ 0 & 0 & \ldots & q_{N-1} & r_{N-1} & p_{N-1} \\ 0 & 0 & 0 & \ldots & 0 & q_N & r_N \end{bmatrix}^{t_2} \quad (1)
$$

This states that whenever the battery enters in to Tx state with a nominal capacity of i, it leaves Tx state with a nominal capacity of $i - 2 * t_1$, with a probability of 1. Here, we assume that the nodes have at least k packets in the data buffer for transmission. We assume the matrix index to start from 0 for the ease of denoting 0th or the dead state. When the battery enters in to Rx state with a remaining nominal capacity of i, the probability that the battery will leave Rx state, after t_2 slots, with a nominal capacity of j, where $j = i$ is given by $Rx_{i,i}$. Hence, $Rx_{i,i}$ is the probability that the battery does not recover any charge after spending t_2 slots in Rx state and $Rx_{i,j}$ represents the probability that the battery, on entering Rx state with a nominal capacity of i leaves Rx state, after spending t_2 basic time units with a nominal capacity of j.

In Figure 1(c), let the battery enter Rx state with a remaining nominal capacity of i units. Hence, at 0th time unit, it remains in state R_{0i}. After t_2 time units, the battery can be in any of the states from $i - t_2$ to $i + t_2$ based on the number of packets received and the probability of recovery; that is, if the node receives data in all the t_2 time units, it goes to $i - t_2$ state and if the node does not receive any packet in Rx state, then it remains idle for t_2 time units and recovers t_2 charges with a probability Rx_{i,t_2}. Figure 1 shows two such instances. Path 1 shows that the battery remains in the same state i even after idling for t_2 time units, which is represented by the probability $Rx_{i,i}$ and Path 2 shows the state transitions of the battery while traversing from state i to state N in t_2 idle slots. Thus the probability for this transition to happen is $Rx_{i,N}$ if $(N - i) \leq t_2$ and zero otherwise. This is because in t_2 idle slots, the maximum recovery of a battery, starting from state i, is $i + t_2$. Path 3 shows that the node with a remaining battery charge of i receives more than i packets and hence goes to the 0th state or the dead state. Similarly, as shown in Path 4, if the node enters Tx state with a nominal capacity of N, it leaves Tx state with a nominal capacity of $N - 2 * t_1$, with probability 1. The states of the battery at different time units in Tx state is shown using Path 4.

In order to understand the battery behavior for a single occurrence of the pattern, we should calculate the probability matrix for the Markov model representing the pattern. The probability matrix is the one which differs from one MAC protocol to another and characterizes them, the calculation of which is explained in detail in the subsequent subsections for RTMAC and BAMAC(k) protocols. For the generalized model shown in Figure 1, the probability matrix for the pattern is given by, $P = [T]^{t_1}[R]^{t_2}[T]^{t_3} \ldots [R]^{t_{n-1}}[T]^{t_n}$. The time duration for which the Markov model remain in transient states is calculated as follows: (a) Given any probability matrix P, calculate matrix $Q=[Q_{(i,j)}]$, where i and j represent only the transient states. Remove the rows and columns corresponding to the absorbing states of the Markov model. In our protocol, state 0 corresponds to the absorption state or the dead state of the model. Hence, $Q_{(i+1,j+1)} = P_{(i,j)}$. The index of matrix Q starts from index 1 to represent the absence of the state 0. (b) Calculate matrix $M = (I - Q)^{-1}$, where I is the identity matrix. (c) Now $M_{(i,j)} \times \Delta t$ represents the total number of times the battery enters state j if the starting state is i and Δt is the time duration the Markov model spends in state j. Based on the above steps M is calculated as follows,

$$
M = \begin{bmatrix}
1 & 0 & \cdots & 0 & 0 \\
0 & 1 & \cdots & 0 & 0 \\
\vdots & \vdots & & \vdots & \vdots \\
0 & 0 & \cdots & 0 & 0 \\
z_{0,0} & z_{0,1} & \cdots & z_{0,N-1} & z_{0,N} \\
z_{1,0} & z_{1,1} & \cdots & z_{1,N-1} & z_{1,N} \\
\vdots & \vdots & & \vdots & \vdots \\
z_{N-k-2,0} & z_{N-k-2,1} & \cdots & z_{N-k-2,N-1} & z_{N-k-2,N}
\end{bmatrix}^{-1}
\qquad
Z_{(i,j)} = \begin{cases} 1 - Rx_{i,j} & if\ i=j \\ -Rx_{i,j} & otherwise \end{cases}
$$

Here, since state 0 was removed, matrix starts from index 1 (representing state 1). Hence, M is an $N \times N$ matrix starting from state 1 (index 1) whereas, P is an $(N +$

1) $\times$ ($N + 1$) matrix starting from state 0 (index 0). Let, T_{active} of a battery model give the total active time of batteries. Here we assume that the starting state is state N, that is, we start with a fully charged battery system. Now the time for which it remains active or the lifetime of the battery can be given by, $T_{active} = \sum_{i=1}^{N} M_{N,i}$. Hence, the total number of discharges is given by,

$$T_{life} = \sum_{i=1}^{N} M_{N,i} \times \sum_{j=1}^{i-1} P_{i,j} \tag{2}$$

where $P_{i,j}$ corresponds to the entry at the ith row and jth column. T_{life} corresponds to the total number of left transitions of the model which is the total number of discharges and hence, represents the lifetime of the network. Discharge of a node's battery occurs due to packet transmissions or receptions. Lifetime of a node is proportional to the number of packets transmitted before it enters into the dead state. In order to calculate the total number of packets transmitted, the value T_{life}, which corresponds to both transmissions and receptions of packets, has to be calculated. Then, based on the traffic pattern, the total number of discharges caused due to transmissions alone can be derived. T_{life} cannot exceed the theoretical capacity, even if the recovery of the battery is very high. A lesser value for T_{life} than the theoretical capacity of the battery shows the inefficient battery consumption introduced by the MAC protocol.

3 Verification of the Analytical Model

3.1 For TDMA Protocols

We use RTMAC protocol to verify our analytical model. The real-time medium access control protocol (RTMAC) [2] provides a bandwidth reservation mechanism for supporting real-time traffic in ad hoc wireless networks. We make the following assumptions in the implementation of RTMAC protocol. We assume all the flows to be real-time with a very long deadline. We also assume the abortion of a flow to occur only due to the death of a node and not because of the missing of a flow deadline. We assume one basic time unit Δt as the time taken for one packet transmission. Δt consists of several superframe slots.

Analysis of RTMAC Protocol. As mentioned earlier, the proposed analytical model varies with different MAC protocols, only in their probability matrices. The value of t_i, $1 \leq t_i \leq S_{slots}$ and $\sum_{i=1}^{i=3} t_i < S_{slots}$ varies depending on the nature of the protocol. Thus we first calculate, from the behavior of the RTMAC protocol, the unit probability matrix of the analytical model for a single occurrence of the pattern. In RTMAC protocol, this time duration is the superframe time duration. The cycle time in RTMAC protocol is also the superframe time. In one superframe, each node reserves a set of slots and then transmit its packets. Once the reservation is made in the first superframe, the same reservation is carried for the subsequent superframes. The Markov model representing the behavior of RTMAC protocol is shown in Figure 2. The probability matrix for each of the superframes remains the same. In RTMAC protocol for every node, the

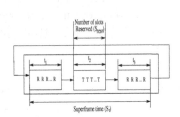

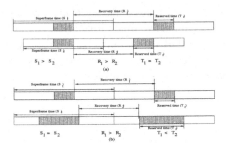

Fig. 2. Illustration of RTMAC protocol   **Fig. 3.** Superframes in RTMAC protocol

probability matrix is given by $P = [R]^{t_1}[T]^{t_2}[R]^{t_3}$, where $0 \leq t_1, t_2, t_3 \leq S_{slots}$ and S_{resv} is the total number of slots reserved by that node. Here S_{slots} is the total number of superframe slots and Tx and Rx are given in Equation 1, respectively. Once the probability matrix is calculated, we use Equation 2 to calculate T_{life}. Based on the traffic in the network, we calculate the maximum number of packets transmitted by the node, which is the parameter used to define the lifetime of the network. We now discuss the lifetime calculation of the network, for the following cases.

(1) In general the total number of discharges by a battery (both due to transmissions and receptions) correspond to its lifetime. We assume that the set of receivers and transmitters of the network form a disjoint set and all the nodes reserve equal number of slots per superframe because of the uniform traffic characteristics. We also assume that more than one node does not transmit to the same receiver. Thus, the transmitters of these networks are the first ones to die, because they discharge twice the charge as the receivers. Thus the lifetime of the transmitters become the lifetime of the network. If a node encounters T_{life} discharges, and if we assume each packet transmission (reception) to correspond to two (one) units battery discharges the total number of packets transmitted by the node is given by, $T_{pkts} = \frac{T_{life}}{2}$. If the node is the receiver, the total number of packets received by the nodes is given by, $T_{pkts} = T_{life}$.

(2) If we assume each node in the network to transmit and receive equal number of packet per superframe, the total number of packets transmitted and received by the node is given by $T_{pkts} = \frac{T_{life}}{3}$.

Comparative Study of Theoretical and Simulation Results. We now provide a comparative study of the analytical results with those of simulations. All the protocols discussed were implemented using GloMoSim simulator. All the nodes were assumed to be sending packets with the same transmission power. We used the following simulation parameters: simulation area - 2000m × 2000m, number of nodes - 10 to 40, transmission power - 12dB, channel bandwidth - 2Mbps, C - 2000, N - 250, and battery parameter g - 0.05. The routing protocol used was Dynamic Source Routing (DSR) protocol. In the following discussion, capacity of a battery refers to nominal battery capacity unless otherwise specified. We have assumed the data packet size to be 512 bytes. In the discussions that follow, we assume a static ad hoc wireless networks. All the results in this section have been obtained at 95% confidence level. We make the following assumptions in the implementation of RTMAC protocol. (a) We assume only one flow in the network. In the discussions of the results, we also show that the relaxation of this assumption has no significance on the performance of the protocol, in terms of lifetime. (b) We assume listening to the channel by a node as idling of the node and thus to consume zero power. (c) As mentioned in the analysis, we assume that the set of receivers and transmitters of the network form a disjoint set and all the nodes reserve equal number of slots per superframe because of the uniform traffic characteristics. (d) Similarly a node does not receive packets from more than one source node. (e) We assume the nodes' data buffer to have at least k packets.

In the RTMAC protocol, every node reserves a set of slots in the superframe. The time between two transmissions and recoveries is shown in Figure 3. An increase in the number of nodes in the network does not affect the network lifetime. Thus the time durations T and R are affected only by duration of superframe and number of slots reserved per node per superframe. Figure 4 shows the theoretical estimation and

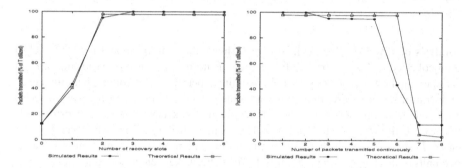

Fig. 4. Effect of duration of superframe on network lifetime

Fig. 5. Effect of k on network lifetime

simulation results of the lifetime of the network for varying recovery slots, which is introduced by varying the superframe duration. As shown in the figure, the lifetime increases as the superframe time (number of recovery slots) increases, which can be observed from Figure 3(a) where $R_1 > R_2$. Thus, as the superframe time increases, the nodes get more time for recovery, which increases the nominal capacity of their batteries. In addition, recovery capabilities of the battery increases with the increase in the nominal capacity. Hence, a burst of recovery slots is favorable than an equivalent amount of discrete recovery slots. The maximum lifetime is attained when whole of the theoretical capacity is utilized. Similarly, Figure 5 shows the lifetime of the network for varying values of k, where k is the number of packets transmitted continuously or the number of slots reserved by the node for packet transmission. It is clear from Figure 3(b) that as k increases, the lifetime of the network decreases because of the increase in the number of slots. A continuous discharge decreases both the theoretical and nominal capacities of the battery, which reduces the recovery capacity effect drastically. Hence, a continuous burst of transmission is highly unfavorable for batteries. The discrepancies between the theoretical and the simulation results are in due to the randomness involved in the probabilistic recovery of the batteries of the nodes.

3.2   For Pattern-Based CSMA Protocols

We use BAMAC(k) protocol [1], a round-robin MAC scheduling protocol, to verify our analytical model. Battery Aware MAC (BAMAC(k)) protocol, an energy-efficient contention-based MAC protocol, tries to increase the lifetime of the nodes by exploiting the recovery capacity effect of the battery.

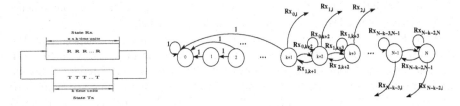

Fig. 6. Illustration of BAMAC Protocol   **Fig. 7.** Markov model representing battery behavior

Analysis of BAMAC Protocol. Figure 6 shows the Markov model for BAMAC(k) protocol. Here, R is the probability matrix of the battery for one basic time unit in Rx state. In the BAMAC(k) protocol, assuming a perfect round-robin scheduling, each node transmits for k basic time units and remains in the receiving or recovery state for $n \times k$ basic time units where n is the number of neighbors. Here, $(nk+k)\Delta t$ corresponds to one cycle time for BAMAC protocol. Hence, in the battery model, whenever the node enters Tx state, it remains there for k units of time and in each basic time unit discharges two units of its charge with probability 1. The battery, then, enters Rx state and remains there for $n \times k$ units of time. The transition probability matrix for Rx state is given by $Rx = R^{nk}$. As in RTMAC, we assume the matrix index to start from 0 for the ease of denoting 0th or the dead state. When the battery enters in to Rx state with a remaining nominal capacity of i, the probability that the battery will leave Rx state, after nk slots, with a nominal capacity of j, where $j = i$ is given by $Rx_{i,i}$. Hence, $Rx_{i,i}$ is the probability that the battery does not recover any charge after spending nk slots in Rx state and $Rx_{i,j}$ represents the probability that the battery, on entering Rx state with a nominal capacity of i leaves Rx state, after spending nk basic time units with a nominal capacity of j. The transition probability matrix for $Rx = R^{nk}$ and $Tx = T^k$ states are given as follows.

$$Rx = \begin{bmatrix} 1 & 0 & 0 & 0 & \dots & 0 & 0 \\ q_1 & r_1 & p_1 & 0 & \dots & 0 & 0 \\ 0 & q_2 & r_2 & p_2 & 0 & \dots & 0 \\ \vdots & & & & \ddots & & \vdots \\ 0 & 0 & 0 & \dots & 0 & q_N & r_N \end{bmatrix}^{nk} \quad Tx = \begin{bmatrix} 1 & 0 & 0 & \dots & 0 & 0 \\ 1 & 0 & 0 & \dots & 0 & 0 \\ 1 & 0 & 0 & \dots & 0 & 0 \\ 0 & 1 & 0 & \dots & 0 & 0 \\ \vdots & & & & & \vdots \\ 0 & \dots & 0 & 1 & 0 & 0 \end{bmatrix}^{k} = \begin{bmatrix} 1 & 0 & 0 & \dots & 0 & 0 \\ \vdots & & & & & \vdots \\ 1 & 0 & 0 & \dots & 0 & 0 \\ 0 & 0 & 1 & \dots & 0 & 0 \\ \vdots & & & & & \vdots \\ 0 & \dots & 0 & 0 & 1 & \dots \end{bmatrix}$$

where, T is the probability matrix for one basic time unit in Tx state. This states that whenever the battery enters in to Tx state with a nominal capacity of i, it leaves Tx state with a nominal capacity of $i - 2k$, with a probability of 1. Here we assume that the data buffer for all the nodes remains always full. That is the nodes always have packets for transmission. Hence, the one-step transition probability matrix for the Markov model for one cycle time of BAMAC(k) is given by

$$P = Tx \times Rx = T^k \times R^{nk} = \begin{bmatrix} 1 & 0 & \cdots & 0 & 0 \\ \vdots & \vdots & \cdots & \vdots & \vdots \\ 1 & 0 & \cdots & 0 & 0 \\ Rx_{0,0} & Rx_{0,1} & \cdots & Rx_{0,N-1} & Rx_{0,N} \\ \vdots & \vdots & \cdots & \vdots & \vdots \\ Rx_{N-k-2,0} & Rx_{N-k-2,1} & \cdots & Rx_{N-k-2,N-1} & Rx_{N-k-2,N} \end{bmatrix}$$

Hence, the final Markov model, for one cycle time unit $((nk+k)\Delta t)$ is shown in Figure 7. Figure 7 shows that a battery in state i, at the end of 1 cycle, can be in any of the states from $i - 2k - nk$ (after discharging $2k$ charge units in the transmission of k packets and discharging nk units in the reception nk packets from all the n neighbors) to $i - 2k + nk$ (after discharging $2k$ charge units and recovering for the whole nk time units). The probability value $Rx_{i-k-2,j}$ ($P_{i,j}$) refers to the probability that the battery goes to state j from state i. Once the probability matrix is calculated, the steps discussed in Section 2.1 are used to calculate the values of T_{active} and T_{life}. In general, T_{life} represents the lifetime of the network. However, as in RTMAC, we assume the lifetime of the transmitter as the lifetime of the network and thus the total number of packets transmitted represent the lifetime of the network. As mentioned earlier, based on the traffic pattern, the total number of discharges caused due to transmissions alone can be derived. We now explain a method to calculate the number of packets transmitted, for one such traffic pattern. In our theoretical model, two discharges of a battery correspond to either a packet transmission or reception of 2 packets. Calculation of total number of packet transmission depends on the value of $q_{\Delta t}$. For example, if we assume that in nk time units spent by the battery in Rx state, each node receives k packets from one out of n neighbors, the probability that a packet is received in one time unit of Rx state is given by $q_{\Delta t} = \frac{k}{nk} = \frac{1}{n}$. Thus, the total number of packets transmitted, in this case, is given by, $Total\ number\ of\ transmissions = \frac{2 \times T_{life}}{3}$. Similarly, if a node does not receive any packet in nk time units, the total number of packets transmitted is equal to $\frac{T_{life}}{2}$. Hence, the total number of packets transmitted can be calculated based on the value of T_{life} and the traffic pattern.

Comparative Study of the Simulation and Theoretical Results. We now discuss the effect of k value on the performance on the system. Figures 8 and 9 show the number of packets transmitted as k value increases from 1 to 20 and 1 to 250, respectively. The corresponding graphs obtained using theoretical analysis are provided in Figures 10 and 11. As shown in the Figure 8-11, an increase in the number of neighbors (n) corresponds to an increase in the number of recovery slots (nk). This ultimately increases the number of packets transmitted because of an increase in the recovery probability. Thus, if longer battery lifetime and higher number of packet transmissions are favored, a smaller value of k, that is $k = 1$, is preferred. Whereas, if higher throughput is preferred, higher values of k is chosen. The discrepancy between the theoretical and the simulation results is mainly because, in the theoretical analysis, a perfect round-robin scheduling of the nodes is assumed. Hence, there exists exactly nk recovery slots and k transmission slots, whereas, in the simulations, we assume a random back-off.

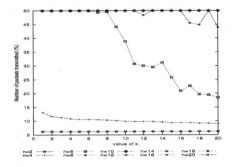

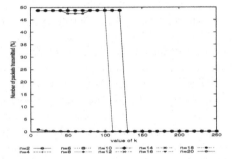

Fig. 8. Number of packets transmitted calculated through simulations for k values 1 to 20

Fig. 9. Number of packets transmitted calculated through simulations for k values 1 to 250

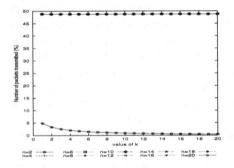

Fig. 10. Theoretical analysis of number of packets transmitted for k values 1 to 20

Fig. 11. Theoretical analysis of number of packets transmitted for k values 1 to 250

4   Limitations of the Model and Directions for Future Work

The model discussed so far assumes that all the nodes in the network possess similar battery and traffic characteristics. However, in real scenario, end users (mobile nodes) of the ad hoc wireless network may have heterogeneous nodes, which may vary in voltages, recovery capacity effect, current rating, lifetime, operational environment, and weight. Hence, heterogeneity has to be taken into consideration in the proposed model.

As mentioned in Section 3.1, we assume that the nodes to consume zero power when they listen to the channel. However, listening consumes a sizable amount of power, which is almost equal to that for packet reception, in the case of sensor networks. The power consumption in the listen state of a node can easily be incorporated in the proposed analytical model, by assuming the listening of node to consume battery charge. In future, we plan to propose a generalized analytical model to analyze the lifetime of both homogeneous and heterogeneous ad hoc wireless networks, for all types of MAC protocols. The problems with CSMA protocols is that they do not follow a specific pattern in transmitting their packets because they follow a random back-off scheme. The behavior of a node's battery recovery and discharge is affected by the activities at all the higher layer protocols. However, since this paper is the first attempt towards constructing a model for estimating lifetime of ad hoc networks, we assumed only the effect of MAC layer to be present on the discharge and recovery of the underlying nodes' batteries.

5 Conclusions

Ad hoc wireless networks are characterized by highly constrained energy resources. All the layers of the protocol stack should possess energy efficient protocols. Thus, there exists a need for analyzing the lifetime of ad hoc networks, in the presence of these protocols. We, in this paper, proposed a general analytical model using discrete-time Markov chain with probabilistic recovery to analyze the lifetime of the homogeneous ad hoc wireless networks, in the presence of TDMA and pattern-based CSMA-based MAC protocols. We also verified the model using two MAC protocols, RTMAC and BAMAC(k). Finally, we have discussed a general mechanism to extend the proposed basic model to estimate the efficiency of all CSMA protocols.

References

1. S. Jayashree, B. S. Manoj, C. Siva Ram Murthy, "On Using Battery State for Medium Access Control in Ad hoc Wireless Networks", *Proceedings of ACM MOBICOM 2004*, pp. 360-373, September 2004.
2. B. S. Manoj and C. Siva Ram Murthy, "Real-Time Traffic Support for Ad Hoc Wireless Networks," *Proceedings of IEEE ICON 2002*, pp. 335-340, August 2002.
3. C. F. Chiasserini and R. R. Rao, "A Model for Battery Pulsed Discharge with Recovery Effect," *Proceedings of IEEE WCNC 1999*, vol. 2, pp. 636-639, September 1999.
4. C. F. Chiasserini and R. R. Rao, "Energy Efficient Battery Management", *Proceedings of IEEE INFOCOM 2000*, vol. 2, pp. 396-403, March 2000.
5. P. Rong and M. Pedram, "Battery-Aware Power Management Based on Markovian Decision Processes", *Proceedings of ICCAD 2002*, pp. 707-713, November 2002.
6. R. Wattenhofer, L. Li, P. Bahl, and Y. M. Wang, "Distributed Topology Control for Power-Efficient Operation in Multi-Hop Wireless Ad Hoc Networks," *Proceedings of IEEE INFOCOM 2001*, vol. 3, pp. 1388-1397, April 2001.
7. A. Srinivas and E. Modiano, "Minimum Energy Disjoint Path Routing in Wireless Ad Hoc Networks", *Proceedings of ACM MOBICOM 2003*, pp. 122-133, September 2003.
8. C. Siva Ram Murthy and B.S. Manoj, "Ad Hoc Wireless Networks: Architectures and Protocols", *Prentice Hall, NJ, USA, 2004*.

A Proxy Based Efficient Checkpointing Scheme for Fault Recovery in Mobile Grid System

Imran Rao[1], Nomica Imran[1], PilWoo Lee[2], Eui-Nam Huh[1,*], and TaeChoong Chung[1]

[1] Department of Computer Engineering, Kyung Hee University,
Yongin-si, Gyeonggi-do, 449-701 South Korea
{imran, nomica}@oslab.khu.ac.kr, {johnhuh, tcchung}@khu.ac.kr
[2] Korea Institute of Science Technology and Information (KISTI)
pwlee@kisti.re.kr

Abstract. Mobile Grid is an emerging and prospering field of distributed computing where mobile devices are enjoying the benefits of Grid. Challenges faced by mobile Grid are unpredictable network quality, lower trust, limited resources (battery power, network bandwidth, storage, processing power, etc) and extended periods of disconnections which may result in lost of the work done by these devices. We, therefore, need a proper fault tolerance scheme for these mobile hosts. A major issue is the appropriate handling of failures with minimal processing and storage overhead on mobile hosts. To meet these goals, we propose a proxy-based coordinated checkpointing scheme for our mobile to Grid middleware, Mobile Access to Grid Infrastructure (MAGi). In this scheme mobile hosts seamlessly store checkpoints on their respective proxies running on the middleware. Together with the central coordinator component, these proxies act as a centralized checkpointing store. This approach makes it efficient to rollback to the latest consistent global snapshot, without direct involvement of the mobile hosts, which results in less processing and storage overhead on mobile device as compared to existing schemes.

Keywords: Checkpointing, Fault Tolerance, Mobile Grid.

1 Introduction

Grid computing is based on an open set of standards and protocols that enable coordinated resource sharing and problem solving in dynamic, multi-institutional virtual organizations [2]. With Grid computing, organizations can optimally utilize computing and data storage by pooling them for large capacity workloads, sharing across networks and enabling collaboration across enterprise boundaries. Though the concept of Grid computing is still evolving, yet there have been a number of achievements in the arena of scientific applications, such as EU Grid, Particle Physics Grid and Bio Grid. Extending this potential of the Grid to a wider audience, particularly to users of wireless mobile devices, who are the prospective users of this technology, promises increase in productivity.

---

* Eui-Nam Huh is the corresponding author for this paper.

Y. Robert et al. (Eds.): HiPC 2006, LNCS 4297, pp. 448–459, 2006.
© Springer-Verlag Berlin Heidelberg 2006

Mobile devices promote mobile communication and flexible usage, yet they bring along problems such as unpredictable network quality, lower trust, limited resources (power, network bandwidth etc) and extended periods of disconnections [3]. If such resource limited mobile devices could access and utilize the Grid's resources then they could implicitly obtain results from resource intensive tasks never thought of before. This bridge between the mobile devices and Grid computing is named as Mobile Grid in literatures, can be thought of in two possible ways [4]. First, the mobile devices can act as a Grid resource consumer and can initiate the use of Grid resources, monitor the jobs being executed remotely, and take any results from the Grid. Secondly and more interestingly, mobile devices can also be assumed to participate in Grid as a resource provider and not just a resource recipient. For an example, consider mobile health care system which provides immediate, inexpensive and ubiquitous medical solutions to the remote patients and saves the time of healthcare professionals. These health care services include diagnoses of diseases by symptoms matching, analysis and visualization of patient's health records. Thanks to Mobile Grid technologies, now medical professionals and scientists can access these patients record with their hand-held mobile devices and monitor and schedule these resource intensive tasks on the Grid. Doctors and nurses can access medical information, prescribe treatment and send patient updates from anywhere in the hospital, clinic, or patient's home, without needing to be physically at the site.

Although Mobile Grid promise a lot good to the wireless mobile community and is an intrinsic choice to delegate resource intensive tasks to Grid, the mobility, unpredictable network quality, lower trust and security vulnerabilities (such as device theft or lost) of mobile devices hinder to make this envisage a reality. To breach this gap between resourceful highly available Grid resource and resource constrained unreliable wireless mobile world and make Mobile Grid reality, we need to overcome the inherited fault prone nature of the mobile devices. Traditional fault tolerance techniques are inadequate and unfeasible to meet the mobile Grid challenges, as explained in next Section. Special measures are required to be taken to ensure fault tolerance of the mobile devices and once the fault occurred, to roll back to the last correct state.

In this paper, we present a fault tolerant scheme for our mobile to Grid middleware MAGi [6], enabling heterogeneous mobile devices to access Grid services in a fault tolerant fashion. We extend the work done by [5] and propose a proxy based extension of this checkpointing approach. Our proxy-based coordinated checkpointing scheme takes storage and processing overhead from low-power mobile devices and delegates it to their respective proxies running on mobile service station (MSS).

Rest of this paper is organized as follow: In section 2, we analyze the problem in more details and lay down the requirements for our proposed systems. Section 3 elaborates our proposed proxy-based checkpointing scheme in details. In section 4, we present the simulation results and analyze our work. Section 5 surveys the existing work in the field of fault tolerance, in general and in mobile Grid in specific, and gives a comparison with our work. In section 6, we conclude our work and highlight the future directions.

2 Problem Statement and Research Issues

Unlike wired distributed Grid systems, availability of resources changes dynamically in mobile Grid system. The node may suddenly disappear due to disconnections or exhausted power and can lead to unpredictable results and even failure of the system which can further cause the loss of opportunity, financial loss or even loss of human lives. Since the failure probability increases with unpredictable network quality, fault tolerance is an essential characteristic of such mobile Grid systems. These mobile Grid systems must provide redundancy and mechanisms to detect and localize errors as well as to reconfigure the system and to recover from error states.

Traditional fault tolerance techniques are inadequate and unfeasible to meet the mobile Grid challenges. To elaborate these challenges, we first explain the checkpoint in the conventional wireless systems and then highlight then their short comings which set a layout for our proposed solution.

2.1 Overview of Checkpointing

Checkpointing is the saving of program state, usually to stable storage, so that it may be reconstructed later in time. Checkpointing provides the backbone for rollback recovery (fault-tolerance). To be recoverable, the processing node must save its system states and log incoming and outgoing messages from time to time on a stable storage so that later it can be re-inserted by reading its most recent saved state and message logs. Because of possible non-determinism, a recoverable node cannot send the same messages it sent before its failure, other processes might need to rollback to ensure consistency. In this scenario, a local checkpoint is then a snapshot of the local state of a process and a global checkpoint is a set of all local checkpoints saved on nonvolatile storage. For constructing a consistent snapshot of the system, local snapshot is selected for each node at the time of recovery. A set of all local snapshot points, that together can achieve a consistent snapshot of the system, is known as recovery line. An important issue to recover from node failure is collection of consistent global snapshot. A correct and consistent global snapshot must not have lost or orphan messages. The problem of ensuring that the system recovers to a consistent state after transient failures has two components, checkpoint creation and rollback-recovery.

There are two approaches to create checkpoints, coordinated and non-coordinated. With the first approach, processes coordinates their checkpointing actions such that each processes only its most recent checkpoint, and the set of checkpoints in the system is guaranteed to be consistent. When a failure occurs, they system restarts from these checkpoints [7]. In the second approach, processes takes checkpoints independently and save all checkpoints on stable storage, Upon a failure, processes must find a consistent set of checkpoints among the saved ones [8]. Many algorithms have been proposed to efficiently calculate the consistent global snapshot. If algorithm is non-intrusive and efficient, it is considered as a good snapshot collection algorithm. Non-intrusive algorithm doesn't force system nodes to freeze there computations during snapshot collection and hence are non-blocking. Efficient algorithm keeps efforts required for collecting a consistent snapshot to a minimum.

2.2 Fault Recovery Issues in Mobile Grid

Mobile Grid is composed of a fixed network and a wireless network that interact with each other. The wireless network consists of mobile hosts (MHs) which have the capacity to exchange messages with a mobile support station (MSS). MSS is defined as the infrastructure machine (usually thought to be on a fixed and wired network) that is rich in resources and communicates with MHs within its range. The MH may move from one cell to another while communicating with the fixed network (or with another MH through the fixed network). Cell is a logical or geographical area covered by an MSS. Checkpointing schemes for wired networks rely on active participation of nodes to process, calculate and store a local checkpoint. But mobile devices being light in nature cannot optimally calculate and store the checkpoints locally. The critical and challenging problem of mobile computing is how to cope with the special characteristics of the mobile wireless environment, to make balanced usage of computation and communication and support the user's mobility. Fault tolerance schemes proposed for wired distributed systems are not suitable for distributed mobile Grid environment because of following reasons;

- Checkpointing includes the time and processing to trace the dependability tree and to save the states of the processes on the stable storage. Mobile devices are mostly limited in there resources like low processing power, short life battery and limited storage capacity. These limitations provide obstacles in implementing traditional checkpointing schemes on MHs. Any checkpointing scheme designed for mobile networks should consider these vulnerabilities.
- Distributed checkpointing schemes, in which every node can initiate a checkpointing process, leads to another performance issue for the resource limited mobile devices. In order to keep the checkpointing sequence number updated, any time a process takes a checkpoint; it has to notify all the processes in the system, which will consequently result in network over flooding with control messages.
- MHs are more prone to catastrophic failures like theft, physical damage and loss. To ensure security, checkpointing schemes must rely on some other alternative stable repositories.
- If the network connection goes down (due to power loss, being out of range etc) or some other error occur in the normal execution of the system, user would lose his connection and would have to start mediating with the Grid service from scratch which would result in loss of precious time and resources.

3 Proposed Solution

Due to the intrinsic limitations of the non-coordinated checkpointing, such as considerably large amount of time, processing and stable storage is required to calculate and save frequent checkpoints; coordinated checkpointing is an attractive approach for transparently adding fault tolerance to the wireless mobile devices. We present an enhanced version of [5] technique, which purposed that the checkpoint can be stored on any storage media, either on the MH or on MSS. But mobile devices, being light in processing and having limited storage, cannot efficiently store and bear the calculation overhead of constructing the dependability checkpointing tree as suggested by [5]. To solve these issues, we propose a proxy based coordinated checkpointing extension of this approach. Our scheme is based upon the assumption that in our system, all the interaction of concern is only between a MH and MSS, that is no two MH can communicate with each other directly. This assumption is very practical and acceptable in a middleware based mobile Grid environment. Please note, we are limiting our discussion to one wireless mobile cell. In a system with multiple wireless cells and MSS, the scheme can easily be extended as explained in our previous work [6]. We also assume that a single MH initiates the algorithm to take the permanent checkpointing. Again as all communication is through MSS, if there are concurrent invocations, it can easily be synchronized.

In our proposed coordinated checkpointing scheme, we introduce the concept of mobile host proxy (MHP), which seamlessly communicates to their respective MH and takes storage and processing overhead from low-power mobile device. This MHP is the only interface to communicate to MH from MSS and vice versa. MSS coordinates among all the MHPs to calculate globally consistent snapshot. We propose that rather than calculating and storing the checkpoints at the mobile device, MH delegates this task to its respective MHP. MH offline stores its checkpoints on its respective proxy seamlessly, which subsequently participates in the global checkpointing calculation algorithm without direct involvement of the MH. As MHP is a static host and resides on the MAGi middleware, which is rich in resources and have access to unlimited power of Grid, at least in theory, this delegation results in better performance and reliability as compared to existing techniques. Fig. 1 elaborates this scenario.

Our proxy based checkpointing algorithm uses enhancement two-phase commit (2PC) protocol in which we saves two types of checkpoints on MHP; permanent and tentative. A permanent checkpoint can not be undone, while a tentative checkpoint can be undone or changed to permanent. In first phase MHP takes its latest local snapshots and mark it as tentative. In tentative phase, MHP refrains from communication with other processes. Moreover, after receiving tentative checkpoints from all relevant processes, MSS will convert the tentative checkpoints to permanent and store in the stable storage to rollback for fault recovery. If any relevant process refuses to take tentative checkpoint, it will notify back to MSS which will notify all the participating processes to rollback the checkpointing activity and discard their tentative checkpoints. By relevant process we mean a process which received or sent a message from/to the checkpoint initiator after taking its last permanent checkpoint. So only affected processes are involved in the checkpointing process, which save the considerable overhead on the system. In addition to the tentative and permanent checkpoints, we employ the concept of mutable checkpoints [5] in our proxy based checkpointing scheme to avoid the checkpointing inconsistencies [16], [1]. Mutable checkpoints are neither tentative and nor permanent. When a MH takes a mutable checkpoints it doesn't send the checkpoint requests to other MHs and don't need to save the checkpoint on stable storage.

MHP being the gateway to MH, log and number all the messages sent and received by MH. Moreover, MHP after a certain time quanta, request its corresponding MH to take a local snapshot of its processes (which includes process states, function stack).

This time quanta can be adjusted by the administrator and depends upon the network availability. Moreover every MH may have its own snapshot frequency. This process of taking local snapshot is an offline activity and is not synchronized with the global checkpointing process. After taking its local snapshot, MH sends it back to MHP. Subsequently MHP stores this local checkpoint in its personal storage which is readily available to it. After time quanta expire, MHP will repeat these steps and will update the local snapshots. This latest local of snapshot will enable MHP to rollback to an appropriate global consistent state of the system without direct involvement of the MH.

It is important to note that some of the MHPs will just mark the latest as tentative which will save the time to connect to MH and receive the snapshot. It means that if the MHP got the latest snapshot, it needs not to ask MH to send again. If the snapshot MHP has in its personal stable storage is expired, only than it will ask MH to send the latest snapshot. Fig. 2 explains the process. We will use terms MHP and MH interchangeably when there is no ambiguity.

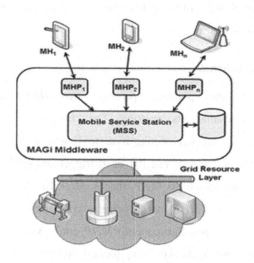

Fig. 1. Proxy based Checkpointing Scheme for Mobile Grid

3.1 System Model

Any MH_p in the system can initiate the checkpointing process by sending a request to through its MHP_p. If there are n mobile hosts, system can be modeled as;

$$\{(MH_p, MHP_p), MSS\} \forall p \in \{1,2,3,...,n\}$$

Let MH_p has its mobile host proxy MHP_p residing on the resourceful fixed MSS. MHP_p maintain two array R_p and csn_p of n bits where $R_p[j]=1$ means that MHP_p receive a message from MHP_j in the current checkpoint interval and $csn_p[j]$ represents the checkpoint sequence number of MHP_j known at MHP_p. Besides that

MHP_p maintains two boolean variables cp_state and $sent_p$ to indicate if MHP_p is in checkpointing state and if it has sent a message during its current checkpointing interval respectively. MHP_p uses variables old_csn_p to save the csn of the current tentative checkpoint. It also uses a tuple $trigger(pid,inum)$ of checkpointing initiator identifier pid and csn number at pid. In our algorithm we use a non-negative variable ω which is used to detect the termination of checkpointing algorithm. Initially array $csn_p[j]$, cp_state, ω are all initialized to 0s. $trigger(pid,inum)$ is initialized to $(p,0)$ at MHP_p. When MHP_p sends a message, it appends its $csn_p[p]$ to the message. Also MHP_p checks if cp_state is equal to one. If so, MHP_p also piggybacks its trigger with the message. If MHP_p receives a message from MHP_q, MHP_p takes a tentative checkpoint if and only if $old\_csn_p \leq req\_csn$ (where req_csn is appended with the checkpointing request). Note that old_csn_p is literally used to instead of $csn_p[p]$.

3.2 Checkpointing Scheme

Let MH_q initiate the checkpointing process proxy MHP_q at time t_2 such that $t_2 < t_1 + \Delta_p$ and resumes its working. The checkpointing initiation process includes incrementing its $csn_q[q]$, setting its ω to 1, setting its cp_state_q to 1 and storing its own identifier, the new $csn_q[q]$ in its trigger and sending this information along with its latest snapshot to its MHP, that is MHP_q in this case. After receiving this information from MH_q, MHP_q marks the received snapshot as tentative and saves it on its personal local storage. Subsequently MHP_q notify MSS for its willingness to take a checkpoint by sending checkpointing request. Request carries the trigger of the initiator, R_q and ω. MSS multicasts this request to all MHP_i connected such that $R_q[i] = 1$ and where $i \neq q$.

Upon receipt of the checkpointing request from MHP_q, MHP_p will decide its willingness to take the checkpoint by evaluating if this checkpoint is relevant by comparing req_csn with its old_csn, that is if and only if $old\_csn_p \leq req\_csn$. If any of MHP_p is not dependant upon the initiator, it will simply discard the request and will continue working as normal. Otherwise, MHP_p, on behalf of MH_p, updates its csn and cp_state and compares checkpointing request message trigger with its own trigger. If message trigger is equal to its own trigger (implying that MHP_p has already taken a checkpoint for this checkpointing), MHP_p checks if it has a mutable checkpoint which has a trigger identical to message trigger. If not, MHP_p sends the appended ω to the initiator; otherwise, MHP_p turns the status of mutable checkpoint as tentative and then propagates the request.

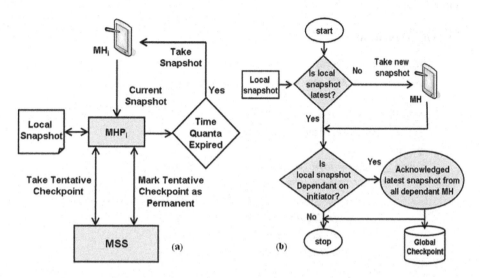

Fig. 2. (a) Calculating local snapshot (b) Calculating global consistent checkpoint

If MHP_p propagates the request to all MHs on which it depends, it may result in a large number of redundant system messages since some MH on which MHP_p depends may have received the request from other MH. The [8] algorithm uses this approach and its system message overhead can be as large as $O(N^2)$, where N is the number of MH in the system. On the other hand, only propagating the request to MHs on which MHP_p depends, but MHP_q (the sender) does not, may not work since receiving a request does not necessarily mean that the MH inherits the request. We solve this problem by attaching some information (csn and R which are saved in a structure called MR in the algorithm) to the request as does [9]. MHP_p only propagates the request to each MHP_k on which MHP_p depends, but MHP_k may not have inherited the request; that is, if MHP_p knows (by MR) some other process has sent the request to MHP_k with $req\_csn \geq scn_p[k]$ (req_csn is appended with the request and saved in $MR[k].csn$), it does not need to send the request to MHP_k ; otherwise, it has to send the request since MHP_k may inherit the request from MHP_p . Also, MHP_p appends the initiator's trigger and a portion of the received weight to all those requests. Then, MHP_p sends a *reply* to the initiator with the weight equal to the remaining weight and resumes its underlying computation.

If $msg\_trigger \neq own\_trigger$, MHP_p takes a tentative checkpoint, increases its csn_p, and propagates the request as above. At last, MHP_p clears R_p and $sent_p$, sends a reply to the initiator with the remaining weight, and then resumes its underlying computation.

3.3 The Algorithm

MHP_p receives checkpointing request from MHP_q :
$(MH_q, request, MR, recv\_csn, msg\_trigger_q, req\_csn, \omega_q)$
$csn_p[q] := recv\_csn$
if old $\_csn > req\_csn$
        send $(MH_p, reply, \omega_q)$ to the initiator and return.
$cp\_state_p := 1$
if $msg\_trigger = own\_trigger$
    if $CP_p.trigger = msg.trigger$
           $prop\_cp(CP_p.R, MR, MH_p, msg\_trigger, recv\_weight)$
           save $CP_p.mutable$ on stable storage
           old $\_csn_p := csn_p[p]; CP_p := NULL;$
           send $(MH_p, reply, \omega_p)$ to the initiator
    else
           send $(MH_p, reply, \omega_q)$ to the initiator
else
    $increment(csn_p[p]); own\_trigger := msg\_trigger;$
    $prop\_cp(R_p, MR, MH_p, msg\_trigger, \omega_q)$
    Take a local checkpoint (on the stable storage at MHP_p);
    old $\_csn_p := csn_p[p];$
    send $(MH_p, reply, \omega_p)$ to the initiator
    $sent_p := 0; reset R_p;$

4 Simulation and Analysis of Results

We simulated our enhancement on [5] to gauge the performance effects due to the mobile host-proxies inclusions in the systems. After every fixed time interval Δ_p, MH_p takes its local snapshot, describing current state of MH_p, and sends to MHP_p. We define $S_p(t_1)$ as local snapshot of MH_p at time t_1. Let φ is the time to calculate local snapshot at MH_p. If λ is the wireless network bandwidth between MH_p and MHP_p and θ is the data load associated with $S_p(t_1)$ then Γ, time taken to transfer snapshot from MH_p to MHP_p, can be calculated as follows;

$$\Gamma = \theta / \lambda$$

Note that Δ_p must be set greater then Γ. Let later at some time t_{cp}, MHP_p receive a checkpointing request from MH_q. The total time to take a checkpoint is;

$$\Theta = \Gamma + \varphi + \phi$$

Where ϕ is the time taken to participate in the checkpoint process.

Now note that MHP_p will only request a new local snapshot from MH_p if there was some message exchange to and from MH_p otherwise MHP_p will use locally stored snapshot of MH_p which will save considerable time. As shown in the Fig. 3. Every time the local snapshot is used from the personal storage of the MHP_p, the total time taken to calculate the snapshot is less as compared. When the, locally stored snapshot is invalid, MHP_p requests a new snapshot from MH_p. In that case the time taken to take a checkpoint is equal to the system without proxies as seen in the case 6, 11 and 18.

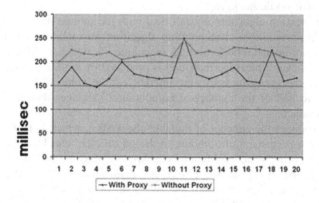

Fig. 3. Time comparison to take a checkpoint using proxies and without proxies

5 Related Work

The characteristics of MH and its specific errors prevent traditional error recovery techniques to be directly applicable and efficient for building fault-tolerant mobile applications. We surveyed existing fault-recovery techniques proposed for both traditional distributed systems and mobile Grid systems.

First coordinated checkpointing algorithm was presented by [10]. They assume that all communications between processes are atomic, which is too restrictive. [8] algorithm relaxes there assumption by allowing only those processes need to take the checkpoint that have communicated with the checkpoint initiator either directly or indirectly, thus reducing the number of synchronization messages and the number of checkpoints. In this algorithm if any of the involved processes is not able to or not willing to take a checkpoint, the entire checkpointing process is aborted.

All of the above coordinated checkpointing algorithms are blocking in nature as they require processes to be blocked during checkpointing. Checkpointing includes the time to trace the dependency tree and to save the state of processes on the stable storage, which may be long. Therefore, blocking algorithms may dramatically reduce the performance of the system and are not good for mobile devices.

One of the earliest works done in the category of non-blocking checkpointing was by [11], which deals with static nodes and system messages are sent along all the links in network during snapshot collection. This leads to a message complexity of $O(n^2)$. Authors in [12] use the checkpoint sequence number to identify orphan messages. This sequence number avoids the need for processes to be blocked during checkpointing. However, this approach is centralized in nature as it requires the initiator to communicate with all processes in the computation. The algorithm proposed by [13] uses a similar idea as [14] with an exception that the processes which did not communicate with others during the previous checkpoint interval change do not need to take new checkpoints. Both algorithms [13] and [14] suffers because of there centralized nature which assume that a distinguished initiator decides when to take a checkpoint. Therefore, they suffer from the disadvantages such as one-site failure, traffic bottle-neck, etc.

The above mention algorithm doesn't deal with the mobile devices. Ref [15] was the pioneers in presenting a checkpointing algorithm for mobile computing systems. They use uncoordinated checkpointing technique in which a MH takes a local checkpoint whenever a message reception is preceded by a message sent at that MH. If the send and receive of messages are interleaved, the number of local checkpoints will be equal to half of the number of computation messages this may degrade the system performance.

Ref [5] introduces the concept of mutable checkpoint, which is neither a tentative checkpoint nor a permanent checkpoint, to design efficient checkpointing algorithms for mobile computing systems. Mutable checkpoints need not be saved on the stable storage and can be saved anywhere, e.g., the main memory or local disk of MHs. Taking a mutable checkpoint avoids the overhead of transferring large amounts of data to the stable storage at MSSs over the wireless network. This scheme, however, fails to overcome the storage overhead on mobile devices. Our scheme overcomes this drawback by delegating resource intensive tasks to the MHPs residing on the resource rich MSS.

6 Conclusion and Future Work

In this paper, we presented a proxy based checkpointing scheme for mobile Grid environment. First we highlighting the checkpointing in general and then we discussed how traditional checkpointing scheme are not suitable for mobile Grid environment. We also listed the main research issues faced by mobile Grid in the field of fault tolerance and fault recovery. In this paper, we extend the work done by [5] and propose a proxy based extension of this checkpointing approach. Our proxy-based coordinated checkpointing scheme takes storage and processing overhead from low-power mobile devices and delegates it to their respective proxies running on mobile service station (MSS).

In future we plan to implement our proposed scheme and accumulate the experiment result to find correctness of our scheme. We also intend to investigate the performance and storage overhead our scheme as compared to the other existing solutions.

References

[1] G. Cao, M. Singhal. "On the Impossibility of Min-Process Non-Blocking Checkpointing and An Efficient Checkpointing Algorithm for Mobile Computing Systems", pp. 37-44, In Proc. of Int'l Conf. on Parallel Processing, Aug. 1998

[2] Foster, C. Kesselman, and S. Tuecke, "The Anatomy of the Grid", International Journal of Supercomputing Applications, Vol.15, No.3, pp.200-222, 2001.

[3] Forman, G. and Zahorjan, J.: The Challenges of Mobile Computing, IEEE Computer, vol. 27, no. 4, (April 1994)

[4] Sang-Min Park, Young-Bae Ko, Jai-Hoon Kim: "Disconnected Operation Service in Mobile Grid Computing", pp. 499-513, in Proc. of Intl. Conf. Service-Oriented Ccomputing, ICSOC 2003, Trento, Italy, Dec. 2003

[5] Guohong Cao, Mukesh Singhal, "Checkpointing with mutable checkpoints", Theoretical Computer Science, Volume 290, Issue 2, Jan 2003

[6] Ali Sajjad, et. al., "MAGI - Mobile Access to Grid Infrastructure: Bringing the gifts of Grid to Mobile Computing", pp 311-322, NODe/GSEM 2005

[7] Y. Tamir, C.H.Sequin, "Error Recovery in Multicomputers using global checkpoints", in Proc. 13th Intl. conf. Parallel Processing, Aug, 1984.

[8] Richard Koo , Sam Toueg, "Checkpointing and rollback-recovery for Distributed Systems", IEEE Transactions on Software Engineering, v.13 n.1, p.23-31, Jan. 1987

[9] Guohong Cao, Mukesh Singhal, "Mutable Checkpoints: A New Checkpointing Approach for Mobile Computing Systems," IEEE Transactions on Parallel and Distributed Systems, vol. 12, no. 2, pp. 157-172, Feb., 2001.

[10] G. Barigazzi and L. Strigini. "Application-transparent setting of recovery points.", pp. 48 - 55, In Proc. of the 13th Intl. Symposium on Fault-Tolerant Computing Systems, 1983

[11] K. M. Chandy and L. Lamport, "Distributed Snapshots: Determining Global States of Distributed Systems", pp 63-75, ACM Transactions on Computer Systems,1985

[12] E.N. Elnozahy, D.B. Johnson, and W. Zwaenepoel, "The Performance of Consistent Checkpointing", pp 39-47, In Proc of the 11th Symposium on Reliable Distributed Systems, October 1992

[13] L.M. Silva and J.G. Silva, "Global Checkpointing for Distributed Programs", pp. 155-162, In Proc. of IEEE Symposium on Reliable Distributed Systems, Oct. 1992

[14] E.W. Dijkstra, " Self-stabilizing Systems in Spite of Distributed Control", pp 643-644, Communications of the ACM vol.17, 1974

[15] Acharya, B. R. Barinath, "Checkpointing Distributed Applications on Mobile Computing", pp. 73-80, In Proc. of the 3rd International Conference on Parallel and Distributed Information Systems, Sep. 1994

[16] G. Cao, M. Singhal, "On Coordinated Checkpointing in Distributed Systems", pp. 213-225, IEEE Transactions on Parallel and Distributed Systems, vol. 9, no. 12, Dec. 1998

Impact of Noise on Scaling of Collectives:
An Empirical Evaluation

Rahul Garg and Pradipta De

IBM India Research Laboratory, Hauz Khas, New Delhi 110 016
grahul@in.ibm.com, pradipta.de@in.ibm.com

Abstract. It is increasingly becoming evident that operating system interference in the form of daemon activity and interrupts contribute significantly to performance degradation of parallel applications in large clusters. An earlier theoretical study has evaluated the impact of system noise on application performance for different noise distributions [1]. Our work complements the theoretical analysis by presenting an empirical study of noise in production clusters. We designed a parallel benchmark that was used on large clusters at SanDeigo Supercomputing Center for collecting noise related data. This data was fed to a simulator that predicts the performance of collective operations using the model of [1]. We report our comparison of the predicted and the observed performance. Additionally, the tools developed in the process have been instrumental in identifying anomalous nodes that could potentially be affecting application performance if undetected.

1 Introduction

Scaling of parallel applications on large high-performance computing systems is a well known problem [2–4]. Prevalence of large clusters, that uses processors in order of thousands, makes it challenging to guarantee consistent and sustained high performance. To overcome variabilities in cluster performance and provide generic methods for tuning clusters for sustained high performance, it is essential to understand theoretically, as well as using empirical data, the behavior of production mode clusters. A known source of performance degradation in large clusters is the *noise in the system* in the form of daemons and interrupts [2, 3]. Impact of OS interference in the form of interrupts and daemons can even cause an order of magnitude performance degradation in certain operations [2, 5].

A formal approach to study the impact of noise in these large systems was initiated by Agarwal et al. [1]. The parallel application studied was a typical class of kernel that appears in most scientific applications. Here, each node in the cluster is repetitively involved in a computation stage, followed by a collective operation, such as barrier. This scenario was modeled theoretically, and impact of noise on the performance of the parallel applications was studied for three different types of noise distributions.

In this paper, our goal is to validate the theoretical model with data collected from large production clusters. Details revealed through empirical study helps in fine-tuning the model. This allows us to establish a methodology for predicting the performance of large clusters. Our main contributions are: (i) We have designed a parallel benchmark that measures the noise distribution in the cluster; (ii) Using data collected from the

Y. Robert et al. (Eds.): HiPC 2006, LNCS 4297, pp. 460–471, 2006.

production clusters we make performance predictions made by the theoretical model proposed earlier. This validation step enables us to predict the performance of large clusters. We report the prediction accuracy against measurements at the SanDiego Supercomputing Center (SDSC). We discovered that measurements of noise distributions also help in identification of misbehaving nodes or processors.

In addition to making performance predictions, our study could be useful in performance improvements. Traditional techniques for performance improvement either fall in the category of *noise reduction* or *noise synchronization*. Noise reduction is achieved by removing several system daemons, dedicating a spare processor to absorb noise, and reducing the frequency of daemons. Noise synchronization is achieved by explicit co-scheduling or gang scheduling [6–8]. Most of these implementations require changing the scheduling policies. Our work gives insight into another technique for improving performance, that can be called *noise smoothing*. If the model predicts the actual performance reasonably well, then the systems can be tuned to ensure that the noise does not have heavy tail (i.e. infrequent interruptions that take long time). This technique may complement the other approaches currently used in large high-performance systems.

The rest of the paper is organized as follows. The theoretical model for capturing the impact of noise on cluster performance, based on [1], is presented in Section 2. In Section 3, we present the details of the parallel benchmark that we have designed. Section 4 presents the analysis of the data collected on the SDSC clusters. Finally, we conclude in Section 5.

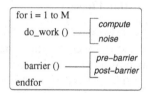

Fig. 1. Typical code block in a parallel scientific application

2   Theoretical Modeling of System Noise

In this section, we briefly introduce the theoretical model of collective communication, as described earlier in [1]. In this model, a parallel program consists of a sequence of iterations of a *compute phase* followed by a *communicate phase*, as shown in Figure 1. In the compute phase, all the threads of the program locally carry out the work assigned to them. There is no message exchange or I/O activity during the compute phase. The communicate phase consists of a collective operation such as a barrier. We are interested in understanding the time it takes to perform the collective operation as a function of the number of threads in the system.

Consider a parallel program with N threads running on a system that has N processors. We assume, for simplicity of analysis, that $N = 2^k - 1$ for some positive

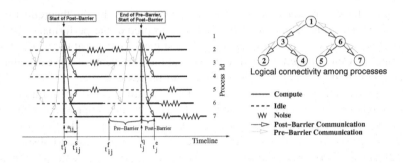

Fig. 2. The diagram shows a typical computation-barrier cycle, along with pre-barrier and post-barrier phases for a barrier call, and interruptions in compute phase due to system noise

integer k. Figure 2 shows the sequence of events involving one iteration of the loop in Figure 1. In this figure, we used 7 processors which are logically organized as a binary tree to demonstrate the operation of one iteration of the parallel code. Figure 2 shows the decomposition of the communicate phase into pre-barrier and post-barrier stages, and the interruptions introduced during compute phase by the triggering of noise. In the figure, t_{ij}^s denotes the start time of the compute phase, t_{ij}^f denotes the finish time of the compute phase, and t_{ij}^e denotes the end of the barrier phase of the iteration.

2.1 Modeling the Communication Phase

The barrier operation comprises of two stages: a *pre-barrier* stage succeeded by a *post-barrier* stage. We assume that these stages are implemented using message passing along a complete binary tree, as shown in Figure 2. The basic structure does not change for any implementation based on a fixed tree of bounded degree, such as a k-ary tree. A process is associated to each node of the binary tree. A special process, called the *root* (Process 1 in Figure 2) initiates the *post-barrier* stage and concludes the *pre-barrier* stage. In the post-barrier stage, a start-compute-phase message initiated by root is percolated to all leaf nodes. At the end of computation each node notifies its parents of completion of work. This stage ends when the root finishes its computation and receives a message from both its children indicating the same. An iteration of the loop in Figure 1 would thus consist of a compute stage, followed by a pre-barrier and a post barrier stage. Let t_j^p denote the start of the post-barrier stage just preceding the j-th iteration, and t_j^q denotes the time at which the post-barrier stage of the j-th iteration concludes, as shown in Figure 2. Following t_j^q, the iteration can continue with other book-keeping operations before beginning the next iteration. Also, the time taken by a barrier message to reach node i in the post-barrier phase is denoted by a_{ij}.

For simplicity, we assume that each message transmission between a parent and a child node takes time τ, which is referred to as the *one-way latency*. Thus, for the root process (process id 1) this value is zero, i.e. $a_{1j} = 0$ in Fig 2. For all the leaves i,

$a_{ij} = \tau(\log(N+1)-1)^1$. Thus, for the case of $N = 7$ in the figure, $a_{2j} = 2\tau$, $a_{3j} = \tau$, $a_{4j} = 2\tau$, $a_{5j} = 2\tau$, $a_{6j} = \tau$, and $a_{7j} = 2\tau$, for all j.

2.2 Modeling the Compute Phase

Let W_{ij} represent the amount of work, in terms of number of operations, carried out by thread i in the compute phase of j-th iteration. If the system is *noiseless*, time required by all processors to finish the assigned work will be constant, i.e. time to complete work W_{ij} is w_{ij}. The value of the constant typically depends on the characteristics of the processor, such as clock frequency, architectural parameters, and the state of the node, such as cache contents. Therefore, in a noiseless system the time taken to finish the computation is, $t_{ij}^f - t_{ij}^s = w_{ij}$.

Due to presence of system level daemons that are scheduled arbitrarily, the wall-clock time taken by processor i to finish work W_{ij} in an iteration varies. The time consumed to service daemons and other asynchronous events, like network interrupts, can be captured using a variable component δ_{ij} for each thread i in j-th iteration. Thus, the time spent in computation in an iteration can be accurately represented as,

$$t_{ij}^f - t_{ij}^s = w_{ij} + \delta_{ij}$$

where δ_{ij} is a random variable that captures the overhead incurred by processor i in servicing the daemons and other asynchronous events. Note that δ_{ij} also includes context switching overheads, as well as, time required to handle additional cache or TLB misses that arise due to cache pollution by background processes. The characteristics of the random variable δ_{ij} depends on the work W_{ij}, and the system load on processor i during the computation. The random variable δ_{ij} models the *noise* on processor i for j-th iteration, shown as the noise component in Figure 2.

2.3 Theoretical Results

The theoretical analysis of [1] provides a method to estimate the time spent at the barrier call. We will show here that the time spent at the barrier can be evaluated indirectly by measuring the total time spent by a process in an iteration, that consists of the compute and communicate phases. The time spent by a process in an iteration may be estimated by the amount of work, noise distributions and the network latencies.

In this analysis we make two key assumptions of stationarity and spatial independence of noise. Since we assume that our benchmark is run in isolation, therefore only noise present is due to system activity. This should stay constant over time, giving stationarity of the distribution. Secondly, the model in [1] assumes that the noise across processors is independent (i.e. δ_{ij} and δ_{kj} are independent for all i, j, k). Thus there cannot be any co-ordinated scheduling policy to synchronize processes across different nodes.

The time spent idling at the barrier call is given by,

$$b_{ij} = t_{ij}^e - t_{ij}^f \tag{1}$$

---

[1] From here on, *log* refers to log_2 and *ln* refers to log_e.

We first derive the distribution of $(t_j^q - t_j^p)$, and then use it to derive the distribution of b_{ij}. From the figure it can be noted that, $(t_j^q - t_j^p)$ depends on a_{ij}, w_{ij}, and the instances of the random variable δ_{ij}. Now, if the network latencies are constant (τ), it is easy to verify that, if $a_{ij} \leq \tau \log((N+1)/2), \forall(i,j)$,. Thus we have,

Lemma 1. For all j, $\max_{i \in [1...N]} t_{ij}^f - t_{ij}^s \leq t_j^q - t_j^p \leq \max_{i \in [1...N]}(t_{ij}^f - t_{ij}^s) + 2\tau \log((N+1)/2)$.

We model $(t_j^q - t_j^p)$ as another random variable θ_j. Now, Lemma 1 yields,

Theorem 1. For all iterations j, the random variable θ_j may be bounded as[2],

$$\max_{i \in [1...N]}(w_{ij} + \delta_{ij}) \leq \theta_j \leq \max_{i \in [1...N]}(w_{ij} + \delta_{ij}) + 2\tau \log((N+1)/2).$$

For a given j, all δ_{ij} are independent for all i. Thus if we know the values of w_{ij} and the distributions of δ_{ij}, then the expectations as well as the distributions of θ_j may be approximately computed as given by Theorem 1. For this, we independently sample from the distributions of $w_{ij} + \delta_{ij}$ for all i and take the maximum value to generate a sample. Repeating this step a large number of times gives the distribution of $\max_{i \in [1...N]}(w_{ij} + \delta_{ij})$. Now, b_{ij} can be decomposed as (see Figure 2)

$$b_{ij} = (t_j^q - t_j^p) - (t_{ij}^f - t_{ij}^s) - (a_{i(j+1)} - a_{ij}) \qquad (2)$$

Since we have assumed a fixed one-way latency τ, $a_{ij} = a_{i(j+1)} = \tau$, therefore distribution of barrier time b_{ij} is given by,

$$\theta_j - (w_{ij} + \delta_{ij})$$

Using Theorem 1, θ_j can be approximately computed to within $2\tau \log((N+1)/2)$ just by using w_{ij} and δ_{ij}. Therefore, the barrier time distribution can be computed just by using noise distribution and w_{ij}. If w_{ij} are set to be equal for i, then w_{ij} cancels out, and barrier time distribution can be approximated just by using δ_{ij}.

In this paper we attempt to validate the above model by comparing the measured and predicted performance of the barrier operations on real systems. We evaluate if Theorem 1 can be used to give a reliable estimate of collectives performance on a variety of system. For this, we designed a micro-benchmark that measures the noise distributions $\delta_i(w)$. The benchmark also measures the distribution of θ_j, by measuring $t_{ij}^e - t_{ij}^s$. We implemented a simulator that takes the distributions of $w_{ij} + \delta_{ij}$ as inputs, and outputs the distribution of $\max_{i \in [1...N]}(w_{ij} + \delta_{ij})$. We compare the simulation output with the actual distribution of θ_j obtained by running the micro-benchmark. We carry out this comparison on the Power 4 cluster at SDSC with different values of work quanta, w_i.

3   Methodology for Empirical Validation

Techniques to measure noise accurately is critical for our empirical study. This section presents the micro-benchmark kernel used to measure the distributions. We first tested this benchmark on a testbed cluster and then used it to collect data on the SDSC cluster.

---

[2] For random variables, X and Y, we say that $X \leq Y$ if $P(X \leq t) \geq P(Y \leq t), \forall t$.

Algorithm 1 The Parallel Benchmark kernel

 1: **while** elapsed_time < period **do**
 2:    busy-wait for a randomly chosen period
 3:    MPI_Barrier
 4:    t_{ij}^s = get_cycle_accurate_time
 5:    do_work (iteration_count)
 6:    t_{ij}^f = get_cycle_accurate_time
 7:    MPI_Barrier
 8:    t_{ij}^e = get_cycle_accurate_time
 9:    store $(t_{ij}^f - t_{ij}^s), (t_{ij}^e - t_{ij}^f), (t_{ij}^e - t_{ij}^s)$
10:    MPI_Bcast (elapsed_time);
11: **end while**

3.1 Parallel Benchmark (PB)

The Parallel Benchmark (PB)[3] aims to capture the compute-barrier sequence of Figure 1. The kernel of PB is shown in Algorithm-1. The PB executes a compute process (Line 5) on multiple processors assigning one process to each processor. The *do_work* can be any operation. We have chosen it to be a Linear Congruential Generator (LCG) operation defined by the recurrence relation,

$$x_{j+1} = (a * x_j + b) \bmod p. \tag{3}$$

A barrier synchronization call (Line 7) follows the fixed work (W_i). The time spent in different operations are collected using cycle accurate timers and stored in Line 9. In the broadcast call in Line 10 rank zero process sends the current elapsed time to all other nodes ensuring that all processes terminate simultaneously. Daemons are usually invoked with a fixed periodicity which may lead to correlation of noise across iterations. The random wait (Line 2) is intended to reduce this correlation. The barrier synchronization in Line 3 ensures that all the processes of the PB commence simultaneously.

Since the benchmark measures the distributions $\delta_i(W)$ for a fixed W we omit the subscript j that corresponds to the iteration number in the subsequent discussion.

3.2 Testing the Parallel Benchmark

We first study the PB on a testbed cluster. The testbed cluster has 4 nodes with 8 processors on each node. It uses identical Power-4 CPUs on all nodes. IBM SP switch is used to connect the nodes. The operating system is AIX version 5.3. Parallel jobs are submitted using the LoadLeveler[4] and uses IBM's Parallel Operating Environment (POE). The benchmark uses MPI libraries for communicating messages across processes.

---

[3] We refer this benchmark as PB, acronym for Parallel Benchmark, in the rest of the paper.

[4] The LoadLeveler is a batch job scheduling application and a product of IBM.

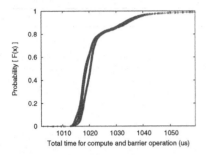

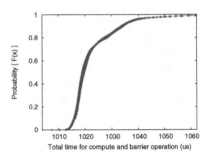

Fig. 3. Distribution of time taken in an iteration for computation and barrier using the Parallel Benchmark code shown in Fig 1

Fig. 4. Distribution of time taken in an iteration for computation and barrier operation using random wait, but *no broadcast* during execution

The goal in running on the testbed cluster was to fine-tune the PB for use on larger production clusters. We first execute the code block as shown in Algorithm 1 on the testbed cluster, and record the distributions of $(t_i^f - t_i^s) \sim w_i + \delta_i$, $t_i^e - t_i^f = b_i$, and $(t_i^e - t_i^s) \sim \theta$. Figure 3 shows the distribution of $\theta \sim (t_j^e - t_j^s)$ (Section 2.3) for PB. Ideally, the distributions for all processes (i.e. for all i) should exactly match, assuming $a_{i(j+1)} = a_{ij}, \forall i, j$. Interestingly, whenever broadcast was enabled in our experiment, the distributions were different. There were 4 bands formed, which we correspond to the 4 different nodes of the cluster. Figure 4 shows the distributions are identical when the broadcast in Line 10 of Figure 1 was omitted.

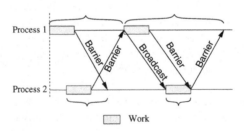

Fig. 5. Sequence of message exchange in a work-barrier-broadcast loop

In order to explain this discrepancy we closely looked at the implementation of barrier. Barrier is implemented in two steps: a shared memory barrier which synchronizes all the processes on a node, followed by message passing implementation that synchronizes across nodes. The message passing implementation across the nodes is not based on binary tree as assumed in the model described earlier. We illustrate the message passing mechanism for barrier synchronization in Figure 5 for a case with two processes. As soon as a barrier is called on a process, a message is sent to the other process. The barrier call ends when the process receives a message from the other process. Broadcast is implemented by sending a single message. Figure 5 shows the message exchange between the two processes, and the calculation of total time for an iteration. It can be seen that if process 2 starts its work after process 1 then it consistently measures smaller time per iteration. Adding random wait, desynchronizes the start time and mitigates the above anomaly.

Finally, we also verified the stationarity assumption by executing PB multiple times at different times of the day. As long as the PB is executed in isolation without any other user process interrupting it the results stay unchanged.

3.3   Predicting Cluster Performance

We implemented a simulator that repeatedly collects independent samples from N distributions of $w_i + \delta_i$ (as measured by the PB), and computes the distribution of $\max_{i \in [1...N]}(w_i + \delta_i)$.

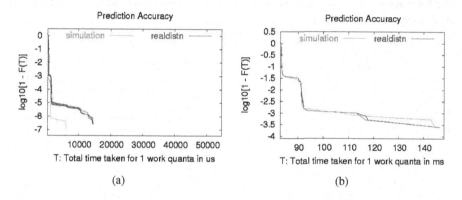

(a)   (b)

Fig. 6. Comparing distribution of $\max_{i \in [1...N]}(w_i + \delta_i)$ against distributions of $t_i^e - t_i^s$ computed from empirical data on 32 processors of a testbed cluster, for (a) $w = 300\mu s$ and (b) $w = 83ms$

Figure 6(a) and Figure 6(b) show the distributions of the time taken by an iteration (θ) on 32 processors in our testbed cluster, along with the output of the simulator ($\max_{i \in [1...N]}(w_i + \delta_i)$). Two different quanta values (w) of $300\mu s$ and $83ms$, were used to model small and large choice of work respectively. Since, in the simulation the communication latency involved in the collective call is not accounted, hence the output of the simulator is lower than the real distribution.

Interestingly, the accuracy of the prediction with larger quanta values is better even without accounting for the communication latency. This is because when the quanta value (W) is large, the noise component, $\delta_i(W)$ is also large thereby masking the communication latency part.

4   Benchmark Results on SanDiego Supercomputing Center (SDSC)

This section presents our observations from the data collected on the SanDiego Super-computing Center's DataStar cluster [9]. The DataStar compute resource at SDSC [9] offers 15.6 TFlops of compute power. For our experiments, we used two combination of nodes: (a) 32 8-way nodes with 1.5 GHz CPU, giving a total of 256 processors, and (b) 128 8-way nodes with a mix of 1.5 GHz and 1.7 GHz processors, giving a total of 1024 processors. The nodes are IBM Power-4 and Power-4+ processors. Each experiment is executed for about 1 hour in order to collect a large number of samples. However, in order to avoid storage of this large time series, we convert the data into discrete probability distributions with fixed number of bins. The distributions are used to compute different statistics related to each experiment.

4.1   Benchmark Results on SDSC (256 Processors on 32 Nodes)

The PB was run on 256 processors spawning 32 nodes. All nodes and processors in this experiment were identical.

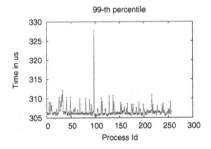

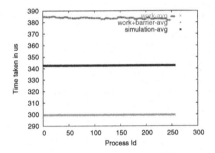

Fig. 7. This plot shows the values corresponding to 99-th percentile in the distribution of $w_i + \delta_i$ for $w = 300\mu s$

Fig. 8. The plot shows the mean for three distributions for each process: $w_i + \delta_i$, which is the work-avg, $t_i^e - t_i^s$, which is the total time for iterations including barrier wait, and $\max_{i \in [1...N]}(w_i + \delta_i)$, for a quanta of $300\mu s$

The value corresponding to 99-th percentile in the distribution of $w_i + \delta_i$ with $w = 300\mu s$ is plotted in Figure 7. Next, we plotted the distribution of $t_i^e - t_i^s$ for all the processes in Figure 9, along with the (predicted) distribution of $\max_{i \in [1...N]}(w_i + \delta_i)$ obtained using the simulator. It is seen that the simulation predicts the real distribution very closely on this production cluster, except in the tail part of the distribution.

Further insight is revealed in Figure 8, which shows the average time for the distributions of $(w_i + \delta_i)$, $t_i^e - t_i^s$, and $\max_{i \in [1...N]}(w_i + \delta_i)$. It shows that the mean value of $\max_{i \in [1...N]}(w_i + \delta_i)$ is about $50\mu s$ less than the mean of $t_i^e - t_i^s$ distributions. This

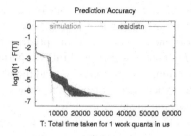

Fig. 9. Comparing distribution of $\max_{i \in [1...N]}(w_i + \delta_i)$ against distributions of $t_i^e - t_i^s$ for 256 processes, for $w_i = 300\mu s$

is accounted by the communication latency, $2 * \tau * log(N + 1)/2$, which is calculated to be $2 * 5\mu s * log(32)/2 = 40\mu s$ for the 32 node cluster in DataStar.

4.2 Benchmark Results on SDSC (1024 Processors on 128 Nodes)

We repeated the experiments on a larger cluster of 128 nodes with 1024 processors. However, in this experiment there were 2 different sets of processor types.

In Figure 10(a), we have plotted the 99-th percentile of $(w_i + \delta_i)$ distribution for each processor. It shows a spike around processor id 100. A zoom-in of the region between processor id 75 and 100 is shown in Figure 10(b). There are a set of 8 processors starting from id 89 to 96 which takes significantly longer to complete its workload. All these processors belong to a single node. This indicates that one node is anomalous and slowing down rest of the processes in this cluster. We discussed this with the SDSC system administrator who independently discovered problems with the same

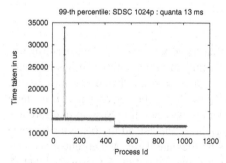

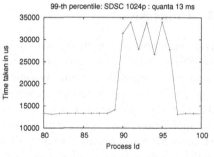

(a) The graph shows the 99-th percentiles from the distributions of $(w_i + \delta_i)$ for 1024 processors. There is a noticeable spike indicating that some processors take significantly longer to complete the computation phase.

(b) Zooming into the spiked area of Figure 10(a) shows that there are 8 processes on 1 particular node that is taking up significantly longer to finish the work.

Fig. 10.

node (possibly after receiving user complaint). Our run on the 256 processor system had the same problem (see Figure 7) due to the same node.

Finally, in Figures 11(a) and Figures 11(b) the prediction made by the simulator is compared against the observed distribution. In this experiment, the match between the predicted distribution of $\max_{i \in [1...N]}(w_i + \delta_i)$ and the observed distribution is not as good as in the previous experiment (for both the values of $w = 300\mu s$ and $w = 13ms$). For the $300\mu s$ case, the mean of the $\max_{i \in [1...N]}(w_i + \delta_i)$ was found to be $1.93ms$, while the mean of the $t_i^e - t_i^s$ was in the range $2.54ms$ to $2.93ms$ (for different processes i); while, for the $13ms$ case, the mean for $\max_{i \in [1...N]}(w_i + \delta_i)$ distribution was $23.191ms$ and the mean of the $t_i^e - t_i^s$ ranged from $24ms$ to $28.36ms$. At present, we are unable to explain this anomaly. We are conducting more experiments on different systems to pinpoint the cause of this.

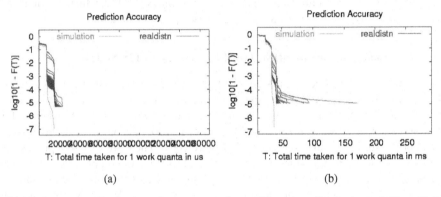

Fig. 11. Comparing distribution of $\max_{i \in [1...N]}(w_i + \delta_i)$ against distributions of $(t_i^e - t_i^s)$ for 1024 processes for $w_i = 300\mu s$ and $w_i = 13ms$ on SDSC cluster

5 Conclusion

High performance computing systems are often faced with the problem performance variability and lower sustained performance compared to the optimal. It has been noticed that system activities, like periodic daemons and interrupts, behave as noise for the applications running on the large clusters and slows down the performance. If a single thread of a parallel application is slowed down by the Operating System interference, the application slows down. Hence it is important to understand the behavior of noise in large clusters in order to devise techniques to alleviate them. A theoretical analysis of the impact of noise on cluster performance was carried out by Agarwal et al. [1]. A model for the behavior of noise was designed to predict the performance of collective operations in cluster systems. In this paper, we have attempted to validate the model using empirical data from a production cluster at SanDiego Supercomputing Center. We have designed a benchmark for collecting performance statistics from clusters. Besides providing the means to validate the model, the measurements from the benchmark proved useful in identifying system anomalies, as shown in the the case of the SDSC cluster.

Acknowledgment

Firstly, we would like to thank SanDeigo Supercomputing Center (SDSC) who provided us substantial time on their busy system for our experiments. We would like to thank Marcus Wagner for helping us in collecting the data from the SanDiego Supercomputing Center. Thanks to Rama Govindaraju and Bill Tuel for providing us with insights on the testbed cluster we have used for fine-tuning the parallel benchmark and helping us in collecting the data.

References

1. S. Agarwal, R. Garg, and N. K. Vishnoi, "The Impact of Noise on the Scaling of Collectives," in *High Performance Computing (HiPC)*, 2005.
2. T. Jones, L. Brenner, and J. Fier, "Impacts of Operating Systems on the Scalability of Parallel Applications," Lawrence Livermore National Laboratory, Tech. Rep. UCRL-MI-202629, Mar 2003.
3. R. Giosa, F. Petrini, K. Davis, and F. Lebaillif-Delamare, "Analysis of System Overhead on Parallel Computers," in *IEEE International Symposium on Signal Processing and Information Technology (ISSPIT)*, 2004.
4. F. Petrini, D. J. Kerbyson, and S. Pakin, "The Case of the Missing Supercomputer Performance: Achieving Optimal Performance on the 8192 Processors of ASCI Q," in *ACM Supercomputing*, 2003.
5. D. Tsafrir, Y. Etsion, D. G. Feitelson, and S. Kirkpatrick, "System Noise, OS Clock Ticks, and Fine-grained Parallel Applications," in *ICS*, 2005.
6. J. Moreira, H. Franke, W. Chan, L. Fong, M. Jette, and A. Yoo, "A Gang-Scheduling System for ASCI Blue-Pacific," in *International Conference on High performance Computing and Networking*, 1999.
7. A. Hori and H. Tezuka and Y. Ishikawa, "Highly Efficient Gang Scheduling Implementations," in *ACM/IEEE Conference on Supercomputing*, 1998.
8. E. Frachtenberg, F. Petrini, J. Fernandez, S. Pakin, and S. Coll, "STORM: Lightning-Fast Resource Management," in *ACM/IEEE Conference on Supercomputing*, 2002.
9. "DataStar Compute Resource at SDSC." [Online]. Available: http://www.sdsc.edu/user_services/datastar/

DDSS: A Low-Overhead Distributed Data Sharing Substrate for Cluster-Based Data-Centers over Modern Interconnects

Karthikeyan Vaidyanathan, Sundeep Narravula, and Dhabaleswar K. Panda

Department of Computer Science and Engineering
The Ohio State University
{vaidyana, narravul, panda}@cse.ohio-state.edu

Abstract. Information-sharing is a key aspect of distributed applications such as database servers and web servers. Information-sharing also assists services such as caching, reconfiguration, etc. In the past, information-sharing has been implemented using ad-hoc messaging protocols which often incur high overheads and are not very scalable. This paper presents a new design for a scalable and a low-overhead *Distributed Data Sharing Substrate* (DDSS). DDSS is designed to support efficient data management and coherence models by leveraging the features of modern interconnects. It is implemented over the OpenFabrics interface and portable across multiple interconnects including iWARP-capable networks in LAN/WAN environments. Experimental evaluations with networks like Infini-Band and iWARP-capable Ammasso through data-center services show an order of magnitude performance improvement and the load resilient nature of the substrate. Application-level evaluations with Distributed STORM achieves close to 19% performance improvement over traditional implementation, while evaluations with check-pointing application suggest that DDSS is highly scalable.

1 Introduction

Distributed applications in the fields of nuclear research, biomedical informatics, satellite weather image analysis etc., are increasingly getting deployed in cluster environments due to their high computing demands. Advances in technology have facilitated storing and sharing of the large datasets that these applications generate, typically through a web interface forming web data-centers [1]. A web data-center environment (Figure 1) comprises of multiple tiers; the first tier consists of front-end servers such as the proxy servers that provide services like web messaging, caching, load balancing, etc. to clients; the middle tier comprises of application servers that handle transaction processing and implement business logic, while the back-end tier consists of database servers that hold a persistent state of the databases and other data repositories. In order to efficiently host these distributed applications, current data-centers also need scalable support for intelligent services like dynamic caching of documents, resource management, load-balancing, etc. Apart from communication and synchronization, these applications and services exchange some key information at multiple sites (e.g, timestamps of cached copies, coherency and consistency information, current system load). However, for the sake of availability, high-performance and low-latency, programmers use

Y. Robert et al. (Eds.): HiPC 2006, LNCS 4297, pp. 472–484, 2006.

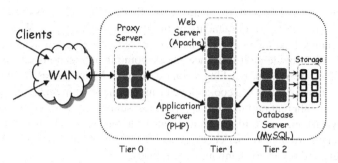

Fig. 1. Web data-centers

ad-hoc messaging protocols for maintaining this shared information. Unfortunately, as mentioned in [2], the code devoted to these protocols accounts for a significant fraction of overall application size and complexity. As system sizes increase, this fraction is likely to increase and cause significant overheads.

On the other hand, System Area Network (SAN) technology has been making rapid progress during the recent years. SAN interconnects such as InfiniBand (IBA) [3] and 10-Gigabit Ethernet (10GigE) have been introduced and are currently gaining momentum for designing high-end computing systems and data-centers. Besides high performance, these modern interconnects provide a range of novel features and their support in hardware, e.g., Remote Direct Memory Access (RDMA), Atomic Operations, Offloaded Protocol support and several others. Recently OpenFabrics [4] has been proposed as the standard interface that allows protable implementations over several modern interconnects like IBA, and iWARP capable ethernet interconnects including [5] Chelsio, Ammasso [6], etc., both in LAN/WAN environments.

In this paper, we design and develop a low-overhead distributed data sharing substrate (DDSS) that allows efficient sharing of data among independently deployed servers in data-centers by leveraging the features of the SAN interconnects. DDSS is designed to support efficient data management and coherence models by leveraging the features like one-sided communication and atomic operations. Specifically, DDSS offers several coherency models ranging from null coherency to strict coherency.

Experimental evaluations with IBA and iWARP-capable Ammasso networks through micro-benchmarks and data-center services such as reconfiguration and active caching not only show an order of magnitude performance improvement over traditional implementations but also show the load resilient nature of the substrate. Application-level evaluations with Distributed STORM using DataCutter achieves close to 19% performance improvement over traditional implementation, while evaluations with checkpointing application suggest that DDSS is scalable and has a low overhead. The proposed substrate is implemented over the OpenFabrics standard interface and hence is portable across multiple modern interconnects.

2   Constraints of Data-Center Applications

Existing data-center applications such as Apache, MySQL, etc., implement their own data management mechanisms for state sharing and synchronization. Databases

communicate and synchronize frequently with other database servers to satisfy the coherency and consistency requirements of the data being managed. Web servers implement complex load-balancing mechanisms based on current system load, request
patterns, etc. To provide fault-tolerance, check-pointing applications save the program
state at regular intervals for reaching a consistent state. Many of these mechanisms are
performed at multiple sites in a cooperative fashion. Since communication and synchronization are an inherent part of these applications, support for basic operations to read,
write and synchronize are critical requirements from the DDSS. Further, as the nodes in
a data-center environment experience fluctuating CPU load conditions the DDSS needs
to be resilient and robust to changing system loads.

Higher-level data-center services are intelligent services that are critical for the efficient functioning of data-centers. Such services require sharing of some state information. For example, caching services such as active caching [7] and cooperative
caching [8], [9] require the need for maintaining versions of cached copies of data and
locking mechanisms for supporting cache coherency and consistency. Active resource
adaptation service requires the need for advanced locking mechanism in order to reconfigure nodes serving one website to another in a transparent manner and needs simple
mechanism for data sharing. Resource monitoring services, on the other hand, require
efficient, low overhead access to the load information on the nodes. The DDSS has to
be designed in a manner that meets all of the above requirements.

3   Design Goals of DDSS

To effectively manage information-sharing in a data-center environment, the DDSS
must understand in totality, the properties and the needs of data-center applications
and services and must cater to these in an efficient manner.

Caching dynamic content at various tiers of a multi-tier data-center is a well known
method to reduce the computation and communication overheads. Since the cached
data is stored at multiple sites, there is a need to maintain cache coherency and consistency. Broadly, to accommodate the diverse coherency requirements of data-center
applications and services, DDSS supports a range of coherency models. The six basic
coherency models [10] to be supported are: 1) *Strict* Coherence, which obtains the most
recent version and excludes concurrent writes and reads. Database transactions require
strict coherence to support atomicity. 2) *Write* Coherence, which obtains the most recent
version and excludes concurrent writes. Resource monitoring services [11] need such a
coherence model so that the server can update the system load and other load-balancers
can read this information concurrently. 3) *Read* Coherence is similar to write coherence except that it excludes concurrent readers. Services such as reconfiguration [14]
are usually performed at many nodes and such services dynamically move applications
to serve other websites to maximize the resource utilization. Though all nodes perform
the same function, such services can benefit from a read coherence model to avoid two
nodes looking at the same system information and performing a reconfiguration. 4)
Null Coherence, which accepts the current cached version. Proxy servers that perform
caching on data that does not change in time usually require such a coherence model. 5)
Delta coherence guarantees that the data is no more than x versions stale. This model is

particularly useful if a writer has currently locked the shared segment and there are several readers waiting to the read the shared segment. 6) *Temporal Coherence* guarantees that the data is no more than t time units stale.

Secondly, to meet the consistency needs of applications, DDSS should support versioning of cached data and ensure that requests from multiple sites view the data in a consistent manner. Thirdly, services such as resource monitoring require the state information to be maintained locally since the data is updated frequently. On the other hand, services such as caching and resource adaptation can be cpu-intensive and hence require the data to be maintained at remote nodes distributed over the cluster. Hence, DDSS should support both local and remote allocation in the shared state. Due to the presence of multiple threads on each of these applications at each node in the data-center environment, DDSS should support the access, update and deletion of the shared data for all threads in a transparent manner. DDSS should also provide asynchronous interfaces for reading and writing the shared information Further, as mentioned in Section 2, DDSS must be designed to be robust and resilient to load imbalances and should have minimal overheads and provide high performance access to data. Finally, DDSS must provide an interface that clearly defines the mechanism to allocate, read, write and synchronize the data being managed.

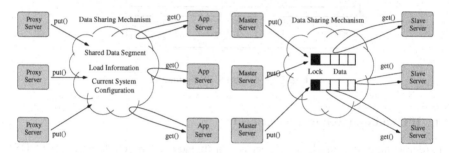

Fig. 2. DDSS using the proposed Framework (a) Non Coherent Distributed Data Sharing Mechanism (b) Coherent Distributed Data Sharing Mechanism

4   Proposed DDSS Framework and Implementation Issues

The basic idea of DDSS is to allow efficient sharing of information across the cluster by creating a logical shared memory region. It supports two basic operations, *get* operation to read the shared data segment and *put* operation to write onto the shared data segment. Figure 2a shows a simple distributed data sharing scenario with several processes (proxy servers) writing and several application servers reading certain information from the shared environment simultaneously. Figure 2b shows a mechanism where coherency becomes a requirement. In this figure, a set of master and slave servers access different portions of the shared data. All master processes waits for the lock since the shared data is currently being read by multiple slave servers.

In order to efficiently implement distributed data sharing, several components need to be built. Figure 3 shows the various components of DDSS that help in satisfying the needs of the current and next generation data-center applications. Broadly, in the figure, all the colored boxes are the components which exist today. The white boxes are the ones which need to be designed to efficiently support next-generation data-center applications. In this paper, we concentrate on the boxes with the *dashed lines* by providing either complete or partial solutions. In this section, we describe how these components take advantage of advanced networks in providing efficient services.

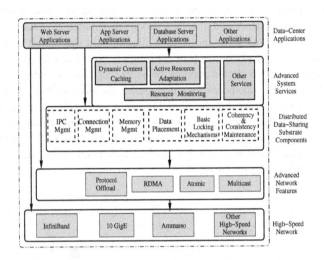

Fig. 3. Proposed DDSS Framework

4.1 IPC and Connection Management

In order to support multiple user processes or threads in a system to access the DDSS, we optionally provide a run-time daemon to handle the requests from multiple processes. We use shared memory channels and semaphores for communication and synchronization purposes between the user process and the daemon. The daemon establishes connections with other data sharing daemons and forms the distributed data sharing framework. Any service which is multi-threaded or the presence of multiple services need to utilize this component for efficient communication. Connection management takes care of establishing connections to all the nodes participating in either accessing or sharing its address space with other nodes in the system. It allows for new connections to be established and existing connections to be terminated.

4.2 Memory Management and Data Access

Each node in the system allocates a large pool of memory to be shared with DDSS. We perform the allocation and release operations inside this distributed memory pool. One way to implement the memory allocation is to inform all the nodes about an allocation. However, informing all the nodes may lead to large latencies. Another approach is to assign one node for each allocation (similar to home-node based approach but the node can maintain only the metadata and the actual data can be present elsewhere). This approach reduces the allocation latency. The nodes maintain a list of free blocks available within the memory pool. During a *release_ss()* operation, we inform the designated remote node for that allocation. During the next allocation, the remote node searches through the free block list and informs the free block which can fit the allocation unit. While searching for the free block, for high-performance, we get the first fit free block which can accommodate the allocation unit. High-speed networks provide one-sided operations (like RDMA read and RDMA write) that allow access to remote memory without interrupting the remote node. In our implementation, we use these operations to perform the read and write. All the applications and services mentioned in Figure 3 will need this interface in order access/update the shared data.

4.3   Data Placement Techniques

Though DDSS hides the placement of shared data segments, it also exposes interfaces to the application to explicitly mention the location of the shared data segment (e.g. local or remote node). For the remote nodes, the interface also allows the application to choose a specific node. In our implementation, each time a data segment is allocated, the next data segment is automatically allocated on a different node. This design allows the shared data segments to get well-distributed among the nodes in the system and accordingly help in distributing the load in accessing the shared data segments for data-center environments. This is particularly useful in reducing the contention at the NIC in the case where all the shared segment resides in one single node and several nodes needs to access different data segment residing on the same node. In addition, distributed shared segments also help in improving the performance for applications which use asynchronous operations on multiple segments distributed over the network.

4.4   Basic Locking Mechanisms

Locking mechanisms are provided using the atomic operations which is completely handled by modern network adapters. The atomic operations such as Fetch-and-Add and Compare-and-Swap operate on 64-bit data. The Fetch-and-Add operation performs an atomic addition at a remote node, while the Compare-and-Swap compares two 64-bit values and swaps the remote value with the data provided if the comparison succeeds. In our implementation, every allocation unit is associated with a 64-bit data which serves as a lock to access the shared data and we use the Compare-and-Swap atomic operation for acquiring and checking the status of locks. If the locks are implicit based on the coherence model, then DDSS automatically unlocks the shared segment after successful completion of *get()* and *put()* operations. Each shared data segment has an associated lock. Though we maintain the lock for each shared segment, the design allows for maintaining these locks separately. Similar to the distributed data sharing, the locks can also be distributed which can help in reducing the contention at the NIC if too many processes try to acquire different locks on the same node.

4.5   Coherency and Consistency Maintenance

As mentioned earlier, we support six different coherence models. We implement these models by utilizing the RDMA and atomic operations of advanced networks. However, for networks which lack atomic operations, we can easily build software-based solutions using the send/receive communication model. In the case of Null coherence model, since there is no explicit requirement of any locks, applications can directly read and write on the shared data segment. For strict, read, write coherence models, we maintain locks and *get()* and *put()* operations internally acquire locks to DDSS before accessing or modifying the shared data. The locks are acquired and released only when the application does not currently hold the lock for a particular shared segment. In the case of version-based coherence model, we maintain a 64-bit integer and use *IBV_WR_ATOMIC_FETCH_AND_ADD* atomic operation to update the version for every *put()* operation. For *get()* operation, we perform the actual data transfer only if the current version does not match with the version maintained at the remote end. In delta coherence model, we split the shared segment into memory hierarchies and support up to x versions. Accordingly, applications can ask for up to x previous versions of the data using the *get()* and *put()* interface. Basic consistency is achieved through maintaining versions of the shared segment and applications can get a consistent view of the shared data segment by reading the most recently updated version. We plan to provide several consistency models as a part of future work.

4.6   DDSS Interface

Table 1 shows the current interface that is available to the end-user applications or services. The interface essentially supports six main operations for gaining access to DDSS: *allocate_ss(), get(), put(), release_ss(), acquire_lock_ss(), release_lock_ss()* operations. The *allocate_ss()* operation allows the application to allocate a chunk of memory in the shared state. This function returns a unique shared state key which can be shared among other nodes in the system for accessing the shared data. *get() and put()* operations allow applications to read and write data to the shared state and *release_ss()* operation allows the shared state framework to reuse the memory chunk for future allocations. *acquire_lock_ss() and release_lock_ss()* operations allow end-user application to gain exclusive access to the data to support user-defined coherency and consistency requirements. In addition, we also support asynchronous operations such as *async_get()*,

Table 1. DDSS Interface

DDSS Operation	Description
int allocate_ss(nbytes, type, ...)	allocate a block of size nbytes in the shared state
int release_ss(key)	free the shared data segment
int get(key, data, nbytes, ...)	read nbytes from the shared state and place it in data
int put(key, data, nbytes, ...)	write nbytes of memory to the shared state from data
int acquire_lock_ss(key)	lock the shared data segment
int release_lock_ss(key)	unlock the shared data segment

async_put( ), wait_ss( ) and additional locking operations such as *try_lock( )* operation to support a wide range of applications to use such features.

DDSS is built as a library which can be easily integrated into distributed applications such as checkpointing, DataCutter [12], web servers, database servers, etc. For applications such as datacutter, several data sharing components can be replaced directly using the DDSS. Further, for easy sharing of keys, i.e., the key to an allocated data segment, DDSS allows special identifiers to be specified while creating the data sharing segment. Applications can create the data sharing segment using this identifier and DDSS will make sure that only one process creates the data segment and the remaining processes will get a handle to this data segment. For applications such as web servers and database servers, DDSS can be integrated as a dynamic module and all other modules can make use of the interface appropriately. In addition, DDSS can also be used to replace traditional communication such as TCP/IP. In our earlier work, cooperative caching [9], we have demonstrated the capabilities of high-performance networks for data-centers with respect to utilizing the remote memory and support caching of varying file sizes. DDSS can also be utilized in such environments. However, for very large file sizes which cannot fit in a cluster memory, applications will need to rely on the file system to store and retrieve the data. Another aspect of DDSS that is currently not supported is fault-tolerance. This is especially required for applications such as databases. If applications can explicitly inform DDSS for taking frequent snapshots, this feature can be implemented as a part of DDSS. We plan to implement this as a part of future work.

5   Experimental Results

In this section, we analyze the applicability of DDSS with services such as reconfiguration and active caching and with applications such as Distributed STORM and checkpointing. We evaluate our DDSS framework on two interconnects IBA and Ammasso using the OpenFabrics implementation. The iWARP implementation of OpenFabrics over Ammasso was available only at the kernel space. We wrote a wrapper for user applications which in turn calls the kernel module to fire appropriate iWARP functions. Our experimental testbed consists of a 12 node cluster with dual Intel Xeon 3.4 GHz CPU-based EM64T systems. Each node is equipped with 1 GB of DDR400 memory. The nodes were connected with MT25128 Mellanox HCAs (SDK v1.8.0) connected through a InfiniScale MT43132 24-port completely non-blocking switch. For Ammasso experiments, we use two node dual Intel Xeon 3.0 GHz processors with a 512 kB L2 cache and a 533 MHz front side bus and 512 MB of main memory.

5.1   Microbenchmark

Measuring Access Latency: The latency test is conducted in a ping-pong fashion and the latency is derived from round-trip time. For the measuring the latency of *put( )* operation, we run the test performing several *put( )* operations on the same shared segment and average it over the number of iterations. Figure 4a shows the latencies of different coherence models by using the *put( )* operation of DDSS using OpenFabrics over IBA through a daemon process. We observe that the 1-byte latency achieved by null and read coherence model is only 20μs and 23μs. We observed that the overhead of communicating with the daemon process is close to 10-12μs which explains the large latencies with null and read coherence models. For write and strict coherency model, the latencies are 54.3μs and 54.8μs respectively. This is due to the fact both write and strict coherency models use atomic operations to acquire the lock before updating the shared data. Version-based and delta coherence models report a latency of 37μs and 41μs respectively, since they both need to update the version status maintained at the remote node. Also, as the message size increases, we observe that the latency increases for all coherence models. We see similar trends for *get()* operations with the basic 1-byte latency of get being 25μs. Figure 4b shows the performance of *get()* operation with several clients accessing different portions from a single node. We observe that DDSS is highly scalable in such scenarios and the performance is not affected for increasing number of clients. Figure 4c shows the performance of *get()* operation with several clients accessing the same portion from a single node. Here, we observe that for relatively lesser contention-levels of up to 40%, the performance of *get()* and *put()* operations do not seem to be affected. However, for contention-levels more than 40%, the performance of clients degrades significantly in the case of strict and write coherence model mainly due to the waiting time for acquiring the lock. We see similar trends in the performance of latencies using OpenFabrics over Ammasso. We have included these results in [13].

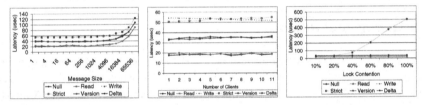

Fig. 4. Basic Performance using OpenFabrics over IBA: (a) *put()* operation (b) Increasing Clients accessing different portions (*get()*) (c) Contention accessing the same shared segment (*put()*)

Measuring Substrate Overhead: One of the critical issues to address on supporting DDSS is to minimize the overhead of the middle-ware layer for applications. We measure the overhead for different configurations (i) Direct scheme allows application to directly communicate with underlying network through DDSS library, (ii) Thread-based scheme allows application to communicate through a daemon process for accessing DDSS and (iii) Thread-based asynchronous scheme is same as thread-based scheme except that applications use asynchronous calls. We see that the overhead is less than a microsecond (0.35μs) through the direct scheme. If the run-time system needs to support multiple threads, we observe that the overhead jumps to 10μs using the thread-based scheme. The reason being the overhead of round-trip communication between the application thread and the DDSS daemon consumes close to 10μs. If the application uses asynchronous operations (thread-based asynchronous scheme), this overhead can be significantly reduced for large message transfers. However, in the worst case scenario, for small message sizes and scheduling of asynchronous operations followed by a wait operation can lead to an overhead of $12\mu s$. The average synchronization time observed in all the schemes is around $20\mu s$.

5.2 Data-Center Service Evaluation

Dynamic Reconfiguration: In our previous work [14] we have shown the strong potential of using the advanced features of high-speed networks in designing reconfiguration techniques. In this section, we use this technique to illustrate the overhead of using DDSS for such a service in comparison with implementations using native protocols. We modified our code base to use the DDSS and compared it with the previous implementation. Also, we emulate the loaded conditions of a real data-center scenario by firing client requests to the respective servers. As shown in Figure 5a, we see that the average reconfiguration time is only $133\mu s$ for increasing loaded servers. The x-axes indicates the number of servers that are currently heavily loaded. The y-axes shows the reconfiguration time using the native protocol (white bar) and using DDSS (white bar + black bar). We observe that the DDSS overhead (black bar) is only around $3\mu s$ for varying load on the servers. Also, as the number of loaded servers increase, we see no change in the reconfiguration time. This indicates that the service is highly resilient to the loaded conditions in the data-center environment. Further, we see that the number of reconfigurations increase linearly as the number of loaded servers increase from 5% to 40%. Increasing the loaded servers further does not seem to affect the reconfiguration time and when this reaches 80%, the number of reconfiguration decreases mainly due to insufficient number of free servers for performing the reconfiguration. Also, for increasing number of reconfigurations, several servers get locked and unlocked to perform efficient reconfiguration. The figure clearly shows that the contention for acquiring locks on loaded servers does not affect the total reconfiguration time showing the scalable nature of this service. In this experiment, since we have only one process per node performing the reconfiguration, we use the direct model for integrating with the DDSS.

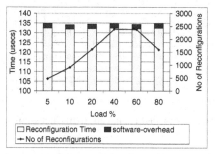
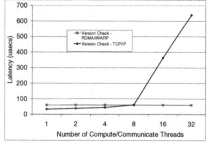

Fig. 5. Software Overhead on Data-Center Services (a) Active Resource Adaptation using Open-Fabrics over IBA (b) Dynamic Content Caching using OpenFabrics over Ammasso

Strong Cache Coherence: In our previous work [7], we have shown the strong potential of using the features of modern interconnects in alleviating the issues of providing strong cache coherence with traditional implementations. In this section, we show the load resilient nature of the one-sided communication in providing such a service using DDSS over Ammasso. Figure 5b, we observe that as we increase the number of server compute threads, the time taken to check for the version increases exponentially for a two-sided communication protocol such as TCP/IP. However, since DDSS is based on one-sided operations (RDMA over iWARP in this case), we observe that the time taken for version check remains constant for increasing number of compute threads.

5.3 Application-Level Evaluation

STORM with DataCutter: STORM [12] is a middle-ware service layer developed by the Department of Biomedical Informatics at The Ohio State University. It is designed to support SQL-like select queries on datasets primarily to select the data of interest and transfer the data from storage nodes to compute nodes for processing in a cluster computing environment. In distributed environments, it is common to have several STORM applications running which can act on same or different datasets serving the queries of different clients. If the same dataset is processed by multiple STORM nodes and multiple compute nodes, DDSS can help in sharing this dataset in a cluster environment so that multiple nodes can get direct access to this shared data. In our experiment, we modified the STORM application code to use DDSS in maintaining the dataset so that all nodes have direct access to the shared information. We vary the dataset size in terms of number of records and show the performance of STORM with and without DDSS. Since larger datasets showed inconsistent values, we performed the experiments on small datasets and we flush the file system cache to show the benefits of maintaining this dataset on other nodes memory. As shown in Figure 6a, we observe that the performance of STORM is improved by around 19% for 1K, 10K and 100K record dataset sizes using DDSS in comparison with the traditional implementation.

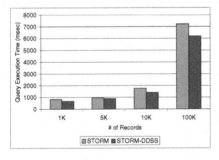
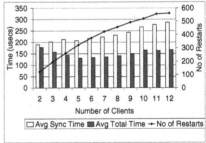

Fig. 6. Application Performance over IBA (a) Distributed STORM application (b) Check-pointing

Check-pointing: We use a check-pointing benchmark to show the scalability and the performance of using DDSS. In this experiment, every process attempts to checkpoint a particular application at random time intervals. Also, every process simulates the application restart, by attempting to reach a consistent check-point and informing all other processes to revert back to the consistent check-point at other random intervals. In

Figure 6b, we observe that the average time taken for check-pointing is only around $150\mu s$ for increasing number of processes. As this value remains almost constant with increasing number of clients and application restarts, it suggests that the application scales well using DDSS. Also, we see that the average application restart time to reach a consistent checkpoint increases with the increase in the number of clients. This is expected as each process needs to get the current checkpoint version from all other processes to decide the most recent consistent checkpoint. Further, we noticed that the DDSS overhead for checkpointing in comparison with native implementation is only around $2.5\mu s$.

6   Related Work

Several distributed data sharing models have been proposed in the past for a variety of environments. The key aspects that distinguish DDSS from previous work is its ability to exploit features of high-performance networks, its portability over multiple interconnects, its support for relaxed coherence protocols and its minimal overhead. Further, our work is mainly targeted for real data-center environment on very large scale clusters.

Run-time data sharing models such as InterWeave [15], Khazana [16], InterAct [17] offer benefits to applications in terms of relaxed coherency and consistency protocols. Khazana proposes the use of several consistency models. InterWeave allows users to define application-specific coherence models. Many of these models are implemented using traditional two-sided communication protocols targeting the WAN environment addressing issues such as heterogeneity, endianness, etc. Such protocols have been shown to have significant overheads in a real cluster-based data-center environment under heavy loaded conditions. Also, none of these models take advantage of high-performance networks for communication, synchronization and efficient data management. Though many of the features of high-performance networks are applicable only in a cluster environment, with the advent of advanced protocols such as iWARP included in the OpenFabrics standard, DDSS can also work well in WAN environments.

7   Conclusion and Future Work

This paper proposes and evaluates a low-overhead distributed data sharing substrate (DDSS) for data-center environments. Traditional data-charing implementations using ad-hoc messaging often incur high overheads and are not very scalable. DDSS on the other hand, is designed to support efficient data management and coherence models while minimizing overheads by leveraging the features of modern interconnects. DDSS is implemented over the OpenFabrics interface and is portable across multiple interconnects including iWARP-capable networks both in LAN/WAN environments. Application-level evaluations with Distributed STORM using DDSS show close to 19% performance benefit over traditional implementation, while evaluations with check-pointing application suggest that DDSS is scalable and has a low overhead.

We plan to enhance DDSS to support advanced locking mechanisms and study the benefits of DDSS for services and applications like meta-data management, storage of BTree data structures in database servers and advanced caching techniques.

Funding Acknowledgment: This research is supported in part by Department of Energy's Grant #DE-FC02-01ER25506, and National Science Foundation's grants #CNS-0403342 and# CNS-0509452; grants from Intel, Mellanox, Sun Microsystems and Linux Networx; and equipment donations from Intel, Mellanox and Silverstorm.

References

1. Shah, H.V., Minturn, D.B., Foong, A., McAlpine, G.L., Madukkarumukumana, R.S., Regnier, G.J.: CSP: A Novel System Architecture for Scalable Internet and Communication Services. 3rd USENIX Symposium on Internet Technologies and Systems, (2001)
2. Tang, C., Chen, D., Dwarkadas, S., Scott, M.: Integrating Remote Invocation and Distributed Shared State (2004)
3. InfiniBand Trade Association. (http://www.infinibandta.com)
4. OpenFabrics Alliance: OpenFabrics. (http://www.openfabrics.org/)
5. Shah, H.V., Pinkerton, J., Recio, R., Culley, P.: DDP over Reliable Transports (2002)
6. Ammasso, inc. (http://www.ammasso.com)
7. Narravula, S., Balaji, P., Vaidyanathan, K., Krishnamoorthy, S., Wu, J., Panda, D.K.: Supporting Strong Coherency for Active Caches in Data-Centers in InfiniBand. SAN. (2004)
8. Zhang, Y., Zheng, W.: User-level communication based cooperative caching. In ACM SIGOPS Operating Systems. (2003)
9. Narravula, S., Jin, H.W., Vaidyanathan, K., Panda, D.K.: Designing Efficient Cooperative Caching Schemes for Data-Centers over RDMA-enabled Networks. In CCGRID). (2005)
10. Chen, D., Tang, C., Sanders, B., Dwarkadas, S., Scott, M.: Exploiting high-level coherence information to optimize distributed shared state. In Proc. of the 9th ACM Symp. on Principles and Practice of Parallel Programming. (2003)
11. Vaidyanathan, K., Jin, H.W., Panda, D.K.: Exploiting RDMA Operations for Providing Efficient Fine-Grained Resource Monitoring in Cluster-based Servers. In Workshop on Remote Direct Memory Access (RDMA): RAIT, Barcelona, Spain (2006)
12. The STORM Project at OSU BMI. (http://storm.bmi.ohio-state.edu/index.php)
13. Vaidyanathan, K., Narravula, S., Panda, D.K.: Soft Shared State Primitives for Multi-Tier Data-Center Services. Technical Report OSU-CISRC-1/06-TR06, OSU (2006)
14. Balaji, P., Vaidyanathan, K., Narravula, S., Savitha, K., Jin, H.W., Panda, D.K.: Exploiting Remote Memory Operations to Design Efficient Reconfiguration for Shared Data-Centers. In Workshop on RAIT, San Diego, CA (2004)
15. Chen, D., Dwarkadas, S., Parthasarathy, S., Pinheiro, E., Scott, M.L.: InterWeave: A Middleware System for Distributed Shared State. In LCR. (2000)
16. Carter, J., Ranganathan, A., Susarla, S.: Khazana: An Infrastructure for Building Distributed Services. In ICDCS. (1998)
17. Parthasarathy, S., Dwarkadas, S.: InterAct: Virtual Sharing for Interactive Client-Server Application. Workshop on Languages, Compilers, and Systems for Computers. (1998)

Proactive Fault Tolerance in MPI Applications Via Task Migration

Sayantan Chakravorty, Celso L. Mendes, and Laxmikant V. Kalé

Department of Computer Science, University of Illinois at Urbana-Champaign
{schkrvrt, cmendes, kale}@uiuc.edu

Abstract. Failures are likely to be more frequent in systems with thousands of processors. Therefore, schemes for dealing with faults become increasingly important. In this paper, we present a fault tolerance solution for parallel applications that proactively migrates execution from processors where failure is imminent. Our approach assumes that some failures are predictable, and leverages the features in current hardware devices supporting early indication of faults. We use the concepts of processor virtualization and dynamic task migration, provided by Charm++ and Adaptive MPI (AMPI), to implement a mechanism that migrates tasks away from processors which are expected to fail. To demonstrate the feasibility of our approach, we present performance data from experiments with existing MPI applications. Our results show that proactive task migration is an effective technique to tolerate faults in MPI applications.

1 Introduction

High-performance systems with thousands of processors have been introduced in the recent past, and the current trends indicate that systems with hundreds of thousands of processors should become available in the next few years. In systems of this scale, reliability becomes a major concern, because the overall system reliability decreases with a growing number of system components. Hence, large systems are more likely to incur a failure during execution of a long-running application.

Many production-level scientific applications are currently written using the MPI paradigm [1]. However, the original MPI standards specify very limited features related to reliability or fault tolerance [2]. In traditional MPI implementations, the entire application has to be shutdown when one of the executing processors experiences a failure. Given the practical problem of ensuring application progress despite the occurrence of failures in the underlying environment, some alternative MPI implementations have been recently proposed (we discuss representative examples in Section 5). Most of these solutions implement some form of redundancy, forcing the application to periodically save part of its execution state. In our previous work, we have demonstrated solutions following this general scheme, using either checkpointing/restart mechanisms [3, 4] or message-logging approaches [5].

In this paper, we present a new solution that goes one significant step further: instead of waiting for failures to occur and reacting to them, we proactively migrate the execution from a processor where a failure is imminent, without requiring the use of spare processors. We build on our recent work on fault-driven task migration [6], and develop

Y. Robert et al. (Eds.): HiPC 2006, LNCS 4297, pp. 485–496, 2006.

a scheme that supports MPI applications transparently to the user. Based on processor virtualization and on the migration capabilities of AMPI, our runtime system handles imminent faults by migrating the MPI tasks to other processors.

To be effective, this approach requires that failures be predictable. We leverage the features in current hardware devices supporting early fault indication. As an example, most motherboards contain temperature sensors, which can be accessed via interfaces like ACPI [7]. Meanwhile, recent studies have demonstrated the feasibility of predicting the occurrence of faults in large-scale systems [8] and of using these predictions in system management strategies [9]. Hence, it is possible, under current technology, to act appropriately before a system fault becomes catastrophic to an application. In this paper we focus on handling warnings for imminent faults and not on the prediction of faults. For unpredictable faults, we can revert back to traditional recovery schemes, like checkpointing [3] or message logging [5].

Our strategy is entirely software based and does not require any special hardware. However, it makes some reasonable assumptions about the system. The application is warned of an impending fault through a signal to the application process on the processor that is about to crash. The processor, memory and interconnect subsystems on a warned node continue to work correctly for some period of time after the warning. This gives us an opportunity to react to a warning and adapt the runtime system to survive a crash of that node. The application continues to run on the remaining processors, even if one processor crashes.

We decided on a set of requirements before setting out to design a solution. The time taken by the runtime system to change (response time), so that it can survive the processor's crash, should be minimized. Our strategy should not require the start up of a new "spare" [4, 5] process on either a new processor or any of the existing ones. When an application loses a processor due to a warning, we expect the application to slow down in proportion to the fraction of computing power lost. Our strategy should not require any change to the user code. We verify in Section 4 how well our protocol meets these specifications.

2   Processor Virtualization

Processor virtualization is the key idea behind our strategy for proactive fault tolerance. The user breaks up his computation into a large number of objects without caring about the number of physical processors available. These objects are referred to as virtual processors. The user views the application in terms of these virtual processors and their interactions. Charm++[10] and Adaptive-MPI (AMPI) [11] are based on this concept of processor virtualization. The Charm++ runtime system is responsible for mapping the virtual processors to physical processors. It can also migrate a virtual processor from one physical processor to another at runtime. Charm++ supports message delivery to and creation, deletion, migration of the virtual processors in a scalable and efficient manner. It also allows reductions and broadcasts in the presence of migrations.

Coupling the capability of migration with the fact that for most applications the computation loads and communication patterns exhibited by the virtual processors tend to persist over time, one can now build measurement based runtime load balancing techniques. Dynamic load balancing in Charm++ has been used to scale diverse applications such as cosmology[12] and molecular dynamics [13] to thousands of processors.

Adaptive MPI (AMPI) [11] is an implementation of the Message Passing Interface (MPI) on top of Charm++. Each MPI process is a user-level thread bound to a Charm++ virtual processor. The MPI communication primitives are implemented as communication between the Charm++ objects associated with each MPI process. Traditional MPI codes can be used with AMPI after no or slight modifications. These codes can also take advantage of automatic migration, automatic load balancing and adaptive overlap of communication and computation.

3   Fault Tolerance Strategy

We now describe our technique to migrate tasks from processors where failures are imminent. Our solution has three major parts. The first part migrates the Charm++ objects off the warned processor and ensures that point-to-point message delivery continues to function even after a crash. The second part deals with allowing collective operations to cope with the possibility of the loss of a processor. It also helps to ensure that the runtime system can balance the application load among the remaining processors after a crash. The third part migrates AMPI processes away from the warned processor. The three parts are interdependent, but for the sake of clarity we describe them separately.

3.1   Evacuation of Charm++ Objects

Each migratable object in Charm++ is identified by a globally unique index which is used by other objects to communicate with it. We use a scalable algorithm for point-to-point message delivery in the face of asynchronous object migration, as described in [14]. The system maps each object to a *home* processor, which always knows where that object can be reached. An object need not reside on its home processor. As an example, an object on processor A wants to send a message to an object (say X) that has its home on processor B but currently resides on processor C. If processor A has no idea where X resides, it sends the message to B, which then forwards it to C. Since forwarding is inefficient, C sends a routing update to A, advising it to send future X-bound messages directly to C.

The situation is complicated slightly by migration. If a processor receives a message for an object that has migrated away from it, the message is forwarded to the object's last known location. Figure 1 illustrates this case as object X migrates from C to another processor D. X's home processor (B) may not yet have the correct address for X when it forwards the message to C. However, C forwards it to D and then D sends a routing update to A. B also receives the migration update from C and forwards any future messages to D. [14] describes the protocol in much greater detail.

When a processor (say E) detects that a fault is imminent, it is possible in Charm++ to migrate away the objects residing there. However, this crash would disrupt message delivery to objects which have their homes on E, due to the lack of updated routing information. We solve that problem by changing the index-to-home mapping such that all objects mapped to E now map to some other processor F. This mapping needs to

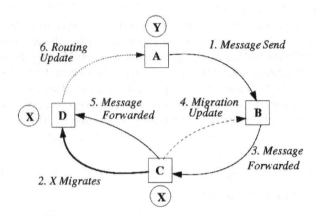

Fig. 1. Message from object Y to X while X migrates from processor C to D

change on all processors in such a way that they stop sending messages to E as soon as possible. The messages exchanged for this distributed protocol are shown in Figure 2. As previously described, objects which had their home on E now have their home on F.

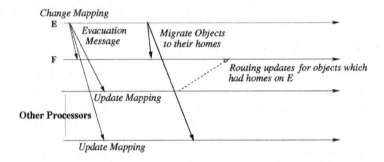

Fig. 2. Messages exchanged when processor E is being evacuated

The index-to-home mapping is a function that maps an object index and the set of valid processors to a valid processor. If the set of valid processors is given by the bitmap *isValidProcessor*, the initial number of processors is *numberProcessors* and *sizeOfNode* is the number of processors in a node, then an index-to-home mapping is given by:

$start \leftarrow possible \leftarrow (index$ **mod** $numberProcessors)$
while $!isValidProcessor[possible]$**do**
$\quad possible \leftarrow (possible + $**sizeOfNode**$)$**mod**$numberProcessors$
$\quad$ **if inSameNode**$(start, possible)$**then**
$\quad\quad$ **abort**$(``No\ valid\ node\ left")$
$\quad$ **end**
end
return $possible;$

For efficiency, we derive the mapping once and store it in a hashtable for subsequent accesses. When an *evacuation* message is received, we repopulate the hashtable. An analysis of the protocol (omitted here due to space constraints) shows that the only messages that E needs to process after being warned were sent to it by processors which had not yet received the *evacuation* message from E. Once all processors receive the *evacuation* message, no messages destined for Charm++ objects will be sent to E.

This protocol is robust enough to deal with multiple simultaneous fault warnings. The distributed nature of the algorithm, without any centralized arbitrator or even a collective operation, makes it robust. The only way two warned processors can interfere with each other's evacuation is if one of them (say H) is the home for an object existing on the other (say J). This might cause J to evacuate some objects to H. Even in this case once J receives the *evacuation* message from H, it changes its index-to-home mapping and does not evacuate objects to H. Only objects that J evacuates before receiving an *evacuation* message from H are received by H. Though H can of course deal with these by forwarding them to their new home, this increases the evacuation time. This case might occur if H receives J's *evacuation* message before it receives its own warning and so does not send an evacuation message to J. We reduce the chances of this by forcing a processor to send an *evacuation* message to not only all valid processors but also processors that started their evacuation recently.

3.2 Support for Collective Operations in the Presence of Fault Warnings

Collective operations are important primitives for parallel programs. It is essential that they continue to operate correctly even after a crash. Asynchronous reductions are implemented in Charm++ by reducing the values from all objects residing in a processor and then reducing these partial results across all processors [14]. The processors are arranged in a k-ary reduction tree. Each processor reduces the values from its local objects and the values from the processors that are its children, and passes the result along to its parent. Reductions occur in the same sequence on all objects and are identified by a sequence number. If a processor were to crash, the tree could become disconnected. Therefore, we try to rearrange the tree around the tree node corresponding to the warned processor. If such node is a leaf, then the rearranging involves just deleting it from its parent's list of children. In the case of an internal tree node, the transformation is shown in Figure 3. Though this rearrangement increases the number of children for some nodes in the tree, the number of nodes whose parent or children change is limited to the node associated to the warned processor, its parent and its children.

Since rearranging a reduction tree while reductions are in progress is very complicated, we adopt a simpler solution. The node representing the warned processor polls its parent, children and itself for the highest reduction that any of them has started. Because the rearranging affects only these nodes, each of them shifts to using the new tree when it has finished the highest reduction started on the old tree by one of these nodes. If there are warnings on a node and on one of its children at the same time, we let the parent modify the tree first and then let the child change the modified tree. Other changes to the tree can go on simultaneously in no specific order.

The Charm++ runtime provides support for asynchronous broadcasts to its objects [14]. It simplifies the semantics of using broadcasts by guaranteeing that all objects

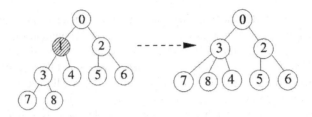

Fig. 3. Rearranging of the reduction tree, when processor 1 receives a fault warning

receive broadcasts in the same sequence. All broadcasts are forwarded to an appointed serializer. This processor allots a number to a broadcast and sends it down the broadcast tree to all other processors. Each processor delivers the broadcast messages to the resident objects in order of the broadcast number. Contrary to intuition, this does not create a *hotspot* since the number of messages received and sent by each processor during a broadcast is unchanged.

We can change the broadcast tree in a way similar to the reduction tree. However, if the serializer receives a warning we piggyback the current broadcast number along with the *evacuation* message. Each processor changes the serializer according to a predetermined function depending on the set of valid processors. The processor that becomes the new serializer stores the piggybacked broadcast count. Any broadcast messages received by the old serializer are forwarded to the new one.

It is evident from the protocol that evacuating a processor might lead to severe load imbalance. Therefore, it is necessary that the runtime system be able to balance the load after a migration caused by fault warning. Minor changes to the already existing Charm++ load balancing framework allow us to map the objects to the remaining subset of valid processors. As we show in Section 4, this capability has a major effect on performance of an application.

3.3   Processor Evacuation in AMPI

We modified the implementation of AMPI to allow the runtime system to migrate AMPI threads even when messages are in flight, i.e. when there are outstanding MPI requests or receives. This is done by treating outstanding requests and receives as part of the state of an AMPI thread. When a thread migrates from processor A to B, the queue of requests is also packed on A and sent to processor B. At the destination processor B, the queue is unpacked and the AMPI thread restarts waiting on the queued requests. However, just packing the requests along with the thread is not sufficient. Almost all outstanding requests and receives are associated with a user-allocated buffer where the received data should be placed. Packing and moving the buffer from A to B might cause the buffer to have a different address on B's memory. Hence the outstanding request that was copied over to the destination would point to a wrong memory address on B.

We solve this problem by using the concept of *isomalloc* proposed in PM^2 [15]. AMPI already uses this to implement thread migration. We divide the virtual address space equally among all the processors. Each processor allocates memory for the user only in the portion of the virtual address space alloted to it. This means that no two buffers allocated by the user code on different processors will overlap. This allows all user buffers in a thread to be recreated at the same address on B as on A. Thus, the buffer addresses in the requests of the migrated thread point to a valid address on B as well. This method has the disadvantage of restricting the amount of virtual address space available to the user on each processor. However, this is a drawback only for 32-bit machines. In the case of 64-bit machines, even dividing up the virtual address space leaves more than sufficient virtual address space for each processor.

4   Experimental Results

We conducted a series of experiments to assess the effectiveness of our task migration technique under imminent faults. We measured both the response time after a fault is predicted and the overall impact of the migrations on application performance. In our tests, we used a 5-point stencil code, written in C and MPI, and the *Sweep3d* code, which is written in Fortran and MPI. The 5-point stencil code allows a better control of memory usage and computation granularity than a more complex application. Sweep3d is the kernel of a real ASCI application; it solves a 3D Cartesian geometry neutron transport problem using a two-dimensional processor configuration.

We executed our tests on NCSA's Tungsten system, a cluster of 3.2 GHz dual-Xeon nodes, with 3 GBytes of RAM per node, and two kinds of interconnects, Myrinet and Gigabit-Ethernet. Each node runs Linux kernel 2.4.20-31.9. We compiled the stencil program with GNU GCC version 3.2.2, and the Sweep3d program with Intel's Fortran compiler version 8.0.066. For both programs, we used AMPI and Charm++ over the Myrinet and Gigabit interconnects. We simulated a fault warning by sending the USR1 signal to an application process on a computation node.

4.1   Response Time Assessment

We wanted to evaluate how fast our protocol is able to morph the runtime system such that if the warned processor crashes, the runtime system remains unaffected. We call this the *processor evacuation* time. We estimate the processor evacuation time as the maximum of the time taken to receive acknowledgments that all evacuated objects have been received at the destination processor and the last message processed by the warned processor. It should be noted that these acknowledgment messages are not necessary for the protocol; they are needed solely for evaluation. The measured value is, of course, a pessimistic estimate of the actual processor evacuation time, because it includes the overhead of those extra messages.

The processor evacuation time for the 5-point stencil program on 8 and 64 processors, for different problem sizes and for both interconnects, is shown in Figure 4(a). The evacuation time increases linearly with the total problem size until at least 512 MB. This shows that it is dominated by the time to transmit the data out from the warned processor. Thus, the evacuation time in Myrinet is significantly smaller than in Gigabit.

Figure 4(b) presents the processor evacuation time for two problem sizes, 32 MB and 512 MB, of the 5-point stencil calculation on different numbers of processors. For both interconnects, the evacuation time decreases more or less linearly with the data volume

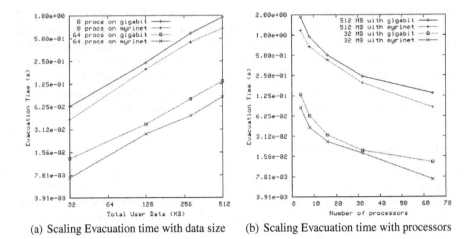

(a) Scaling Evacuation time with data size   (b) Scaling Evacuation time with processors

Fig. 4. Processor evacuation time for MPI 5-point stencil calculation

per processor. Myrinet has a significantly faster response time than Gigabit. Table 1 shows similar data corresponding to the evacuation time for Sweep3d, for a problem size of $150 \times 150 \times 150$. These experiments reveal that the response to a fault warning is constrained only by the amount of data on the warned processor and the speed of the interconnect. In all cases, the evacuation time is under 2 seconds, which is much less than the time interval demanded by fault prediction as reported by other studies [8]. The observed results show that our protocol scales to at least 256 processors. In fact, the only part in our protocol that is dependent on the number of processors is the initial *evacuate* message sent out to all processors. The other parts of the protocol scale linearly with either the size of objects or the number of objects on each processor.

4.2   Overall Application Performance

We evaluated the overall performance of the 5-point stencil and Sweep3d under our task migration scheme in our second set of experiments. We were particularly interested in observing how the presence of warnings and subsequent task migrations affect application behavior. For two executions of the 5-point stencil on 8 processors and a dataset size of 288 MB, we observed the execution profiles shown in Figure 5(a). We generated one warning in both executions. In the first execution, the evacuation prompted by the warning at iteration 85 forces the tasks in the warned processor to be sent to other processors. The destination processors become more loaded than the others, resulting in much larger iteration times for the application. In the second execution, we

Table 1. Evacuation time for a 150^3 Sweep3d problem on different numbers of processors

Number of Processors	4	8	16	32	64	128	256
Evacuation Time (s)	1.125	0.471	0.253	0.141	0.098	0.035	0.025

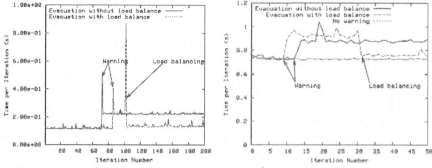

(a) 5-point stencil with 288MB of data on 8 pro-   (b) 150^3 Sweep3d problem on 32 processors
cessors

Fig. 5. Time per iteration for different applications in the presence of warnings

used AMPI's dynamic load balancing capabilities to re-balance the load among the remaining processors after an evacuation at iteration 70. After the re-balancing occurs (at
iteration 100), the remaining processors have nearly the same load, and performance
improves significantly. In this phase, the performance drop relative to the beginning
of the execution is exactly proportional to the computational capability that was lost
(namely, one processor in the set of eight original processors).

We evaluated the effects of evacuation and load balancing on Sweep3d executions
on 32 processors that solved a 150^3 sized problem. Figure 5(b) shows the application
behavior, in terms of iteration durations, for the cases when no processor is evacuated,
one processor is evacuated and when an evacuation is followed by a load balancing
phase. The evacuations are triggered by externally inserted fault warnings. The performance of the application deteriorates after a warning in both cases. The performance
hit of around 10% increase in time per iteration is far more than the loss in computation power of 3%. This is probably caused by computation and communication load
imbalance among processors. After load balancing the performance improves significantly and the increase in time per iteration relative to the warning-free case is around
4%. Thus the loss in performance is very similar to the loss in computation power once
AMPI has performed load balancing.

The Projections analysis tool processes and displays trace data collected during application execution. We use it to assess how parallel processor utilization changes across
a Sweep3d execution of a 150^3 problem on 32 processors. We trigger warnings on Node
3 which contains two processors: 4 and 5. This tests the case of multiple simultaneous
warnings by evacuating processors 4 and 5 at the same time. Before the warnings occur, processors have nearly uniform load and similar utilization (Figure 6(a)). After
evacuation takes place, processors 4 and 5 quit the execution and their objects get distributed among the other processors. However, this creates a load imbalance among the
processors, with some taking longer than others to finish iterations. The redistribution
of objects can also increase communication load by placing objects that communicate

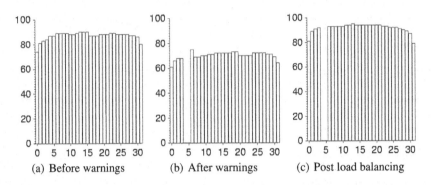

(a) Before warnings   (b) After warnings   (c) Post load balancing

Fig. 6. Utilization per processor for the 150^3 Sweep3d on 32 processors

frequently on different processors. This reduces the utilization significantly on all processors (Figure 6(b)). Finally, after load balancing, the remaining processors divide the load more fairly among themselves and objects that communicate frequently are placed on the same processor, resulting in a much higher utilization (Figure 6(c)). These experiments verify that our protocol matches the goals laid out in Section 1.

5   Related Work

The techniques for fault tolerance in message-passing environments can be broadly divided in two classes: checkpointing schemes and message-logging schemes. In checkpoint based techniques, the application status is periodically saved to stable storage, and recovered when a failure occurs. The checkpointing can be coordinated or independent among the processors. However, due to the possible rollback effects in independent schemes, most implementations use coordinated checkpointing. Representatives of this class are CoCheck [16], Starfish [17] and Clip [18].

In message-logging techniques, the central idea is to retransmit one or more messages when a system failure is detected. Message-logging can be optimistic, pessimistic or causal. Because of the complex rollback protocol, optimistic logging [19] is rarely used; instead, pessimistic logging schemes are more frequently adopted, like in FT-MPI [20], MPI/FT [21], MPI-FT [22] and MPICH-V [23]. Causal logging (such as in [24]) attempts to strike a balance between optimistic and pessimistic logging; however, its restart is also non-trivial.

In all of these proposed fault-tolerant solutions, some corrective action is taken in reaction to a detected failure. In contrast, with the proactive approach that we present in this paper, fault handling consists in migrating a task from a processor where failures are imminent. Thus, no recovery is needed. In addition, both checkpointing and message-logging impose some execution overhead even in the case of no faults, whereas our technique incurs no overhead if faults are not present. Other studies have proposed proactive fault-tolerant schemes for distributed systems [25], but no previous study has considered MPI applications.

6   Conclusion and Future Work

We have presented a new technique for proactive fault tolerance in MPI applications, based on the task migration and load balancing capabilities of Charm++ and AMPI. When a fault is imminent, our runtime system proactively attempts to migrate execution off that processor before a crash actually happens. This processor evacuation is implemented transparently to the application programmer. Our experimental results with existing MPI applications show that the processor evacuation time is close to the limits allowed by the amount of data in a processor and the kind of interconnect. The migration performance scales well with the dataset size. Hence, the fault response time is minimized, as required in our specifications in Section 1. Our experiments also demonstrated that MPI applications can continue execution despite the presence of successive failures in the underlying system. Load balancing is an important step to improve parallel efficiency after an evacuation. By using processor virtualization combined with load balancing, our runtime system was able to divide the load among the remaining fault-free processors, and application execution proceeded with optimized system utilization.

We are currently working to enhance and further extend our technique. We plan to bolster our protocol so that in the case of false positives it can expand the execution back to wrongly evacuated processors. We will also extend our protocol to allow recreating the reduction tree from scratch. We plan to investigate the associated costs and benefits and the correct moment to recreate the reduction tree. In addition, we will generate our fault warnings with information derived from temperature sensors in current systems.

Acknowledgments. This work was supported in part by the US Dep. of Energy under grant W-7405-ENG-48 (HPC-Colony) and utilized parallel machines of NCSA.

References

1. Gropp, W., Lusk, E., Skjellum, A.: Using MPI. Second edn. MIT Press (1999)
2. Gropp, W., Lusk, E.: Fault tolerance in message passing interface programs. International Journal of High Performance Computing Applications **18**(3) (2004) 363–372
3. Huang, C.: System support for checkpoint and restart of Charm++ and AMPI applications. Master's thesis, Dep. of Computer Science, University of Illinois, Urbana, IL (2004) Available at http://charm.cs.uiuc.edu/papers/CheckpointThesis.html.
4. Zheng, G., Shi, L., Kalé, L.V.: FTC-Charm++: An in-memory checkpoint-based fault tolerant runtime for Charm++ and MPI. In: 2004 IEEE International Conference on Cluster Computing, San Diego, CA (2004)
5. Chakravorty, S., Kalé, L.V.: A fault tolerant protocol for massively parallel machines. In: FTPDS Workshop at IPDPS'2004, Santa Fe, NM, IEEE Press (2004)
6. S. Chakravorty, C. L. Mendes and L. V. Kale: Proactive fault tolerance in large systems. In: HPCRI Workshop in conjunction with HPCA 2005. (2005)
7. Hewlett-Packard, Intel, Microsoft, Phoenix, Toshiba: Advanced configuration and power interface specification. ACPI Specification Document, Revision 3.0 (2004) Available from http://www.acpi.info.
8. Sahoo, R.K., Oliner, A.J., Rish, I., Gupta, M., Moreira, J.E., Ma, S., Vilalta, R., Sivasubramaniam, A.: Critical event prediction for proactive management in large-scale computer clusters. In: Proceedings og the ACM SIGKDD, Intl. Conf. on Knowledge Discovery Data Mining. (2003) 426–435

9. Oliner, A.J., Sahoo, R.K., Moreira, J.E., Gupta, M., Sivasubramaniam, A.: Fault-aware job scheduling for BlueGene/L systems. Technical Report RC23077, IBM Research ((2004))

10. Kalé, L.V., Krishnan, S.: Charm++: Parallel programming with message-driven objects. In Wilson, G.V., Lu, P., eds.: Parallel Programming using C++. MIT Press (1996) 175–213

11. Huang, C., Lawlor, O., Kalé, L.V.: Adaptive MPI. In: Proceedings of the 16th International Workshop on Languages and Compilers for Parallel Computing (LCPC 03), College Station, TX (2003)

12. Gioachin, F., Sharma, A., Chackravorty, S., Mendes, C., Kale, L.V., Quinn, T.R.: Scalable cosmology simulations on parallel machines. In: 7th International Meeting on High Performance Computing for Computational Science (VECPAR). (2006)

13. Kalé, L.V., Kumar, S., Zheng, G., Lee, C.W.: Scaling molecular dynamics to 3000 processors with projections: A performance analysis case study. In: Terascale Performance Analysis Workshop, International Conference on Computational Science(ICCS), Melbourne, Australia (2003)

14. Lawlor, O.S., Kalé, L.V.: Supporting dynamic parallel object arrays. Concurrency and Computation: Practice and Experience 15 (2003) 371–393

15. Antoniu, G., Bouge, L., Namyst, R.: An efficient and transparent thread migration scheme in the PM^2 runtime system. In: Proc. 3rd Workshop on Runtime Systems for Parallel Programming (RTSPP) San Juan, Puerto Rico. Lecture Notes in Computer Science 1586, Springer-Verlag (1999) 496–510

16. Stellner, G.: CoCheck: Checkpointing and process migration for MPI. In: Proceedings of the 10th International Parallel Processing Symposium. (1996) 526–531

17. Agbaria, A., Friedman, R.: Starfish: Fault-tolerant dynamic MPI programs on clusters of workstations. Cluster Computing 6(3) (2003) 227–236

18. Chen, Y., Plank, J.S., Li, K.: Clip: A checkpointing tool for message-passing parallel programs. In: Proceedings of the 1997 ACM/IEEE conference on Supercomputing (CDROM). (1997) 1–11

19. Strom, R., Yemini, S.: Optimistic recovery in distributed systems. ACM Transactions on Computer Systems 3(3) (1985) 204–226

20. Fagg, G.E., Dongarra, J.J.: Building and using a fault-tolerant MPI implementation. International Journal of High Performance Computing Applications 18(3) (2004) 353–361

21. Batchu, R., Skjellum, A., Cui, Z., Beddhu, M., Neelamegam, J.P., Dandass, Y., Apte, M.: Mpi/fttm: Architecture and taxonomies for fault-tolerant, message-passing middleware for performance-portable parallel computing. In: Proceedings of the 1st International Symposium on Cluster Computing and the Grid, IEEE Computer Society (2001) 26

22. Louca, S., Neophytou, N., Lachanas, A., Evripidou, P.: MPI-FT: Portable fault tolerance scheme for MPI. Parallel Processing Letters 10(4) (2000) 371–382

23. Bouteiller, A., Cappello, F., Hérault, T., Krawezik, G., Lemarinier, P., Magniette, F.: MPICH-V2: A fault tolerant MPI for volatile nodes based on the pessimistic sender based message logging programming via processor virtualization. In: Proceedings of Supercomputing'03, Phoenix, AZ (2003)

24. Elnozahy, E.N., Zwaenepoel, W.: Manetho: Transparent rollback-recovery with low overhead, limited rollback, and fast output commit. IEEE Transactions on Computers 41(5) (1992) 526–531

25. Pertet, S., Narasimhan, P.: Proactive recovery in distributed CORBA applications. In: Proceedings of the International Conference on Dependable Systems and Networks. (2004) 357–366

Exploring Energy-Performance Trade-Offs for Heterogeneous Interconnect Clustered VLIW Processors

Rahul Nagpal and Y.N. Srikant

Department of Computer Science and Automation,
Indian Institute of Science, Bangalore, India
{rahul, srikant}@csa.iisc.ernet.in

Abstract. Clustered architecture processors are preferred for embedded systems because centralized register file architectures scale poorly in terms of clock rate, chip area, and power consumption. Although clustering helps by improving clock speed, reducing energy consumption of the logic, and making the design simpler, it introduces extra overheads by way of inter-cluster communication. This communication happens over long global wires which leads to delay in execution and significantly high energy consumption.

In this paper, we propose a new instruction scheduling algorithm that exploits scheduling slacks of instructions and communication slacks of data values together to achieve better energy-performance trade-offs for clustered architectures with heterogeneous interconnect. Our instruction scheduling algorithm achieves 35% and 40% reduction in communication energy, whereas the overall energy-delay product improves by 4.5% and 6.5% respectively for 2 cluster and 4 cluster machines with marginal increase (1.6% and 1.1%) in execution time. Our test bed uses the Trimaran compiler infrastructure.

1 Introduction

ILP architectures have been developed to meet the need for high performance in embedded and other applications. Two major ILP design philosophies are superscalar architecture and VLIW architecture. Superscalar processors have dedicated hardware responsible for scheduling instructions at runtime to improve the performance. The high power consumption, chip area, and cost of these architectures make them less suitable for embedded systems. Another design philosophy is the VLIW architecture, where the compiler is responsible for scheduling. This simplifies the hardware but in order to exploit the ILP in emerging embedded applications, more functional units that can operate in parallel are required. This in turn requires more number of read and write ports and hence increased chip area, cycle time, and power consumption.

A clustered VLIW architecture [5] has more than one register file and connects only a subset of functional units to a register file. Groups of small computation clusters can be fully or partially connected using either a *point-to-point* network

Y. Robert et al. (Eds.): HiPC 2006, LNCS 4297, pp. 497–508, 2006.

or a *bus-based* network. Clustering avoids area and power consumption problems of centralized register file architectures while retaining high clock speed, and can be leveraged to get better performance. Texas Instrument's VelociTI, HP/ST's Lx, Analog's TigerSHARC, and BOPS' ManArray are examples of the recent commercial clustered micro-architectures. IBM's eLite is a research proposal for a novel clustered architecture. A compiler for these architectures is responsible for distributing the operations to the resources in different clusters.

Communication of data values in clustered VLIW architectures happens over long wires having high load capacitance which in effect takes more time and consumes more energy. Earlier Studies report that a very high percentage (30% to 40%) of total processor energy consumption is attributed to interconnects [14]. Clearly, clustered architectures are attractive only if their benefits outweigh the performance and energy penalties due to interconnections. Thus efficient means of using interconnects are important for clustered VLIW architectures. The primary goal so far has been reduction in the latency of communication to minimize communication delays [6]. It has been shown that using 50nm technology, it is possible to design wires consuming 1/5 the energy but having twice the delay [2]. Though VLSI technology enables design of interconnects with wires having different energy characteristics, to the best of our knowledge, there has been no effort in the direction of using energy-efficient interconnects for clustered VLIW architectures.

In this paper, we propose and evaluate a new energy-aware instruction scheduling algorithm which exploits multiple interconnects of different energy and delay characteristics in the context of clustered VLIW architectures. The proposed algorithm takes into consideration the interconnect characteristics, and communication slacks of data values together with the scheduling slacks of instructions while steering the communication to an appropriate interconnect, thereby reducing energy consumption without much performance degradation. We consider different flavors of homogeneous interconnects such as latency-optimized and energy-optimized as well as heterogeneous interconnects together with the variation in degree of clustering (no clustering, 2-clustered and 4-clustered) to perform a detailed performance evaluation. Our evaluation uses the Trimaran compiler infrastructure.

The rest of the paper is organized as follows. In section 2, we present the motivation for this work and section 3 gives a detailed description of our energy-aware instruction scheduling algorithm. Section 4 presents a detailed performance evaluation of proposed algorithm and section 5 presents the related work. We conclude in section 6 with pointers to future directions.

2   Motivation

Previous studies have reported that the performance degrades by approximately 12% when the latency of communication is doubled for a four clustered architecture, and that increasing the interconnection bandwidth from one to two improves the performance by as much as 10% [7]. A high speed path for com-

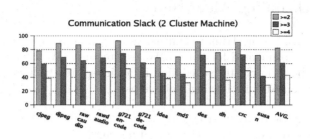

Fig. 1. Communication Slack for 2 Cluster Machine Model

munication of data values among clusters indeed enables better performance. But we argue that not all data values need to be communicated on a high speed path, and that many communications are non-critical and can still happen on a slow path without affecting performance. *We define the communication slack of a data value on clustered architectures as the number of cycles between the time when the data value to be communicated becomes available (due to completion of execution of the producing instruction) and when the instruction that requires the data value is actually scheduled.* The available communication slack of a data value on clustered architecture is affected by data dependencies among instructions, limitation on the available number of functional units, and the limitations on the number of available cross-paths, their bandwidth, and the latency of cross-path communication.

Figure 1 presents quantitative results to substantiate our arguments. This figure presents the percentage of required communication that has a slack of three cycles (two cycles and four cycles) or more for a two-cluster machine with two high speed bidirectional cross-paths between clusters. We observe that all the benchmarks have many communications with high slack values. In particular *djpeg, g721encode, des, and crc* have 70% to 75% of communications with slack value of three cycles or higher. On an average, we observe that 60.88% (82.51% and 43.16%) of communications can sustain a latency of three cycles (two cycles and four cycles respectively) or higher. Thus, having both the cross-path wires optimized for low latency (resulting in high energy consumption) is an overkill. This is because improving the latency of communication channel requires closely spaced repeaters which increase the area and energy overheads of repeaters [2]. A more suitable option that we propose is to design interconnects between clusters with some paths optimized for latency and others for energy. Thus, critical communication can take place over the fast but more energy-consuming wires, and the other not-so-critical communication can happen over the slower but energy-efficient wires. Further mechanisms (software or hardware) that can steer the communications to the appropriate cross-path depending on the communication slack of the data value should be available. Our instruction scheduler is one such software mechanism.

3   The Scheduling Algorithm

The Elcor backend of the Trimaran infrastructure has a list scheduling algorithm designed and implemented for flat VLIW architectures. We have extended this algorithm to perform cluster scheduling in an integrated fashion and it (Refer Algorithm 1) consists of following main steps :

1. Prioritizing the ready instructions
2. Assignment of a cluster to the selected instruction
3. Assignment of cross-paths for transferring data values to the target cluster.

3.1   Prioritizing the Ready Instructions

Instructions in the ReadyList are prioritized using a priority function that uses instruction slack and number of consumers of the instruction respectively. Instructions with less slack should be scheduled early and are given higher priority over instruction with more slack to avoid unnecessary stretching of the schedule. Instructions with the same slack values are further ordered in decreasing order of the number of consumers. An instruction with more successors is more constrained in the sense that its spatial and temporal placement affects scheduling of more instructions and hence should be given higher priority. Giving preference to an instruction with more dependent instructions also enables better scheduling decisions by uncovering a larger portion of the graph.

3.2   Cluster Assignment

Once an instruction has been selected for scheduling, we make a cluster assignment decision. The primary constraints are :

– The chosen cluster should have at least one free resource of the type needed to perform this operation
– Given the bandwidth of the channels among clusters and their usage, it should be possible to satisfy the communication needs of the operands of this instruction on the cluster by scheduling these communications in the earlier cycles

Selection of a cluster from the set of the feasible clusters is done as follows. We determine the earliest time when we can schedule the operation under consideration on each of the clusters in the feasible cluster list while adhering to all the resource and communication constraints. The operation is primarily assigned to that cluster where it can be scheduled at the earliest after accommodating all the communications. In the case of a tie, the operation is assigned to the cluster that minimizes communication requirements. The communication cost is computed by determining the number and type of communications needed by a binding in the earlier cycles as well as the communication that will happen in the future. Future communications are determined by considering the successors of

---

Algorithm 1. The Main Scheduling Loop

Initialize Early Cycle, Late Cycle and Priorities of the operations
ReadyList=Start Operation
while (CurrentOperations =UnSchedList.pop()) **do**
  Compute EarlyCycle of the CurrentOperation
  Initialize MinCycle, MinCommCost, and MinCommOption
  for (CurrentCluster ranging from FirstCluster through LastCluster) **do**
    for (CurrentClusterCycle ranging from EarlyCycle through MaxScheduledCycle) **do**
      Compute the Cross-path Requirements in CurrentCommOption
      Compute the Communication Cost in CurrentCommCost
      if (FU and Cross-paths required by CurrentOperation are available in CurrentCycle for
      CurrentCluster) **then**
        break
      end if
    end for
    if (($CurrentClusterCycle < MinCycle$) || ($CurrentClusterCycle ==$
    $MinCycle$ && $CurrentCommCost <= MinCommCost$)) **then**
      MinCycle=CurrentClusterCycle
      MinCommCost=CurrentCommCost
      MinCommOption=CurrentCommOption
      TargetCluster=CurrentCluster
    end if
  end for
  while (CurrentComm=CurrentCommOption.pop()) **do**
    Determine the EarlyCommCycle, LateCommCycle and the CommSlack for CurrentComm
    Schedule the CurrentComm using minimum energy consuming cross-path between Early-
    CommCycle and LateCommCycle
  end while
  Schedule CurrentOperation on TargetCluster in MinCycle
end while

---

this instruction which have one of their parents bound on a cluster different from the cluster under consideration. This is due to the fact that if the instruction is bound to the cluster under consideration, it will surely lead to communication(s) in the future while scheduling the successors of the instructions.

3.3 Cross-Path Binding

The cross-path assignment scheme is designed to minimize the energy consumption due to inter-cluster communication without affecting runtime performance. To schedule a communication, its earliest scheduling cycle, latest scheduling cycle, and slack values are determined first. The earliest scheduling cycle for a communication is the cycle in which the data value to be communicated is produced in the source cluster, plus one. The latest scheduling time for communication is the scheduling cycle of first consuming instruction, minus one. The difference between the earliest scheduling cycle and the latest scheduling cycle is the communication slack. In order to avoid delaying the consuming instruction and the consequent possible stretch of the schedule, a communication is assigned to a least energy consuming cross-path that can transfer the data value within the available slack for communication. Thus the cross-path assignment scheme maximizes the usage of low power cross-paths subject to the availability of slack in the communication, and thus, as far as possible, performance degradation is minimized and energy saving is maximized.

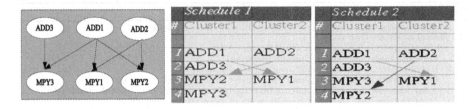

Fig. 2. An Example (a) Data Dependency Graph (b) Schedule 1 (c) Schedule 2

3.4   An Example

Figure 2 shows a portion of a data dependency graph and two possible schedules for this dependency graph. We assume a two-clustered machine with each cluster having an adder, a multiplier, and a fast communication bus. Schedule 1 has ADD1 and ADD2 scheduled on adders of cluster 1 and cluster 2 respectively in cycle 1. To perform multiplication, the results of these operations are transferred to the other cluster in cycle 2. The remaining addition operation ADD3 is also initiated in cycle 2 on cluster 1. The results of ADD1 and ADD2 can be used in cycle 3 on cluster 1 and cluster 2 respectively to perform MPY2 and MPY1 on multipliers. Though MPY3 does not require any inter-cluster communication, it is still executed in cluster 1 at cycle 4 because of non-availability of a multiplier in cycle 3. The scheduler decides to schedule MPY2 ahead of MPY3 in schedule 1 assuming that MPY2 is on the critical path. However, MPY3 gets preference if it is on the critical path as shown in schedule 2. Note that in this case, MPY2 needs to be scheduled in cycle 4 on cluster 1 again because cluster 1 has only one multiplier. *The important point to note here is that the scheduler when scheduling MPY2 in cycle 4 in cluster 2 has the knowledge that it can take two cycles to transfer the result of ADD2 over the communication channel without stretching the schedule.* In such a situation if a slow but more energy-efficient bus is available, our schedulers decide to steer communication to such a bus (as shown with darker arrow in schedule 2).

Notably, even though three additions are ready to be scheduled in the first cycle only two of them can be scheduled (only two adders are available in this case). Similarly though the addition operations finish in opposite clusters in cycle one the results can not be utilized for multiplications in cycle 2 because it takes at least one cycle to transfer the results to the other clusters. This shows how contention among computation and communication resources in clustered architectures manifests itself in the form of greater computation and communication slack. The contention for resources is more in clustered architectures as compared to flat architectures because of distribution of resources. Our scheduling algorithm leverages this increased slack and takes into consideration the criticality of an instruction and the available cycles to communicate requisite data values while scheduling an instruction in a given cycle. Accordingly, communication is assigned to the most energy-efficient cross-path that can transfer the value in the available communication cycles.

4 Experimental Evaluation

4.1 Setup

We have modified the Trimaran suite[1] to generate and simulate code for a variety of clustered VLIW configurations. We have used twelve benchmarks out of which nine are from mediabench *(viz. cjpeg, djpeg, rawcaudio, rawdaudio, g721encode, g721decode, md5, des, and idea)*, two from netbench*(viz. crc, and dh)*, and one *(susan)* is from MiBench. We have adapted the Epic-Explorer[2] to determine the energy consumption in the different components of the data-path. Epic-Explorer is a collection of well known activity based power models. The energy parameters used for heterogeneity in interconnects are the same as the one used in [1] which are based on the circuit level analysis presented in [2] [3]. It takes into account wires with different latency and energy profiles to determine the overall energy consumption of the interconnect. *In our simulations, we consider a latency of one cycle and three cycles for latency optimized (L) bus and energy optimized (P) bus respectively. The dynamic and the leakage energies of the L bus are 2.80 times and 2.64 times the dynamic and the leakage energies of the P bus respectively* [1] [2] [3]. A detailed description of our experimental setup and energy models used is available in the associated technical report [11].

We present results for a two-cluster machine and a four-cluster machine with each machine having one integer unit, one floating unit, one branch unit, and one load store unit in each cluster. To ensure a fair comparison and to determine the exact trade-offs, results are presented in comparison with an equivalent BASE machine with no clustering but with the same number of functional units and registers. Thus, the BASE VLIW machine has two functional units and four functional units of each type for 2-clustered and 4-clustered VLIW. Both the BASE machines have 64 integer and 64 floating point registers. These registers are evenly divided among clusters. Thus the 2-clustered machine has 32 registers of each type and the 4-clustered machine has 16 registers of each type in each cluster. Each configuration has two bidirectional buses between each pair of clusters. The first configuration called LL, has both the buses implemented with delay-optimized wires (one cycle transfer time). The second configuration called PP, has both the buses implemented with the energy-optimized wires (three cycles transfer time). The third configuration represents a heterogeneous interconnect called LP, and has one L bus (one cycle transfer time) and one P bus (three cycle transfer time).

4.2 Results

We have performed a detailed experimental evaluation of the proposed scheme in terms of run-time performance, energy, and energy-delay product. These results are discussed in detail in the following subsections.

---

[1] http://www.trimaran.org/

[2] http://epic-explorer.sourceforge.net/

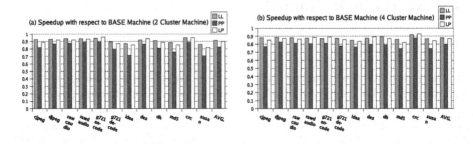

Fig. 3. Speedup w.r.t. the BASE Machine

Performance. We compare the number of cycles taken to execute the program on different configurations. Figure 3(a) shows speedup for a two-cluster machine with different interconnect configurations with respect to the corresponding BASE machine. We observe that the LL configuration achieves the best performance among all the clustered configurations as expected. The average performance degradation while going from a BASE machine to a 2-clustered machine with the LL configuration is 8.65% whereas the average performance degradation for the PP and the LP configurations is 17.74% and 10.11% with respect to the BASE configuration. The results for a 4-cluster configuration (Figure 3(b)) show similar trends.

It is clear that heterogeneous interconnects (LP-with a fast and a slow bus) offer nearly the same performance as that of a homogeneous interconnect (LL-with two fast buses) as the performance degradation is only marginal (1.64% and 1.11% for two-cluster and four-cluster machines respectively). However, an energy optimized homogeneous interconnect (having two slow buses) suffers an intolerably high performance degradation of 10.08% and 9.56% for two and four cluster machines respectively.

Energy Consumption. We observe that the PP configuration shows about 64.66% reduction in communication energy as compared to the LL. The average communication energy reduction for the LP configuration is 35.54% with respect to the LL configuration (Refer Figure 4 (a)). Unlike PP configuration, there is much more variation here in communication energy saved for different benchmarks depending upon the available communication slack for the benchmark and the effectiveness of the proposed scheme in terms of mapping the communication to appropriate bus. We observed 69.74% (PP over LL) and 39.98% (LP over LL) reduction in communication energy while going from the LL to the PP and the LP configuration respectively for a 4-clustered configuration.

The simplification of components and a high degree of communication in the context of clustered VLIW architectures lead to a higher contribution of communication energy to the overall processor energy consumption. An earlier study attributes 36% of energy consumption in interconnects [14]. Conservatively assuming interconnect energy consumption as 20% of the overall processor energy

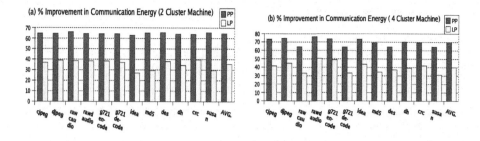

Fig. 4. % Reduction in Communication Energy w.r.t. the LL Configuration

consumption [3], Figures 5(a) presents total energy consumption for two cluster machine. We observe 25.44%, 22.14%, and 20.81% increase in energy for the LL, PP, and the LP configuration compared to the BASE configuration. Energy consumptions of the PP and the LP configurations are 4.32% and 6.15% less than the LL configuration respectively. The corresponding figures for a four cluster machine are slightly higher but show similar trends. There are two reasons of high energy consumption in clustered architectures apart from energy consumption due to interconnects. Firstly, the communication delays prolong the execution on clustered architectures which in turn increases the leakage energy consumption in components. Secondly, clustering causes execution of extra move instructions due to inter-cluster communication which cause extra dynamic energy consumption. Huge saving in communication energy in the PP configuration offsets the increase in processing energy in all the benchmarks except *idea, md5, and susan*. These benchmarks have exceptionally high increase in processing energy because they have fewer communications with high slack values compared to other benchmarks. The LP configuration performs the best in terms of energy consumption. This is because of significant savings in communication energy in the LP configuration with only marginal increase in the processing energy compared to the LL configuration. Clustering reduces the complexity of components and gives additional benefits in terms of energy consumption. However, these benefits can not be attributed to our scheduling algorithm. *Therefore, to fairly quantify the benefits due to scheduling algorithm alone, we have conservatively used the same technology parameters for determining the energy consumption in flat BASE and clustered configurations. In reality, the energy benefits of clustering will be more because of reduction in processing energy due to simplification of components.*

Energy-Delay Product. Figure 6(a) presents the total energy-delay product for different configurations of a 2-cluster machine as compared to the BASE machine. We observe that the total energy-delay product increases by 31.83%, 35.59%, 28.67% for the LL, PP, and the LP configurations respectively. The average increase in total energy-delay product while going form the LL to the

---

[3] Refer to associated technical report [11] for a detailed sensitivity analysis.

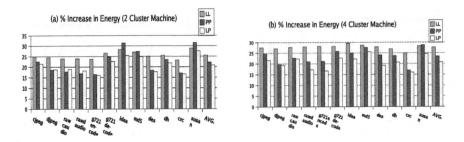

Fig. 5. % Increase in Energy w.r.t. the BASE Machine

PP and the LP configurations is 5.96% and -4.45% (actually decrease for the LP) respectively. Notably the LP configuration provides an improvement over the PP configuration by 10.12%. The results for 4-cluster machine is depicted in Figure 6(b).

The PP configuration, having both interconnects optimized for energy, achieves huge reduction in communication energy. However, the performance degradation due to slow interconnects leads to large performance penalties and the resulting increase in processing energy annuls the benefits obtained due to reduction in communication energy. On the other hand, the LL configuration offers the best performance but at the cost of high energy penalty of delay-optimized interconnects. The LP configuration performs extremely well in terms of energy-delay product. The proposed selective scheduling steers only critical communications to the high speed interconnect. Thus, it maximizes the usage of the low energy interconnect. As a result, it incurs only slight performance and energy penalties as compared to the delay-optimized LL configuration, but is still able to obtain a significant reduction in communication energy. Programs in which more communications have high communication slacks, *viz., djpeg, g721encode, des, and crc* suffer less performance degradation and consequently less increase in processing energy in the LP configuration as compared to the LL configuration. These programs also achieve significant reduction in energy in the LP configuration because of usage of the P bus whenever possible. As a result, the energy-delay product for these programs is significantly better in the LP configuration. Even programs in which moderate number of communications have high slack values yield significant benefit in energy-delay product in the LP configuration as compared to the LL configuration because of a selective choice of the bus. Reader is referred to the associated technical report [11] for detailed analysis of results.

5   Related Work

Earlier proposals for scheduling on clustered VLIW architectures can be classified into two main categories, viz., phase-decoupled approaches [9] [4] and phase-coupled approaches [12] [7] [10]. A phase-decoupled approach partitions a data

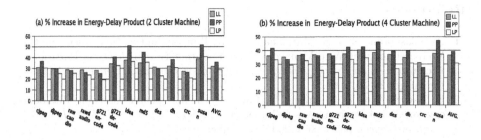

Fig. 6. % Increase in Energy-Delay Product w.r.t the BASE Machine

flow graph (DFG) into clusters to reduce inter-cluster communication while approximately balancing the load among clusters. The annotated DFG is then scheduled using a traditional list scheduler. However, the phase-decoupled approach is known to suffer from the phase ordering problem. An integrated approach to scheduling combats the phase-ordering problem by combining spatial and temporal scheduling decisions in a single phase.

Zhang et al., [15] have proposed a scheme to reduce dynamic and leakage energy in the functional units of VLIW processor, that exploits the slacks in already scheduled code to remap the functional units. Kim et al., [8] have proposed a leakage energy management scheme for VLIW processors that determines the ILP available in the program at a loop level granularity and keeps only the canonical set of functional units sufficient to exploit this ILP in active mode. Andrei et al., [13] have proposed various inter-cluster communication model for clustered architecture and performed a quantitative analysis to compare their benefits. Gonzalez et al., [6] have evaluated different kinds of interconnects from performance perspective and concluded that a point-to-point interconnect with an effective latency-aware steering scheme is more efficient than a bus-based interconnect. Balasubramonian et al., [1] have evaluated techniques such as cache pipelining, exploiting narrow bit-width operands, and interconnect load balancing in the context of superscalar architectures with heterogeneous interconnect.

6   Conclusions and Future Directions

In this work, we have proposed a new energy-aware instruction scheduling algorithm for clustered VLIW architectures that is capable of exploiting interconnect characteristics to get energy benefits without showing much performance degradation. The major conclusion that we draw form this work is that clustered architecture with heterogeneous interconnect offers better energy-performance trade-offs when used with an effective scheduling algorithm as compared to a cluster VLIW architecture with homogeneous interconnect (which is either optimized or latency or power). In future, we would like to develop an integrated algorithm for reducing energy consumption in functional units as well as interconnects.

References

1. R. Balasubramonian, N. Muralimanohar, K. Ramani, and V. Venkatachalapathy. Microarchitectural Wire Management for Performance and Power in Partitioned Architectures. In *Proc. of Intl. Symp. on High-Performance Computer Architecture*, pages 28–39, 2005.
2. K. Banerjee and A. Mehrotra. A Power-Optimal Repeater Insertion Methodology for Global Interconnects in Nanometer Designs. In *Proc. of IEEE Trans. on Electron Devices*, pages 2001–2007, November 2002.
3. M. L. Mui, K. Banerjee, and A. Mehrotra. A Global Interconnect Optimization Scheme for Nanometer Scale VLSI with Implications for Latency, Bandwidth and Power Dissipation. In *IEEE Trans. on Electron Devices*, pages 195–203, 2004.
4. M. Chu, K. Fan, and S. Mahlke. Region-based Hierarchical Operation Partitioning for Multicluster Processors. *SIGPLAN Notices*, pages 300–311, 2003.
5. P. Faraboschi, G. Brown, J. A. Fisher, and G. Desoli. Clustered Instruction-level Parallel Processors. Technical report, Hewlett-Packard, 1998.
6. A. G. Joan-Manuel Parcerisa, Julio Sahuquillo and J. Duato. Efficient Interconnects for Clustered Microarchitectures. In *Proc. of Int. Conf. on Parallel Architectures and Compilation Techniques*, pages 291–300, 2002.
7. K. Kailas, A. Agrawala, and K. Ebcioglu. CARS: A New Code Generation Framework for Clustered ILP Processors. In *Proc. of Intl. Symp. on High-Performance Computer Architecture*, page 133, 2001.
8. H. S. Kim, N. Vijaykrishnan, M. Kandemir, and M. J. Irwin. Adapting Instruction Level Parallelism for Optimizing Leakage in VLIW Architectures. In *Proc. of Conf. on Language, Compiler, and Tool for Embedded Systems*, pages 275–283, 2003.
9. V. S. Lapinskii, M. F. Jacome, and G. A. De Veciana. Cluster Assignment for High-Performance Embedded VLIW processors. *ACM Trans. on Design and Automation of Electronic Systems*, pages 430–454, 2002.
10. R. Nagpal and Y. N. Srikant. Integrated Temporal and Spatial Scheduling for Extended Operand Clustered VLIW Processors. In *Proc. of Conf. on Computing Frontiers*, pages 457–470, 2004.
11. R. Nagpal and Y. N. Srikant. Exploring Energy-Performance Trade-offs for Heterogeneous Interconnect Clustered VLIW Processors. Technical Report, Dept. of CSA, Indian Institute of Science (http://www.archive.csa.iisc.ernet.in/TR), 2005.
12. E. Ozer, S. Banerjia, and T. M. Conte. Unified Assign and Schedule: A New Approach to Scheduling for Clustered Register File Microarchitectures. In *Proc. of Intl. Symp. on Microarchitecture*, pages 308–315, 1998.
13. A. Terechko, E. L. Thenaff, M. Garg, J. V. Eijndhoven, and H. Corporaal. Inter-Cluster Communication Models for Clustered VLIW Processors. In *Proc. of Intl. Symp. on High-Performance Computer Architecture*, page 354, 2003.
14. H. Wang, L.-S. Peh, and S. Malik. Power-driven Design of Router Microarchitectures in On-chip Networks. In *Proc. of Symp. on Microarchitecture*, page 105, 2003.
15. W. Zhang, N. Vijaykrishnan, M. Kandemir, M. J. Irwin, D. Duarte, and Y.-F. Tsai. Exploiting VLIW Schedule Slacks for Dynamic and Leakage Energy Reduction. In *Proc. of Intl. Symp. on Microarchitecture*, pages 102–113, 2001.

Distributed Anemone: Transparent Low-Latency Access to Remote Memory

Michael R. Hines, Jian Wang, and Kartik Gopalan

Computer Science Department, Binghamton University
{mhines, jianwang, kartik}@cs.binghamton.edu

Abstract. Performance of large memory applications degrades rapidly once the system hits the physical memory limit and starts paging to local disk. We present the design, implementation and evaluation of *Distributed Anemone* (Adaptive Network Memory Engine) – a lightweight and distributed system that pools together the collective memory resources of multiple machines across a gigabit Ethernet LAN. Anemone treats remote memory as another level in the memory hierarchy between very fast local memory and very slow local disks. Anemone enables applications to access potentially "unlimited" network memory without any application or operating system modifications. Our kernel-level prototype features fully distributed resource management, low-latency paging, resource discovery, load balancing, soft-state refresh, and support for 'jumbo' Ethernet frames. Anemone achieves low page-fault latencies of $160\mu s$ average, application speedups of up to 4 times for single process and up to 14 times for multiple concurrent processes, when compared against disk-based paging.

1   Introduction

Performance of large-memory applications (LMAs) can suffer from large disk access latencies when the system hits the physical memory limit and starts paging to local disk. At the same time, affordable, low-latency, gigabit Ethernet is becoming commonplace with support for jumbo frames (packets larger than 1500 bytes). Consequently, instead of paging to a slow local disk, one could page over a gigabit Ethernet to the unused memory of remote machines and use the disk only when remote memory is exhausted. Thus, remote memory can be viewed as another level in the traditional memory hierarchy, filling the widening performance gap between low-latency RAM and high-latency disk. In fact, remote memory paging latencies of about $160\mu s$ or less can be easily achieved whereas disk read latencies range anywhere between 6 to 13ms. A natural goal is to enable unmodified LMAs to transparently utilize the collective remote memory of nodes across a gigabit Ethernet LAN. Several prior efforts [1, 2, 3, 4, 5, 6, 7, 8] have addressed this problem by relying upon expensive interconnect hardware (ATM or Myrinet switches), slow bandwidth limited LANs (10Mbps/100Mbps), or heavyweight software Distributed Shared Memory (DSM) [9, 10] systems that require intricate consistency/coherence techniques and, often, customized application

Y. Robert et al. (Eds.): HiPC 2006, LNCS 4297, pp. 509–521, 2006.
© Springer-Verlag Berlin Heidelberg 2006

programming interfaces. Additionally, extensive changes were often required to the LMAs or the OS kernel or both.

Our earlier work [11] addressed the above problem through an initial prototype, called the *Adaptive Network Memory Engine (Anemone)* – the first attempt at demonstrating the feasibility of transparent remote memory access for LMAs over commodity gigabit Ethernet LAN. This was done without requiring any OS changes or recompilation, and relied upon a central node to map and exchange pages between nodes in the cluster. Here we describe the implementation and evaluation of a *fully distributed* Anemone architecture. Like the centralized version, distributed Anemone uses lightweight, pluggable Linux kernel modules and does not require any OS changes. Additionally, it achieves the following significant improvements over centralized Anemone. (1) Memory resource management is distributed across the whole cluster. There is no single control node. (2) Paging latency is reduced by over a factor of 3 – from around $500\mu s$ in the to less than $160\mu s$. (3) Clients can perform load-balancing across multiple memory servers, taking into account their memory usage and paging load. (4) A *distributed resource discovery* mechanism enables clients to discover newly available servers and track memory usage across the cluster. (5) A *soft-state refresh* mechanism enables memory servers to track the liveness of clients and their pages. (6) The distributed version incorporates the flexibility of whether or not 'jumbo' frames should be used, allowing Anemone to operate in networks with any MTU size. (7) We are currently incorporating reliability into the paging process, where clients can replicate pages to protect against server failures. We evaluated our prototype using unmodified LMAs such as ray-tracing, network simulations, in-memory sorting, and k-nearest neighbor search. Results show that average page-fault latencies reduce from 8.3ms to $160\mu s$, single-process applications speed up by up to a factor of 4, and multiple concurrent processes by up to a factor of 14, when compared against disk-based paging.

2   Design and Implementation

Distributed Anemone has two major software components: the *client module* on low memory machines and the *server module* on machines with unused memory. The client module appears to the client system simply as a block device that can be configured as the primary swap device. Whenever an LMA needs more virtual memory, the pager (swap daemon) in the client swaps out pages from the client to other server machines. As far as the pager is concerned, the client module is just a block device not unlike a hard disk partition. Internally, however, the client module maps swapped out pages to remote memory servers.

The servers themselves are also commodity machines, but have unused memory to contribute, and can in fact switch between the roles of client and server at different times, depending on their memory requirements. Client machines discover available servers by using a simple distributed resource discovery mechanism. Servers provide regular feedback about their load information to clients, both as a part of the resource discovery process and as a part of regular paging

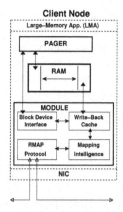

Fig. 1. The components of a client

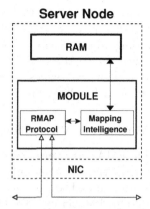

Fig. 2. The components of a server

process. Clients use this information to schedule page-out requests. For instance, a client can simply choose the least loaded server node to send a new page. Also, both the clients and servers use a soft-state refresh protocol to maintain the liveness of pages stored at the servers. The earlier Anemone prototype differed in that the page-to-server mapping logic was maintained at a central *Memory Engine*, instead of individual client nodes. Although simpler to implement, this centralized architecture incurred two extra round trip times on every request besides forcing *all traffic* to go through the central Memory Engine, which can become a single point of failure and a significant bottleneck.

2.1   Client and Server Modules

Figure 1 illustrates the client module that handles paging operations. It has four major components: (1) The Block Device Interface (BDI), (2) a basic LRU-based write-back cache, (3) mapping logic for server location of swapped-out pages, and (4) a Remote Memory Access Protocol (RMAP) layer. The pager issues read and write requests to the BDI in 4KB data blocks. The BDI, in turn, performs read and write operations to our write-back cache (for which pages do not get transmitted until eviction). When the cache is full, a page is evicted to a server using RMAP. Figure 2 illustrates the two major components of the server module: (1) a hash table that stores client pages along with client's identity (layer-2 MAC address) and (2) the RMAP layer. The server module can store/retrieve pages for any client machine. Once the server reaches capacity, it responds to the requesting client with a negative acknowledgment. It is then the client's responsibility to select another server, if available, or to page to disk if necessary. Page-to-server mappings are kept in small hashtables whose buckets are allocated using the `get_free_pages()` call. Linked-lists contained within each bucket hold 64-byte entries that are managed using the Linux slab allocator (which performs fine-grained management of small, equal-sized memory objects). Standard disk block devices interact with the kernel through a *request queue* mechanism, which permits the kernel to group spatially consecutive block I/Os (BIO) together into one "request" and schedule them using an elevator algorithm for seek-time minimization. Unlike disks, Anemone is essentially random access with a fixed read/write latency. Thus BDI does not need to group sequential BIOs. It can bypass request queues, perform out-of-order transmissions, and asynchronously handle un-acknowledged, outstanding RMAP messages.

2.2   Transparent Virtualization

To enable LMAs to transparently access remote memory (no relinking or re-compilation), the client module exports a BDI to the pager. Additionally, no changes are required to the core OS kernel because the BDI is implemented as a self-contained kernel module. One can invoke the standard open, read, and write system calls on the BDI like any other block device. Although our paper does not focus on this aspect, the BDI can be used as a low-latency store for temporary files and can even be memory-mapped by applications aware of the remote memory. The system also performs two types of multiplexing in the presence of multiple clients and servers: (a) Any single client can transparently access memory from multiple servers as one pool via the BDI, and (b) Any single server can share its unused memory pool among multiple clients simultaneously. This provides the maximum flexibility in efficiently utilizing the global memory pool and avoids resource fragmentation.

2.3   Remote Memory Access Protocol (RMAP)

RMAP is a tailor-made, low-overhead communication protocol for remote memory access within the same subnet. It implements the following features: (1) Reliable Packet Delivery, (2) Flow-Control, and (3) Fragmentation and Reassembly. While one could technically communicate over TCP, UDP, or even the IP protocol layers, this choice comes burdened with unwanted protocol processing. Instead RMAP takes an integrated, faster approach by communicating directly with the network device driver, sending frames and handling reliability issues in a manner that suites the needs of the Anemone system. Every RMAP message is acknowledged except for soft-state and dynamic discovery messages. Timers trigger retransmissions when necessary (which is extremely rare) to guarantee reliable delivery. We cannot allow a paging request to be lost, or the application that depends on that page will fail altogether. RMAP also implements flow control to ensure that it does not overwhelm either the receiver or the intermediate network card and switches. However, RMAP does not require TCP's features such as byte-stream abstraction, in-order delivery, or congestion control. Hence we chose to implement RMAP as a light-weight window-based reliable datagram protocol. All client nodes keep a static-size window to control the transmission rate, which works very well for purely in-cluster communication.

The last design consideration in RMAP is that while the standard memory page size is 4KB (or sometimes 8KB), the maximum transmission unit (MTU) in traditional Ethernet networks is limited to 1500 bytes. RMAP implements dynamic fragmentation/reassembly for paging traffic. Additionally, RMAP also has the flexibility to use *Jumbo frames*, which are packets with sizes greater than 1500 bytes (typically between 8KB to 16KB). Jumbo frames enable RMAP to transmit complete 4KB pages to servers using a single packet, without fragmentation. Our testbed includes an 8-port switch that supports Jumbo Frames (9KB packet size). We observe a 6% speed up in RMAP throughput by using Jumbo Frames. In this paper, we conduct all experiments with 1500 byte MTU with fragmentation/reassembly performed by RMAP.

2.4   Distributed Resource Discovery

As servers constantly join or leave the network, Anemone can (a) seamlessly absorb the increase/decrease in cluster-wide memory capacity, insulating LMAs from resource fluctuations and (b) allow any server to reclaim part or all of its contributed memory. This objective is achieved through *distributed resource discovery* described below, and *soft-state refresh* described next in Section 2.5. Clients can discover newly available remote memory in the cluster and the servers can announce their memory availability. Each server periodically broadcasts a *Resource Announcement* (RA) message (1 message every 10 seconds in our prototype) to advertise its identity and the amount of memory it is willing to contribute. Besides RAs, servers also piggyback their memory availability information in their page-in/page-out replies to individual clients. This distributed mechanism permits any new server in the network to dynamically announce its presence and allows existing servers to announce their up-to-date memory availability information to clients.

2.5   Soft-State Refresh

Distributed Anemone also includes soft-state refresh mechanisms (keep-alives) to permit clients to track the liveness of servers and vice-versa. Firstly, the RA message serves an additional purpose of informing the client that the server is alive and accepting paging requests. In the absence of any paging activity, if a client does not receive the server's RA for three consecutive periods, it assumes that the server is offline and deletes the server's entries from its hashtables. If the client also had pages stored on that server that went offline, it needs to recover the corresponding pages from a copy stored either on the local disk on on another server's memory. Soft-state also permits servers to track the liveness of clients whose pages they store. Each client periodically transmits a *Session Refresh* message to each server that hosts its pages (1 message every 10 seconds in our prototype), which carries a client-specific session ID. The client module generates a different and unique ID each time the client restarts. If a server does not receive refresh messages with matching session IDs from a client for three consecutive periods, it concludes that the client has failed or rebooted and frees up any pages stored on that client's behalf.

2.6 Server Load Balancing

Memory servers themselves are commodity nodes in the network that have their own processing and memory requirements. Hence another design goal of Anemone is to a avoid overloading any one server node as far as possible by transparently distributing the paging load evenly. In the earlier centralized architecture, this function was performed by the memory engine which kept track of server utilization levels. Distributed Anemone implements additional coordination among servers and clients to exchange accurate load information. Section 2.4 described the mechanism to perform resource discovery. Clients utilize the server load information gathered from resource discovery to decide the server to which they should send new page-out requests. This decision process is based upon one of two different criteria: (1) The number of pages stored at each active server and (2) The number of paging requests serviced by each active server. While (1) attempts to balance the memory usage at each server, (2) attempts to balance the request processing overhead.

2.7 Fault-Tolerance

The ultimate consequence of failure in swapping to remote memory is no worse than failure in swapping to local disk. However, the probability of failure is greater in a LAN environment because of multiple components involved in the process, such as network cards, connectors, switches etc. Although RMAP provides reliable packet delivery as described in Section 2.3 at the protocol level, we are currently implementing two alternatives for tolerating server failures: (1) To maintain a local disk-based copy of every memory page swapped out over the network. This provides same level of reliability as disk-based paging, but risks performance interference from local disk activity. (2) To keep redundant copies of each page on multiple remote servers. This approach avoids disk activity and reduces recovery-time, but consumes bandwidth, reduces the global memory pool and is susceptible to network failures.

3 Performance

The Anemone testbed consists of one 64-bit low-memory AMD 2.0 GHz client machine containing 256 MB of main memory and nine remote-memory servers. The DRAM on these servers consist of: four 512 MB machines, three 1 GB machines, one 2 GB machine, one 3 GB machine, totaling to almost 9 gigabytes of remote memory. The 512 MB servers range from 1.7 GHz to 800 MHz Intel processors. The other 5 machines are all 2.7 GHz and above Intel Xeons, with mixed PCI and PCI express motherboards. For disk based tests, we used a Western Digital WD800JD 80 GB SATA disk, with 7200 RPM speed, 8 MB of cache and $8.9ms$ average seek time, (which is consistent with our results). This disk has a 10 GB swap partition reserved on it to match the equivalent amount of remote memory available in the cluster, which we use exclusively when

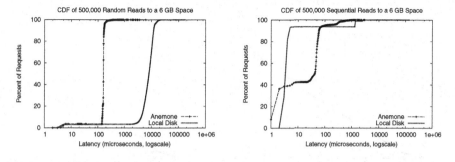

Fig. 3. Comparison of latency distributions for random and sequential reads

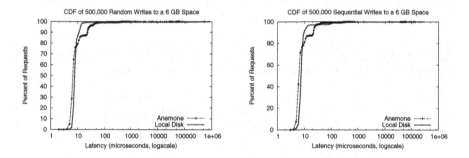

Fig. 4. Comparison of latency distributions for random and sequential writes

comparing our system against the disk. Each machine is equipped with an Intel PRO/1000 gigabit Ethernet card connected to one of two 8-port gigabit switches, one from Netgear and one from SMC. The performance results presented below can be summarized as follows. Distributed Anemone reduces read latencies to an average $160\mu s$ compared to $8.3ms$ average for disk and $500\mu s$ average for centralized Anemone. For writes, both disk and Anemone deliver similar latencies due to write caching. In our experiments, Anemone delivers a factor of 1.5 to 4 speedup for single process LMAs, and delivers up to a factor of 14 speedup for multiple concurrent LMAs. Our system can successfully operate with both multiple clients and multiple servers. Due to space constraints, we omit the speedup results for multiple clients, which emulates the single-client speedups.

3.1 Paging Latency

Figures 3 and 4 show the distribution of observed read and write latencies for sequential and random access patterns with both Anemone and disk. Though real-world applications rarely generate purely sequential or completely random memory access patterns, these graphs provide a useful measure to understand the underlying factors that impact application execution times. Most random

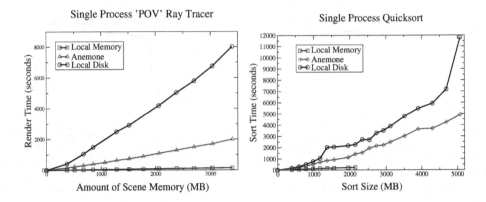

Single Process 'POV' Ray Tracer

Single Process Quicksort

Fig. 5. Execution times of POV-ray for increasing problem sizes

Fig. 6. Execution times of STL Quicksort for increasing problem sizes

Table 1. Average application execution times and speedups for local memory, Distributed Anemone, and Disk. N/A indicates insufficient local memory.

	Size (GB)	Local Mem	Distr. Anemone	Disk	Speedup $\frac{Disk}{Anemone}$
Povray	3.4	145	1996	8018	4.02
Quicksort	5	N/A	4913	11793	2.40
NS2	1	102	846	3962	4.08
KNN	1.5	62	721	2667	3.7

read requests to disk experience a latency between 5 to 10 milliseconds. On the other hand most requests in Anemone experience only around $160\mu s$ latency. Most sequential read requests to disk are serviced by the on-board disk cache within 3 to $5\mu s$ because sequential read accesses fit well with the motion of disk head. In contrast, Anemone delivers a range of latency values, most below $100\mu s$. This is because network communication latency dominates in Anemone even for sequential requests, though it is masked to some extent by the prefetching performed by the pager and the file-system. The write latency distributions for both disk and Anemone are comparable, with most latencies being close to $9\mu s$ because writes typically return after writing to the buffer cache.

3.2 Application Speedup

Single-Process LMAs: Table 1 summarizes the performance improvements seen by unmodified single-process LMAs using the Anemone system. The first application is a **ray-tracing program called POV-Ray**. The memory consumption of POV-Ray was varied by rendering different scenes with increasing number of colored spheres. Figure 5 shows the completion times of these increasingly large renderings up to 3.4 GB of memory versus the disk using an equal amount of

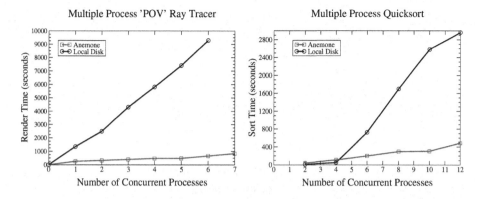

Multiple Process 'POV' Ray Tracer   Multiple Process Quicksort

Fig. 7. Execution times of multiple con- **Fig. 8.** Execution times of multiple concurrent processes executing POV-ray   current processes executing STL Quick-
                                          sort

local swap space. The figure clearly shows that Anemone delivers increasing application speedups with increasing memory usage and is able to improve the
execution time of a single-process POV-ray by a factor of 4 for 3.4 GB memory usage. The second application is a **large in-memory Quicksort program**
that uses a C++ STL-based implementation, with a complexity of $O(N \log N)$
comparisons. We sorted randomly populated large in-memory arrays of integers.
Figure 6 shows that Anemone delivers a factor 2.4 speedup for a single-process
Quicksort using 5 GB of memory. The third application is the popular **NS2 network simulator**. We simulated a delay partitioning algorithm [12] on a 6-hop
wide-area network path using voice-over-IP traffic traces. Table 1 shows that,
with NS2 requiring 1GB memory, Anemone speeds up the simulation by a factor
of 4 compared to disk based paging. The fourth application is the **k-nearest
neighbor (KNN)** search algorithm on large 3D datasets, which are useful in
applications such as medical imaging, molecular biology, CAD/CAM, and multimedia databases. Table 1 shows that, when executing KNN search algorithm
over a dataset of 2 million points consuming 1.5GB memory, Anemone speeds
up the simulation by a factor of 3.7 over disk based paging.

Multiple Concurrent LMAs: In this section, we test the performance of Anemone
under varying levels of concurrent application execution. Multiple concurrently
executing LMAs tend to stress the system by competing for computation, memory and I/O resources and by disrupting any sequentiality in paging activity.
Figures 7 and 8 show the execution time comparison of Anemone and disk
as the number of POV-ray and Quicksort processes increases. The execution
time measures the time interval between the start of the execution and the
completion of last process in the set. We try to keep each process at around
100 MB of memory. The figures show that the execution times using disk-
based swap increases steeply with number of processes. Paging activity loses
sequentiality with an increasing number of processes, making the disk seek and

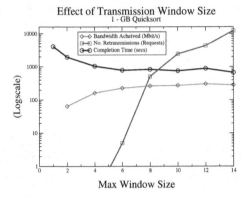

Fig. 9. Effects of varying the transmission window using Quicksort

rotational overheads dominant. On the other hand, Anemone reacts very well as execution time increases very slowly, due to the fact that network latencies are mostly constant, regardless of sequentiality. With 12–18 concurrent LMAs, Anemone achieves speedups of a factor of 14 for POV-ray and a factor of 6.0 for Quicksort.

3.3   Tuning the Client RMAP Protocol

One of the important knobs in RMAP's flow control mechanism is the client's transmission window size. Using a 1 GB Quicksort, Figure 9 shows the effect of changing this window size on three characteristics of the Anemone's performance: (1) the number of retransmissions, (2) paging bandwidth, which is represented in terms of "goodput", i.e. the amount of bandwidth obtained after excluding retransmitted bytes and header bytes, and (3) completion time. As the window size increases, the number of retransmissions increases because the number of packets that can potentially be delivered back-to-back also increases. For larger window sizes, the paging bandwidth is also seen to increase and saturates because the transmission link remains busy more often, delivering higher goodput in spite of an initial increase in the number of retransmissions. However, if driven too high, the window size will cause the paging bandwidth to decline considerably due to increasing number packet drops and retransmissions. The application completion times depend upon the paging bandwidth. Initially, an increase in window size increases the paging bandwidth and lowers the completion times. Similarly, if driven too high, the window size causes more packet drops, more retransmissions, lower paging bandwidth and higher completion times.

3.4   Control Message Overhead

To measure the control traffic overhead due to RMAP, we measured the percentage of control bytes generated by RMAP compared to the amount of data bytes transferred while executing a 1GB POVRay application. Control traffic refers to the page headers, acknowledgments, resource announcement messages, and soft-state refresh messages. We first varied the number of servers from 1 to 6, with a single client executing the POV-Ray application. Next, we varied the number of clients from 1 to 4 (each executing one instance of POV-Ray), with 3 memory servers. The percentage control traffic overhead was consistently measured at 1.74% – a very small percentage of the total paging traffic.

4   Related Work

To the best of our knowledge, Anemone is the first system that provides unmodified LMAs with a completely transparent and virtualized access to cluster-wide remote memory over commodity gigabit Ethernet LANs. The earliest efforts [13, 14] in using remote memory aimed to improve memory management, recovery, concurrency control, and read/write performance for in-memory database and transaction processing systems. The first two remote paging systems [1, 2] incorporated extensive OS changes to both the client and the memory servers and operated upon 10Mbps Ethernet. The Global Memory System (GMS) [3] was designed to provide network-wide memory management support for paging, memory mapped files, and file caching. This system was also closely built into the end-host operating system and operated upon a 155Mbps DEC Alpha ATM Network. The Dodo project [4, 15] provides a user-level library based interface that a programmer can use to coordinate all data transfers to and from a remote memory cache, requiring legacy applications to be modified. Work in [5] implements a remote memory paging system in the DEC OSF/1 operating system as a customized device driver over 10Mbps Ethernet. A remote paging mechanism [7] specific to the Nemesis operating system was designed to permit application-specific remote memory access and paging. The Network RamDisk [6] offers remote paging with data replication and adaptive parity caching by means of a device driver based implementation. Other remote memory efforts include software distributed shared memory (DSM) systems [9, 10], which allow a set of independent nodes to behave as a large shared memory multiprocessor, often requiring customized programming to share common data across nodes. This goal is different from that of Anemone which allows unmodified application binaries to execute and use remote memory transparently. Samson [8] is a dedicated memory server with a highly modified OS over a Myrinet interconnect that actively attempts to predict client page requirements and delivers the pages just-in-time to hide the paging latencies. The NOW project [16] performs cooperative caching via a global file cache in the xFS file system, while [17] attempts to avoid inclusiveness within the cache hierarchy.

5   Conclusions

In this paper, we presented Distributed Anemone – a system that enables unmodified large memory applications to transparently utilize the unused memory of nodes across a gigabit Ethernet LAN. Unlike its centralized predecessor,

Distributed Anemone features fully distributed memory resource management, low-latency remote memory paging, distributed resource discovery, load balancing, soft-state refresh to track liveness of nodes, and the flexibility to use Jumbo Ethernet frames. We presented the architectural design and implementation details of a fully operational Anemone prototype. Evaluations using multiple real-world applications, include ray-tracing, large in-memory sorting, network simulations, and nearest neighbor search, show that Anemone speeds up single process application by up to a factor of 4 and multiple concurrent processes by up to a factor of 14, compared to disk-based paging. Average page-fault latencies are reduced from 8.3ms with disk based paging to $160\mu s$ with Anemone. There are several exciting avenues for further research in Anemone. We are incorporating fault-tolerance mechanisms into Anemone using page replication across servers as well as local disk. Additionally, compression of pages holds the potential to further reduce communication overhead and increase the effective storage capacity.

References

1. Comer, D., Griffoen, J.: A new design for distributed systems: the remote memory model. In Proc. of the USENIX 1991 Summer Technical Conference (1991) 127–135
2. Felten, E., Zahorjan, J.: Issues in the implementation of a remote paging system. Tech. Report 91-03-09, Comp. Science Dept., University of Washington (1991)
3. Feeley, M., Morgan, W., Pighin, F., Karlin, A., Levy, H.: Implementing global memory management in a workstation cluster. Operating Systems Review, 15th ACM Symposium on Operating Systems Principles 29(5) (1995) 201–212
4. Koussih, S., Acharya, A., Setia, S.: Dodo: A user-level system for exploiting idle memory in workstation clusters. In: Proc. of the Eighth IEEE Intl. Symp. on High Performance Distributed Computing (HPDC-8). (1999)
5. Markatos, E., Dramitinos, G.: Implementation of a reliable remote memory pager. In: USENIX Annual Technical Conference. (1996) 177–190
6. Flouris, M., Markatos, E.: The network RamDisk: Using remote memory on heterogeneous NOWs. Cluster Computing 2(4) (1999) 281–293
7. McDonald, I.: Remote paging in a single address space operating system supporting quality of service. Tech. Report, Dept. of Comp. Science, Univ. of Glasgow (1999)
8. Stark, E.: SAMSON: A scalable active memory server on a network (Aug. 2003)
9. Dwarkadas, S., Hardavellas, N., Kontothanassis, L., Nikhil, R., Stets, R.: Cashmere-VLM: Remote memory paging for software distributed shared memory. In: Proc. of Intl. Parallel Processing Symposium, San Juan, Puerto Rico. (April 1999) 153–159
10. Amza, C., et. al., A.C.: Treadmarks: Shared memory computing on networks of workstations. IEEE Computer 29(2) (Feb. 1996) 18–28
11. Hines, M., Lewandowski, M., Gopalan, K.: Implementation experiences in transparently harnessing cluster-wide memory. In: Proc. of the International Symposium on Performance Evaluation of Computer and Telecommunication Systems (SPECTS), Calgary, Canada (Aug. 2006)
12. Gopalan, K., Chiueh, T.: Delay budget partitioning to maximize network resource usage efficiency. In: Proc. IEEE INFOCOM'04, Hong Kong, China. (March 2004)

13. Garcia-Molina, H., Lipton, R., Valdes, J.: A massive memory machine. IEEE Transactions on Computers **C-33 (5)** (1984) 391–399

14. Bohannon, P., Rastogi, R., Silberschatz, A., Sudarshan, S.: The architecture of the Dali main memory storage manager. Bell Labs Technical Journal **2**(1) (1997) 36–47

15. Acharya, A., Setia, S.: Availability and utility of idle memory in workstation clusters. In: Measurement and Modeling of Computer Systems. (1999) 35–46

16. Anderson, T., Culler, D., Patterson, D.: A case for NOW (Networks of Workstations). IEEE Micro **15**(1) (1995) 54–64

17. Wong, T., Wilkes, J.: My cache or yours? Making storage more exclusive. In: Proc. of the USENIX Annual Technical Conference. (2002) 161–175

Scalable Localization in Wireless Sensor Networks

Muralidhar Medidi, Roger A. Slaaen, Yuanyuan Zhou,
Christopher J. Mallery, and Sirisha Medidi*

School of Electrical Engineering and Computer Science
Washington State University
Pullman, WA 99164-2752

Abstract. Localization, an important challenge in wireless sensor networks, is the process of sensor nodes self-determining their position. The difficulty encountered is in cost-effectively providing acceptable accuracy in localization. The potential for the deployment of high density networks in the near future makes scalability a critical issue in localization. In this paper we propose Cluster-based Localization (CBL), which provides effective localization suitable for large and highly-dense networks. CBL utilizes both a computationally-intensive localization technique (non-metric multidimensional scaling (MDS)) and a less intensive trilateration to achieve balance between performance and cost. Clustering is utilized to select a subset of nodes to perform MDS and then extend their localization to the remaining network. Besides providing scalability clustering overcomes local irregularities and provides good accuracy even in irregular networks with or without obstacles. Simulation results illustrate that CBL reduces both computation and communication, while still yielding acceptable accuracy.

1 Introduction

Wireless sensor networks (WSN) are typically densely populated ad-hoc networks composed of small, resource-constrained, immobile nodes. The ability for a sensor to self-determine its own position, enabling the node to correlate its data with a location, is critical in many domains [1]. Minimizing the cost of nodes is a critical consideration, which makes equipping all sensor nodes with GPS capabilities infeasible [2]. Furthermore, the number of sensor nodes deployed in a sensor network is typically high, e.g., on the order of thousands and possibly even millions, and the network density can reach a few hundred nodes per square meter [3],[4]. The need for position information in high-density WSN motivates the need for scalable localization.

In this paper we propose a scalable hop-based localization technique called Cluster-based Localization (CBL). CBL's use of clustering is motivated by the need for scalability and efficiency. Scalability is gained by, first using an expensive but accurate localization technique on a small subset of the nodes in the

---

* This work was supported in part by NSF grant CNS-0454416.

Y. Robert et al. (Eds.): HiPC 2006, LNCS 4297, pp. 522–533, 2006.

network. The derived position estimates for the chosen subset will be used as references when localizing the remaining nodes. This approach leads to a significant reduction in the computation required to localize the entire network.

Although scalability was the primary goal for performing the localization in different stages in CBL, employing clustering has also provided the benefit of smoothing over local variations within the network. This smoothing effect prevents the global localization results from being skewed by the affects of local aberrations. In particular, CBL is able to maintain accuracy even in the presence of RF-opaque obstacles and irregular network topologies. Although our CBL implementation uses non-metric MDS in localization of the representative nodes, this technique could be replaced so that CBL could benefit from other localization ones.

The rest of this paper is organized as follows. Section 2 presents related work in the area of sensor network localization. CBL is described in Section 3. Section 4 contains performance evaluation and comparisons of CBL and Section 5 provides some concluding remarks.

2   Related Work

Previous attempts at localization in sensor networks can be categorized into two groups: range-aware and hop-based. In range-aware techniques a distance measure between neighboring nodes is used to estimate node positions. In hop-based ones ranging equipment is not necessary, and the estimated distances between nodes are typically approximated to the hops in the shortest path.

Range-aware localization techniques typically derive inter-node distances based on received signal strength. APS [5], a distributed localization technique, extends both distance vector routing and GPS positioning. Similiar distance estimates are used in [6]. In [7], a similiar, but more coarse estimation process is described, in which nodes adjacent to at least one anchor are localized first. Received signal strength measurements of broadcasts from a single mobile beacon node are used in [8]. The beacon's current position, will place constraints on the possible position of a node. Another technique using a mobile beacon node, utilizing time-of-arrival measurements and probablistic estimation, is proposed in [9]. In [10], four mobile beacons are used, creating a rectangle with the "un-localized" node in the middle, allowing for trilateration. PRI [11] provides improved performance by augmenting hop information with any available ranging information. In [12], known peer-to-peer communications are modeled as a set of geometric constraints on nodes' positions. In [13], the Approximate Maximum-Likelihood method and Direction of Arrival estimation are reviewed. In [14], sensing constraints caused by mobile objects at several nodes are utilized to improve the accuracy of localization. RangeQ determines node positions by means of a distributed range quantization technique which is similar to quantization in image processing [11]. Ji and Zha present a distributed localization technique based on the estimation-comparison-correction paradigm. It applies multidimensional scaling (MDS) [15] to merge each individual node's map of the network topology into one global map of the network

[16]. In [17], previous work from rigidity theory is extended to networks where not all nodes are localizable. Chan, Luk and Perrig [1] propose a scalable localization, where a complex clustering is used to produce a highly regular structure. This regularity can then be utilized, along with arbitrarily positioned anchor nodes, in the localization.

The problem with calculating distances by means of signal strength measurements is that since all possible sources of signal interference cannot be accurately anticipated prior to sensor deployment, the estimated distances can become inaccurate due to multi-path interference, line-of-sight obstructions, etc. Hop-based localization techniques aim to remove the dependency on any form of ranging equipment. A theoretical analysis of network connectivity for node self-localization is presented in [18]. An extension of the algorithm proposed in [5] is the differential Ad-Hoc Positioning System [19], which describes a differential error correction scheme designed to reduce the cumulative distances and positioning error over multiple hops. HOP-TERRAIN [20], similar to [5], utilizes an additional refinement phase to improve the localization accuracy. A low-power-dedicated hardware localization technique utilizing the HOP-TERRAIN is proposed in [21]. SHARP [22] technique adopts a hybrid hop-based and range-aware approach for localization. SHARP attempts to perform localization on a subset of the nodes in a network using inter-node distance estimates. This subset of nodes will be localized using MDS [2], and are then used to localize the remaining nodes with APS [5]. Shang *et al.* proposed MDS-MAP [23], an algorithm that utilizes MDS to perform global localization. Since MDS-MAP, there have been notable extensions of this technique: MDS-MAP(P)[2] and MDS-MAP(R)[24]. MDS-MAP(P) is a distributed algorithm that uses patches of relative maps, that can be computed in parallel, to estimate the absolute positions of nodes. On the otherhand, the distributed MDS-MAP(R) estimates the relative positions. A similar algorithm based on the estimation-comparison-correction paradigm and MDS [15] is described in [16].

3   Cluster-Based Localization

To localize all sensor nodes accurately, the localization process usually involves computationally-intensive procedures, e.g., MDS that has $O(n^3)$ computational complexity [2],[25]. Since sensor nodes are expected to be resource-constrained and networks to be large and dense, light-weight localization algorithms that achieve decent accuracy and high scalability are desirable. Therefore, we attempt to address the trade-offs between accuracy and scalability, so as to make our localization algorithm, CBL, a practical solution for sensor network applications.

CBL adopts a hierarchical idea to perform the whole localization procedure, i.e., instead of regarding the whole network as a flat topology in which all nodes share the same localization procedure as in most existing localization algorithms, we deliberately separate the nodes into two types, and apply different localization approaches on them to improve scalability as well as ensure accuracy. The first type of nodes are *representatives* that account for a small portion of the nodes in the whole network. Representatives will be selected to reflect a good abstraction of the network and localized by using complex but accurate localization algorithms, which essentially improves the overall localization accuracy. The remaining nodes, which account for the majority of the network, can employ a lighter-weight localization process that uses the representatives as reference nodes. By applying this mechanism, CBL achieves decent accuracy with relatively low computation and communication overhead.

3.1 Representative Selection

Since representatives will be localized first and used as reference nodes for estimating the remaining nodes' positions, they should provide a good abstraction of the whole network. Further, the representative selection process itself should be light-weight to ensure the overall scalability. Therefore, in CBL we apply a simple single-hop and size-bounded clustering algorithm to select representatives, or *Cluster Heads* (CHs), from the network. The single-hop feature of the clustering algorithm ensures the algorithm's efficiency, and the size-bounded feature enables each cluster to be formed fairly uniformly and makes cluster heads represent a good abstraction of the whole network.

We assume each node has a unique ID. To prepare for clustering, initially each node performs neighbor discovery to become aware of its single-hop neighborhood information. After that, a node either assumes to be a CH if it has the largest ID among its neighbors, or waits to be contacted if any of its neighbors have larger IDs. A CH will start to contact its neighbors to build its own cluster whose size is bounded by a threshold θ. A node becomes a *Cluster Member* (CM) of the first neighboring CH that contacts it, or becomes a CH if all its larger-ID neighbors have already spoken by either becoming a clusterhead with full size-bounded cluster or becoming a CM in some other cluster. To enable the network formed by CHs to provide a good abstraction of the whole network, a CH c will attempt avoiding CMs that cover overlapping transmission areas with c's existing CMs. After deciding its own role (CH/CM), a node informs its neighbors about its decision. Eventually, within a few rounds each node will finish determining its role (CH/CM) and each cluster forms a star topology with a CH at the center. In Fig. 1(a), we illustrate a random network and the connectivity between nodes. In Fig. 1(b), we show the resulting clusters in which a CM is represented as a line from its CH to itself, and in Fig. 1(c) we show the connectivity for each cluster. It can be observed that the CMs of each cluster are reasonably well distributed around the CH; and these CHs, or representatives, provide a good abstraction of the original topology.

3.2 Cluster Head Localization

CHs are taken as representatives and their localization accuracies have great impact on the overall localization accuracy. Since CHs usually account for a small portion of the whole network, it is affordable to apply complex localization algorithms to estimate CHs' positions. Further, we want to apply a hop-based

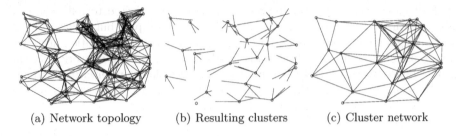

(a) Network topology   (b) Resulting clusters   (c) Cluster network

Fig. 1. Example clustering

algorithm to localize CHs so as to improve CBL's applicability. Based on these three considerations, we chose MDS as the basic technique to localize CHs, since hop-based MDS localization algorithm MDS-MAP(P) [2] was shown to achieve a good accuracy. CBL uses similar approaches to localize CHs, however, in CBL only CHs are involved in the MDS computation, therefore the computation overhead is contained. CBL does not involve any pre-installed anchor nodes as opposed to MDS-MAP(P), which reduces the cost of sensor hardware. Further, in CBL we employ the non-metric MDS technique that has a weaker requirement on the input data than the classical MDS used in MDS-MAP(P).

To apply MDS for local map construction, each CH c needs to collect the distance information from its "two-step" clusters' CHs (we call neighboring clusters "one-step" clusters). After that, c computes shortest distances (in hops) between each pair of the involved CHs, which will be taken as the input for the MDS algorithm to estimate positions of the involved CHs and to create a local map. MDS is a set of well-known data analysis techniques for geometrical position estimation and information visualization; see [15] for details. In our localization, we utilize the non-metric MDS, which assumes a less-stringent monotonicity constraint than the classical metric MDS deployed in [2].

After obtaining a local map through MDS calculation, each CH will attempt to merge its local map with its neighboring CHs'. Similar to MDS-MAP(P), the merging is a completely distributed process, in which the CHs with larger IDs will have higher priorities to choose one of their neighboring CHs to merge. Further, neighboring maps will be merged based on their common nodes, i.e. the maps with the highest number of common nodes should be merged first. When merging, we apply the *best linear transform* technique to transform one map onto another, and the new coordinates are computed based on the average of the common nodes' coordinates.

3.3   Cluster Member Localization

After determining each CH's coordinates, we will take CHs as reference nodes to estimate the remaining nodes', or CMs', positions. We choose the least-square triangulation technique for CMs' localization to obtain decent accuracy with low computation overhead. In particular, each CH first calculates the euclidean distances to the CHs in its neighboring clusters, then estimates the average hop distance by using these euclidean distances and corresponding hop length. After that, each CH broadcasts a LOCATION message, which contains its coordinates and the estimated hop distance, to the CMs within α hops (α should be chosen to balance the message overhead and the localization accuracy). Each CM waits for at least three LOCATION messages from different CHs, and then perform a triangulation to determine its own coordinates. If a CM receives more than three LOCATION messages, a *least square error* technique is applied.

4   Performance Evaluation

CBL was implemented in ns-2 [26], and our main goals when evaluating its performance were to determine the accuracy, scalability, and how irregularities in a network topology will affect performance. The simulation environment parameters are: ns2 version 2.29; area $= 100 \times 100 m^2$; 100 nodes; transmission range $= 15m$ and cluster size varied from 1 to 10. We utilize a commonly used metric, *accuracy*, to measure how much estimated positions deviate from actual positions. Two maps, *estimated* and *actual*, are compared by finding the best linear transformation (rotating, shifting and/or scaling) of the estimated map onto the actual and calculating the average Euclidean difference between each corresponding point. In the simulations the node-density is controlled by varying the enclosing network area while maintaining the same number of nodes. We also varied the upper-bound for cluster-size θ to determine how changes in cluster-sizes affect the performance of CBL. For comparison, we have included both MDS-MAP(P) and DV-hop results at similar densities for reference. The data for MDS-MAP(P) is obtained from [2] and are based on MATLAB simulations; a CAML implementation of DV-hop was used to obtain results for comparison.

During cluster-head (CH) localization, all hop values used are perturbed by adding noise σ $(0 \leq \sigma \leq 10^{-5})$ before the non-metric MDS is employed to eliminate ties and help the non-metric MDS converge, because the non-metric MDS technique is dependent on the differences in the distance estimations.

4.1   Random Network

The topologies in our random networks are square-shaped $n \times n$ (n controls density) networks, with 100 randomly placed nodes. As a key step, the clustering we deployed directly affects the quality of localization as well as the computation overhead. We observed that the average cluster-size stays close to the given upper-bound, which reflects the clusters' high quality in spite of our lightweight clustering. Further, as expected, cluster-density drops significantly given a larger upper bound.

In Figure 2 we show the resulting localization accuracy for random networks. Because we extend cluster-head (CH) localizations to obtain member (CM) localizations, we present CHs' accuracy results in Figure 2(a) and all nodes' accuracy results in Figure 2(b). Our results show that increasing the cluster size does

not lead to significant performance degradation. Although MDS-MAP(P) and
DV-hop achieve slightly better accuracy than CBL, CBL does not utilize nor
depend on any anchor nodes or extra refinement of the estimates as DV-hop and
MDS-MAP(P) require. Furthermore, it should also be observed that the difference between the accuracy of all nodes and that of only cluster-heads is very
small. This reflects the effectiveness of the clustering technique we applied, and
shows that if the CH localization is accurate enough the low cost trilateration
technique provides good results.

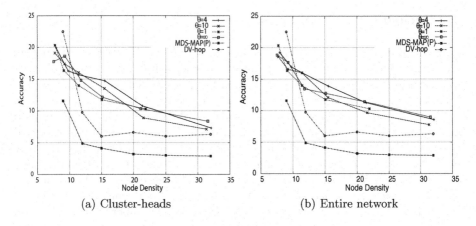

(a) Cluster-heads   (b) Entire network

Fig. 2. Accuracy for localization random topologies

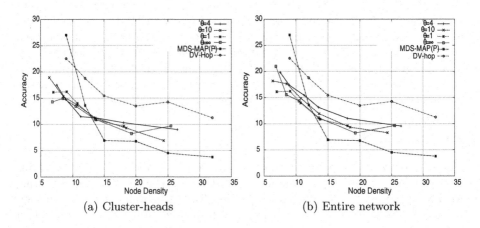

(a) Cluster-heads   (b) Entire network

Fig. 3. Accuracy for localization in C-shaped topologies

4.2 C-Shaped Network

C-shaped topologies are commonly employed to stress any localization technique
[2],[5]. In this irregular topology the estimated distances between nodes can

deviate greatly from the actual Euclidean distances. The resulting cluster-densities and average cluster-size are very similar to those for random networks.

In Figure 3, we show the resulting accuracy for both the cluster-heads Fig. 3(a) and all nodes Fig. 3(b). As in the random topologies, changes in cluster-sizes do not affect the resulting accuracy. Furthermore, compared to MDS-MAP(P), CBL's accuracy is only slightly worse, while providing the same lowered computation as in the random networks. Compared to DV-hop, CBL provides better accuracy for irregular topologies. This is mostly because of DV-hop's dependence on the uniformity and regularity of the network.

4.3   Irregular Node Densities

To test how well CBL will perform in topologies with changing node densities, we created an irregular topology. The irregular topologies, also referred to as biased in the literature, are another fairly common test. An illustration of an irregular topology can be found in Figure 5(a), where the area containing the higher node-density is indicated by the square in the lower left corner. We can see from Figure 4 that, as expected, CBL is not greatly affected by the irregularity in node-densities; verifying that in CBL innacurate estimates do not propagate and affect a significant part of a network.

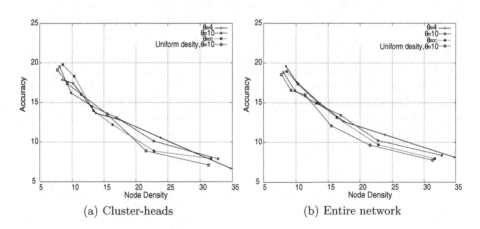

(a) Cluster-heads   (b) Entire network

Fig. 4. Accuracy for localization in irregular node-density topologies

4.4   Obstacles

Obstacles usually create great difficulty for any localization technique because any network with obstacles will have a much more irregular structure and the inter-node distance estimates are likely to be more inaccurate than those without obstacles. We performed simulations on random topologies with two different types of RF-opaque obstacles; four line-shaped (Figure 5(b)) and one H-shaped (Figure 5(c)) obstacle similar to the ones used in literature. As shown in Figure 5(c) the nodes close to the horizontal line of the H-shaped obstacle will be ones most affected and these nodes will derive severely overestimated inter-node distances between themselves and nodes on the other side of the horizontal line.

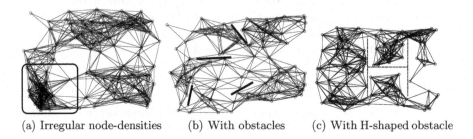

(a) Irregular node-densities   (b) With obstacles   (c) With H-shaped obstacle

Fig. 5. Irregular network topologies

Figure 6 shows the resulting accuracy for the entire network with both four obstacles Fig. 6(a) and one H-shaped obstacle Fig. 6(b). We omit the graphs for the cluster-heads due to space constraints, but the same trend seen in other topologies is also evident for networks with obstacles. The localization accuracy achieved is very similar to that of networks without obstacles. This is because the use of clusters produces a technique that is less sensitive to any type of irregularities, including obstacles, and by covering a larger geographical area the clusters can in effect reach around many of the obstacles. Although the inter-cluster distances at times might be overestimated the overall accuracy is not significantly affected. The same trend as in the random networks is observed, where increasing node-densities improves the accuracy and there is a small degradation in accuracy when comparing that of all nodes and that of CHs.

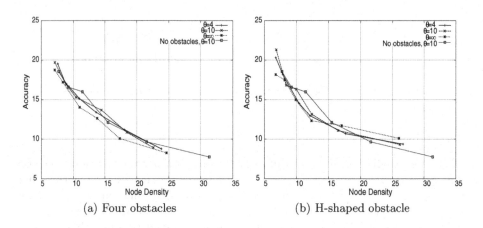

(a) Four obstacles   (b) H-shaped obstacle

Fig. 6. Accuracy for localization in random topologies with obstacles

4.5   Scalability and Overhead

CBL is explicitly designed to improve scalability by reducing computation overhead. We show its time and message overhead, which are two important metrics that reflect its scalability, in Figure 7. In Figure 7(a), we show the running-time in seconds for ns-2 simulations with different-sized clusters. As expected, CBL significantly reduces the running time if clustering is applied. The running-time is reduced to 6% when using clusters of size five. The reason for this reduction is mostly that the computation overhead of MDS techniques greatly depends on the network density, and this density is reduced by the clustering. As shown in Figure 7(b), we see a significant decrease in the messages needed to perform the localization as the cluster size is increased from 1. This is mainly because the messages exchanged when merging local maps are greatly reduced, and the complexity of each merge is less as each local map is smaller. Because of the very

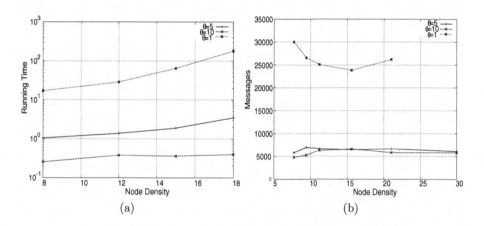

(a)   (b)

Fig. 7. Computation and communication overhead

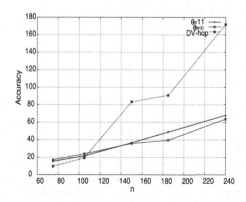

Fig. 8. Accuracy in $n \times n$ networks

high running-time of ns-2, when using single-node clusters, we were not able to complete simulations at the higher densities.

In Figure 8, we plotted the accuracy of CBL as the network size increases: we maintain the same density and increase network area and nodes correspondingly. We have included DV-hop as a reference. For all DV-hop simulations we used 4 anchor nodes, and we can see that CBL (which uses no anchors) degrades more gracefully in accuracy as the number of nodes and network size increases, than DV-hop.

5   Conclusions and Future Work

In CBL, we utilized clustering to obtain an abstraction of the network, differentiating between the localization of cluster-heads and that of cluster-members. This reduces the computation and communication cost while providing decent localization accuracy. The use of clustering also makes the localization more resilient to irregularities like obstacles: CBL performed well in regular networks, irregular networks and in networks with irregular node-densities. In all our simulations we used a simplified transmission model with no irregularities. This model may not be reflective of all realistic scenarios and the irregular transmission model may create more complex topologies, which might reduce the accuracy of any localization technique. We have shown that localizing only a small subset of nodes and using this to quickly and efficiently localize the remaining nodes is feasible and produces good results with lower computation complexity. Our goal was to create a scalable and more feasible localization algorithm; CBL achieves this scalability, even when using a high-cost technique such as non-metric MDS for localizing cluster-heads. CBL can provide a parameterized (by controlling the upper bound of cluster size) abstraction of networks. Although CBL utilizes non-metric MDS, a computationally-expensive but accurate localization, other localization techniques can also be used in CBL instead; the effect of changing the technique needs further investigation. Feasible localization is achieved by not introducing any new constraints on the network or sensor nodes: CBL does not rely on any kind of specialized, extra capable nodes such as anchors or beacons, or predetermined information about distributions, size, etc.

References

1. Chan, H., Luk, M., Perrig, A.: Using clustering information for sensor networks localization. In: DCOSS. (2005)
2. Shang, Y., Ruml, W.: Improved MDS-based localization. In: INFOCOM. Volume 4. (2004) 2640–2651
3. Akyildiz, I., Su, W., Sankarasubramaniam, Y., Cayirci, E.: A survey on sensor networks. IEEE Communications Magazine (2002) 102–114
4. Shih, E., S.Cho, Ickes, N., Min, R., Sinha, A., Wang, A., Chandrakasan, A.: Physical layer driven protocol and algorithm design for energy-efficient wireless sensor networks. In: MobiCom. (2001)

5. Niculescu, D., Nath, B.: Ad hoc positioning system (APS). In: GLOBECOM. Volume 5. (2001) 2926–2931
6. Savarese, C., Rabaey, J., Beutel, J.: Locationing in distributed ad-hoc wireless sensor networks. In: ICASSP. Volume 4. (2001) 2037–2040
7. Yao, Q., Tan, S., Ge, Y., Yeo, B., Yin, Q.: An area localization scheme for large wireless sensor networks. In: VTC. Volume 2. (2005) 2835–2839
8. Sichitiu, M., Ramadurai, V.: Localization of wireless sensor networks with a mobile beacon. In: MSN. (2004) 174–183
9. Sun, G., Guo, W.: Comparison of distributed localization algorithms for sensor networks with a mobile beacon. In: ICNSC. Volume 1. (2004) 536–540
10. Patro, R.: Localization in wireless sensor network with mobile beacons. In: IEEE Convention of Electrical and Electronics Engineers in Israel. (2004) 22–24
11. Li, X., Shi, H., Shang, Y.: A partial-range-aware localization algorithm for ad hoc wireless sensor networks. In: LCN. (2004) 77–83
12. Doherty, L., Pister, K., El Ghaoui, L.: Convex position estimation in wireless sensor networks. In: INFOCOM. Volume 3. (2001) 1655–1663
13. Bergamo, P., Asgari, S., Wang, H., Maniezzo, D., Yip, L., Hudson, R., Yao, K., Estrin, D.: Collaborative sensor networking towards real-time acoustical beamforming in free-space and limited reverberance. TMC **3**(3) (2004) 211–224
14. Galstyan, A., Krishnamachari, B., Lerman, K., Pattem, S.: Distributed online localization in wireless sensor networks using a moving target. In: IPSN. (2004) 61–70
15. Coxon, A.: The Users Guide to Multi Dimensional Scaling. Heinemann Educational Books (1982)
16. Ji, X., Zha, H.: Robust sensor localization algorithm in wireless ad hoc sensor networks. In: INFOCOM. (2003) 527–532
17. Eren, T., Whiteley, W., Belhumeur, P.: Further results on sensor network localization using rigidity. In: EWSN. (2005) 405–409
18. Liu, K., Wang, S., Ji, Y., Yang, X., Hu, F.: On connectivity for wireless sensor networks localization. In: IWCCC. Volume 2. (2005) 879–882
19. Perkins, D., Tumati, R.: Reducing localization errors in sensor ad hoc networks. In: IPCCC. (2004) 723–729
20. Savarese, C., Rabaey, J., Langendoen, K.: Robust positioning algorithms for distributed ad-hoc wireless sensor networks. In: USENIX Annual Technical Conference. (2002) 317–327
21. Karalar, T., Yamashita, S., Sheets, M., Rabaey, J.: A low power localization architecture and system for wireless sensor networks. In: SIPS. (2004) 89–94
22. Ahmed, A., Hongchi, S., Shang, Y.: Sharp: A new approach to relative localization in wireless sensor network. In: ICDCS. (2005) 892–898
23. Shang, Y., Ruml, W., Zang, Y., Fromhertz, M.: Localization from mere connectivity. In: MobiHoc, Annapolis, MD (2003) 202–212
24. Shang, Y., Meng, J., Shi, H.: A new algortihm for relative localization in wireless sensor networks. In: IPDPS. (2004)
25. Morrison, A., Ross, G., Chalmers, M.: Fast multidimensional scaling through sampling, springs and interpolation. Information Visualization **2** (2003) 68–77
26. URL: ns-2, discrete event simulator. http://www.isi.edu/nsnam/ns/ (2006)

A Novel Real-Time MAC Protocol Exploiting Spatial and Temporal Channel Diversity in Wireless Industrial Networks

Kavitha Balasubramanian, G.S. Anil Kumar, G. Manimaran, and Z. Wang

Dept. of Electrical and Computer Engineering
Iowa State University, Ames, IA 50011, USA
{kavitha, anil, gmani, zhengdao}@iastate.edu

Abstract. Wireless technology is increasingly finding its way into industrial communication because of the tremendous advantages it is capable of offering. However, the high bit error rate characteristics of wireless channel due to conditions like attenuation, noise, channel fading and interference seriously impact the timeliness and guarantee that need to be provided for real-time traffic. Existing wired protocols including the popular PROFIBUS perform unfavorably when extended or adapted to the wireless context. Other wireless protocols proposed either do not adapt well to erroneous channel conditions or do not provide real-time guarantees. In this paper, we present a novel real-time MAC (Medium Access Control) protocol that is specifically tailored to the message characteristics and requirements of the industrial environments. The protocol exploits both the spatial and temporal diversity of the wireless channel to effectively schedule real-time messages in the presence of bursty channel error conditions. Simulation results show that the proposed protocol achieves much better loss rate compared to baseline protocols under bursty channel conditions.

1 Introduction

The term industrial communication denotes the interaction between various classes of devices in setups such as production control, control of chemical plants, air control, communication systems in cars, planes and trains, power station control and so on. The applications in these setups are very complex, therefore their functionality needs to be distributed to a number of systems or devices, which communicate with each other. In this paper, we are concerned mainly with the traffic generated on a network operating at the device level of factory communication systems which includes various controllers, sensors and actuators.

Industrial networks differ significantly from traditional LANs due to special requirements of their applications like the need for hard timing and bandwidth guarantees and supporting priorities. Predictable inter-task communication is extremely critical in such industrial real-time systems because unpredictable delays in the delivery of messages can affect the completion time of the tasks participating in message communication, resulting in deadline misses and eventually

Y. Robert et al. (Eds.): HiPC 2006, LNCS 4297, pp. 534–546, 2006.
© Springer-Verlag Berlin Heidelberg 2006

performance losses, halts/resets of manufacturing pipelines or defects in products. Several wired protocols like PROFIBUS are being used in industries and factories that meet such stringent timing requirements.

Recently, the growing popularity of wireless communication in numerous fields has led to its increased dependability, performance improvement and cost reduction. Hence wireless networks are beginning to represent a viable choice for industrial applications because they can offer several attractive features like reduced cost of cabling, ease of configuration and maintenance, extended mechanical freedom and mobility and preventing losses arising due to potential damage of cabling caused by mechanical moving parts, high temperatures and other hostile conditions. Thus, it is very likely in the near future, there will be a proliferation of wireless implementations of factory communication systems.

In spite of having such clear benefits, wireless technology has its own drawbacks arising due to the unreliable characteristics of the wireless medium which makes it, in its current state, unsuitable for supporting real-time communication. Effects due to fading, interference from other users and shadowing from objects degrade the channel performance. In addition, distance dependent path loss and co-/adjacent channel interference influence the channel. Hence the wave propagation environment (number of propagation paths, their respective losses) and its time varying nature (moving people, moving machines and metal surfaces) play a dominant role in constituting channel characteristics [1]. Also due to heavy obstruction, the wireless medium of industrial environments are known to suffer more serious large-scale path loss and fading than other indoor environments [2]. Consequently, the wireless link exhibits both bit errors and packet losses (change in bit values in a packets data part) which vary strongly over time and tend to occur in bursts.

Since wireless networks are substantially different from their wired counterparts with respect to the channel conditions, technologies developed for wired networks cannot be directly adopted. In most wired network models for real-time systems, the communication links are assumed to have a fixed capacity over time. This assumption may be invalid in wireless environments, where link capacities can be temporarily degraded due to fading, attenuation, and path blockage [1]. In addition, existing wireless standards such as IEEE 802.15.1 (Bluetooth) and IEEE 802.15.4 (Zigbee) also provide no mechanisms for supporting real-time messages. Hence, there arises a need to design and develop special MAC protocols and techniques which take both the channel characteristics and the hard real-time requirements of the messages into account. In the next section, we present the related work in this area.

2 Related Work and Motivation

A number of measurement studies[1, 3, 4]reveal the time-variable and high error rates of the wireless channel. Results published by Willig et. al.[1] indicate that the popular Gilbert Elliot model with some modifications is a useful tool for simulating bit errors on a wireless link, which we use in the present work.

Several proposals have been made that extend the wired protocols used for industrial communication over to a wireless medium. In [5], the authors explore the use of IEEE 802.11 for industrial communication by analyzing the possibility of implementing protocols based on master-slave architecture of traditional field buses on a IEEE 802.11 PHY. In [6], the adaptive-intervals MAC protocol has been proposed that uses a polling-based approach combined with group testing feature for improving the delay in low load conditions. In [7], the authors discuss different architectures that make use of a spread spectrum repeater to integrate distant wireless stations with a wired segment.

In addition, many MAC protocols and schemes have been proposed to increase the reliability offered by wireless links. In [8] and [9], the authors make use of channel conditions while making packet dispatching decisions. However, the traffic considered in [9] is best effort. In [8], a technique that estimates the channel state beforehand and uses a centralized priority queue based scheduling mechanism is proposed. However, accurate estimation techniques that predict the exact future channel state is unfeasible. In [10], the authors investigate schemes to support combined scheduling of periodic and aperiodic real-time traffic over master-slave Bluetooth networks. In [2], the authors explore the use of Direct sequence spread spectrum(DSSS) CDMA technology to build Industrial Control Wireless LAN with enhanced robustness. In [11] and [12], the authors introduce the concept of antenna redundancy and compare it with modifications made to the Automatic Repeat Request (ARQ) protocol. The ARQ schemes proposed do not work well at high error rates and antenna redundancy requires additional hardware in all communicating devices if any-to-any communication need to be implemented.

The rest of the paper is organized as follows. In Section 3, details about the system model are provided following which we introduce our basic framework in section 4. In section 5, we explain in detail about the Exchange Protocol and present the findings of the simulation studies in section 6. We conclude in section 7 providing directions for our future work.

3   Network and Channel Model

We study a single-hop industrial environment consisting predominantly of real-time periodic message with occasional aperiodic messages/alarms being generated due to faulty or abnormal outcome of some process which require higher priority. The communication medium is wireless characterized by high bit error rate due to phenomena like noise, attenuation, fading and interference. We assume that messages destination is a node in the single hop.

The bursty error characteristics of the wireless environment in a typical industrial setup can be captured by the Discrete-Time Gilbert-Elliot Channel Model [1, 13, 14]. Time in the super-frame is divided into slots and the model works with slotted time where state transitions happen at the end of each slot. The state space of the Gilbert-Elliot model contains the following two states: GOOD

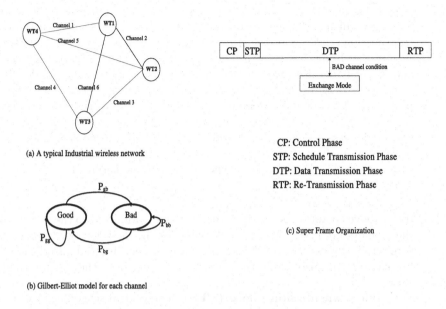

(a) A typical Industrial wireless network

(b) Gilbert-Elliot model for each channel

CP: Control Phase
STP: Schedule Transmission Phase
DTP: Data Transmission Phase
RTP: Re-Transmission Phase

(c) Super Frame Organization

Fig. 1. Channel model and super-frame format

and BAD. When in the GOOD state, no bit errors occur in the data sent in the corresponding slot. Hence the transmission succeeds when done in an exclusive manner. On the other hand, when the channel is in BAD state certain bit errors occur in the received data unit and the data transmission is considered erroneous since an Acknowledgement (ACK) is not received. Fig. 1(b) shows the state diagram along with the transition probabilities. We assume that each channel between a given source-destination pair is statistically independent. In Fig. 1(a), each solid line between two wireless nodes represent an independent channel over which the Gilbert-Elliot channel model is applied.

4   Basic Framework

The medium is shared by all the wireless nodes and transmissions follow a super-frame structure that repeats itself. The super-frame is divided into slots and each message would occupy several slots. In a slot, a sender is able to transmit a unit of the message and receive the corresponding acknowledgment(ACK). The absence of an ACK indicates that the channel between the source an destination is in a bad state and the unit is marked for re-transmission.

The basic framework consists of a centralized scheduler that collects all the messages available in the system before every super-frame. The scheduler then prepares a schedule that is followed by all nodes in the system. To facilitate such an approach, every super-frame is divided into the following four phases (see Fig. 1(c)):

- **Control phase (CP):** All the messages in the system are sent to the central scheduler which performs an admission test and constructs a non-overlapping transmission schedule for the admitted messages. The admission test checks if the super-frame has enough free slots to accommodate the next message and its recurring instances (for periodic messages only) before its deadline. Consider a periodic message of size M_i occupying N_i slots of the super-frame. Let N_{data} denote the number of slots of the data transmission phase, $N_{admitted}$ denote the number of slots of the super-frame occupied by already admitted messages; $N_{transfer}$ denote the number of transfer slots and N_{exchg} denote the number of exchange slots (more details about the usage of these slots are provided in Section 5). The admission test checks if

$$N_i \leq N_{data} - N_{admitted} - N_{transfer} - N_{exchg} \qquad (1)$$

If the above condition is satisfied, the message is admitted to the system and the scheduler reserves N_i slots exclusively for the message; else the message is rejected. However, aperiodic messages are always admitted into the system by removing an instance of the periodic message, since they require higher priority.
- **Schedule transmission phase (STP):** The central scheduler broadcasts the above constructed schedule to all the nodes in the network.
- **Data transmission phase (DTP):** Each wireless node begins its transmission in its scheduled slot. We assume that all the messages that need to be transmitted during the data phase become ready at the beginning of this phase and every message needs to complete before the end of the super-frame. In spite of allocating enough time slots in an exclusive manner, not all messages will reach the destination without errors because of the erroneous channel condition. Therefore, some messages might miss the deadline. The number of deadline misses will depend on the exact data transmission protocol. We present two basic schemes here which would be used for transmitting messages in this phase. However, our main contribution is the Exchange protocol which we present in Section 5 and compare it against the following two basic schemes.

In Time Division Multiplexing with Variable number of Retransmissions (**TDMVR**), when the channel is underloaded, all the unutilized slots towards the end of the super frame are used for re-transmission. In Time Division Multiplexing with Constant number of Retransmissions (**TDMCR**), the schedule is formed in such a way that all the unutilized slots are equally distributed between the transmitting nodes. Although these schemes enable full utilization of the channel in case of of an underloaded system by increasing the attempts available for existing message transmissions, they do not adapt to the bursty error conditions of the channel. The exchange protocol presented in the next section adapts to the channel conditions thereby decreasing the number of deadline misses and increasing the effective system utilization.
- **Re-Transmission phase (RTP):** All wireless nodes which could not successfully transmit all their messages during the DTP in the first attempt

contend for channel access (CSMA) and employ a backoff algorithm on collision. A fixed percentage of slots in DTP is allocated for re-transmission. At the end of the superframe, the slots that were unable to be successfully transmitted are declared as deadline misses.

5   Slot Exchange Protocol

We now present the Slot exchange protocol that comes into effect during the DTP as shown in Fig. 1(c). The exchange protocol dynamically adapts to adverse channel conditions and enables effective scheduling of real-time messages in addition to preserving the schedulability guarantees provided to existing messages. Schedulability guarantee implies the fact that when a message is admitted into the system, it is given a certain number of slots (as is occupied by the message) exclusively for data transmission. The scheme caches on two characteristic features - spatial and temporal diversity of the wireless channel; temporal diversity signifies the fact that when a channel is the bad state, it would eventually move to the good state and spatial diversity indicates the condition that if one channel is in bad state, it is possible that a different channel would be in good state.

5.1   Basic Idea and Illustrative Example

During the DTP, each wireless node begins its transmission in its scheduled slot. When a channel between a source destination pair is bad, transmissions begin to fail. During this state, the Exchange protocol is used that works around the occurrences of error bursts. The primary intuition behind the scheme is to postpone the transmission on a channel in a bad state to a later time and schedule transmissions on a channel in a good state with the hope that the channel in the bad state would change into good state in the meantime. This protocol forms its basis on the wireless channel characteristic of correlated packet losses i.e. on a channel which is erroneous, a single packet loss would be followed by back-to-back packet losses. Hence the exchange protocol takes advantage of this characteristic feature to perform efficient scheduling of real-time messages.

Consider a simple network with three wireless nodes shown in Fig 2(a). Let the messages that need to be transmitted be: 12, 23 and 13 where the first number indicates the source and second number indicates the destination. Figure 2(b) shows the channel condition variation with time. The shaded slots indicate that the channel is in bad state. The original schedule given by the central scheduler is shown in Fig. 2(c) and the schedule of the basic schemes is given in Fig. 2(d) which would lead to 6 slots being unsuccessful.

In the exchange protocol, once a node (exchange initiator), notices that its channel to the destination is in bad state, it exchanges its slots (as many as possible) with a different node(exchanged sender). As a result of the exchange, the exchange initiator performs its transmissions in the slots of the exchanged sender and vice versa. This basic idea is depicted in Fig. 2(e), where the exchange initiator, node 2 exchanges its 6 slots with the exchanged sender 1. The final schedule due to the exchange is shown in Fig. 2(f) where only 1 slot is unsuccessful.

Several different heuristics can be applied for a choice of the exchanged sender, based on channel correlation, estimation of the burst length and priority. In this paper, for simplicity, we use the next node in the trasnmission schedule which has a message to transmit for exchange.

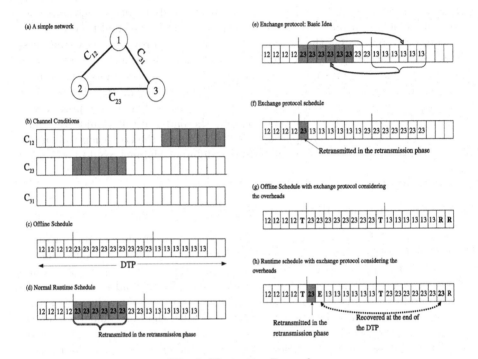

Fig. 2. Illustrative Example

5.2   Protocol Details

The basic idea of the exchange protocol is to avoid transmissions on a channel in the bad state by passing control to a different transmitter-receiver pair whose channel is in good state. In order to preserve the schedulability guarantee, the exchange protocol incurs some control overhead.

When a exchange initiator wants to exchange its slots with an exchanged sender, a slot called the exchange slot (slot 7 indicated as E in Fig. 2(h)) is used in which a two way handshake is performed. The exchange initiator sends an exchange request (N_{req}) along with the maximum slots it want to exchange which is typically till the end of its data transmission phase and the exchanged sender replies with an ACK that denotes the actual number of slots it has available for exchange ($N_{available}$). In the example, $N_{exchg} = N_{available} = 6$.

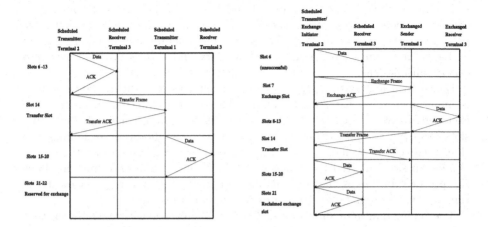

Fig. 3. (a) Timing diagram for the offline schedule   (b) Timing diagram with exchange

Since for every exchange initiated, an exchange slot is being consumed, the number of exchanges that can be performed is limited to N_{exchg} in every super-frame. The scheduler broadcasts the N_{exchg} value to all nodes during the STP. To compensate for the exchange slot (to maintain the schedulability guarantee) which are being used by the exchange initiator from the scheduled slots that it has been allocated for transmission, N_{exchg} number of slots are reserved at the end of the super-frame (indicated by R in Fig. 2g). From this pool of reserved slots, every exchange initiator exclusively gets a slot for every exchange it has performed. To enable these functions, an exchange counter, N_{ctr}, is maintained that denotes the number of exchanges that has been performed in the super-frame until the current time. This exchange counter is passed on between the transmitting nodes by means of the transfer slot (indicated by T in Fig. 2g) occurring at the end of every message transmission. Therefore at the beginning of the transmission, each node knows how many more exchanges can be performed. Each time an exchange is performed, the exchange counter is decremented by the exchange initiator and the value of the exchange counter is passed onto the exchanged sender in the exchange slot. In this way, the exchanged sender knows how many more exchanges it can perform during the exchange period. After its exchange period, it passes on the value of the exchange counter to the next transmitting station in the transfer slot. If the exchange counter becomes 0, no more exchanges are performed. If any of the transfer or exchange slots are completed the exchange counter is reset to zero and the transmissions proceed as per the offline schedule.

Let N_{ctr} denote the current value of the exchange counter. When a node uses up an exchange slot for performing exchange, it decrements the exchange counter to $N_{ctr} - 1$ and $N_{exchg} - N_{ctr}$th slot is used by this exchange initiator from the reserved slots. In the above example, assume that $N_{exchg} = 2$. Therefore node 2 has the exchange counter of 2 before performing the exchange. During exchange, it decrements the exchange counter to 1 and uses the (2-1) = 1st slot from among the slots reserved for exchange(slot 21 in the example) since it is the first node performing the exchange. Note that when an exchange initiator performs an exchange, it is limited to its message boundary and it does not spill over into other transmissions.

Hence, by using the transfer slots and the exchange slots, the exchange counter is maintained in a distributed manner. This enables limiting the number of exchanges in every super-frame and thus enables controlling the number of actual slots available for data transmission. In addition, it allows for reclaiming the slots used up for exchange in a exclusive manner; thus preserving the actual number of slots allotted to each node for performing data transmission. Thus the protocol preserves the schedulability guarantee given for messages at the time of admission and effectively uses the channel resources.

The timing diagrams shown in the Fig. 3 explain the exact transmissions that take place for the above example during the working of the Exchange protocol.

6   Simulation Studies

We simulated a single hop wireless network with 10 nodes over a 1Mbps channel with periodic messages of size 1050 bits and aperiodic of size 450 bits. Each slot has a time duration equal to the transmission time of 150 bits. Approximately 10% of slots in every super-frame is allocated for re-transmission. We simulated the different channel conditions using the Gilbert-Elliot model for different values of the model parameters. In our simulation studies we compared the performance of the above proposed protocols. The performance metric for all our simulation studies is the loss rate defined as the ratio of number of deadline violated to the number of messages admitted. P_{bb} represents the probability that the channel remains in a bad state given that it is in a bad state. P_{gg} represents the probability that the channel remains in the good state, given that the channel is in a good condition. N_e denotes the number of exchanges that can be performed in a given super-frame. M_s is the number of slots required for the complete transmission of a message. Total number of messages per super-frame is given by N_m.

6.1   Results and Discussions

Effect of bad state probability (P_{bb}): Figure 4(a) compares the loss rates incurred by the above three protocols by varying P_{bb}. The other parameters are assumed as follows: $P_{gg} = 0.9$, $S_l = 1$, $N_e = 11$, $M_s = 7$, $N_m = 10$. The graph has two distinct regions of interest corresponding to $P_{bb} < 0.8$ (small burst region) and $P_{bb} \geq 0.8$ (large burst region). In the small burst region, with low values of P_{bb} the channel quickly switches to the good state and the benefits of the exchange protocol are not very significant. In fact the overhead due to the exchange scheme overshadows the benefits of the protocol. On the other hand, in the large burst region (shown enlarged in Fig. 4(b))which depicts the typical

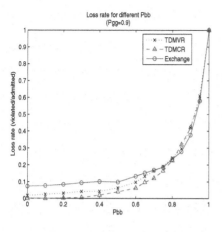
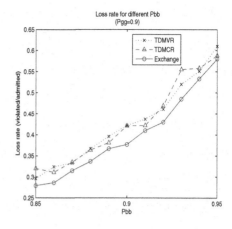

Fig. 4. (a) Effect of P_{bb} (b) Effect of the high P_{bb}

industrial environment, the exchange protocol performs better than the basic schemes due to the fact that the exchange protocol *exchanges* the slots of a bad channel at the beginning of the bad burst with a good channel which is not noisy. At $P_{bb} = 0.9$, the exchange protocol gives an improvement of 10.6% over TDMCR and 10.5 % over TDMCR. Interestingly, towards the end of the large burst region where $P_{bb} \geq 0.96$, exchange protocols behave similar to the basic protocols due to the fact that the channel experiences significantly long bursts that deferred transmissions also encounter the erroneous channel condition.

Effect of good state probability (P_{gg}): Fig. 5(a) compares the loss rates incurred by the three protocols by varying P_{gg}. The other parameters are assumed as follows: $P_{bb} = 0.9$, $S_l = 1$, $N_e = 11$, $M_s = 7$, $N_m = 10$. The graph has two distinct regions of interest corresponding to $P_{gg} < 0.8$ and $P_{gg} \geq 0.8$. At low values of P_{gg} the channel quickly switches to the bad state and hence experiences frequent bad state bursts whose size is depicted by the P_{bb} value. This results in an exchange being performed from a bad channel to another channel that also moves into bad state frequently; hence the benefits of the exchange protocol are not very significant. At high P_{gg} (shown enlarged in Fig. 5(b)), which is the typical scenario in an industrial environment, the exchange protocol performs better than the basic schemes because the channels are in good state for a longer time and when the channel is erroneous,the exchange protocol *exchanges* its slots with a good channel. At $P_{gg} = 0.91$, the exchange protocol gives an improvement of 14% over the basic schemes.Therefore, at very large values of P_{gg} the exchange protocol performs better than the others and at $P_{gg} = 1$, all the schemes show similar results.

Effect of number of exchange slots (N_e): We study the effect of the N_e by varying the message sizes and number of messages per super-frame. We have chosen $P_{gg} = P_{bb} = 0.9$ for these simulations.

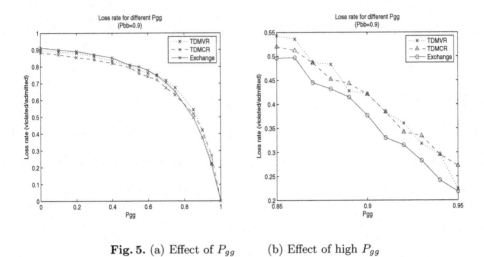

Fig. 5. (a) Effect of P_{gg} (b) Effect of high P_{gg}

- Effect of message size (M_s): Fig. 6(a) shows the effect of the N_e for different M_s values keeping N_m fixed at 10 . With the increasing N_e, the loss rate for all message sizes decrease due to the benefits of the exchange protocol. However, after a point, N_e becomes more than the maximum number of exchanges that need to be performed and hence the loss rate saturates beyond that point. The saturation point depends on the message size, number of messages and channel parameters. For large message sizes, the saturation point is higher (12 for message of size 10 while it is 8 for message of size 4) since more exchanges can be performed.
- Effect of number of messages (N_m): Fig. 6(b) shows the effect of the N_e for different N_m values keeping M_s fixed at 7 . With increasing N_e, the loss

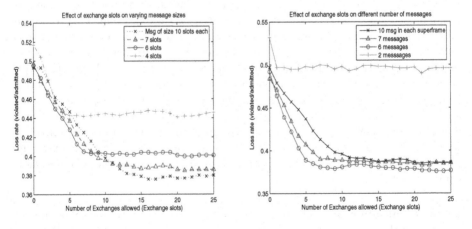

Fig. 6. (a) Effect of N_e for different M_s (b) Effect of N_e for different N_m

rate for all N_m values decrease due to the benefits of the exchange protocol. However, after a point, N_e becomes more than the maximum number of exchanges that need to be performed and hence the loss rate saturates beyond that point. As in the previous case, the saturation point is higher for large number of messages (2 for $N_m = 2$ while it is 12 for $N_m = 10$) since the number of exchanges that can be performed is more when the number of messages increase.

7   Conclusions

In this paper, we proposed a novel MAC protocol for real-time message scheduling which adapts to the channel conditions by exploiting spatial and temporal channel diversity characteristics of the wireless medium. Our simulation results show that the proposed exchange protocol provides better loss rate as compared to the generic protocols. In our future work, we would like to make the protocol distributed and extend it to multi-hop networks. We also plan to improve the protocol through channel estimation techniques.

References

1. A. Willig, M. Kubisch, C. Hoene, and A. Wolisz, "Measurements of a Wireless Link in an Industrial Environment using an IEEE 802.11-Compliant Physical Layer," *IEEE Trans. on Industrial Electronics*, vol. 43, no. 6, pp. 1265-1282, Dec. 2002.
2. Q. Wang, X. Liu, W. Chen, W. He and M. Caccamo,"Building Robust Wireless LAN for Industrial Control with DSSS-CDMA Cellphone Network Paradigm," *in Proc. of the IEEE Real-Time Systems Symposium*, Miami, Dec. 2005.
3. D. Eckhard and P. Steenkiste, "Measurement and analysis of the error characteristics of an in-building wireless network," *in Proc. Of ACM SIGCOMM 96 Conference*, pp. 243-254, California, Aug. 1996.
4. D. Duchamp and N. Reynolds, "Measured performance of wireless LAN," *in Proc. of 17th Conf. on Local Computer Networks*,Minneapolis, 1992.
5. D. Miorandi and S. Vitturi, "Analysis of Master-Slave Protocols for Real-Time Industrial Communications over IEEE802.11 WLANs," *in Proc. of INDIN04*, Berlin, June 2004.
6. A. Willig and A. Kpke, "The Adaptive-Intervals MAC protocol for a wireless PROFIBUS," *In Proc. of 2002 IEEE Intl. Symp. on Industrial Electronics*, L'Aquila, Italy, July 2002.
7. P. Morel, A. Croisier, and J.-D. Decotignie, "Requirements for wireless extensions of a FIP fieldbus," *in Proc. IEEE International Conference on Emerging Technologies and Factory Automation*, pp. 116-122, Nov. 1996.
8. P. Bhagwat, P. Bhattacharya, A. Krishna, and S. Tripathi, "Enhancing throughput over wireless LANs using Channel State Dependent Packet Scheduling," *Proc. of the IEEE INFOCOM*, vol. 3, pp. 1133-1140, March 1996.
9. A. Willig, "A MAC protocol and a scheduling approach as elements of a lower layers architecture in wireless industrial LANs," *2nd IEEE Int. Workshop on Factory Communication Systems*, Barcelona, 1997.

10. K. Sairam, N. Gunasekaran, S.R. Redd, "Bluetooth in Wireless Communication," *IEEE Communications*, Vol. 40, No. 6, June 2002.
11. A. Willig, "Antenna Redundancy for Increasing Transmission Reliability in Wireless Industrial LANs," *in Proc. 9th IEEE International Conference on Emerging Technologies and Factory Automation*, pp. 7-14, 2003.
12. A. Willig, "Exploiting redundancy concepts to increase transmission reliability in wireless industrial LANS," *IEEE Trans. on Industrial Electronics*, vol. 50 no. 11, Nov. 2003.
13. E.O. Elliot, "Estimates of error rates for codes on burst-noise channels," *Bell Syst. Tech. J* 42(9): 1977-1997, Sept. 1963.
14. E.N. Gilbert, "Capacity of a burst-noise channel," *Bell Syst. Tech. J.* vol. 39, no. 8: 1253-1265, Sept. 1960.

Collective Communication Costs Analysis over Gigabit Ethernet and InfiniBand

Hyacinthe Nzigou Mamadou[1,2], Takeshi Nanri[3], and Kazuaki Murakami[1,2,3]

[1] Department of Informatics, Kyushu University
6-1 Kasuga park, Kasuga-shi 816-8580, Japan
[2] Institute of Systems & Information Technologies / KYUSHU
1-22 Momochihama, Fukuoka-shi 814-0001, Japan
[3] Computing and Communications Center, Kyushu University
10-1 Hakozaki, Fukuoka-shi 812-8581, Japan

Abstract. Users of parallel machines need to have a good grasp for how different communication patterns and styles affect the performance of message-passing applications. MPI Collective communications involve multiple processors, and their performance prediction is a tricky task to perform. In order to evaluate the performance of collective communications, we attempt to extend LogGP and P-LogP standard point-to-point models. Our objective is to compare these models with the empirical data, and identify the most suitable for performance characterization of collective communications. The models proposed are related with the implemented algorithms in MPICH. The experimental results performed on clusters of 16 and 64 processors connected by Infiniband and Gigabit Ethernet networks respectively, show the same trend. For any collective operation, given a number of processors and a range of message sizes, there is at least one model that predicts the performance precisely.

1 Introduction

MPI (Message Passing Interface) is a set of specifications of functions for communications and managements on parallel computing. It has become one of the most popular method of programming many type of parallel computers. The communication routines of MPI can be classified into point-to-point communications and collective communications. The last one involves multiple processes to collaborate on solving a problem, and their performance prediction is a tricky task to perform.

Moreover, since the library routines are implemented before the topology information is known, it is impossible for the library to utilize topology specific algorithms. Various forms of software adaptability are supported in most MPI implementations. For example in MPICH-1.2.5, an implementation of MPI, the algorithm for the Broadcast operation with large message sizes is different from that of the operation with small message sizes. Still, the implementation algorithms continue to be adapted in order to realize the communication efficiently and achieve high performance. These adaptations lead to new models for performance estimation.

Y. Robert et al. (Eds.): HiPC 2006, LNCS 4297, pp. 547–559, 2006.
© Springer-Verlag Berlin Heidelberg 2006

In addition, for any communication algorithm, there are many system parameters that can affect the performance of the algorithm. These parameters, which include operating system context switching overheads, the ratio between the network and the processor speeds, the synchronization of the processor's clock, the number of processes involved in the communication, the switch design, the amount of buffer memory in switches, and the network topology, are very difficult to combine, in order to make efficient performance prediction with models.

MPI developers have long recognized the need for the communication routines to be adaptable to the system architecture and the application communication pattern in order to achieve high performance on a system. However the software adaptability supported in the current MPI libraries is insufficient and these libraries are not able to achieve high performance on many platforms. Furthermore, the existing performance prediction models of collective communications are not enough accurate. Thus, the problem to choose the best model that can accurately predict the collective operations within a wide area network still provide opportunities to be improved.

This paper gives a simple approach that is directly based on the algorithms used in the MPI implementation evaluated, that is MPICH-1.2.5. Using these algorithms, we attempt to extend standard point-to-point models that also are message size dependent, such as LogGP [2] and P-LogP [3], in order to predict the performance of collective communications. We compare these extended models with the empirical run-time data of respective operation, collected by using Intel MPI Benchmarks 2.3 [7].

The rest of this paper proceeds as follows. Section 2 discusses the related work. Section 3 gives an overview of the standard models evaluated, LogGP and P-LogP performance models. Section 4 describes our method of performance modeling, with all the algorithms used. Section 5 presents and evaluate the experimental results. Section 6 discusses some potential communication problems on performance prediction. Section 7 relates the conclusions and future work.

2   Related Work

The message passing interface still provides critical issues of research regarding the performance of collective communication area. The communication models optimization follows a real understanding of implemented algorithms and the network topology. Several approaches to model the communication performance of a multicomputer have been proposed in the literature.

R. Thakur et al. [4] and R. Rabenseifner et al. [5] used Hockney model to analyze the performance of different collective operation algorithms. Vadhiyar et al. used a modified LogP model which took into account the number of pending requests that had been queued [6].

More recently, Dongara et al. [8] have evaluated Hockney [9], LogP [1], LogGP [2] and P-LogP [3] standard models applied to MPI collective operations. Their approach consists of modeling each collective communication using many different algorithms, and try to give an optimized algorithm on different communication models.

Instead of developing a new implementation with its own library, or using many algorithms to predict the performance, this paper gives another estimation approach based directly on the algorithms of the MPI implementation evaluated. We extend the models most frequently used by the message passing community,and that also are message size dependent, such as LogGP and P-LogP. Our approach assumes that the first two performance model parameters are derived from P-LogP parameters as explained in [3].

3   Characteristics of LogGP and P-LogP

LogGP [2] and P-LogP [3] are standard point-to-point performance models that reflect important parameters required to estimate the network communication performance for parallel computers. They are frequently used to predict the run-time of parallel algorithms. Both also are message size dependent. Their parameters calculations are in some cases time consuming. This section provides a brief summary characteristics of all these models.

3.1   LogGP

The LogGP model addresses the performance of a parallel machine in terms of five parameters, as follows : (1) L = latency or the upper bound on the time to transmit a message from its source to destination, (2) o = overhead or the time period during which the processor is busy sending or receiving a message, (3) g = gap per message or the minimum time interval between consecutive sends and receives, (4) P = the number of processors involved in the operation, (5) the G parameter introduces the Gap per byte or the time per byte. Its inverse characterizes the available per processor communication bandwidth for long messages. LogGP model predicts the time to send a message of size m between two nodes as $L + (m-1)G + 2 * o$.

3.2   P-LogP

The P-LogP model, or Parameterized-LogP, is also based on five parameters; the number of processor P, the process-to-process latency L, overhead send $O_s(m)$, overhead receive $O_r(m)$ and the gap per message g(m). This model distinguishes between the send overhead, $O_s(m)$, and the receive overhead $O_r(m)$. Furthermore, these overheads and the gap g(m) are all defined as a function of the message size m.

For sufficiently long messages, receiving may already start while the sender is still busy, thus, O_s and O_r may overlap. g(m) is the reciprocal value of the end-to-end bandwidth from process to process for messages of a given size m. P-LogP model predicts the time elapsed to send a message of size m between two distinct nodes as $L + g(m)$.

4   Implementation and Performance Modeling

MPICH-1.2.5 implements collective communications using several algorithms based on MPI point-to-point functions. After checking the source code of this implementation, we have applied and extended LogGP and P-LogP performance models on each collective routine, in order to estimate the cost of the collective communication algorithms in terms of latency and bandwidth usage. This section introduces the methodology used, and presents the models we have built for some collective operations; MPI_Bcast, MPI_Reduce, MPI_Allgather and MPI_Allreduce.

4.1   General Assumptions

We assume that the time taken to send a message between any two distinct nodes can be modeled as $\alpha + m\beta$, where α is the latency (or startup time) per message, independent of message size, β is the transfer time per byte, and m is the total number of bytes transferred. Thus, this time is the same as modeled by LogGP and P-LogP standard models between two distincts nodes in the previous section. We assume further that the time taken is independent of how many pairs of processes are communicating with each other, independent of the distance between the communicating nodes , and that the communication links are bidirectional; that is, a message can be transfered in both directions on the link in the same times as in one direction. The node of the network interface is assumed to be single ported; that is, at most one message can be sent and one message can be received simultaneously. In the case of reduction operations, we will use t_a to represent the computation cost per byte for performing the reduction operation locally on any process. This experiment uses single precision floating point data (4 bytes) for the computation operations. In this work we also assume that LogGP model basic parameters are derived from P-LogP parameters as explained in [3].

4.2   Broadcast Models

MPI_Bcast is an operation that sends a message from the root process to all processes of a group communicator. Since all nodes need to receive the same data, node that already received the data can be used as a new data source. After the broadcast is completed, the data of the communication buffer of the root is copied to all other processes. MPICH-1.2.5 implements MPI_Bcast with two different algorithms according to the message sizes.

For short message size, it uses a minimum spanning tree (MST) algorithm, also called binary tree. This uses a fairly basic recursive subdivision algorithm. The root process sends to the process size/2 away; the receiver becomes a root for a subtree and applies the same process. With also handling for subtrees that are not a power of two in size, the cost of this algorithm is
$T_{tree} = \lceil \log_2 P \rceil (\alpha + m\beta).$

For long message size, it does a scatter followed by an allgather. it first scatter the buffer into $\frac{m}{P}$ data, using an MST algorithm. This costs $T_{tree} = \lceil \log_2 P \rceil \alpha + \frac{P-1}{P} m\beta$.

For the allgather, it uses a recursive doubling algorithm. This takes also $\lceil \log_2 P \rceil$ steps. In each step, pairs of processes exchange all the data they have, and we take care of non-power-of-two situations by using the ceiling function. This costs approximately $T_{double} = \lceil \log_2 P \rceil \alpha + \frac{P-1}{P} m\beta$.

Therefore for long message, the total cost is : $T_{total} = 2\lceil \log_2 P \rceil \alpha + 2\frac{P-1}{P} m\beta$.

This algorithm has twice the latency as the MST algorithm used for short message, but requires lower bandwidth : $2 \cdot m \cdot \beta$ versus $m \cdot \log_2 P \cdot \beta$. Therefore, for long messages and when $\log_2 P > 2$, this algorithm will perform better.

Hence, applied to LogGP and P-LogP performance models with respect of the latency and the bandwidth parts, the result costs are given in Table 1.

Table 1. Broadcast Cost Estimation

Standard models	Broadcast	
	Short message size (Binary tree)	Large message size (Binary tree + Recursive doubling)
LogGP	$\lceil \log_2 P \rceil (L + 2o + (m-1)G)$	$2\lceil \log_2 P \rceil (L + 2o)$ $+\frac{2(P-1)}{P}(m-1)G$
P-LogP	$\lceil \log_2 P \rceil (L + g(m))$	$2\lceil \log_2 P \rceil L + 2(P-1)g(\lceil \frac{m}{P} \rceil)$

4.3 Reduce Models

This function distinguishes a communication and a local computation part. The reduction operation combines the data stored in the input buffer of each process in the group communicator, using a specified operation, and returns the combined value in the output buffer of the root process. As in the case of the short message size part of MPI_Bcast, MPI_Reduce is implemented using the binary tree algorithm for both short and long size messages. We consider single precision floating point datatype, and t_a stands for the time cost of an arithmetic operation. By the same analogy and logic used with Broadcast function, we have derived the resulting cost of reduce, and the extended models are discribed in Table 2.

4.4 Allgather Models

MPI_Allgather is a gather operation in which the data contributed by each process is gathered on all processes, instead of just the root process as in MPI_Gather. Allgather is implemented in MPICH using the recursive doubling algorithm. The full description of this algorithm can be found in [4]. As with MPI_Bcast logic, we have derived the cost of Allgather, and the resulted models are given in Table 3.

Table 2. Reduce Cost Estimation

Standard models	Reduce (Binary tree)
LogGP	$\lceil \log_2 P \rceil (L + 2o + (m-1)G + \frac{m}{4} \times t_a)$
P-LogP	$\lceil \log_2 P \rceil (L + g(m) + \frac{m}{4} \times t_a)$

Table 3. Allgather Cost Estimation

Standard models	Allgather (Recursive doubling)
LogGP	$\lceil \log_2 P \rceil (L + 2o) + \dfrac{P-1}{P} \times (m-1) \times G$
P-LogP	$\lceil \log_2 P \rceil L + \dfrac{P-1}{P} \times g(m)$

4.5   Allreduce Models

This function also distinguishes a communication and a local computation part. It applies a reduction operation that combines the data stored in the input buffer of each process in the communicator group, using a specified operation, and returns the combined value in the output buffer of all the processes in that communicator.

MPI_Allreduce is implemented using recursive doubling algorithm for both, short and long size messages. Using the same logic as the previous functions, the resulting models of Allreduce can be found in Table 4.

Table 4. Allreduce Cost Estimation

Standard models	Allreduce (Recursive doubling)
LogGP	$\lceil \log_2 P \rceil (L + 2o + (m-1)G + \dfrac{m}{4} \times t_a)$
P-LogP	$\lceil \log_2 P \rceil (L + g(m) + \dfrac{m}{4} \times t_a)$

5   Results Evaluation

This section presents the environment and tools of the experiments, then evaluates the empirical results of the collective communication operations.

5.1   System Configuration and Tools of the Experiments

Our experiment was carried out on two different platforms. One is a GbE(Gigabit Ethernet) Cluster at the Center for GRID Research and Development of the

National Institute of Informatics in Tokyo. Another one is an IB(InfiniBand) Cluster at the Computing and Communications Center of Kyushu University. Table 5 illustrates their respective system configurations.

Table 5. GbE and IB System Configurations

	GbE System Configuration	IB System Configuration
Number of nodes	64	16
CPU	Intel Xeon 3.06GHz (2CPUs per node)	Intel Xeon 3.06GHz (2CPUs per node)
RAM	1 GB	7 GB
O.S	RedHat Linux 9	RedHat Enterprise $Linux2.4.21 + SCore5.8$
Compiler	gcc 3.2.2	Fujitsu C compiler 5.0
Interconnect	Gigabit Ethernet Switches	Infiniband Switch
MPI Implementation	Mpich-1.2.5	Fujitsu MPI

Our methodology assumes that the parameters of LogGP standard model are derived from P-LogP basic parameters. Hence, one of the key points of this experiment is to accurately measure the parameters of P-LogP standard model. For this reason, we have used the procedure called MPI LogP Benchmark [3]. We have also used the Intel MPI Benchmarks version 2.3 [7] to measure the algorithm run-time of all collective communications implemented in MPICH-1.2.5. These programs consist of taking the time between conventional pairs of sending and receiving messages. We achieved the experiments using only one CPU per node.

5.2   Gigabit Ethernet vs InfiniBand

Gigabit Ethernet and InfiniBand are two distincts network technologies in high performance computing environment that we have evaluated in these experiments. The latencies obtained after using the MPI LogP Benchmark are 16.52 μs and 3.75 μs respectively. Table 6 describes the data of the parameters of P-LogP standard model for both networks, also gathered with the same Benchmark.

5.3   Empirical Data Evaluation

The following are the empirical results of the collective communication we have evaluated.

Reduce. The experimental results of MPI_Reduce using 16 processors on the GbE cluster, are illustrated by the Fig.1. This graph shows that the related P-LogP model performs better on prediction than other models in many cases of the message sizes.

Table 6. P-LogP parameters

Message Size (Bytes)	GbE Cluster			IB Cluster		
	Overhead send (μs)	Overhead recv (μs)	gap (μs)	Overhead send (μs)	Overhead recv (μs)	gap (μs)
1	4.70	5.57	6.37	1.30	1.50	7.30
4	4.83	5.60	6.67	1.40	1.50	7.30
16	4.74	5.70	7.32	1.30	1.50	7.20
64	4.92	5.34	8.60	1.30	1.45	7.40
256	4.95	6.36	17.82	1.60	1.45	7.30
1024	5.32	7.14	53.87	2.00	2.20	10.97
4096	21.90	18.32	100.54	4.50	3.62	20.67
16384	115.20	266.98	259.00	30.55	58.97	64.00
65536	295.29	690.45	694.27	54.14	136.99	139.80
262144	1567.61	2389.61	2395.21	136.15	462.41	463.33
524288	3806.80	4712.00	4628.50	291.90	899.60	896.80

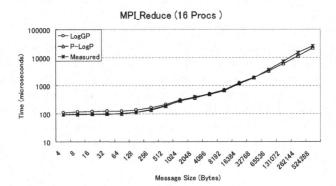

Fig. 1. Reduce using 16 processors of GbE Cluster

Moreover, the percentage of the relative gap is given in the graph of Fig.2. This gap represents the ratio of the difference between the real run-time and the best prediction time of all related new models, to the measured runtime. The figure presents that we have achieved a relative gap average around 5% for a large interval of the messages. We can also notice in this figure that, when the message size increases, the gap between the prediction and the measured time also increases.

Allreduce. The evaluation results of MPI_Allreduce executed with 8 processors on the GbE and IB clusters, are presented in Fig.3 and Fig.4 respectively. The two figures show all the models predict quite well the real performance of this collective communication operation. With a deep observation of the graphs, we can assess that P-LogP has the best performance prediction on a large intervalle of the message sizes in IB system; while LogGP predicts the best on almost all cases of the message sizes in GbE system.

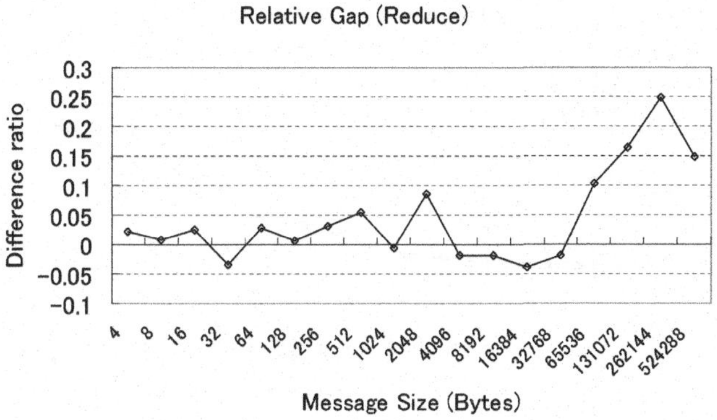

Fig. 2. Relative Gap Ratio on 16 Processors of GbE Cluster

Broadcast and Allgather. Fig.5 and Fig.6 show respectively, the MPI_Bcast and MPI_Allgather experiment results using 64 and 16 processors on the GbE and IB Clusters, which is the maximum. number of nodes of each cluster.

We see that, in Fig.5, the related P-LogP model predicts better than the extended LogGP. Although that prediction is fairly well for short message sizes up to 512 bytes; from 1 KB the gap between the prediction and the measured time increases according to the message size.

On the other hand, LogGP has the best performance prediction for short message sizes up to 256 bytes in Fig.6.

These two Figures illustrate clearly the fact that, when the message sizes or the number of processors increase, the gap between the prediction datas and the real performance runtimes increases accordingly. In all cases, the measured runtime is larger than the prediction time. This can be caused by many factors that we attempt to discuss some in the next section.

6 Discussion

This section discusses important points of the standard models evaluated. It also presents some potential problems that could significantly affect the performance estimation.

6.1 Performance Models

The results of the experiments show that, for small message sizes, there is at least one model that precisely predicts the performance of the collective communications. However, it is difficult to predict accurately the run-time of all the routines with only one model. For example the related P-LogP predicts better than the extended LogGP model using 16 processors on the reduction operation. For all the routines, when the number of processors is less than 16, the

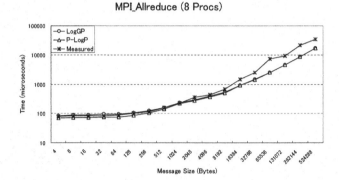

Fig. 3. Allreduce using 8 processors of GbE Cluster

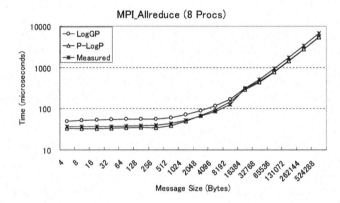

Fig. 4. Allreduce using 8 processors on IB Cluster

gap between the predicted time with the most precise model and the empirical run-time was under 15%.

The accurate measurements of P-LogP model parameters are also very important in this work. As written earlier, we used a procedure described in [3], and which presents an efficient method to determine these parameters on message passing platforms. That MPI Benchmark first measures the parameters of P-LogP model, then derives LogGP model parameters. Although the related P-LogP model predicts better in many cases, the measurement of its parameters is more complicated and time consuming.

From these facts we can advice two important ways to predict well the performance of collective routines.

Firstly, choose only one standard model to estimate the time cost of each collective communication, which is P-LogP performance model, since it is the one that predicts well in most of the cases compared to LogGP model. The demerit of this model is its complexity. P-LogP model is defined as (L, P, Os(m), Or(m), g(m)) five tuples network, where three are message size dependent, and the

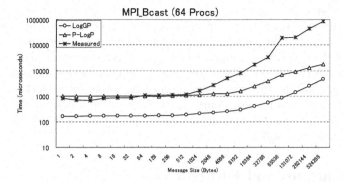

Fig. 5. Broadcast with 64 processors of GbE Cluster

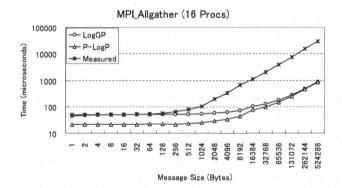

Fig. 6. Allgather with 16 processors of IB Cluster

measurements of these parameters are also more complicated and time consuming compared to LogGP.

Secondly, we can choose to change the model within LogGP or P-LogP, depending on the situations. In this case, for the collective routine to evaluate, the number of processors involved and the message size will represent the variable parameters, to determine the optimized situation. Obviously, the number of models represents the drawback of this solution.

6.2 Potential Performance Prediction Problems

So far we have developed a theoretical underpinning for performance prediction of collective communication operations. Our method of modeling collective communications is simple, since it focuses only on the algorithms used to implement those functions. Also using a single Benchmark to determine the parameters of all the standard models leads to a reliable experimental environment.

A general observation is that when the message size or the number of processors involved in the communication increases, the gap between the prediction and the real performance run-time also increases. This can be caused by many factors of the evaluated systems. For any communication and computation system, there are many system parameters that can affect the performance estimation algorithms of collective communications. These parameters, which include operating system context switching overheads, the ratio between the network and the processor speeds, the synchronization of the processor's clock, the number of processes involved in the communication, the switch design, the amount of buffer memory in switches and the network topology, are not so easily remedied and are difficult to model.

7   Conclusions and Future Work

Using some variants of LogP model, such as LogGP and P-LogP, we constructed a set of models, based on the implemented algorithms, to predict the performance of MPI collective communications. Their comparison with the empirical data on Gigabit Ethernet and Infiniband clusters has provided useful informations. All the models have given satisfaction into various aspects of different algorithms, and their relative performance. For a given number of processors and an interval of message size, it is always possible to identify a performance model that estimates precisely the collective operation. We could achieve a relative gap percentage up to 10% in many cases. However, for large message size using large number of processors, the relative gap percentage is still very high. Furthermore, neither of the models was able to completely handle all the situations, in terms of the routine evaluated, the number of processors and the message size.

The experimental results have introduced many insights. There are a lot of directions in which the work in this paper should be extended. The result of the evaluation clearly shows that, for a relatively small number of processors intercommunicating with each other using short messages, the prediction time and the real performance run-time of collective operations, are close enough. When the number of nodes are increasing, the relative gap also increases. Hence, the main future work include a deep investigation of the factors that lead to such variations especially the network contention problem. For collective communications with large size messages, the network contention problem can significantly affect the communication performance. This is particularly true when the number of nodes is large enough that not all nodes are connected by a single switch.

References

1. D. Culler, R. Karp, D. Patterson, A. Sahay, K. E. Schauser, E. Santos, R. Subramonian, and T. von Eiken. LogP : Towards a realistic model of parallel computation. In *Proceedings of the 4th ACM SIGPLAN symposium on principles and pratice of parallel programming*,ACM 1993.
2. A. Alexandrov, M. F. Ionescu, K. E. Schauser, and C.Scheiman. LogGP : Incorporating long messages into the LogP model - one step closer towards a realistic model for parallel computation. In *Proceedings of the seventh annual ACM symposium on Parallel algorithms and architectures*, pages 95-105. ACM Press, 1995.

3. T. Kielmann, H. Bal, and K. Verstoep. Fast measurement of LogP parameters for message passing platforms. In J. D. P. Rolim, editor, *IPDPS Workshops*, volume 1800 of *Lecture Notes in Computer Science*, pages 1176-1183, Cancun, Mexico, May 2000. Springer-Verlag. http://www.cs.vu.nl/albatross/.
4. R. Thakur and W. Gropp. Improving the performance of collective operations in MPICH. In J. Dongara, D. Laforenza, and S. Orlando, editors, *Recent Advances in Parallel Virtual Machine and Message Passing Interface*, number LNCS2840 in Lecture Notes in Computer Science, pages 257-267. Springer Verlag, 2003. 10th European PVM/MPI User's Group Meeting, Venice, Italy.
5. R. Rabenseifner and J. L. Traff. More efficient reduction algorithms for non-power-of-two number of processors in message-passing parallel system. In *Proceedings of EuroPVM/MPI*, Lecture Notes in Computer Science. Springer-Verlag, 2004.
6. S.S. Vadhiyar, G.E. Fagg, and J. J. Dongara. Automatically tuned collective communications. In *Proceedings of the 2000 ACM/IEEE conference on Supercomputing (CDROM)*, page 3. IEEE Computer Society, 2000.
7. Intel MPI Benchmarks version 2.3. *http://www.intel.com/cd/software/products/asmo-na/eng/cluster/mpi/219848.htm*
8. Jelena P. Sivac-Grbovi'c, T. Angskun, G. Bosilca, G. E. Fagg, E.Gabriel Jack J. Dongarra. Performance Analysis of MPI Collective Operations. *4th International Workshop on Performance Modeling Evaluation and Optimization*, April 2005.
9. R.Hockney. The communication challenge for MPP: Intel Paragon and Meiko CS-2. *Parallel computing*, 20(3):389-398, March 1994.

An Efficient MAP Classifier for Sensornets

Zille Huma Kamal[1], Ajay Gupta[1], Leszek Lilien[1], and Ashfaq Khokhar[2]

[1] Dept. of Computer Science, Western Michigan University, Kalamazoo, MI
{zkamal, gupta, llilien}@cs.wmich.edu
[2] Dept. Computer Science, Dept. Electrical & Computer Eng., University of Illinois, Chicago
ashfaq@uic.edu

Abstract. The classification phase is computationally intensive and frequently recurs in tracking applications in sensor networks. Most related work uses traditional signal processing classifiers, such as Maximum A Posterior (MAP) classifier. Naïve formulations of MAP are not feasible for resource constraint sensornet nodes. In this paper, we study computationally efficient methods for classification. We propose to use one-sided Jacobi iterations for eigen value decomposition of the covariance matrices, the inverse of which are needed in MAP classifier. We show that this technique greatly simplifies the execution of MAP classifier and makes it a feasible and efficient choice for sensornet applications.

Keywords: Classification, Jacobi Iterations, Sensor Networks, Collaborative Processing.

1 Introduction

Sensornets have been envisioned as a cost effective paradigm for monitoring, controlling or sensing in military, agricultural, or commercial applications. In any of these applications, the fundamental criterion for detecting the presence or absence of an object or an environmental condition requires 'sensing' of the features or modalities present in the environment. After detecting a stimulus, classification algorithms process the sensed data to categorically determine and identify the event (stimulus), results of this classification phase can be used to decide whether a reaction/response is warranted. In monitoring/surveillance applications—[1], [2], [3], [4], to mention a few, the classification phase is followed by a tracking phase, where the sensornet nodes work collaboratively to track the stimulus, which could be an enemy combat vehicle or the spread of hazardous chemicals in the environment.

The classification process is a computationally intensive process, repeatedly initiated for every measurable disturbance in the environment that the sensornet is trained to observe. A traditional approach to classification in sensornets is to use classifiers that are commonly employed in signal-processing applications, such as Linear Vector Quantization, Support Vector Machines (SVM), k–Nearest Neighbor (kNN) and Maximum Likelihood (ML), among which ML is considered most feasible for sensornets due to its minimal storage and computation requirements [4].

Y. Robert et al. (Eds.): HiPC 2006, LNCS 4297, pp. 560–571, 2006.
© Springer-Verlag Berlin Heidelberg 2006

When all categories are equally probable, the Maximum A Posterior (MAP) classifier is the same as ML ([22], [23]). In most cases, we can easily hold and justify this assumption. Thus, we critically review the computations and communications required in a distributed MAP classifier and its feasibility in resource constraint sensornet applications.

Computation of the MAP classifier requires the inverse of the covariance matrices that represent the pre-determined categories. Often inverse computation is avoided, due to the numerical instability [18]. Instead, techniques that approximate the inverse, decompose the inverse computation, or perform orthogonal transformations, are preferred [16]. Therefore, we instigate the use of stable, parallelizable and efficient one-sided Jacobi iterations and transform the problem of computing the inverse of the covariance, into computing its Eigen Value Decomposition (EVD). We will show that the substitution of EVD of covariance matrix in place of its inverse, in MAP can greatly reduce the cost of this classifier. To the best of our knowledge, using Jacobi in MAP has not been considered previously.

The main contributions of this paper are to show how to use one-sided Jacobi iterations in sensornets and consequently minimize the cost of executing the MAP classifier in sensor networks.

We begin by summarizing related work in this area in section 2. In section 3 we give an overview of the MAP classifier. In Section 4, we review the one-sided Jacobi iterations and its computation and communication costs. We present computational and communicational costs of MAP with Jacobi and MAP with LU decomposition (previous work [21]] in Section 5 and compare the power consumption in Section 6. We conclude with our findings and contributions in Section 7.

2 Related Work

Authors in [1], [3] and [4] have proposed the use of classifiers such as kNN and MAP in classification applications of sensornets. However, their test-bed and sensing platform is composed of WINS 3.0 Sensoria [9], [10] nodes that are more powerful in terms of energy, processor speed and power, and memory capacity, compared to the more constraint Crossbow's MICA2 Motes [11], Intel Motes [14] or the envisioned dust size COTS Dust [12], [13]. In this paper, we use *sensors* or *nodes* interchangeably to generalize small resource constrained motes, like MICA2 or Intel Motes. An analytical study conducted in [5] concluded that there was an acute trade-off between accuracy and efficiency of classifiers in sensornets and it became apparent that traditional classifiers were computationally and communicationally intensive and unpractical for today's resource constraint sensornet (mote-like) nodes.

Some alternate approaches to conventional classification include novel, computationally efficient influence field patterns [7] to classify objects, but its accuracy is shown to be directly dependent on the reliability of the underlying sensornet, a tough characteristic to build into a *dynamic* sensornet. Gu et. al. [24] considers hierarchical classification, first on node level, then group level followed by a base level, which has been shown to be feasible and accurate for sensornets. In [6], power conservation is built into sensornet surveillance application by differentiating the sensing coverage of the region monitored based on the proximity to the highly sensitive area, e.g. the army base station in a military application.

Other alternate techniques [8] use mobile agents (robots or humans) to gather and transfer data from sensornet to a more powerful command/base station for processing. While this approach may be appropriate in certain situations, it seems infeasible for applications that are envisioned to be deployed in monitoring impermeable, insecure regions.

Though, it seems that traditional signal processing classifiers, like MAP, are inappropriate for sensornets that are resource constraint [24], in terms of processing power, small storage and absence of floating point hardware [24], [25], [29]. Our goal is to avoid reinventing the wheel and adapt these existing tested classifiers, specifically, the MAP classifier, and try to make it feasible for these sensornets.

3   Background

3.1   Classification

Classification consists of three phases: training, testing and deployment. In the *training phase*, the sensors in the sensornet are exposed to N *known* sample event feature matrices from each of the k predetermined categories. The user interactively classifies the events for the sensors. An *event* is an $f{\times}d$ feature matrix, where f is the number of features monitored in the environment and d is the temporal dimension. Based on these sample feature matrices, each sensor can locally compute the *mean* (μ) and *covariance* (δ) matrix for a category as in Eq. 1 and 2. $M_{i,j}^{\alpha}$ is the event feature matrix from the training phase (indicated by α) at sensor i for category j, for $1 \le i \le n_0$, where n_0 is the total number of sensors in the sensornet and $1 \le j \le k$.

$$\mu_{i,j} = \frac{1}{N} \sum_{\alpha=1}^{N} M_{i,j}^{\alpha} \ . \tag{1}$$

$$\delta_{i,j} = \frac{1}{N-1} \sum_{\alpha=1}^{N} \left(M_{i,j}^{\alpha} - \mu_{i,j} \right) \left(M_{i,j}^{\alpha} - \mu_{i,j} \right)^{T} \ . \tag{2}$$

In the *testing phase*, probability of false positives and false negatives are computed to determine a belief probability for the sensor's accuracy, by allowing the *sensors to classify the test sample events*. In the deployment phase, sensors are deployed in the region to be monitored, unmonitored themselves, and repeat the classification process to decide whether a detected object is of interest or not.

3.2   MAP Classifier

In this paper we study the Maximum A Posterior (MAP) classifier, the computation of which may be complex for resource constraint sensornets. However, by removing terms that are constant across the categories, computation of MAP can be simplified ([5], [22], [23]), to the computation of the *Mahalanobis distance* [22] between the mean of a category and the event/object detected (Eq. 3). The detected object is classified into the category that minimizes the Forbenius norm [26] of the Mahalanobis distance (Eq. 4), where $M_{i,j}$ represents event recorded at sensor i for some *unknown* category j, and $\left\|MAP_{i,j}\right\|_F$ represents the Frobenius norm of the $d{\times}d$ MAP matrix. Here, β is used to indicate event feature matrices that are collected during the testing phase, however, same computations are performed during deployment phase.

$$MAP_{i,j}^{\beta} = \left(M_{i,j}^{\beta} - \mu_{i,j}\right)^T \delta_{i,j}^{-1}\left(M_{i,j}^{\beta} - \mu_{i,j}\right). \tag{3}$$

$$\left\|MAP_{i,j}^{\beta}\right\|_F = \sum_{p=1}^{d}\sum_{q=1}^{d} MAP_{i,j}^{\beta}(p,q). \tag{4}$$

Generally, sensornet 'surveillance' applications decompose the region monitored into logically or physically disparate cells, clusters or sub regions. We design clusters of size d sensors where d is the larger dimension of the events. These dense clusters enable us to exploit parallelism to run the computationally-intensive MAP quickly, efficiently and in a load balanced manner.

We assume for brevity that the distributed FFT ([27], [34]) for sensornets can be extended to perform efficient d-point temporal processing of the $f{\times}d$ matrix, distributively across the d sensors. So, each sensor holds a column of the mean and covariance matrices, as in Fig. 1, which illustrates the clustering and data distribution.

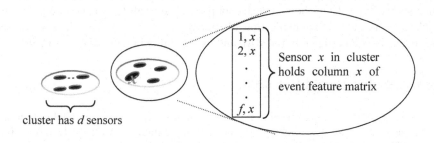

Fig. 1. Data distribution in a cluster node

The most complex computation in Eq. 3, is finding the inverse of $\delta_{i,j}$ for all k categories. One might consider offloading the computation of $\delta_{i,j}^{-1}$ to a more powerful node such as a personal computer. However, this would inhibit the sensornet from dynamically updating $\delta_{i,j}$ and its inverse that could have increased the accuracy of the classification. To make MAP feasible for sensornets, we must find fast, stable and efficient algorithms for the computation or decompose it into smaller less computationally intensive parts.

Traditionally, computation of inverse of an $f{\times}f$ matrix requires $O\!\left(f^3\right)$ operations [28], and is slow, inefficient, unbalanced, serial and tedious. Instead, we consider the eigen value decomposition (EVD) of $\delta_{i,j}$, such that it is decomposed into its eigenvectors and eigen values (Eq. 5), where Λ is a diagonal matrix, with eigen values on the diagonal and E is the matrix, whose columns are eigen vectors. By substituting Eq. 5 in Eq. 3, we get Eq. 6.

$$\delta_{i,j} = E\Lambda E^T \Rightarrow \delta_{i,j}^{-1} = E\Lambda^{-1}E^T . \tag{5}$$

$$MAP_{i,j} = \left(M_{i,j} - \mu_{i,j}\right)^T E\Lambda^{-1}E^T \left(M_{i,j} - \mu_{i,j}\right)$$
$$\Rightarrow MAP'_{i,j} = \left(M'_{i,j} - \mu'_{i,j}\right)^T \Lambda^{-1}\left(M'_{i,j} - \mu'_{i,j}\right) . \tag{6}$$

$$MAP_{i,j} = \frac{\left(m'_{11} - \mu'_{11}\right)^2}{\lambda_1} + \frac{\left(m'_{21} - \mu'_{21}\right)^2}{\lambda_2} + ... + \frac{\left(m'_{f1} - \mu'_{f1}\right)^2}{\lambda_f} + ... + \frac{\left(m'_{fd} - \mu'_{fd}\right)^2}{\lambda_f} . \tag{7}$$

Matrix algebra says, $\left(M'_{i,j} - \mu'_{i,j}\right) = \left(\left(M_{i,j} - \mu_{i,j}\right)^T E\right)^T = E^T\left(M_{i,j} - \mu_{i,j}\right)$ and multiplication of a matrix with Λ^{-1} is just a scalar multiplication of column x of the matrix with $\frac{1}{\lambda_x}$, which is the reciprocal of the diagonal entry in column x of Λ. Using these rules, Eq. 6 becomes the equation of a hyper-ellipse (Eq. 7), centered at μ' [22], and $m'_{p,q}$, $\mu'_{p,q}$ are the p,q entry in matrix $M'_{i,j}$ and $\mu'_{i,j}$ and λ_p is the pth eigen value in Λ. Eq. 7 is simpler than Eq. 6, as matrix multiplication is eliminated.

We use the inherently parallel one-sided Jacobi iterations for EVD of the covariance matrix. It has been shown that Jacobi parallelizes well [20], hence we do not include any speedup or parallelization efficiency analysis. We show how one-sided Jacobi can be used efficiently for MAP and its feasibility in mote-like sensornets.

4   One-Sided Jacobi Iterations

Jacobi Iterations gained renewed popularity with advancements in parallel computing [19] and are used for eigen (or spectral or singular) valued decomposition (EVD/SVD) of a symmetric covariance matrix. The goal is to apply similarity transformations to zero-off (eliminate) the off-diagonal entries. The multipliers used for the transformations are accumulated, and yield the eigen vectors of the covariance matrix and the remaining diagonal entries represent the eigen values. The inherent parallelism of one-sided Jacobi iterations makes it a better choice than its two-sided Jacobi counterpart [15], [20], for our distributed sensornets. This parallelism will enable a faster and more robust application.

The similarity transformation to eliminate entry p,q of the matrix, termed a *step* [20], only affects columns p and q of the matrix, and can be performed in parallel with the step for entry a,b, $\forall \ a \neq p$ and $b \neq q$. This implies $f/2$ steps can be performed in parallel, to eliminate the off-diagonal entries of an $f{\times}f$ matrix. A *sweep* [20] contains $f{-}1$, if f is even, or f, if f is odd, steps that are required to eliminate all off-diagonal entries. After every step, an exchange of columns (or rows) must occur, termed *transition* [20].

Generally, more than one sweep may be required, since a step may introduce a non-zero entry in a previously zeroed off entry. Typically, $\log_2 f$ sweeps converge a symmetric matrix, faster convergence can be achieved with a threshold ε, ([15], [20]), that determines whether an entry is considered for elimination.

The order in which the transformations are applied also affects the rate of convergence [20]. Various orderings, such as row-cyclic [15] and block recursive (BR) [20] have been shown to achieve high convergence. Without loss of generality, we will use BR ordering, whose efficient implementation on hypercube has been shown in [20]. Therefore, for simplicity, we abstract a hypercube topology from the sensornet for each cluster, such that for the $f{\times}f$ covariance matrix, we have a hypercube with $\log_2 f$ dimensions and $f/2$ sensors.

4.1 Block Recursive (BR) Ordering

Initially in BR ordering, we distribute two columns of $\delta_{i,j}$ to each sensor, the smaller index value column is called *top* and the larger becomes the *bottom* column, succeeded by three phases delineated as: (i) Rotation phase, a series of parallel steps and transitions, (ii) Top and Bottom Exchange (TBX) phase, a step followed by a transition, and (iii) Last Exchange (X) phase, the last step and transition of a sweep. The Jacobi procedure is presented in the program code, One_Sided_Jacobi and the computations required for a step are delineated in the program code, Step.

Initially, we let, $S_0'=\delta_{i,j}$, $S_0=\delta_{i,j}$ and $U_0=I$ (the identity matrix). If we assume that the iterations converge in $\log_2 f$ sweeps, the eigen vector matrix E will be stored in the U_i's, i.e. $E = U_{\log_2 f}$ and similarly $\Lambda = S_{\log_2 f}$ [20]. The step computations required in the Rotation, TBX and X phases, iteratively update the initial matrices, S_0' and $U_0=I$ and consequentially S_k is updated as in line 10, in One_Sided_Jacobi procedure.

```
One_Sided_Jacobi(δ_{i,j}, ε)        Step(S_k, S'_k, U_k, p, q, ε)
(1) S'_0 = δ_{i,j}; U_0 = I;         (1) θ = [S_k(q,q) − S_k(p,p)] / S_k(p,q)
(2) for k = 1 to log₂f
(3)     //perform sweep              (2) c_{p,q} = cos θ; s_{p,q} = sin θ
(4)     while(p,q in S_k > ε)        (3) S'_{k+1}(•,p) = c_{p,q}*S'_k(•,p)+s_{p,q}*S'_k(•,q)
(5)         //compute S'_k&U_k→Step  (4) S'_{k+1}(•,q) = −s_{p,q}*S'_k(•,p)+c_{p,q}*S'_k(•,q)
(6)         Rotation; TBX; X         (5) U_{k+1}(•,p) = c_{p,q}*U_k(•,p)+s_{p,q}*U_k(•,q)
(9)     end while                    (6) U_{k+1}(•,q) = −s_{p,q}*U_k(•,p)+c_{p,q}*U_k(•,q)
(10) S_k = U_k^T S'_k                end procedure
(11) end for
end procedure
```

The dimensions used to exchange the columns in a transition are generated as $D_1^{BR} = <0>$ and $D_r^{BR} = <D_{r-1}^{BR}, r-1, D_{r-1}^{BR}>$ [20], where, D_r^{BR} is the BR sequence of dimension(s) used for transitions in rotation phase r, D denotes the dimensions used for column exchange and is different from d used earlier to denote the points of temporal processing of each feature. The dimension used for transitions of successive sweep, j, is a permutation of the dimensions used in the respective rotation of the first sweep, simply generated as, $(k - j)\mathrm{mod}(\log_2 f)$, for every dimension k, in the sequence D_r^{BR}, for rotation r in the first sweep [20].

4.2 Adapting One-Sided Jacobi to Sensornets

The matrix multiplication required to update S_k (line 10, procedure One_Sided_Jacobi) may not be feasible for our sensornet, since the columns of S_k are distributed and communication is expensive. Furthermore, if all we need is $E = U_{\log_2 f}$ and $\Lambda = S_{\log_2 f}$, then we should avoid updating S_i' (lines 3–4, procedure Step). Then, the computation of θ (line 1, procedure Step), can be formulated as the inner product of the columns of U and S (Eq. 8) ([30],[31]), where u_p^k and s_p^k are

the pth columns of U_k and S_k, respectively, and $\left\langle u^p, s^p \right\rangle = \sum_{i=1}^{f} u_i^p \cdot s_i^p$.

$$\tan 2\theta = \frac{2\left\langle u_p^k, s_q^k \right\rangle}{\left(\left\langle u_p^k, s_p^k \right\rangle - \left\langle u_q^k, s_q^k \right\rangle\right)} = \Psi . \tag{8}$$

$$\theta = \frac{\tan^{-1}(\Psi)}{2} . \tag{9}$$

Then, by letting $t_{p,q} = \tan \theta$, we have $c_{p,q} = \cos \theta = \dfrac{1}{\sqrt{1 + t_{p,q}^2}}$ and

$s_{p,q} = \sin \theta = c_{p,q} * t_{p,q}$, which will replace lines 1–2 in procedure Step and eliminate line 10 in procedure One_Sided_Jacobi. This way, the entire step can be computed locally, as the sensors hold the necessary columns. Moreover, to overcome the absence of hardware support for floating point operations, and to conserve processor cycles we can maintain a small finite table that stores values of θ and its trigonometric cosine and sine values. This approach has been shown to achieve accurate results in [17] and we refer readers to it for a detailed study of this method.

4.3 Computational and Communicational Cost of One-Sided Jacobi Iterations

The preceding sections review classification and one-sided Jacobi iterations, indicating the adaptations made to them to make them feasible for sensornets. We now derive computational and communicational costs of one-Sided Jacobi and MAP based on these discussions.

4.3.1 Cost of a Sweep - Steps and Transitions in Rotation, TBX and X Phases
The computation costs incurred in a step are: computation of Ψ (Eq. 8), table lookup operation, scalar-vector multiplications and vector-vector additions (lines 5–6,

procedure Step), that incur a total cost $12f + 6f + O(\log_2 m) \approx 18f$ *local* operations without *any* additional communication, where m is the number of θ values stored in the table for lookup.

There are log_2f rotations in a sweep and the number of steps and transitions in a Rotation phase, r, are $2^r - 1$ [20]. Then, the total number of steps and transitions in a sweep from the Rotations are $\sum_{r=1}^{\log_2 f} 2^r - 1 = \dfrac{2^{\log_2 f+1} - 1}{2 - 1} - \log_2 f = 2f - 1 - \log_2 f$.

There is a TBX phase after every Rotation phase, with one step and transition in it, so there are log_2f steps and transitions from the TBX phases of a sweep. Since, there is only one X phase in a sweep, the *total number of steps and transitions* in a sweep are $2f - 1 - \log_2 f + \log_2 f + 1 = 2f$.

4.3.2 Total Cost for One-Sided Jacobi

The step is the only and purely local computation required in the one-sided Jacobi iterations. The cost of *a* sweep is $(2f)(18f) = 36f^2$ at a sensor. So, the total *computational* cost incurred after log_2f sweeps is $(36f^2)(\log_2 f) = 36f^2 \log_2 f$ across $f/2$ sensors of the d sensors in a cluster of the sensornet. The local computation minimizes communication costs thereby optimizing overall MAP computation costs. Each of the $f/2$ sensors involved in the Jacobi computation do an equal amount of work and hence expend equal energy which increases network lifetime.

In a transition, each sensor incurs communicational costs to send and receive a message, which is a column of the matrix. A TinyOS message payload is 29 bytes long [32], therefore, a message with f elements, assuming an element is stored in 4 bytes, requires $\left\lceil \dfrac{4f}{29} \right\rceil$ messages to be sent and received. Thus, the total *communicational* cost incurred across the sensors is the total number of messages sent and received in log_2f sweeps i.e. $2f\left\lceil \dfrac{4f}{29} \right\rceil \log_2 f = 2f \log_2 f\left(\dfrac{4f}{29} + 1\right)$.

5 Cost of Maximum a Posterior (MAP) Classifier

5.1 MAP with Jacobi (MAP-J)

Once $\delta_{i,j}$ converges, MAP can be computed as presented in Eq. 7, for which we first redistribute the columns held by the $f/2$ sensor nodes that were participating in the one-sided Jacobi iterations. This way, sensor node i holding column i of $M_{i,j}$, now also holds column i of S_k^{-1} $(=\Lambda^{-1})$ and U_k, where $k = \log_2 f$. Then, each of the f sensors can concurrently compute one of the fraction entities of Eq. 7 locally. This requires 3 arithmetic operations for each of the f entries in a column of the matrix, followed by summation of all local fractions, which implies a total of $4f–1$ (i.e. $3f+f–1$) operations at each of the d sensors in a cluster.

Thus, the MAP classifier incurs a total *computational* cost of $(4f-1)d = (4fd-d)$. To efficiently aggregate these distributed fractions, we can use a tree-like structure so that all entries can be aggregated in $log_2 d$ levels, where each level requires transmission and reception of one message. Thus, the total messages sent and received are d, the total *communicational* cost of MAP with Jacobi (MAP-J).

5.2 MAP Using LU (MAP-LU)

Recall the inverse of the covariance matrix in MAP can also be computed using the popular LU decomposition technique. Our earlier work in [21] presents this approach in detail. In this section, we just restate the results briefly to facilitate comparison with MAP classifier using one-sided Jacobi. The LU decomposition is used to decompose the covariance matrix into its Lower (L) and Upper (U) triangular matrices. Eq. 3 can now be formulated as Eq. 10, where, D is the diagonal entries of U.

$$MAP_{i,j} = \left(M_{i,j} - \mu_{i,j}\right)^T LD^{-1}L^T \left(M_{i,j} - \mu_{i,j}\right)$$
$$\Rightarrow MAP'_{i,j} = \left(M'_{i,j} - \mu'_{i,j}\right)^T D^{-1} \left(M'_{i,j} - \mu'_{i,j}\right) \tag{10}$$

However, this cannot be further simplified like we did in Eq. 7. This is because, LU decomposition is not a similarity transformation, like one-sided Jacobi that can preserve the eigen values [20] of the original matrix, and allow the simplification of Eq. 7.

Therefore, the cost of MAP with LU (MAP-LU) can be stated as follows: let $Z = \left(M_{i,j} - \mu_{i,j}\right)^T L$ and $\left(M_{i,j} - \mu_{i,j}\right)L^T = Z^T$, so computing Z incurs $2f^2 d$ operations (i.e. fd subtractions and fd dot products) and $\left\lceil \frac{4f}{29} \right\rceil = O(f)$ message exchanges for interchanging columns of L. The cost of computing the norm of MAP (Eq. 4) is $2d^2 f - d^2$ for the d^2 dot products and the communicational cost for the norm is $\left\lceil \frac{4d}{29} \right\rceil = O(d)$.

Thus, over the d sensor nodes of a cluster, the total computational cost of computing MAP-LU is $O(2d^2 f + 2df^2 - d^2) = O(d^2 f)$ with a communicational cost of $O(d+f) = O(d)$, since $d \gg f$.

6 Power Consumption Comparison

We now compare the power consumption of MAP-J with MAP-LU. Various authors, ([29], [33]), to mention a few have studied the average power consumption of a Mica mote-type sensor node in various modes. We use the power ratings presented in [29], where the power consumption for transmitting, receiving and computing (only CPU operational) is approximately 10 mA, 7 mA and 8mA, respectively. From Section 5.1, MAP-J expends a total of $(4fd - d) \times 8 + d \times 10 + d \times 7 = 32fd + 9d$ mA and similarly, MAP-LU, consumes $16d^2 f - 8d^2 + 16f^2 d + \frac{68(f+d)}{29} + 34$ mA. These power consumptions are plotted in Fig. 2.

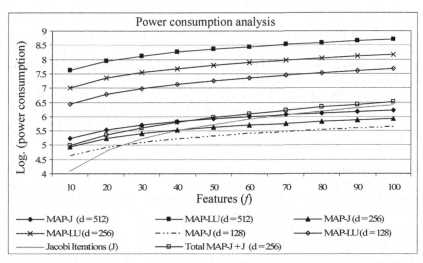

Fig. 2. Comparing power consumed by MAP-J, MAP-LU, with varying dimensions of temporal processing (d), cost of one-sided Jacobi with BR (J) and total cost of MAP-J + J

It is evident that there are orders of magnitude saving in power consumption of MAP-J with MAP-LU, irrespective of temporal processing points (d). This is because EVD allows the simplifications of Eq. 7 that enables faster computation. Principal Component Analysis (PCA) [26] type of techniques can also be used in MAP-J so that only the first few principal or *significant* eigen values are computed and used and not all f. Although this is not further explored in this paper, it can be seen that PCA will only speedup the MAP-J computation more but at the possible risk of reduced accuracy.

Fig. 2 also shows that cost of computing Jacobi (J) is not drastically expensive, as the total cost of MAP-J and cost of Jacobi is still magnitudes lower than MAP-LU. Furthermore, typically, 20-40 features are monitored, in which range, cost of Jacobi (J) does not exceed MAP-J, for $d \geq 256$. Most importantly, note total power consumption of MAP and Jacobi Iterations (MAP-J + J) is still magnitudes lower than cost of MAP-LU alone, irrespective of the temporal processing dimension, and not taking into account cost of LU, which can only increase costs.

It is also interesting to note that we were able to overcome floating point hardware limitation without any explicit software modules with MAP-J. This may not be the case for MAP-LU, as at the least floating point emulation software would be needed.

The significant difference of MAP-LU and MAP-J is the reduction in total cost by a factor of $O(d)$, which is typically in the order of 512. Also, the computations have been simplified to vector additions, rather than the more complex vector multiplications of MAP-LU.

7 Conclusion

We presented a feasible, practical and efficient method for executing the traditional Maximum A Posterior (MAP) classifier distributively, in resource constraint sensornets. We do so by simplifying the computation of the MAP classifier into an equation of a hyper-ellipse (Eq. 7) ([23],[20]). This greatly reduces the computations required and significantly cuts down the communication cost, which occur when the MAP classifier uses the traditional approach of LU decomposition to overcome computation of inverse of covariance matrix. Not only are the one-sided Jacobi iterations efficient but they are also more stable than LU decomposition [28]. Furthermore, we do not have to compromise with any hardware limitations, with the one-sided Jacobi iterations.

We have thus shown feasibility of executing traditional signal processing classifiers such as MAP on resource constraint sensornets. Our future work includes, analyzing the accuracies of these methods for classification and effect of PCA on the accuracy of one-sided Jacobi iterations.

Acknowledgements. This research was supported in part by the NSF, under grants ACI-0000442, ACI-0203776, IIS-0242840 and MRI-0215356, by the Dept. of Education grant R215K020362, and a Congressional Award, administered by the US Dept. of Education, Fund for the Improvement of Education. The authors would also like to acknowledge Western Michigan University for its support and contributions to the WiSe (Wireless Sensornets) Laboratory, Computational Science Center and Information Technology and Image Analysis (ITIA) Center. Any opinions, findings, and conclusions or recommendations expressed in this material are those of the author(s) and do not necessarily reflect the views of the funding agencies or institutions.

References

[1] D. Li, K. Wong, Y. Hu, and A. Sayeed, Detection, Classification, tracking of targets in micro-sensor networks. *IEEE Signal Processing Magazine*, March 2002.

[2] A. D'Costa and A. M. Sayeed, Collaborative signal processing for distributed classification in sensor networks. *Lecture Notes in Computer Science* (Proceedings of IPSN'03), (Springer-Verlag, Berlin Heidelberg), pp. 193–208, (F. Zhao and L. Guibas (eds.), Apr'03.

[3] M. Duarte and Y. H. Hu, Distance Based Decision Fusion in a Distributed Sensor Network. *International Symposium on Information Processing in Sensor Networks* (IPSN) 2003.

[4] Marco Duarte and Yu-Hen Hu, Vehicle Classification in Distributed Sensor Networks. *Journal of Parallel and Distributed Computing*, Vol. 64 No. 7, 2004.

[5] Z. H. Kamal, M. A. Salahuddin, A. Gupta, M. Terwilliger, V. Bhuse, and B. Beckmann, Analytical Analysis in Decision and Data Fusion. *Proc. of Conference on Embedded Systems and Applications*, June 2004.

[6] T. Yan, T. He, and J. Stankovic, Differentiated Surveillance for Sensor Networks. *In Proc. of 1st ACM Conference on Embedded Networked Sensor Systems*, L.A., CA, Dec 2003.

[7] A. Arora, P. Dutta, S. Bapat, V. Kulathumani, H. Zhang, V. Naik, V. Mittal, H. Cao, M. Demirbas, M. Gouda, Y-R. Choi, T. Herman, S. S. Kulkarni, U. Arumugam, M. Nesterenko, A. Vora and M. Miyashita, A Line in the sand: A wireless sensor network for target detection, classification, and tracking. *Computer Networks (Elsevier)*, 2004.

[8] Y. Xu, and H. Qi, Distributed Computing Paradigm for Collaborative Signal and Information Processing in Sensor Networks. *J. of Parallel and Distributed Computing*, Aug 2004.

[9] Sensoria Corporation, last accessed May 12, 2006, sensoria.com

[10] WINS 3.0 Sensing Platform, sensoria.com/pdf/WINS-3.0-Wireless-Sensing-Platform.pdf

[11] Crossbow Technology Incorporated, last access May 12, 2006, xbow.com

[12] Dust Networks, last accessed May 12, 2006 www-bsac.eecs.berkeley.edu/archive/users/warneke-brett/index.html

[13] B. Warneke, M. Last, B. Leibowitz, K.S.J. Pister, Smart Dust: Communicating with a Cubic-Millimeter Computer. *IEEE Computer*, Computer Society, Piscataway, NJ, Jan. 2001.

[14] Intel Corporation, 2006, last accessed May 12, 2006.intel.com/research/exploratory/motes.htm

[15] D. Gimenez, V. Hernandez, R. Geijn, A. Vidal, A Jacobi Method by Blocks on a Mesh of Processors. *Concurrency: Practice and Experience*, Vol 9 Issue 5, 1997 John Wiley & Sons.

[16] R. Brent, Parallel Algorithms in Linear Algebra. *Proc. 2nd NEC Research Symp., Aug'91.*

[17] T.J. Herron, K.M. Reddy, R. Garg, and K. Devanahalli, Eigen Decompositions of Covariance Matrices on a Fixed Point DSP. *National Conference on Communication*, Jan 31 - Feb 2, 2003, Indian Institute of Technology, Madras, India

[18] D. Bertsekas, and J. Tsitsiklis, Parallel and Distributed Computation – Numerical Methods. Prentice-Hall, Inc., Englewood, New Jersey, 1989.

[19] B. B. Zhou, R. P. Brent and M. H. Kahn, Efficient one-sided Jacobi algorithms for singular value decomposition and the symmetric eigen problem. *Proc. IEEE First International Conference on Algorithms and Architectures for Parallel Proc.*, IEEE Press, 1995, 256-262.

[20] D. Royo, A. Gonzalez, and M. Valero-Garcia, Low Communication Overhead Jacobi Algorithms for Eigenvalues Computation on Hypercubes. *J. Supercomputing*, 14(2), Sept'99.

[21] Z. H. Kamal, A. Gupta, L. Lilien, and A. Khokhar, Classification Using Efficient LU Decomposition in Sensornets. *Proceedings of WSN 2006*, July 2006.

[22] C. W. Therrien, Decision Estimation and Classification. An Introduction to Pattern Recognition and Related Topics. John Wiley & Sons, NYC, NY, 1989

[23] Online Resource, Last Accessed May 9, 2006 profc.udec.cl/~gabriel/tutoriales/rsnote/cp11/cp11-7.htm

[24] L. Gu, D. Jia, P. Vicaire, T. Yan, L. Luo, A. Tirumala, Q. Cao, T. He, J. Stankovic, T. Abdelzaher, B. Krogh, Lightweight Detection and Classification for Wireless Sensor Networks in Realistic Environments. *In Proc. of the 3rd ACM Conf. SenSys'05*, November.

[25] S. Mohan, F. Mueller, D. Whalley and C. Healy, Timing Analysis for Sensor Network Nodes of the Atmega Processor Family, Real-Time and Embedded Technology and Applications Symposium, March 2005.

[26] S. J. Leon, Linear Algebra with Applications 5th Ed. Prentice Hall, NJ, 1998.

[27] T. Canli, M. Terwilliger, A. Gupta, and A. Khokhar, Power-Time Efficient Algorithm for Computing FFT in Sensor Networks, *In Proc. of the 2nd ACM SenSys'04*, November.

[28] A. Grama, A. Gupta, G. Karypis, and B. Kumar, Introduction to Parallel Computing, 2nd Ed. Pearson Education Limited, 2003

[29] V. Shnayder, M. Hempstead, B. Chen,G. W. Allen, and M. Welsh, Simulating the Power Consumption of Large Scale Sensor Network Applications. *Proc. 2nd ACM SenSys'04*, Nov.

[30] P. J. Eberlein and H. Park, Efficient Implementation of Jacobi Algorithms and Jacobi Sets on Distributed Memory Architectures. *Journal Parallel Distributed Computing* 8(4), (1990)

[31] P. J. Eberlein, On the Schur Decomposition of a Matrix for Parallel Computation. *IEEE Transaction Computers* 36(2), 1987

[32] TinyOS Tutorial-Lesson 4: Component Composition and Radio Communication, 2003. Last accessed May 10, 2006. www.tinyos.net/tinyos-1.x/doc/tutorial/lesson4.html

[33] M. Srivastava, Sensor node platforms and energy issues. *Mobicom* 2002 Tutorial. Available online and last accessed May 12, 2006 nesl.ee.ucla.edu/tutorials/mobicom02/slides/Mobicom-Tutorial-2-MS.pdf

[34] T. Canli, M. Terwilliger, A. Gupta and A. Khokhar, Power Efficient Algorithms for Computing Fast Fourier Transform over Wireless Sensor Networks. *The 4th ACS/IEEE Conf. Computer Systems and Applications*, Dubai, UAE, March 2006.

Performance Evaluation of a Chip-MultiThreading Server for High Performance Computing Applications

Myungho Lee[1], Yeonseung Ryu[1], Tae-Sun Chung[2], and Neungsoo Park[3,*]

[1] Department of Computer Software, MyongJi University,
Yong-In, Gyung Gi Do, Korea
{myungho1, ysryu}@mju.ac.kr
[2] Department of Information & Computer Engineering, Ajou University
Su-Won, Gyung Gi Do, Korea
tschung@ajou.ac.kr
[3] Department of Computer Science & Engineering, Konkuk University
Seoul, Korea
neungsoo@konkuk.ac.kr

Abstract. Shared Memory Multiprocessor (SMP) systems based on processors with Chip-level MultiThreading (CMT) technology are becoming mainstream servers in High Performance Computing (HPC) applications and commercial business applications as well. With multiple threads executing on a processor chip at the same time, CMT servers promise to deliver higher aggregate performance than servers without CMT technology. However, resource sharing among the threads executing on the same processor chip can cause conflicts and hurt the performance. Thus in order to obtain high performance and scalability on CMT servers, it is crucial to understand the performance impact that the CMT processors have on the target applications. In this paper, we evaluate the performance of an example high-end CMT server, Sun Fire E25K, using HPC applications parallelized with OpenMP standard, SPEC OMPL (standard OpenMP benchmark suite). We also study the performance impact of the resource conflicts on the CMT processor for each benchmark program.

Keywords: Chip-MultiThreading, SMP, High Performance Computing, OpenMP, Scalability.

1 Introduction

Recently, microprocessor designers have been considering many design choices to effectively utilize the ever increasing effective silicon area with the increase of transistor density. Instead of employing a complicated processor pipeline with an emphasis on improving a single thread's performance, incorporating multiple processor cores on a single chip (or Chip Multi-Processor) has become a main stream

---

* Corresponding author.

design trend. As a Chip Multi-Processor (CMP), it can execute multiple software threads on a single chip at the same time. Thus it supports Chip-level MultiThreading (CMT) and provides a larger capacity of computations performed per chip for a given time interval (or throughput). Examples are Dual-Core Intel Xeon [3], UltraSPARC IV, IV+, T1 microprocessors from Sun Microsystems [11, 13]. Intel Pentium IV with Hyperthreading technology [4] is another form of a CMT processor. Although it is not a CMP, it has support for two software threads executing on a single processor chip. Other similar products from vendors such as IBM [5], AMD [1], and Fujitsu are also being developed or already in the market.

Shared-Memory Multiprocessor (SMP) servers based on CMT processors are already introduced in the market (e.g., Sun Fire E25K [11] from Sun Microsystems based on UltraSPARC IV processors) and will become more popular. They are rapidly being adopted in High Performance Computing (HPC) applications as well as commercial business applications. CMT servers promise to deliver higher throughput performance than the servers based on single core processors. However, resources on the CMT processors such as cache, cache/memory bus, etc., are shared among the cores on the same chip, which can cause thrashing and hurt performance. For example, the pressure on the memory bus can be significantly increased as all the cores on the same chip generate their memory traffic on the shared bus. Thus exploiting the full performance potential of a CMT server is a challenging task. In order to obtain optimal performance on the CMT server, it is important to understand the performance impact that the CMT processors have on the target applications, in particular the effects of the resource conflicts.

In this paper, we evaluate the performance of a high-end CMT server, Sun Fire E25K, using a suite of HPC applications, SPEC OMPL [9], written using OpenMP standard for SMP [7]. We use the Sun Studio 9 compiler suite [12] to generate fairly high optimized executables for SPEC OMPL programs. We use features of Solaris 10 Operating System friendly to HPC applications and run the optimized executables of the SPEC OMPL on E25K. When running the SPEC OMPL, we use many threads (143 and 72) to evaluate the overall performance and scalability of E25K. Furthermore, in order to evaluate the performance impact of resource conflicts on the CMT, we run the programs using 72 threads with and without the CMT feature (using both cores of 36 processors vs. using only one core of 72 processors). The experimental results show a decent overall scalability of 1.66 from 72 threads to 143 threads. Also, the results show the performance impact of the resource conflicts for each test program.

The rest of the paper is organized as follows: Section 2 describes the architecture of a generic CMT processor and an example CMT server, Sun Fire E25K. Section 3 describes the OpenMP programming model and our test benchmark suite, SPEC OMPL. Section 4 shows our methodologies to generate optimized executables for SPEC OMPL and to exploit other compiler, OS features for obtaining high performance for SPEC OMPL. Section 5 shows the experimental results on E25K. Section 6 wraps up the paper with conclusions.

2 CMT Server

We first describe the architecture of the state-of-the-art CMT processor. We then describe an example high-end CMT server, Sun Fire E25K, which we used for our performance experiments in this paper.

The current main design for CMT processors is based on CMP's such as Dual-Core Intel Xeon [3], Sun Microsystems' UltraSPARC IV, IV+, among others. Some CMT processors even incorporate Simultaneous MultiThreading (SMT) [14] or similar technologies on a core. Examples are IBM Power5 [5] and UltraSPARC T1 microprocessor [13] from Sun. Figure 1 shows the architecture of a generic CMT processor [2]. On each processor chip, there are N-processor cores, with each core having its own cache on chip. The N-cores share a larger cache on or off the processor chip. Each core also has M hardware threads performing SMT or similar features. For example, the UltraSPARC T1 from Sun includes 8 cores, with each core supporting 4 hardware threads. In total, 32 (= 8 x 4) threads can execute on a chip at the same time. Each core has 8KB private data cache. The level-2 unified cache is 3MB in size.

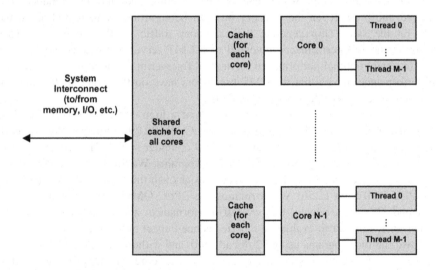

Fig. 1. Architecture of a generic CMT processor

The Sun Fire E25K server is the first generation throughput computing server from Sun Microsystems which aims to dramatically increase the application throughput via CMT technology. The server is based on the UltraSPARC IV processor and can scale up to 72 processors executing 144 threads (two threads per processor) simultaneously. The system offers up to twice the compute power of the UltraSPARC III Cu (predecessor to UltraSPARC IV processor) based high-end systems.

The UltraSPARC IV is a dual-threaded CMT (see Figure 2). It contains two enhanced UltraSPARC III Cu cores (or Thread Execution Engines: TEE's), a memory controller, and the necessary cache tag for 8 MB of external L2 cache per core. The off-chip L2 cache is 16 MB in size (8 MB per core). The two cores share the Fireplane System Interconnect, as well as the L2 cache bus.

The basic computational component of the Sun Fire E25K server is the UniBoard [11]. Each UniBoard consists of up to four UltraSPARC IV processors, their L2 caches, and associated main memory. Sun Fire E25K can contain up to 18 UniBoards, thus at maximum 72 UltraSPARC IV processors. In order to maintain cache coherency system wide, the snoopy cache coherency protocol is used within the UniBoard and directory-based cache coherency protocol is used among different UniBoards. The memory latency, measured using lat_mem_rd ( ) routine of lmbench, to the memory within the same UniBoard is 240nsec and 455nsec to the memory in different Uniboard (or remote memory).

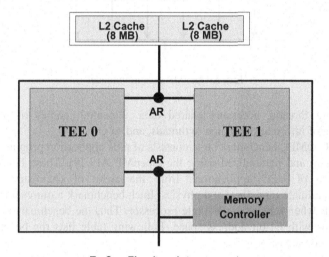

To Sun Fireplane Interconnect

Fig. 2. UltraSPARC IV processor

3 SPEC OMPL Benchmarks

We first explain the underlying execution model for OpenMP. We, then, introduce the SPEC OMPL benchmark suite used for evaluating the performance of an example CMT server Sun Fire E25K.

The underlying execution model for OpenMP is fork-join (see Figure 3) [7]. A *master* thread executes sequentially until a parallel region of code is encountered. At that point, the *master* thread forks a team of *worker* threads. All threads participate in executing the parallel region concurrently. At the end of the parallel region (the *join* point), the team of worker threads and the master synchronize. After then the *master* thread alone continues sequential execution. OpenMP parallelization incurs an overhead cost that does not exist in sequential programs: cost of creating threads,

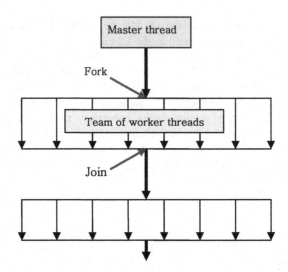

Fig. 3. OpenMP execution model

synchronizing threads, accessing shared data, allocating copies of private data, bookkeeping of information related to threads, and so on.

The SPEC OMPL benchmark suite consists of nine application programs written in C and Fortran, and parallelized using the OpenMP API [9]. These benchmarks are representative of HPC applications from the areas of chemistry, mechanical engineering, climate modeling, and physics. Each benchmark requires a memory size up to 6.4 GB when running on a single processor. Thus the benchmarks target large-scale systems with 64-bit address space. Following table lists the benchmarks and their application areas.

Table 1. SPEC OMPL Benchmarks

Benchmark Programs	Application Areas	Programming Languages
311.wupwise_l	Quantum chromodynamics	Fortran
313.swim_l	Shallow water modeling	Fortran
315.mgrid_l	Multi-grid solver	Fortran
317.applu_l	Partial differential equations	Fortran
321.equake_l	Earthquake modeling	C
325.apsi_l	Air pollutants	Fortran
327.gafort_l	Genetic algorithm	Fortran
329.fma3d_l	Crash simulation	Fortran
331.art_l	Neural network simulation	C

4 Preparing Executables and System Environments

Using Sun Studio 9 compiler suite [12], we've generated executables for the benchmarks in SPEC OMPL suite. By using a common set of compiler options provided by the Sun Studio 9, very high level of compiler optimizations are applied to the benchmarks. The common set of compiler flags are –fast –openmp –xipo=2 – autopar –xprofile –xarch=v9a. These options provide many common and also advanced optimizations such as scalar optimizations, loop transformations, data prefetching, memory hierarchy optimizations, interprocedural optimizations, profile feedback optimizations, among others. The –openmp option processes openmp directives and generate parallel code for execution on multiprocessors. The –autopar option provides automatic parallelization by the compiler beyond user-specified parallelization. This can further improve the performance.

The Solaris 10 Operating System provides features which help improve performance of HPC applications. They are Memory Placement Optimization (MPO) and Multiple Page Size Support (MPSS). For programs with intensive data accesses to localized regions of memory, efficiently utilizing MPO feature in Solaris 10 can significantly improve the performance. With the default MPO policy called first-touch, memory accesses can be kept on the local board most of the time, whereas, without MPO, those accesses would be distributed all over the boards (both local and remote) which can become very expensive. For programs which use a large amount of memory, using large size pages (supported by MPSS feature) can significantly reduce the number of TLB entries needed for the program and the number of TLB misses, thus significantly improve the performance [8]. We are enabling both MPO and MPSS for our runs of SPEC OMPL executables.

OpenMP threads can be bound to processors using the environment variable SUNW_MP_PROCBIND which is supported by thread library in Solaris 10. Processor binding, when used along with the static scheduling, benefits applications that exhibit a certain data reuse pattern where data accessed by a thread in a parallel region will either be in the local cache from a previous invocation of a parallel region, or in local memory due to the OS's first-touch memory allocation policy.

5 Performance Results

Using the compiler flags in Section 4, we've generated fairly high optimized executables for SPEC OMPL. We also enabled MPO and MPSS features in Solaris 10. We then run the SPEC OMPL programs on Sun Fire E25K while using the processor binding. In the following subsections, we present the performance results.

5.1 Overall Performance and Scalability

While running SPEC OMPL suite on Sun Fire E25K, we've used both 72 threads and 143 threads. In both runs, both cores of the UltraSPARC IV processors (clock rate = 1050 Mhz) were used. Thus 36 UltraSPARC IV processors were used in the 72 threads run and 72 processors were used for the 143 threads run. One core is left idle to take care of system processes or daemons.

The performance of SPEC OMPL is computed as follows [9]:

- For each benchmark, the run time (wall clock time) is measured. Then we divide the run time with the reference run time given for each benchmark. The resulting number is multiplied with 16,000. This is the score for each benchmark.
- After computing scores for the all 9 benchmarks, we then compute their geometric mean. This mean is the overall performance of the benchmark suite.

Table 2. Performance Comparisons for SPEC OMPL—143 threads vs. 72 threads

Benchmark Programs	72 threads	143 threads	Scalability:143/72
311.wupwise_l	282508	504121	1.78
313.swim_l	151927	287973	1.9
315.mgrid_l	178496	302214	1.69
317.applu_l	156009	278397	1.78
321.equake_l	96142	120950	1.26
325.apsi_l	106463	172362	1.62
327.gafort_l	147356	222660	1.51
329.fma3d_l	155068	288431	1.86
331.art_l	1001238	1653466	1.65
Geometric Mean	186995	310772	1.66

On 143 threads run, we've obtained an overall performance of 310,772. Then we've conducted 72 threads run which results in the overall performance of 186,995. The scalability is computed as 1.66 (310772/186995) when the number of threads is increased from 72 to 143. Table 2 summarizes the performance results and scalability. Four of the benchmarks show scalabilities greater than 1.78 (311.wupwise, 313.swim_l, 317.applu_l, 329.fma3d_l). In the case of 313.swim_l which consumes a lot of memory bandwidth, the high peak aggregate memory bandwidth available on E25K helps obtain the high scalability. The L2 cache miss rate is high for 329.fma3d_l in 72 threads run. When the number of threads gets doubled from 72 to 143, the amount of data loaded onto L2 cache gets diminished significantly. This results in a significant reduction in L2 cache misses, due to effectively reduced working set sizes.

The scalabilities for 321.equake_l and 327.gafort_l are noticeably low: 1.26, 1.51. In the case of 321.equake_l, the synchronization overhead increased significantly as the number of threads increased from 72 to 143. 327.gafort_l has two hot loops with critical sections inside. The critical section loops take a long time to execute, while performing intensive load and store operations from/to the memory. As they are sequential portions, they act as the "serial bottleneck" in Amdahl's law and leads to low scalability. Figure 4 shows the scalability bar graph for each benchmark for better illustrations of the results.

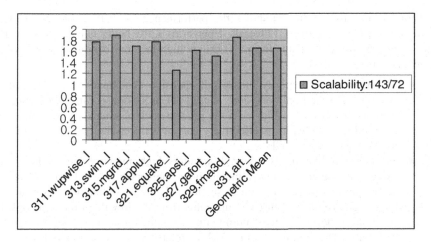

Fig. 4. Scalability bar graph per each benchmark program

5.2 Performance Impact of Resource Conflicts on CMT

In order to measure the performance impact of resource conflicts on CMT (level-2 cache bus, memory bus) on SPEC OMPL, we've measured the performance of 72 threads runs in two ways:

1. Using 36 UltraSPARC IV processors (experiments conducted in section 5.1), thus using both cores of the processor
2. Using 72 UltraSPARC IV processors, thus using only one core per processor. There are no resource conflicts between the two cores on the same processor chip.

Table 3. 72 threads case—36x2 vs. 72x1

Benchmark Programs	36 chip X 2 cores	72 chip x 1 core	1 core vs. 2 cores
311.wupwise_l	282508	299246	1.06
313.swim_l	151927	211008	1.39
315.mgrid_l	178496	213833	1.20
317.applu_l	156009	160670	1.03
321.equake_l	96142	98853	1.03
325.apsi_l	106463	130035	1.22
327.gafort_l	147356	187597	1.27
329.fma3d_l	155068	155060	1.00
331.art_l	1001238	1293218	1.29
Geometric Mean	186995	216592	1.16

Table 3 (and also Figure 5) shows the performance for both 1 and 2, and also shows the speedup of 2 over 1. Overall, 2 performs 1.16x better than 1. Benchmarks with greater performance gains from 2 consume high memory bandwidths and/or use large amounts of memory:

- 313.swim_l is a memory bandwidth-intensive benchmark as mentioned in section 5.1. When only one core is used per processor, it can fully utilize the memory bandwidth available on the processor chip, whereas when two cores are used the bandwidth is effectively halved between the two cores. This led to 1.39x gain of 2 over 1.
- 315.mgrid_l, like in 313.swim_l, requires high memory bandwidth. Thus using only one core can have much higher memory bandwidth which leads to 1.20x gain. However, as seen in section 5.1, the low scalability from the 72 threads run to the 143 threads run stems from the low iteration counts of the two hottest loops. The maximum iteration counts for the two hottest loops are 512 only. The iteration count 512 is relatively small compared with the number of threads used (143, 72). Thus the portion for parallelization overheads such as synchronization also increases as the number of threads is increased. This prevented the benchmark from achieving high scalability in Section 5.1.

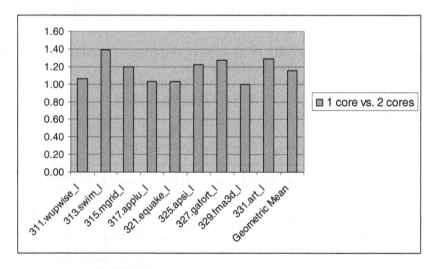

Fig. 5. Performance comparisons of 1 core vs. 2-cores

- 325.apsi_l and 331.art_l allocate large amount of memory per thread at run-time. Thus, instead of allowing 8 threads to allocate large memory on the same UniBoard's memory, allowing only 4 threads can have significant performance benefit.
- 327.gafort_l suffers from intensive memory loads and stores generated from the critical section loops. Allocating 8 threads on two different UniBoards reduces the pressure on the memory bandwidth significantly compared with allocating 8 threads on the same UniBoard.

Benchmarks other than the above (311.wupwise_l, 317.applu_l, 321.equake_l, 329.fma3d_l) relatively give less pressure on the memory bandwidth and/or consume smaller amount of memory. Thus both 1 and 2 result in the same level of performance. These benchmarks are not heavily affected by the resource conflicts and more suitable for execution on CMT servers.

6 Conclusions

In this paper, we first described the architecture of a generic CMT server and an example CMT server, Sun Fire E25K, in detail. Then we introduced the OpenMP execution model along with the SPEC OMPL benchmark suite used to evaluate the example CMT server Sun Fire E25K. We showed our methodology to generate highly optimized executables for SPEC OMPL benchmarks using the options of Sun Studio 9 compiler. We also described the features in Solaris 10 OS (MPO, MPSS) which help improve HPC application performance. Using the executables for SPEC OMPL generated by the Sun Studio 9 compilers, and also the MPO and MPSS features in Solaris 10, we've obtained high scalability on E25K. We've also measured the performance impact of the resource conflicts on CMT processor for SPEC OMPL using either one core or both cores of a CMT processor. It turned out that the benchmarks which require large memory bandwidths and/or consume large amounts of memory suffer when both cores are used due to the saturations on the memory bus and the main memory.

Acknowledgements

The authors would like extend their thanks to the Center for Computing and Communications of the RWTH Aachen University for allowing the access of Sun Fire E25K server. This research was partially supported by the MIC (Ministry of Information and Communication), Korea, under the ITRC (Information Technology Research Center) support program supervised by the IITA(Institute of Information Technology Assessment).

References

1. *AMD Multi-Core: Introducing x86 Multi-Core Technology & Dual-Core Processors*, http://multicore.amd.com/2005
2. Shailender Chaudhry, Paul Caprioli, Sherman Yip, and Marc Tremblay, *High-Performance Throughput Computing*, IEEE Micro, May-June, 2005.
3. Intel Dual-Core Server Processor, http://www.intel.com/business/bss/products/server/dual-core.htm
4. Intel Hyperthreading Technology, http://www.intel.com/technology/hyperthread/index.htm
5. R. Kalla, B. Sinharoy, and J. Tendler, *IBM POWER5 chip: a dual core multithreaded processor*, IEEE Micro, March-April 2004.
6. K. Olukotun et. al., *The Case for a single Chip-Multiprocessor*, International Conference on Architectural Support for Programming Languages and Operating Systems, 1996.
7. *OpenMP Architecture Review Board*, http://www.openmp.org

8. *Solaris 10 Operating System*, http://www.sun.com/software/solaris
9. *The SPEC OMP benchmark suite*, http://www.spec.org/omp
10. L. Spracklen and S. Abraham, *Chip MultiThreading: Opportunities and Challenges*, 11th International Symposium on High-Performance Computer Architecture (HPCA-11), pp 248-252, 2005.
11. *Sun Fire E25K server*, http://www.sun.com/servers/highend/sunfire_e25k/index.xml
12. *Sun Studio 9 Software*, http://www.sun.com/software/products/studio/index.html
13. *Sun UltraSPARC T1 microprocessor*, http://www.sun.com/processors/UltraSPARC-T1
14. D. Tullsen, S. Eggers, and H. Levy, *Simultaneous MultiThreading: Maximizing On-Chip Parallelism*, International Symposium on Computer Architecture, 1995.

A Study on the Locality Behavior of Minimum Spanning Tree Algorithms

Guojing Cong and Simone Sbaraglia

IBM T.J. Watson Research Center
Yorktown Heights, NY, 10598
{gcong, ssbarag}@us.ibm.com

Abstract. Locality behavior study is crucial for achieving good performance for irregular problems. Graph algorithms with large, sparse inputs, for example, oftentimes achieve only a tiny fraction of the potential peak performance on current architectures. Compared with most numerical algorithms graph algorithms lay higher pressure on the memory system. In this paper, using the minimum spanning tree problem as an example, we study the locality behavior of graph algorithms, both sequential and parallel, for arbitrary, sparse instances. We show that the inherent locality of graph algorithms may not be favored by the current architecture, and parallel graph algorithms tend to have significantly poorer locality behaviors than their sequential counterparts. As memory hierarchy gets deeper and processors start to contain multi-cores, our study suggests that architectural support and new parallel algorithm designs are necessary for achieving good performance for irregular graph problems.

Keywords: memory locality, graph algorithm, minimum spanning tree.

1 Introduction

Graph abstractions are used in many science and engineering problems, for example, data mining, determining gene function, clustering in semantic webs, and security applications. Graph problems with large arbitrary, sparse instances are challenging to solve on current architectures (e.g., see [3, 4]). For dense linear algebra packages near peak performances are repeatedly reported. Yet we have not seen similar performance results for graph problems. Graph algorithms tend to lay higher pressure on the memory system. For architectures with deep memory hierarchy, locality features are crucial to the performance of the algorithms. In this paper, using the minimum spanning tree (MST) problem as an example, we study the locality behaviors of graph algorithms and compare their performances with different cache configurations.

The MST problem finds a spanning tree of a connected graph G with the minimum sum of edge weights. MST is one of the most studied combinatorial problems with practical applications in VLSI layout, wireless communication, distributed networks [15, 24, 26], and recent problems in biology and medicine [5, 13], and national security [7]. MST is also often a key step in other graph problems [16, 17, 23, 25].

Moret and Shapiro give an empirical analysis of MST algorithms in [18]. Implementations of Prim's, Kruskal's and Cheriton-Tarjan's algorithms on several architectures

Y. Robert et al. (Eds.): HiPC 2006, LNCS 4297, pp. 583–594, 2006.

are compared. Through extensive comparisons, Prim's algorithm is found to be the best candidate. Computer architectures have since evolved, and Prim's algorithm may no longer be the fastest on current platforms. Moreover, running times alone are generally not sufficient to estimate the relative performance of algorithms on new architectural configurations. As memory hierarchy gets deeper, cache performance becomes crucial to an application. Whether the locality behavior of an algorithm fits well with the cache configuration affects the overall performance. Understanding the locality behavior helps the adaptation of algorithms to target platforms and dynamic configurations (e.g., shutdown of a cache bank to reduce power consumption).

In this paper we study the locality behavior of three MST algorithms, that is, Prim's, Borůvka's, and Kruskal's, and show how cache configurations (e.g., cache size and line size) affect their performance. We include Borůvka's algorithm as it can be easily parallelized to run in poly-log time under the PRAM model. As processors increasingly adopt multi-core designs, solving a problem in parallel is important for performance. The locality behavior of a parallel graph algorithm can be very different from the sequential counterpart as the designs are drastically different. Comparison of their locality behavior brings insight to efficient parallel algorithm design and better architectural support.

Cache-friendly algorithms, for example, external memory algorithms and cache-oblivious algorithms, abound in the literature. These algorithms assume some memory hierarchy models, and minimize the number of block transfers between hierarchy levels. Common design techniques include divide-and-conquer and sequential scan, for which the I/O complexity (number of blocks transfered) is relatively easy to analyze. For other algorithms that do not employ these techniques, however, it is hard to analyze for I/O complexity under these hierarchy models. Also the locality behavior of an algorithm is an inherent property that should not depend on the memory hierarchy of a target platform (while its performance certainly depends on how well the locality behavior fits with the cache configuration). In our study we do not analyze the MST algorithms under these existing memory models. Instead we characterize locality through *Least-Recently-Used* (LRU) stack distance analysis that is discussed in Section 2.

2   Characterizing Locality Behavior

LRU stack distance was first used in the "stack processing" technique proposed by Mattson *et al.* for evaluating cost-performance of storage hierarchies [14]. LRU stack distance is also referred to as *reuse distance*, and the two names are used interchangeably in the literature. Locality of a program can be studied by computing the LRU stack distance histogram (e.g., see [21]).

Consider a trace of k memory accesses, $T = T_1, T_2, \ldots, T_k$, that access a set of c addresses. For a storage system with *Least-Recently-Used* replacement policy, access T_i is a hit if the size of the fast memory is larger than the stack distance $\Delta(T_i)$. A histogram can be derived if we compute for each $\Delta \in [0 : c]$, the total number of accesses that have reuse distance Δ. The LRU stack distance histogram has been used as a machine-independent metric of locality (e.g., see [6]). With LRU stack distance analysis it is possible to perform various optimizations on a program (e.g., see [21, 27]).

In our study we use the binary rewriting approach to get a memory access trace. We intercept each *load* and *store* instruction using SIGMA [10], and compute the reuse distance histogram on the fly to avoid dumping huge traces.

3   Comparison of Three MST Algorithms

In this section we compare the locality behavior of the three MST algorithms, that is, Prim's, Kruskal's and Borůvka's. For each algorithm the exact process of constructing an MST is influenced by the topology, edge density, and weight distribution of the input graph. We focus on sparse random graphs with randomly-assigned edge weights. We choose random graphs because they are the most challenging to solve on parallel computers. As memory access pattern is highly dependent on the inputs, the study of arbitrary graphs can expose regular memory patterns of the algorithms. A random graph of n vertices and m edges is generated by randomly picking a pair of vertices and connecting them with an edge until m edges are generated.

For Prim's algorithm (denoted as Prim), we use the implicit binary heap described [9]. For Kruskal's algorithm (denoted as Kruskal), we use non-recursive merge sort as the sorting routine. The union-find data structure is used to maintain the disjoint sets of elements. Borůvka's algorithm (denoted as Borůvka) is composed of Borůvka iterations that have three steps: *find-min*, *connect-components*, and *compact-graph*. The algorithm iterates until only isolated vertices are left. All MST implementations run in $O(m \log n)$ time. Fig. 1 shows the LRU stack distance histograms for each algorithm with an input graph of $1K$ vertices and $4K$ edges (we use $1K$ to denote 1024).

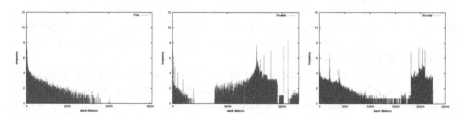

Fig. 1. Histograms of stack distances for three MST implementations

One common feature of the three plots in Fig. 1 is the blanks in the histogram. The ratio of the number of observed distinct stack distances over the memory footprint size is 40% for Prim, 54% for Kruskal, and 73% for Boruvka. In each plot, the minimum reuse distance is 0, and the maximum is c, the size of the footprint. Large concentrations of distribution are observed around certain distances. For example, there are concentrations around small reuse distances in all plots. Each plot has a different shape. For Prim the histogram monotonously decreases with the reuse distance. For Borůvka and Kruskal, there are concentrations of distribution around large reuse distances.

The plot on the left of Fig. 2 presents a different view of the same histogram data. The x axis is the stack distance. The y value shows the percentage of accesses with stack distance no bigger than x. Alternatively, y can also be viewed as a cache hit ratio

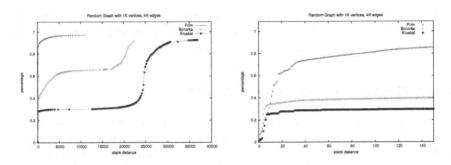

Fig. 2. The ratio plots for three implementation of MST algorithms with an input of $1K$ vertices, $4K$ edges

for a fully associative cache of size x with the LRU replacement policy. In the rest of the paper we refer to such plots as ratio plots. For each of the three algorithms, the shape of the line in the ratio plots is different. Prim and Kruskal achieve fairly good cache hit ratios with small cache sizes. The curve of Borůvka remains flat at low ratios for a range of reuse distances, and jump abruptly at relatively large distances. The plot on the right of Fig. 2 is a zoomed-in view for reuse distances in the range of [0:150]. Prim achieves a hit ratio of over 80% at a cache size of only 120 words.

3.1   Locality of Prim

Starting from a single vertex, Prim grows an MST one edge at a time. Prim maintains a heap to retrieve the lightest-weight edge. During the execution, for sparse inputs, most memory accesses occur around accessing the heap data structure. Each heap operation incurs $O(\log h)$ memory accesses, where $h = O(n)$ is the size of the heap. We focus our analysis on the heap operations.

In our experiments, for all graphs of different sizes, ranging from $1K$ vertices to $100K$ vertices, a hit ratio of more than 70% is achieved with fewer than 20 words. In fact hit ratios of more than 80% are achieved with around 120 words (integers), for all input sizes. Within the reuse distance range of $[0, 50]$, the curves are nearly identical and the hit ratio appear to be independent of the input size.

The magic numbers observed (i.e., 70% and 50 words) are dependent not only the topology, edge density, and weight distribution of the input, but also the actual programming of the algorithm. Instead of modeling the tree construction process and giving rough bounds, we show that a significant percentage of memory accesses incur short reuse distances.

Recall that a reuse distance is associated with each memory access. We now consider the reuse distance for the memory accesses incurred by *Extract_Min*, *Insert* and *Decrease_Key*. *Extract_Min* removes the top element of the heap, and places the last element as the new top. It then iteratively inspects a node and its two children starting from the top. If the parent has larger weight, it then is swapped with one of the children. During each iteration there are three reads (reading the weights) and two writes (swapping). The parent and one of the children are accessed twice, and the second access has a distance of $O(1)$. More exactly, the distance will be 1 or 2 depending on whether the left or right child gets swapped. Here we do not consider the interference of other auxiliary data structures, for example, a temporary location to facilitate the swap. So at least $\frac{2}{5}$ of the accesses generated by *Extract_Min* are within a constant distance. *Insert* appends an element to the end of the heap and then compares iteratively from the end whether an element is larger than its parent. If the parent is larger, it then gets sifted down. Successive sifting incurs constant reuse distance, and $\frac{1}{2}$ of the accesses have distance $O(1)$. *Decrease_Key* works similarly as *Extract_Min*, and about $\frac{1}{2}$ of the accesses have distance $O(1)$. Note that although the distances are constant, in practice they can take a range of values due to book-keeping activities. For example, to enable *Decrease_Key*, the positions of each vertex in the heap are recorded in an array. Updating the positions increases reuse distances (still $O(1)$ though) for heap accesses in *Decrease_Key*. According to our analysis, an estimate of 40 to 50 percent of accesses have constant reuse distances (disregarding book-keeping activities).

In addition to constant reuse distances, some operation incur $O(\log n)$ distance. For example, to maintain the size of the heap, after each *Extract_Min* or *Insert*, a counter is either incremented or decremented. Access to the counter generates reuse distance of $O(\log n)$ as *Extract_Min* and *Insert* incur $O(\log n)$ accesses to different memory locations. The top of the heap is accessed every time in *Extract_Min*, and the largest reuse distance incurred by *Extract_Min* is $O(d \log n)$, where d is the largest degree of all vertices. The distribution of reuse distances for the rest of memory operations is governed by the random process of constructing an MST. It is easy to construct scenarios that incur large (e.g., $O(n)$) reuse distances.

As the ratio plots show good locality of the simple binary heap, it is then interesting to compare with other more sophisticated implementations of heaps. Heaps (and priority queues) have been studied extensively, and quite a few data structures are proposed, for example, Fibonacci heap, pairing heap, and splay trees. Sanders presents a data structure called *sequence heap* [20] and shows that for a cache configuration with size M and block size B, I insertions and I deletions can be performed with $I(2R/B + O(1/k + (\log k)/m))$ I/Os and $I(\log I + \log R + \log m + O(1))$ comparisons, where $m = \Theta(M)$, $k = \Theta(M/B)$, $R = \lceil \log_k \frac{I}{m} \rceil$. The motivation of *sequence heap* is based on the fact that merging k sequences is I/O efficient under the external memory model. Arge *et al.* designed cache-oblivious priority queues based on similar observations [2]. In his study Sanders has four heap implementations, denoted as *hslow* (implicit binary heap), *h2* (binary heap with the "bounce" heuristic [12]), *h4*(4-ary heap), and *knh* (the *sequence heap*), respectively.

In Fig. 3 are the ratio plots for the four different heap implementations. Surprisingly, the textbook binary heap (*hslow*) has the best locality behavior in terms of reuse distances. At each distance, the ratio for *hslow* is consistently higher than the ratios for other implementations. In practice, however, *hslow* is found to be the slowest for most inputs on current architectures, for example, SUN SPARC V9 and IBM Power 4. In fact, *knh* are four times faster than *hslow* for many inputs. Although *hslow* tends to make more memory accesses (about 1.5 times as many as *knh*), the difference does not fully explain the observed poor performance of *hslow*, especially considering its good locality behavior. The fastest implementation is *knh*. As it mostly works with sorted

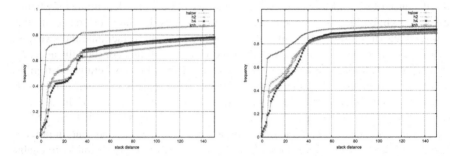

Fig. 3. The ratio plots for four heap implementations. The plot is for stack distance in range [0:150]. On the left are the plots for 1000 *Insert* followed by 1000 *Extract_Min*. On the right are the plots for 1000 *Insert*, *Extract_Min*, and *Insert* followed by 1000 *Extract_Min*, *Insert*, *Extract_Min*.

sequences, it exhibits good spatial locality. Current architectures that typically have long cache lines and long latency to main memory impose the requirement of spatial locality for good performance. Unfortunately, spatial locality is scarce in *hslow*.

All heap operations start with a certain node v, and inspect v's parent and/or children. Due to the layout of the implicit binary heap in memory, whenever a block is brought into the cache, except for node v that is currently being accessed, it is unlikely that the rest of the block contains v's parent or children unless v is near the top of the heap. In this case, long cache line causes fetching data that is not used in the near future and wastes memory bandwidth.

There is no machine-independent metric in the literature to measure the spatial locality of a program. Recently Snir and Yu studied the theoretical aspects of temporal and spatial locality [22]. While they acknowledge that LRU stack distance analysis captures well temporal locality, they also point out that in terms of predicting cache miss bandwidth, temporal locality and spatial locality can not be studied in isolation. We present further experimental results in Section 4.

3.2   Locality of Kruskal

For Kruskal, sorting dominates the execution time, and dictates the shape of the plot. For the implementation with merge sort, the hit ratio remains low until the distance and hence cache size becomes very large. In fact only at a size that can hold all the data structures used for sorting does the hit ratio reach above 90%. Fig. 4 shows the ratio plots for Kruskal with three different inputs. The vertical line in each plot is $\Delta = 6m$. Recall that non-recursive merge sort employs an auxiliary buffer. For an input with m edges, as each edge in the data structure has three elements (two vertices and the weight), the size of the total memory usage is $2m * 3 = 6m$ words. The plots show that a cache has to be of size at least $6m$ words in order to have reasonably good hit ratios. Otherwise the hit ratio is as low as 30%, even for cache size $6m - 1$. Unfortunately, $6m$ is in direct proportion to the input size, and the algorithm exhibits poor temporal locality behavior.

In practice, for many inputs, Kruskal with merge sort is the fastest among all implementations. As long as the the data structure fits in main memory, our implementation with merge sort beats the version with quick sort for large inputs on all tested platforms. This is largely due to the fact that merge sort has very good spatial locality that are especially advantageous for long cache lines. For n (assuming $n = 2^k$, $k \in N$) elements, merge sort takes k iterations. In iteration $1 \leq i < k$, $\frac{n}{2^i}$ pairs of consecutive sequences (each of length 2^i) are merged. Whenever a block is brought into cache, it contains data that is soon to be used. We further presents experimental results in Section. 4.

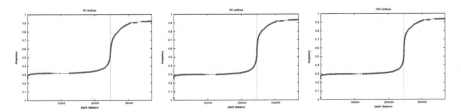

Fig. 4. The ratio plots for Kruskal with three inputs. The input sizes from left to right are $1K$, $5K$, and $10K$ vertices. $m = 4n$.

3.3 Locality of Borůvka

With Borůvka, the surges in the ratio plots are at distances in direct proportion to the input size, as shown in Fig. 5. In Fig. 5 we show the ratio plots for three different input sizes, that is, random graphs with $1K$, $5K$, and $10K$ vertices, and $m = 4n$ edges. The vertical line in each plot is $\Delta = n * 3 + m * 4$. That is exactly the size of the input. With our adjacency list representation, for each vertex there are three data fields. Each data field takes a word of memory. For each edge incident to vertex v, there are two elements: u, the other vertex, and w, the edge weight. Each edge appears twice in the adjacency list. The size of the input is thus $3n + 4m$.

Algorithms with such reuse behavior as shown in Fig. 5 generally scans through the data structures repeatedly for multiple runs, and each run can be considered as an algorithmic phase that may have similar or different characteristics. These algorithms generally lend themselves to parallelization as in the case of Borůvka's algorithm. In fact the Borůvka iteration (*find-min*, *connected-components* and *compact-graph*) is employed in other parallel MST algorithms (e.g., see [8, 11]).

As shown in Fig. 5, even with a fully associative cache, the cache needs to be at least of the size of the input in order to have good hit ratios. Otherwise, the hit ratio is well below 90%. The vertical line on the left of each plot ($\Delta = 4n$) crosses the curve at about *frequency*=60%. The line corresponds nicely to the size of the four auxiliary data structures used in the algorithm, that is *Min*, *Min_ind*, *D*, *Alive*. Consistently, with a cache size of $4n$ words, the hit ratio is around 60%. In order not to contract the graph which is costly as it involves memory allocation and copying, we use the D and *Min* arrays for each vertex (and supervertex) to record the component it belongs to and the smallest weight of the adjacent edges. *Min_ind* records the other vertex (or supervertex) that is incident to the edge with smallest weight. The *Alive* array shows whether a vertex should be considered in the Borůvka iteration. With a cache size of $4n$, most accesses to D in our implementation are cache hits. We refer interested readers to [4] for details of the algorithm.

There are two cache sizes for Borůvka that can achieve reasonable hit ratios. More specifically, one is $4n$ and the second is $3n + 4m$. The effectiveness of caching is highly dependent on the input size. In contrast to Prim and similar to Kruskal, Borůvka does not exhibit good temporal locality. What is worse is that Borůvka does not exhibit good spatial locality either, and most accesses to the arrays are irregular.

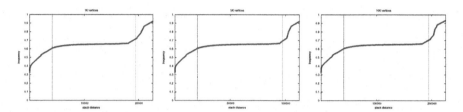

Fig. 5. Ratio plots for Borůvka with three different inputs

The reuse distance analysis of Borůvka suggests poor temporal locality for parallel graph algorithms (we mainly focus on PRAM algorithms since most interesting parallel graph algorithms are based on PRAM) due to the inherent phase behavior. In addition, the irregular nature of the input dictates poor spatial locality behavior.

3.4 Locality of Parallel MST Algorithms

In this section we consider the locality of parallel Borůvka's algorithm. The locality of parallel Borůvka's algorithm is representative for at least some stages in the more complex MST algorithms. In fact, the graft-and-shortcut approach used in parallel Borůvka's algorithm is also frequently used in other parallel graph algorithms, for this class of algorithms, we expect to see similar locality behavior.

The parallel implementations of *find-min* and *connect-components* are straightforward. We have two implementations of the *compact-graph* step. One of them contracts the graph using parallel sorting routines, while the other adopts a data structure called *flexible adjacency list* that avoids large scale sorting. We refer interested readers to [4] for details of the implementations.

Fig. 6 shows the ratio plots for two implementations of parallel Borůvka's algorithm. Again the input is a random graph with $1K$ vertices and $4K$ edges. We emulate the parallel algorithm with one thread. The locality behavior for each thread with multiple threads should be similar. The two ratio plots of Fig. 6 look roughly like the plots in Fig. 5, and show poor locality in terms of reuse distance. At a large distance about $130,000$ words, the hit ratio reaches above 80%. The range of the reuse distance is significantly larger than that of the sequential implementation. This is due to the fact in both implementations, after each iteration new instances of the graph (either fully or partially compacted) are generated and the next iteration works on the new instances. Fig. 6 partly explains why it is difficult to achieve good parallel speedup for sparse arbitrary instances on current parallel computers. The parallel algorithms have poorer locality than the sequential algorithms, and as far as we are aware of, there are no mature techniques for improving the locality behavior of parallel graph algorithms. As cache performance becomes even more crucial, the gap between theoretical results and actual performances can be increasing.

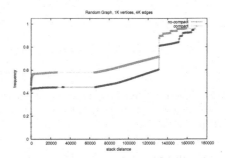

Fig. 6. The ratio plots for two implementations of parallel Borůvka's algorithm. The implementation labeled as *compact* compacts the graph using parallel sorting routines. The *no-compact* version uses the *flexible adjacency list* representation.

4   Simulation Results

We next present our experimental results with different cache configurations that support our analysis in the prior section. We run the algorithms on the RSIM simulator [19] that simulates modern processors and memory sub-system. Instead of giving pages of specifications for the processor, we use similar settings as in prior studies (e.g., see [1]). The important features include instruction-level parallelism, out-of-order scheduling, non-blocking reads and speculative execution. As we only run sequential algorithms, we do not use any of the multiprocessor features such as memory consistency protocols. In our study we use directly-mapped L1 cache and 2-way set associative L2 cache, and the input is a random graph with $1K$ vertices and $4K$ edges.

First we vary the cache line size, and measure the performance. As the cache line size increases, each transfer brings more data into cache, and the spatial locality of an algorithm becomes more important for performance.

In Fig. 7, the plot on the left shows how the performance varies with different cache line sizes. The size of the cache is kept constant ($1KB$ L1 and $4KB$ L2) in the experiments. The smallest cache line size that can be simulated is 16 bytes. With the increase of line size, the performance of both Prim and Borůvka decreases while that of Kruskal improves. The results support our analysis that both Prim and Borůvka do not have good spatial locality and is not favored by long cache lines.

The plot on the right of Fig. 7 shows the performance of the algorithms with different cache sizes, from $1KB$ L1 and $4KB$ L2 to $128KB$ L1 and $512KB$ L2. The performance, measured as instruction per cycle (IPC), improve as the cache size increases. The performance curves in this plot are correlated with the ratio plots in Fig. 2. Yet it is not straightforward to predict the performance with real cache configurations from

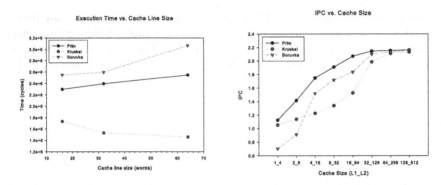

Fig. 7. Performance of MST algorithms with different cache configurations. For the plot on the left, we experiment with cache line sizes of 16 bytes, 32 bytes and 64 bytes. The plot on the right shows performance for different cache sizes.

the ratio plots. According to the ratio plot we would expect the performance curve for Prim's algorithm rise sharply at a small cache size and remain somewhat flat afterwards. This is obviously not true in the performance plot. The discrepancy is mostly due to the associativity of the cache and the cache line size. For Kruskal the IPC increases sharply at 32K bytes (L1 cache size), and the whole input (of size roughly 24K bytes) fits in L1. For Borůvka, there are two sharp increases with the performance curve. The increases correspond roughly to the sharp increases in the ratio plots.

5   Conclusion and Future Work

In this paper we studied the locality behavior of MST algorithms. As memory hierarchy deepens, locality is becoming even more important to the performance. We show that Prim with implicit binary heap has better temporal locality than the cache-aware implementations in our study. A significant percentage of the memory accesses incurred by the heap operation have $O(1)$ or $O(\log n)$ reuse distances. However, architectures with long cache lines impose the requirement of spatial locality for good performance, and penalize the performance of Prim with implicit binary heap. Kruskal (with non-recursive merge sort) exhibits poor temporal locality as many reuse distances are in the order of $O(n)$. Due to its good spatial locality, it runs fast on current architectures. Increasing cache line size in general improves its performance. Comparing Prim and Kruskal, it seems that good spatial locality fits better with current cache organizations.

Both the sequential and parallel implementations of Borůvka show poor temporal and spatial locality. In future work we will further investigate the locality behaviors of parallel graph algorithms. This is especially meaningful as many processors adopt multi-core designs. Our study of Borůvka's algorithm hints that poor locality might be inherent in the PRAM algorithms. In order to verify, we will need to find a metric for measuring spatial locality. On one hand, it is important to design parallel algorithms with reasonable locality behavior. On the other hand, special architectural support, for example, multi-threaded architecture, is necessary to tolerate the memory access latency for parallel algorithms. We will also investigate the impact of locality enhancing techniques such as vertex reordering on the performance of parallel algorithms. For Prim and Kruskal in our study, from the analysis of the algorithms, we do not expect to see too big a difference in the stack distance distribution. For Borůvka, however, there can be interesting findings, and we expect similar results with many other parallel algorithms.

References

1. S.V. Adve, V.S. P, and P. Ranganathan. Recent advances in memory consistency models for hardware shared-memory systems. In *proceedings of the IEEE, special issue on distributed shared-memory*, pages 445–455, 1999.
2. G. Aloupis, P. Bose, E.D. Demaine, S. Langerman, H. Meijer, M. Overmars, and G.T. Toussaint. Computing signed permutations of polygons. In *Proc. of the 14th Canadian Conf. on Computational Geometry (CCCG)*, pages 68–71, Lethbridge, Alberta, Canada, August 2002.
3. D. A. Bader and G. Cong. A fast, parallel spanning tree algorithm for symmetric multiprocessors (SMPs). In *Proceedings of the 18th International Parallel and Distributed Processing Symposium (IPDPS 2004)*, Santa Fe, New Mexico, Apr 2004.
4. D. A. Bader and G. Cong. Fast shared-memory algorithms for computing the minimum spanning forest of sparse graphs. In *Proc. 18th Int'l Parallel and Distributed Processing Symp. (IPDPS 2004)*, Santa Fe, New Mexico, 2004.
5. M. Brinkhuis, G.A. Meijer, P.J. van Diest, L.T. Schuurmans, and J.P. Baak. Minimum spanning tree analysis in advanced ovarian carcinoma. *Anal. Quant. Cytol. Histol.*, 19(3):194–201, 1997.
6. C. CaΒcaval and D.A. Padua. Estimating cache misses and locality using stack distances. In *Proceedings of the 17th annual international conference on Supercomputing*, pages 150–159, San Francisco, CA, 2003.
7. C. Chen and S. Morris. Visualizing evolving networks: Minimum spanning trees versus pathfinder networks. In *IEEE Symp. on Information Visualization*, Seattle, WA, October 2003. to appear.
8. R. Cole, P.N. Klein, and R.E. Tarjan. A linear-work parallel algorithm for finding minimum spanning trees. In *Proceedings of the 6th Annual ACM Symposium on Parallel Algorithms and Architectures*, pages 11–15, Cape May, NJ, 1994.
9. T. H. Cormen, C. E. Leiserson, and R. L. Rivest. *Introduction to Algorithms*. MIT Press, Inc., Cambridge, MA, 1990.
10. L. DeRose, K. Ekanadham, J.K. Hollingsworth, and S. Sbaraglia. Sigma: a simulator infrastructure to guide memory analysis. In *Proceedings of the 2002 ACM/IEEE conference on Supercomputing*, pages 1–13, 2002.
11. D.R. Karger, P.N. Klein, and R.E. Tarjan. A randomized linear-time algorithm to find minimum spanning trees. *J. ACM*, 42(2):321–328, 1995.
12. D.E. Knuth. *The Art of Computer Programming: Sorting and Searching*, volume 3. Addison-Wesley Publishing Company, Reading, MA, 1973.
13. M. Matos, B.N. Raby, J.M. Zahm, M. Polette, P. Birembaut, and N. Bonnet. Cell migration and proliferation are not discriminatory factors in the in vitro sociologic behavior of bronchial epithelial cell lines. *Cell Motility and the Cytoskeleton*, 53(1):53–65, 2002.
14. R.L. Mattson, J. Gecsei, D.R. Slutz, and I.L. Traiger. Evaluation techniques for storage hierarchies. *IBM Systems Journal*, 9:78–117, 1970.
15. S. Meguerdichian, F. Koushanfar, M. Potkonjak, and M. Srivastava. Coverage problems in wireless ad-hoc sensor networks. In *Proc. INFOCOM '01*, pages 1380–1387, Anchorage, AK, April 2001. IEEE Press.
16. G. L. Miller and V. Ramachandran. Efficient parallel ear decomposition with applications. Manuscript, UC Berkeley, MSRI, January 1986.

17. Y. Moan, B. Schieber, and U. Vishkin. Parallel ear decomposition search (EDS) and st-numbering in graphs. *Theoretical Computer Science*, 47(3):277–296, 1986.
18. B.M.E. Moret and H.D. Shapiro. An empirical assessment of algorithms for constructing a minimal spanning tree. In *DIMACS Monographs in Discrete Mathematics and Theoretical Computer Science: Computational Support for Discrete Mathematics 15*, pages 99–117. American Mathematical Society, 1994.
19. V.S. Pai, P. Ranganathan, and S.V. Adve. RSIM: an execution-driven simulator for ILP-based shared-memory multiprocessors and uniprocessor. In *Proceedings of the 3rd workshop on computer architecture education*, 1997.
20. P. Sanders. Fast priority queues for cached memory. *ACM J. Experimental Algorithmics*, 5(7), 2000. www.jea.acm.org/2000/SandersPriority/.
21. X. Shen, Y. Zhong, and C. Ding. Locality phase prediction. In *Proceedings of the 11th international conference on Architectural support for programming languages and operating systems*, pages 165–176, Bostaon, MA, 2004.
22. M. Snir and J. Yu. On the theory of spatial and temporal locality. Technical Report UIUCDCS-R-2005-2611, University of Illinois at Urbana-Champaign, 2005.
23. R.E. Tarjan and U. Vishkin. An efficient parallel biconnectivity algorithm. *SIAM J. Computing*, 14(4):862–874, 1985.
24. Y.-C. Tseng, T.T.-Y. Juang, and M.-C. Du. Building a multicasting tree in a high-speed network. *IEEE Concurrency*, 6(4):57–67, 1998.
25. U. Vishkin. On efficient parallel strong orientation. *Information Processing Letters*, 20(5):235–240, 1985.
26. S.Q. Zheng, J.S. Lim, and S.S. Iyengar. Routing using implicit connection graphs. In *9th Int'l Conf. on VLSI Design: VLSI in Mobile Communication*, Bangalore, India, January 1996. IEEE Computer Society Press.
27. Y. Zhong, M. Orlovich, X. Shen, and C. Ding. Array regrouping and structure splitting using whole-program reference affinity. In *Proceedings of the ACM SIGPLAN 2004 conference on Programming language design and implementation*, pages 255–266, Washington, DC, 2004.

A Precomputation-Based Scheme for QoS Routing and Fair Bandwidth Allocation

Chia-Hung Wang[1] and Hsing Luh[2]

Department of Mathematical Sciences, National Chengchi University,
Wen-Shan, Taipei 11623, Taiwan
{[1]93751502, [2]slu}@nccu.edu.tw

Abstract. We propose a precomputation-based scheme offering Pareto optimal solutions to the network optimization problem. This scheme precomputes bandwidth allocation (rate vector) and end-to-end paths with QoS guarantees. It prepares a routing database, identifying an optimal path upon each connection request. We propose a mixed-integer optimization model with nonlinear utility functions. The purpose of this work is to choose the optimal solutions and to provide decision makers a set of solutions satisfying users' preferences with fairness. It prepares for a proportionally fair treatment of all competing connections. Numerical results show that the proposed model can provide each connection with its fair share of the bandwidth which is proportional to the target rate.

Keywords: Communication Networks, QoS Routing, Proportional Fairness.

1 Introduction

Quality of Service (QoS) has been the major issue for telecom providers [1], [2], and [13]. Packet-switched networks have been proposed to offer the QoS guarantees in integrated-services networks. The Universal Mobile Telecommunications System (UMTS) offers multiple services, which provide the capability for information transfer between access points [11]. UMTS network services have different QoS classes while connection oriented and connectionless services are offered for point-to-point communication. The main function of the core network with UMTS provisioning is to provide switching, routing and transit for user traffic. A core network also contains the databases and network management functions.

QoS routing concerns the selection of a path satisfying the QoS requirements of a connection [1], [9], [13], etc. The path selection process involves the knowledge of the connection's QoS requirements and information on the availability of bandwidth [2]. QoS routing poses major challenges in algorithmic design [3]. Depending on the specifics and the number of QoS metrics involved, computation in real time required for path selection become prohibitively expensive as the network size grows.

Precomputation-based methods are performed by means of a two-phase procedure [9]. The first phase is executed in advance, which is to precompute solutions summarized in a database for later usage. When an event arrives, the

Y. Robert et al. (Eds.): HiPC 2006, LNCS 4297, pp. 595–606, 2006.

second phase is activated to promptly provide an adequate solution of event's parameters. The key idea of precomputation is to effectively reduce the time needed to handle an event by performing a certain amount of computations in advance. We present an approach for the fair resource allocation problem and QoS routing in communication networks offering multiple services for users. The objective of the optimization problem is to determine the amount of bandwidth for each class to maximize the sum of the users' satisfaction.

The remainder of this paper is organized as follows. In the next section, we describe the network optimization problem. In Section 3, a precomputation-based scheme for bandwidth allocation and QoS routing are formally introduced. Numerical results are given in Section 4. Finally, we remark results of our work.

2   Network Optimization Problem

2.1   Problem Definition

Consider a directed network topology $G = (V, E)$ as shown in Figure 1, where V and E denote the set of nodes and the set of links in the network respectively. There are m different QoS classes of connections in the network. Let $I = \{1, \ldots, m\}$ be an index set which consists of m different QoS classes. The specific QoS requirements, for each class i, include minimal bandwidth requirement b_i and maximal end-to-end delay constraint D_i. We denote the total number of connections, for each class i, by K_i. Let J_i, for each class i, be an index set consisting of K_i connections, that is, $J_i = \{1, \ldots, K_i\}$. All connections are delivered between the same source o and destination d in this (core) network. Every connection in the same class i expects the same bandwidth θ_i and has the same QoS requirement.

Definition 1. *A bandwidth allocation* $(\theta_1, \ldots, \theta_m)$ *is called a **feasible bandwidth allocation** if the bandwidth requirements are satisfied for all classes.*

Suppose, for each link e, we have a mean delay ℓ_e related to the link's speed, propagation delay, and maximal transfer unit. The maximal possible link capacity is U_e on each link $e \in E$, and the link cost is κ_e of using one unit bandwidth. A connection j in each class i should be routed through a path $p_{i,j}$ between o and d. Under a limited available budget B, we plan to allocate the bandwidth in order to provide each class with maximal possible QoS and determine the optimal path from o to d under guaranteed service. Decision variables are listed as follows: $A_{i,j}(e)$ is the bandwidth allocated to link $e \in E$ for connection j in class i, θ_i denotes the bandwidth allocated to each connection in class i, and $\chi_{i,j}(e)$ is a binary variable which determines whether the link e is chosen for connection j in class i. The purpose of this work is to present an mathematical model that provides the decision maker to explore a set of solutions satisfying users' preferences with fairness.

Definition 2. *A feasible path* $p_{i,j} = \{e \in E \mid \chi_{i,j}(e) = 1\}$ *is called a **Pareto optimal path**, for a connection j of class i, if no other feasible path is as less*

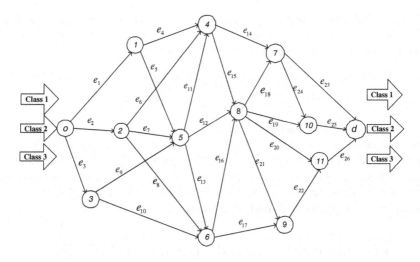

Fig. 1. A sample network topology

as $p_{i,j}$ with respect to two evaluations, path cost and o-to-d delay, and strictly less than $p_{i,j}$ with respect to at least one evaluation.

Definition 3. *The set of all Pareto optimal paths is called the* **routing database** *P. That is, $P = \{p_{i,j} \mid p_{i,j}$ is the Pareto optimal path from o to d, $\forall\, j \in J_i,\ i \in I\}$.*

Definition 4. *A link e is called* **bottleneck link** *if the usage of bandwidth achieves its link capacity, i.e., $\sum_{i \in I} \sum_{j \in J_i} A_{i,j}(e) = U_e$.*

2.2   Proportional Fairness

Kelly et al. [4] advocated proportional fairness characterized by $\log(\theta_i)$. This log utility function is strictly concave. The proportional fair bandwidth allocation is determined by the following objective function:

$$\max\ \sum_{i \in I} K_i \log(K_i \theta_i). \tag{1}$$

Determining the maximizer of (1) can be done explicitly for simple networks. A proportionally fair bandwidth allocation (rate vector) is defined in [7] as follows:

Definition 5. *A feasible bandwidth allocation $(\theta_1, \ldots, \theta_m)$ is called* **proportionally fair** *if and only if, for any other feasible allocation $(\hat\theta_1, \ldots, \hat\theta_m)$, we have:*

$$\sum_{i=1}^{m} \frac{\hat\theta_i - \theta_i}{\theta_i} \leq 0. \tag{2}$$

In this case Kelly et al. [4] have shown that the maximizer of (1) corresponds to a proportionally fair bandwidth allocation.

Mo and Walrand [8] characterized the class of (w, α)- proportionally fair bandwidth allocation, for any given number α $(\alpha > 0, \; \alpha \neq 1)$, as the following objective function:

$$\max \sum_{i \in I} w_i K_i^\alpha \frac{(K_i \theta_i)^{1-\alpha}}{1 - \alpha},$$ (3)

where w_i is a fixed parameter.

In equilibrium, connections that share the same links do not necessarily equally share the available bandwidth [5], [14]. Their shares reflect how they value the bandwidth as expressed by their utility functions and how their use of the bandwidth implies a cost on others. This could be a basis to provide differentiated services in terms of different bandwidth allocations.

3 A Precomputation-Based Scheme for Network Management

In this section we propose a precomputation-based scheme offering optimal solutions to the network optimization problem, which precomputes paths under network constraints.

3.1 Network Optimization Model

We transform the different measurements onto a normalized scale by using achievement functions [12]. Since pages are limited, proofs of the following results are skipped and will be provided under request. First, we give the following lemma on the limits of the achievement function.

Lemma 1. *Let κ be the cheapest cost per unit bandwidth given in an end-to-end path. Suppose the total budget is B. There exists a finite number $M_i \leq B/\kappa K_i$ such that $\theta_i \leq M_i$, $\forall \; i$, where K_i is the number of connections in class i.*

Depending on the specified aspiration and reservation levels, a_i and r_i, respectively, we construct our achievement function of θ_i as follows [12]:

$$f_i(\theta_i) = \log_{\alpha_i} \frac{\theta_i}{r_i},$$ (4)

where $\alpha_i = a_i/r_i$. Formally, we define $f_i(\cdot)$ over the range $[0, M_i]$, with $f_i(0) = -\infty$ and $f_i'(0) = \infty$. It is a strictly increasing function of θ_i, having value 1 if $\theta_i = a_i$, and value 0 if $\theta_i = r_i$. Next, we present an appealing property of the achievement function (4), which holds in the bandwidth allocation problem we are studying.

Proposition 1. *The achievement function $f_i(\theta_i)$ is continuous, increasing, and concave.*

These results are entirely consistent with those assumptions on the utility functions for end-to-end flow control in [6], where the objective is to maximize the aggregated source utility over their transmission rates. Obviously, the maximizer of (4) corresponds to a proportionally fair bandwidth allocation.

Let $E_o \subseteq E$ and $E_d \subseteq E$ be subsets of links connected with the source o and destination d respectively. We denote $E_\nu^{in} \subseteq E$ a subset of incoming links to the node $\nu \in V$, and we also denote $E_\nu^{out} \subseteq E$ a subset of outgoing links from the node $\nu \in V$. When adopting the achievement function (4) interpreted as a measure of QoS on networks, we may formulate the mathematical model of the fair bandwidth allocation. The precomputation-based maximization model is formulated as follows:

$$\max \ \sum_{i \in I} w_i f_i(\theta_i) \tag{5}$$

$$s.t. \ \sum_{e \in E} \sum_{i \in I} \sum_{j \in J_i} \kappa_e A_{i,j}(e) \leq B \tag{6}$$

$$\sum_{i \in I} \sum_{j \in J_i} A_{i,j}(e) \leq U_e, \ \forall e \in E \tag{7}$$

$$\sum_{e \in E} \ell_e \chi_{i,j}(e) \leq D_i, \ \forall \, j \in J_i, \ i \in I \tag{8}$$

$$A_{i,j}(e) - M \cdot \chi_{i,j}(e) \leq 0, \ \forall e \in E, \ j \in J_i, \ i \in I \tag{9}$$

$$\theta_i - A_{i,j}(e) \leq M(1 - \chi_{i,j}(e)), \ \forall e \in E, \ j \in J_i, \ i \in I \tag{10}$$

$$A_{i,j}(e) - \theta_i \leq M(1 - \chi_{i,j}(e)), \ \forall e \in E, \ j \in J_i, \ i \in I \tag{11}$$

$$\theta_i \geq b_i, \ \forall i \in I \tag{12}$$

$$\sum_{e \in E_o} A_{i,j}(e) = \theta_i, \ \forall \, j \in J_i, \ i \in I \tag{13}$$

$$\sum_{e \in E_\nu^{in}} A_{i,j}(e) = \sum_{e \in E_\nu^{out}} A_{i,j}(e), \ \forall \nu \in V', \ j \in J_i, \ i \in I \tag{14}$$

$$\sum_{e \in E_d} A_{i,j}(e) = \theta_i, \ \forall \, j \in J_i, \ i \in I \tag{15}$$

$$A_{i,j}(e) \geq 0, \ \forall e \in E, \ j \in J_i, \ i \in I \tag{16}$$

$$\theta_i \geq 0, \ \forall i \in I \tag{17}$$

$$\chi_{i,j}(e) = 0 \ or \ 1, \ \forall e \in E, \ j \in J_i, \ i \in I, \tag{18}$$

where M is a sufficiently large positive number, $w_i \in (0,1)$ is given for each class i, and $\sum_{i \in I} w_i = 1$.

We have the budget constraint (6) due to the limited budget on network planning. The constraint (7) says that the aggregate bandwidth of all connections at any link does not exceed the capacity. We have the end-to-end delay constraint (8) since every connection has the maximal end-to-end delay constraint. The inclusion of constraints (9)-(11) is equivalent to at least one of $A_{i,j}(e) = 0$ and $A_{i,j}(e) = \theta_i$ being satisfied by either $\chi_{i,j}(e) = 0$ or $\chi_{i,j}(e) = 1$. Constraints

(10), (11), and (12) show that every connection in the same class uses the same bandwidth and has the same bandwidth requirement. Constraints (13), (14), and (15) express the node conservation relations indicating that flow in equals flow out for every connection j in class i. Although $A_{i,j}(e)$ are continuous variables, constraints (13)-(15) are standard flow conservation constraints with the help of constraints (9)-(11). Continuous decision variables and binary variables must be nonnegative in constraints (16)-(18). Moreover, we have the following properties.

Theorem 1. *This precomputation-based maximization model is bounded.*

This result follows because each achievement function is bounded by Lemma 1.

Theorem 2. *This precomputation-based model is NP-complete.*

3.2   Pareto Optimal Solutions

After the optimization, we have the optimal solutions $A_{i,j}^*(e)$ and θ_i^*, which represent the optimal bandwidth allocation for each link e and for each connection of class i. We also determine the optimal choices of links, $\chi_{i,j}^*(e)$. The optimal solution θ_i^* is unique, providing the proportional fairness to each class. We find the bandwidths allocated to each class i, $K_i\theta_i^*$, and the maximal bandwidth, $R_{i,e}$, by which the link e can offer for class i, i.e.,

$$R_{i,e} = \sum_{j \in J_i} A_{i,j}^*(e). \tag{19}$$

Proposition 2. *A link e is the bottleneck link if $\sum_{i \in I} R_{i,e} = U_e$.*

Bandwidth are allocated along less expensive paths that connect the origin o and the destination d. From the optimization of these precomputation schemes, we have the Pareto optimal path and a routing database, P.

Proposition 3. *If $p_{i,j} = \{e \in E \mid \chi_{i,j}^*(e) = 1\}$ for connection j in class i, then path $p_{i,j}$ is the Pareto optimal path from the source o to the destination d. Moreover, the Pareto optimal path $p_{i,j}$ is unique for connection j in class i.*

Proposition 4. *The unit cost for bandwidth along the optimal path $p_{i,j}$ is*

$$\sum_{e \in p_{i,j}} \kappa_e \chi_{i,j}^*(e)$$

for connection j in class i.

Proposition 5. *If link e belongs to the optimal path $p_{i,j}$, then the bandwidth by which the link e can offer for connection j in class i is the same. That is, $A_{i,j}^*(e) = A_{i,j}^*(e')$ for all $e, e' \in p_{i,j}$.*

Proposition 6. *Let* $\theta_{i,p} \geq 0$, *for each class* i, *be the bandwidth allocated to each optimal path* $p \in P$. *Then we have*

$$\sum_{p \in P} \theta_{i,p} = K_i \theta_i^* \qquad (20)$$

and

$$0 \leq \sum_{i \in I} \theta_{i,p} \leq \min_{e \in p} U_e. \qquad (21)$$

Next, we study the sensitivity to the maximal number K_i of connections for each class i. Let

$$c_i = \frac{\sum_{j \in J_i} \sum_{e \in E} \kappa_e A_{i,j}^*(e)}{K_i} \qquad (22)$$

be a mean budget allocated to each connection in class i. If π_i denotes the reserved budget for each class i, then the budget constraint (6) is represented as

$$\sum_{i \in I} (K_i c_i + \pi_i) = B, \qquad (23)$$

where $\pi_i \geq 0, \forall\, i \in I$.

Definition 6. *The ratio* $\sum_{j=1}^{K_i} \sum_{e \in E} \kappa_e A_{i,j}^*(e)/B$ *is called a* **budget ratio** *allocated to class* i.

Budget ratio is defined by proportional share of total budget with respect to each class. Each class is given a percentage, budget ratio, of the total budget B. It must further be noted that the budget ratio can be computed from $c_i K_i/B$.

Theorem 3. *Let* c_p *be the unit path cost along the Pareto optimal path* $p \in P$, *i.e.,* $c_p = \sum_{e \in p} \kappa_e$. *If the budget* B *satisfies*

$$B \geq \sum_{i \in I} \pi_i + \sum_{p \in P} c_p \min_{e \in p} U_e, \qquad (24)$$

then there exists one Pareto optimal path p *which contains at least one bottleneck link. Moreover, link* e *is the bottleneck link if* $U_e = \sum_{p \ni e} \sum_{i \in I} \theta_{i,p}$.

3.3 A Routing Scheme with End-to-End QoS Guarantees

After applying the optimization model, we obtain a network $G = (V, E')$, where V is the original set of nodes and $E' \subseteq E$ is the subset of links belonging to each end-to-end path $p \in P$. Each link $e \in E'$ is characterized by delay ℓ_e. Let $n(p_{i,j})$ be the number of links along path $p_{i,j}$ and σ_i be the mean packet size for each class i, $i \in I$. When a connection j in class i is routed along a path $p_{i,j}$, the following end-to-end delay $D(p_{i,j})$ applies (Atov et al. [1]):

$$D(p_{i,j}) = \frac{n(p_{i,j}) \cdot \sigma_i}{\theta_i^*} + \sum_{e \in p_{i,j}} \ell_e. \qquad (25)$$

For each connection of class i, we find a feasible path p such that $D(p) \leq D_i$.

To balance the loads over the network, we seek a path for which the residual bandwidth of its bottleneck link is maximal. For a path p, we denote by A_p the residual bandwidth of its bottleneck link, that is,

$$A_p = \min_{e \in p}(R_{i,e} - \theta_i^*). \tag{26}$$

The problem is to find an optimal path p maximizing A_p from the routing database P. For each class i, we give the following scheme with end-to-end QoS guarantees.

$$\begin{aligned} \max \quad & A_p \\ \text{s.t.} \quad & A_p \leq R_{i,e} - \theta_i^*, \quad \forall e \in p \\ & D(p) \leq D_i \\ & p \in P. \end{aligned} \tag{27}$$

This routing scheme distributes the connection among the paths so as to avoid overloaded links. The goal of this scheme is to enhance the performance of IP traffic while utilizing the bandwidth on All-IP networks economically. This QoS routing is to make more efficient use of bandwidth on the network, and its concept is consistent with the shortest remaining processing time discipline.

4   Numerical Results

Consider a sample network (as shown in Figure 1) where $V =\{$node o, node 1, ..., node $d\}$ and $E = \{e_k,\ k = 1, 2, \ldots, 26\}$ denote the set of nodes and the set of links in the network respectively. Each connection is delivered from node o to node d. Table 1 shows the capacity U_e, mean delay ℓ_e, and the link cost κ_e of bandwidth for each link $e \in E$. In Table 2, three different QoS classes are given, where class 1 has the highest priority and class 3 has the lowest priority.

Table 1. Characteristics of each link

Characteristics	e_1	e_2	e_3	e_4	e_5	e_6	e_7	e_8	e_9
Capacity (kbps)	35,000	45,000	55,000	53,000	47,000	36,000	37,000	45,000	40,000
Cost ($)	7	6	5	14	11	14	7	13	8
Delay (sec)	0.03	0.032	0.035	0.012	0.02	0.012	0.03	0.015	0.027
	e_{10}	e_{11}	e_{12}	e_{13}	e_{14}	e_{15}	e_{16}	e_{17}	e_{18}
Capacity (kbps)	50,000	45,000	46,000	45,000	44,000	46,000	36,000	35,000	54,000
Cost ($)	14	7	11	5	5	10	5	7	5
Delay (sec)	0.012	0.03	0.02	0.035	0.035	0.022	0.035	0.03	0.035
	e_{19}	e_{20}	e_{21}	e_{22}	e_{23}	e_{24}	e_{25}	e_{26}	
Capacity (kbps)	40,000	53,000	41,000	40,000	52,000	44,000	42,000	50,000	
Cost ($)	7	9	6	8	13	6	8	6	
Delay (sec)	0.03	0.025	0.032	0.027	0.015	0.032	0.027	0.032	

Table 2. Characteristics of each QoS class

Class i	Bandwidth Requirement b_i (kbps)	Asp. Level a_i (kbps)	Res. Level r_i (kbps)	Mean Packet Size σ_i (kb)	Max. Delay D_i (sec)
1	160	334	167	35	0.89
2	80	166	83	16.6	1.02
3	25	56	28	12.5	2.34

Table 3. A database as $(K_1, K_2, K_3) = (80, 120, 150)$ and $(w_1, w_2, w_3) = (0.6, 0.3, 0.1)$

Class i	Optimal Path p	Path Flow $\theta_{i,p}$	No. of Connect.	No. of Links $n(p)$	Unit Path Cost	Delay $D(p)$
	$e_1 - e_4 - e_{14} - e_{23}$	5010	15	4	39	0.511
	$e_2 - e_6 - e_{14} - e_{23}$	7346	22	4	38	0.513
1	$e_2 - e_7 - e_{11} - e_{14} - e_{23}$	6009	18	5	38	0.666
	$e_2 - e_7 - e_{13} - e_{16} - e_{19} - e_{25}$	7015	21	6	38	0.818
	$e_2 - e_7 - e_{13} - e_{16} - e_{20} - e_{26}$	1334	4	6	38	0.818
	$e_1 - e_4 - e_{14} - e_{23}$	1826	11	4	39	0.511
	$e_2 - e_6 - e_{14} - e_{23}$	7301	44	4	38	0.494
2	$e_2 - e_7 - e_{11} - e_{14} - e_{23}$	3816	23	5	38	0.642
	$e_2 - e_7 - e_{13} - e_{16} - e_{19} - e_{25}$	4646	28	6	38	0.789
	$e_2 - e_7 - e_{13} - e_{16} - e_{20} - e_{26}$	2321	14	6	38	0.789
	$e_1 - e_4 - e_{14} - e_{23}$	624	16	4	39	1.387
	$e_2 - e_6 - e_{14} - e_{23}$	1830	47	4	38	1.389
3	$e_2 - e_7 - e_{11} - e_{14} - e_{23}$	1012	26	5	38	1.761
	$e_2 - e_7 - e_{13} - e_{16} - e_{19} - e_{25}$	1752	45	6	38	2.131
	$e_2 - e_7 - e_{13} - e_{16} - e_{20} - e_{26}$	622	16	6	38	2.131

Table 4. Change in the weight with $(K_1, K_2, K_3) = (80, 120, 150)$ and $B = 2000, 000$

Weight	Bandwidth	Utility	Total Utility	Budget Ratio
$(\frac{1}{3}, \frac{1}{3}, \frac{1}{3})$	(301,166,56)	(0.852,1,1)	0.951	(0.459,0.381,0.160)
$(0.4, 0.3, 0.3)$	(301,166,56)	(0.852,1,1)	0.941	(0.460,0.380,0.160)
$(0.4, 0.4, 0.2)$	(301,166,56)	(0.852,1,1)	0.941	(0.459,0.380,0.160)
$(0.5, 0.3, 0.2)$	(334,144,56)	(1,0.798,1)	0.939	(0.510,0.329,0.160)
$(0.5, 0.4, 0.1)$	(334,166,39)	(1,1,0.464)	0.946	(0.509,0.380,0.110)
$(0.6, 0.2, 0.2)$	(334,144,56)	(1,0.798,1)	0.960	(0.509,0.331,0.160)
$(0.6, 0.3, 0.1)$	(334,166,39)	(1,1,0.464)	0.946	(0.510,0.379,0.110)
$(0.7, 0.2, 0.1)$	(334,144,56)	(1,0.798,1)	0.960	(0.510,0.331,0.160)
$(0.8, 0.1, 0.1)$	(334,143,56)	(1,0.789,1)	0.979	(0.510,0.328,0.160)

We assume every connection in class i, for $i = 1, 2, 3$, has the same aspiration level a_i kbps (i.e. kilobits/sec), reservation level r_i kbps, mean packet size σ_i kb, maximal end-to-end delay D_i, and bandwidth requirement b_i kbps.

We assume the number of connections, K_i, are independently and identically distributed as Poisson distribution with parameter λ_i, where $\lambda_1 = 80$, $\lambda_2 = 120$,

Table 5. Numerical results of 35 samples from the Poisson arrivals with mean $\lambda_1 = 80$, $\lambda_2 = 120$, $\lambda_3 = 150$, and $(w_1, w_2, w_3) = (0.6, 0.3, 0.1)$

No. of Connect.	Bandwidth (kbps)	Utility	Total Utility	Budget Ratio
$(95, 115, 141)$	$(334,134,37)$	$(1,0.69,0.42)$	0.85	$(0.605,0.295,0.100)$
$(78, 128, 133)$	$(334,158,47)$	$(1,0.93,0.74)$	0.95	$(0.497,0.384,0.118)$
$(89, 124, 159)$	$(334,138,35)$	$(1,0.74,0.32)$	0.85	$(0.567,0.327,0.106)$
$(68, 106, 161)$	$(352,166,56)$	$(1.07,1,1)$	1.04	$(0.464,0.337,0.173)$
$(85, 128, 153)$	$(334,143,37)$	$(1,0.79,0.42)$	0.88	$(0.541,0.350,0.109)$
$(79, 119, 146)$	$(334,166,43)$	$(1,1,0.62)$	0.96	$(0.504,0.375,0.120)$
$(78, 123, 163)$	$(334,165,37)$	$(1,0.99,0.42)$	0.94	$(0.497,0.387,0.116)$
$(83, 131, 162)$	$(334,142,37)$	$(1,0.78,0.42)$	0.88	$(0.528,0.356,0.116)$
$(96, 126, 148)$	$(334,118,37)$	$(1,0.51,0.42)$	0.79	$(0.612,0.282,0.105)$
$(74, 121, 145)$	$(334,166,53)$	$(1,1,0.91)$	0.99	$(0.471,0.384,0.146)$
$(86, 140, 142)$	$(334,131,37)$	$(1,0.66,0.42)$	0.84	$(0.547,0.351,0.102)$
$(81, 112, 158)$	$(334,166,43)$	$(1,1,0.62)$	0.96	$(0.516,0.354,0.130)$
$(78, 111, 160)$	$(334,166,50)$	$(1,1,0.83)$	0.98	$(0.497,0.351,0.152)$
$(66, 128, 150)$	$(335,166,56)$	$(1.01,1,1)$	1.00	$(0.423,0.404,0.160)$
$(75, 118, 151)$	$(334,166,52)$	$(1,1,0.88)$	0.99	$(0.477,0.374,0.149)$
$(82, 133, 131)$	$(334,142,47)$	$(1,0.78,0.74)$	0.91	$(0.523,0.361,0.117)$
$(81, 147, 162)$	$(334,132,37)$	$(1,0.66,0.42)$	0.84	$(0.516,0.369,0.116)$
$(84, 114, 163)$	$(334,160,37)$	$(1,0.95,0.42)$	0.93	$(0.535,0.349,0.116)$
$(80, 122, 148)$	$(334,165,37)$	$(1,1,0.42)$	0.94	$(0.511,0.384,0.105)$
$(79, 114, 149)$	$(334,166,48)$	$(1,1,0.77)$	0.98	$(0.502,0.361,0.136)$
$(79, 115, 147)$	$(334,166,47)$	$(1,1,0.76)$	0.98	$(0.504,0.363,0.133)$
$(81, 118, 145)$	$(334,166,40)$	$(1,1,0.51)$	0.95	$(0.516,0.373,0.111)$
$(71, 136, 143)$	$(334,162,47)$	$(1,0.97,0.74)$	0.96	$(0.452,0.420,0.127)$
$(98, 123, 150)$	$(334,115,37)$	$(1,0.47,0.42)$	0.78	$(0.62,0.27,0.11)$
$(76, 123, 152)$	$(334,166,44)$	$(1,1,0.64)$	0.96	$(0.484,0.389,0.127)$
$(72, 123, 161)$	$(334,166,49)$	$(1,1,0.82)$	0.98	$(0.489,0.389,0.152)$
$(75, 123, 150)$	$(334,166,46)$	$(1,1,0.73)$	0.97	$(0.477,0.390,0.133)$
$(74, 103, 131)$	$(351,166,56)$	$(1.07,1,1)$	1.04	$(0.505,0.329,0.140)$
$(78, 120, 140)$	$(334,166,46)$	$(1,1,0.72)$	0.97	$(0.496,0.380,0.123)$
$(73, 138, 176)$	$(334,156,37)$	$(1,0.91,0.42)$	0.91	$(0.465,0.409,0.125)$
$(70, 125, 155)$	$(334,166,54)$	$(1,1,0.94)$	0.99	$(0.446,0.395,0.158)$
$(87, 132, 152)$	$(334,134,37)$	$(1,0.69,0.42)$	0.85	$(0.554,0.338,0.108)$
$(78, 132, 165)$	$(334,153,37)$	$(1,0.88,0.42)$	0.91	$(0.496,0.386,0.118)$
$(81, 127, 160)$	$(334,153,37)$	$(1,0.88,0.42)$	0.91	$(0.515,0.370,0.114)$
$(92, 109, 151)$	$(334,147,37)$	$(1,0.83,0.42)$	0.89	$(0.584,0.309,0.107)$
$(80.1, 123.1, 151.5)$	$(335.0,154.4,43.3)$	$(1,0.89,0.62)$	0.93	$(0.512,0.361,0.125)$

and $\lambda_3 = 150$. Under the budget $B = \$2,000,000$, we plan to allocate the bandwidths in order to provide each class with maximal utility (4). By linearizing the achievement function (4) [12], the mathematical model is programmed in a Mixed Integer Programming (MIP) form and ready to be solved by the popular modelling packages CPLEX. This example led to an MIP with 31,203 variables and 31,880 constraints. The computation time is about 809 seconds, including 10,003 iterations, and 194 nodes of the branch-and-bound tree.

After executing the precomputation-based scheme, we provide a database as shown in Table 3 for given parameters. In Table 3, it gives, for each path p in the routing database P, the path flow $\theta_{i,p}$ which is computed by (20) in Proposition 6. Moreover, it gives the number of connections and number of links $n(p)$ along path p. These paths are the candidates for the adequate solution in a routing scheme (27) with end-to-end QoS guarantees. By Proposition 4, we determine the unit path cost, $\sum_{e \in p} \kappa_e$, for using one unit bandwidth along the path $p \in P$. These paths are Pareto optimal solutions with end-to-end QoS guarantees. The path flow $\theta_{i,p}$ in (20) and (21), for each class i, is the aggregated bandwidth of connections along path p. The number of connections, for each class, along path $p \in P$ is also determined. We also find that, by Theorem 3, link e_2 is the bottleneck link since $\sum_{p \ni e_2} \sum_{i=1}^{3} \theta_{i,p} = 45,000 = U_2$. By the computation of (25), we can list (in Table 3) the approximate ene-to-end delay $D(p)$ along the Pareto optimal path $p \in P$ for all classes. The results suggest that as paths traverse a larger number of links, the end-to-end delay becomes large. A path $p_{i,j}$ between o and d is guaranteed if $D(p_{i,j}) \leq D_i$ for a connection j in class i.

We now explore how changes in the model's parameters, w_i and K_i, affect the optimal allocation. First, we observe the sensitivity to the weight w_i assigned to each class. Given mean arrival numbers, we compare it by changing the weight assigned to each class. Table 4 shows the computational result. Observe that enlarging the difference between classes 1 and 3 will increase the total satisfaction. The increase is mainly contributed by class 1 since w_1, the weight assigned to class 1, is larger than the others. Next, as the weight is fixed, we analyze it by collecting 35 random samples from the Poisson arrivals with means $\lambda_1 = 80$, $\lambda_2 = 120$, and $\lambda_3 = 150$. The results are shown in Table 5. It shows that increasing number of connections of classes 1, 2, and 3 will decrease the optimal value in (5). The sample means are also listed in the last line of Table 5. Computational experiences show a different topology does not change the validity of the model. Note, according to Definition 5, that the bandwidth allocation $(\theta_1^*, \theta_2^*, \theta_3^*)$ in tables are proportionally fair by our approach.

5 Conclusions

We present an optimization model for balancing resources with proportional fairness and providing a routing scheme on networks. The precomputation-based scheme is taken in advance, which is to precompute solutions for a routing database. It enables decision makers to identify an optimal path upon each connection request through a simple routing procedure. This scheme determines the amount of required bandwidth for each class to maximize the sum of the users' utilities. We find the Pareto optimal bandwidth allocation under a limited budget, and this allocation can provide proportional fairness to every class. Numerical results show that this scheme can provide each connection with its fair share of the bandwidth which is proportional to the target rate. To design on-line routing algorithms for dynamic traffic, an additional computation may be activated when connections arrive, which could select one of the optimal solutions from the routing database.

Acknowledgments. The authors are grateful to the referees for helpful remarks and suggestions. This work was supported by NSC 94-2213-E-004-003.

References

1. Atov, I., Tran, H. T., and Harris, R. J.: OPQR-G: Algorithm for efficient QoS partition and routing in multiservice IP Networks. Computer Communications **28** (2005) 1987–1996
2. Bai, Y. and Ito, M. R.: Class-Based Packet Scheduling to Improve QoS for IP Video. Telecommunication Systems **29** (**1**) (2005) 47–60
3. van Hoesel, S.: Optimization in telecommunication networks. Statistica Neerlandica **59** (**2**) (2005) 180–205
4. Kelly, F. P., Maulloo, A. K., Tan, D. K. H.: Rate control for communication networks: Shadow prices, proportional fairness and stability. Journal of the Operational Research Society **49** (1998) 237–252
5. Kelly, F. P.: Fairness and stability of end-to-end congestion control. European Journal of Control **9** (2003) 159–176
6. Low, S. H., Lapsley, D. E.: Optimization flow control–Part I: Basic algorithm and convergence. IEEE/ACM Transactions on Networking **7** (**6**) (1999) 861–874
7. Massoulié, L., Roberts, J.: Bandwidth sharing: objectives and algorithms. IEEE/ACM Transactions on Networking **10** (**3**) (2002) 320–328
8. Mo, J., Walrand, J.: Fair end-to-end window-based congestion control. IEEE/ACM Transactions on Networking **8** (**5**) (2000) 556–567
9. Orda, A., Sprintson, A.: Precomputation schemes for QoS routing. IEEE/ACM Transactions on Networking **11** (**4**) (2003) 578-591
10. Paganini, F., Wang, Z., Doyle, J. C., Low, S. H.: Congestion control for high performance, stability, and fairness in general networks. IEEE/ACM Transactions on Networking **13** (**1**) (2005) 43–56
11. The UMTS Forun: Enabling UMTS/Third Generation Services and Applications. UMTS Forun Report **11** (2000)
12. Wang, C. H. and Luh, H.: Network Dimensioning Problems of Applying Achievement Functions. Lecture Notes in Operations Research: Operations Research and Its Applications **6** (2006) 35–59
13. Wu, H., Jia, X., He, Y., Huang, C.: Bandwidth-guaranteed QoS routing of multiple parallel paths in CDMA/TDMA *ad hoc* wireless networks. International Journal of Communication Systems **18** (2005) 803–816
14. Ye, H. Q., Qu, J.: Stability of data networks: Stationary and bursty models. Operations Research **53** (**1**) (2005) 107-125

Heuristics for Flash-Dissemination in Heterogenous Networks*

Mayur Deshpande, Nalini Venkatasubramanian, and Sharad Mehrotra

School of Information and Computer Science, University of California, Irvine
{mayur, nalini, sharad}@ics.uci.edu

Abstract. Flash Dissemination is a particularly useful form of data broadcast that arises in many mission-critical applications. The goal is rapid distribution of medium amounts of data in as short a time period as possible. While optimal algorithms are available for a highly constrained case (all nodes having the same bandwidth and latency), there is relatively little work in the context of heterogenous networks. Most systems and protocols today either use trees or randomized mesh-based techniques to deal with heterogeneity and work with local knowledge. We argue that a protocol with global knowledge can perform much better. In this paper, we propose two centralized heuristics – DIM-Rank and DIM-Time that use global knowledge to schedule data transfer between nodes. The heuristics are based upon insights from broadcast theory. We perform experimental evaluation of these two heuristics with decentralized randomized approaches and show that DIM-Rank achieves faster dissemination than decentralized approaches across a range of heterogeneity metrics.

1 Introduction

Fast distribution of data to multiple receivers is a basic primitive and required functionality in several application domains. In this paper, we study a particularly useful form of dissemination that arises in mission-critical applications which we term as *flash dissemination*. Such a scenario consists of rapid dissemination of medium amounts of data to a large number of recipients in a very short period of time. Consider, for example, an organization that has geographically distributed data-centers located at various ISP points. Periodically, the data centers need to be synchronized with a global master list (or latest security patches). Fast delivery of this information to all centers is critical to avoid loss of downtime or observable 'glitch' by users. As another example, from the emergency management domain, consider a service such as "Shakecast" [2] of the USGS (United States Geographical Survey). Earthquake information sent out by Shakecast is a Shake-Map" (image-file) of 100-300KB. This information is sent to various city and county emergency management organizations that subscribe to the USGS. The goal is to provide accurate and timely data and information about seismic events as quickly as possible.

At an abstract level, these applications fall under the network wide broadcast problem where a particular node wants to broadcast some data to all other recipient nodes as fast

---

* This work was supported by NSF Grant, Award Numbers 0331707 and 0331690.

Y. Robert et al. (Eds.): HiPC 2006, LNCS 4297, pp. 607–618, 2006.

as possible. While network wide broadcast is a mature area with more than 20 years of research, the problem of high-speed dissemination in heterogenous networks is a new problem. In the highly constrained case of all nodes with homogenous bandwidth and latency, an optimal solution to the achievable lower bound was proposed in 1980 [6]. This was rediscovered again in 2005 [8] (in the context of overlay P2P (Peer-to-Peer) networks) and the authors also proposed an alternative approach using a hypercube to achieve the lower bound. However, when nodes are allowed to have varying bandwidths and/or latency, the problem becomes NP-hard [10]. In [10], the authors only consider a case where the data to be distributed is one single piece (or chunk). Multi-chunk distribution in heterogenous networks adds further complexity to this scenario but leads to faster dissemination.

Current systems either use overlay trees (Narada [12], Splitstream [4]), or more recently, meshes (Bit-torrent [1], CREW [5]) or a hybrid of both (Bullet [11], Bimodal-Multicast [3]) to deal with multi chunk distribution in heterogenous networks. Though not mathematically proven, randomized approaches perform quite well in real world settings and much empirical work substantiates this [8, 11]. However, many of these systems [1, 11] are either tailored towards streaming or large amounts (GBs) of data or small size events. In the scenario of interest to us, data size is usually in the middle range of hundreds of KBs to tens of MBs. As we explain later, fast dissemination of medium size data requires a protocol to do both *ramp-up* and *sustained-throughput* very well. Ramp-up is the time needed for each node to start participating in the dissemination process. In sustained-throughput, a node is able to sustain high transfer rates. Currently, CREW is a protocol that addresses this special data range.

However, all these systems use some form of randomization in their protocols and work with mostly local knowledge. We argue that when low dissemination time is of utmost importance, centralized approaches with global knowledge can make a crucial difference in performance. We propose two heuristics for multi-chunk dissemination in heterogenous networks that we call DIM-Time and DIM-Rank. The heuristics need global knowledge and a centralized 'scheduler' to orchestrate data transfer between nodes. By global knowledge, we mean pair-wise bandwidth and latency measures among all participating nodes. The heuristics are based upon the insights obtained from the original optimal solution to homogenous data broadcast [6]. We show via experiments that DIM-Rank achieves lower dissemination time as compared to randomized approaches across a range of heterogeneity metrics. The rest of the paper is as follows. In Sec-2, we formalize the problem of flash dissemination. In Sec-3 we present our centralized heuristics, and situate them in a taxonomy of related research. We compare the heuristics to randomized mesh based approaches in Sec-4 and conclude in Sec-5

2   Problem Formalization

In this section, we first define the problem of flash dissemination more concretely. Let ν be a set of N nodes $\nu = N_1, N_2, ... N_N$ connected by an underlying fully connected network. Let $N_{seed} \in \nu$ be the seeder node with the data item D to be disseminated. The objective is to get D to all non-seeder nodes in ν as fast as possible.

2.1 Chunk Based Representation of Data

We view the data item D to be disseminated as a sequence of M equal sized chunks. A chunk is an aggregation of one or more bytes of data. Meta-information regarding the chunks contains details on how received chunks must be 'stitched' to get back the original information. A node has to receive (and verify) the whole chunk before it can transfer it further. Chunk dissemination is advantageous and flexible because chunks can be disseminated asynchronously, be received out-of-order and then finally assembled. Furthermore, it supports increased concurrency in the dissemination process since multiple nodes can start propagating the chunks they have received so far. In fact, [6] showed how such a chunk-based dissemination leads to an optimal solution in a homogenous network and how an optimal chunksize for a given dissemination can be found.

2.2 Chunk Based Dissemination over Heterogenous Networks

Each node in the network has a certain *capacity* or rate at which it can transmit (or receive) data to (from) another node (also called it's *bandwidth*). Additionally, there is certain *delay* (or *latency*) defined as the time it takes for one byte of data to be transmitted from the sender to the receiver. In a homogenous network all nodes have the same bandwidth and equal inter-node latency, so the time to transmit a message between any two peers is equal. However, this is not an accurate model to capture dissemination in the Internet where peers have different bandwidths (T3, T1, DSL, etc.) and inter-peer latencies vary considerably (from 1-1000 millisecs).

We use the following characterization to describe chunk-based dissemination in heterogenous networks. Let the maximum capacity/bandwidth of a node n be $MaxBW(n)$. Different nodes can have different Max-bandwidths. Any pair of nodes, (x, y) has a latency denoted as $Lat(x, y)$. We assume $Lat(x, x)$ can be approximated to 0 and $Lat(x, y) = Lat(y, x)$. A node's bandwidth may be partially reserved for an ongoing transfer and its leftover (or available) bandwidth is denoted as $AvailBW(x)$. When a pair of nodes initiates a chunk-transfer, the sustained bandwidth for the transfer is denoted as $SusBW(x, y)$ and it has the following property: $SusBW(x, y) \leq Min(AvailBW(x), AvailBW(y))$. The time required to transfer a chunk of size D between $x \rightarrow y$ is then $Lat(x, y) + \frac{D}{SusBW(x,y)}$. This time can also be higher, if for example, a connection needs to be established first between the two nodes. For instance, in TCP, a 3-way handshake is needed to establish the connection and hence the time required can be approximated as $3 * Lat(x, y) + \frac{D}{SusBW(x,y)}$. Nodes can 'split' their bandwidth into any combination of uploads and downloads. Thus a node can be engaged in multiple transfers, some upload and some download. Next, we present our heuristics for flash dissemination over heterogeneous networks.

3   Centralized Flash Dissemination Heuristics

In this section, we present our heuristics, DIM-Time and Dim-Rank, for flash dissemination in heterogeneous networks. However, we first provide a brief summary of the theoretical basis for these heuristics. We conclude with a taxonomy of research on data broadcast and situate our heuristics in it.

3.1   Theoretical Background

The theoretical base consists of two results for scheduling data broadcast: a multi-chunk broadcast in a homogenous network and a single-chunk broadcast in heterogenous bandwidth network.

Homogenous Network. In a homogenous network, the optimal solution for broadcast [6, 8], can be broken down into two main parts: **Phase 1: Ramp-up phase** - which ensures that every node receives at least one chunk and **Phase 2: Sustained-throughput phase** - which ensures that the total available capacity is used to transfer chunks.

The lower bound for the first phase is $Log(N)$ since the number of nodes that have atleast one chunk can be doubled in every unit of time. This is achieved quite simply if in each time unit, a node that has a chunk picks as a receiver, a node that does not have any chunk. The lower bound for the second phase is $2M - 1$ and achieving it involves realizing some clever insights. We do not delve into the full details but only present the main intuition here.

After the ramp-up phase, the whole set of nodes can be partitioned into two equal sets of of $\{givers\}$ and $\{receivers\}$. Then, each $giver$ gives one chunk to one $receiver$. If the right chunks are transferred, then the set of nodes can again be partitioned and this can be continued until all nodes have all chunks. The crux of the problem, however, is in deciding how to partition and what chunks to transfer. The intuition employed by the optimal algorithm(s) is to make 'rare' chunks more 'popular' so that there are no bottlenecks (where there are more receivers who want chunks than there are givers of those chunks). The rarity or popularity of a chunk is defined by how many nodes possess that chunk; with the crossover point being half the nodes.

Heterogeneous Network. The optimal solution in a homogenous network works because all nodes have the same bandwidth and equal inter-node latency, so that the time to transmit a message between any two peers is equal. This is not true in practice since individual link bandwidths and inter-peer latencies vary significantly. A recent result [10] shows that the problem of minimizing the time for broadcasting a single message in a heterogeneous network is a NP-hard problem; the authors also show that the Fastest-Node-First (FNF) heuristic is optimal in many cases for single-message broadcast. The FNF broadcast tree problem is restricted to one message and it is not entirely clear if it is also a good heuristic for broadcast of multiple messages. For example, when there is only one message to transmit, then different peers are picked for reception of message at consecutive steps from the seeder. However, if there are multiple messages, then it is not entirely clear whether the fastest node should get all the messages first or another scheme should be followed.

3.2   Our Heuristics: DIM-RANK and DIM-Time

Using metrics that capture the key insights of the optimal solution (in the homogenous case), we derive two heuristics, that are better than a simple adaptation of FNF. An elegant property of these heuristics is that in the case of a homogenous network, they work close to the optimal solution and in the case of single-message dissemination in heterogenous network they work like FNF. We then embed these heuristics into a demand-driven framework, thus realizing a dynamically adaptive system for flash-dissemination system in a heterogenous network.

Any solution that addresses the key challenge of optimal partitioning must determine the following at each decision point: (1)the set of transmitters, (2)the set of receivers and (3)chunks that transmitters must send to receivers. Note that(1) and (3) are intertwined since what chunks a node possess factors into deciding the node's role. To aid in optimized partitioning, we define the following metrics:

Chunk Spread: The *spread* of a chunk is, c_i, is the total number of nodes that have c_i. More formally, $spread(c_i) = |\{P_i\}|$ where $\{P_i\} = \{x : x \text{ contains } c_i\}$. Spread of a chunk thus quantifies how rare of popular a chunk is[1].
Node Rank: The node's rank is defined as $Rank(n) = \sum_i \frac{1}{spread(c_i)}$ i.e. the summed inverse of spread of all chunks that it contains. Thus, a node's rank is higher if it either contains rare chunks or many chunks. Conversely, if a node only contains popular chunks, it's rank is low.

The choice of the above metrics is not arbitrary; it captures key insights of the optimal solution. In the optimal solutions, the receivers were nodes that either did not have any chunks or had chunks that were in the *majority*, i.e. half of the nodes already had the same chunk. Nodes which had rare chunks were the transmitters. Thus, when deciding which chunk to transmit among a set of chunks, a node should transmit the lower-spread chunk. Similarly, if the rank of a node is high, it should be considered for transmitting a chunk and if it's rank is lower, it should instead receive a chunk. We derive two heuristics (DIM-Time and DIM-Rank) using the metrics defined. We assume that the heuristic is to be applied to a set of nodes that have spare capacity (called *AvailNodes* henceforth) and decisions have to be made on how to split them into transmitters and receivers and what chunks should be transferred. The operational flow of both DIM-Time and DIM-Rank is shown in Fig-1.

The algorithms are run by a central scheduler and nodes report their initial capacity, any change to capacity and what chunk they received to it. The scheduler, therefore has full knowledge (about both chunks and spare bandwidth) of the network. One point in the implementation of the scheduler is how often it should be run. If it is run too often, then the number of nodes in *availNodes* may be too small at any given point of time. If the scheduler is run too infrequently, then nodes will waste time just waiting for the scheduler to tell them what to do. Thus, there exists an optimal periodicity of the scheduler. We do not address this issue here though one could use a policy where the scheduler comes into effect when the spare capacity of the system reaches a certain threshold or a certain fraction of nodes have spare capacity.

Fig. 2 depicts an sample dissemination for 3 chunks (from Node 1) among 10 nodes (*Nodes 1 and 2 have twice the bandwidth of the remaining nodes*) for 3 centralized protocols - (i) a naive-FNF adaptation, (b) DIM-Time and (c) DIM-Rank. Transfer of a chunk takes one unit of time between node-1 and node 2; and two units of time between

---

[1] If so desired, spread can be normalized by the total number of nodes.

INPUT: $\{AvailNodes\}$: List of nodes with spare capacity
BEGIN
```
1)     While |{AvailNodes}| > 0
2)         giver = getHighestRankedNode({AvailNodes})
3)         {Receivers} = {AvailNodes} − giver
4)         ForEach recvr ∈ {Receivers}
5)             If {giver.chunks} ⊂ {recvr.chunks} # nothing to give
6)                 {Receivers} − recvr
7)         If {Receivers} = ∅ # no receivers possible for this giver
8)             {AvailNodes} − giver
9)             continue # go back and pick another giver
10)        While AvailableBandwidth(giver) > 0 AND |{Receivers}| > 0
11)            If DIM_TIME
12)                recvr = getHighestSpareCapacityNode({Receivers})
13)            If DIM_RANK
14)                recvr = getLowestRankedNode({Receivers})
15)            {PossibleChunks} = {giver.chunks} − {recvr.chunks}
16)            chunkToTransfer = getRarestChunk ({PossibleChunks})
17)            StartTransfer : giver --→ recvr With chunk chunkToTransfer
18)            {Receivers} − recvr
19)            If AvailableBandwidth(giver) <= 0
20)                {AvailNodes} − giver
21)            If AvailableBandwidth(recvr) <= 0
22)                {AvailNodes} − recvr
END
```

Fig. 1. DIM_RANK and DIM_TIME Psuedocode

any other pair of nodes. The Naive-FNF adaptation for multiple chunks works as follows: (a) the highest capacity node is the transmitter, (b)the next highest capacity node with missing chunks is the receiver and (c) at every time step, the receiver node obtains the next missing chunk. As Fig. 2 shows, DIM-Rank achieves the fastest dissemination with lowest dissemination time, 11 time units. DIM-Time takes 13 units of time and Naive-FNF takes 15 units of time.

Discussion: So far we have ignored the cost of computing the schedule in the central coordinator. In the best case scenario, the central scheduler has to sort the list of nodes just once, thus incurring a cost of $O(NLogN)$. However, it is possible that the scheduler has to sort the list for the transmitters and then sort the list again for receivers for each of the transmitters. This can happen for all M chunks, leading to a worst case computing cost of $O(M*NLogN)^2$. As M and N increase, therefore, the worst case computing cost increases exponentially. We have also not considered the effect of node dynamicity. Handling new nodes is straightforward – they report directly to the scheduler. However, for node leaves, we can assume that either the scheduler maintains a heart-beat to each node or other nodes detect dead nodes and report them to the scheduler.

3.3 Taxonomy of Broadcast and Our Contributions

We can view solutions to the dissemination problem along multiple dimensions as illustrated in Figure-3.

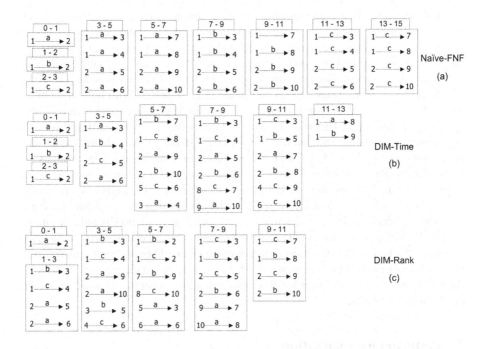

Fig. 2. Steps in Dissemination for Three Greedy Heuristics

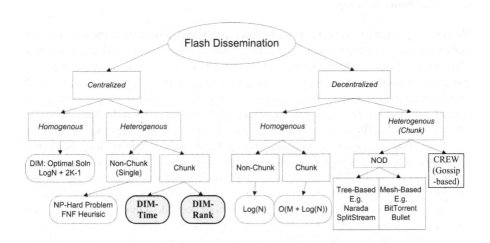

Fig. 3. Scope of our Contributions

1. Centralized versus decentralized decision making for dissemination
2. Dissemination for Homogenous versus heterogenous networks
3. Non-chunk (single message) versus chunk-based dissemination

We illustrate the best solutions (that we are aware of), using the taxonomy in Fig-1. We situate our heuristics (shown in bold) in the figure. For the centralized side of the tree, we have already described the related work. For the decentralized side, we start with results in the homogenous case. The results obtained in decentralized case are stochastic in nature. Gossip-based broadcast offers bounds on how long it would take for a single chunk (message) to be broadcast to N nodes. With high probability, it would take ($O(LogN)$) rounds to disseminate a single message (with high probability) to all N nodes [9]. Theoretical results for multi-chunk gossip starting from only one seeder are not yet proven but for M nodes each starting with a chunk, the dissemination time $O(LogN + M)$ [7]. In heterogenous networks, decentralized systems use an overlay topology to spread chunks. Within overlay based schemes, one can divide the systems into whether they are neighbor-oriented or not. In neighbor-oriented diffusion (NOD) schemes, nodes keep track of the chunks that the neighbors have and then setup exchanges. In case of a tree overlay, the flow of chunks is only in one direction. Splitstream [4], Bullet [11] and Bit-torrent [1] are all examples of NOD based systems. CREW, on the other hand does not maintain neighbor state but uses the overlay as a membership management service for its gossip-based mechanism. In [5], we show CREW to be much faster for flash dissemination as compared to other overlay dissemination systems.

4   Performance Evaluation

4.1   Experiments Setup

We have implemented all four approaches on top of a middleware platform that we built, called RapID. For scalability testing of thousands of peers, we needed to run multiple RapidPeers on a single host. Further, since we wanted to control the delay and throughput between peers for experiments, we developed an emulation layer that intercepts all peer-peer communications. A call to transfer a chunk to a target-peer would therefore not actually transfer the chunk but only emulate the time taken for it; both on the sender and receiver side. The emulation layer is quite detailed and provides the peer with all the details that it normally would ask from the Operating System, such as the current rate of transfer of all ongoing chunks, available bandwidth, (TCP)connection-cache of open connections, etc. When a chunk-transfer is initiated between two peers, the emulation layer on the sender side emulates an upload and the emulation layer on the receiver emulates a download. The emulation layers on both peers do a quick handshake to determine at what rate the transfer will progress. This is arrived by following the formula of $SusBW$ as noted in Sec-2. When the appropriate time as elapsed, the emulation layers, readjust the $availBW$ and send appropriate events to the Peers that the transfer is complete. Control messages exchanged between peers are similarly emulated to reflect the latency between the peers. TCP-connection setup time is also emulated if two peers communicate for the first time. Since, there is no actual data transfer between peers, multiple peers can be run a host without hitting the maximum NIC bandwidth. Experiments are run in both homogenous and heterogenous settings. Default values for experiments are shown in Figure-4. For the decentralized protocols (CREW and NOD),

High-BW Peers (10 Mbps)	2	Med-BW Peers (1Mbps)	10	Low-BW Peers (100Kbps)	200

Total File Size	128 KB	Chunk Size	4 KB

Fig. 4. Default Values Used in Experiments

we construct uniform-random overlays between the peers with average node degree of 4, since these protocols seem particularly suited to sparse random graphs.

Total time for dissemination is our primary metric and this is calculated in an experiment as follows. The required instances of peers are started up and they all contact a known server to obtain the emulation parameters. The server assigns each peer its capacity (bandwidth) and also gives it a latency-vector to other peers. This vector contains all the latencies for this peer to contact to any other peer in the system. The latency vector is used by the emulation layer to appropriately delay inter-peer communications. The server then tags one of the peers with all the chunks. At this point the server records the $startTime$. In the centralized heuristics, the peers contact the central scheduler who then schedules the peers. As peers receive all chunks, they report the event to the server. When all peers get all chunks, the server notes the ($stopTime$). The total dissemination time is calculated as $stopTime - startTime$.

4.2   Performance Results

We now show the dissemination time of the four approaches under various scenarios. We start with a homogenous case and then progressively relax the constraints, showing it's impact on the four protocols.

Performance in Ideal Homogenous Network. We begin with a 'baseline' ideal-world that is homogenous; all nodes have the same bandwidth (1Mbps) and inter-node latency (2ms). The file to be disseminated is of size 128K Bytes and split into the ideal number of chunks. (as calculated from the optimal solution). The goal is to asses how well the protocols scale in a homogenous world. One expects the total time to complete is linear in $LogN$, where N is the total number of nodes. Figure5(a) shows the total time each protocol takes to disseminate the file to all the nodes. Note that the x-axis is a log-scale of number of nodes and hence straight lines indicates scalability in $LogN$. Fig-5(b) shows CDF plots for disseminating the file to 1000 nodes.

DIM-Rank performs the best while DIM-Time performs the worst (though all of them scale linearly with $LogN$). DIM-Time performs the worst since it optimizes for lower transfer time. In a homogenous network with equal latencies, the seeder sends all chunks to one node at first. The deciding factor is whether a node has an open connection to another node or not, since the latency is longer when a new connection has to be opened. This trend continues so that nodes get all chunks before they start transmitting to others. Figure5(b) shows this clearly. The number of nodes that complete in DIM-Time (rightmost plot) doubles with time but since nodes get all chunks before

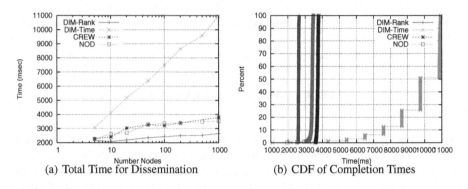

(a) Total Time for Dissemination   (b) CDF of Completion Times

Fig. 5. Scalability Results

disseminating, DIM-Time scales as $O(M * Log(N))$ (hence the steep slope). The other protocols scale closer to $O(M + Log(N))$. This case, therefore, shows how real-world situations can affect heuristics even in the simple case of a homogenous network.

Effect of Bandwidth Heterogeneity. To the baseline model, we now introduce bandwidth heterogeneity. The settings for this experiment are as follows. A network of 10 nodes with medium-bandwidth (1Mbps) is the initial baseline. To this, we first add a varying number of low-bandwidth (100Kbps) nodes and study it's effects (Fig-6(a)). We now move the baseline to a network with 10 medium-bandwidth (med-bw) nodes and 200 low-bandwidth (low-bw) nodes. We then study the effect of introducing high-bandwidth (high-bw) nodes of 10Mbps into this network (Fig-6(b)).

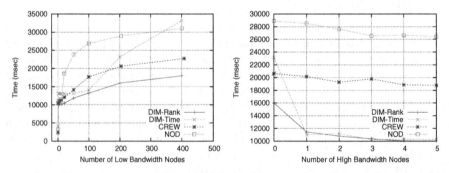

(a) Effect of Increasing Low Bandwidth Nodes (b) Effect of Introducing High Bandwidth Nodes

Fig. 6. Performance in Bandwidth Heterogenous Networks

Intuitively, one would expect that introducing low-bandwidth nodes increases the dissemination time and conversely, introducing high-bandwidth nodes decreases the dissemination time. Fig-6(a) shows that the introduction of the first low-bw node causes a significant jump in total dissemination time. This is because the total dissemination time is dictated by when the low-bw node finishes. After this data point, the slope is more linear for all four protocols and they scale linearly in $LogN$, where N is now the number of low-bw nodes. When the first high-bw node is introduced (Fig-6(a)), there is a dramatic reduction in dissemination time for the centralized protocols. It is as if the addition of one high-bw node compensated for the addition of 200 low-bw nodes. Dim-Time and Dim-Rank are fully able to exploit high-bandwidth nodes whereas the effect is more limited for the decentralized protocols. Further introduction of high-bw servers has only marginal effect. Thus, for a given network, introducing high-bw nodes can have a significant impact initially and almost no value for later additions.

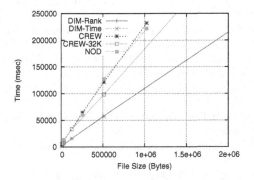

Fig. 7. Performance with Increasing Data Size

Effect of File Size. Fig-7 shows the effect when the total data to be disseminated is increased. DIM-Rank has a clear superior performance over all the other protocols. Moreover, as the data size is increased, the gap widens between DIM-Rank and the other protocols making it very desirable. However, note that when data size increases, M increases and this in turn increases the computing need on the central scheduler. In our experiments, we ran the scheduler on a powerful machine so that it could always finish its computation before the next cycle. With increasing node size and file size, the computation load can increase dramatically. Thus, while DIM-Rank may be a very good heuristic for disseminating medium amounts of data to hundreds (or even thousands) of peers, the cost justification for DIM-Rank has to be evaluated carefully for particular application needs.

5   Concluding Remarks

In this paper, we presented two new heuristics, DIM-time and DIM-Rank for the flash dissemination problem in heterogeneous networks. These heuristics were developed using broadcast theory. Of the two, DIM-Rank offers much lower dissemination times across a variety of metrics. In general, the centralized approaches fare better than the randomized, local-knowledge, decentralized protocols. DIM-Rank will find most use when low dissemination time is of utmost necessity and a centralized coordinator who has global knowledge can be used. The cost of computing the schedule in the central coordinator, however, is non-trivial. It is atleast $O(N LogN)$ and can be as bad as $O(M * N LogN)^2$ in the worst case. An interesting course of future work would be to investigate approximation techniques that achieve the same effect as DIM-Rank without the high computation costs.

References

1. Bittorrent: http://bitconjurer.org/bittorrent/.
2. Shakecast: http://earthquake.usgs.gov/resources/software/shakecast//.
3. K. P. Birman, M. Hayden, O. Ozkasap, Z. Xiao, M. Budiu, and Y. Minsky. Bimodal multicast. In *ACM TOCS*, 1999.
4. M. Castro, P. Druschel, A.-M. Kermarrec, A. Nandi, A. Rowstron, and A. Singh. Splitstream: High-bandwidth multicast in a cooperative environment. In *SOSP*, 2003.
5. M. Deshpande, B. Xing, I. Lazardis, B. Hore, N. Venkatasubramanian, and S. Mehrotra. Crew: A gossip-based flash-dissemination system. In *ICDCS*, 2006.
6. A. M. Farley. Broadcast time in communication networks. In *SIAM Journal on Applied Mathematics*, volume 39, 1980.
7. C. Fernandess and D. Malkhi. On collaborative content distribution using multi-message gossip. In *IPDPS*, 2006.
8. P. Ganesan and M. Seshadri. On cooperative content distribution and the price of barter. In *ICDCS*, 2005.
9. R. Karp, C. Schindelhauer, S.Shenker, and B. Vocking. Randomized rumor spreading. In *IEEE Symposium on Foundations of Computer Science (FOCS) 2000.*, 2000.
10. S. Khuller and Y.-A. Kim. On broadcasting in heterogeneous networks. In *ACM-SIAM Symposium on Discrete algorithms*, 2004.
11. D. Kostic, A. Rodriguez, J. Albrecht, and A. Vahdat. Bullet: High bandwidth data dissemination using an overlay mesh. In *Usenix Symposium on Operating Systems Principles (SOSP)*, 2003.
12. S. S. Yang-hua Chu, Sanjay G. Rao and H. Zhang. A case for end system multicast. In *Measurement and Modeling of Computer Systems*, 2001.

B-PIC: A Novel Caching Scheme for Multimedia Streaming Servers*

Ohhoon Kwon[1], Taeseok Kim[1], Hyokyung Bahn[2], and Kern Koh[1]

[1] School of Computer Science and Engineering, Seoul National University
{ohkwon, tskim, kernkoh}@oslab.snu.ac.kr
[2] Department of Computer Science and Engineering, Ewha University
bahn@ewha.ac.kr

Abstract. With the recent proliferation of video-on-demand services, caching in a multimedia streaming server is becoming increasingly important. Previous studies have shown that request interval based caching and its extension for considering different video popularity performs well for various streaming environments. In this paper, we show that block level refinement of this existing scheme can further improve the performance of streaming servers. Trace driven simulations with real world VOD traces have shown that the proposed scheme improves the cache hit rate and the startup latency.

Keywords: caching, multimedia, streaming, VOD (Video-on-Demand).

1 Introduction

Caching in a multimedia streaming server is an effective way to improve the performance of server systems and reduce the service latency. Due to the large volume of multimedia objects and the strictly sequential access pattern, traditional buffer cache management schemes such as LRU (Least Recently Used) will not work well for multimedia server systems. To address this problem, request interval based caching schemes have been proposed [1-6]. By caching only the data in the interval between two successive requests on the same object, the following request can be serviced directly from the buffer cache without I/O operations. Kim et al. proposed the Popularity-aware Interval Caching (PIC) scheme that extends the interval caching by considering different popularity of multimedia objects [7, 11]. PIC estimates the popularity of multimedia objects based on the request intervals of each object and exploits the estimated popularity in predicting future request times. Based on this information, PIC extends the original interval caching by including predicted intervals in the candidate of caching. However, in the PIC scheme, the sliding window of a predicted interval proceeds as time progresses. As a result, the prefix of popular objects may not be cached though prefix caching is effective in reducing the startup latency.

---

* This work is supported by IGI (institute for Graphic Interfaces) grant 2006. The email address of correspondence is bahn@ewha.ac.kr.

Y. Roberts et al. (Eds.): HiPC 2006, LNCS 4297, pp. 619–628, 2006.

In this paper, we show that the block level refinement of PIC can resolve the afore-mentioned problem and can further improve the performance of streaming servers. Trace-driven simulations with real world VOD traces show that the proposed scheme performs better than the PIC, the IC (interval caching), the LRU (Least Recently Used), and the MRU (Most Recently Used) algorithms in terms of the cache hit rate and the start block misses.

The remainder of this paper is organized as follows. We review some existing works on caching algorithms in multimedia streaming environments in Section 2. Section 3 presents the system architecture of the multimedia streaming server environments. Section 4 presents a new caching scheme for multimedia streaming servers. We evaluate the performance of the proposed scheme in Section 5. Finally, we conclude this paper in Section 6.

2 Related Works

A variety of studies on the caching of multimedia streaming objects have recently been studied. Dan and Sitaram proposed a caching scheme for video-on-demand servers named interval caching that exploits the short term temporal locality of accessing the same multimedia object consecutively [1, 2]. The interval caching scheme organizes all consecutive request pairs by the increasing order of memory requirements. It then allocates memory space to as many of the consecutive pairs as possible. When an interval is cached, the following stream does not need any disk access since it could be serviced directly from the memory buffer cache.

Ozden et al. proposed a cache replacement algorithm named distance caching which is similar to the interval caching scheme [5, 6]. It assigns a priority value to each request based on its distance from the previous request and always replaces the block consumed by the request with the lowest priority value (longest distance from the previous request) over all streams.

However, the interval caching and the distance caching schemes exploit only the short term temporal locality of two consecutive requests on an identical object and do not consider the popularity of objects. Consequently, when the size of a multimedia object is not sufficiently large or when the inter-arrival time of stream requests is too long, there is little opportunity to obtain the effectiveness of caching.

Kim et al. proposed the Popularity-aware Interval Caching (PIC) scheme to resolve these problems [7, 11]. PIC estimates the popularity of multimedia objects based on the request intervals of each object and exploits the estimated popularity in predicting future request times. Based on this information, PIC incorporates the caching of predicted intervals into the original interval caching scheme.

Some recent studies have extended the original interval caching scheme for various caching environments. Sarhan and Das proposed the distributed interval caching (DIC) scheme that extends the original interval caching for network attached disk (NAD) architectures [3]. Almeida et al. considered the two-level caching architecture for streaming objects at proxy servers. They employed the interval caching scheme at the buffer cache layer and the LFU (Least Frequently Used) algorithm at the disk cache layer [4]. They showed that this two-level caching scheme performs better than previous approaches. Cho et al. presented the Hybrid Buffer cache Management (HBM) scheme for VOD servers [12]. In their HBM scheme, a video stream is not just assumed to be accessed sequentially but can form a looping pattern in some applications such as online education servers. Based on this assumption, HBM detects the access pattern of each multimedia file and then employs the distance caching or LRU algorithm appropriate for file accesses. Lee et al. showed that improving the hit ratio alone is not sufficient to guarantee the hiccup-free service and efficient disk bandwidth utilization in multimedia systems [13]. They proposed a new caching scheme, namely the Preemptive but Safe Interval Caching (PSIC), and showed that PSIC provides services to additional streams with the saved disk bandwidth. Recently, Fernandez et al. proposed the Iteration Set Caching (ISC) scheme that evolved from the original interval caching scheme to obtain a better performance for variable bit-rate streams [9]. While the relative ordering of intervals in the original interval caching scheme is statically determined, ISC dynamically changes the ordering of caching blocks to support variable bit rates.

3   System Architecture

In this section, we present the system architecture of the multimedia streaming servers. Our multimedia server consists of an I/O manager, a buffer manager, and a network manager as shown in Fig. 1. The buffer manager divides the memory buffer into

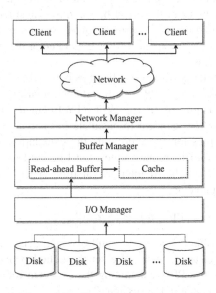

Fig. 1. The multimedia streaming server architecture

the *cache* and the *read-ahead buffer*. The read-ahead buffer stores data to be sent immediately to clients while the cache stores data already sent to clients which can be reused when requests for the same object arrive. Note that data in the memory buffer do not actually move their physical positions (from the read-ahead buffer to the cache) but just a cache flag is used to indicate whether it is in the cache. For each stream request, when the requested block is not in the cache, the I/O manager acquires a free block, inserts it into the read-ahead buffer, and starts disk I/O. On the other hand, if the requested block is in the cache, the cached block is serviced directly without I/O operations. Finally, the network manager reads necessary data blocks from the memory buffer and sends them to the client through the network.

4 Block Level Refinement of Popularity-Aware Interval Caching

In the interval caching scheme, for two consecutive requests for the identical streaming object, the later stream will read the data brought into the memory buffer by the earlier stream if the data is retained in the buffer until it is read by the later stream. Understanding such dependencies makes it possible to guarantee the continuous delivery of the later stream with a small amount of buffer space. An *interval* denotes the distance of the offsets between two consecutive requests on an identical object. The interval caching scheme aims to maximize the number of concurrent streams serviced from the memory buffer. With a given buffer space, therefore, the interval caching scheme sorts the intervals based on their size and caches from the shortest interval. Kim et al. incorporated the virtual interval concept into the original interval caching scheme to consider the different popularity of multimedia objects [7]. A *virtual interval* is defined as the distance of the offset between the latest request on an object and the *virtual request* on that object. A virtual request is not a real request from a client but a predicted request that is expected to be generated at that time based on the past requests on an object [7, 11].

We estimate the popularity of multimedia objects based on past reference behaviors similar to PIC. However, unlike PIC, we use this popularity information in deciding the targets of prefix caching by the block level refinement of PIC. We use the *expected reference probability* concept to calculate the potential benefit of each block when it is cached. Expression (1) represents the calculation of the predicted inter-arrival time (PI) based on past request times. Let I be the latest real inter-arrival time and PI_{k-1} be the $(k-1)$th predicted inter-arrival time. Then, the kth predicted inter-arrival time PI_k is computed as

$$PI_k = \alpha I + (1-\alpha) PI_{k-1} \tag{1}$$

where α is a constant value between zero and one, and determines how much weight is put on the latest inter-arrival time. We set the default value of α as 0.6 through empirical analysis. We use PI in calculating the expected reference probability p_i of multimedia object i as shown in Expression (2).

$$p_i = 1 / PI \tag{2}$$

Finally, the profit of each prefix block b of object i is calculated by

$$\text{Profit}_i(b) = p_i \cdot 1 / t_b \tag{3}$$

where t_b is the remaining time until block b will be referenced assuming object i will be referenced from now on. For example, the first three blocks of object i have the t_b value of 1, 2, and 3, respectively. Similarly, the profit of each block b in the real interval i can be calculated as

$$\text{Profit}_i(b) = 1 / t_b \tag{4}$$

where t_b is the remaining time until block b will be referenced. p_i is omitted in this expression because reference probability p_i is equal to 1 for real intervals. Our scheme calculates the profit of all blocks and allocates the buffer space to block b (including prefix blocks) by the decreasing order of Profit(b). Fig. 2 shows an example of the profit calculations. In this simple example, we assume that clients consume a block per each time unit.

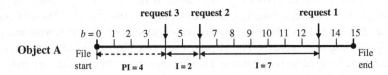

Fig. 2. In this example, $p_A = 1/\text{PI} = 0.25$; Hence, the profits of prefix blocks are Profit(0) = 0.25/1, Profit(1) = 0.25/2, Profit(2) = 0.25/3, and Profit(3) = 0.25/4. Similarly, the profits of interval blocks are Profit(4) = 1/1, Profit(5) = 1/2, Profit(6) = 1/1, Profit(7) = 1/2, Profit(8) = 1/3, Profit(9) = 1/4, etc. The caching order of blocks is determined by the profit value.

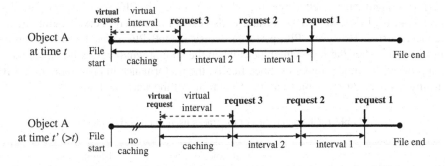

Fig. 3. When the time is t, PIC caches the virtual interval that contains the prefix of object A. However, as time progresses, the sliding window of the virtual interval proceeds, and the prefix of object A is not included in the virtual interval at time t'. As a result, the prefix of object A is not cached, and if there comes a request on object A after time t, the prefix of this stream will be a miss.

In the original PIC scheme, as time progresses, the sliding window of the virtual interval also proceeds. As a result, the prefix of popular objects may not be cached in PIC though prefix caching is effective in reducing the startup latency. Fig. 3 shows an example of this problem.

In contrast to PIC, our scheme retains the prefix blocks of popular objects in the memory buffer even though time progresses. This allows start blocks of popular objects to be hits from the memory buffer. Our scheme could maximize the benefits of caching by the block-level refinement of PIC. At the same time, through the prefix caching concept, it caches the prefix of popular objects before they are actually requested. This could eventually reduce the startup latency of popular streams perceived by users.

5 Performance Evaluation

In this section, we present the performance evaluation results for various caching algorithms to assess the effectiveness of our scheme namely the Block-level Popularity-aware Interval Caching (B-PIC). We gathered real world VOD traces from two commercial VOD servers, namely OnGameNet and Hanmir [8, 10]. The OnGameNet trace has 293 video files whose average playback time is 883 seconds with an average inter-arrival time of 6 seconds. The Hanmir trace has 1266 video files whose average playback time is 1078 seconds with an average inter-arrival time of 21 seconds. Table 1 summarizes the characteristics of the traces.

We conducted extensive simulations to compare the performance of our scheme with those of PIC (Popularity-aware Interval Caching), IC (Interval Caching), LRU (Least Recently Used), and MRU (Most Recently Used). Fig. 4 shows the number of start block misses for the five schemes as a function of the cache size. Since B-PIC caches the prefix blocks of popular stream objects before actual requests arrive, it performs significantly better than the other four schemes. PIC shows better performance than IC, LRU, and MRU because it also caches predicted intervals before actual requests arrive. However, PIC performs worse than B-PIC since PIC evicts prefix blocks and caches the following blocks when the sliding window of a predicted interval proceeds as time progresses. Specifically, the performance improvement of B-PIC against original PIC is as much as 46.7% in terms of the start block misses.

Table 1. Characteristics of the traces used in our experiments

	OnGameNet trace	Hanmir trace
number of requests	3091	6938
average object length	883 seconds	1078 seconds
average inter-arrival time	6 seconds	21 seconds
distinct number of objects	293	1266

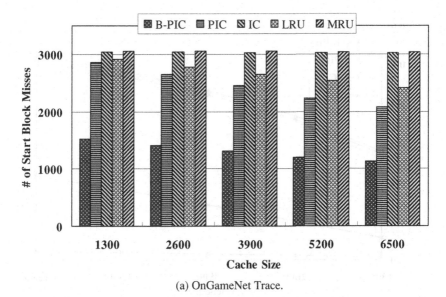

(a) OnGameNet Trace.

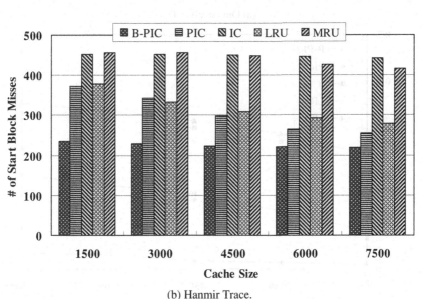

(b) Hanmir Trace.

Fig. 4. Comparison of the number of start block misses for the B-PIC, PIC, IC, LRU, and MRU as a function of the cache size

Fig. 5 shows the hit rate of the five schemes as a function of the cache size. As mentioned in Section 1, traditional buffer management schemes such as LRU and MRU do not perform well in our experiments. For both of the traces, B-PIC shows consistently the best performance in terms of the hit rate irrespective of the cache size.

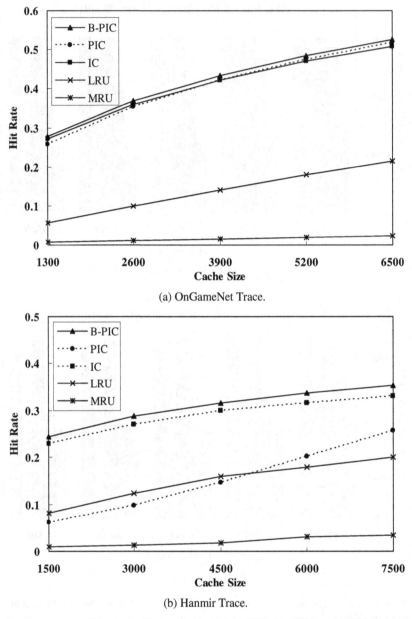

(a) OnGameNet Trace.

(b) Hanmir Trace.

Fig. 5. Comparison of the cache hit rate for the B-PIC, PIC, IC, LRU, and MRU as a function of the cache size

IC also shows good performance for both of the traces though it performs slightly worse than B-PIC for all cases. PIC performs similarly to IC for the OnGameNet trace, but it performs worse than B-PIC and IC by a large margin for the Hanmir trace. This is because the inter-arrival times of the Hanmir trace are relatively long and their variations are also large. As a result, the accuracy of the predicted intervals in PIC degrades severely. Unlike PIC, however, B-PIC shows consistently the best performance because it evaluates the profit of each block based on the expected reference probability, so the priority of a real interval block in B-PIC is relatively higher when compared with PIC.

6 Conclusion

In this paper, we presented the block-level refinement of the Popularity-aware Interval Caching (B-PIC) scheme for multimedia streaming servers. By caching the prefix of popular streaming objects as well as the request interval in the block level, B-PIC performs better than PIC, IC, LRU, and MRU in terms of the cache hit rate and start block misses for the real world VOD traces we considered. Specifically, we have shown that the performance improvement of B-PIC against original PIC in terms of start block misses is as much as 46.7%.

Acknowledgment

The authors thank OnGameNet and Hanmir for making their traces available.

References

1. A. Dan and D. Sitaram, "Buffer Management Policy for an On-Demand Video Server," IBM Research Report RC19347, T.J. Watson Research Center, Yorktown Heights, NY.
2. A. Dan and D. Sitaram, "A Generalized Interval Caching Policy for Mixed Interactive and Long Video Environments," *Proceedings of SPIE Multimedia Computing and Networking Conference*, San Jose, CA, 1996.
3. N. J. Sarhan and C. R. Das, "Caching and Scheduling in NAD-Based Multimedia Servers," *IEEE Transactions on Parallel and Distributed Systems*, Vol.15, No.10, pp.921-933, Oct. 2004.
4. J. M. Almeida, D. L. Eager, M. K. Vernon, "A Hybrid Caching Strategy for Streaming Media Files," *Proceedings of the SPIE/ACM Conference on Multimedia Computing and Networking*, 2001.
5. B. Ozden, R. Rastogi and A. Silberschatz, "Buffer Replacement Algorithms for Multimedia Storage Systems," *Proceedings of the 3rd IEEE International Conference on Multimedia Computing and Systems*, Hiroshima, Japan, pp. 172-180, 1996.
6. B. Ozden, R. Rastogi and A. Silberschatz, "Disk Striping in Video Server Environments," *Proceedings of the 3rd IEEE International Conference on Multimedia Computing and Systems, Hiroshima*, Japan, pp. 580-589, 1996.
7. T. Kim, H. Bahn, and K. Koh, "Popularity-Aware Interval Caching for Multimedia Streaming Servers," *IEE Electronics Letters*, Vol.39, No.21, pp. 1555-1557, Oct. 2003.
8. OnGameNet Co. Ltd, http://www.ongamenet.com.
9. J. Fernandez, J. Carretero, F. Garcia-Carballeira, A. Calderon, and J. Perez-Menor, "New stream caching schemas for multimedia systems," *IEEE Int'l Conf. Automated Production of Cross Media Content for Multi-Channel Dist.*, 2005.
10. Hanmir, Co. Ltd, http://www.hanmir.net.

628   O. Kwon et al.

11. T. Kim, H. Bahn, and K. Koh, "Efficient Cache Management for QoS Adaptive Multimedia Streaming Services," *Lecture Notes in Computer Science*, Springer-Verlag, Vol.3768, pp.1-11, Oct. 2005.
12. K. Cho, Y. Ryu, Y. Won and K. Koh, "A hybrid buffer cache management scheme for VOD server," *IEEE Int'l Conf. Multimedia and Expo (ICME 2002)*, Vol. 1, pp. 241- 244, 2002.
13. K. Lee, Y. Y. Park, H. Y. Yeom, "Pre-emptive but safe interval caching for real-time multimedia system," *Int'l Journal of Computer Systems Science and Engineering*, Vol. 18, No. 2, pp. 87-94, 2003.

Gvu: A View-Oriented Framework for Data Management in Grid Environments

Pradeep Padala and Kang Shin

EECS, University of Michigan
{ppadala, kgshin}@umich.edu

Abstract. In a grid, data is stored in geographically-dispersed virtual organizations with varying administrative policies and structures. Current grid middleware provide basic data-management services including data access, transfer and simple replica management. Grid applications often require much more sophisticated and flexible mechanisms for manipulating data than these, including logical hierarchical namespace, automatic replica management and automatic latency management. We propose a view-oriented framework that builds on top of existing middleware and provides global and application-specific logical hierarchical views. Specifically, we developed mechanisms to create, maintain, and update these views. The views are synchronized using an efficient group communication protocol. Gvu (pronounced G-view) is built as a distributed set of synchronized servers and scales much better than the existing grid services. We conducted experiments to measure various aspects of Gvu and report on the results, showing Gvu to outperform existing grid services, thanks to its distributed nature.

1   Introduction

Grids[1] have become the favorite choice for executing data-intensive scientific applications. Scientific applications in domains, such as high energy physics, bio-informatics, medical image processing and earth observations, often analyze and produce massive amounts of data (sometimes of the order of petabytes). The applications access and manipulate data stored in various sites on the grid. They also have to distribute and publish the derived data.

Let us consider how a typical scientific application interacts with the data grid.

1. A physicist participating in a high-energy physics experiment would like to execute a CMS (Compact Muon Solenoid) application.
2. The application requires various input files. It has to find the location of files using a catalog, index or database system where information about the location of the file is stored. The application usually uses a logical file name (LFN) to index into the catalog and find the physical location.
3. The files may have to be replicated at various sites in the grid for the application to find a nearby location to quickly access the file.

Y. Robert et al. (Eds.): HiPC 2006, LNCS 4297, pp. 629–640, 2006.
© Springer-Verlag Berlin Heidelberg 2006

4. The physicist, having gathered all the information about the input files, runs the jobs on various sites. If a site does not have the required input data, the data is pre-fetched before the job starts.
5. The jobs execute using the input files and produce derived data.
6. The derived data needs to be distributed to various sites on the grid for usage by other scientists and for archival purposes.
7. Finally, the output data locations have to be published in a catalog so that other scientists can locate the data.

Using current middleware and grid file systems as they currently exist, the above scenario requires the application to perform complex interactions with grid services. Globus [2] middleware, one of the most popular grid toolkits, provides data management mechanisms including GridFTP [3] and RLS (Replica Location Service) [4]. GridFTP is an enhanced version of the popular File Transfer Protocol (FTP) that provides high performance using parallel streams, parallel file transfers, command pipelining, etc. To realize the above scenario, ad-hoc mechanisms using GridFTP, RLS, and metadata catalog services (MCS) [5] can be developed. Unfortunately, these mechanisms lack flexibility and power.

Therefore, the key research question is: *What are the data management requirements of typical workloads in grid environments and how do we provide flexible and powerful mechanisms for manipulating data?* Thain *et al.* [6] surveyed six scientific application workloads run in grid environments and concluded that traditional distributed file systems are inefficient for the *batch-pipelined* nature of these workloads.

Currently there are three different data management mechanisms available in grid environments. On one hand, data-management facilities like replica location and metadata management can be provided by different services that can be combined in various ways depending on the application. On the other hand, one can develop a unified grid file system that provides a consistent file-system-like interface to the application. Researchers [7–9] have worked on providing a file system-like interface to the data on the grid (a detailed comparison is provided in the next section). Although there is no consensus on the grid file system interfaces, these efforts have succeeded in providing uniform access to heterogeneous storage systems distributed over a grid. Certain key features that are missing are global hierarchical name space and application- and user-specific views of the data.

Why do we need a global hierarchical name space? If we consider the scenario explained earlier, jobs of an application running on different sites can see others' files as they are created in a global hierarchical tree. *Why do we need a logical name space?* Data in a grid is stored in various sites at different physical locations. It would be more flexible for an application to refer to the data using a logical name instead of a complicated physical name that might change over time. In a single administrative domain, creating this logical hierarchical name space is easy. NFS (Network File System) [10] already provides a simple, though inflexible, mechanism for doing this. In a grid, the data is scattered in different virtual organizations (VOs). Consider a grid file system that provides logical global hierarchical name space. Would that solve all the problems in the above scenario? Not completely. Consider the situation where an application manipulates thousands of files and produces many more files. How do we allow flexible access by other applications which want to use the same data? With the existing tools, this would be a nightmare. The Virtual Data System (VDS) [11] provides a convenient way of maintaining and querying *recipes* for data derivations [11], but it does not provide a way of finding and creating files. A view that contains only the files manipulated by a particular application will solve this problem.

In this paper, we develop mechanisms for creating the global hierarchical namespace and application-specific views on top of it. We first review the existing mechanisms for manipulating data in a grid or distributed system. We next describe the architecture of Gvu. We then provide the details of the implementation. We conclude with experiments demonstrating the usage and performance impact of Gvu.

2   Related Work

There is a vast volume of literature on distributed file systems solving various problems that occur in distributed data sharing. CIFS and NFS (v2 and v3) [10] provide a global namespace, but the naming is only at a local domain level. They also have security weaknesses that are not suitable in wide-area grids. NFSv4 has many enhancements and provides a global physical view of the system. An effort called GridNFS, taken up by CITI at the University of Michigan, to customize NFSv4 for grids is still in its infancy.

Other distributed file systems including AFS [12], Coda [13], and GFS [14] are distributed file systems that are designed for multiple clients to access files by using file caches, and do not perform very well in the data-intensive computing environments that are commonly seen in grids. It is interesting to note that AFS provides a global physical view of the distributed system. The physical view is quite inflexible and does not allow sites to export application-specific views.

In the grid realm, the focus has been on providing high-performance data access. Grid-specific data access mechanisms including GridFTP [3], LegionFS [9], and Gfarm [7] succeed in this respect. Gfarm provides highly scalable and high-bandwidth read/write operations by integrating process and data scheduling. It also provides replica management and supports file fragments, but creates a static view of the global namespace similar to AFS and leaves it to the user to handle it. The centralized metadata database used in Gfarm might become a bottleneck. It is also unclear how Gfarm servers interact with each other.

The Storage Resource Broker (SRB) [8] developed by SDSC provides some interesting capabilities to grid data management. SRB provides a uniform interface to heterogeneous data resources and provides replica management. The metadata catalog (MCAT), which is a part of SRB, provides a way of accessing the data sets using attributes. The key feature of this system with respect to our work is the usage of logical names. SRB fails, though, in providing a hierarchical view of the logical names, and has no concept of application-specific views.

In [15], we provided a detailed comparison of grid file system features in a survey submitted to the GFS-WG (Grid File Systems Working Group). This work is in progress and currently compares Gfarm and SRB, two of the most popular grid data management mechanisms. GFS-WG recently released RNS (Resource Namespace Service) specification, which is still in draft form. It describes many of the features that we envisioned earlier in this work.

3 Design of Gvu

We have considered the following issues in designing Gvu.

- *Distributed vs. Centralized*: Since the metadata and files are distributed over the grid, Gvu should not be centralized, but use a set of distributed servers that are synchronized.
- *View ownership*: The logical view exported by a site is owned by the site administrator, but the application and user-specific views created on top of the global view are owned by the respective applications or users.
- *Fault-tolerance*: Gvu should tolerate faults in a Gvu server. Currently, Gvu handles the crash failure of any number of Gvu servers. If a Gvu server goes down, the user will still be able to access the metadata related to the files on the corresponding site, but will not be able to access the files.
- *Performance*: Since the Gvu servers are synchronized over a wide-area network, it is important to keep the communication among the Gvu servers to minimum. We have implemented batching of metadata updates to improve performance.
- *Scalability*: Gvu should be scalable with the number of clients. Currently, Gvu provides better scalability than RLS and MCS combined because of its distributed nature.
- *Consistent, flexible and powerful API*: Gvu provides a familiar file-system-like interface. Once the view is created, the interaction with the view is very similar to the interaction with a traditional file system.

The following subsections detail design of Gvu.

3.1 Gvu Servers

Each site on the grid runs a Gvu server that maintains the local logical view (explained in the next section) for that site. The Gvu server usually runs on the gatekeeper machine, which has access to local schedulers, local clusters and local file systems. The servers communicate with each other using a reliable, fault-tolerant group communication protocol. There has been a substantial volume of research on providing reliable, fault-tolerant group multicast. We use a toolkit called *distview* developed by Litiu and Prakash, in which a server pool (called Corona) maintains the shared information. The publishers (clients) can submit data to the server pool and subscribers can receive the data either in synchronous or asynchronous mode. The communication protocol provided has all the properties that we require for synchronizing the Gvu servers. We have decided to use `distview`, because of its features, support for distributed collaboration and source availability.

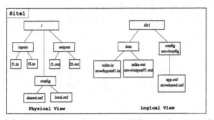

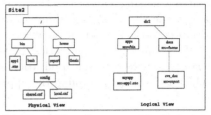

(a) Physical and exported logical view of site1

(b) Physical and exported logical view of site2

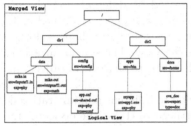

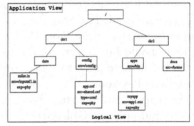

(c) The merged logical view

(d) Application-specific view

Fig. 1. Gvu views

3.2   Logical Views

Each administrator of the site creates a local logical view that may not necessarily correspond to the physical view. The logical view is specified using a configuration file written in XML. Note that the files and directories in the logical view can correspond to arbitrary places in the physical view. The attributes for the files are read from the extended attributes stored by the on-disk file system. The logical views are synchronized among the servers using `distview`.

Figures 1(a) and 1(b) show two sites and how their logical views are formed. Figure 1(a) shows the physical and exported logical view of `site1`. Note that only a few directories and files are exported to the grid. The administrator can also specify attributes in the logical view. The attributes are not shown in the first two figures for clarity. Note that the original `outputs` directory is exported as `data` in the logical view. Similarly, `site2` exports `bin` directory as `apps`. A few of the features provided by Gvu for the creation of views are worth mentioning. The user can create a logical directory without any corresponding physical directory, but the files in the directory have to be specified with fully qualified physical paths. In Figure 1(a), the data directory is a logical directory. If a directory has a corresponding src attribute, meaning it has a physical directory, then the files under the directory are assumed to be under the corresponding physical directory, unless its absolute path is specified.

After the Gvu servers are initialized with respective configuration files, the views are merged and a global view is formed. Figure 1(c) shows the merged global view. For clarity, src attribute is not shown in the figure. One important question while merging is: *What should Gvu do when name conflicts occur?* There is no single answer to this question. Gvu can either provide unique names automatically or ask the administrator to change the logical views. We leave the decision to the site administrator.

When an application queries the Gvu to *get all the files related to experiment* *"phy"*, Gvu returns an application-specific view as shown in the Figure 1(d), which is formed by running the appropriate query on the global view.

3.3   View Synchronization and Security

To support distributed collaboration, views created by different users have to be synchronized. For example, when two users run the query explained in the previous section, they should both see the same view and any changes done by a user should be seen by all the users sharing the view. The synchronization is achieved using distview which provides mechanisms to share Java objects in a distributed system.

Security is an important issue on the grid due to different administrative domains and policies. GSI (Grid Security Infrastructure) [18] is the de-facto standard for providing security on the grid. *How do the authentication mechanisms affect Gvu views?* There are two issues related to Gvu security: access control of files and access control of views. We have implemented security by wrapping Gvu calls with GSI. GSI can map an identity to a local user and Gvu can check the permissions of files to see whether a user has enough privileges to access the file. Adding access control to views is tricky and complicated. Some of the issues are: How do we set the access control list for a local logical view? How can we add access control for application or user-specific views? Where do we store them? One possible solution is to create a security configuration file similar to gridmap-file that specifies access control for local logical views. Access control lists for user or application specific views can be maintained by the local Gvu server. We leave more detailed analysis and implementation as future work.

4   Implementation and Experimental Results

We have implemented Gvu in Java, and it uses various mechanisms provided by distview toolkit to share GvuTree objects. The GvuTree object is a self-sufficient data structure that can identify the global tree, application, user-specific views and respective mappings using a global hash table.

We have conducted experiments measuring various aspects of Gvu. We have set up two separate testbeds for the experiments. The first testbed is a small grid created in the RTCL (Real Time Computing Laboratory) at the University of Michigan. We used this testbed for debugging and for conducting experiments that did not depend on the wide-area nature of a real grid. We used Grid3

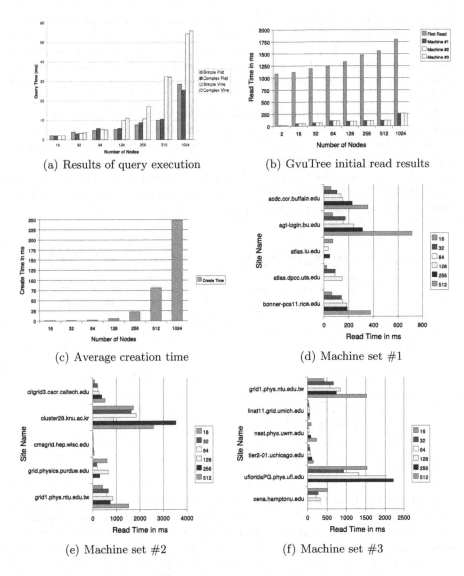

(a) Results of query execution

(b) GvuTree initial read results

(c) Average creation time

(d) Machine set #1

(e) Machine set #2

(f) Machine set #3

Fig. 2. Experimental results

production grid for our real-world scenarios and for understanding the impact of wide-area network on Gvu.

4.1 Experiments on the RTCL Grid

Query Execution. Figure 2(a) shows the execution time of running queries on established trees. The X-axis shows the number of nodes (files) in the GvuTree represented in XML. Four scenarios were run with two types of queries and two types of tree structure: Simple queries that check only one attribute per node, and complex queries that check two attributes. Flat trees had one directory filled with many nodes, and vines had many directories of increasing depth. Ten queries were executed for each data point and the time represents only the time needed to run the query on the server before the results are sent to the application.

As expected, the search times were linear in the number of nodes because every node must be checked in the current implementation. This could be improved by caching the results and only regenerating the parts of tree that have changed since the last query. The difference between simple and complex queries was negligible. The vines took up twice as long to execute because of the overhead involved in the function calls. The total time to execute a query on 1K nodes is under 60 ms, which is an acceptable cost.

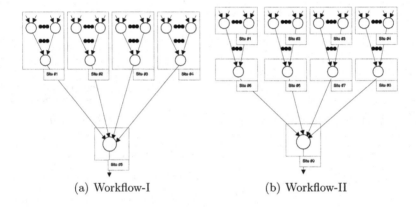

(a) Workflow-I   (b) Workflow-II

Fig. 3. Application workflows

Read Time for GvuTree. Figure 2(b) shows the read time for GvuTree when the server is initialized. One server is initialized with logical view containing files of sizes 16, 32, ..., and the time taken to read the GvuTree on a separate server is recorded. The experiment is conducted on three different machines after restarting the first server and Corona. An average of 20 runs is taken on each of the machines.

The very first read takes a large amount of time due to Java's initialization and serialization of the GvuTree object. Once this is done, the read time for subsequent GvuTree reads is minimal.

Create and Delete Time of Files. Creation and deletion time for single files are 9.78ms and 1.34ms respectively. The low times are due to the in-memory updates. We also ran macro benchmarks by creating a different number of files in a single directory and results can be seen in Figure 2(c). The delay increases as we create more files, because more time is needed to send the updated (bigger) tree to all the Gvu servers. This can be improved in various ways. For example, one can send only updated parts of the tree to other Gvu servers. Another interesting mechanism would be batching of commands. We have implemented batching for the experiments done on Grid3.

4.2   Experiments on Grid3

Grid3 project developed under the auspices of iVDGL (International Virtual Data Grid Laboratory) is a data grid consisting of more than 25 sites with thousands of processors. Grid3 is used by various scientific communities including high-energy physics, bio-chemistry, astrophysics and astronomy.

Read Time for GvuTree. We have run the GvuTree read experiment with different sites for different number of files. Figures 2(d), 2(e) and 2(f) show the read times for various sites. Certain sites were down during a few periods of running the experiments. Note the high latency experienced on the Korean and Taiwanese sites (cluster28.knu.ac.kr, grid1.phys.ntu.edu.tw) and low latencies at the Michigan and Wisconsin sites (linat11.grid.umich.edu, cmsgrid.hep.wisc.edu). The low latency is due to the proximity of these sites to the Corona server in the RTCL. You can see certain anomalies in read times at the Florida site (ufloridapg.phys.ufl.edu). This is due to the high load on the site at the time of our experiments.

Real-World Scenarios. To better understand the behavior of Gvu, we have run two workflows that are similar to CMS workflows in various scenarios. The workflows we used for our experiments are shown in Figures 3(a) and 3(b). The circles represent the jobs and the arrows show the data dependencies between the jobs. All the jobs are similar except that they are run with different inputs and produce different outputs. The workflows have three levels of jobs. Experiments are conducted for a different number of first level jobs. The scheduling of the jobs is done using a simple load-balancing mechanism with an equal number of jobs running on each site. The third level job is run on a separate site.

The workflows are run with Gvu and MCS + RLS and the execution time of the workflow is compared. A pseudo scientific application is written to test the two systems. The application checks with Gvu or MCS for the existence of an input file and if it is available it requests either Gvu or RLS for the location of the file. It produces the output as soon as all the inputs are available.

Both the workflows are run with two different sets of sites. The first set of sites are connected by a wide-area network with latencies on the order of 60ms. The second set of sites are connected by a wide-area network with latencies on the order of 20ms. The two different sets of sites are chosen to demonstrate how Gvu copes up with the high latencies experienced in wide-area networks.

Performance of Workflow-I. As one can see in Figure 3(a), workflow I has better data locality, since the level-1 and level-2 jobs are run on the same site.

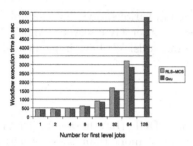

(a) Performance of workflow-I with high latency network

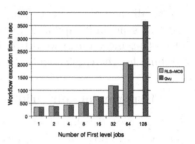

(b) Performance of workflow-I with low latency network

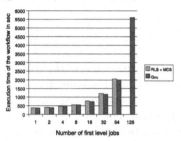

(c) Performance of workflow-II

Fig. 4. Grid3 experimental results

This is the common mechanism for submitting jobs on the grid. The execution time of the workflow with high latency network is shown in Figure 4(a). Note that for 128 jobs, the workflow didn't finish when RLS and MCS are used. This is because of the limit (100) on the number of connections to RLS. This also shows an important aspect of Gvu with respect to the scalability. Though Gvu servers were also highly-loaded for 128 first-level jobs, the performance degraded smoothly, because of the distributed nature of the Gvu.

Performance of the workflow with a low latency network can be seen in Figure 4(b). As expected, the performance improvement is small with Gvu, because of the fast response times from RLS and MCS (due to the low latency network). However, Gvu still performs better than RLS and MCS, because most of the `stat file` requests are handled locally.

Performance of Workflow-II. Workflow-II shown in Figure 3(b) is similar to workflow I except that the level-2 jobs are submitted to different sites. This destroys the data locality and yields poor performance. However, Gvu still performs better than RLS and MCS, due to its distributed nature. Note that the performance improvement is less than that with workflow I. Figure 4(c) shows the performance of workflow II with sites connected with an average latency (30ms) network. We did not run workflow II with a low latency network as we could not find enough site that are near us.

5 Future Work

Gvu framework raises interesting questions for sharing data on a grid. *How do we synchronize views that share files?* An efficient synchronization algorithm is needed to update all the views that have a file when the file is updated. This is not trivial and requires careful design.

Various optimizations can be done to improve the query execution performance. XML databases may provide clues on how to implement the queries. Caching can be used on the client and local Gvu servers to improve performance. More work is needed to implement access control lists for and views. Policies need to be developed to resolve conflicts in merging views. Real-world scenarios have to be explored to understand the effect of conflicts and when they occur. More work is also needed for achieving better fault-tolerance of Gvu and Corona servers.

References

1. Foster, I., Kesselman, C., eds.: The Grid: Blueprint for a Future Computing Infrastructure. Morgan Kaufmann Publishers, San Francisco, California (1999)
2. Foster, I., Kesselman, C.: The Globus project: A status report. In: Proceedings of the Heterogeneous Computing Workshop, IEEE Computer Society Press (1998) 4–18
3. Allcock, B., Bester, J., Bresnahan, J., Chervenak, A.L., Foster, I., Kesselman, C., Meder, S., Nefedova, V., Quesnel, D., Tuecke, S.: Data management and transfer in high-performance computational grid environments. Parallel Computing **28**(5) (2002) 749–771
4. Chervenak, A.L., Deelman, E., Foster, I., Guy, L., Hoschek, W., Iamnitchi, A., Kesselman, C., Kunst, P., Ripeanu, M., Schwartzkopf, B., Stockinger, H., Stockinger, K., Tierney, B.: Giggle: A framework for constructing scalable replica location services. In: SC'2002 Conference CD, Baltimore, MD, IEEE/ACM SIGARCH (2002) pap239.
5. Singh, G., Bharathi, S., Chervenak, A., Deelman, E., Kesselman, C., Manohar, M., Patil, S., Pearlman, L.: A metadata catalog service for data intensive applications. In ACM, ed.: SC2003: Igniting Innovation. Phoenix, AZ, November 15–21, 2003, New York, NY 10036, USA and 1109 Spring Street, Suite 300, Silver Spring, MD 20910, USA, ACM Press and IEEE Computer Society Press (2003) ??–??
6. Thain, D., Bent, J., Arpaci-Dusseau, A., Arpaci-Dusseau, R., Livny, M.: Pipeline and batch sharing in grid workloads. In: Proceedings of the Twelfth IEEE Symposium on High Performance Distributed Computing. (2003)
7. Tatebe, O., Morita, Y., Matsuoka, S., Soda, N., Sekiguchi, S.: Grid datafarm architecture for petascale data intensive computing. In Bal, H.E., Löhr, K.P., Reinefeld, A., eds.: Proceedings of the Second IEEE/ACM International Symposium on Cluster Computing and the Grid (CCGrid2002), Berlin, Germany, IEEE, IEEE Computer Society (2002) 102–110
8. Baru, C., Moore, R., Rajasekar, A., Wan, M.: The sdsc storage resource broker. In: Proceedings of IBM Centers for Advanced Studies Conference, IBM (1998)
9. White, B.S., Walker, M., Humphrey, M., Grimshaw, A.S.: LegionFS: A secure and scalable file system supporting cross-domain high-performance applications. In: SC'2001 Conference CD, Denver, ACM SIGARCH/IEEE (2001) U. of VA.

10. Sun Microsystems, I.: NFS: Network file system protocol specification. Internet Request for Comments (1094) (1989)

11. Foster, I., Voeckler, J., Wilde, M., Zhao, Y.: Chimera: A virtual data system for representing, querying and automating data derivation. In: Proceedings of the 14th Conference on Scientific and Statistical Database Management, Edinburgh, Scotland (2002)

12. Howard, J.H.: An overview of the andrew file system. In: Proceedings of the USENIX Winter Conference, Berkeley, CA, USA, USENIX Association (1988) 23–26

13. Satyanarayanan, M.: Coda: a highly available file system for a distributed workstation environment. In IEEE, ed.: Proceedings of the Second Workshop on Workstation Operating Systems (WWOS-II), September 27–29, 1989, IEEE Computer Society Press (1989) 114–116

14. Soltis, S.R., Ruwart, T.M., Erickson, E., Preslan, K.W., O'Keefe, O.: The global file system. In Jin, H., Cortes, T., Buyya, R., eds.: High Performance Mass Storage and Parallel I/O: Technologies and Applications. IEEE/Wiley Press, New York (2001) chap. 23.

15. Padala, P.: A survey of the grid file systems. GGF Informational Document, Grid File System Working Group, Global Grid Forum (2003)

16. Pazandak, P., Vesudevan, V.: Sematic file systems (1997)

17. Gifford, D.K., Jouvelot, P., Sheldon, M.A., Jr., J.W.O.: Sematic file systems. Proceedings of the 13th ACM Symposium on Operating Systems Principles (October 1991) 16–25

18. Foster, I., Kesselman, C., Tsudik, G., Tuecke, S.: A security architecture for computational grids. In: ACM Conference on Computers and Security. ACM Press (1998) 83–91

Author Index

Lecture Notes in Computer Science

For information about Vols. 1–4245

please contact your bookseller or Springer